SYNTHETIC APERTURE RADAR SIGNAL PROCESSING

SYNTHETIC APERTURE RADAR SIGNAL PROCESSING

with MATLAB Algorithms

Mehrdad Soumekh

A Wiley-Interscience Publication

JOHN WILEY & SONS, INC.

New York / Chichester / Weinheim / Brisbane / Singapore / Toronto

This book is printed on acid-free paper. ♾

Published simultaneously in Canada.

To order books or for customer service please, call 1(800)-CALL-WILEY (225-5945).

Library of Congress Cataloging-in-Publication Data:

Soumekh, Mehrdad.

Synthetic aperture radar signal processing with MATLAB algorithms / Mehrdad Soumekh

p. cm.

"A Wiley-Interscience publication."

Includes bibliographical references and index.

ISBN 0-471-29706-2 (cloth)

1. Synthetic aperture radar. 2. Signal processing—Mathematics. 3. MATLAB. I. Title.

TK6592.S95S68 1999

621.3848—dc21 98-29322

Printed in the United States of America

10 9 8 7 6

To My Beloved Parents
Mrs. Malkah and Dr. Eliahou Soumekh
and
Hertzel, Flora, Lyda, Fereshteh, Benhoor, and Shiaoyan

CONTENTS

Preface xv

Introduction xix

Synthetic Aperture Radar / xix
The Book / xxi
Organization / xxii
SAR and ISAR Databases / xxiv

1 Range Imaging 1

1.1 System Model / 7
1.2 Reconstruction via Matched Filtering / 14
1.3 Range Solution / 17
1.4 Data Acquisition and Signal Processing / 17
Time Domain Sampling / 18
Time Interval of Sampling / 19
Number of Time Samples / 19
1.5 Reconstruction Algorithm / 20
1.6 Reconstruction via Pulse Compression for Chirp Signals / 23
Signal Model / 23
Reconstruction / 25
Range Resolution / 28
Time Domain Sampling / 28
Residual Video Phase Error / 31
Upsampling to Recover Alias-Free Echoed Signal / 31
Electronic Counter-Countermeasure (ECCM) via Amplitude Modulation of Chirp Signals / 34

1.7 Frequency-Dependent Target Reflectivity / 39
Reconstruction via Target Signature Matched Filtering / 44

1.8 MATLAB Algorithms / 45

2 Cross-range Imaging **47**

2.1 System Model / 54

2.2 Spherical PM Signal within an Infinite Aperture / 60

2.3 Reconstruction via Matched Filtering: Infinite Aperture / 64
Synthetic Aperture (Slow-time) Sampling / 66
Cross-range Resolution / 66

2.4 Spherical PM Signal within a Finite Aperture / 67
Instantaneous Frequency / 67
Slow-time Fourier Transform / 68
Slow-time Angular Doppler Spectrum / 71

2.5 Reconstruction via Matched Filtering: Finite Aperture / 72

2.6 Cross-range Resolution / 75

2.7 Data Acquisition and Signal Processing / 75
Synthetic Aperture Sampling for a Broadside Target Area / 75
Synthetic Aperture Sampling for a Squint Target Area / 78
Reducing PRF via Slow-time Compression / 80
Cross-range Gating via Slow-time Compression / 88

2.8 Reconstruction Algorithm / 89
Baseband Conversion of Target Area / 90
Zero-padding in Synthetic Aperture Domain / 91
Slow-time Doppler Domain Subsampling / 93
Reducing Bandwidth of Reconstructed Image / 94

2.9 Synthetic Aperture-Dependent Target Reflectivity / 95
AM-PM Signal Model / 96
Slow-time Fourier Transform of AM-PM Signal / 98
Reconstruction / 114
Representation in Slow-time Angular Doppler Domain / 121

2.10 Reconstruction via Target Signature Slow-time Matched Filtering / 127
Type 1: Generalization of Spotlight SAR / 127
Type 2: Generalization of Stripmap SAR / 130
Type 3: Partial Observability / 135

2.11 MATLAB Algorithms / 135

3 SAR Radiation Pattern **137**

3.1 Transmit Mode Radar Radiation Pattern / 142
Synthetic Aperture (Slow-time) Dependence / 156

3.2 Radiation Pattern in Three-Dimensional Spatial Domain / 157
Radar Footprint / 157
Slant-Range / 162

3.3 Transmit-Receive Mode Radar Radiation Pattern / 163

3.4 Transmit-Receive Mode Radar-Target Radiation Pattern / 165

3.5 Polarization / 171

3.6 MATLAB Algorithms / 174

4 Generic Synthetic Aperture Radar **176**

4.1 System Model / 186

4.2 Fast-time Fourier Transform / 195

4.3 Slow-time Fourier Transform / 196

4.4 Reconstruction / 196

4.5 Digital Reconstruction via Spatial Frequency Interpolation / 198
Baseband Conversion of Target Area / 201
Interpolation from Evenly Spaced Data / 203
Interpolation from Unevenly Spaced Data / 204

4.6 Digital Reconstruction via Range Stacking / 206

4.7 Digital Reconstruction via Time Domain Correlation and Backprojection / 212
Time Domain Correlation Algorithm / 212
Backprojection Algorithm / 214

4.8 Frequency and Synthetic Aperture Dependent Target Reflectivity / 215

4.9 Motion Compensation Using Global Positioning System / 217
Spatial Frequency Modeling of Motion Errors / 218
Narrow-Beamwidth Motion Compensation / 219
Wide-Beamwidth Motion Compensation / 223
Three-Dimensional Wide-Beamwidth Motion Compensation / 226
Motion Compensation for Backprojection / 230

4.10 Motion Compensation Using In-Scene Targets / 230
Narrow-Beamwidth Motion Compensation / 230

Wide-Beamwidth Motion Compensation / 231
Three-Dimensional Wide-Beamwidth Motion Compensation / 234

4.11 Polar Format Processing / 234
Plane Wave Approximation-Based Reconstruction / 234
Narrow-Beamwidth Approximation / 238
Narrow-Bandwidth and Narrow-Beamwidth Approximation / 239
Wavefront Curvature Compensation / 242
Motion Compensation Using Global Positioning System / 243

4.12 Conventional ISAR Modeling and Imaging / 244
ISAR Model / 246
Slow-time Compression or Motion Compensation / 246
Polar Format Processing / 247

4.13 Range-Doppler Imaging / 249
Fresnel Approximation-Based Reconstruction / 251
Narrow-Bandwidth and Narrow-Beamwidth Approximation / 255

4.14 Three-Dimensional Imaging with Two-Dimensional Azimuth and Elevation Synthetic Apertures / 256
System Model / 257
Reconstruction / 258

4.15 Electronic Counter-Countermeasure via Pulse Diversity / 259

5 Spotlight Synthetic Aperture Radar **262**

5.1 Mechanically Beam-Steered Spotlight SAR / 270
Mechanical Beam Steering / 270
System Model / 274
Reconstruction / 274

5.2 Electronically Beam-Steered Spotlight SAR / 277
Electronic Beam Steering / 277
System Model / 280
Reconstruction / 281

5.3 Bandwidth of Spotlight SAR Signal / 282
Single Target / 283
Target Area / 286

5.4 Resolution and Point Spread Function / 290

5.5 Data Acquisition and Signal Processing / 293
Fast-time Domain Sampling and Processing / 294

Slow-time Domain Sampling and Processing / 297
Reducing PRF via Slow-time Compression / 302
Digital Spotlighting / 306
Subaperture Digital Spotlighting / 317

5.6 Reconstruction Algorithms and SAR Image Processing / 319
Digital Reconstruction via Spatial Frequency Interpolation / 319
Reconstruction in Squint Spatial Coordinates / 329
Slow-time Doppler Domain Subsampling / 335
Reducing Bandwidth of Reconstructed Image / 336
Digital Reconstruction via Range Stacking / 349
Digital Reconstruction via Time Domain Correlation and Backprojection / 354
Effect of Slow-time Doppler Filtering / 362
Effect of Motion Errors in Slow-time Doppler Spectrum / 369

5.7 MATLAB Algorithms / 370

6 Stripmap Synthetic Aperture Radar **373**

6.1 System Model / 381
Radar Radiation Pattern / 382
Stripmap SAR Signal Model / 384

6.2 Reconstruction / 387

6.3 Bandwidth of Stripmap SAR Signal / 388
Planar Radar Aperture / 389
Curved Radar Aperture / 391

6.4 Resolution and Point Spread Function / 391

6.5 Data Acquisition and Signal Processing / 397
Fast-time Domain Sampling and Processing / 398
Slow-time Domain Sampling and Processing / 402
Slow-time Compression and Processing / 404
Subaperture Digital Spotlighting / 409
Reducing Side Lobes Doppler Aliasing via Slow-time Upsampling / 415

6.6 Reconstruction Algorithms and SAR Image Processing / 418
Digital Reconstruction via Spatial Frequency Interpolation / 418
Slow-time Doppler Domain Subsampling / 425
Reducing Bandwidth of Reconstructed Image / 428
Digital Reconstruction via Range Stacking / 445

Digital Reconstruction via Time Domain Correlation and Backprojection / 446
Effect of Beamwidth (Slow-time Doppler) Filtering / 451
Effect of Motion Errors in Slow-time Doppler Spectrum / 453
Subpatch "Mosaic" Digital Reconstruction with Subaperture Data / 456

6.7 Moving Target Detection and Imaging / 465
SAR Signal Model for a Moving Target with a Constant Velocity / 467
Three-Dimensional Imaging in Motion-Transformed Spatial Domain and Relative Speed Domain / 472
Moving Target Indicator: SAR Ambiguity Function / 474

6.8 MATLAB Algorithms / 477

7 Circular Synthetic Aperture Radar **486**

7.1 System Model / 491
CSAR Signal Model / 493
Fourier Properties of Slant Plane Green's Function / 494

7.2 Reconstruction / 497
Slant Plane to Ground Plane Transformation / 498
Ground Plane CSAR Reconstruction / 500

7.3 Bandwidth of CSAR Signal / 507

7.4 Resolution and Point Spread Function / 509
Full Rotation Aspect Angle Measurement / 509
Partial Rotation Aspect Angle Measurement / 512

7.5 Data Acquisition and Signal Processing / 515
Fast-time Domain Sampling and Processing / 515
Slow-time Domain Sampling and Processing / 516
Digital Spotlighting and Clutter Filtering / 517

7.6 Reconstruction Algorithms and CSAR Image Processing / 518
Digital Reconstruction via Spatial Frequency Interpolation / 518
Reducing Bandwidth of Reconstructed Image / 521
Digital Reconstruction via Time Domain Correlation and Backprojection / 522

7.7 Three-Dimensional Imaging / 525
Target Resolvability from Single-Tone Fringe Patterns / 533

7.8 Three-Dimensional Imaging with Two-Dimensional Circular and Elevation Synthetic Apertures / 539

System Model / 543
Reconstruction / 545
Digital Reconstruction / 551

8 Monopulse Synthetic Aperture Radar **553**

8.1 Along-Track Moving Target Detector Monopulse SAR / 559
Along-Track Monopulse SAR System Geometry / 561
Monostatic SAR Signal Model / 562
Bistatic SAR Signal Model / 564
Synthesis of Monostatic SAR Signal from Bistatic SAR Signal / 565
Moving Target Indicator / 567
Effect of Variations in Altitude and Nonlinear Motion / 569

8.2 Effect of Uncalibrated and Unstable Radars / 570
Amplitude Patterns of Monopulse Radars / 571
Instability of Monopulse Radars / 574
Wide-Beamwidth Monopulse Radars / 574

8.3 Signal Subspace Registration of Uncalibrated SAR Images / 575
System Model / 576
Signal Subspace Processing / 578
Estimating Calibration Error Impulse Function / 583
Application in MTD Monopulse SAR / 584
Application in Automatic Target Recognition SAR / 585

8.4 Slant Plane Topographic Mapper Monopulse SAR / 587
Slant Plane Monopulse SAR System Geometry / 588
Monostatic and Bistatic SAR Signal Models / 588
Narrow-Bandwidth and Narrow-Beamwidth Approximation: Interferometric SAR (IF-SAR) / 589
Wide-Bandwidth and Wide-Beamwidth Model / 592
Estimating Slant-range Shift via Signal Subspace Processing / 594

8.5 Multistatic Monopulse ISAR / 596
Multistatic ISAR Model / 596
Motion Tracking via Signal Subspace Processing / 598

8.6 MATLAB Algorithms / 603

Bibliography **605**

Index **611**

PREFACE

Synthetic aperture radar (SAR) is one of the most advanced engineering inventions of the twentieth century. SAR is an airborne or satellite-borne radar system that provides high-resolution maps of remote targets on a terrain, a planet, and so forth. The E-3 AWACS (Airborne Warning and Control System) airplanes that are used extensively in surveillance missions in the Persian Gulf region to detect and track maritime and airborne targets, the E-8C Joint STARS (Surveillance Target Attack Radar System) airplanes that were used to detect and locate ground targets in the Gulf War, and the NASA space shuttles are equipped with this radar system. SAR systems are a highly developed combination of precision hardware and electronic design for data acquisition, and advanced theoretical principles of mathematics and physics to convert the acquired data to high-resolution images. The origin of the SAR theoretical principles can be traced back to Gabor's theory of *wavefront reconstruction* [gab]; this theory is also the foundation of many other *coherent* imaging systems in diverse fields such as geophysical exploration and diagnostic medicine [s94].

The utility of the wavefront reconstruction theory in SAR was recognized during the inception of this imaging system in the 1950s and 1960s. However, the lack of fast computing machines and advanced digital information (signal) processing algorithms at that time prevented the development of wavefront reconstruction-based SAR imaging methods. The early SAR systems were based on optical (analog) processing of the measured echoed signal using the Fresnel approximation for image formation [cut; wil]; this SAR processor, in the analog form or its digital version which was introduced for the spaceborne SAR in the late 1970s [cur], is also known as *range-Doppler imaging* [br69].

In the 1970s another SAR imaging method, known as *polar format processing*, was introduced for the high-resolution spotlight SAR systems [aus; wal]; this method was based on what was known as the *plane wave* approximation. An extensive theoretical and practical knowledge base has been shaped over the past 40 years by the approximations inherent in these methods; some of these concepts are either incorrect or not applicable when viewed in the framework of modern high-resolution SAR systems.

The first digital signal processing methods for SAR image formation via the wavefront reconstruction theory were introduced in the late 1980s. SAR wavefront reconstruction theory not only provides a tool for SAR image formation but also reveals functional properties of the SAR signal that contradict or were not predicted in the approximation-based theoretical foundation of either range-Doppler imaging or polar format processing.

A prominent example involves the slow-time Doppler bandwidth of the spotlight SAR system. Polar format processing theory states that this bandwidth is proportional to the beamwidth angle of the physical radar aperture; this also supports the classical SAR theory principles for the stripmap SAR signal. This slow-time Doppler bandwidth dictates the minimum required pulse repetition frequency (PRF) of the radar.

However, SAR wavefront reconstruction theory shows that the slow-time Doppler bandwidth of the spotlight SAR signal is proportional to the beamwidth angle of the physical radar aperture *plus* the beamwidth look angle of the *synthetic aperture* (Chapter 5). Unfortunately, most of the users of polar format processing who tried to understand and implement the SAR wavefront reconstruction missed this subtle but very important issue.

The processing of aliased slow-time Doppler data in X band spotlight SAR systems has been a key reason for the failure to exploit *coherent* (complex) target signatures for automatic target recognition (ATR). Filtering out the higher slow-time Doppler data is a commonly used scheme to deal with the Doppler aliasing problem; this turns out to be a premature and simplistic approach that removes a significant amount of useful information.

The wavefront reconstruction theory reveals unique functional properties of and redundancies in the SAR signal that cannot be found in any other information processing modality. These redundancies could be exploited to recover uncorrupted data from aliased data.

Another classic example of how wavefront reconstruction theory has changed the way SAR signals are viewed involves *filtering* of the acquired SAR data. Conventional frequency domain filtering, for example, applying a Hamming window, is known to improve the quality of data in speech processing. Such filters have also been used in range-Doppler imaging or polar format processing to suppress clutter and/or improve the shape of the point spread function.

However, these filters mainly reduce the errors that are due to the approximations in range-Doppler imaging or polar format processing; these errors turn out to be more prominent in the higher slow-time Doppler frequencies where the filter attenuates the SAR data. Based on the SAR wavefront reconstruction theory, applying these filters on SAR data results in loss of *useful* high Doppler frequency information and, consequently, in loss of resolution.

The main message of the above examples is that one cannot view the SAR signal and its properties via the range-Doppler or polar format processing theory and then use the SAR wavefront reconstruction theory for image formation. Unfortunately, most of the SAR hardware systems that have been developed since the 1970s result in *aliased* SAR data when their acquired data are viewed via the SAR wavefront reconstruction theory. Does this mean that the wealth of the data collected by these SAR systems are contaminated and that these SAR systems must be rebuilt? As we will discover, the answer to this question is no.

This book provides a foundation in signal processing and includes digital algorithms for the SAR wavefront reconstruction theory. It both establishes the constraints for acquiring the SAR data from the system design point of view and provides digital signal and image processing algorithms for proper and alias-free implementation of the SAR wavefront reconstruction. As will become clear in this book, the unique properties of the SAR signal cannot be found in any other information processing system, such as speech processing, or the conventional beamforming methods for array processing.

For instance, the *unconventional* SAR wavefront reconstruction theory shows that under certain conditions, an *aliased* SAR database can be used to recover its corresponding *alias-free* SAR data (Chapters 2 and 5). This scheme is used to recover the alias-free SAR data of a SAR system that was built based on the polar format processing theory.

The SAR wavefront reconstruction theory has also brought new tools for addressing classical SAR processing issues such as motion compensation and radar calibration (Chapter 4). The motion and calibration errors in SAR can be shown to amplitude modulate the SAR signal which itself is a phase-modulated signal (Chapters 2 and 3). The combined AM-PM signal then dictates the point spread function of the SAR system. This concept has been used by Papoulis to analyze the Fourier properties of optical AM-PM signals [pap]; in the 1960s O'Neal also predicted such properties for the SAR signals.

A major issue associated with using SAR for automatic target recognition (ATR) is the variations of the radar radiation pattern over time or the calibration problem. Owing to the calibration errors, the point spread function of a target in a SAR image varies from one experiment to another. The AM-PM modeling of uncalibrated SAR signals enables the user to address the calibration problem via the *signal subspace* least squares methods which are used for blind deconvolution [s98a; wid] (Chapter 8).

A similar calibration problem is also encountered in moving target detection (MTD) using along-track monopulse SARs. In recent years attempts to use displaced phase center arrays (DPCAs) to emulate along-track monopulse MTD-SARs over a relatively long slow-time have been unsuccessful due to the calibration errors. Signal subspace processing of uncalibrated along-track monopulse MTD-SARs is also a digital signal processing approach used to encounter calibration errors (Chapter 8).

In addition to the traditional *linear* motion stripmap and spotlight SAR systems (Chapters 5 and 6), newer SAR modalities are also being investigated. One of these new SAR systems involves a circular flight path over a desired (spotlighted) target area. The objective in the circular SAR (CSAR) systems is to acquire the radar signature of the targets in the spotlighted scene over all possible *aspect angles* (Chapter 7). Such a database is useful for target detection and classification in ATR-SAR problems. CSAR also provides some form of *three-dimensional* imaging capability that the linear SAR systems do not possess. Meanwhile, a CSAR signal has functional properties and, as a result, digital signal processing problems that are different from those of the linear SAR signal.

The theoretical principles in this book are accompanied by results from four sets of realistic SAR, ISAR, and CSAR (turntable) databases. MATLAB* algorithms are also provided for simulation, digital signal processing, and reconstruction of various one-dimensional radar and multidimensional SAR imaging modalities. These algorithms are available at the ftp site ftp://ftp.mathworks.com/pub/books/soumekh.

The key to the success of a high-resolution SAR system is 90 percent due to good radar hardware, and 10 percent due to *intelligent* digital signal processing, though some radar scientists might argue that the ratio is likely to be 99 to 1. It can be argued that the development of each of these two aspects of the SAR system based on mature and concrete foundations will enhance the success of the other.

Acknowledgments

I am grateful for my collaboration with Rolan Bloomfield, S. I. Chou, Robert Dinger, Robert DiPietro, Ronald Fante, Lawrence Hoff, Liz Jones, Scott Jones, William Kendall, Eric Nelson, David Nobles, Richard Perry, Michael Pollock, Edward Rupp, Roger Rysdyk, Alan Stocker, Steve Tran, Jay Trischman, Michael Wicks, Christopher Yerkes, and William Zwolinski. I am thankful to S. I. Chou, Ira Ekhaus, Joseph Fitzgerald, Robert Nelson, Thierry Rastello, and Christopher Yerkes for their comments and suggestions in the course of this project.

I am also grateful to the members of the radar groups who collected the ISAR, SAR, and CSAR (turntable) databases cited in this book; these groups are identified in the introductory chapter. I also wish to extend my gratitude to those who provided the data. Many thanks to Betty Brown, Rosalyn Farkas, and Donald DeLand for helping me in editing and preparing this book.

The work reported in this book was supported by the following government agencies and private industries: National Science Foundation, Grant MIP-9004996, P. Ramamoorthy, Program Director; Bell Aerospace (Textron), William Zwolinski, Technical Coordinator; Space Computer Corporation, Alan Stocker, Technical Coordinator; Summer Faculty Fellowships at Naval Research and Development (SPAWAR Systems Center), Robert Dinger (1992-95) and Lawrence Hoff (1995), Technical Coordinators; Summer Faculty Fellowship at Rome Laboratory, Air Force Office of Scientific Research, Michael Wicks, Technical Coordinator; Naval Research and Development, Grants N66001-95-M-1383 and N66001-7052-7595, Michael Pollock, Technical Coordinator; MITRE Corporation, Richard Perry, Technical Coordinator; Office of Naval Research, Grants N00014-96-1-0586 and N00014-97-1-0966, William Miceli, Program Officer; and Air Force Office of Scientific Research, Grant 99-NM-080, John Sjogren, Program Officer. Their generous support of this work is greatly appreciated.

MEHRDAD SOUMEKH

Buffalo, NY
October 1998

*MATLAB® is a registered trademark of the MathWorks, Inc. For MATLAB product information, please contact: The MathWorks, Inc., 3 Apple Hill Drive, Natick, MA 01760-2098 USA.

INTRODUCTION

SYNTHETIC APERTURE RADAR

Vision is perhaps the most critical component of the human sensory system. The lens in the human eye collects the optical waves that are being reflected from the objects in its surrounding medium, which are then interpreted as visual information in the brain.

Human-made vision systems imitate the function of the eye–the sensor—and that of the brain—the processor. Human-made vision systems are primarily built to improve on our ability to resolve targets, for example, binoculars, radars, and sonars, or to capture the image of a scene, for example, cameras. In either case the ability to view or capture a scene improves with a larger lens aperture (in a binocular or camera), a larger radar antenna aperture, or larger acoustic transducer aperture; the key to *better vision* is a *larger aperture*. Unfortunately, it is extremely difficult to either build or maintain a *physically* large aperture vision system. (Recall the Hubbel telescope.)

In the 1950s an invention in radar by Wiley revolutionized the way human-made vision systems are constructed. This invention was called *synthetic aperture radar*, or *SAR*. The principal idea behind SAR is to synthesize the effect of a large-aperture *physical* radar, whose construction is infeasible. *What is the significance of a larger aperture?* This becomes evident in the following example.

The lateral or *cross-range* resolution of a $D = 1$-meter diameter radar antenna with wavelength $\lambda = 1$ meter at the range $R = 1000$ meters is

$$\text{Lateral resolution} = \frac{R\lambda}{D} = 1000 \text{ meters},$$

which is very poor. Yet, based on the SAR theory and signal processing, if we move this small 1-meter aperture radar along an *imaginary* aperture with length $D_{\text{eff}} = 1000$ meters (D_{eff} stands for *effective* radar diameter or aperture), then the lateral resolution with wavelength $\lambda = 1$ meter at the range $R = 1000$ meters becomes

$$\text{Lateral resolution} = \frac{R\lambda}{D_{\text{eff}}} = 1 \text{ meter};$$

this is a tremendous improvement over the 1000-meter lateral resolution of the small 1-meter physical radar antenna. The imaginary aperture of $D_{\text{eff}} = 1000$ meters is called a *synthetic aperture*. In recent years the principle of synthetic aperture has also been utilized in other human-made vision systems, such as sonar.

Is synthetic aperture imaging a novel human invention? Not quite. While this has not been clearly demonstrated due to the complexity of biological structures, it is believed that the acoustic or ultrasonic sensory systems in dolphins and bats perform some form of synthetic aperture processing. A dolphin or a bat uses lateral motion to create a synthetic aperture. The lateral motion-based synthetic aperture formation and processing is referred to as *Doppler* processing in the SAR literature.

Since synthetic aperture imaging was originally introduced for radar systems, there has been a common misconception that synthetic aperture radar and sonar are exclusively for military purposes. However, synthetic aperture imaging has commercial as well as military applications. Some of the commercial and military applications of synthetic aperture imaging are:

- Topographic imaging of the surface of planets
- Topographic imaging of the ocean floors
- Assessing the condition of crops
- Airborne reconnaissance
- Underground resources (e.g., oil) exploration
- Surveillance and air traffic control
- Automatic aircraft landing
- Mine detection
- Concealed target detection in foliage
- Interior imaging of buildings in a rescue or hostage crisis
- Detecting buried historical objects
- Missile detection and tracking
- Ground moving target detection and tracking
- Medical ultrasonic imaging of obstructed biological structures
- Nondestructive testing of obstructed mechanical structures

Some of the above applications, such as topographic imaging, are heavily investigated and even commercialized. Some, such as medical ultrasonic imaging, are behind in development due to misunderstandings and the unconventional nature of synthetic aperture imaging. For instance, one of the areas of interest in diagnostic medicine echo imaging is to obtain quantitative as well as qualitative information regarding the condition of the cardiovascular system using ultrasonic sources. Due to the low signal to clutter power ratio in these problems, some form of synthetic aperture phased array data collection strategy is required [s97] (see also the introduction to Chapter 2).

A medical ultrasonic imaging group has investigated and developed a circular synthetic aperture phased array imaging system for this problem.[†] However, this is achieved by dropping *one* element (or subaperture) and turning on another in the successive transmit/receive modes of the array; this would make the acquired data susceptible to various system noise sources (thermal, quantization, nonlinear scattering, etc.). The reconstruction algorithm for this system, which does not exploit wavefront reconstruction theory, is also based on a time-consuming array beamformer which makes the approach susceptible to motion artifacts.

Meanwhile, the radar community, where synthetic aperture imaging got its first start, has been more willing than the medical imaging community to accept and adapt modern synthetic aperture imaging systems and their wavefront reconstruction theory-based signal and image processing. This might be due to the fact that the radar engineers are more capable of comprehending a mathematically oriented concept that requires extensive digital signal processing for its implementation.

This book is intended to provide a basic understanding of the wavefront reconstruction signal theory principles behind SAR imaging, and their digital implementation. This is done by not only presenting a theoretical foundation for the multidimensional signals of a SAR system but also providing a physical meaning for these signals. The analysis and wording in the book are for synthetic aperture *radar*. Yet the general treatment of the synthetic aperture imaging problem, which is applicable in near-range as well as far-range and wideband sources as well as narrowband sources at any carrier frequency, makes the analysis suitable for the synthetic aperture-based imaging systems of sonar, diagnostic medicine, geophysical exploration, nondestructive testing, and other applications of this type.

It is anticipated that this approach would remove the myth that SAR signal theory and digital signal processing is a raw and difficult mathematical concept. Synthetic aperture-based imaging systems have the potential to dominate most human-made vision systems. With a better understanding of SAR signal processing, the hidden power and applications of synthetic aperture imaging would be more readily captured and pursued, and put to a much wider use than they have been up to the present time.

THE BOOK

With the advent of powerful digital signal processing algorithms, both multidimensional signal analysis and reconstruction in imaging systems may now be formulated via more concrete *theoretical* principles and foundations. These theoretical foundations, which were deemed to be impractical for implementation in realistic imaging systems, not only can be used to develop accurate and computationally manageable reconstruction algorithms, but they also provide a new understanding of the information content of the signals that are utilized for image formation.

[†]M. O'Donnell and L. Thomas, "Efficient synthetic aperture imaging from a circular aperture with possible application to catheter-based imaging," *IEEE Trans. Ultrason. Ferroelectr. Freq. Control*, **39**: 366–380, 1992. See also papers by the first author in the May 1997 and January 1998 issues of the same journal.

This is particularly true in SAR imaging. High-resolution SAR offers one of the prime examples of the role of digital signal processing and signal theory in advancing this field. Digital image formation in SAR is heavily dependent on the *discrete Fourier analysis* of the SAR signal and the target function. This discrete-time and discrete-space analysis imposes restrictions on the parameters used to acquire SAR data, such as the radar pulse repetition frequency (PRF), the fast-time A/D conversion rate, and the size of the imaging area for its alias-free implementation.

The successful implementation of the advanced SAR multidimensional digital signal processing algorithms requires a thorough understanding of the SAR signal theory and its sampling constraints. The digital signal processing issues associated with the new high-resolution SAR systems have brought new complexities and also new misunderstandings for those familiar with the traditional SAR systems.

The purpose of this book is to present a multidimensional signal processing framework for high-resolution SAR data acquisition, image formation, and analysis. In addition to the development of the theoretical principles, special emphasis is given to constraints on system parameters, digital signal processing, and digital image reconstruction problems of SAR; MATLAB algorithms and numerical examples are provided to assist the reader in this endeavor.

The book is intended as a research and teaching monograph for radar signal and image processors, and for individuals interested in applications of multidimensional signal processing. It is written from the point of view of a signal processor. The style is tailored for the members of the signal processing community who are interested in learning about the signal processing theory and algorithms for SAR. The book is also of interest to radar engineers who are eager to understand the principles of emerging advanced multidimensional signal processing algorithms for SAR.

As background for this book, a thorough understanding of *Fourier analysis* and *phase modulation* is needed. This requirement is certain to be met by signal processors and radar engineers, though someone with a background in wave theory would also be able to follow the development in the text. The analytical development is presented in the framework of signal processing and basic communication theory, and on some occasions, a brief physical interpretation of a topic is also provided.

ORGANIZATION

Synthetic aperture radar is an echo-mode array imaging system. Similar to the other array imaging systems, SAR provides a multidimensional database (i.e., the acquired data) that can be manipulated via signal processing means to obtain a multidimensional image that carries information on the target area under study (i.e., the target reflectivity function). The signal (information) subspace of the acquired data is formed by varying the radar frequency and radiation pattern in the target area.

The simplest way to look at two-dimensional SAR imaging is via the principle that range imaging, the x domain information, is constructed via the variations of the radar signal frequency (bandwidth), and that the cross-range imaging, the y domain information, is formed through the variations of the radar radiation pattern (synthetic aperture or radar aspect angle). However, the mapping of the two-dimensional signal subspace of the acquired SAR data into the two-dimensional target function is more complicated than *two separable one-dimensional mappings*. In fact the theory of high-resolution SAR image formation is based on a *nonseparable two-dimensional inverse problem* [s94].

Our discussion in this book begins with the basic principles behind *one-dimensional* range imaging via utilization of a radar bandwidth (Chapter 1) and cross-range imaging via utilization of a synthetic aperture (Chapter 2). These basic principles, and their associated data acquisition and digital signal processing issues, will assist us in developing more concrete high-resolution SAR imaging algorithms in the later chapters. In particular, spectral (Fourier) decomposition of propagating waves and the *wavefront* reconstruction theory, which are the theoretical foundation for formulating SAR signal theory and imaging in this book, are presented in Chapter 2. Chapter 1 also includes a discussion on the design and processing of radar signals for electronic counter countermeasure (ECCM), for example, phase-modulated (phase-perturbed) chirp signals.

The discussion in Chapter 3 brings out the special role of the radar radiation pattern and its Fourier properties using the wavefront reconstruction theory. These are critical for the reader in comprehending the origin of the SAR signal and the *multidimensional signal subspace* that it occupies. Once this is established, the formulation of the SAR digital signal processing and imaging algorithms is based on an efficient manipulation of this signal subspace.

The material in Chapter 4 provides the reader's first exposure to the principle behind SAR imaging. In this chapter a SAR scenario is considered, called *generic SAR*, that lacks many practical components of a realistic SAR system. Yet the generic SAR provides a simple way to view the multidimensional SAR signals and their Fourier properties. The problem of interpolation from unevenly spaced data for digital reconstruction of the SAR image is discussed in this chapter.

An alternative approach, which is referred to as *range stacking*, is outlined that does not require any interpolation. The chapter also contains a discussion on two other digital reconstruction methods for SAR which are known as the *time domain correlation* (TDC) and *backprojection* imaging.

Modeling and imaging airborne targets using *inverse* SAR (ISAR) principles are formulated. Motion compensation using global positioning system (GPS) data, and in-scene targets are examined. Application of radar signaling methods of Chapter 1 and *pulse diversity* for ECCM, and their associated signal processing in SAR, are outlined. The chapter concludes with a discussion on three-dimensional imaging using a two-dimensional planar synthetic aperture in the azimuth and elevation domains.

Chapters 5 and 6 discuss system modeling and imaging for broadside and squint spotlight SAR, and stripmap SAR within the framework of the spectral properties of the SAR radiation pattern that is developed in Chapter 3. This provides the reader with an understanding of the information contents of the two SAR systems, and the limitations of their corresponding acquired data signal subspaces.

Chapter 7 is concerned with a SAR geometry in which a circular aperture is formed by the motion of the radar-carrying aircraft. This system is a spotlight SAR system which has the capability to provide the imaging scene signature at all aspect angles. The system also provides some form of three-dimensional imaging capability. A discussion on three-dimensional SAR imaging using a combination of circular and elevation synthetic apertures is also provided.

The final chapter of the book, Chapter 8, provides a discussion on a specialized SAR system that uses two or more receiving radars, with varying spatial coordinates, to record echoed data due to transmission from a single radar. This system, which is called *monopulse* SAR, has utility in detecting moving targets in an imaging scene (along-track monopulse SAR), and topographic terrain mapping (slant plane monopulse SAR which is also called interferometric SAR, or IF-SAR). Signal subspace processing methods for registering uncalibrated SAR images and motion tracking, and its applications in monopulse SAR systems, are also examined.

The signal processing concepts for spotlight SAR in Chapter 5 are accompanied by the real data from a spotlight inverse SAR (ISAR) system; the ISAR data are from an airborne commercial aircraft. The signal processing principles and properties for stripmap SAR in Chapter 6 are accompanied with the real data from an experimental foliage penetrating (FOPEN) ultra wideband UHF stripmap SAR system; this database is also studied in the framework of a spotlight SAR system in Chapter 5. The analytical study of circular SAR in Chapter 7 is accompanied by the real turntable data of a T-72 tank. Chapter 8 also uses the real ISAR data to demonstrate registration and tracking applications of the signal subspace processing methods.

Most chapters of the book (except for Chapters 4 and 7) conclude with a discussion on MATLAB algorithms for the signal and image processing principles that are developed in that chapter; for these programs, output figures are identified by *P*. Throughout each chapter these program figures are referenced in addition to the figures and examples of that chapter.

SAR AND ISAR DATABASES

As we mentioned earlier, a set of realistic SAR and ISAR databases are used in this book to assist the reader in visualizing and comprehending the SAR signal theory principles. In this section we identify the origin of these databases, and the parameters that were used to acquire them.

X Band ISAR Data of a Commercial Aircraft

This ISAR database was collected by the Radar Group at the SPAWAR Systems Center (formerly Naval Command, Control and Ocean Surveillance Center) at San Diego. The radar signal was a chirp burst. The following are the parameters of this ISAR database:

- Fast-time (A/D converter) sample spacing: $\Delta_t = 2$ ns
- Pulse repetition frequency of the radar: PRF $= 200$ Hz

- Fast-time frequency bandwidth of the radar: $[\omega_{\min}/2\pi, \omega_{\max}/2\pi] \approx [9, 9.5]$ GHz
- Speed of the aircraft (estimated): $v_t = 92.22$ m/s
- Slant range and cross-range: $(X_c, Y_c) = (5.23, 2.08)$ km
- Number of fast-time samples: 256
- Number of slow-time samples: 256
- Start point of fast-time sampling: $T_s = 37.27$ μs

We will see in our discussion on spotlight SAR in Chapter 5 that the above parameters translate into the following:

- Carrier frequency of the radar: $\omega_c/2\pi = 9.25$ GHz
- Sample spacing in the synthetic aperture domain: $\Delta_u = v_t/\text{PRF} = 0.23$ m
- Synthetic aperture: $[-L, L] = [-59, 59]$ m
- Slow-time data acquisition period: $2L/v_t = 1.28$ s
- Squint angle: $\theta_c = \arctan(Y_c/X_c) = 21.66°$

An interesting feature of this ISAR database is that the aircraft possessed *nonlinear* motion components (pitch, yaw, roll, acceleration, etc.) that make the processing and interpretation of the data more complicated. We will show the effects of these nonlinear motion components in the reconstructed ISAR image in Chapter 5. This database will also be examined in the framework of ISAR tracking in Chapter 8. The other interesting feature of this ISAR database is its nonzero squint angle θ_c which has an important impact on the point spread function of the ISAR imaging system.

UHF Band Stripmap SAR Data

This SAR database is part of a set of stripmap SAR databases collected by the radar group at SRI International to study foliage penetrating (FOPEN) characteristics and potential of the UHF band SAR for reconnaissance applications. An impulse-type radar signal was used to radiate the target area.

The imaging area was composed of a near-range foliage region and clear land behind it. Various stationary trucks and corner reflectors were positioned in the foliage and clear land regions. As we will see, the SAR data also contain contributions from a surrounding farm and vehicles that were in motion in the area (though these were not the intended targets). Also there exists a foliage area at the far-range that shows up in the reconstructed SAR images.

We will present results for two sets (i.e., two different runs of the radar-carrying aircraft) of these SAR databases. We will use one of the *stripmap* SAR databases (aircraft runs) within a relatively *small* synthetic aperture in Chapter 5 (spotlight SAR) to show its similarity to the spotlight SAR data. In Chapter 6 (stripmap SAR), we will examine the SAR data for the same run of the radar-carrying aircraft within a larger synthetic aperture interval. We will also present results for the SAR data which were collected during the other run of the radar-carrying aircraft.

The following are the parameters of the SAR data for the two runs of the radar-carrying aircraft. The parameters that are different in the two runs are labeled accordingly; the parameters that are common are not labeled.

- Fast-time (A/D converter) sample spacing: $\Delta_t = 4$ ns
- Pulse repetition frequency of the radar: PRF $= 200$ Hz
- Fast-time frequency bandwidth of the radar: $[\omega_{\min}/2\pi, \omega_{\max}/2\pi] \approx [200, 400]$ MHz
- Speed of the radar-carrying aircraft: $v_r = 77.73$ m/s (first run), and 76.38 m/s (second run)
- Number of fast-time samples: 1024 (first run), and 512 after fast-time zero-padding (second run)
- Number of slow-time samples: 1024 (first run), and 2048 (second run)
- Start point of fast-time sampling: $T_s = 3.4$ μs (first run), and 4.06 μs after fast-time zero-padding (second run)

In the case of the second run, the actual number of the fast-time samples is 256, and the start point of fast-time sampling is at $T_s = 4.74$ μs. Due to the relatively small number of the fast-time samples, the FFT leakage (wrap-around) errors are significant; this phenomenon is particularly noticeable because the *strong* foliage signature is at the start point (boundary) of the fast-time sampling. The data in the fast-time domain are zero-padded to reduce these leakage errors.

The above parameters can be used to obtain the following additional SAR system specifications:

- Carrier frequency of the radar: $\omega_c/2\pi = 300$ MHz
- Mean range: $X_c = 816$ m (first run), 787.2 m (second run)
- Sample spacing in the synthetic aperture domain: $\Delta_u = v_r/\text{PRF} = 0.39$ m (first run), and 0.38 m (second run)
- Synthetic aperture: $[-L, L] = [-199, 199]$ m (first run), and $[-391, 391]$ m (second run)
- Slow-time data acquisition period: $2L/v_r = 5.12$ s (first run), and 10.24 s (second run)

The SAR data for the first run appear to be *cleaner*; they came with a reference signal for fast-time matched-filtering. The first run data also contained more fast-time samples (1024), which carry the SAR signature of the farm area and the moving targets around it. The SAR images formed from the two runs show variations in the SAR signature of the imaging scene that are mainly due to the calibration errors of the radar. We should point out that the relative aspect angle and the coordinates of the radar are different for the two runs; that is, the mean range X_c corresponds to two different points in the spatial domain for the two runs.

Chapter 6 also contains a few examples on processing another UHF band SAR database (P-3 data). An analysis of the P-3 SAR data, based on the principles that are developed in this book, can be found in [s98s].

X Band Circular SAR (Turntable) Data of an T-72 Tank

This turntable SAR or circular SAR (CSAR) database is part of a set of databases collected by the radar group at Georgia Tech Research Institute (GTRI). The data were provided by Algorithm Integration Branch (WL/AACI) WPAFB and Veda Incorporated.

The radar that was used for the CSAR data collection was a *stepped frequency* type. This database was converted into *pulsed* radar data (see Chapter 1). The following are the parameters of the resultant pulsed CSAR data:

- Fast-time sample spacing: $\Delta_t = 0.825$ ns
- Fast-time frequency bandwidth of the radar: $[\omega_{\min}/2\pi, \omega_{\max}/2\pi] \approx [9, 10.2]$ GHz
- Number of fast-time samples: 98
- Aspect angle sample spacing: $\Delta_\theta = 0.05°$
- Mean slant range of the target area: $R_c = 52.88$ m
- Fast-time samples are calibrated at the mean slant range: $T_c = 2R_c/c = 0.353\ \mu$s
- Depression (slant) angle: $\theta_z = 30°$

Some of the resultant parameters of interest for the CSAR system are as follows:

- Carrier frequency of the radar: $\omega_c/2\pi = 9.6$ GHz
- Number of aspect angle (slow-time) samples for $\theta \in [0, 2\pi)$: $360/\Delta_\theta = 7200$
- Radius of circular synthetic aperture on the ground plane; $R_g = R_c \cos\theta_z = 45.8$ m
- Fast-time interval of sampling: $[T_s, T_f] = [0.312, 0.392]\ \mu$s

The CSAR database was originally intended to be used to determine the radar cross section of targets at various aspect angles and fast-time frequencies. The conventional processing of these CSAR data is over a few degrees (e.g., $\pm 2°$) of the aspect angle domain; the backprojection algorithm or the approximation-based polar format processing is commonly used for the reconstruction. However, in Chapter 7, we will use the turntable data over the full 360 degrees of aspect angles to show the *three-dimensional* potential of CSAR data. We will also show some of the *spotlight* SAR signal properties of the CSAR system.

1

RANGE IMAGING

Introduction

In this chapter we explore the general concept of *range imaging* and its associated digital signal processing issues. The principle behind range imaging or *echo location* was developed early in the twentieth century [bal; edd; sko; woo]. The objective in range imaging is to measure the distance or range of targets in a scene and, perhaps, identify their structural type (e.g., a bird, a small airplane, or a large airplane). In some ways our sensory system provides us with a primitive tool for echo location. For instance, if a person shouts (transmits a burst of acoustic waves) in a mountain area, then multiple echoes can be heard (received) by that person. The timing of these echoes depends on the relative location (range) of the reflectors with respect to that person (transmitter and receiver). Moreover a bigger reflector is likely to produce a stronger echo. (This is not always true.) The latter measure is referred to as a target *reflectivity*.

A human-made echo location or range imaging system is an advanced version of the mountain experiment. The early echo location systems were based on some form of analog circuitry designed to accurately *quantify* the location of a target and its reflectivity. In the radar community a target distance from the radar transmitter/receiver is called the target *range*, and a target reflectivity is called the target radar cross section (RCS). The basic approach in the early radar (or sonar) range imaging systems was to *ping* a target area with a burst of microwave (or acoustic) energy, and then record and display the resultant echoes.

As the time duration of the original ping (transmitted radar signal) is decreased, the user's ability to resolve targets in the display is improved; a smaller ping time duration reduces the overlap between two echoes in the time domain. Unfortunately, the power of the transmitted radar signal decreases when the ping duration is decreased. In this case the echo from a target with a weak RCS (e.g., a small airplane) may not be observable in a noisy environment or in the presence of targets with much stronger RCS values (e.g., larger airplanes or clouds; clouds are also called *clutter*).

The introduction of frequency modulated (FM) radar transmission, for example, *chirp*, solved the above-mentioned resolution versus power problem [edd; kla; sko; s94; woo]. The early FM chirp radars also used analog circuitry for their processing; certain digital signal processing can be performed to reduce errors [bar; carr].

Digital or discrete processing of information brought flexibility and new tools for radar signal processors. The basic approach is to convert analog information, such as a radar-echoed signal, into *sampled* data via a device called an analog to digital (A/D) converter. The sampled data can be stored in a computer. The computer can then perform complex operations, for example, narrowband filtering or Fourier transformation, on the sampled data which cannot be easily implemented via analog circuitries.

Outline

Our discussion in this chapter is on digital signal processing of echoed signals for range imaging. Our first task in range imaging, or in any other imaging problem, is to establish a mathematical relationship between what we measure (echoed signal) and what we desire to retrieve from it (target range and reflectivity). This is called *system modeling* of an imaging problem, and is treated in Section 1.1 for range imaging. Note that the mathematical relationship should be a *good* respresentative of the interaction between the radiating energy (radar waves) and the target in the imaging scene. One should not develop a system model that involves too many approximations in the above-mentioned physical interaction. On the other hand, an overcomplicated physical model and its mathematical interpretation could tie up our analysis with redundant issues.

The next step is to develop a procedure for extracting the desired information, that is, the target range and reflectivity, from the measurement; this is called target *reconstruction*, and is based on a manipulation of the system model of the imaging problem. Reconstruction is also called *inversion* of the system model [s94]. A given imaging problem might have a unique *theoretical* inversion or reconstruction. As we will see, theoretical inverse solutions are not *practical*. Throughout this book we develop reconstruction methods that are based on the classical communication principle of *matched filtering* [car; gru; hel; van]; the matched filtered-based reconstruction methods provide practical inverse solutions that are not sensitive to the additive noise [s94]. For range imaging, the matched filtered reconstruction is outlined in Section 1.2.

The performance of any imaging system is measured by its ability to *resolve* targets; such a measure is called the *resolution* of the imaging system. Resolution in range imaging and its dependence on the radar signal parameters are discussed in Section 1.3.

We treat digital signal processing issues of range imaging systems in Section 1.4. The main questions here are *when* we should start and stop measuring the echoed signal and *how often* we should sample it. We will show that these issues are also related to the radar signal parameters. Once we establish some kind of discrete or digital sense of the measured information base, we are then ready to outline a step-by-step algorithm for the reconstruction; this is done in Section 1.5.

As we pointed out earlier, our approach to the inverse problems of radar in this book is based on matched filtering, and Section 1.2 contains this method for the problem of range imaging. However, there are other practical reconstruction methods. A well-known reconstruction method for range imaging when the radar signal is an FM chirp signal is called time domain *pulse compression*; this method is described in Section 1.6. This method was originally implemented in analog range imaging systems; its digital form is perhaps the most popular range reconstruction method.

Yet, as we will see, the pulse compression-based reconstruction suffers from certain problems in the modern high-resolution radar systems. We bring up this method not for the purpose of reconstruction; it turns out that the principle behind pulse compression has a certain utility in our future discussions. Electronic counter-countermeasure (ECCM) via amplitude modulation of chirp signals are also examined.

The analytical development in the chapter is concluded with a discussion of a more complicated physical system model for the interaction of the radar signal with a target; this is done in Section 1.7. In the new model, targets that exhibit frequency-dependent reflectivity are examined. The implication of this in the reconstructed range image is discussed.

The last section of the chapter, Section 1.8, the Matlab codes for the digital range imaging algorithms of Chapter 1 are discussed. Two sets of examples are considered, and their outputs are enumerated as Figures P1.1–P1.8. In case 1 (Figures P1.1*a*–P1.8*a*), the radar pulse time duration is 0.1 μs; in case 2 (Figures P1.1*b*–P1.8*b*), the radar pulse time duration is 10 μs. The Matlab code also includes a section for frequency-dependent target reflectivity function (Figures P1.9–1.10). The Matlab code outputs, Figures P1.1–P1.10, are cited throughout Sections 1.1 to 1.7 where the analytical development of the range imaging problem is presented.

Mathematical Notations and Symbols

In this chapter we deal with functions of time t and its Fourier counterpart temporal frequency ω. The units of t and ω are *seconds* and *radians* (or *radians/second*), respectively; the unit of $\omega/2\pi$ is *Hertz*. To identify the temporal Fourier transform of a time domain signal, for example, $s(t)$, we use the conventional uppercase form of $S(\omega)$. The forward Fourier transform of a time domain signal with respect to time t is identified via

$$\begin{aligned} S(\omega) &= \mathcal{F}_{(t)}\Big[s(t)\Big] \\ &= \int_{-\infty}^{\infty} s(t)\exp(-j\omega t)\,dt. \end{aligned}$$

The inverse Fourier transform of a frequency domain signal with respect to temporal frequency ω is identified via

$$\begin{aligned} s(t) &= \mathcal{F}^{-1}_{(\omega)}\Big[S(\omega)\Big] \\ &= \frac{1}{2\pi}\int_{-\infty}^{\infty} S(\omega)\exp(j\omega t)\,d\omega. \end{aligned}$$

We also encounter functions of a one-dimensional spatial domain x, which we call the range domain; the unit of x is *meters*. The Fourier counterpart domain for the range x is denoted with k_x which is called the range *spatial frequency* or *wavenumber* domain; the unit of k_x is *radians/meter*. The forward Fourier transform of a spatial domain signal, for example, $f(x)$, with respect to the range x is expressed via

$$\begin{aligned} F(k_x) &= \mathcal{F}_{(x)}\Big[f(x)\Big] \\ &= \int_{-\infty}^{\infty} f(x)\exp(-jk_x x)\,dx. \end{aligned}$$

The inverse Fourier transform of a spatial frequency domain signal with respect to spatial frequency k_x is identified via

$$\begin{aligned} f(x) &= \mathcal{F}^{-1}_{(k_x)}\Big[F(k_x)\Big] \\ &= \frac{1}{2\pi}\int_{-\infty}^{\infty} F(k_x)\exp(jk_x x)\,dk_x. \end{aligned}$$

We will also see scenarios in which the range domain is a *scale* transformation of the time domain; for example,

$$x = at,$$

where a is a constant (i.e., it does not vary with time t). Note that

$$t = \frac{x}{a}.$$

Suppose that the functional mapping of a time domain signal $s(t)$ into the range domain is

$$\begin{aligned} g(x) &= s(t) \\ &= s\left(\frac{x}{a}\right). \end{aligned}$$

Then the Fourier transform of $s(t)$ with respect to t, that is, $S(\omega)$, and the Fourier transform of $g(x)$ with respect to x, that is, $G(k_x)$, are related via the following functional mapping:

$$G(k_x) = |a|S(\omega),$$

where

$$k_x = \frac{\omega}{a}.$$

We can also write the functional mapping in the frequency domain via

$$G(k_x) = |a| S\,(ak_x)\,.$$

Throughout this book, we often encounter *bandpass* signals in the various domains. A signal is said to be a bandpass signal if its forward or inverse Fourier transform is nonzero within a *finite* interval. For example, if the time domain signal $s(t)$ is a bandpass signal, then

$$S(\omega) \begin{cases} \neq 0 & \text{for } \omega \in [\omega_c - \omega_0, \omega_c + \omega_0], \\ = 0 & \text{otherwise.} \end{cases}$$

The constants (ω_c, ω_0) identify the extent of the region or the frequency band in the ω domain over which the signal $S(\omega)$ is nonzero; ω_c is the *center* point of the band or the *carrier* frequency, and $\pm\omega_0$ is the *size* of the band or the *baseband bandwidth*. The symbol

$$\omega \in [\omega_c - \omega_0, \omega_c + \omega_0]$$

is used to identify this frequency band, which is also called the *support* band of $S(\omega)$.

We also encounter examples of bandpass signals in the frequency domain. For instance, if the signal $F(k_x)$ is a bandpass signal, then

$$f(x) \begin{cases} \neq 0 & \text{for } x \in [X_c - X_0, X_c + X_0], \\ = 0 & \text{otherwise.} \end{cases}$$

The constants (X_c, X_0) identify the support region of $f(x)$ in the range domain; X_c is the center point of the support, and $2X_0$ is the size of the support region.

The following is a list of mathematical symbols used in this chapter, and their definitions:

$a(t)$	Amplitude signal of transmitted LFM chirp radar signal
B_0	Radar signal half bandwidth in Hertz; radar signal baseband bandwidth in Hertz is $\pm B_0$ $(B_0 = \omega_0/2\pi)$
c	Wave propagation speed $c = 3 \times 10^8$ m/s for radar waves $c = 1500$ m/s for acoustic waves in water $c = 340$ m/s for acoustic waves in air
$f(x)$	Target function in range domain
$F(k_x)$	Fourier transform of target function
$f_m(x)$	Mine matched-filtered target function
$F_m(k_x)$	Fourier transform of mine matched-filtered target function

$f_0(x)$	Ideal target function in range domain
$F_0(k_x)$	Fourier transform of ideal target function
k	Wavenumber: $k = \omega/c$
k_x	Spatial frequency or wavenumber domain for range x
n	Index representing a specific target [used with (σ_n, x_n)]
N	Number of measurement time bins
N_a	Number of phase harmonics in amplitude signal of transmitted chirp radar signal
$p(t)$	Transmitted radar signal
$P(\omega)$	Fourier transform of transmitted radar signal
$\mathrm{psf}(x)$	Point spread function in range domain
$\mathrm{psf}_t(t)$	Point spread function in time domain for matched filtering
$\mathrm{psf}_n(x)$	Point spread function of nth target in range domain
$\mathrm{psf}_\omega(\omega)$	Point spread function in frequency domain for pulse compression
$s(t)$	Echoed signal from target area
$S(\omega)$	Fourier transform of echoed signal from target area
$s_c(t)$	Pulse-compressed signal
$S_c(\omega)$	Fourier transform of pulse-compressed signal
$s_{cb}(t)$	Baseband pulse-compressed signal
$S_{cb}(\omega)$	Fourier transform of baseband pulse-compressed signal
$s_b(t)$	Baseband-echoed signal
$S_b(\omega)$	Fourier transform of baseband-echoed signal
$s_M(t)$	Matched-filtered echoed signal
$S_M(\omega)$	Fourier transform of matched-filtered echoed signal
$s_{Mb}(t)$	Baseband matched-filtered echoed signal
$S_{Mb}(\omega)$	Fourier transform of baseband matched-filtered echoed signal
$s_n(t)$	Echoed signal from nth target
$S_n(\omega)$	Fourier transform of echoed signal from nth target
$s_0(t)$	Reference-echoed signal
$S_0(\omega)$	Fourier transform of reference-echoed signal
$s_{0b}(t)$	Baseband reference-echoed signal
$S_{0b}(\omega)$	Fourier transform of baseband reference-echoed signal
$\mathrm{sinc}(\cdot)$	sinc function: $\mathrm{sinc}(a) = \sin(\pi a)/\pi a$
t	Time domain
t_i	Measurement time bins for baseband-echoed signal
t_{ic}	Measurement time bins for baseband pulse-compressed signal
t_n	Round trip delay of echoed signal from nth target
T_f	Ending point of time sampling for echoed signal
T_p	Radar signal pulse duration

T_s	Starting point of time sampling for echoed signal
T_x	Range swath echo time period
x	Range domain
x_i	Range bins
x_n	Range of nth target
X_c	Midrange or center point of target area
X_0	Half-size of target area in range domain; size of target area in range domain is $2X_0$
y	Cross-range domain
α	Quadratic coefficient of chirp signal phase
β	Linear coefficient of chirp signal phase
$\delta(\cdot)$	Delta or impulse function
Δ_t	Sample spacing of echoed signal
Δ_{tc}	Sample spacing of pulse-compressed signal
Δ_x	Range resolution
Δ_ω	Sample spacing of Fourier transform of echoed signal
η_n	Resonating cylinder parameter (related to its physical property)
$\mathcal{F}$	Forward Fourier transform operator
$\mathcal{F}^{-1}$	Inverse Fourier transform operator
γ_n	Resonating cylinder parameter (related to its diameter)
ω	Temporal frequency domain for time t
ω_c	Radar signal carrier or center frequency
ω_{cc}	Pulse-compressed signal carrier or center frequency
ω_i	Temporal frequency bins
$\omega_{ip}(t)$	Instantaneous frequency of transmitted chirp pulse
$\omega_{in}(t)$	Instantaneous frequency of echoed signal from nth target
ω_0	Radar signal half bandwidth in radians; radar signal baseband bandwidth is $\pm\omega_0$
ω_{0c}	Pulse-compressed signal half bandwidth in radians; pulse-compressed signal baseband bandwidth is $\pm\omega_{0c}$
ρ_ℓ	Resonating cylinder echo signature
$\sigma_m^*(\omega)$	Mine-matched filter
σ_n	Reflectivity of nth target (invariant in radar frequency)
$\sigma_n(\omega)$	Reflectivity of nth target (radar frequency-dependent)

1.1 SYSTEM MODEL

Consider the imaging scenario in a two-dimensional spatial domain which is depicted in Figure 1.1. In this problem, a set of targets are located at a *fixed known* cross-range

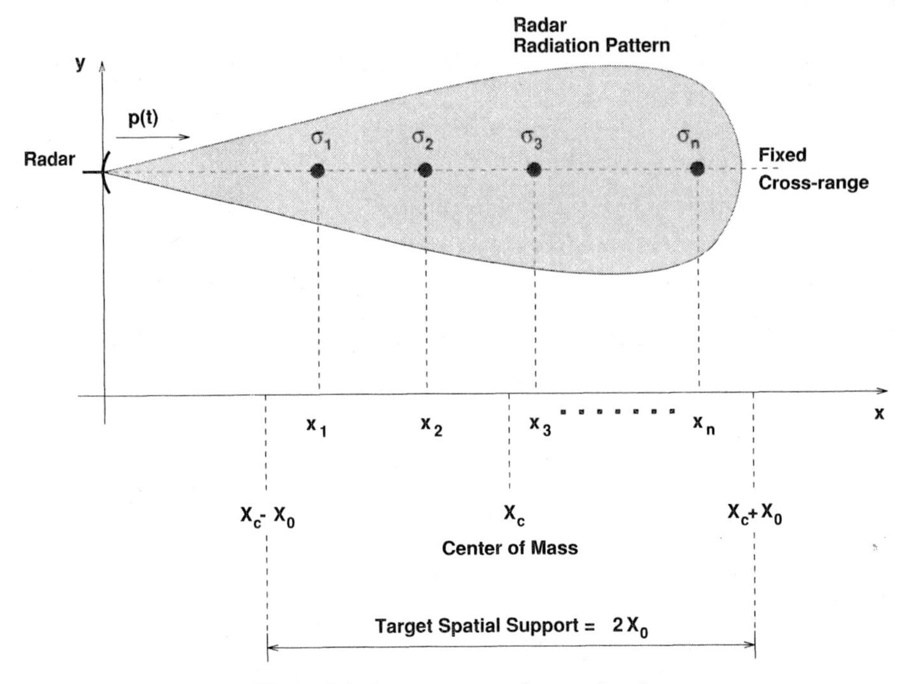

Figure 1.1. System geometry for range imaging.

position, for example, y. The range and reflectivity values of these targets, labeled (x_n, σ_n), $n = 1, 2, \ldots$, are *unknown*. The targets are assumed to reside within a finite spatial region $[X_c - X_0, X_c + X_0]$; X_c is the center of mass or the mean range of the target area, and $2X_0$ is the size of the target area.

- For a practical radar system in the three-dimensional spatial domain, the target range domain area $[X_c - X_0, X_c + X_0]$ is called the *radar swath*. The radar swath is dictated by what we refer to as the radar *radiation pattern*. A radar radiation pattern is analogous to the *field of view* of a camera, that is, the target area which is *observable* to the radar. The radiation pattern of a radar system depends on the coordinates of the radar as well as its type and dimensions [edd; sko; s94]. This will be discussed in Chapter 3.

A transmitting/receiving radar is located at the same known target area cross-range, y. We want to utilize the variations of the radar signal frequency for range imaging, that is, by identifying the range x_n and reflectivity σ_n of the targets in the imaging scene. Clearly the best we can do to identify the range and reflectivity of the targets is to form the following range domain function:

$$f_0(x) = \sum_n \sigma_n \delta(x - x_n),$$

which we refer to as the *ideal* target function; an example is shown in Figure 1.2*a*. The reason for calling this target function *ideal* will become obvious in our future discussion.

- One could define the center of mass of the target area in the range domain X_c for a continuous target function $f_0(x)$ via

$$X_c = \frac{\int_x x|f_0(x)|^2\, dx}{\int_x |f_0(x)|^2\, dx},$$

and for a discrete target model via

$$X_c = \frac{\sum_n x_n|\sigma_n|^2}{\sum_n |\sigma_n|^2}.$$

In general, the radar mean swath is not the same as the target center of mass. In practice, the user has some approximate a priori knowledge of the target center of mass in the cross-range domain, that is, X_c.

There are many hardware- or software-based methods to form the database that corresponds to the variations of the radar signal frequency. One way to tap into this signal subspace (radar band) is to utilize a large bandwidth-pulsed signal, labeled $p(t)$. We denote the duration of this pulsed signal by T_p; an example of a radar-pulsed signal is shown in Figure 1.3*a*. Another method for range imaging is based on *frequency stepping* [s94; weh]. Each method has certain advantages and disadvantages. For instance, the frequency stepping method is useful when the user does not have access to a high-speed A/D converter to record the incoming large-bandwidth echoed signals. We formulate the range imaging problem via illumination with a large-bandwidth radar signal.

- The information content of range imaging via utilizing a large-bandwidth radar signal is equivalent to the information content of range imaging via frequency stepping. In fact one database can be *digitally synthesized* from the other.

The principle of *range imaging* or *echo location* through varying the radar frequency (bandwidth) is well established [bla; edd; sko; s94; woo]. The idea is that if we illuminate a one-dimensional target area in the x (range) domain with $p(t)$, the received echoed signal has the following form:

$$s(t) = \sum_n \sigma_n p\left(t - \frac{2x_n}{c}\right),$$

where σ_n and x_n are, respectively, the nth target reflectivity and range and c is the propagation speed; $2x_n/c$ is the round-trip delay for the radar signal to travel from the radar to the nth target and back to the radar. If we define the range domain x to

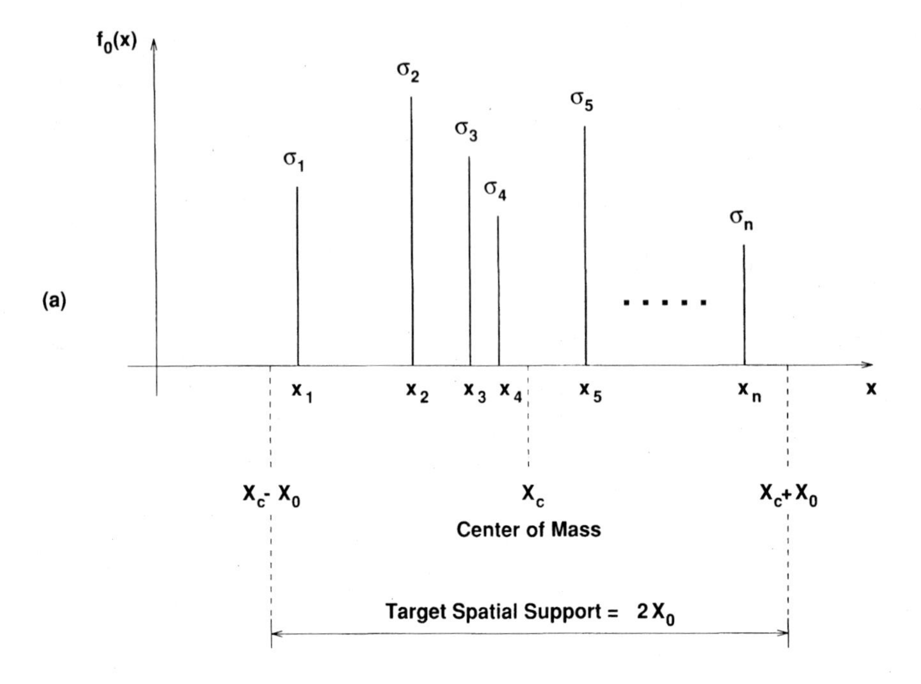

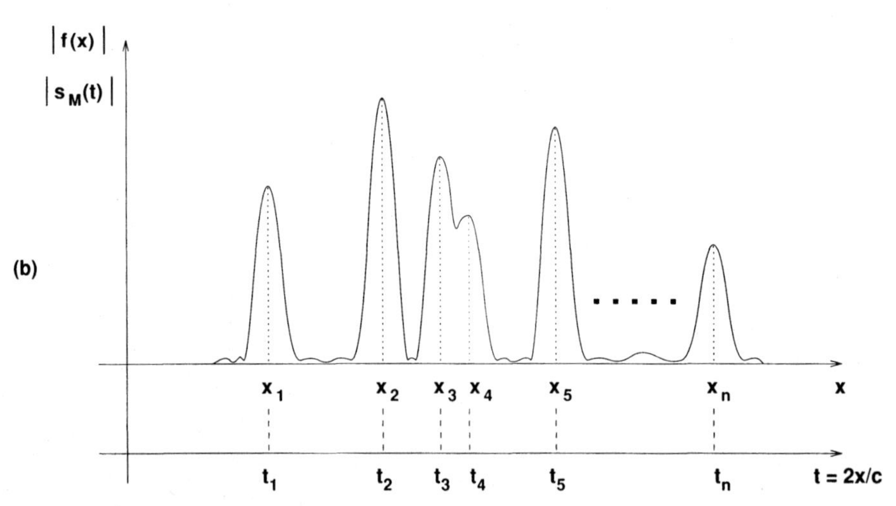

Figure 1.2. (a) Ideal target function $f_0(x)$; (b) matched-filtered signal $s_M(t)$ and reconstructed target function $f(x)$.

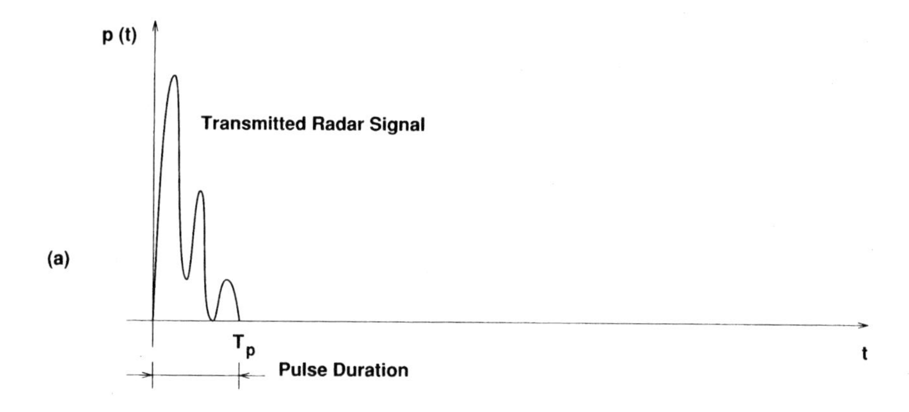

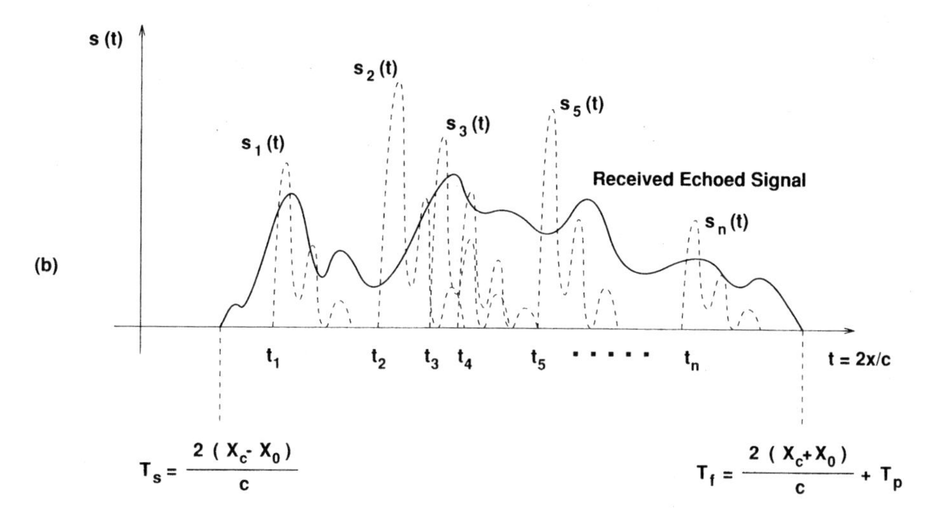

Figure 1.3. (a) Transmitted radar pulsed signal $p(t)$; (b) measured echoed signal $s(t)$.

be the linear transformation of the time axis,

$$x = \frac{ct}{2},$$

then using the expression for the ideal target function $f_0(x)$ in the echoed signal model, we can write

$$\begin{aligned} s(t) &= f_0(x) * p(t) \\ &= f_0\left(\frac{ct}{2}\right) * p(t) \end{aligned}$$

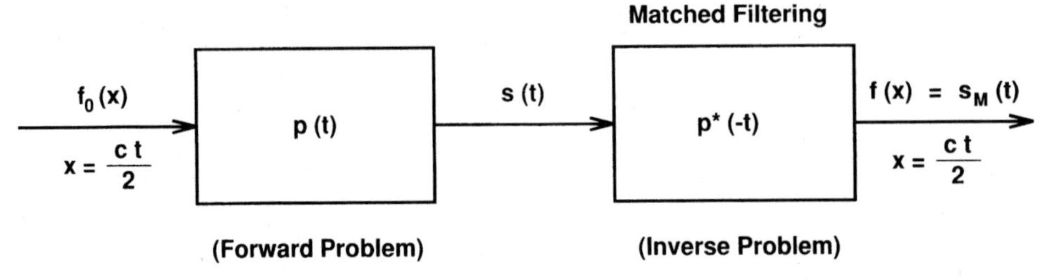

Figure 1.4. Forward and inverse system models for range imaging.

where $*$ denotes convolution in the time domain. (Constants are suppressed in the above equation.) This functional property is depicted in Figure 1.4, and is identified as the *forward* problem. The forward problem of a system is to determine its output from the knowledge of its impulse response and the applied input.

An example of the echoed signal $s(t)$ is shown in Figure 1.3*b* (Figure P1.1). In this figure, the dashed lines represent the echoed signal from the individual reflectors, that is,

$$s_n(t) = \sigma_n p\left(t - \frac{2x_n}{c}\right), \qquad n = 1, 2, \ldots.$$

The received echoed signal is the sum of all these individual returns:

$$s(t) = \sum_n s_n(t).$$

In Figure 1.3*b* the signal $s(t)$ is shown as an arbitrary solid line. Also in this figure the time point

$$t_n = \frac{2x_n}{c}$$

is the round-trip delay for the echoed signal from the nth target.

In practice, the number of reflectors is infinite, since the target area, such as a forest region with human-made targets in it, is composed of a continuum of reflectors. Thus a more appropriate system model for the echoed signal is

$$\begin{aligned} s(t) &= \int_x f_0(x) p\left(t - \frac{2x}{c}\right) dx \\ &= f_0\left(\frac{ct}{2}\right) * p(t), \end{aligned}$$

where $f_0(x)$ is the target reflectivity function in the continuous range domain. It is not difficult to show that a discrete or continuous system model represents the same phenomenon (for engineers, not mathematicians) and that the analysis that follows is suitable for both models. *Why are we using a discrete model?* This is just for notational simplicity. It turns out that it is easier to employ a single *sum* instead of an *integral* (one-dimensional range or cross-range imaging), or *integrals* (multidimensional SAR imaging), to represent the target region.

The true physics of scattering of a radar signal by a target would indicate that the above model (discrete or continuous) is too simplistic. In fact many assumptions must be made in order to end up with the above system model. For instance, one should associate an amplitude factor x_n^{-2} with $s_n(t)$. (The origin of this amplitude factor can be traced to the divergence of the radar wave in the *three-dimensional* spatial domain Green function [mor; s94]; the one-dimensional Green function does not have such a factor.) However, in most practical radar systems, the variations of this amplitude factor within the radar range swath, that is, $x \in [X_c - X_0, X_c + X_0]$ where $X_0 \ll X_c$, is negligible compared with the radar system errors such as the quantization noise, thermal additive noise, and multiplicative noise (instability of the radar). In our discussion we suppress the amplitude factor x_n^{-2}.

There is one assumption in the system model which will be removed later. The assumption is that the target reflectivity σ_n [or the continuous form $f_0(x)$] is invariant in radar frequency. This is not the case in practice. In particular, this issue becomes significant in wideband radar systems. At this point in our discussion, we will put this on the back burner, and examine more basic principles of range imaging. We will bring up the issue of radar frequency-dependent target reflectivity function later in this chapter and in Chapter 3.

We now continue with the analysis of the above simplified discrete system model. The Fourier transform of the echoed signal is

$$S(\omega) = P(\omega) \sum_n \sigma_n \exp\left(-j\omega \frac{2x_n}{c}\right).$$

Provided that $p(t)$ has nonzero components at all ω (frequency), that is, its bandwidth is *infinite*, we can form the ideal target function via

$$\begin{aligned} \mathcal{F}^{-1}_{(\omega)}\left[\frac{S(\omega)}{P(\omega)}\right] &= \sum_n \sigma_n \delta\left(t - \frac{2x_n}{c}\right) \\ &= f_0\left(\frac{ct}{2}\right), \end{aligned}$$

which carries information on the location and reflectivity of the targets. The reconstructed ideal target function is a linear mapping of the time axis of the above function, that is,

$$f_0(x) = \sum_n \sigma_n \delta(x - x_n),$$

where

$$x = \frac{ct}{2}.$$

We will show that $f_0(x)$, the ideal target function, cannot be formed in practice due to the practical limitations on the radar signal; that is, the radar signal has a *finite* bandwidth. In fact the subscript of *zero* in $f_0(x)$ is used to showcase the fact that the range resolution for a radar with an infinite bandwidth is zero.

1.2 RECONSTRUCTION VIA MATCHED FILTERING

The division of $S(\omega)$ by $P(\omega)$, which resulted in the target information, is called *source deconvolution*, where the source is $P(\omega)$. In practice, a signal with *infinite* bandwidth does not exist. In fact all radar signals are *bandpass* signals. An example of a practical $P(\omega)$ is shown in Figure 1.5*a* (Figure P1.4); the midpoint of the radar signal band, ω_c, is called the *carrier* or *center* frequency; and the size of the radar signal band, $\pm\omega_0$, is called the *baseband bandwidth* of the radar signal.

One of the practical methods for range imaging is based on an operation called *matched filtering*. Matched filtering also has certain desirable properties when the measured signal is corrupted with an additive white noise [car; hel; s94; van]. The matched filtering operation is formed via

$$\begin{aligned} s_M(t) &= \mathcal{F}^{-1}_{(\omega)}\Big[S(\omega)P^*(\omega)\Big] \\ &= \mathcal{F}^{-1}_{(\omega)}\left[\sum_n \sigma_n |P(\omega)|^2 \exp\left(-j\omega\frac{2x_n}{c}\right)\right] \\ &= \sum_n \sigma_n \text{psf}_t\left(t - \frac{2x_n}{c}\right), \end{aligned}$$

where P^* is the complex conjugate of P, and

$$\text{psf}_t(t) = \mathcal{F}^{-1}_{(\omega)}\Big[|P(\omega)|^2\Big],$$

is called the *point spread function* of the range imaging system in the time domain for matched filtering. An example of the magnitude of the matched-filtered signal $s_M(t)$ is shown in Figure 1.2*b* (Figure P1.6); the Fourier spectrum of the matched filtered signal is shown in in Figure 1.5*c* (Figure P1.5). Note that the matched-filtered signal can also be expressed as

$$\begin{aligned} s_M(t) &= s(t) * p^*(-t) \\ &= f_0\left(\frac{ct}{2}\right) * \text{psf}_t(t), \end{aligned}$$

where $*$ denotes time domain convolution. The signal $p^*(-t)$, which is the inverse Fourier transform of $P^*(\omega)$, is called the *matched filter*.

The matched filtering operation is depicted in Figure 1.4 and is identified as the *inverse* problem. *The inverse problem of a system is to determine its input (or impulse response) from the knowledge of its impulse response (or input) and the output.*

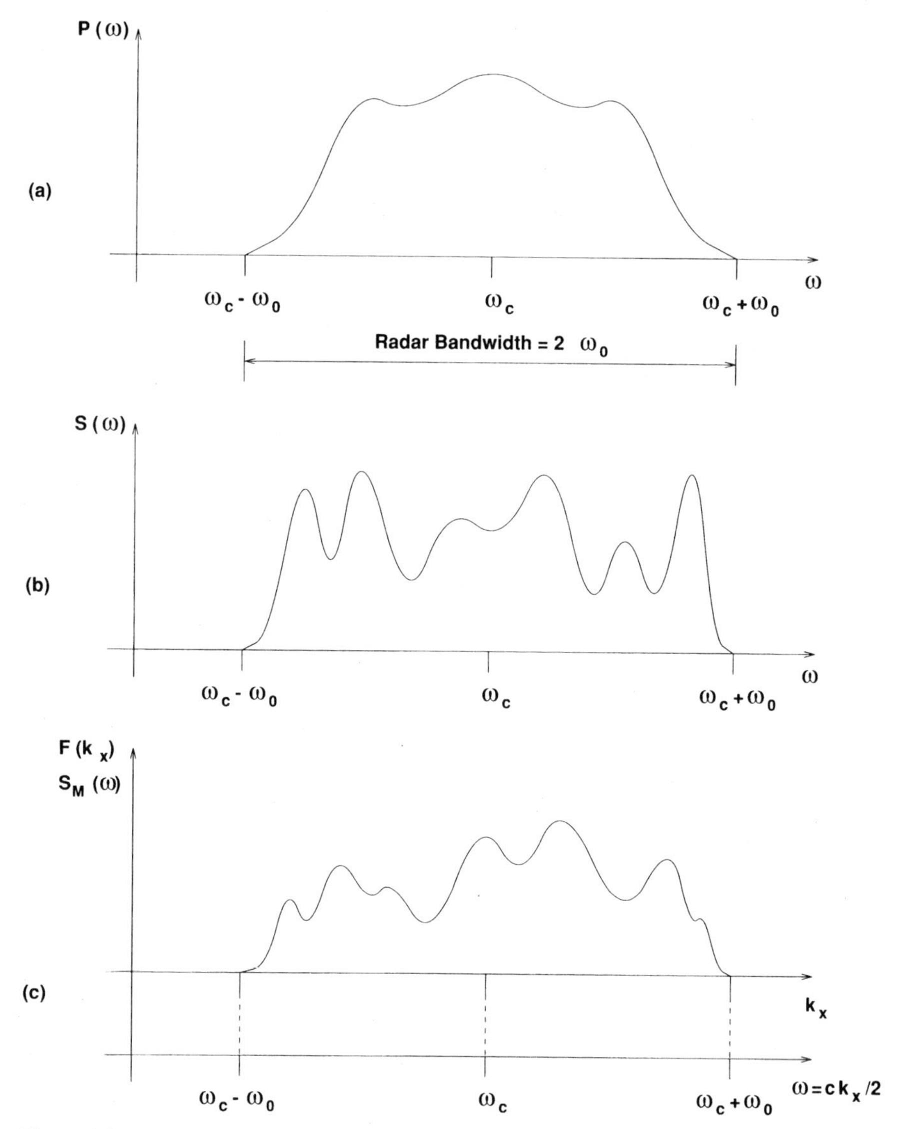

Figure 1.5. (a) Fourier transform of transmitted radar signal $P(\omega)$; (b) Fourier transform of measured echoed signal $S(\omega)$; (c) Fourier transform of reconstructed target function $F(k_x)$.

The point spread function depends on the spectral shape of the radar signal. For instance, when $|P(\omega)| = 1$ within the radar bandwidth, that is, $\omega \in [\omega_c - \omega_0, \omega_c + \omega_0]$, the point spread function becomes

$$\text{psf}_t(t) = \exp(j\omega_c t)\text{sinc}\left(\frac{\omega_0 t}{\pi}\right),$$

where

$$\text{sinc}(a) = \frac{\sin(\pi a)}{\pi a},$$

is the sinc function.

The target function is formed from the matched-filtered signal $s_M(t)$ via a scale transformation of the time domain as shown in Figure 1.2*b* (Figure P1.6):

$$\begin{aligned} f(x) &= s_M(t) \\ &= \sum_n \sigma_n \text{psf}_t \left[\frac{2}{c}(x - x_n) \right]. \end{aligned}$$

Here range mapping and its spatial frequency domain are defined by

$$\begin{aligned} x &= \frac{c}{2}t, \\ k_x &= \frac{2}{c}\omega = 2k, \end{aligned}$$

where $k = \omega/c$ is called the wavenumber. Since we are concerned with target information in the range x domain, it is more convenient to express the point spread function in the x domain instead of the t domain; that is,

$$\text{psf}(x) = \mathcal{F}^{-1}_{(\omega)} \left[|P(\omega)|^2 \right],$$

where $x = ct/2$. We call psf(x) the point spread function in the range domain. In this case the target function formed via matched filtering becomes

$$f(x) = \sum_n \sigma_n \,\text{psf}(x - x_n).$$

- For display purposes, the magnitude of the complex signal $f(x)$ is used; an example is shown in Figure 1.2*b*. In this case the effect of the carrier term $\exp(j\omega_c t)$ in the point spread function becomes "almost" transparent to the user. However, the complex version (real and imaginary parts) of the signal $f(x)$ carries useful *coherent* (magnitude and phase) information that can be used to identify the type of a target. This will be discussed in SAR imaging. The phase of the complex signal $f(x)$ is also used in monopulse radar systems for height finding, moving target detection, and so on.

Note that

$$f(x) = f_0(x) * \text{psf}(x),$$

where $*$ denotes convolution in the range (or time) domain, and

$$\begin{aligned} F(k_x) &= S_M(\omega) \\ &= F_0(k_x)|P(\omega)|^2. \end{aligned}$$

Examples of $S(\omega)$ and $F(k_x)$ are shown in Figure 1.5*b* and *c*, respectively (Figure P1.5). Note that as the bandwidth of $p(t)$ increases, the psf of the imaging system gets *sharper*; that is, the spread of psf(x) in the range (or time) domain gets smaller. In this case a sharper psf improves the ability of the user to *resolve* targets in the range domain, that is, the range resolution. This measure of resolvability will be quantified next.

1.3 RANGE SOLUTION

Range resolution is a measure of the resolvability of the targets in the range domain by a given radar signal. As we mentioned before, the spread of the point spread function, psf(x), determines this measure. Since psf(x) is the inverse Fourier transform of $|P(\omega)|^2$, with $x = ct/2$, and the baseband support of $P(\omega)$ is $2\omega_0$, a simple (crude) radar rule of thumb says that the support of psf(x) in the time domain is approximately within a time interval of π/ω_0. (Recall that the forward or inverse Fourier transform of a rectangular pulse function is a sinc function; the size of the sinc function main lobe is *inversely* proportional to the rectangular pulse function duration.)

Then, with $x = ct/2$, this translates into the following range resolution:

$$\begin{aligned} \Delta_x &= \frac{c}{2}\frac{\pi}{\omega_0} \\ &= \frac{c}{4B_0}, \end{aligned}$$

where $B_0 = \omega_0/2\pi$ is the radar signal baseband bandwidth in Hertz. If $\pm\omega_0$ is the 3 dB baseband bandwidth of the radar signal, then Δ_x represents the *nominal* range resolution of the radar system. In practice, this range resolution measure is not achievable due to uncertainties in the knowledge of the transmitted radar signal.

1.4 DATA ACQUISITION AND SIGNAL PROCESSING

Our signal processing is based on digital storage and manipulation of the echoed signal $s(t)$. Due to limitations of computer storage, we have to translate the analog $s(t)$ into a finite-length array, that is, a finite number of samples of $s(t)$, which can then be handled by a digital computer. The choice of these finite samples are dictated by two factors: the bandwidth of the radar signal [support band of $P(\omega)$ in the frequency ω domain], and the support of the target area or $f(x)$ in the range domain. This is described next.

- Note that $F(k_x)$ has a finite support in the $k_x = 2\omega/c$ domain if $P(\omega)$ has a finite support in the ω domain. Thus, to state that both $P(\omega)$ and $f(x)$ have finite support regions in their respective domains is contradictory. This goes back to the classical problem of representing an analog time or spatial domain signal via a finite sequence, and its Fourier transform via another finite sequence (the discrete Fourier transform, DFT, of the time/space sequence), and *tolerable* aliasing errors which we have to live with [s94].

Time Domain Sampling

We first examine the role of the radar bandwidth. As we pointed out earlier, the transmitted radar signal has energy within a finite band in the frequency domain (bandlimited). We denote the frequency domain support band of radar signal via

$$\omega \in [\omega_c - \omega_0, \omega_c + \omega_0],$$

where ω_c is the radar carrier frequency and $\pm\omega_0$ is the radar baseband bandwidth; see Figure 1.5*a* (Fig. P1.4).

The echoed signal $s(t)$ is a linear combination of shifted versions of the transmitted radar signal $p(t)$; that is,

$$s(t) = \sum_n \sigma_n p\left(t - \frac{2x_n}{c}\right),$$

where σ_n's are the coefficients of the linear model and $2x_n/c$'s are the shifts. Thus the support of its Fourier transform, $S(\omega)$, is also within the radar band $[\omega_c - \omega_0, \omega_c + \omega_0]$, with the baseband bandwidth of $\pm\omega_0$; see Figure 1.5*b* (Figure P1.3). To sample the echoed signal $s(t)$ without aliasing, the following Nyquist sampling criterion should be satisfied:

$$\Delta_t \le \frac{\pi}{\omega_0}.$$

In practice, $\pm\omega_0$ is the 3-dB band of the radar signal around its carrier. In this case a finer sample spacing may be used to provide a *guard* band against the small energy of the radar signal which falls outside $[\omega_c - \omega_0, \omega_c + \omega_0]$. It is worth noting that the range resolution and the time domain sampling are related via

$$\begin{aligned}\Delta_x &= \frac{c}{2}\max(\Delta_t)\\ &= \frac{c\pi}{2\omega_0}.\end{aligned}$$

Time Interval of Sampling

Next we bring the target support area into the picture. The spatial domain support band of the target function $f_0(x)$ is (see Figure 1.2)

$$x \in [X_c - X_0, X_c + X_0],$$

where X_c is the target area mean range and $2X_0$ is the size of the target area. The echoed signal from the closest reflector at $x = X_c - X_0$ arrives at the receiver at the time (see Figure 1.3*b*)

$$T_s = \frac{2(X_c - X_0)}{c}.$$

While the echoed signal from the farthest reflector at $x = X_c + X_0$ arrives at the receiver at the time $2(X_c + X_0)/c$, it lasts until the time (see Figure 1.3*b*)

$$T_f = \frac{2(X_c + X_0)}{c} + T_p,$$

where T_p is the duration of the pulsed radar signal. The echoed signals that are due to the reflectors that reside between the closest and farthest reflectors fall between the time points T_s and T_f.

Thus, to capture the echoed signals from all of the reflectors in the target region $x \in [X_c - X_0, X_c + X_0]$, we have to acquire the time samples of the echoed signal $s(t)$ in the following time interval:

$$t \in [T_s, T_f].$$

Note that the length of this time interval is

$$T_f - T_s = \frac{4X_0}{c} + T_p.$$

(The pulse duration T_p is small as compared with $4X_0/c$ in some applications.) In practice, one might use a larger time interval to reduce boundary (wraparound) errors of DFT processing; this results in a time or range domain guard band (zero-padding).

Number of Time Samples

We refer to the time points separated by Δ_t within the time interval $[T_s, T_f]$ as the measurement time bins:

$$t_i = T_s + (i - 1)\Delta_t, \qquad i = 1, 2, \ldots, N,$$

where N, the number of the measured samples, is the smallest integer which is less than

$$\frac{T_f - T_s}{\Delta_t}.$$

Since we perform DFT and inverse DFT on the resultant database, the numerical errors could be reduced by selecting N to be an *even* number. (There are problems with the *fftshift* routine in Matlab when N is an odd number.) Thus one may use the following for determining the required number of time bins:

$$N = 2\left\lceil \frac{T_f - T_s}{2\Delta_t} \right\rceil,$$

where $\lceil a \rceil$ denotes the smallest integer which is larger than a. Using the transformation $x = ct/2$, the measurement time bins translate into the following *range bins* where the target function $f(x)$ is formed:

$$\begin{aligned} x_i &= \frac{ct_i}{2} \\ &= (X_c - X_0) + (i-1)\frac{c\Delta_t}{2}, \qquad i = 1, 2, \ldots, N. \end{aligned}$$

We call these *measurement* range bins. We will identify another set of range bins in our discussion of digital reconstruction.

The temporal frequency bins are separated by Δ_ω, where this frequency spacing can be identified from the DFT sampling equation [s94]:

$$\Delta_\omega = \frac{2\pi}{N\Delta_t}.$$

Then the frequency bins are at

$$\omega_i = \omega_c + \left(i - \frac{N}{2} - 1\right)\Delta_\omega, \qquad i = 1, 2, \ldots, N.$$

1.5 RECONSTRUCTION ALGORITHM

The initial processing of the echoed signal is done at the hardware level via its baseband conversion; that is,

$$s_b(t) = s(t)\exp(-j\omega_c t).$$

This operation converts the bandpass signal $s(t)$ into a lowpass signal which is suitable for analog to digital (A/D) conversion [s94]. After A/D conversion the computer stores a finite number of samples of the baseband-echoed signal $s_b(t)$ within the time bins t_i, $i = 0, 1, 2, \ldots, N-1$.

For the matched filtering, we need to define the matched filter $p^*(-t)$ in discrete form in the computer. Then one can perform discrete convolution on $s_b(t)$ and $p^*(-t)$. While this is a perfectly valid approach for range imaging, we present an implementation of the range reconstruction via a procedure which we refer to as *reference signal matched filtering*. We will also utilize this concept for the cross-range imaging and SAR-imaging problems in future chapters. The reference signal matched filtering provides a *physical* interpretation of the reconstruction problem in these imaging systems and serves as a unifying tool to solve them. The reference signal matched filtering for range imaging is described next.

Suppose that there is a unit reflector at the range $x = 0$. Then the echoed signal from this target has the form $s(t) = p(t)$. The matched filtering of this single echo with $p^*(-t)$ results in a sinc-like *blip* [i.e., $\text{psf}_t(t)$] at the origin $t = 0$, which indicates that there is a target at $x = 0$. Note that matched filtering $s(t)$ with $p^*(-t)$ is equivalent to *correlating* $s(t)$ with $p^*(t)$; the output shows a strong correlation at $t = 0$.

Now, if the unit reflector is moved to X_c (the target area mean range), then the echoed signal will be shifted by $2X_c/c$; that is, the echoed signal becomes $p[t - (2X_c/c)]$. If we incorporate the same shift in the correlator, that is, we use $p^*[t - (2X_c/c)]$ as the *correlating reference signal*, then we should still see a blip (correlation) at $t = 0$. Correlating the echoed signal from a target at, for example, $X_c + x_n$ with the reference signal $p^*[t - (2X_c/c)]$ results in a sinc-like blip at $t = 2x_n/c$.

We now exploit this correlation with the reference signal from the target area mean range in the digital implementation of range imaging. For this, we first compute (in the software) the reference-echoed signal from a unit reflector at the center of the target region, that is (Figure P1.2),

$$s_0(t) = p\left(t - \frac{2X_c}{c}\right)$$

and its baseband version

$$s_{0b}(t) = s_0(t)\exp(-j\omega_c t)$$

at the measurement time bins t_i, $i = 1, 2, \ldots, N$.

Next the correlation (matched filtering) is performed in the software via

$$s_{Mb}(t) = \mathcal{F}^{-1}_{(\omega)}\Big[S_b(\omega)S^*_{0b}(\omega)\Big],$$

where $S^*_{0b}(\omega)$ is the matched filter in the frequency domain, and $s_{Mb}(t)$ is the baseband version of the matched-filtered signal $s_M(t)$ (Figure P1.6).

It is critical to view the above operation as a correlation with the reference signal, which is the echo from the unit reflector at the target area mean range X_c. Due to this, the sample in the middle of the $|s_{Mb}(t_i)|$ array (note that $|s_M(t_i)| = |s_{Mb}(t_i)|$), that is, the sample number $i = (N/2) + 1$, corresponds to the blip signature of the target at the midrange X_c. Thus the target image $f(x)$ is formed at the following range bins:

$$x_i = X_c + \left(i - \frac{N}{2} - 1\right)\frac{c\Delta_t}{2}, \qquad i = 1, 2, \ldots, N,$$

which are different from the range bins which we identified earlier. We call these *reconstruction* range bins.

The measurement and reconstruction range bin will be the same if the user chooses the reference correlating signal to be

$$s_0(t) = p(t - t_{(N/2)+1})$$

Both definitions (or similar definitions) of the reference correlator $s_0(t)$ work for what we intend to extract from the measured echoed signal. However, the user should stay with one, and use its corresponding range bins.

Algorithm Figure 1.6 is a block diagram representation of a range-imaging system from the hardware that is used for data acquisition to the discrete-time reconstruction in the software. The digital reconstruction algorithm (software) for range imaging can be summarized by the following:

Step 1. Compute the discrete Fourier transform (DFT) of the measured baseband-echoed signal to obtain $S_b(\omega)$.

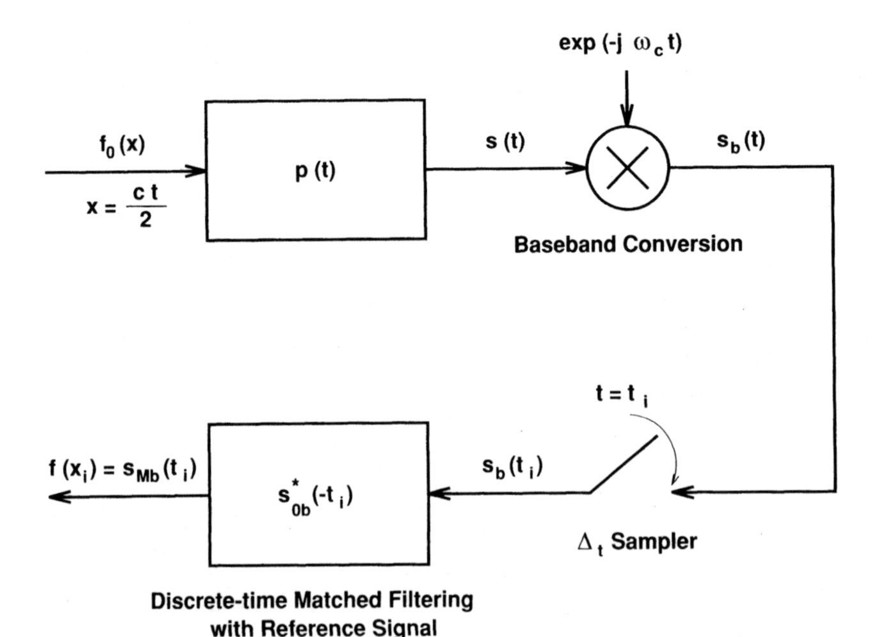

Figure 1.6. Block diagram representation of a range imaging system.

Step 2. Compute the reference baseband-echoed signal $s_{0b}(t)$ *at the same time (range) bins as the measured* $s_b(t)$.

Step 3. Compute the DFT of the reference baseband-echoed signal to obtain $S_{0b}(\omega)$. *(You may use the analytical expression for* $S_{0b}(\omega)$ *if it is available.)*

Step 4. Compute the inverse DFT of the product

$$S_b(\omega) S_{0b}^*(\omega)$$

to obtain the baseband matched-filtered signal $s_{Mb}(t)$.

Step 5. Display $|s_{Mb}(t)|$ *versus appropriate range bins* $x = ct/2$.

1.6 RECONSTRUCTION VIA PULSE COMPRESSION FOR CHIRP SIGNALS

Signal Model

A pulsed chirp or linear frequency modulated (LFM) radar signal is defined as

$$p(t) = a(t) \exp(j\beta t + j\alpha t^2),$$

where $a(t) = 1$ for $0 \le t \le T_p$ and is zero otherwise; T_p is the chirp pulse duration [bar; bla; edd; kla; sko; s94; woo]. In our discussion we assume that both α and β are positive quantities; α or 2α is called the chirp rate. A chirp pulse is a phase-modulated (PM) signal. We will provide a discussion on the Fourier properties of PM signals in the later chapters. The *instantaneous frequency* of the chirp pulse in the interval $0 \le t \le T_p$ is the derivative of its phase function with respect to time:

$$\begin{aligned} \omega_{ip}(t) &= \frac{d}{dt}(\beta t + \alpha t^2) \\ &= \beta + 2\alpha t \end{aligned}$$

Note that with $\alpha > 0$, the instantaneous frequency is an increasing function of time; such a chirp is said to be *upsweep*.

The minimum value of $\omega_{ip}(t)$ is β, and its maximum value is $\beta + 2\alpha T_p$. The spectral support band of a chirp signal is approximately bounded by these minimum and maximum values, that is, $|P(\omega)| > 0$ for

$$\omega \in [\beta, \beta + 2\alpha T_p].$$

Thus the carrier (midfrequency) of a chirp pulse is

$$\omega_c = \beta + \alpha T_p,$$

and its baseband bandwidth is

$$\pm\omega_0 = \pm\alpha T_p.$$

Clearly the baseband bandwidth of a chirp pulse increases with its duration. The same is not true for a rectangular pulsed signal which amplitude-modulates a carrier (i.e., a chirp pulse with zero chirp rate $\alpha = 0$).

In the case of a pulsed chirp radar signal, the echoed signal from the target scene can be rewritten as follows:

$$\begin{aligned} s(t) &= \sum_n \sigma_n a\left(t - \frac{2x_n}{c}\right) \exp\left[j\beta\left(t - \frac{2x_n}{c}\right) + j\alpha\left(t - \frac{2x_n}{c}\right)^2\right] \\ &= \sum_n \sigma_n a\left(t - \frac{2x_n}{c}\right) \exp\left[j\beta\left(t - \frac{2x_n}{c}\right)\right] \exp\left[j\left(\alpha t^2 - \frac{4\alpha x_n}{c}t + \frac{4\alpha x_n^2}{c^2}\right)\right] \\ &= \sum_n \sigma_n a(t - t_n) \exp[j\beta(t - t_n)] \exp\left[j\left(\alpha t^2 - 2\alpha t_n t + \alpha t_n^2\right)\right], \end{aligned}$$

where $t_n = 2x_n/c$ is the round-trip delay for the echoed signal from the nth target. An example of a transmitted chirp pulse and received echoed signals from two targets are shown in Figure 1.7.

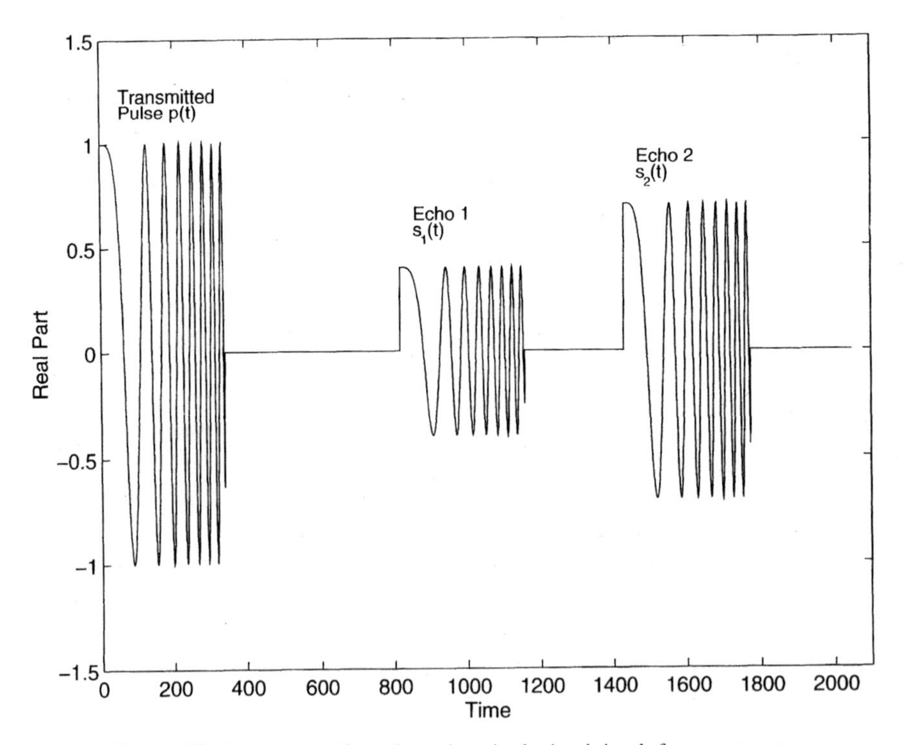

Figure 1.7. Transmitted chirp pulse and received echoed signals from two targets.

Reconstruction

If we *mix* the complex conjugate of this signal with the phase of the transmitted chirp signal, we obtain

$$s_c(t) = s^*(t) \underbrace{\exp(j\beta t + j\alpha t^2)}_{\text{Reference phase}}$$

$$= \sum_n \sigma_n a^*(t - t_n) \exp\left(j\beta t_n - j\alpha t_n^2\right) \underbrace{\exp(j2\alpha t_n t)}_{\text{Sinusoid}}$$

which is called the time domain *compressed* or *deramped* signal [bar; carr; kla] (Figure P1.7). Similar to the mixing of the echoed signal with the carrier for baseband conversion, the deramping or pulse compression is also performed by the hardware; the resultant signal is then lowpass filtered, and sampled (passed through an A/D converter). Figure 1.8 shows the time domain compressed signal for the signals in Figure 1.7.

Note the presence of the sinusoid $\exp(j2\alpha t_n t)$ in the deramped signal. The frequency of this sinusoidal carrier is $2\alpha t_n$ which carries information on the round-trip

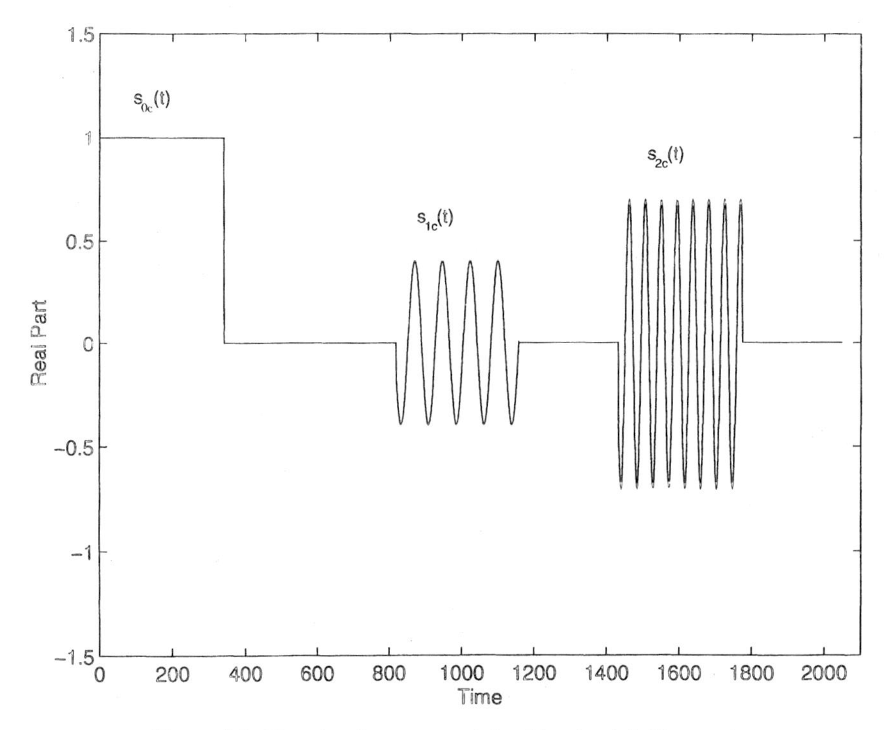

Figure 1.8. Time domain compressed signal for signals in Figure 1.7.

delay t_n or the target range x_n. Thus, by examining the Fourier spectrum of the compressed signal, one should see some kind of energy distribution at the frequency $2\alpha t_n$ which can then be used to identify the target range.

In fact the Fourier transform of the compressed signal is

$$\begin{aligned} S_c(\omega) &= \sum_n \sigma_n \underbrace{\exp\left(j\beta t_n + j\alpha t_n^2 - j\omega t_n\right)}_{\text{Phase function}} \text{psf}_\omega(\omega - 2\alpha t_n) \\ &= \sum_n \sigma_n \exp\left(j\beta t_n + j\alpha t_n^2 - j\omega t_n\right) \text{psf}_\omega\left(\omega - \frac{4\alpha x_n}{c}\right) \end{aligned}$$

where the point spread function in the frequency domain for pulse compression is defined via

$$\text{psf}_\omega(\omega) = \mathcal{F}_{(t)}\left[a^*(t)\right].$$

Note that $a(t)$ is a rectangular pulse, and thus the point spread function is a sinc function.

This indicates that the signature of the nth target appears at the frequency $\omega = 4\alpha x_n/c$, which carries information on the coordinates of the target (Figure P1.8). The inverse (range imaging) equation is a linear transformation of the frequency domain axis into the spatial domain axis via

$$\omega = \frac{4\alpha x}{c}.$$

The *noncoherent* target function is formed via displaying

$$|S_c(\omega)| \approx \sum_n \sigma_n \left|\text{psf}_\omega\left(\omega - \frac{4\alpha x_n}{c}\right)\right|$$

versus

$$x = \frac{c}{4\alpha}\omega.$$

It is more convenient to define the point spread function in the range domain; that is,

$$\text{psf}(x) = \mathcal{F}_{(t)}\left[a^*(t)\right],$$

with $x = c\omega/4\alpha$. The shape of the range-domain point spread function for pulse compression is approximately the same as the shape of the range-domain point spread function for matched filtering, which is

$$\text{psf}(x) = \mathcal{F}^{-1}_{(\omega)}\left[|P(\omega)|^2\right],$$

with $x = ct/2$. This equivalence can be shown by using $|P(\omega)| = 1$ within the radar frequency support band. Using the range-domain point spread function for pulse compression, the noncoherent reconstructed target function via pulse compression becomes

$$|S_c(\omega)| \approx \sum_n \sigma_n |\text{psf}(x - x_n)|.$$

Beyond the mathematical formulation that was shown for range imaging with time domain pulse compression or deramping of chirp signals, the following observations can be made to give this operation a physical meaning. As we pointed out earlier, the instantaneous frequency of the transmitted chirp $p(t)$ is

$$\begin{aligned}\omega_{ip}(t) &= \frac{d}{dt}\left[\beta t + \alpha t^2\right] \\ &= \beta + 2\alpha t,\end{aligned}$$

which is a *linear* function of time. The instantaneous frequency of the echoed signal from the nth reflector $s_n(t) = \sigma_n p[t - (2x_n/c)]$ is

$$\begin{aligned}\omega_{in}(t) &= \frac{d}{dt}\left[\beta\left(t - \frac{2x_n}{c}\right) + \alpha\left(t - \frac{2x_n}{c}\right)^2\right] \\ &= \beta + 2\alpha\left(t - \frac{2x_n}{c}\right),\end{aligned}$$

which is also a linear function of time. The difference between these two instantaneous frequency functions is

$$\omega_{ip}(t) - \omega_{in}(t) = \frac{4\alpha x_n}{c},$$

which results in the removal of the linear (ramp) function of time in the instantaneous frequency (deramping). Figure 1.9 shows the above-mentioned instantaneous frequency deramping or differencing phenomenon for the signals in Figure 1.7.

The above indicates that the *mixing* of the complex conjugate of $s_n(t)$ with the phase of $p(t)$ results in a single-tone sinusoid with frequency $4\alpha x_n/c$. Thus a spectral analysis (Fourier transform) of the resultant signal $s_c(t)$ yields information on the location of the targets. We showed this by obtaining an analytical expression for this Fourier transform, that is, $S_c(\omega)$.

- The magnitude spectrum $|S_c(\omega)|$ contains information on the location of the targets in the one-dimensional range domain. However, the *phase* of the reconstruction plays a critical role in *multidimensional* echo imaging problems such as SAR. This issue will be examined in our discussion on the residual video phase error.

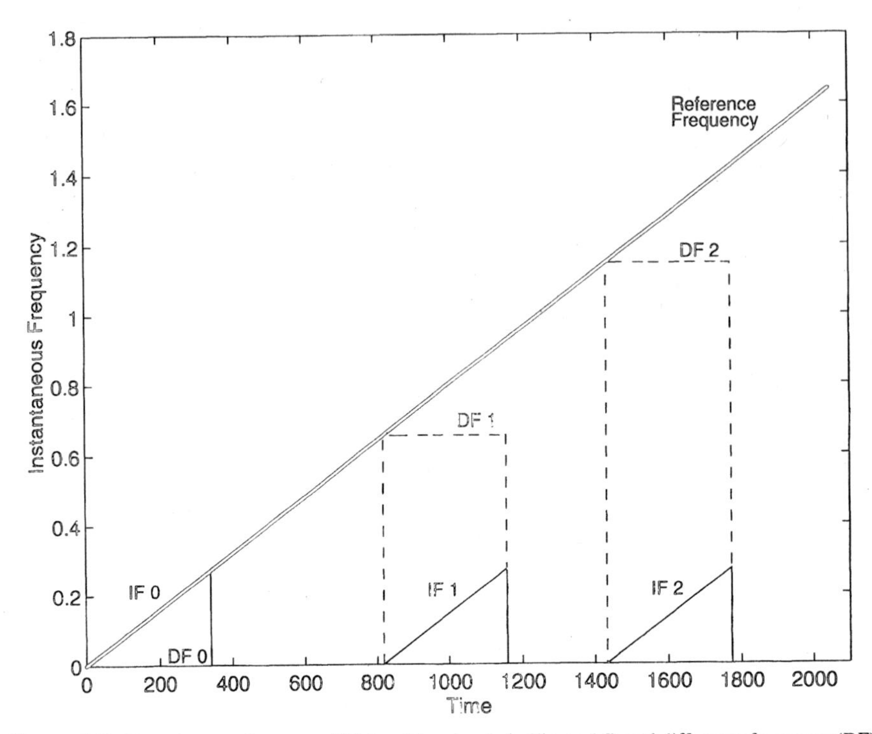

Figure 1.9. Instantaneous frequency (IF) for chirp signals in Figure 1.7, and difference frequency (DF) for deramped signals in Figure 1.8.

Range Resolution

Range resolution for both the matched filtering and time domain chirp pulse compression is the same. This range resolution is

$$\Delta_x = \frac{\pi c}{2\alpha T_p},$$

where T_p is the duration of the chirp pulse in the time domain. The matched filtering can be applied to any type of wideband (FM) transmission. The time domain pulse compression method is applicable for limited types of pulsed signals, such as a chirp. There are other types of wideband signals, such as tone-modulated FM signals, that are easier to generate.

Time Domain Sampling

We showed that the time sampling for the echoed signal $s(t)$ should satisfy the following Nyquist constraint:

$$\Delta_t \leq \frac{\pi}{\omega_0},$$

where $\pm\omega_0$ is the baseband bandwidth of the transmitted radar signal. For a chirp pulse we have

$$\omega_0 = \alpha T_p,$$

where α is the quadratic coefficient of the chirp signal phase and T_p is the chirp pulse duration. Thus the Nyquist sampling constraint for the echoed signal in the case of chirp pulse transmission is

$$\Delta_t \leq \frac{\pi}{\alpha T_p}.$$

This is not the Nyquist time sampling constraint for the pulsed-compressed signal $s_c(t)$. This fact is demonstrated next.

For the pulse compression reconstruction, the mapping of the frequency domain ω into the range domain x was shown to be

$$x = \frac{c}{4\alpha}\omega.$$

The target region is within $x \in [X_c - X_0, X_c, X_0]$. This implies that the compressed signal $s_c(t)$ is a bandpass signal; the support band of $S_c(\omega)$ is

$$\omega \in \left[\frac{4\alpha}{c}(X_c - X_0), \frac{4\alpha}{c}(X_c + X_0)\right].$$

Thus the energy of $S_c(\omega)$ is centered around the carrier

$$\omega_{cc} = \frac{4\alpha}{c}X_c,$$

and its baseband support is $\pm\omega_{0c}$ where

$$\begin{aligned}\omega_{0c} &= \frac{4\alpha}{c}X_0\\ &= \alpha T_x,\end{aligned}$$

and

$$T_x = \frac{4X_0}{c},$$

is called the *range swath echo time period.*

Thus, for alias-free processing of the pulse-compressed signal $s_c(t)$, we should perform baseband conversion on $s_c(t)$ via

$$s_{cb}(t) = s_c(t)\exp(-j\omega_{cc}t).$$

Equivalently, the following is used for pulse compression at the hardware level:

$$\begin{aligned} s_{cb}(t) &= s^*(t)\exp\left(j\beta t + j\alpha t^2 - j\omega_{cc}t\right) \\ &= s^*(t)\exp\left(j\beta t + j\alpha t^2 - j\frac{4\alpha X_c}{c}t\right) \\ &= s^*(t)\text{Phase}\left[p\left(t - \frac{2X_c}{c}\right)\right]\exp\left(j\frac{2\beta X_c}{c} - j\frac{4\alpha X_c^2}{c^2}\right). \end{aligned}$$

Note that to form the baseband version of the compressed signal, the complex conjugate of the echoed signal is mixed with the phase of the reference signal, that is,

$$s_0(t) = p\left(t - \frac{2X_c}{c}\right).$$

For range reconstruction the Fourier transform of this signal, that is, $S_{cb}(\omega)$, is displayed versus $x = c\omega/4\alpha + X_c$ (note the shift by X_c).

Moreover, since the baseband bandwidth of the pulse-compressed signal is $\pm\omega_{0c}$, the following Nyquist time-sampling constraint should be satisfied for the pulse-compressed signal:

$$\begin{aligned} \Delta_{tc} &\le \frac{\pi}{\omega_{0c}} \\ &= \frac{\pi}{\alpha T_x}, \end{aligned}$$

which is different from the Nyquist sampling rate for the echoed signal; that is,

$$\Delta_t \le \frac{\pi}{\alpha T_p}.$$

Thus, depending on the relative values of chirp pulse duration T_p and range swath echo time period T_x, an alias-free set of samples of baseband echoed signal $s_b(t_i)$, $i = 1, \ldots, N$, may yield an aliased set of samples of baseband pulse-compressed signal $s_{cb}(t_i)$, $i = 1, \ldots, N$, and vice versa.

Most pulse compression-based radar systems operate under the mode that $T_x < T_p$, that is, a relatively small range swath. This is clearly a restrictive constraint if one desires to image a large target area (large X_0 or T_x). Also, if the target area is located at a close range, then a large value for the pulse duration T_p would jam the receiver. (The echoed signals arrive while the radar is still transmitting.)

In the case of a large-range swath, the mixing of the received echoed signal is done with a *periodic* version of $p(t)$ and some form of *range-gating* [bar; bla; edd; sko]. This operation introduces another form of error in the pulse-compressed signal which is not described here.

Residual Video Phase Error

Recall the expression for the compressed or deramped echoed signal

$$s_c(t) = \sum_n \sigma_n a^*(t - t_n) \underbrace{\exp\left(j\beta t_n - j\alpha t_n^2\right)}_{\text{Undesirable phase}} \underbrace{\exp(j2\alpha t_n t)}_{\text{Desirable phase}}$$

In the above expression, the term labeled the *desirable phase* is a component that is critical for coherent SAR imaging; the *undesirable phase* term should be removed. In this case one has to compensate for the unknown phase function

$$\exp\left(j\beta t_n - j\alpha t_n^2 - j\omega t_n\right),$$

which is also called *residual video phase (RVP)* [carr], in $S_c(\omega)$ to form a *coherent* target function, that is, a target function with the desirable phase information intact, or

$$f(x) = \sum_n \sigma_n \exp(-j\omega t_n)\text{psf}(x - x_n),$$

with $x = c\omega/4\alpha$.

For this, one has to use an *approximation*, since RVP depends on the round-trip delay t_n or the target range x_n and thus is unknown. Provided that the point spread function is sharp, and thus the energy of $\text{psf}_\omega(\omega - 2\alpha t_n)$ [or $\text{psf}(x - x_n)$] is mainly around $\omega \approx 2\alpha t_n$ (or $x \approx x_n$), the RVP can be approximated via

$$\exp\left(j\beta t_n - j\alpha t_n^2\right) \approx \exp\left(j\frac{2\beta x}{c} - j\frac{4\alpha x^2}{c^2}\right).$$

In this case the coherent target function can be formed via

$$\begin{aligned} f(x) &\approx \exp\left(-j\frac{2\beta x}{c} + j\frac{4\alpha x^2}{c^2}\right) S_c\left(\frac{4\alpha x}{c}\right) \\ &\approx \sum_n \sigma_n \exp(-j\omega t_n)\text{psf}(x - x_n) \end{aligned}$$

We call the above equation the *coherent* inverse equation for time domain pulse compression which includes the addition of a *phase* term to the Fourier spectrum of the compressed signal.

Upsampling to Recover Alias-Free Echoed Signal

It should be noted that a smaller baseband bandwidth in the measured signal is desirable, since this puts less of a burden on the A/D converter and the output signal-to-noise power ratio also improves. Suppose that $T_x < T_p$, and the user selects the less restrictive sampling rate of the baseband pulse-compressed signal $s_{cb}(t)$ for the A/D converter. Clearly the baseband-measured signal $s_b(t)$ is aliased. It turns out that the alias-free version of $s_b(t)$ can be recovered from the alias-free samples of $s_{cb}(t)$. [The reverse is also true: If $T_p < T_x$, the alias-free version of $s_{cb}(t)$ can be recovered from the alias-free samples of $s_b(t)$.]

For this, suppose that the samples of baseband deramped or compressed signal $s_{cb}(t)$ are available at the time points

$$t_{ic} = T_s + (ic - 1)\Delta_{tc}, \qquad ic = 1, 2, \ldots,$$

where

$$\Delta_{tc} = \frac{\pi}{\alpha T_x};$$

these samples are not aliased. However, the equivalent samples of $s_b(t)$, that is, $s_b(t_{ic})$'s, are aliased, since

$$\Delta_t = \frac{\pi}{\alpha T_p},$$

is *smaller* than Δ_{tc}.

However, we could *increase* the sampling rate of available $s_c(t)$ data via *digital signal processing methods*. This can be achieved via sinc interpolation, or *zero-padding* the Fourier transform of the compressed signal, that is, $S_{cb}(\omega)$; the latter method is used in the Matlab code of this chapter. Sufficient zeros are added to $S_{cb}(\omega)$ such that the sample spacing in the time domain becomes Δ_t; we call this operation *upsampling*.

In this case the available alias-free samples of $s_{cb}(t)$ are at the following time points:

$$t_i = T_s + (i - 1)\Delta_t, \qquad i = 1, 2, \ldots.$$

In order to recover alias-free samples of the echoed signal, the following operation, which we call *decompression* or *ramping*, is performed on the new samples of $s_{cb}(t)$:

$$s(t_i) = s^*_{cb}(t_i) \exp\left(j\beta t_i + j\alpha t_i^2 - j\frac{4\alpha X_c}{c} t_i \right).$$

This is followed by a baseband conversion in the software:

$$s_b(t_i) = s(t_i) \exp(-j\omega_c t_i),$$

which yields *alias-free* samples of the baseband echoed signal. The results in Figures P1.1b–P1.8b are generated via this upsampling and decompression method on alias-free samples of $s_{cb}(t)$. The alias-free samples of $s_b(t)$ can then be used to form the error-free coherent matched-filtered range reconstruction.

Algorithm We showed a block diagram representation for a range imaging system that utilized discrete-time matched filtering for reconstruction in Figure 1.6. That system can also be used for range imaging when the transmitted signal is a chirp signal and $T_x > T_p$. In this case $\Delta_t > \Delta_{tc}$, and thus there is no need for upsampling. (The A/D converter time sampling is set at the larger Δ_t.)

Figure 1.10 is a block diagram representation of a chirp-pulsed range-imaging system, in which $T_x < T_p$, from the hardware that is used for data acquisition to the discrete-time reconstruction in the software. In this case $\Delta_t < \Delta_{tc}$ and thus one has to perform upsampling to retrieve alias-free baseband-echoed signal for discrete-time matched filtering. (The A/D converter time sampling is set at the larger Δ_{tc}.) The following steps are used to recover alias-free baseband echoed signal from the pulse-compressed signal and perform matched-filtered reconstruction when $T_x < T_p$:

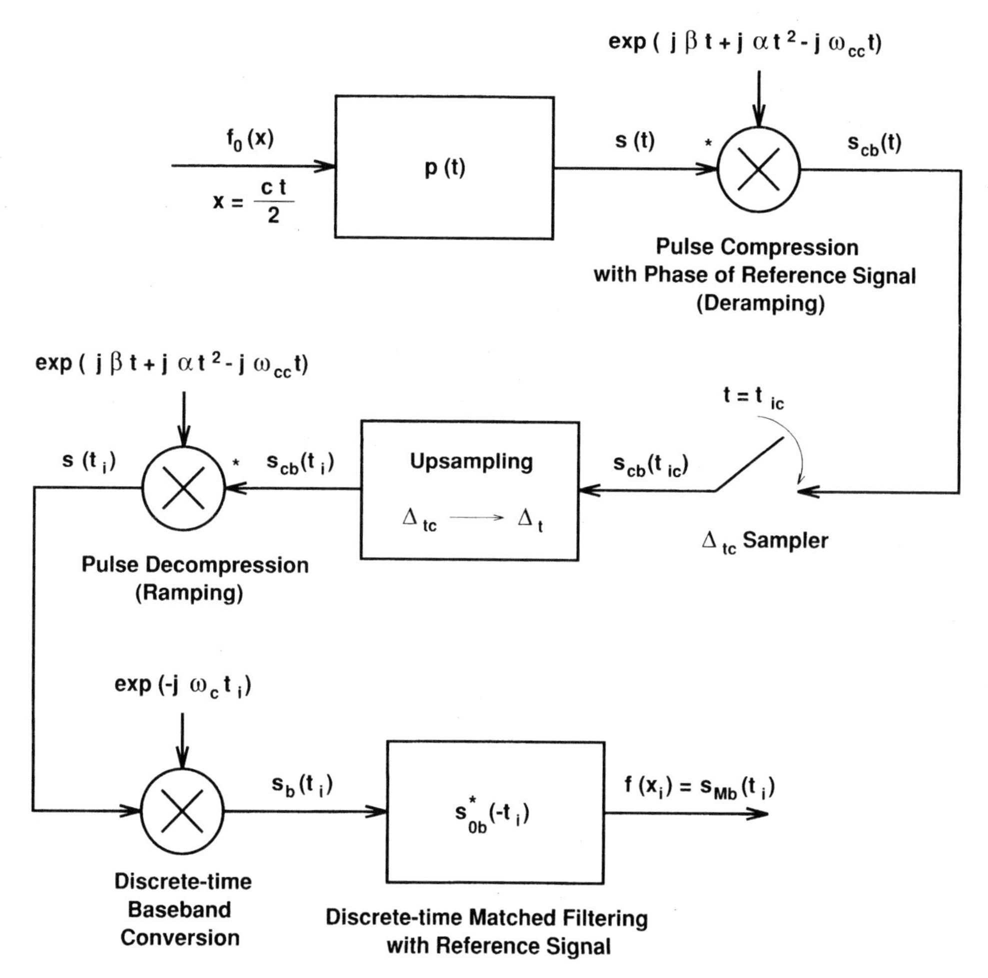

Figure 1.10. Block diagram representation of a chirp-pulsed range imaging system when $T_x < T_p$.

Step 1. Perform pulse compression on the incoming echoed signal with the phase of the reference signal $s_0(t) = p[t - (2X_c/c)]$. *(This is done by the hardware.) The result is the baseband pulse-compressed signal* $s_{cb}(t)$.

Step 2. Sample the baseband pulse-compressed signal $s_{cb}(t)$ *with temporal spacing* Δ_{tc}; *the resultant samples are* $s_{cb}(t_{ic})$, $ic = 1, 2, \ldots$. *The baseband-echoed signal* $s_b(t)$ *would be aliased if it was sampled at the time points* t_{ic}*'s, since* $T_x < T_p$.

Step 3. Obtain the DFT of the baseband pulse-compressed signal sampled data $s_{cb}(t_{ic})$, $ic = 1, 2, \ldots$; *this yields the samples of* $S_{cb}(\omega)$. *Zero-pad this signal in the frequency* ω *domain to achieve the desired aperture sample spacing of* Δ_t *in the time* t *domain (upsampling). The result is a set of samples of* $s_{cb}(t)$ *at the desired (Nyquist) measurement time bins of the baseband-echoed signal, that is,* $s_{cb}(t_i)$, $i = 1, 2, \ldots$.

Note: To reduce wraparound errors due to DFT and IDFT in the above-mentioned upsampling scheme, a "pseudo" discrete Fourier transform (DCT) can be used. This is done via appending $s_{cb}(t)$ *with its mirror image in the time domain. Then the DFT, zero-padding, and the IDFT are performed; the mirror image is discarded in the upsampled data.*

Step 4. Perform pulse decompression *or* ramping *(i.e., the inverse of pulse compression or deramping) to recover the samples of the echoed signal at* t_i*'s, that is,* $s(t_i)$, $i = 1, 2, \ldots$ *Then apply discrete-time (software) baseband conversion on* $s(t_i)$*'s with the carrier to retrieve uncorrupted (alias-free) baseband-echoed signal sampled data* $s_b(t_i)$, $i = 1, 2, \ldots$.

Step 5. Perform discrete-time matched filtering with the reference baseband-echoed signal on $s_b(t_i)$, $i = 1, 2, \ldots$, *to reconstruct the target function in the range domain.*

The Matlab code of this chapter contains the above algorithm; an example is also provided (Figures P1.1b–P1.8b). A similar scheme, which is called *compression, interpolation (or upsampling), and decompression*, will be discussed in the next chapter for processing synthetic aperture signals that exhibit chirp-type properties.

- There are many practical as well as theoretical reasons for using the matched filtering method in SAR imaging. In spite of this, the time domain chirp pulse compression is still being used in practice, and mentioned in various manuscripts. An expert noted that this is a *generational* problem. The fact remains that for $T_x > T_p$ with time sampling Δ_t, or $T_p < T_x$ with time sampling Δ_{tc} in conjunction with *upsampling*, one can retrieve alias-free samples of the baseband echoed signal for coherent matched-filtered range reconstruction.

Electronic Counter-Countermeasure (ECCM) via Amplitude Modulation of Chirp Signals

In some of the military reconnaissance applications of SAR, the user encounters what is referred to as electronic warfare (EW) [goj]. In these scenarios the enemy utilizes a transmitting radar to send out a signal within the band of the SAR system transmitter to confuse and/or saturate the SAR system receiver; this is called electronic counter measure (ECM) to *jam* the SAR system. The ECM jamming signal causes the SAR system to receive and process erroneous information, which results in severe degradations in the output SAR images.

Various types of ECM jamming signals have been used [goj]. For instance, the ECM jamming signal could be composed of regularly or randomly timed replicas of the transmitted chirp radar signal $p(t)$ (repeater jammer), or a powerful random signal in the frequency band of $p(t)$ (noise jammer). The methods used for suppressing the effect of the ECM jamming signal are called electronic counter-countermeasure (ECCM). We now examine a method for ECCM that relies on an amplitude modulation of the LFM chirp signal. The following analysis will also bring out another advantage of range imaging via matched filtering over range imaging via deramping for chirp signals.

As we pointed out earlier in this section, the conventional pulsed chirp radar signal, which is used in the SAR systems, is a *flat-top* chirp that has the following form:

$$p(t) = a(t) \exp(j\beta t + j\alpha t^2),$$

where $a(t) = 1$ for $0 \le t \le T_p$ and is zero otherwise; T_p is the chirp pulse duration. The amplitude signal $a(t)$ is chosen to be flat (or some tapered version of a flat-top signal), since the point spread function for range imaging via pulse compression is

$$\text{psf}_\omega(\omega) = \mathcal{F}_{(t)}\Big[a^*(t)\Big].$$

If one uses a fluctuating signal for the amplitude (phase as well as magnitude) signal $a(t)$, the resultant point spread function will widen, and as a result the range resolution will be degraded. This has been a major impasse for using a fluctuating amplitude signal $a(t)$ for range imaging via deramping of LFM chirp signals [goj].

As we will see in our later discussion of SAR systems, the user relies on multiple (chirp) radar signal transmissions for SAR image formation. Because of this the enemy could use a receiver to construct a good estimate of the LFM chirp parameters; this estimate can then be used to jam the SAR system with erroneous replicas of the transmitted radar signal (repeater jammer). One of the methods that has been suggested to encounter a repeater jammer is to add a small phase to the LFM chirp signal, that is, use an amplitude signal $a(t)$ that is a slowly fluctuating phase function. As we mentioned earlier, this would cause degradations in the point spread function of range imaging via deramping.

Recall from Section 1.2 that the point spread function in range imaging using the matched filtering method is

$$\begin{aligned}\text{psf}_t(t) &= \mathcal{F}^{-1}_{(\omega)}\Big[|P(\omega)|^2\Big]\\ &= p(t) * p^*(-t),\end{aligned}$$

which is the *autocorrelation* of the transmitted radar signal. Thus this point spread function is not changed if the amplitude signal $a(t)$ is chosen to be a highly fluctuating signal, for example, a pseudorandom phase-modulated (PM) signal.

- When the amplitude signal $a(t)$ is a fluctuating signal, the bandwidth of the radar signal $p(t)$ is equal to the bandwidth of the amplitude signal $a(t)$ plus the bandwidth of the LFM chirp signal. The range resolution is still dictated by the bandwidth of the radar signal $p(t)$.

Moreover, to prevent the enemy from estimating this amplitude signal from multiple radar transmissions in a SAR system, the user could alter the amplitude signal $a(t)$ (e.g., using various pseudorandom PM signal) with each pulse transmission; this would confuse the enemy's receiving radar which is trying to estimate the transmitted radar signal of the SAR system. This procedure will be outlined in Chapter 4 (ECCM via pulse diversity).

Note that since a noise jammer is not correlated to the transmitted radar signal, a major portion of a powerful ECM random jamming signal is filtered out by the matched filtering; this is the case for a flat-top or a fluctuating amplitude signal $a(t)$. The same is not true for range imaging via pulse compression (deramping) when $a(t)$ is a fluctuating signal.

As we mentioned earlier, in some of the radar systems the chirp pulse duration is larger than the range swath echo time period T_x; that is, $T_x < T_p$. In these systems it is more practical to sample the pulse-compressed signal to reduce the burden on the A/D converter device. When the amplitude signal $a(t)$ is a fluctuating signal, the pulse-compressed signal should be converted into the echoed signal via the upsampling method which was outlined earlier. Then the range imaging via matched filtering can be performed on the alias-free echoed signal.

- In practice, the quantization level of the A/D converter is automatically adjusted based on the power of the received signal, that is, the desired SAR signal (the baseband-echoed signal or deramped signal) plus the ECM jamming signal. If the power of the ECM jamming signal is significantly higher than the power of the desired SAR signal, then the desired SAR signal may be at the level of the A/D quantization noise. In this case, the A/D is saturated by the ECM jamming signal, so the recovery of the SAR signal is infeasible.

Example We constructed this example by amplitude modulating an LFM chirp signal; see the Matlab code for this chapter. We consider case 2 in the Matlab code where $T_x < T_p$; thus the Matlab code performs upsampling to recover the alias-free echoed signal from the deramped signal. The amplitude signal $a(t)$ is chosen to be a PM signal with a phase composed of N_a harmonics; the amplitude and phase of these harmonics are generated via a uniform random number generator.

The Matlab code also contains a procedure for power equalization of the radar signal spectrum $|P(\omega)|^2$ to improve the point spread function. This is achieved by defining the magnitude of the matched filter to be

$$\begin{cases} \dfrac{P_{\max}^2}{\sqrt{2}\,|P(\omega)|} & \text{for } |P(\omega)| \le \dfrac{P_{\max}}{\sqrt{2}}, \\ |P(\omega)| & \text{otherwise.} \end{cases}$$

One can also use a frequency ω domain window to reduce the *ringing* side lobes in the point spread function; this is done with the help of the *power* window described in Chapter 2.

The case of $N_a = 0$ corresponds to a conventional LFM chirp radar signal with no amplitude modulation; for this case range reconstruction via matched filtering and time-domain compression are, respectively, shown in Figure 1.11*a* and *b*. Fig-

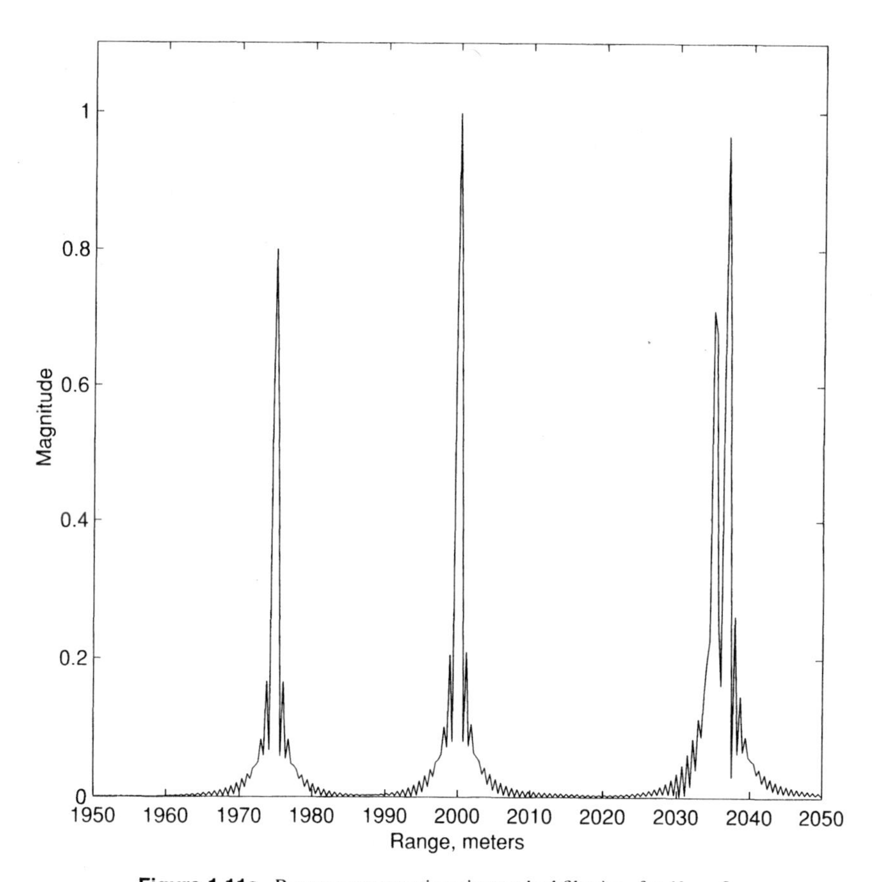

Figure 1.11a. Range reconstruction via matched filtering, for $N_a = 0$.

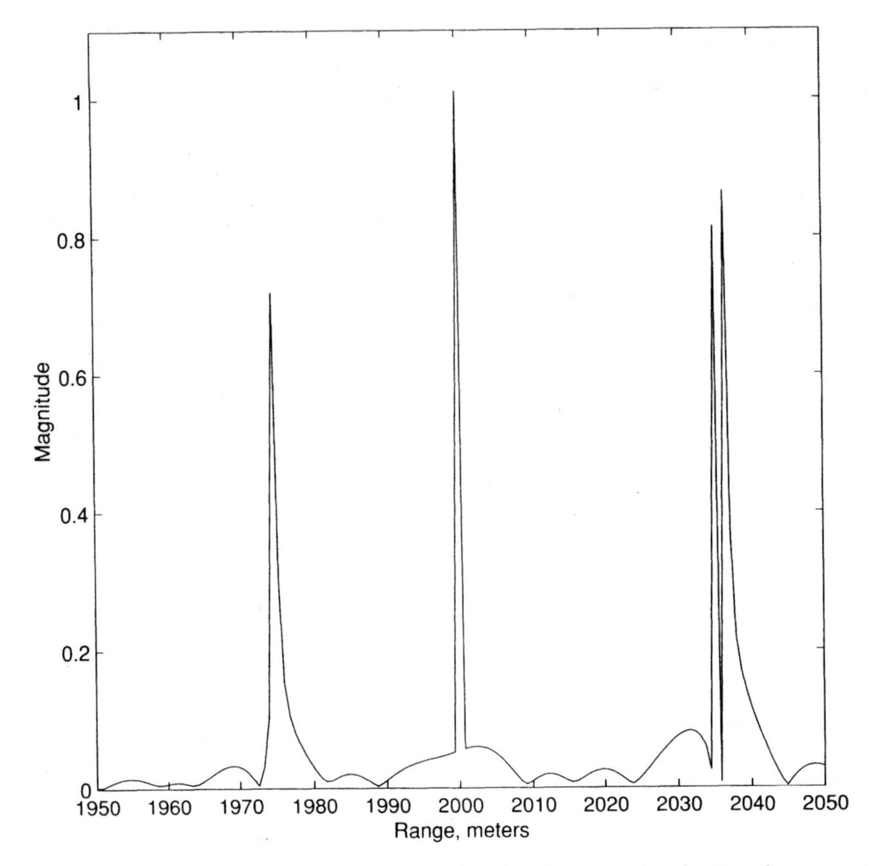

Figure 1.11b. Range reconstruction via time domain compression, for $N_a = 0$.

ures 1.12 and 1.13 show the range reconstructions using matched filtering (*a*), and time-domain compression (*b*) for $N_a = 4$ and 8, respectively. We can observe from Figures 1.12*b* and 1.13*b* that as the number of harmonics N_a is increased, the point spread function for the range reconstruction via time domain compression exhibits more undesirable side lobes, spreading, and target peak power reduction.

Finally we should point out that the radar system circuitry puts limitations on the type of the amplitude signal $a(t)$ that it can generate. A *sharp* set of pulses, such as a pseudorandom sequence, creates discontinuities in the phase of the signal that cannot be supported by the radar system hardware. A practical option would be the use of continuous phase modulation (CPM) [pro] for the amplitude signal $a(t)$; this was used in the above-mentioned example.

Use of continuous phase-modulation of a chirp radar signal has been suggested for ECCM [goj]. However, this has been implemented with deramping (time-domain

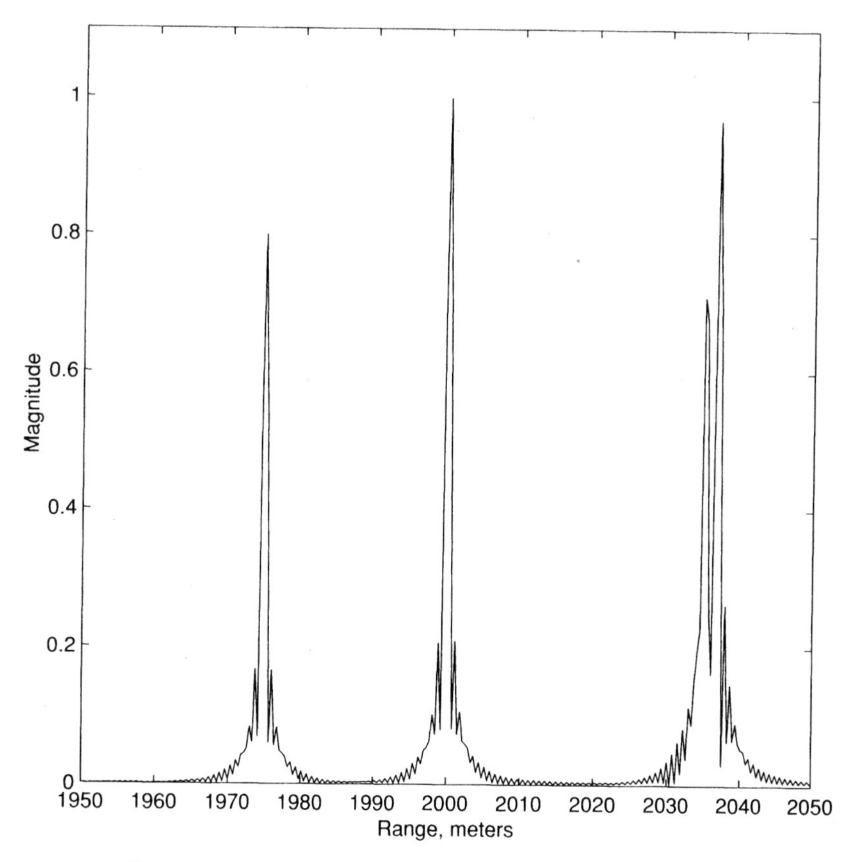

Figure 1.12a. Range reconstruction via matched filtering for $N_a = 4$.

compression) operation for range imaging; as we mentioned earlier, it has undesirable effects on the point spread function of the deramped signal. To rectify this problem, the deramped signal should be converted to the baseband-echoed signal via the procedure that we described earlier; the baseband-echoed signal can then be used for range imaging by matched filtering with the reference signal $p^*(-t)$ which contains information on the amplitude signal $a(t)$.

1.7 FREQUENCY-DEPENDENT TARGET REFLECTIVITY

We now examine the range imaging problem when the target function exhibits a radar frequency-dependent behavior. There are well-established reasons for this phenomenon, for example, resonance in cavities, surface waves, and multipath effects. The physical nature of these is not the subject of our discussion. We are interested

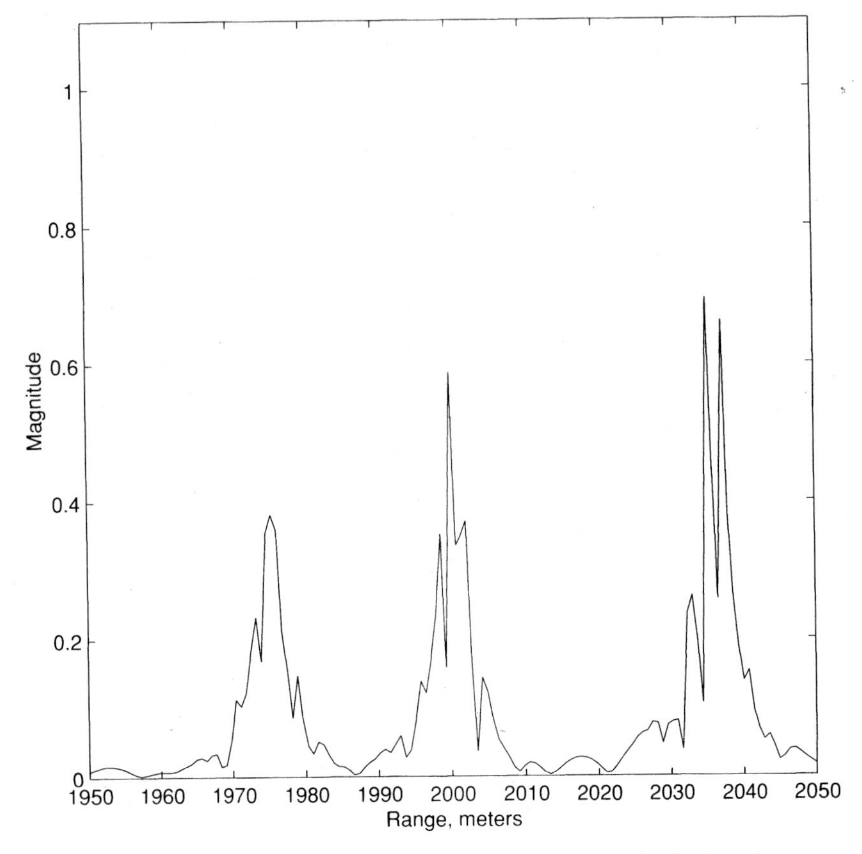

Figure 1.12b. Range reconstruction via time domain compression for $N_a = 4$.

in identifying a signal model for these effects, and we want to trace its effect in the reconstructed image.

We could analyze this by starting with the time domain system model. However, this brings other physical complexities into the picture. As engineers, we take a slight shortcut to avoid these by starting with the system model in the frequency domain, that is,

$$S(\omega) = P(\omega) \sum_n \sigma_n(\omega) \exp\left(-j\omega \frac{2x_n}{c}\right).$$

In the above model, $\sigma_n(\omega)$ represents a frequency-dependent reflectivity for the nth target which, in general, is a complex function; that is, it contains a phase function as well as a magnitude function. We call $\sigma_n(\omega)$ the nth target *radar signature*.

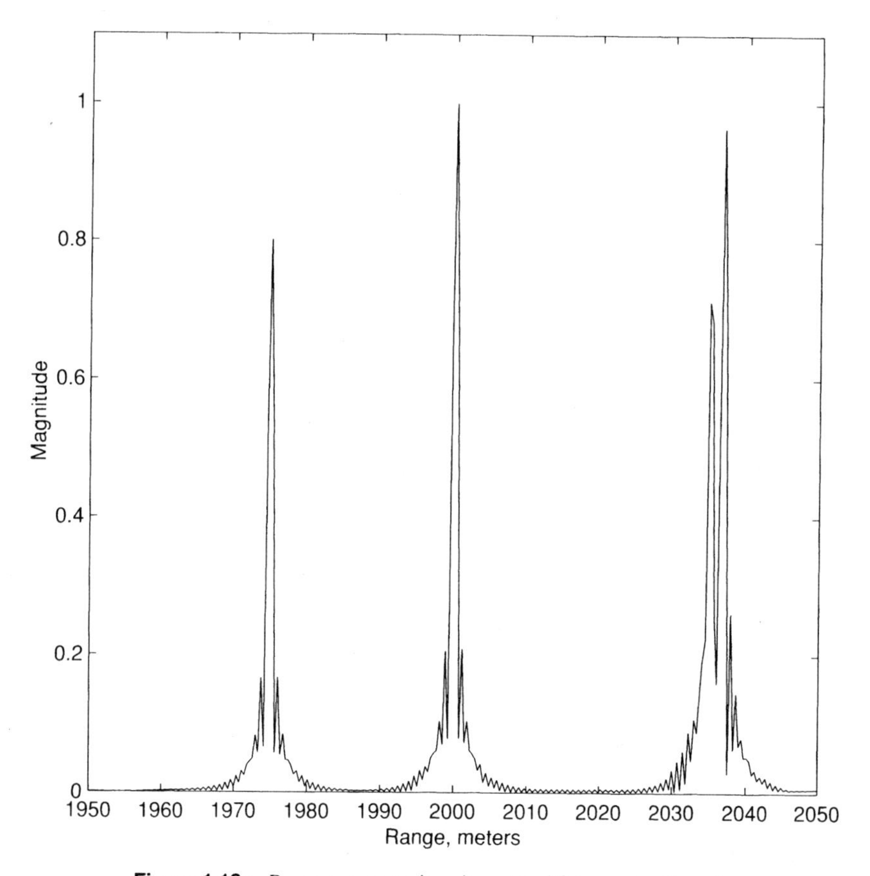

Figure 1.13a. Range reconstruction via matched filtering, for $N_a = 8$.

Note that the contribution of the nth target in the received echoed signal $s(t)$ is the convolution of $p[t - (2x_n/c)]$ and the inverse Fourier transform of $\sigma_n(\omega)$:

$$s_n(t) = p\left(t - \frac{2x_n}{c}\right) * \mathcal{F}^{-1}_{(\omega)}\Big[\sigma_n(\omega)\Big],$$

which is typically a *smeared* version of $p[t - (2x_n/c)]$ (Figure P1.9). Physically this echoed signal $s_n(t)$ is composed of a primary echo from the target, that is, a signal that is proportional to $p[t - (2x_n/c)]$, and secondary multiple echos that have slight delays with respect to the primary echo. The time spacing of these delays are typically *smaller* than $2\Delta_x/c = \pi/\omega_0$ (the Nyquist time sample spacing). As a result these delays are closely packed in the time domain and may not be visually distinguishable.

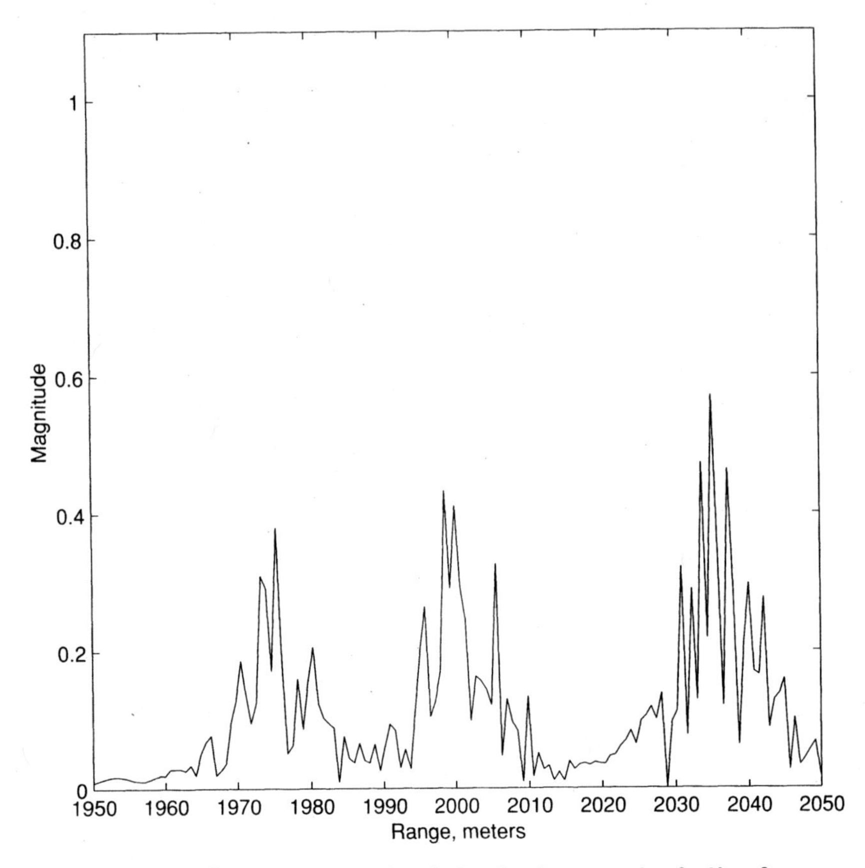

Figure 1.13b. Range reconstruction via time domain compression, for $N_a = 8$.

For the new system model, the matched-filtered target function in the spatial frequency domain becomes

$$\begin{aligned} F(k_x) &= S(\omega)P^*(\omega) \\ &= |P(\omega)|^2 \sum_n \sigma_n(\omega) \exp(jk_x x_n) \end{aligned}$$

with $k_x = 2\omega/c = 2k$. The inverse spatial Fourier transform of this function is the range domain target reconstruction

$$\begin{aligned} f(x) &= \mathcal{F}^{-1}_{(k_x)}[F(k_x)] \\ &= \sum_n \text{psf}_n(x - x_n), \end{aligned}$$

where $\text{psf}_n(x)$ is the point spread function for the nth target, which is

$$\text{psf}_n(x) = \mathcal{F}^{-1}_{(\omega)}\Big[|P(\omega)|^2 \sigma_n(\omega)\Big],$$

with $x = ct/2$. Note that the point spread function now depends on the target (type).

The case in the previous sections corresponds to the scenario when the target radar signature σ_n is invariant in the radar frequencies $\omega \in [\omega_c - \omega_0, \omega_c + \omega_0]$; in that case the point spread function is

$$\text{psf}(x) = \mathcal{F}^{-1}_{(\omega)}\Big[|P(\omega)|^2\Big]$$

with a constant amplitude of σ_n. Now, with variations of σ_n with ω, one should anticipate a point spread function that is less sharp (smeared), since the new point spread function is a convolved version of $\text{psf}(x)$; that is,

$$\text{psf}_n(x) = \text{psf}(x) * \mathcal{F}^{-1}_{(\omega)}[\sigma_n(\omega)].$$

It turns out that the phase of the target radar signature $\sigma_n(\omega)$ is the main source of degradation (smearing) of the point spread function. For instance, the phase of the radar signature of certain human-made metallic targets, such as a cylinder, is sinusoidal:

$$\begin{aligned}\text{phase}[\sigma_n(\omega)] &= \eta_n \sin\left(\frac{2\gamma_n \omega}{c}\right)\\ &= \eta_n \sin(\gamma_n k_x),\end{aligned}$$

where γ_n is related to the size (diameter) of the cylinder and η_n is related to a physical property of the cylinder. (The magnitude of the target radar signature also exhibits a similar sinusoidal behavior.) In this case, the target radar signature $\sigma_n(\omega)$ is a *tone phase-modulated* signal [car]; the inverse Fourier transform of the tone-modulated signal $\sigma_n(\omega)$ in the $x = ct/2$ domain is composed of delta functions separated by γ_n.

We showed earlier that the point spread function $\text{psf}_n(x)$ is the convolution of the inverse Fourier transform of the tone-modulated signal $\sigma_n(\omega)$ with $\text{psf}(x)$. Thus the nth target point spread function is composed of sinc-like blips separated by γ_n:

$$\text{psf}_n(x) = \sum_{\ell} \rho_\ell \, \text{psf}(x - \ell\gamma_n),$$

where ρ_ℓ, $\ell = 0, \pm 1, \pm 2, \ldots$, is typically a decreasing function of $|\ell|$. (Recall the Bessel pattern Fourier spectrum of a tone phase-modulated signal [car].) If γ_n is smaller than the range resolution Δ_x, then the nth target appears like a *smeared* blip around $x = x_n$ (Figure P1.10). We refer to this as a *range-extended* target. *Recall that $s_n(t)$ is composed of a primary echo and secondary echos that are closely packed in the time domain and may not be visually distinguishable.* We have observed the above point spread function in the radar signature of corner reflectors.

This phenomenon is different from what is called *range-spread* targets [edd; sko; van]. A range-spread target refers to a structure, for example, an aircraft, that is composed of multiple reflectors located at different range bins; this corresponds to our earlier model

$$f_0(x) = \sum_n \sigma_n \delta(x - x_n),$$

in which σ_n does not vary with the frequency ω. The above-mentioned range-extended target corresponds to a physical phenomenon, for example, resonance in the cavity of an aircraft engine, which makes that engine reflector appear as a smeared target, or *multiple* targets (see the above discussion on tone phase modulation) in the reconstructed range image.

There exists another phenomenon in range imaging using PM (or FM) radar signals, for example, chirp signals, which results in a point spread function that resembles the above-mentioned resonance tone-modulated spectrum point spread function. This effect was first reported by Klauder et al. [kla] in their original work with chirp radar signals. It turns out that matched filtering of PM signals never yields a *perfect* phase match in the frequency ω domain; this is due to processing (numerical) errors and/or radar signal instability.

Similarly, for the chirp-pulsed signals, the time domain pulse compression does not result in an ideal phase processing due to processing (numerical) errors and/or radar signal instability. In both cases (i.e., matched filtering or time domain pulse compression), the point spread function of a target appears smeared (range-extended) due to the *phase error* in the reference signal used for matched filtering or time-domain pulse compression.

- What is radar signal instability? This refers to the fact that we do not have a precise knowledge of the transmitted radar signal used for matched filtering. It might be due to the fact that the radar signal varies slightly from one transmission to another (e.g., due to the heat in the radar circuitry), or the user just does not have an analytical expression for the radar signal (this is particularly true for impulse-type radars).

In the latter case, when an analytical expression for the radar signal is not available, some use the return from a prominent target, for example, a corner reflector, as the reference signal. We have processed a SAR database for which a corner reflector return was used as the reference signal. After matched filtering, all the targets in the scene (trucks, etc.) exhibited the same range-extended behavior (tone phase-modulated spectrum) of a corner reflector. Clearly the phase information of those targets was *contaminated* by the phase of the reference signal which was extracted from a corner reflector signature.

Reconstruction via Target Signature Matched Filtering

One may use the knowledge of $\sigma_n(\omega)$ for target identification. Suppose that a certain class of targets, for example, buried mines, has a known radar signature $\sigma_m(\omega)$. To identify mines in an imaging scene, the user could construct a target function that is *matched* to the radar signature of a mine:

$$F_m(k_x) = S(\omega)P^*(\omega)\sigma_m^*(\omega);$$

we call $\sigma_m^*(\omega)$ the *mine-matched filter*. This concept is identical to the classical matched filtering which is also used in pattern recognition problems. This operation results in the removal of the phase of the mine radar signatures in $S(\omega)$ (the Fourier transform of the received signal). As a result the reconstructed mines in the range image $f_m(x)$ would exhibit *less* range extension and appear sharper. On the other hand, the other target types, which possess radar signature phase functions that are different from that of a mine radar signature, are likely to appear more widely spread (range extended).

1.8 MATLAB ALGORITHMS

All the Matlab programs in this book use a modified FFT which is described in the code *FFT subroutines*; these codes are listed prior to the codes for range imaging.

The code for range imaging is identified as *pulsed range imaging*. A chirp-pulsed radar signal is used in this program. However, the user could modify the code for other types of radar signals (for the case of matched-filtered reconstruction). Two scenarios are considered in the code. In case 1, the radar pulse time duration is $T_p = 0.1\ \mu$s, and the range swath echo time period is $T_x = 0.67\ \mu$s; thus $T_x > T_p$ and $\Delta_t > \Delta_{tc}$. In this case the measured samples are chosen based on Δ_t which is less restrictive than Δ_{tc}. Clearly this results in an aliased pulse compressed signal as can be seen in the range reconstruction in Figure P1.8a.

In case 2, the radar pulse time duration is increased to $T_p = 10\ \mu$s, while the range swath echo time period is maintained at $T_x = 0.67\ \mu$s; thus $T_x < T_p$ and $\Delta_t < \Delta_{tc}$. In this case the measured samples are chosen based on Δ_{tc}, which is less restrictive than Δ_t. This results in an aliased baseband-echoed signal. However, the Matlab code realizes this problem and performs sufficient upsampling in the compressed signal to convert the time sample spacing from Δ_{tc} to Δ_t. Then the upsampled compressed signal is decompressed to retrieve the alias-free echoed signal which is suitable for matched filtering.

For case 1 in which the compressed signal is aliased, one could develop a similar upsampling code to retrieve the alias-free compressed signal from the alias-free echoed signal. This is not essential, since the coherent matched filtered range reconstruction is preferable to the coherent pulse compressed range reconstruction which contains errors. (See the discussion on processing the deramped P-3 SAR data in Section 6.8.) The code provides the option of making the radar amplitude function a PM signal. The user could select the number of the phase harmonics in $a(t)$ via N_a; the value of $N_a = 0$ is for a conventional LFM chirp signal.

The code provides the following outputs for the two cases:

P1.1 Real part of the baseband echoed signal $s_b(t)$

P1.2 Real part of the baseband reference echoed signal $s_{0b}(t)$

P1.3 Baseband-echoed signal spectrum $|S_b(\omega)|$

P1.4 Baseband reference echoed signal spectrum $|S_{0b}(\omega)|$

P1.5 Baseband matched filtered signal spectrum $|S_{Mb}(\omega)|$. To display $|F(k_x)|$, plot the $|S_{Mb}(\omega)|$ array versus the spatial frequency k_x array

P1.6 Range reconstruction via matched filtering $|f(x)|$. To display $|s_{Mb}(t)|$, plot the $|f(x)|$ array versus the time t array

P1.7 Real part of the time domain compressed signal $s_c(t)$

P1.8 Range reconstruction via time domain compression $|f(x)|$

The end of the program contains a code for frequency-dependent target reflectivity function. Only one target, which is located at the center of target area $x_1 = X_c$, is simulated to see the effect of range extension. For this, the radar signal is a conventional LFM chirp, the radar pulse time duration is $T_p = 0.1$ μs, and the range swath echo time period is $T_x = 0.67$ μs. Results are shown for two cases: (1) $(\eta_1, \gamma_1) = (2, 1)$, and (2) $(\eta_1, \gamma_1) = (2, 5)$. Case 2 shows a slight range extension, while case 1 clearly exhibits the tone phase-modulation spectrum behavior. The following outputs are for the above cases:

P1.9 Real part of the baseband echoed signal $s_b(t)$

P1.10 Range reconstruction via matched filtering $|f(x)|$

2

CROSS-RANGE IMAGING

Introduction

This chapter provides a relatively *modern* view of cross-range imaging from the information obtained by a discrete or *sampled* aperture; this is also referred to as an *array*. The origin of this approach can be found in the wave equation inversion theory, which is also known as *wavefront reconstruction* or *holography* [born; gab; wol]. This theory is the principal foundation of many imaging systems in optics [goo; wol], geophysics [ber80; ber81; hag; loe; mil; pet; rob; sto], and diagnostic medicine [chr; mue]. The basic principle behind wavefront reconstruction is the use of the Fourier decomposition of a Green's function, which is also known as the *spherical phase function* and represents the impulse response of an imaging system [ber80; goo; mor; s94].

The foundation of most *analog* optical imaging systems, which utilize lenses for image formation, is based on approximations of the wavefront reconstruction, for example, the Fresnel or Fraunhofer approximation. The early SAR systems, which utilized some form of analog processing of the received signals, were also based on these approximations [broo; br67; br69; curt; cut; fit; kir751; kir752; kov76; kov77; tom; wil].

The introduction of powerful computers and fast digital signal processing algorithms, for example, FFT, has provided a fresh way to approach array imaging problems [ber81; chr; loe; mue; s94]. A similar revolution has occurred in other information processing systems. For example, one can perform narrowband *digital* filtering on a sampled signal in a computer which cannot be implemented via *analog* filters in practice. In the case of array imaging systems, one can implement the wavefront reconstruction method for image formation in a computer without any need for the Fresnel or Fraunhofer approximation.

A major obstacle in building *digital* information processing systems is the *sampling* system. In the case of time signals, for example, a speech signal in a telephone network or an incoming (echoed) microwave signal to a radar antenna, the challenge has been to build *fast* A/D converters that output a signal with a sufficiently high signal-to-noise (additive, quantization, etc.) *power* ratio. The A/D technology has advanced in many fields, such as speech processing and microwave signal processing.

In the case of spatial signals, for example, a radar, sonar, ultrasonic, or optical aperture, the problem is to construct an array (a sampled aperture) with individual transmitting and/or receiving *elements* (samplers) which are *small* and yet output a sufficiently high signal to noise power ratio. This has also been a rapidly advancing technology. (Perhaps the most difficult ones have been the optical array systems.)

- The size of an array element, or an aperture sampler, is analogous to the integration time of an A/D converter. They both represent the size of a flat-top sampler (or some windowed version of it). A smaller size for the flat-top sampler results in smaller sampling errors. We will see in stripmap SAR systems that in fact the size of the radar antenna (the sampler on the synthetic aperture) dictates the cross-range resolution.

There are two types of arrays: physical and synthesized [chr; s90; s92ju; s94; ste]. *Physical* arrays are composed of individual elements which are distributed on various spatial coordinates; the most common arrays are *linear* ones for which the elements are positioned along a linear path (aperture) in the spatial domain. *Synthesized* arrays, which include SAR, are formed by moving a *single* element along an *imaginary* aperture. In theory, a physical and a synthetic array that have a common aperture length yield the same information, and thus cross-range resolution [s90; s92ju]. In practice, however, a physical array is preferable because it provides a higher signal-to-noise power ratio by allowing the user to *turn on* all the elements simultaneously (big bang).

This is the main reason for using special types of physical arrays which are called *phased* arrays. A phased array allows the user to tap into the information base (linear signal subspace) of a physical array by introducing (relatively slight) delays in the transmitter and/or receiver paths of the elements [chr; s92ju; s94; ste]; the delays are varied to construct the signal subspace. In this case the target is still radiated with the big bang of the physical array. The conventional way to use a phased array for imaging a target is to vary the relative time delays of the elements to focus the radiation pattern of the array at specific points in the spatial domain; this scheme is known as dynamic focusing, which is a variation of *beam steering* [chr; s92ju; s94; ste]. Dynamic focusing is also based on certain approximations. The newer commercial dynamic focusing systems use complicated and time-consuming spatially varying array beam-forming methods. There is a wavefront reconstruction-based algorithm for phased array imaging that does not use approximations [s92ju; s94; s97].

Phased arrays are particularly popular in diagnostic ultrasonic imaging. In these imaging systems the transmitted ultrasonic signals of an element of an array (located outside the patient) need to penetrate the patient's body which has an acoustic impedance that is quite different from that of outside of the body. Most of the ultrasonic energy is bounced back at the boundary (the patient's skin). The resultant penetrated ultrasonic energy, which is much weaker now, has to go through the patient's body which exhibits high *attenuation* at ultrasonic frequency. After being echoed from a reflector inside the patient's body, the resultant ultrasonic energy has to travel the same attenuating path, and penetrate the patient/outside boundary to get to the elements of the array to be recorded. The relatively large acoustic impedance mismatch and attenuating tissue are the main factors for using the *big bang* of a phased array in medical ultrasonic imaging from outside a patient's body. We should point out that a bone has a high attenuation coefficient for ultrasonic sources. Because of this, the exterior-based ultrasonic phased arrays cannot be used to image parts of the human body that are (partially) enclosed in bony structures.

Synthetic arrays use the motion of a single element along an imaginary aperture to synthesize the effect of a physical array. In the Preface we provided a discussion on the utility of this form of data acquisition for radar imaging. This technology also has applications in many other imaging problems. For instance, it is of interest to utilize ultrasonic sources through esophagus to image the human chest cavity (transesophageal ultrasonic echocardiographic systems). The resolution of a (physical) phased array is limited by its aperture size. Meanwhile a large phased array penetrating via the throat is dangerous. Thus it is desirable to develop synthetic aperture array ultrasonic imaging methods that use a *mobile* single element ultrasonic transducer with a dimension that is significantly smaller than the size of a phased array. (The signal-to-noise power ratio is an important issue in using a *single* ultrasonic element; a synthetic aperture ultrasonic imaging system similar to the *stripmap* SAR system, which is discussed in Chapter 6, might provide a viable solution for transesophageal ultrasonic echocardiography.)

Outline

The discussion in this chapter is focused on the *cross-range* imaging capability of a synthetic aperture. The main tool for our approach is the *wavefront* reconstruction principle. As in *range imaging* presented in Chapter 1, our first task is to develop a mathematical relationship between what we measure (aperture signal) and what we desire to retrieve from it (target cross range and reflectivity), that is, the *system modeling*. This is done in Section 2.1 for cross-range imaging. A discussion on what we identify as a *broadside* target area and a *squint* target area is provided; the type of target area plays a role in the manner in which the signals are sampled and digitally processed.

In this section we also encounter a phase-modulated (PM) signal in the system model which we refer to as the *spherical* PM signal. Most of what we identified earlier as a *modern* view of aperture data analysis and wavefront reconstruction is based on the *Fourier* analysis of the spherical PM signal.

For this analysis we begin with the Fourier properties of the spherical PM signal over an *infinite* aperture in Section 2.2. Clearly measurements over an infinite aperture is not practical; however, infinite aperture analysis simplifies our analysis for the time being so that we can bring up more elementary properties of synthetic aperture signals. We use this Fourier analysis to develop a *matched filtered*-based reconstruction for cross-range imaging from synthetic aperture data over an infinite aperture in Section 2.3. Sampling and resolution for this problem are also examined.

Next, we bring up the practical issue of *finite* synthetic aperture measurements in Section 2.4. We show how the spectral band of the measured signal is affected by a finite aperture. In particular, we prove that a *windowing* (finite aperture effect) of the spherical PM signal results in a *windowing* of its Fourier spectrum; a similar phenomenon has been observed in optical processing with a lens [pap]. We revisit reconstruction via matched filtering for a finite synthetic aperture in Section 2.5. Cross-range resolution for this system is examined in Section 2.6.

In Section 2.7 we provide a treatment of the digital signal processing issues for the cross-range imaging with a finite aperture. Synthetic aperture sampling requirements for a broadside and squint target area is outlined. A digital signal processing of synthetic aperture data, which is similar to the pulse compression in range imaging, is introduced to reduce sampling rate in the aperture domain. In some cross-range imaging systems, the synthetic aperture data have to be *zero-padded* prior to any digital signal processing. The reason for this and the minimum size of the synthetic aperture are also discussed.

After establishing the digital signal processing principles, the digital reconstruction algorithm for cross-range imaging is discussed in Section 2.8. A method for reducing (compressing) the bandwidth of the reconstruction is shown.

In Section 2.9 we examine a problem for a synthetic aperture signal that is analogous to the frequency-dependent target reflectivity function in Chapter 1. In this problem we study the effect of a target reflectivity function that varies with the synthetic aperture. This discussion brings up a general analysis of a *windowed* spherical PM signal, which is referred to as the amplitude–modulated PM (AM-PM) signal. A Fourier analysis of the AM-PM signal is also provided. The cross-range reconstruction problem for the AM-PM synthetic aperture signal model is outlined in Section 2.10. The three classes of AM-PM SAR signals will be encountered in spotlight and stripmap SAR systems. The section also includes discussions on sampling and cross-range resolution for these scenarios. Sections 2.9 and 2.10 should be viewed as a summary of the synthetic aperture principles which will be used in the processing of spotlight and stripmap SAR signals in Chapters 5 and 6.

In the final section of Chapter 2, Section 2.11, the Matlab codes for the cross-range imaging problem are examined. Two sets of examples are considered, and their outputs are provided in Figures P2.1–P2.10. In case 1 (Figures P2.1*a*–P2.10*a*), a broadside target area is considered for which synthetic aperture zero-padding is not necessary. In case 2 in Figures P2.1*b*–P2.10*b*, a squint target area is examined; the parameters are chosen such that synthetic aperture zero-padding is essential. Figures P2.1–P2.10 provide an illustration of many modern synthetic aperture concepts of this chapter; these figures are heavily cited in Sections 2.1 through 2.10.

Mathematical Notations and Symbols

For our discussion on cross-range imaging, we deal with functions of a variable which we refer to as the *synthetic aperture* or *slow-time* domain (the reason for this will be discussed). We use the symbol u to represent this variable; the unit of u is meters. The frequency domain for this variable is called synthetic aperture spatial frequency domain or *slow-time Doppler* domain. The symbol k_u is used for this frequency domain variable; the unit of k_u is radians/meter. We also encounter bandpass signals in the u domain which are converted to the lowpass (baseband conversion) for digital processing. The other variables and Fourier integrals are the same as what we were dealing with in Chapter 1.

The following is a list of mathematical symbols used in this chapter, and their definitions:

$a(\omega, x, y)$	Amplitude pattern
$A(\omega, k_u)$	Slow-time Doppler amplitude pattern
$\mathbf{a}(k_u)$	Fourier transform of $a(\omega, x, u)$ with respect to slow-time u
$a_n(\omega, x, y)$	Amplitude pattern for n-target (general case)
$A_n(\omega, k_u)$	Slow-time Doppler amplitude pattern for n-target (general case)
$A_{pn}(\omega, \phi)$	Slow-time angular Doppler amplitude pattern for n-target (general case)
B	Half-beamwidth of amplitude pattern; beamwidth of amplitude pattern is $2B$
B_n	Effective half-beamwidth of amplitude pattern for nth target. Effective beamwidth of amplitude pattern for nth target is $2B_n$
c	Wave propagation speed $c = 3 \times 10^8$ m/s for radar waves $c = 1500$ m/s for acoustic waves in water $c = 340$ m/s for acoustic waves in air
D	Diameter of radar
$f(y)$	Target function in cross-range domain
$F(k_y)$	Fourier transform of target function
$f_c(y)$	Baseband converted target function in cross-range domain
$F_c(k_y)$	Fourier transform of baseband-converted target function
$f_0(y)$	Ideal target function in cross-range domain
$F_0(k_y)$	Fourier transform of ideal target function
$i_n(\omega, u)$	Inverse Fourier transform of slow-time Doppler domain indicator function for nth target; inverse Fourier transform of indicator function $I_n(\omega, k_u)$
$I_n(\omega, k_u)$	Indicator (window) function for slow-time Doppler support band of nth target; indicator (window) function for slow-time Doppler support band of $S_n(\omega, k_u)$

k	Wavenumber: $k = \omega/c$
k_u	Spatial frequency or wavenumber domain for synthetic aperture u; slow-time Doppler (frequency) domain
k_{ui}	Slow-time Doppler sampled points for synthetic aperture signal
k_x	Spatial frequency or wavenumber domain for range x
k_y	Spatial frequency or wavenumber domain for cross-range y
$K_{un}(u)$	Instantaneous frequency of signature of nth target in slow-time domain; instantaneous frequency of $s_n(\omega, u)$
$K_{u0}(u)$	Instantaneous frequency of reference signal in slow-time domain; instantaneous frequency of $s_0(\omega, u)$
L	Half-size of synthetic aperture
$L_{\min}$	Minimum half-length of processed (zero-padded) synthetic aperture; minimum length of processed (zero-padded) synthetic aperture is $2L_{\min}$
M	Number of samples of synthetic aperture signal in slow-time u domain
M_c	Number of samples of slow-time compressed synthetic aperture signal in slow-time u domain
n	Index representing a specific target [used with (σ_n, y_n)]
$p(t)$	Transmitted radar signal; radar signal is a single tone with frequency ω in this chapter
$\text{psf}(y)$	Point spread function in cross-range domain, Type 1
$\text{psf}_n(y)$	Point spread function in cross-range domain for nth target, finite aperture case with no amplitude pattern variations
$\text{psf}_{I_n}(y)$	Point spread function in cross-range domain for nth target, Type 1
PRF	Radar pulse repetition frequency in slow-time
PRI	Radar pulse repetition interval in slow-time
$\text{PSF}_n(k_y)$	Spatial frequency filter function for nth target
r_n	Target radial distance from center of synthetic aperture
R_c	Radial distance of center of target area from center of synthetic aperture
$\mathbf{s}(t, u)$	Echoed signal from target area when radar is at $(0, u)$
$s(\omega, u)$	Fast-time baseband converted version of echoed signal; synthetic aperture signal
$S(\omega, k_u)$	Slow-time Fourier transform of synthetic aperture signal
$s_c(\omega, u)$	Slow-time compressed synthetic aperture signal
$S_c(\omega, k_u)$	Slow-time Fourier transform of slow-time compressed synthetic aperture signal
$s_n(\omega, u)$	Fast-time baseband converted version of echoed signal from nth target; signature of nth target in slow-time domain
$S_n(\omega, k_u)$	Slow-time Fourier transform of signature of nth target
$s_{nc}(\omega, u)$	Slow-time compressed signature of nth target in slow-time domain

$S_{nc}(\omega, k_u)$	Slow-time Fourier transform of slow-time compressed signature of nth target
$s_0(\omega, u)$	Reference synthetic aperture signal
$S_0(\omega, k_u)$	Slow-time Fourier transform of reference synthetic aperture signal
$\text{sinc}(\cdot)$	sinc function: $\text{sinc}(a) = \sin(\pi a)/\pi a$
t	Fast-time domain
u	Synthetic aperture or slow-time domain
u_i	Synthetic aperture sampled points for synthetic aperture signal
u_{ic}	Synthetic aperture sampled points for slow-time compressed synthetic aperture signal
$U_n(k_u)$	Instantaneous frequency of signature of nth target in slow-time Doppler domain; instantaneous frequency of $S_n(\omega, k_u)$
$\mathbf{u}(k_u)$	Phase center of PM signal $\exp\left[-j2k\sqrt{x_n^2 + (y_n - u)^2} - jk_u u\right]$ in u domain for a fixed k_u
v_R	Speed of radar-carrying vehicle (e.g., an aircraft)
x	Range domain
x_n	Range of nth target. $x_n = X_c$ is a constant in this chapter.
X_c	Fixed range of all targets ($x_n = X_c$ for $n = 1, 2, \ldots$)
y	Cross-range domain
y_i	Cross-range sampled points in reconstructed target function
y_n	Cross-range of nth target
Y_c	Center point of target area in cross-range domain; squint cross-range
Y_0	Half-size of target area in cross-range domain; size of target area in cross-range domain is $2Y_0$
$\delta(\cdot)$	Delta or impulse function
Δ_{k_u}	Sample spacing of synthetic aperture signal in slow-time Doppler domain
Δ_{k_y}	Sample spacing of reconstructed target function in spatial frequency domain
Δ_u	Sample spacing of synthetic aperture signal
Δ_{uc}	Sample spacing of slow-time compressed synthetic aperture signal
Δ_y	Cross-range resolution for an infinite synthetic aperture
Δ_{y_n}	Cross-range resolution for a finite synthetic aperture
(η, ζ)	Parameters of power window function
$\mathcal{F}$	Forward Fourier transform operator
$\mathcal{F}^{-1}$	Inverse Fourier transform operator
λ	Wavelength: $\lambda = 2\pi c/\omega$
ω	Single frequency of radar signal
Ω_A	Slow-time Doppler support band of amplitude pattern $A(\omega, k_u)$

Ω_c	Slow-time Doppler support band of slow-time compressed synthetic aperture signal; slow-time Doppler support band of $S_c(\omega, k_u)$
Ω_n	Slow-time Doppler support band of nth target; slow-time Doppler support band of $S_n(\omega, k_u)$
$\lvert\Omega_n\rvert/2$	Baseband Doppler bandwidth of nth target; baseband Doppler bandwidth of $S_n(\omega, k_u)$
Ω_{nc}	Center (carrier) Doppler frequency of nth target; center (carrier) Doppler frequency of $S_n(\omega, k_u)$
Ω_s	Slow-time Doppler support band of synthetic aperture signal; slow-time Doppler support band of $S(\omega, k_u)$
ϕ	Slow-time angular spectral (Doppler) domain
σ_n	Reflectivity of nth target (invariant in synthetic aperture)
θ_c	Squint angle of center of target area
$\theta_{\max}$	Absolute largest aspect angle of targets; largest aspect angle of targets
$\theta_{\min}$	Smallest aspect angle of targets
$\theta_n(u)$	Aspect angle of nth target when radar is at $(0, u)$
ξ_n	Percentage of main lobe of nth target signature in synthetic aperture $[-L, L]$ for Type 3

2.1 SYSTEM MODEL

Consider the imaging scenario in Figure 2.1*a* or 2.1*b* where a group of targets are located at the coordinates (x_n, y_n), $n = 1, 2, \ldots$. In this chapter we consider the special case where all the targets are at a *fixed and known range* value, for example, $x_n = X_c$ $(n = 1, 2, \ldots)$, with X_c a known constant. The cross-range values of these targets, labeled y_n, and their reflectivity σ_n $(n = 1, 2, \ldots)$ are *unknown.*

The case shown in Figure 2.1*a* corresponds to a *finite* target area in the cross-range domain where all the targets fall within the following cross-range interval:

$$y_n \in [-Y_0, Y_0], n = 1, 2, \ldots,$$

where Y_0, the *half-size* of the target area is known; $2Y_0$ is the size of the target area in the cross-range domain. The target area centered around the *origin* in the cross-range y domain is called a *broadside* target area. Figure 2.1*b* depicts a *general* target area, which is called a *squint* target area, where all the targets fall within the following cross-range interval:

$$y_n \in [Y_c - Y_0, Y_c - Y_0], \qquad n = 1, 2, \ldots.$$

Y_c is called the *center of mass* of the target area and is known. Note that the broadside target area is a special case of the squint target area with $Y_c = 0$. In the general squint case the target area is centered at (X_c, Y_c) in the spatial (x, y) domain.

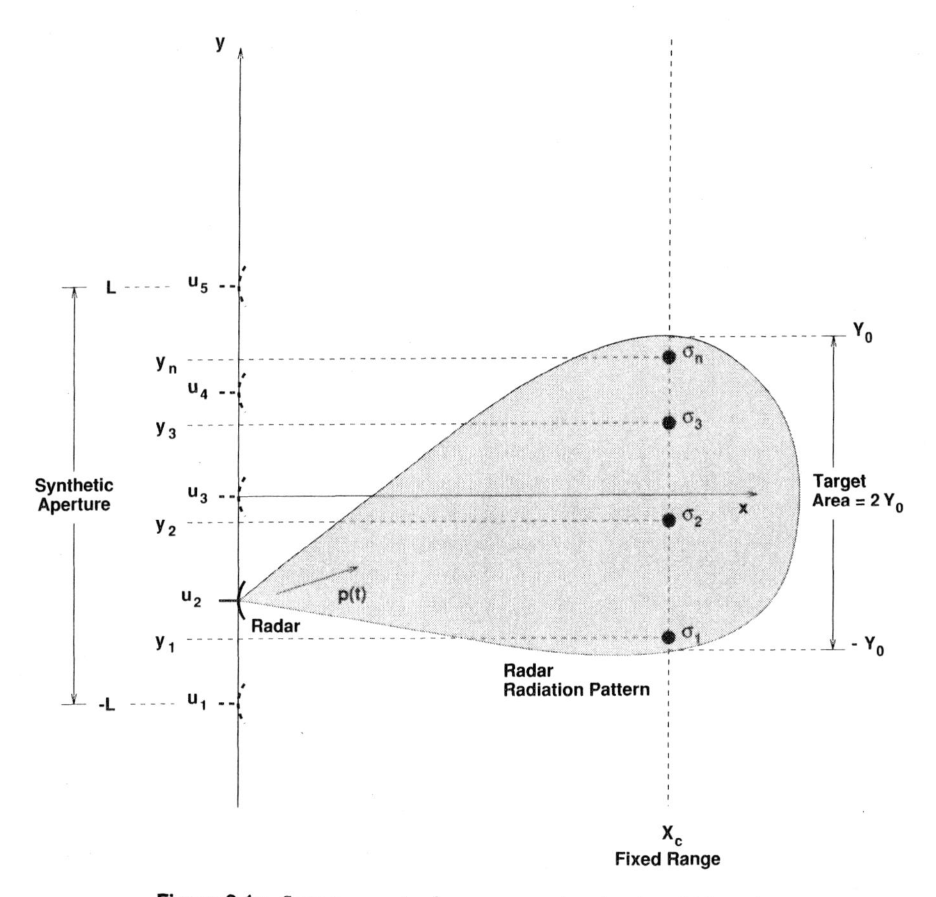

Figure 2.1a. System geometry for cross-range imaging: broadside mode.

Similar to the range imaging problem, we define an *ideal* target function in the cross-range domain via

$$f_0(y) = \sum_n \sigma_n \delta(y - y_n),$$

which represent the best we can do to identify the cross-range coordinates and reflectivities of the targets. The spatial Fourier transform of the ideal target function with respect to the cross-range y is

$$F_0(k_y) = \sum_n \sigma_n \exp(-jk_y y_n).$$

Note that $F_0(k_y)$ is composed of a sum of linear phase functions of y_n's.

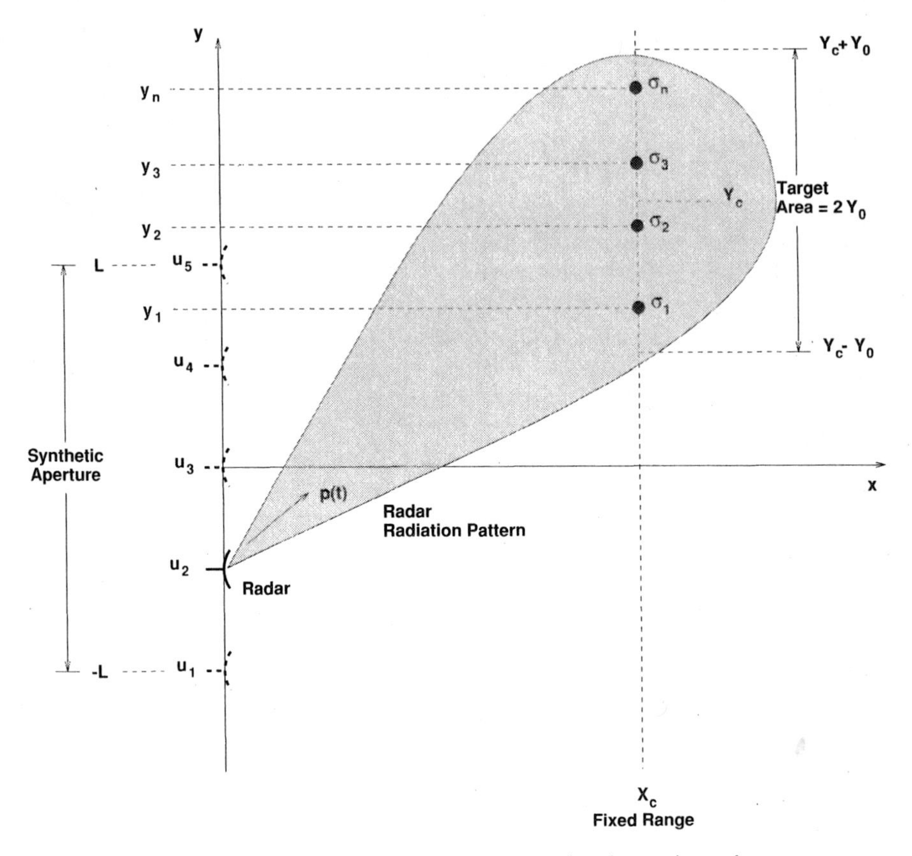

Figure 2.1b. System geometry for cross-range imaging: squint mode.

- One may define the center of mass of the target area in the cross-range domain Y_c for a continuous target function $f_0(y)$ via

$$Y_c = \frac{\int_y y|f_0(y)|^2\,dy}{\int_y |f_0(y)|^2\,dy},$$

  and for a discrete target model via

$$Y_c = \frac{\sum_n y_n|\sigma_n|^2}{\sum_n |\sigma_n|^2}.$$

  In practice, the user has some approximate a priori knowledge of the target center of mass in the cross-range domain, that is, Y_c.

A transmitting/receiving radar is located at $(0, u)$ in the spatial (x, y) domain. *The cross-range location of the radar, that is, u, can be changed.* For instance, the radar is mounted on a vehicle that can move along the cross-range y domain. The u domain is called the *synthetic aperture* domain. For each position of the vehicle on the synthetic aperture, the radar illuminates the target area with its signal $p(t)$ and records the resultant echoed signals.

- The airborne vehicle (aircraft) possesses a continuous motion. For pulsed SAR systems, within a very good approximation, one can assume that the vehicle stops, a radar measurement is made (transmit and receive), and then the vehicle is moved to its next position on the synthetic aperture. This assumption is not valid in airborne synthetic aperture acoustic imaging. One can develop synthetic aperture signal processing principles for a continuous wideband (FM-CW) radar or acoustic system without the stop-and-go assumption for the vehicle. For simplicity, we consider the stop-and-go assumption for the vehicle in this chapter.
- The time domain t is called the *fast-time* domain. This is due to the fact that it corresponds to the time delays associated with the propagation of the echoed signals that move with the fast speed of light, that is, $c = 3 \times 10^8$. The synthetic aperture domain u is referred to as the *slow-time* domain. This term originates from the fact that the synthetic aperture is generated via the motion of a vehicle (e.g., an aircraft) that carries the radar. The motion of the vehicle is much slower than the motion (propagation) of the echoed signals, since the speed of the vehicle is much slower than the speed of light.

Variations of u can be viewed as changing the *aspect angle* of the radar with respect to the target area. For the nth target the aspect angle of the radar when it is located at $(0, u)$ is defined via

$$\theta_n(u) = \arctan\left(\frac{y_n - u}{x_n}\right).$$

This is the angle between the x axis and the line that connects the radar [at $(0, u)$] to the nth target [at (x_n, y_n)]; see Figure 2.2. Note that we use x_n to identify the range of the targets though x_n's are equal to a known constant value X_c. This gives our analysis a more general derivation which is applicable for the cases in which x_n's are unknown and varying with n; these will be studied in future chapters.

Consider the case when the radar transmits a continuous *single frequency* signal

$$p(t) = \exp(j\omega t).$$

Why is a wide-bandwidth radar signal unnecessary? From an information theory point of view, we *might* be able to retrieve *one-dimensional* target information in the cross-range y domain from the *one-dimensional* measurements in the synthetic aperture u domain. We will change the phrase *might* to *could* in the discussion that follows. In our present discussion on one-dimensional cross-range imaging, the trans-

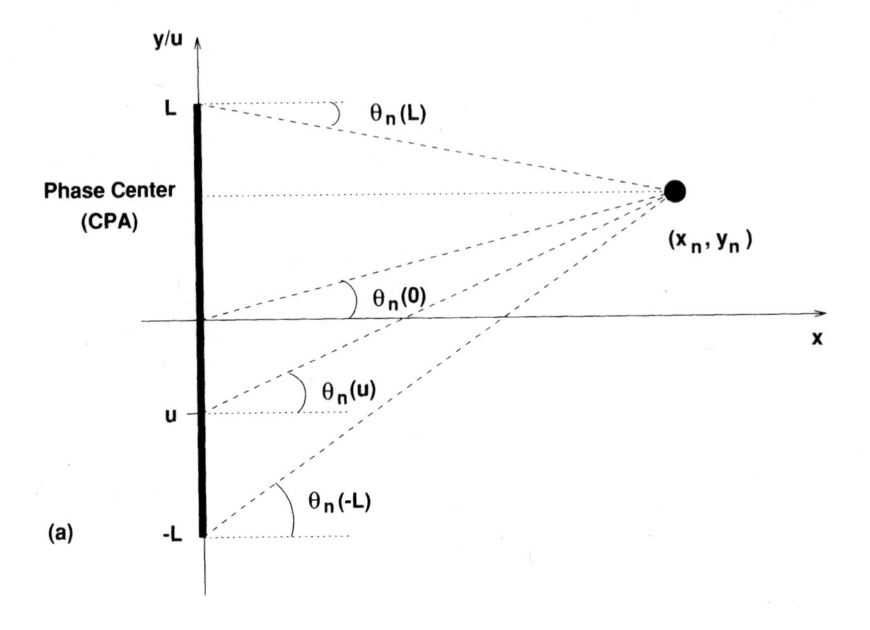

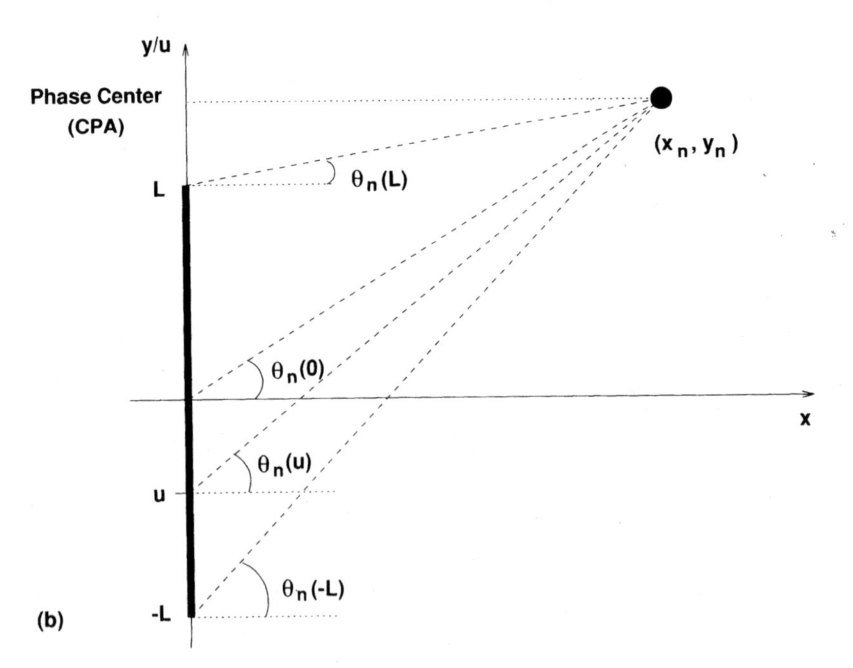

Figure 2.2. Aspect angle of nth target with respect to radar: (a) $|y_n| < L$ and (b) $|y_n| > L$.

mitted radar signal is assumed to be a single-frequency sinusoid with frequency ω; the frequency ω is a constant. In our future discussion on multidimensional SAR imaging, we will deal with *wide-bandwidth* transmitted radar signals where ω becomes a variable that changes within a band, for example, $\omega \in [\omega_c - \omega_0, \omega_c + \omega_0]$. This is the same form of radar signal that we used for range imaging in Chapter 1.

The distance of the radar from the target located at (x_n, y_n) is

$$\sqrt{x_n^2 + (y_n - u)^2}.$$

Thus the recorded echoed signal as u varies can be expressed via the following function of u:

$$\begin{aligned} \mathbf{s}(t, u) &= \sum_n \sigma_n p\left[t - \frac{2\sqrt{x_n^2 + (y_n - u)^2}}{c}\right] \\ &= \exp(j\omega t) \sum_n \sigma_n \exp\left[-j2k\sqrt{x_n^2 + (y_n - u)^2}\right] \end{aligned}$$

where $k = \omega/c$ is called the wavenumber. The point $u = y_n$ on the synthetic aperture is called the closest point of approach (CPA) of the radar for the nth target.

Similar to the range imaging problem of Chapter 1, here we use a discrete model for simplicity of notation; the analysis that follows is also suitable for a continuous target model in the cross-range domain. Moreover, in practice, the target reflectivity σ_n varies with the radar frequency ω and the relative coordinates of the target with respect to the radar, that is, the synthetic aperture position u. These issues will be brought up later in this chapter and in Chapter 3. At this point we wish to present the basic principles behind cross-range imaging via utilizing an aperture.

After fast-time baseband conversion, the recorded signal becomes

$$\begin{aligned} s(\omega, u) &= \mathbf{s}(t, u) \exp(-j\omega t) \\ &= \sum_n \sigma_n \exp\left[-j2k\sqrt{x_n^2 + (y_n - u)^2}\right]. \end{aligned}$$

Note that the fast-time baseband signal $s(\omega, u)$ is identified as a function of the radar frequency ω; this provides a more general symbol that is useful for our future discussion.

We denote the baseband signal for a unit reflector at the center of the *broadside* target area, that is, $(x_n, y_n) = (X_c, 0)$ via

$$s_0(\omega, u) = \exp\left(-j2k\sqrt{X_c^2 + u^2}\right).$$

We call $s_0(\omega, u)$ the *reference* signal. [We will show in our future discussion on reconstruction that the reference signal should be the baseband signal for a unit reflector at the *center* of the general squint target area, that is, $(x_n, y_n) = (X_c, Y_c)$. In

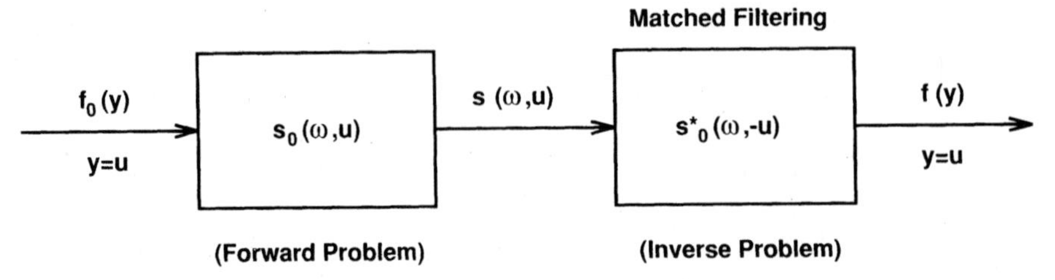

Figure 2.3. Forward and inverse system models for cross-range imaging.

the current discussion it is notationally more convenient to use $Y_c = 0$.] In this case the recorded signal $s(\omega, u)$ can be expressed via

$$s(\omega, u) = f_0(u) * s_0(\omega, u),$$

where $*$ denotes convolution in the synthetic aperture u domain, and

$$f_0(u) = \sum_n \sigma_n \delta(u - y_n)$$

is the ideal target function in the cross-range domain with $y = u$. This convolution model of the recorded signal is shown in Figure 2.3 and is identified as the *forward* problem.

2.2 SPHERICAL PM SIGNAL WITHIN AN INFINITE APERTURE

We can also express the synthetic aperture signal via the sum of the baseband-echoed signals from the individual targets in the imaging scene

$$s(\omega, u) = \sum_n s_n(\omega, u),$$

where

$$s_n(\omega, u) = \sigma_n \exp\left[-j2k\sqrt{x_n^2 + (y_n - u)^2}\right].$$

The phase function $\exp[-j2k\sqrt{x_n^2 + (y_n - u)^2}\,]$ on the right side of the above is a phase-modulated (PM) signal which we refer to as the *spherical* PM signal [mor; sco; s94]. Note that the spherical PM signal is a *nonlinear* phase function of (x_n, y_n) as well as u.

Examples of the spherical PM (synthetic aperture) signal are shown in Figure 2.4*a* and 2.4*b* for $y_n = 50$ m and 150 m, respectively. The dashed line represents the spherical PM signal at all the u values which are shown in this figure. The spherical PM signal over a relatively smaller u region is shown by a solid line; we refer to this as the spherical PM signal over a *finite* aperture.

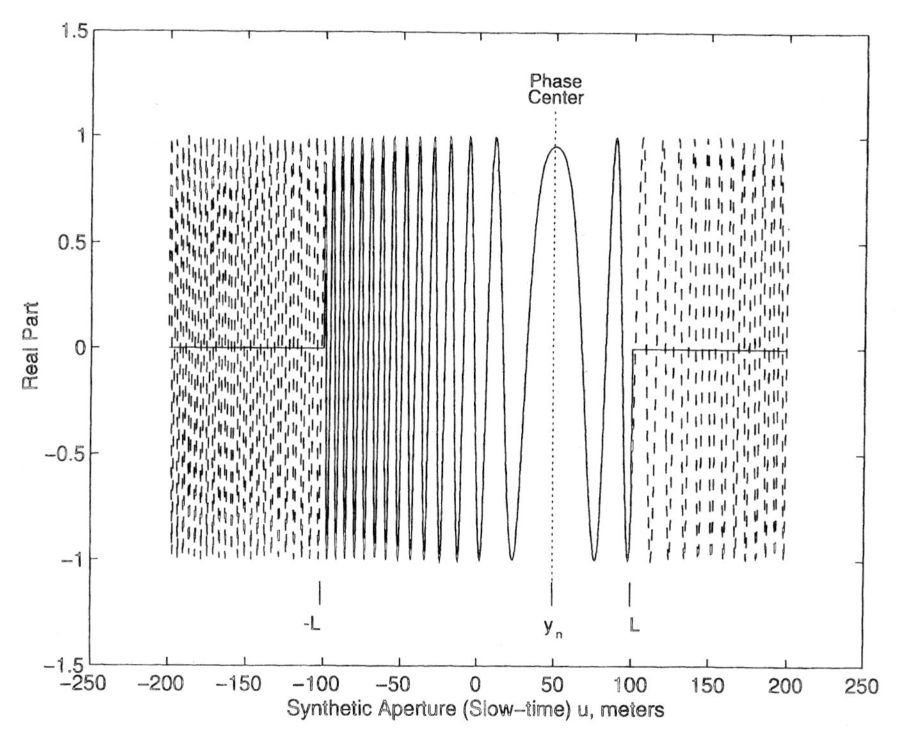

Figure 2.4a. Spherical PM signal $s(\omega, u)$ in synthetic aperture u domain: $y_n = 50$ m.

In this section we examine the Fourier properties of the spherical PM signal when the synthetic aperture measurements are made within an *infinite* aperture, that is, $u \in (-\infty, \infty)$; the effect of a finite aperture will be examined. First, we wish to determine the Fourier transform of the spherical PM signal $s_n(\omega, u)$ with respect to the synthetic aperture u; this is found via the following Fourier integral:

$$S_n(\omega, k_u) = \int_{-\infty}^{\infty} s_n(\omega, u) \exp(-jk_u u)\, du$$
$$= \int_{-\infty}^{\infty} \sigma_n \exp\left[-j2k\sqrt{x_n^2 + (y_n - u)^2}\right] \exp(-jk_u u)\, du,$$

where k_u represents the spatial frequency domain for the synthetic aperture domain u. This spatial frequency variable, that is, k_u, is also referred to as the *slow-time frequency* or *Doppler* domain.

After rearranging the components of the above Fourier integral, we obtain

$$S_n(\omega, k_u) = \sigma_n \int_{-\infty}^{\infty} \exp\left[-j2k\sqrt{x_n^2 + (y_n - u)^2} - jk_u u\right] du$$

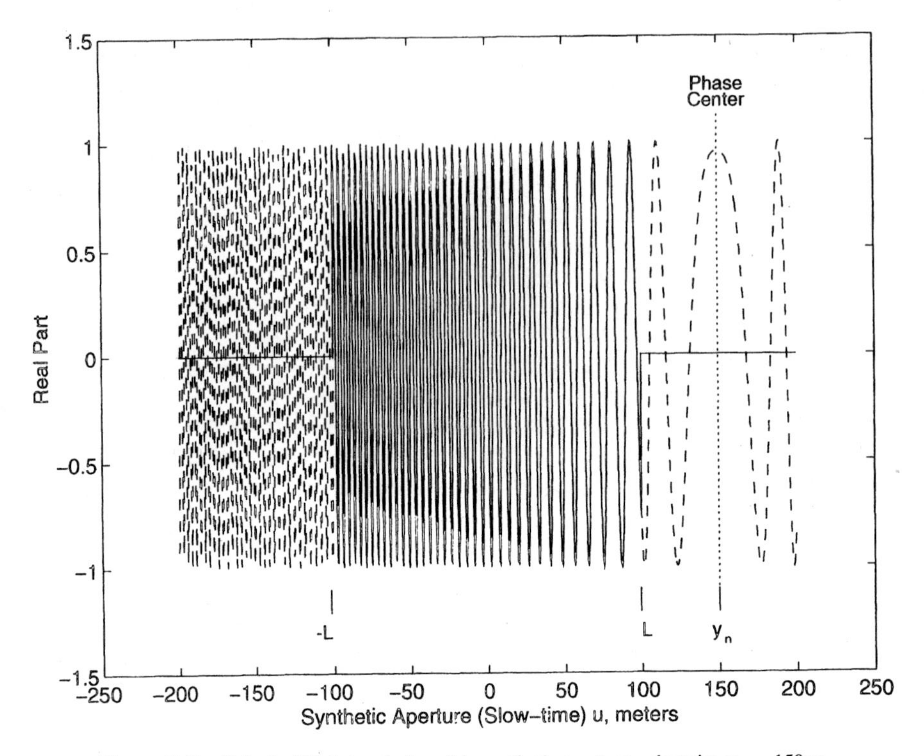

Figure 2.4b. Spherical PM signal $s(\omega, u)$ in synthetic aperture u domain: $y_n = 150$ m.

The above integral can be evaluated using the method of *stationary phase* [mor; pap; sco; s94]. We will go through a more general form of this derivation later. At this point we simply show the outcome, which is

$$S_n(\omega, k_u) = \sigma_n \frac{\exp(-j\pi/4)}{\sqrt{4k^2 - k_u^2}} \exp\left(-j\sqrt{4k^2 - k_u^2}\, x_n - jk_u y_n\right)$$

for $k_u \in [-2k, 2k]$ and zero otherwise. *Note that $S_n(\omega, k_u)$ has a finite support band.* Examples of $S_n(\omega, k_u)$, for the $s_n(\omega, u)$ signals in Figure 2.4*a* and *b* are shown in Figure 2.5*a* and *b* for a *finite* synthetic aperture; the effect of a finite synthetic aperture will be discussed.

The term $\exp(-j\pi/4)/\sqrt{4k^2 - k_u^2}$ is a known and slowly-fluctuating amplitude function that does not play an important role in our analysis. For notational simplicity we suppress this amplitude function to give more prominence to the phase term

$$\exp\left(-j\sqrt{4k^2 - k_u^2}\, x_n - jk_u y_n\right),$$

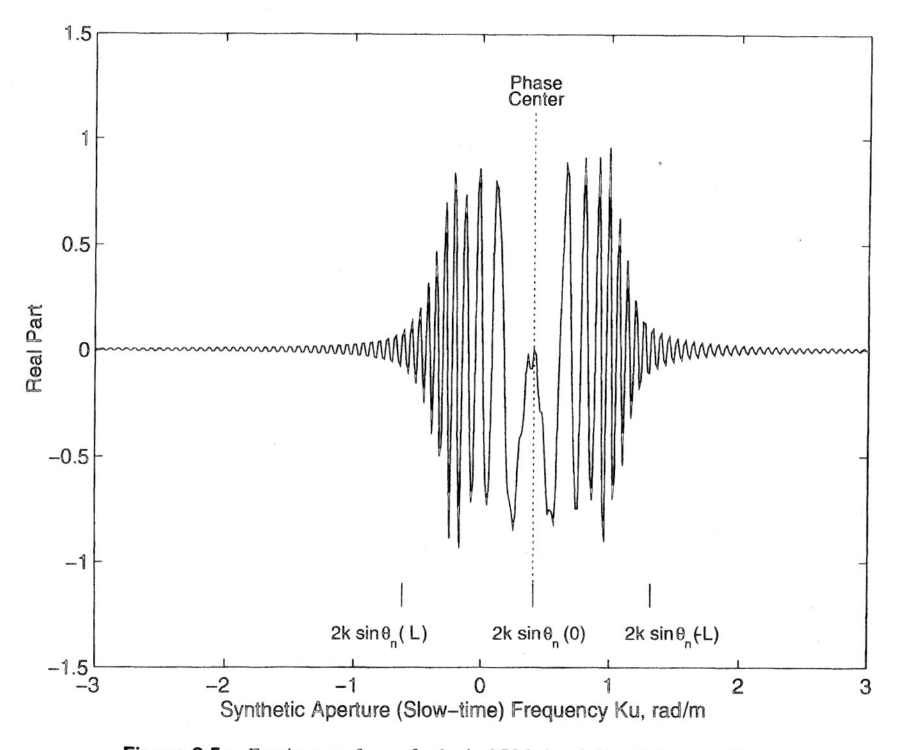

Figure 2.5a. Fourier transform of spherical PM signal $S(\omega, k_u)$: $y_n = 50$ m.

which is the key term in cross-range imaging. Thus we rewrite the spatial Fourier transform of the spherical PM signal via

$$S_n(\omega, k_u) = \sigma_n \exp\left(-j\sqrt{4k^2 - k_u^2}\, x_n - jk_u y_n\right)$$

for $k_u \in [-2k, 2k]$, and zero otherwise. We also denote

$$k_x = \sqrt{4k^2 - k_u^2}.$$

The reason for using the above designation for k_x (range spatial frequency) will become obvious in our discussion of SAR systems in Chapters 4 through 6.

The following observations could be made about the phase function

$$\exp\left(-j\sqrt{4k^2 - k_u^2}\, x_n - jk_u y_n\right)$$

in the slow-time Doppler spectrum of the spherical PM signal $S_n(\omega, k_u)$:

1. This phase function is a *nonlinear* function of k_u. This implies that $S_n(\omega, k_u)$ is also a *PM signal* in the slow-time Doppler k_u domain. (This is simply an

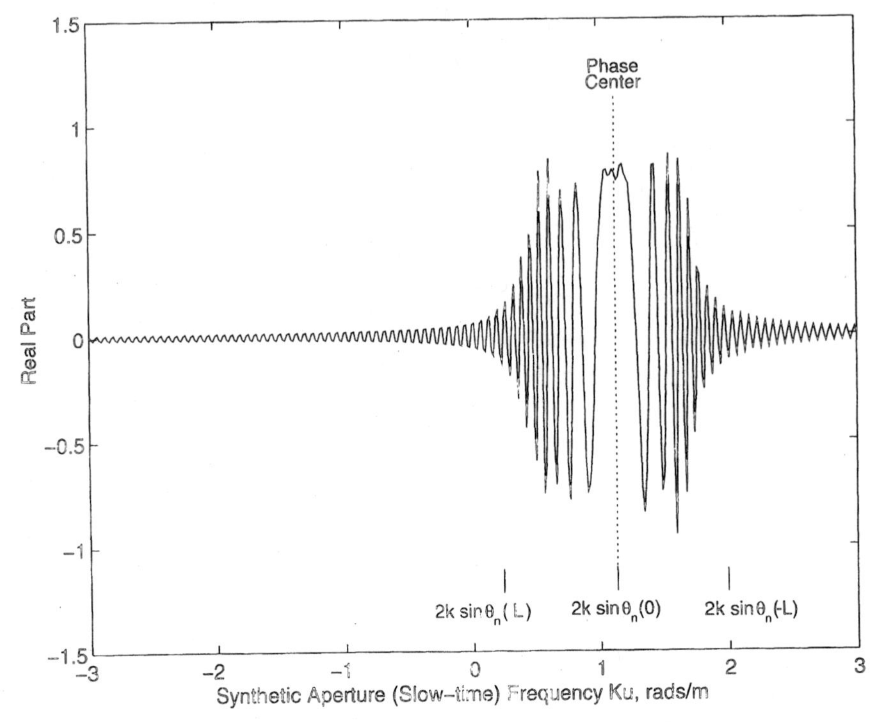

Figure 2.5b. Fourier transform of spherical PM signal $S(\omega, k_u)$: $y_n = 150$ m.

observation; it has no impact on what we are trying to establish for cross-range imaging.)

2. This phase function contains a *linear* function of (x_n, y_n), while the spherical PM signal $s_n(\omega, u)$ is composed of a *nonlinear* phase function of (x_n, y_n). The significance of this in cross-range imaging will become obvious soon.

2.3 RECONSTRUCTION VIA MATCHED FILTERING: INFINITE APERTURE

The spatial Fourier transform of the received baseband signal $s(\omega, u)$ with respect to the synthetic aperture u is the sum of the spatial Fourier transform of the individual spherical PM signals $s_n(\omega, u)$'s; that is,

$$S(\omega, k_u) = \sum_n \sigma_n S_n(\omega, k_u)$$
$$= \sum_n \sigma_n \exp\left(-j\sqrt{4k^2 - k_u^2}\, x_n - jk_u y_n\right)$$

for $k_u \in [-2k, 2k]$, and zero otherwise. Using the fact that all the targets are located at a common range value, that is, $x_n = X_c$ $(n = 1, 2, \ldots)$, we can rewrite the above via

$$S(\omega, k_u) = \exp\left(-j\sqrt{4k^2 - k_u^2}\, X_c\right) \sum_n \sigma_n \exp(-jk_u y_n)$$

for $k_u \in [-2k, 2k]$, and zero otherwise.

Earlier we denoted the baseband signal for a unit reflector at the center of the broadside target area, that is, $(x_n, y_n) = (X_c, 0)$ via

$$s_0(\omega, u) = \exp\left[-j2k\sqrt{X_c^2 + u^2}\right]$$

and called it the *reference* signal. The spatial Fourier transform of the reference signal with respect to the synthetic aperture u is

$$S_0(\omega, k_u) = \exp\left(-j\sqrt{4k^2 - k_u^2}\, X_c\right)$$

for $k_u \in [-2k, 2k]$, and zero otherwise.

Using the expressions for the reference signal $S_0(\omega, k_u)$, and the ideal target function Fourier transform

$$F_0(k_y) = \sum_n \sigma_n \exp(-jk_y y_n)$$

in the expression for $S(\omega, k_u)$, one obtains

$$S(\omega, k_u) = S_0(\omega, k_u) F_0(k_u).$$

Theoretically we could have recovered the ideal target function $F_0(k_u)$ by the following division:

$$\frac{S(\omega, k_u)}{S_0(\omega, k_u)}.$$

This is not a practical option, since both $S(\omega, k_u)$ and $S_0(\omega, k_u)$ have a finite support band which is $k_u \in [-2k, 2k]$.

A practical reconstruction can be achieved via the following matched filtering:

$$\begin{aligned} F(k_y) &= S(\omega, k_u) S_0^*(\omega, k_u) \\ &= \sum_n \sigma_n \exp(-jk_y y_n) \end{aligned}$$

for $k_y \in [-2k, 2k]$, and zero otherwise, where S_0^* is the complex conjugate of S_0, and the cross-range spatial frequency mapping is defined via

$$k_y = k_u.$$

Note that the spatial frequency domain for the cross-range domain is a direct mapping of the spatial frequency domain for the synthetic aperture domain. The inverse Fourier transform of $F(k_y)$ with respect to the spatial frequency k_y yields the cross-range reconstructed image

$$f(y) = \sum_n \sigma_n \operatorname{sinc}\left[\frac{k}{\pi}(y - y_n)\right].$$

Note that this reconstruction may also be expressed as a matched-filtering convolution in the synthetic aperture or slow-time domain:

$$f(y) = s(\omega, u) * s_0^*(\omega, -u),$$

where $*$ denotes convolution in the synthetic aperture u domain, and the cross-range mapping is

$$y = u.$$

This matched filtering operation is shown in Figure 2.3, and is identified as the *inverse* problem.

Synthetic Aperture (Slow-time) Sampling

We showed earlier that the support band of $S(\omega, k_u)$ is $k_u \in [-2k, 2k]$. Thus the received baseband signal $s(\omega, u)$ is bandlimited with baseband bandwidth of $\pm 2k$. In this case, for alias-free sampling of this signal in the synthetic aperture, we should satisfy the following Nyquist rate:

$$\begin{aligned}\Delta_u &\le \frac{\pi}{2k} \\ &= \frac{\lambda}{4},\end{aligned}$$

where $\lambda = 2\pi/k$ is the wavelength at the radar frequency ω.

Cross-range Resolution

The cross-range reconstruction $f(y)$ is composed of sinc-like blips at the cross-range coordinates of the targets. The bandwidth of the observability of the signature of the nth target in the k_u domain, that is, $\sigma_n \exp(-jk_y y_n)$, is within $k_y \in [-2k, 2k]$; this is invariant of n. Thus the cross-range resolution for all the targets is

$$\begin{aligned}\Delta_y &= \frac{\pi}{2k} \\ &= \frac{\lambda}{4}.\end{aligned}$$

It turns out that this is an overoptimistic performance. There is a practical limitation in acquiring synthetic aperture data that greatly affects the cross-range resolution as well as the spectral properties of the received signal in the k_u domain. This is described in the next section.

2.4 SPHERICAL PM SIGNAL WITHIN A FINITE APERTURE

The analysis shown in Sections 2.2 and 2.3 was performed with the assumption that the synthetic aperture measurements were made within an infinite aperture $u \in (-\infty, \infty)$. In practice, the user could only obtain a finite number of measurements in the synthetic aperture u domain within a finite interval $u \in [-L, L]$ as shown with the solid lines in Figures 2.4*a* and *b* with $L = 100$ m; $2L$ is the size of the synthetic aperture. We now examine the effects of finite synthetic aperture in our analysis of spherical PM signal.

Instantaneous Frequency

The spherical PM signal for the nth target is identified via

$$s_n(\omega, u) = \sigma_n \exp\left[-j2k\sqrt{x_n^2 + (y_n - u)^2}\right],$$

which depends on the coordinates of the target in the spatial domain (x_n, y_n), the radar coordinates $(0, u)$, and the wavenumber of the radar signal $k = \omega/c$. The instantaneous frequency of this PM signal in the synthetic aperture domain u is

$$\begin{aligned} K_{un}(u) &= \frac{\partial}{\partial u}\left[-2k\sqrt{x_n^2 + (y_n - u)^2}\right] \\ &= \frac{2k(y_n - u)}{\sqrt{x_n^2 + (y_n - u)^2}} \\ &= 2k \sin\theta_n(u), \end{aligned}$$

where

$$\theta_n(u) = \arctan\left(\frac{y_n - u}{x_n}\right)$$

is the aspect angle of the radar for the nth target when the radar is located at $(0, u)$ on the synthetic aperture; see Figure 2.2.

Figure 2.6*a* and *b* show instantaneous frequency $K_{un}(u)$ functions for the spherical PM signals of Figure 2.4*a* and *b*. The solid line represents this instantaneous frequency over the finite aperture $u \in [-L, L]$. The dashed line is the instantaneous frequency $K_{un}(u)$ function at all the u values shown in these figures.

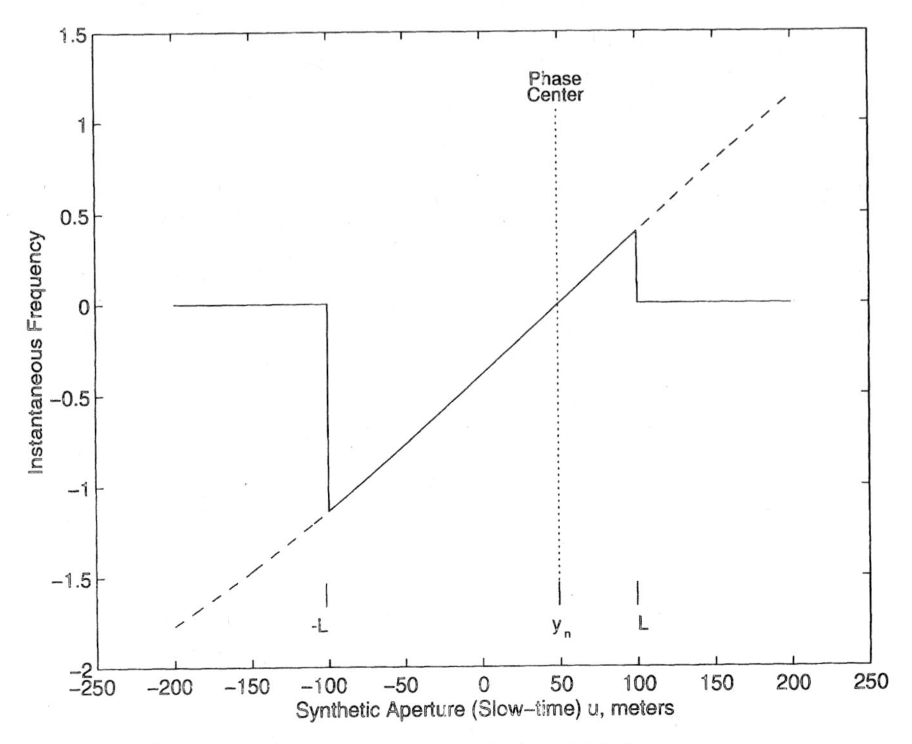

Figure 2.6a. Instantaneous frequency of spherical PM signal: $y_n = 50$ m.

The point $u = y_n$ in the synthetic aperture domain, which we earlier identified as the closest point of approach (CPA) of the radar for the nth target, is called the *phase center* of the spherical signal. At its phase center the instantaneous frequency of the spherical PM signal is zero; that is, $K_{un}(y_n) = 0$ (see Figure 2.6*a* and *b*).

Slow-time Fourier Transform

The Fourier transform of the spherical PM signal within the finite aperture $u \in [-L, L]$ is

$$
\begin{aligned}
S_n(\omega, k_u) &= \int_{-L}^{L} s_n(\omega, u) \exp(-jk_u u)\, du \\
&= \sigma_n \int_{-L}^{L} \exp\left[-j2k\sqrt{x_n^2 + (y_n - u)^2} - jk_u u\right] du.
\end{aligned}
$$

Once again, the above integral can be evaluated using the method of stationary phase which yields (the amplitude function $\exp(-j\pi/4)\sqrt{4k^2 - k_u^2}$ is suppressed in the

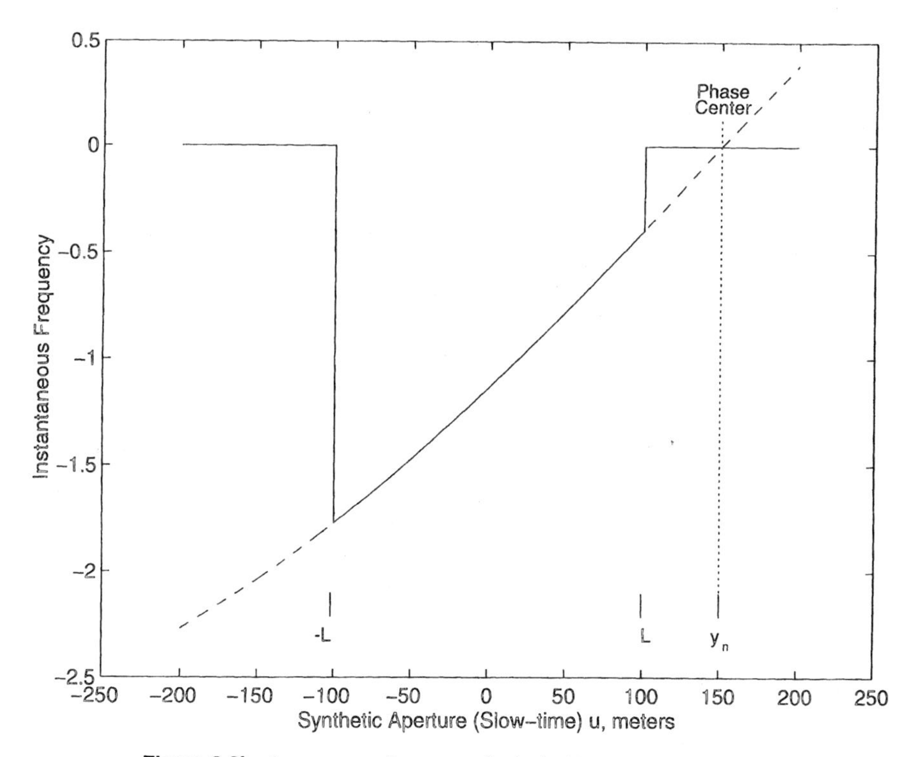

Figure 2.6b. Instantaneous frequency of spherical PM signal: $y_n = 150$ m.

following)

$$S_n(\omega, k_u) = \sigma_n \exp\left(-j\sqrt{4k^2 - k_u^2}\, x_n - jk_u y_n\right)$$

for

$$\begin{aligned} k_u &\in [K_{un}(L), K_{un}(-L)] \\ &= [2k \sin\theta_n(L), 2k \sin\theta_n(-L)], \end{aligned}$$

and zero otherwise. This phenomenon is depicted for two targets (broadside and off-broadside) in Figure 2.7. Examples of $S_n(\omega, k_u)$ are shown in Figure 2.5a and b; the corresponding $s_n(\omega, u)$ are shown in Figure 2.4a and b.

We denote the slow-time Doppler support band of $S_n(\omega, k_u)$ via

$$\Omega_n = [2k \sin\theta_n(L), 2k \sin\theta_n(-L)].$$

In general, the band Ω_n represents a *bandpass* region in the slow-time Doppler domain k_u. The center (carrier Doppler frequency) of this band is approximately located

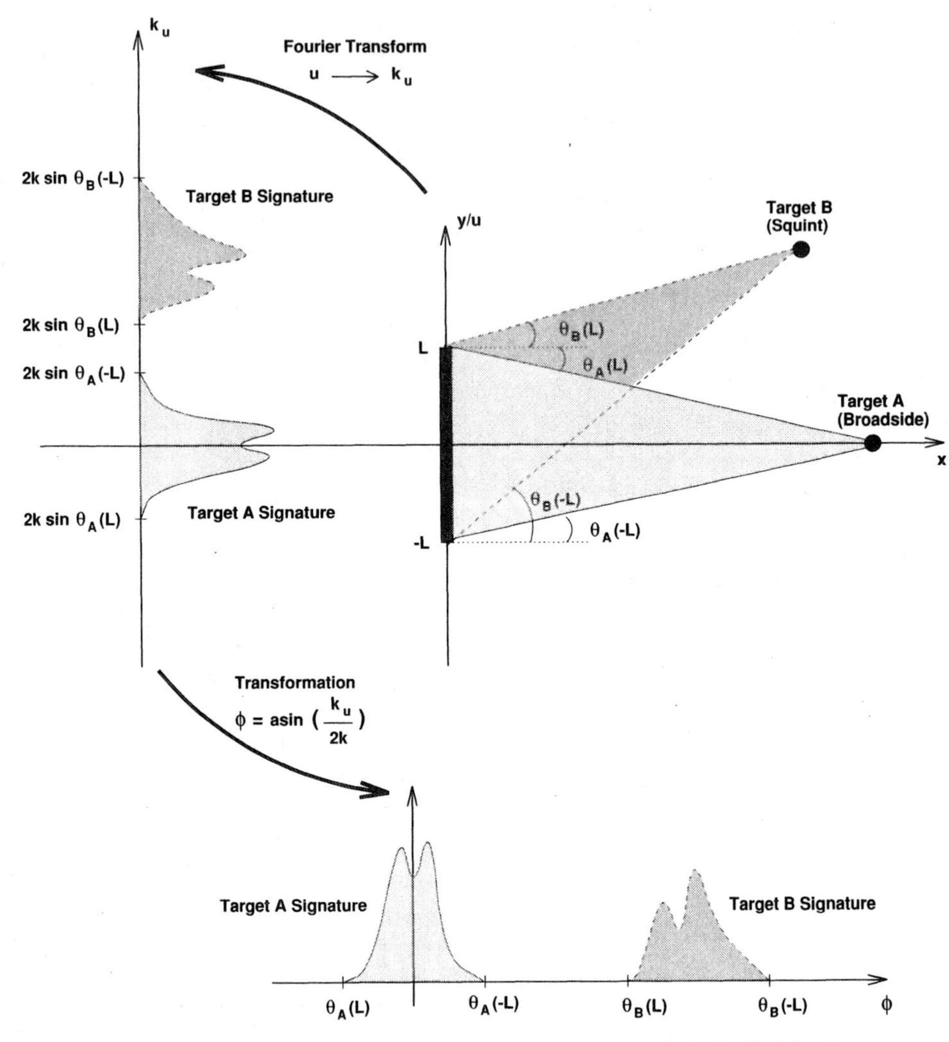

Figure 2.7. Effect of finite aperture in slow-time Doppler spectrum of spherical PM signal.

at

$$\begin{aligned} \Omega_{nc} &\approx K_{un}(0) \\ &= 2k \sin \theta_n(0). \end{aligned}$$

The length of the band Ω_n can be approximated via

$$|\Omega_n| \approx \frac{4kL}{x_n} \cos^2 \theta_n(0);$$

that is, the baseband bandwidth of $S_n(\omega, k_u)$ is $\pm|\Omega_n|/2$. Thus at a fixed range x_n (currently, we are assuming that $x_n = X_c$), the target with $\theta_n(0) = 0$ has the largest baseband Doppler bandwidth; such a target is called a *broadside* or *boresight* target. The targets with $\theta_n(0) \neq 0$ are called *off-broadside* or *squint* targets.

For instance, if the nth target is a broadside one, that is, $y_n = 0$, we have

$$\Omega_n = \left[-2k \sin\left[\arctan\left(\frac{L}{x_n}\right)\right], 2k \sin\left[\arctan\left(\frac{L}{x_n}\right)\right]\right]$$

$$\approx \left[\frac{-2kL}{x_n}, \frac{2kL}{x_n}\right]$$

For $|\theta_n(0)| = \pi/2$ which is equivalent to $|y_n| = \infty$, the baseband slow-time Doppler bandwidth of the target signature is *zero*; that is, the target is not *observable* to the radar.

- The above observations indicate that the finite aperture of measurements in the synthetic aperture u domain or, equivalently, the finite interval of aspect angles over which the target is illuminated, results in a filtering of the Fourier spectrum of the spherical PM signal in the slow-time Doppler k_u domain.
- The bandwidth of this slow-time Doppler filter increases as:

1. the size of the synthetic aperture $2L$ increases,
2. the radar frequency ω increases,
3. the target range $r_n = \sqrt{x_n^2 + y_n^2}$ gets smaller, or
4. the target aspect angle with respect to the origin $\theta_n(0) = \arctan(y_n/x_n)$ gets smaller.

Slow-time Angular Doppler Spectrum

We define the following transformation of the slow-time Doppler domain k_u:

$$\phi = \arcsin\left(\frac{k_u}{2k}\right).$$

The variable ϕ is called the *angular* spectral domain of the slow-time. We may rewrite the spherical PM signal in the (ω, k_u) domain via the following:

$$S_n(\omega, 2k \sin\phi) = \sigma_n \exp(-j2k \cos\phi x_n - j2k \sin\phi y_n)$$

for

$$\phi \in [\theta_n(L), \theta_n(-L)]$$

and zero otherwise. Figure 2.7 shows this coordinate transformation for a broadside target and a squint target.

Note that $S_n(\omega, k_u)$ is also a finite duration PM signal in the slow-time Doppler k_u domain. The instantaneous frequency of this PM signal is

$$\begin{aligned} U_n(k_u) &= \frac{\partial}{\partial k_u}\left[-\sqrt{4k^2 - k_u^2}\,x_n - k_u y_n\right] \\ &= \frac{k_u}{\sqrt{4k^2 - k_u^2}}x_n - y_n \\ &= \tan\phi x_n - y_n. \end{aligned}$$

The phase center of the PM wave $S_n(\omega, k_u)$, that is, the point in the k_u domain where its instantaneous frequency $U_n(k_u)$ is zero, is located at

$$\begin{aligned} k_u &= 2k \sin\left[\arctan\left(\frac{y_n}{x_n}\right)\right] \\ &= 2k \sin\theta_n(0), \end{aligned}$$

or

$$\phi = \theta_n(0),$$

in the slow-time angular spectral domain.

2.5 RECONSTRUCTION VIA MATCHED FILTERING: FINITE APERTURE

We can rewrite the slow-time Fourier transform of the spherical PM signal via

$$S_n(\omega, k_u) = \sigma_n I_n(\omega, k_u) \exp\left(-j\sqrt{4k^2 - k_u^2}\,x_n - jk_u y_n\right),$$

where I_n, the slow-time Doppler domain indicator (support) function for the nth target, is a rectangular window which is defined via

$$I_n(\omega, k_u) = \begin{cases} 1 & \text{for } k_u \in \Omega_n = [2k\sin\theta_n(L), 2k\sin\theta_n(-L)], \\ 0, & \text{otherwise.} \end{cases}$$

Then the slow-time Fourier transform of the received PM signal within $u \in [-L, L]$ is

$$\begin{aligned} S(\omega, k_u) &= \sum_n S_n(\omega, k_u) \\ &= \sum_n \sigma_n I_n(\omega, k_u) \exp\left(-j\sqrt{4k^2 - k_u^2}\,x_n - jk_u y_n\right). \end{aligned}$$

Since all the targets are located at a common range value, that is, $x_n = X_c$ ($n = 1, 2, \ldots$), we can rewrite the above via

$$S(\omega, k_u) = \exp\left(-j\sqrt{4k^2 - k_u^2}\, X_c\right) \sum_n \sigma_n I_n(\omega, k_u) \exp(-jk_u y_n).$$

Recall the slow-time Fourier transform of the reference signal when $u \in (\infty, \infty)$; that is,

$$S_0(\omega, k_u) = \exp\left(-j\sqrt{4k^2 - k_u^2}\, X_c\right)$$

for $k_u \in [-2k, 2k]$, and zero otherwise. *Note that the support of $I_n(\omega, k_u)$ in the slow-time Doppler k_u domain is always a subset of $k_u \in [-2k, 2k]$.*

Consider the matched filtering with reference signal for target reconstruction in the cross-range. For the finite synthetic aperture data, this becomes

$$\begin{aligned} F(k_y) &= S(\omega, k_u) S_0^*(\omega, k_u) \\ &= \sum_n \sigma_n I_n(\omega, k_u) \exp(-jk_y y_n), \end{aligned}$$

where S_0^* is the complex conjugate of S_0, and the cross-range spatial frequency mapping is still defined via $k_y = k_u$. The reconstructed target function in the cross-range domain is formed via the inverse Fourier transform of $F(k_y)$ with respect to the spatial frequency k_y:

$$f(y) = \sum_n \sigma_n i_n(\omega, y - y_n),$$

where

$$i_n(\omega, y) = |\Omega_n| \exp(j\Omega_{nc} y)\, \text{sinc}\left(\frac{1}{2\pi} |\Omega_n| y\right)$$

and

$$\begin{aligned} \Omega_{nc} &\approx 2k \sin\theta_n(0) \\ |\Omega_n| &\approx \frac{4kL}{x_n} \cos^2\theta_n(0) \end{aligned}$$

are, respectively, the center Doppler frequency and length of the slow-time Doppler band Ω_n.

Note that the inverse problem solution in Figure 2.3, that is,

$$f(y) = s(\omega, u) * s_0^*(\omega, -u),$$

where $*$ denotes convolution in the synthetic aperture u domain, and the cross-range mapping $y = u$ is still valid for the *finite* measurements of $s(\omega, u)$. However, for the *analytical* form of the reference signal $s_0(\omega, u)$ and its slow-time Fourier transform $S_0(\omega, k_u)$, one has to assume an *infinite* synthetic aperture. This does not cause any difficulty in the discrete implementation of the matched filtering in the k_u domain, that is, $S(\omega, k_u)S_0^*(\omega, k_u)$. For the u domain implementation of the matched filtering via convolution, $s(\omega, u) * s_0^*(\omega, -u)$, it can be shown that a *finite* aperture reference signal

$$s_0(\omega, u) \qquad \text{for } u \in [Y_c - L - Y_0, Y_c + L + Y_0]$$

is sufficient. (This is not discussed in detail, since we use the slow-time frequency domain implementation of the matched filtering.)

For the display, the magnitude of $f(y)$ is used, which is

$$|f(y)| = \sum_n \sigma_n |\text{psf}_n|(\omega, y - y_n),$$

where the point spread function for the nth target is defined via

$$\text{psf}_n(\omega, y) = i_n(\omega, y),$$

and

$$\begin{aligned} |\text{psf}_n(\omega, y)| &= |i_n(\omega, y)| \\ &= |\Omega_n| \left| \text{sinc}\left(\frac{1}{2\pi}|\Omega_n| y\right) \right|. \end{aligned}$$

Similar to the range imaging and cross-range imaging with an *infinite* synthetic aperture, the cross-range reconstructed image with a finite synthetic aperture is composed of sinc-like blips at the cross-range coordinates of the targets. There exists, however, a major difference: a *shift-varying* point spread function in the cross-range domain. In fact the sharpness (resolution) of the blip for a target depends on the length of the slow-time Doppler band of the target $|\Omega_n|$. For instance, at a fixed range x_n, a broadside target, with $\theta_n(0) = 0$, has the sharpest sinc-like blip because its slow-time baseband Doppler bandwidth is greater than that of the targets with $\theta_n(0) \neq 0$.

Finally, in the above discussion, we defined the reference signal $s_0(\omega, u)$ as the signature of a unit reflector at the center of a *broadside* target area in the cross-range domain $y = 0$. In some applications that involve a *squint* target area, the center of mass of the target area is not located at the origin $y = 0$. We will address this issue in the signal processing and digital implementation of the cross-range imaging algorithm. This will be achieved via redefining the reference signal $s_0(\omega, u)$, and then performing slow-time matched-filtering of the measured signal $s(\omega, u)$ with the new reference signal.

2.6 CROSS-RANGE RESOLUTION

For a given target, the support of its sinc-like point spread function, $\text{psf}_n(y)$, in the cross-range domain is a measure of its resolvability. Again, we use the approximation that the support of a sinc function is approximately equal to the size of its main lobe. In this case the cross-range resolution for the nth target becomes

$$\begin{aligned}\Delta_{y_n} &= \frac{2\pi}{|\Omega_n|} \\ &\approx \frac{x_n \lambda}{4L \cos^2 \theta_n(0)}.\end{aligned}$$

It should be noted that this is a *fair* measure of the cross-range resolution. In practice, one might multiply the above Δ_{y_n} with a number between one and two (fudge factor).

Also recall the classical Rayleigh resolution for a finite aperture [chr; goo; sko; s92ju; s94; ste]: $R\lambda/D$, where R is the range and D is the aperture diameter. In our model the *effective* wavelength is $\lambda/2$ due to the round-trip propagation (not the presence of $2k$ in the spherical PM signal) and the aperture diameter is $2L$. With these, the above cross-range resolution is very similar to the Rayleigh resolution. In fact, for a broadside target with $\theta_n(0) = 0$, it is the same as the Rayleigh resolution.

2.7 DATA ACQUISITION AND SIGNAL PROCESSING

Suppose that the target area in the cross-range domain is limited to the region $y \in [Y_c - Y_0, Y_c + Y_0]$, where Y_c, the center of mass of the target area, and $2Y_0$, the size of the target area, are known constants; see Figure 2.1. We now examine the implications of this condition on y_n's (i.e., a limited target region) in determining the Nyquist sampling rate in the aperture u domain. For this, we start with the special case of $Y_c = 0$, a *broadside* target area; we then examine the general case of $Y_c \neq 0$, a *squint* target area.

Synthetic Aperture Sampling for a Broadside Target Area

The case of $Y_c = 0$ corresponds to the scenario when the center of the synthetic aperture $u = 0$ and the center of mass of the target area are directly facing each other. In this case we have $|y_n| \leq Y_0$ for all n.

We mentioned that the support band of $s_n(\omega, u)$ in the spatial frequency k_u domain is

$$\Omega_n = [2k \sin \theta_n(L), 2k \sin \theta_n(-L)],$$

which is approximately centered at the phase center of $S_n(\omega, k_u)$; that is, $\phi = \theta_n(0)$, where

$$\phi = \arcsin\left(\frac{k_u}{2k}\right).$$

(Recall that the slow-time angular spectrum mapping of $k_u = 2k \sin \phi$.) Consider the measured signal $s(\omega, u)$ which is the *sum* of all $s_n(\omega, u)$'s; that is,

$$s(\omega, u) = \sum_n s_n(\omega, u).$$

The support band of this signal in the k_u domain, Ω_s, is the *union* of the support bands of all $s_n(\omega, u)$'s in the spatial frequency domain. Thus we have

$$\begin{aligned}\Omega_s &= \bigcup_{|y_n| \le Y_0} \Omega_n \\ &= \bigcup_{|y_n| \le Y_0} [2k \sin \theta_n(L), 2k \sin \theta_n(-L)] \\ &= [-2k \sin \theta_{\max}, 2k \sin \theta_{\max}]\end{aligned}$$

where

$$\theta_{\max} = \arctan\left(\frac{Y_0 + L}{X_c}\right),$$

is the *largest* aspect angle of the targets with respect to the radar for $u \in [-L, L]$. If $Y_0 + L \ll X_c$, the above slow-time Doppler band can be approximated via

$$\Omega_s \approx \left[-2k\frac{Y_0 + L}{X_c}, 2k\frac{Y_0 + L}{X_c}\right].$$

An example is shown in Figure 2.8a.

Based on the above slow-time Doppler support band for $S(\omega, k_u)$, the Nyquist sample spacing in the u domain for recording $s(\omega, u)$ is

$$\Delta_u \le \frac{\pi}{2k \sin \theta_{\max}}.$$

When $Y_0 + L \ll X_c$, we have the approximation

$$\sin \theta_{\max} \approx \frac{Y_0 + L}{X_c}.$$

In this case the sampling constraint in the synthetic aperture domain becomes

$$\begin{aligned}\Delta_u &\le \frac{\pi}{-2k[(Y_0 + L)/X_c]} \\ &= \frac{X_c \lambda}{4(Y_0 + L)},\end{aligned}$$

where $\lambda = 2\pi/k$ is the *wavelength* at the radar fast-time frequency ω.

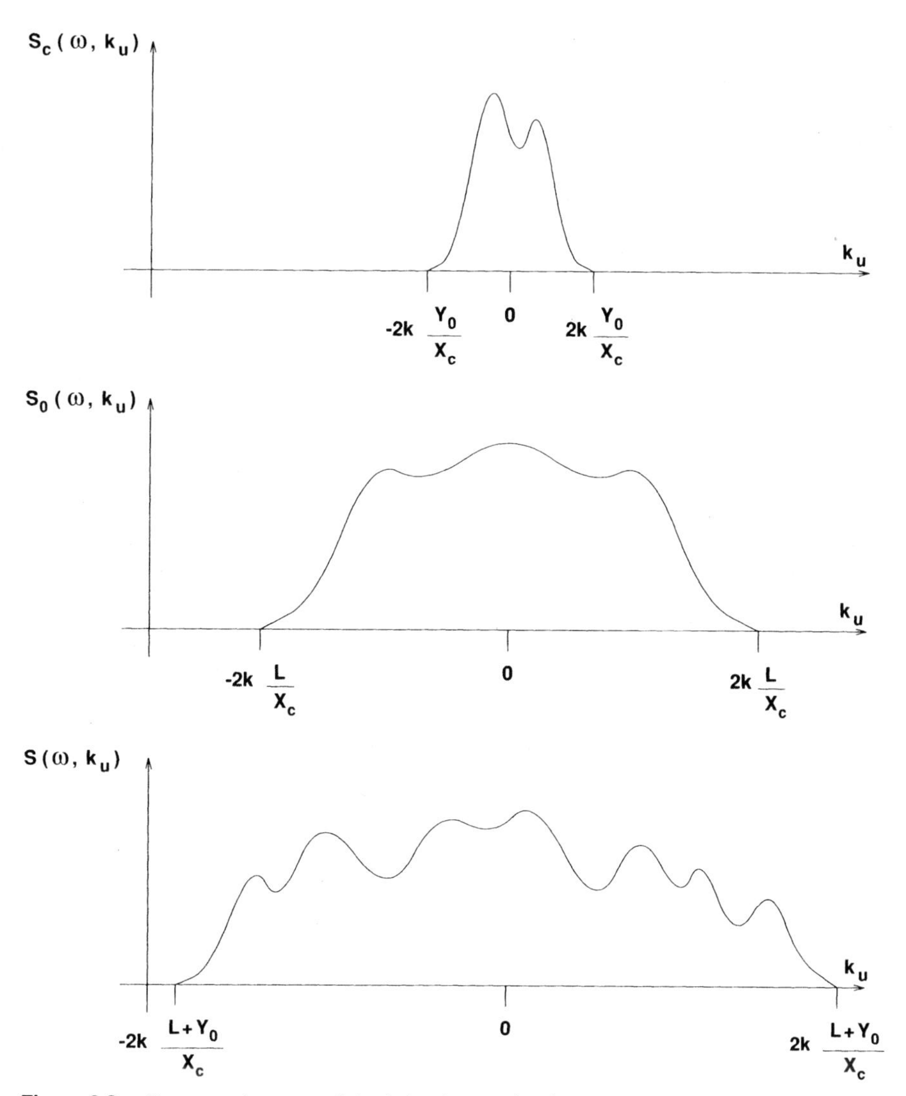

Figure 2.8a. Compressed spectrum $S_c(\omega, k_u)$, reference signal spectrum $S_0(\omega, k_u)$, and synthetic aperture signal spectrum $S(\omega, k_u)$: broadside mode.

- The worst-case scenario corresponds to the case of $L = \infty$ or $Y_0 = \infty$ (infinite target area or synthetic aperture) for which the Nyquist sampling spacing in the synthetic aperture domain is

$$\Delta_u \leq \frac{\pi}{2k} = \frac{\lambda}{4}.$$

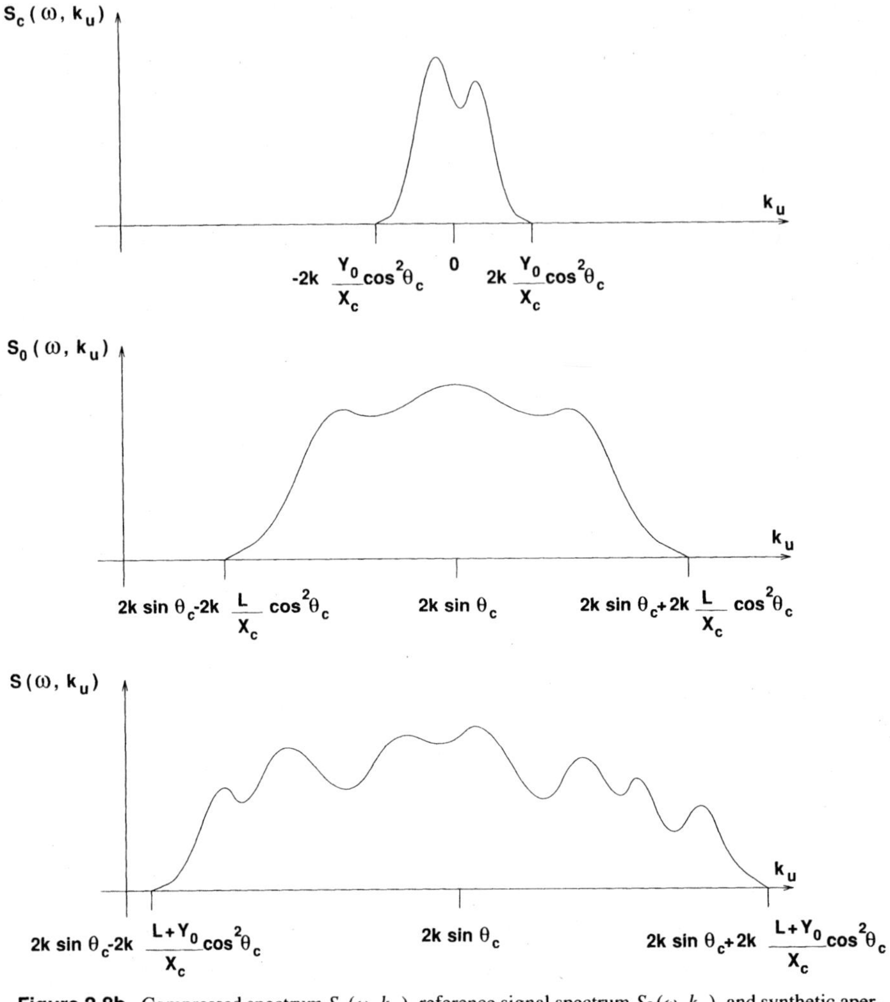

Figure 2.8b. Compressed spectrum $S_c(\omega, k_u)$, reference signal spectrum $S_0(\omega, k_u)$, and synthetic aperture signal spectrum $S(\omega, k_u)$: squint mode.

Synthetic Aperture Sampling for a Squint Target Area

For the general squint case, the target area in the cross-range domain is within the interval $y \in [Y_c - Y_0, Y_c + Y_0]$, where Y_c is the squint cross-range. The squint angle of the center of the target area is defined via

$$\theta_c = \arctan\left(\frac{Y_c}{X_c}\right).$$

Then the slow-time Doppler support band of the squint SAR signal becomes

$$\Omega_s = \bigcup_{y_n \in [Y_c - Y_0, Y_c + Y_0]} \Omega_n$$
$$= [2k \sin \theta_{\min}, 2k \sin \theta_{\max}],$$

where

$$\theta_{\min} = \arctan\left(\frac{Y_c - Y_0 - L}{X_c}\right),$$
$$\theta_{\max} = \arctan\left(\frac{Y_c + Y_0 + L}{X_c}\right),$$

are, respectively, the smallest and largest aspect angle of the targets with respect to the radar for $u \in [-L, L]$. An example of the slow-time spectral distribution of $S(\omega, k_u)$ is shown in Figure 2.8*b*.

Based on the above slow-time Doppler support band for the squint $S(\omega, k_u)$, the Nyquist sample spacing in the u domain for recording $s(\omega, u)$ is

$$\Delta_u \le \frac{\pi}{k(\sin \theta_{\max} - \sin \theta_{\min})}.$$

When $Y_0 + L \ll X_c$, the slow-time Doppler band for the synthetic aperture signal, Ω_s, is approximately centered around

$$k_u \approx 2k \sin \theta_c,$$

and its baseband bandwidth is approximately equal to

$$\pm 2k \frac{Y_0 + L}{X_c} \cos^2 \theta_c.$$

In this case the Nyquist sampling constraint in the synthetic aperture domain can be approximated via

$$\Delta_u \le \frac{\pi}{2k[(Y_0 + L)/X_c] \cos^2 \theta_c}$$
$$= \frac{X_c \lambda \cos^2 \theta_c}{4(Y_0 + L)},$$
$$= \frac{R_c \lambda \cos \theta_c}{4(Y_0 + L)},$$

where $R_c = \sqrt{X_c^2 + Y_c^2}$ is the distance of the center of the synthetic aperture $u = 0$ to the center of mass of the target area; (R_c, θ_c) are the *polar* coordinates for (X_c, Y_c).

For the general squint case, it is crucial to realize that the synthetic aperture signal is a *bandpass* signal; the center of the band is at $2k \sin \theta_c$. Thus one has to perform slow-time baseband conversion via

$$s(\omega, u)\exp(-j2k\sin\theta_c u)$$

before any Fourier (DFT) processing.

Reducing PRF via Slow-time Compression

As we pointed out earlier, the synthetic aperture measurements are generated via the motion of a vehicle that carries the radar. Suppose that the speed of this vehicle is v_R. A quantity that is the slow-time interval between successive radar transmissions (while the vehicle is moved by Δ_u) is called *pulse repetition interval (PRI)*:

$$\text{PRI} = \frac{\Delta_u}{v_R}.$$

Another quantity that represents the number of radar transmissions per unit of the slow-time is called the *pulse repetition frequency (PRF)*:

$$\begin{aligned}\text{PRF} &= \frac{1}{\text{PRI}} \\ &= \frac{v_R}{\Delta_u}.\end{aligned}$$

Clearly the PRF of the radar increases as Δ_u gets smaller (a higher sampling rate on the synthetic aperture). In a practical radar system, it is desirable to use as small a PRF as possible due to hardware limitations of the radar system. In this section we outline a procedure for reducing the PRF of the radar.

For range imaging we utilized fast-time domain pulse compression of chirp signals to formulate the *imaging* problem. We now discuss a procedure which we refer to as *compression of the spherical PM signal in the slow-time domain.* In the following discussion we use a general form for the target range x_n that might have slight deviations from the constant range X_c; that is, $|x_n - X_c| \ll X_c$. This is for the sake of our future analysis where the targets are not located at a fixed range.

Broadside Target Area Provided that the target area size $2Y_0$ and the synthetic aperture length $2L$ are both much smaller than the target area range X_c, we can approximate the instantaneous frequency of the spherical PM signal $s_n(\omega, u)$ via

$$\begin{aligned}K_{un}(u) &= \frac{2k(y_n - u)}{\sqrt{x_n^2 + (y_n - u)^2}} \\ &\approx \frac{2k(y_n - u)}{X_c},\end{aligned}$$

which is a *linear* function of the synthetic aperture u domain; see Figure 2.6*a* and *b*.

- Recall that the instantaneous frequency of a *chirp* signal is also a linear function. The approximation of the spherical PM signal via a chirp signal, that is,

$$\exp\left[-j2k\sqrt{x_n^2+(y_n-u)^2}\right] \approx \exp\left[-j2kx_n - j\frac{k(y_n-u)^2}{X_c}\right],$$

  is called the *Fresnel approximation.*

For a unit reflector located at $(x_n, y_n) = (X_c, 0)$, the spherical PM signal is

$$s_0(\omega, u) = \exp\left(-j2k\sqrt{X_c^2+u^2}\right),$$

which we refer to as the *reference* spherical PM signal. The instantaneous frequency of the reference spherical PM signal is

$$K_{u0}(u) \approx \frac{-2ku}{X_c}.$$

The difference between the instantaneous frequencies for the target at (x_n, y_n) and the target at $(X_c, 0)$ is

$$K_{un}(u) - K_{u0}(u) \approx \frac{2ky_n}{X_c}.$$

Thus the *mixing* of $s_n(\omega, u)$ with the reference signal, that is,

$$\begin{aligned} s_{nc}(\omega, u) &= s_n(\omega, u)s_0^*(\omega, u) \\ &\approx \sigma_n \exp[-j2k(x_n - X_c)] \underbrace{\exp\left(j\frac{2ky_n}{X_c}u\right)}_{\text{Sinusoid}}, \end{aligned}$$

results in a *single-tone sinusoid* in the slow-time u domain with frequency $2ky_n/X_c$. We call the resultant signal, that is, $s_{nc}(\omega, u)$, the *slow-time compressed* version of the spherical PM signal. Provided that the slow-time measurements are made for $u \in [-L, L]$ (finite synthetic aperture), the compressed signal in the slow-time Doppler k_u domain is

$$S_{nc}(\omega, k_u) \approx \sigma_n \exp[-j2k(x_n - X_c)] \operatorname{sinc}\left[\frac{L}{\pi}\left(k_u - \frac{2ky_n}{X_c}\right)\right].$$

For the total measured SAR signal in the (ω, u) domain, that is,

$$s(\omega, u) = \sum_n \sigma_n s_n(\omega, u),$$

the slow-time compression yields

$$\begin{aligned} s_c(\omega, u) &= s(\omega, u)s_0^*(\omega, u) \\ &\approx \sum_n \sigma_n \exp[-j2k(x_n - X_c)] \exp\left(j\frac{2ky_n}{X_c}u\right). \end{aligned}$$

In the slow-time Doppler domain, the compressed signal becomes

$$S_c(\omega, k_u) \approx \sum_n \sigma_n \exp[-j2k(x_n - X_c)] \operatorname{sinc}\left[\frac{L}{\pi}\left(k_u - \frac{2ky_n}{X_c}\right)\right],$$

which carries information on the cross-range location of targets. (Note that the above is an *approximation.*)

Figure 2.8*a* shows an example of slow-time Doppler bandwidth reduction via slow-time compression for a broadside target area. Figure 2.9*a* and *b*, respectively,

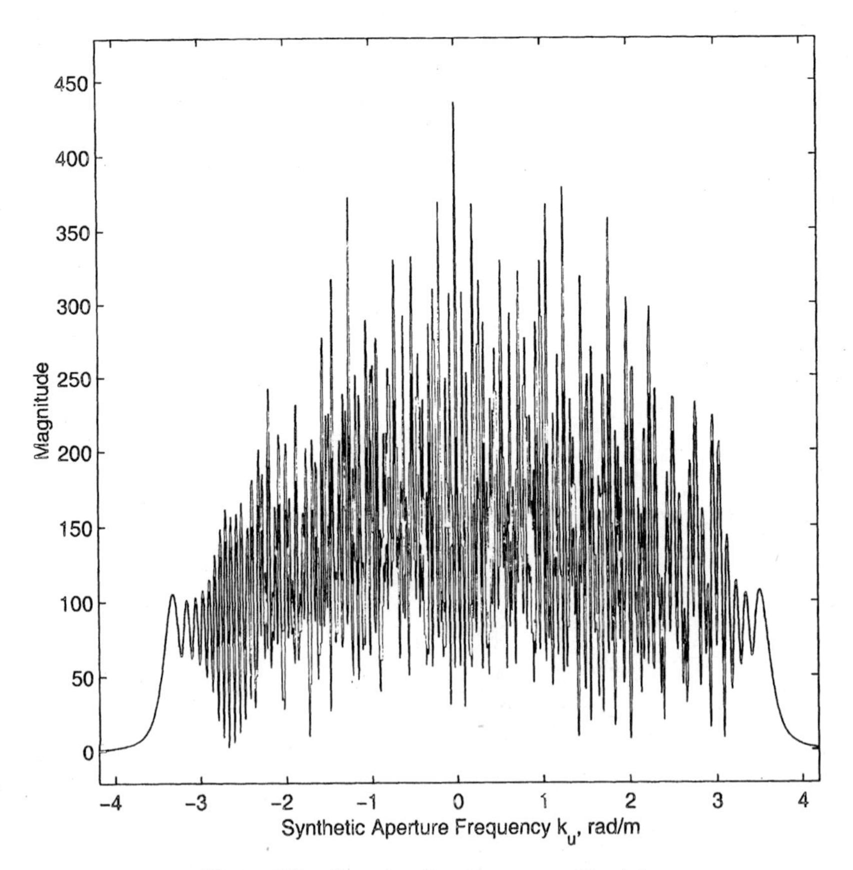

Figure 2.9a. Slow-time Doppler spectrum $S(\omega, k_u)$.

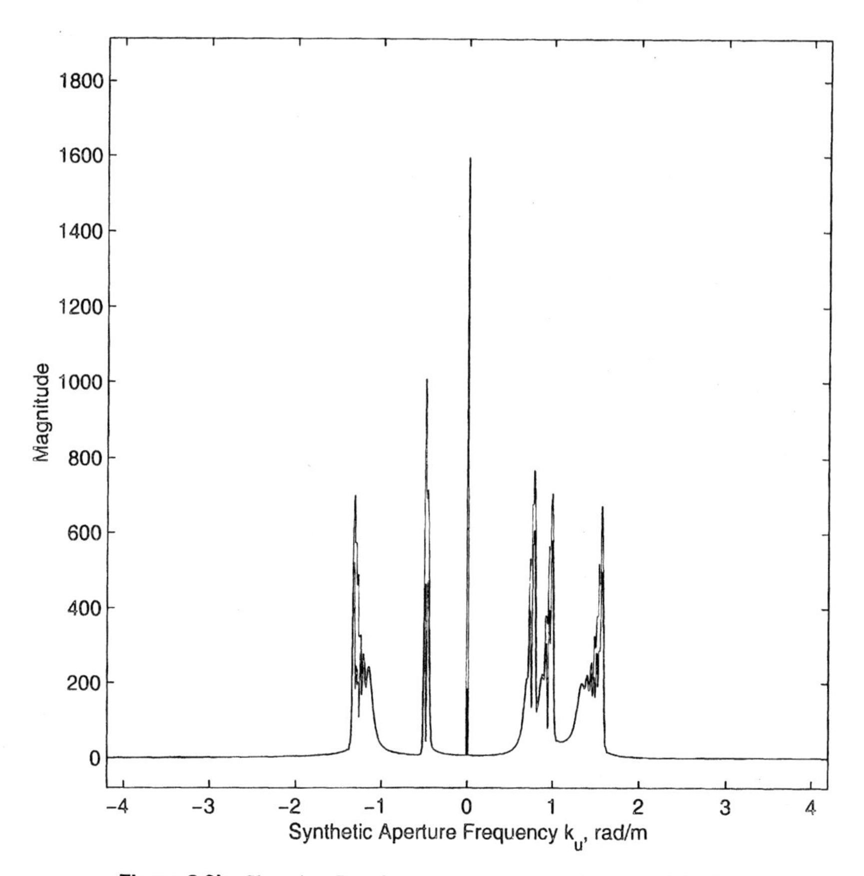

Figure 2.9b. Slow-time Doppler spectrum compressed version $S_c(\omega, k_u)$.

shows a numerical example of the spherical PM signal slow-time Doppler spectrum $S(\omega, k_u)$, and the slow-time Doppler spectrum of its slow-time compressed version $S_c(\omega, k_u)$. The data are generated with six targets in the cross-range domain.

Note that the sinc-like blips in the slow-time compressed signal $S_c(\omega, k_u)$ appear at

$$k_u = \frac{2ky_n}{X_c}.$$

Thus the following scaled version of the slow-time Doppler domain

$$y = \frac{k_u X_c}{2k}$$

could be used to view the signature of the targets in a mapped cross-range domain. In fact there exists a SAR reconstruction method, which is known as polar format pro-

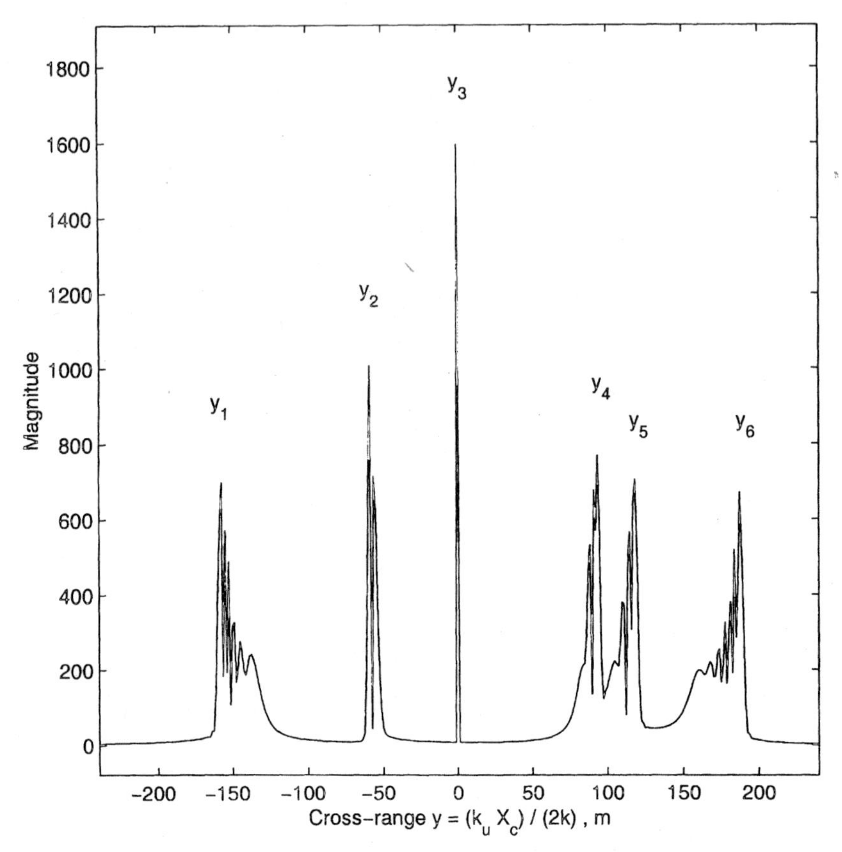

Figure 2.9c. Mapping of compressed signal spectrum into cross-range domain via axis (scale) transformation $y = k_u X_c / 2k$.

cessing, that uses slow-time compression for imaging. Figure 2.9*c* shows $S_c(\omega, k_u)$ versus $y = k_u X_c / 2k$. The cross-range locations of the six targets are marked in Figure 2.9*c*.

Note that the signatures of the targets appear *smeared* in Figure 2.9*c*. This is due to the fact that the expression for $S_c(\omega, k_u)$ in terms of the sinc functions is an *approximation*. Degradations caused by using slow-time compression for SAR imaging (polar format processing) are discussed in [s90; s92ja; s94; s95].

Squint Target Area In the squint case the reference spherical signal is chosen to be the signature of the target located at $(x_n, y_n) = (X_c, Y_c)$; that is,

$$s_0(\omega, u) = \exp\left(-j2k\sqrt{X_c^2 + (Y_c - u)^2}\right).$$

Following the steps we used for the broadside case, we obtain

$$s_c(\omega, u) = s(\omega, u)s_0^*(\omega, u)$$
$$\approx \sum_n \sigma_n \exp[-j2k(r_n - R_c)] \exp\left[j\frac{2k\cos\theta_c(y_n - Y_c)}{R_c}u\right],$$

where

$$r_n = \sqrt{x_n^2 + y_n^2}$$

is the radial range of the nth target and

$$R_c = \sqrt{X_c^2 + Y_c^2},$$

is the radial range of the target which is located at the center of the target area. The compressed signal in the slow-time Doppler k_u domain is

$$S_c(\omega, k_u) \approx \sum_n \sigma_n \exp[-j2k(r_n - R_c)] \operatorname{sinc}\left[\frac{L}{\pi}\left(k_u - \frac{2k\cos\theta_c(y_n - Y_c)}{R_c}\right)\right].$$

This signal also carries information on the cross-range location of targets with respect to the squint cross-range Y_c.

The above analysis indicates that the slow-time Doppler spread of the signature of the nth target in the compressed signal is dictated by the following sinc pattern:

$$\operatorname{sinc}\left[\frac{L}{\pi}\left(k_u - \frac{2k\cos\theta_c(y_n - Y_c)}{R_c}\right)\right];$$

this is true for the general squint case as well as the special broadside case with $\theta_c = Y_c = 0$. Moreover the support band of the above sinc pattern is mainly dictated by its main lobe in the k_u domain which is equal to $2\pi/L$. Thus the slow-time compression reduces the slow-time Doppler spread of the spherical PM signal $s_n(\omega, u)$ into approximately $2\pi/L$; this is much smaller than the slow-time Doppler spread of $s_n(\omega, u)$, that is,

$$\Omega_n = [2k\sin\theta_n(L), 2k\sin\theta_n(-L)]$$

for a practical synthetic aperture imaging system. Figure 2.8*a* and *b*, respectively, shows an example of slow-time Doppler bandwidth reduction via slow-time compression for a broadside and a squint target area.

The slow-time Doppler bandwidth of the compressed signal is the union of the Doppler band generated by the sinusoids

$$\exp\left[j\frac{2k\cos\theta_c(y_n - Y_c)}{R_c}u\right]$$

for $y_n - Y_c \in [-Y_0, Y_0]$ in the slow-time domain. This implies that the compressed slow-time Doppler band is

$$\Omega_c = \left[-2k\cos\theta_c\frac{Y_0}{R_c} - \frac{\pi}{L}, 2k\cos\theta_c\frac{Y_0}{R_c} + \frac{\pi}{L}\right]$$
$$\approx \left[-2k\cos\theta_c\frac{Y_0}{R_c}, 2k\cos\theta_c\frac{Y_0}{R_c}\right].$$

Thus the Nyquist sampling rate in the slow-time domain for the slow-time compressed spherical PM signal $s_c(\omega, u)$ can be obtained via:

$$\Delta_{uc} \le \frac{R_c\lambda}{4Y_0\cos\theta_c}.$$

This is less restrictive that the sampling constraint for the uncompressed synthetic aperture signal $s(\omega, u)$, that is,

$$\Delta_u \le \frac{R_c\lambda}{4(Y_0 + L)\cos\theta_c}.$$

Thus slow-time compression results in a smaller PRF for synthetic aperture cross-range imaging.

Note that the slow-time compression is similar to the fast-time compression for range imaging. However, *slow-time compression of the spherical PM signal is not intended to be used to solve the imaging problem of cross-range imaging.* In fact the Fresnel approximation that was used to develop this principle makes the method unsuitable for high-resolution SAR imaging. We will show that the slow-time domain compression of the spherical PM signal is a signal processing tool for reducing PRF in spotlight SAR systems, and alias-free implementation of the imaging algorithm (via a process which we will call *digital spotlighting*) in stripmap SAR systems.

Algorithm The following steps are performed for reducing PRF via slow-time compression (a block diagram representation of this procedure is included in Figure 2.10 which will be discussed in the section on the digital implementation of the cross-range reconstruction algorithm):

Step 1. Sample the synthetic aperture signal $s(\omega, u)$ with spacing Δ_{uc} in the synthetic aperture (slow-time) u domain. Clearly, this signal is aliased in the slow-time domain, since Δ_{uc} is larger than the Nyquist sample spacing for $s(\omega, u)$, that is, Δ_u.

Step 2. Perform slow-time compression on the aliased synthetic aperture data. This operation is oblivious to the fact that $s(\omega, u)$ is aliased, since it is done by a point-by-point multiplication in the synthetic aperture u domain. (There is no Fourier transform or filtering/convolution operation that cannot be performed with aliased data.)

Step 3. The resultant slow-time compressed sampled data are not *aliased.*

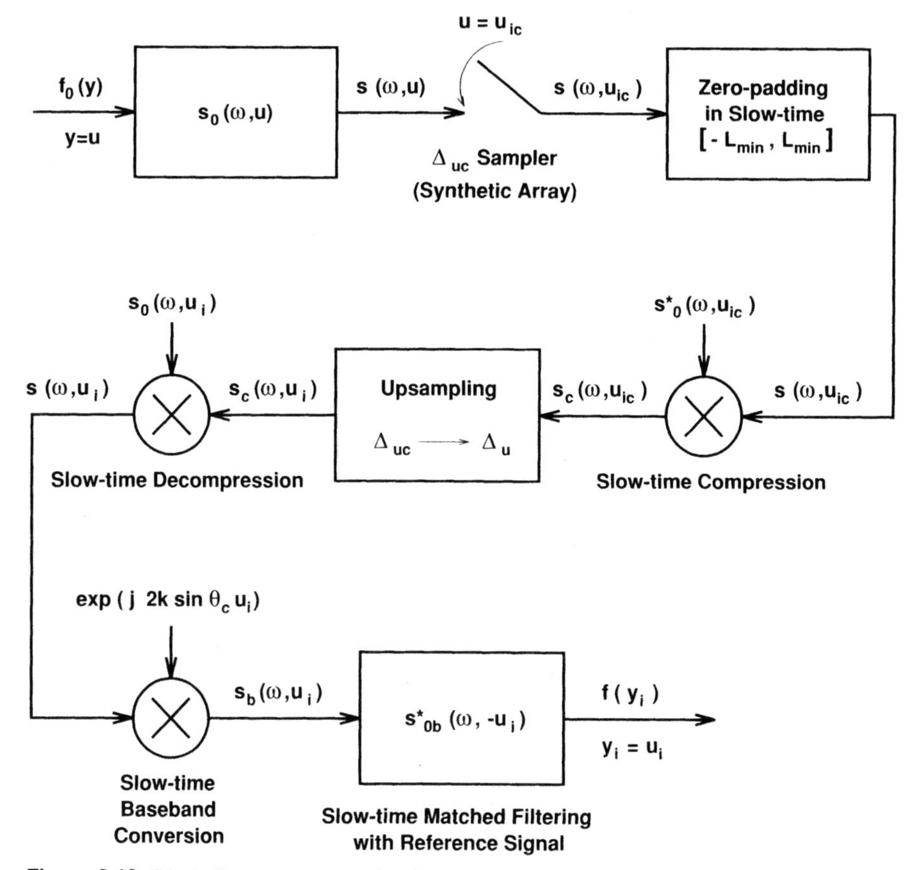

Figure 2.10. Block diagram representation for compression, upsampling, and decompression in synthetic aperture domain.

Step 4. Zero-pad the compressed signal in the slow-time Doppler k_u domain to achieve the desired aperture sample spacing of Δ_u in the u domain (upsampling).

Note: *To reduce wraparound errors due to DFT and IDFT, a "pseudo" discrete cosine transform (DCT) can be used. This is done via appending $s_c(\omega, u)$ with its mirror image in the slow-time domain. Then the DFT, zero-padding, and the IDFT are performed; the mirror image is discarded in the upsampled data.*

If the pseudo DCT procedure was used, then the sinc-like signature of a target at $y = y_n$ appears around two mirror image points in the slow-time Doppler domain:

$$k_u = \frac{\pm 2ky_n}{X_c}.$$

As a result the signature of the targets on the positive (negative) side of the y axis would overlap with the signature of the targets on the negative (positive) side of the y axis. In this case the individual target signatures would not be visible.

To generate Figure 2.9a and c, the pseudo DCT procedure is not used, and thus the individual target signatures are distinguishable. However, the Matlab code for this chapter uses the pseudo DCT procedure. Thus Figures P2.4 and P2.5 exhibit the above-mentioned overlapping of the target signatures in the slow-time Doppler domain distribution of the slow-time compressed signal $S_c(\omega, k_u)$.

Step 5. Perform slow-time decompression *(i.e., inverse of the slow-time compression) via*

$$\begin{aligned} s(\omega, u) &= s_c(\omega, u) s_0(\omega, u) \\ &= s_c(\omega, u) \exp\left[-j2k\sqrt{X_c^2 + (Y_c - u)^2}\right] \end{aligned}$$

to retrieve uncorrupted (unaliased) synthetic aperture sampled data.

Once again, we should emphasize that Δ_u is selected based on the *baseband* bandwidth of $s(\omega, u)$ which, in the general squint case, is a bandpass signal. After the above operations are performed to *artificially* or *digitally* increase the PRF to the amount dictated by Δ_u, the user should still perform the slow-time bandpass conversion

$$s(\omega, u) \exp(j2k \sin \theta_c u)$$

prior to any slow-time Fourier processing or filtering.

Cross-range Gating via Slow-time Compression

Broadside Target Area Consider the approximations developed in the previous section for the slow-time compressed SAR signal:

$$s_c(\omega, u) \approx \sum_n \sigma_n \exp\left[-j2k(x_n - X_c)\right] \exp\left(j\frac{2ky_n}{X_c}u\right)$$

and its Fourier transform with respect to the slow-time,

$$S_c(\omega, k_u) \approx \sum_n \sigma_n \exp[-j2k(x_n - X_c)] \operatorname{sinc}\left[\frac{L}{\pi}\left(k_u - \frac{2ky_n}{X_c}\right)\right].$$

We mentioned that the sinc-like blips in the slow-time compressed signal $S_c(\omega, k_u)$ which appear at

$$k_u = \frac{2ky_n}{X_c},$$

carry information on the cross-range location of targets; see Figure 2.9*b* and *c*.

If the user is interested in imaging a specific portion of the target area, this property of the slow-time compressed signal $S_c(\omega, k_u)$ in the slow-time Doppler k_u domain can be used for extracting or filtering the signature of the desired target area from $S_c(\omega, k_u)$. For instance, suppose that the target area is not finite in the cross-range domain. However, we are interested in imaging the cross-range area $y \in [-Y_0, Y_0]$, where Y_0 is a chosen constant. In this case the cross-range gating is performed by setting $S_c(\omega, k_u)$ equal to zero in the Doppler support band of the targets with $|y| > Y_0$, that is,

$$S_c(\omega, k_u) = 0, \qquad \text{for } |k_u| > \frac{2kY_0}{X_c}$$

We refer to this scheme as *cross-range gating*.

As can be seen from Figure 2.9c, the nth target sinc-like blip signature in $S_c(\omega, k_u)$ has an *inward spreading* from $k_u = 2ky_n/X_c$ toward the zero Doppler $k_u = 0$. Thus, despite the above-mentioned cross-range filtering of targets with $|y| > Y_0$, the filtered signal still contains some contribution (leakage) from these targets. It turns out that the leakage results in a signature that resides outside the target area, that is, $|y| > Y_0$, in the reconstructed image.

Squint Target Area As we showed earlier, for the squint target area the compressed signal in the slow-time Doppler k_u domain is

$$S_c(\omega, k_u) \approx \sum_n \sigma_n \exp[-j2k(r_n - R_c)] \operatorname{sinc}\left[\frac{L}{\pi}\left(k_u - \frac{2k\cos\theta_c(y_n - Y_c)}{R_c}\right)\right],$$

which is composed of sinc-like blips at

$$\begin{aligned} k_u &= \frac{2k\cos\theta_c(y_n - Y_c)}{R_c} \\ &= \frac{2k\cos^2\theta_c(y_n - Y_c)}{X_c}, \qquad n = 1, 2, \ldots. \end{aligned}$$

Thus to extract the signature of the target area within $y \in [Y_c - Y_0, Y_c + Y_0]$, where Y_0 is a chosen constant, the following filtering is performed on the slow-time compressed signal:

$$S_c(\omega, k_u) = 0 \qquad \text{for } |k_u| > \frac{2k\cos\theta_c Y_0}{R_c}.$$

Cross-range gating will be used in our discussion on digital spotlighting in SAR systems.

2.8 RECONSTRUCTION ALGORITHM

Before outlining the digital reconstruction algorithm for cross-range imaging, we must bring out two digital signal processing issues that are associated with the formation of the discrete cross-range image. After outlining the digital reconstruction algorithm, we examine other digital signal processing methods for reducing the size of the discrete reconstructed cross-range image.

Baseband Conversion of Target Area

The problem addressed in this section is the baseband processing associated with the general squint case. As we pointed out earlier, the target area in the squint case is identified via $y \in [Y_c - Y_0, Y_c + Y_0]$. This implies that the target function in the spatial frequency (slow-time Doppler) k_u domain is a *bandpass* signal, that its center frequency is Y_c and its baseband is $\pm Y_0$. To bring this signal to baseband which is suitable for DFT processing, we can make the following simple modification in identifying the *origin* in the spatial (x, y) domain:

Suppose that we move the origin to the center of the target area; thus the unknown cross-range points are within $y_n \in [-Y_0, Y_0]$ and $x_n = 0$ for $n = 1, 2, \ldots$. In this case the center element of the synthetic aperture array moves to (X_c, Y_c), and the synthetic aperture locations of the radar in the spatial domain are

$$(x, y) = (X_c, u - Y_c), \qquad u \in [-L, L].$$

The measured synthetic aperture signal becomes

$$s(\omega, u) = \sum_n \sigma_n \exp\left[-j2k\sqrt{X_c^2 + (y_n + Y_c - u)^2}\right]$$

for $u \in [-L, L]$ and zero otherwise. The slow-time Fourier transform of this signal is

$$S(\omega, k_u) = \sum_n \sigma_n I_n(\omega, k_u) \exp\left[-j\sqrt{4k^2 - k_u^2}\, X_c - jk_u(y_n + Y_c)\right],$$

which is the same as what we had before except for the addition of the phase shift $\exp(-jk_u Y_c)$ due to our new definition of the origin in the cross-range domain.

We also redefine the reference signal with the new origin

$$s_0(\omega, u) = \exp\left[-j2k\sqrt{X_c^2 + (Y_c - u)^2}\right]$$

for $u \in (-\infty, \infty)$ (infinite synthetic aperture). The slow-time Fourier transform of the reference signal is

$$S_0(\omega, k_u) = \exp\left(-j\sqrt{4k^2 - k_u^2} X_c - jk_u Y_c\right)$$

for $k_u \in [-2k, 2k]$, and zero otherwise. Again, the only change in $S_0(\omega, k_u)$ is the addition of the phase function $\exp(-jk_u Y_c)$.

The slow-time matched filtering is performed with this newly defined reference signal in the slow-time Doppler k_u domain

$$F(k_y) = S(\omega, k_u) S_0^*(\omega, k_u),$$

with $k_y = k_u$. The target function $f(y)$ is then obtained from the inverse DFT of the above. Note that all the imaging operations are the same as before (which is only suitable for the special case of a broadside target area); the only change is in the manner that the reference signal is defined in the k_u domain (i.e., the addition of the phase function $\exp(-jk_u Y_c)$).

Zero-padding in Synthetic Aperture Domain

Since the size of the synthetic aperture is $2L$, the sample spacing of the DFT of $s(\omega, u)$ with respect to u in the slow-time Doppler k_u domain is

$$\Delta_{k_u} = \frac{\pi}{L}.$$

Moreover the slow-time matched filtering is achieved via

$$F(k_y) = S(\omega, k_u) S_0^*(\omega, k_u),$$

where the cross-range mapping in the spatial frequency domain is defined by $k_y = k_u$. The resultant samples of $F(k_y)$ in the k_y domain are also separated by

$$\Delta_{k_y} = \frac{\pi}{L}.$$

On the other hand, the size of the target area in the cross-range domain is $2Y_0$ (for either a broadside or a squint target area). To avoid aliasing in the y domain, the sample spacing in the k_y domain should satisfy the following Nyquist rate:

$$\Delta_{k_y} \le \frac{\pi}{Y_0}.$$

If the size of the target area is greater than the synthetic aperture length, that is, $Y_0 > L$, then the matched-filtered data in the k_y domain with sample spacing π/L are aliased. There is a simple remedy for this problem: Zero-pad the synthetic aperture signal $s(\omega, u)$ to create an effective aperture of $[-Y_0, Y_0]$ prior to the DFT processing and slow-time matched filtering. In the general case ($Y_0 > L$ or $Y_0 \le L$), the minimum half-length of the processed (zero-padded) synthetic aperture is chosen to be

$$L_{\min} = \max(L, Y_0).$$

There exists another digital signal processing–oriented rationale for the zero-padding operation when $Y_0 > L$. This comes from the fact that one has to avoid u-domain (or y-domain) circular convolution aliasing when the slow-time matched filtering; that is,

$$s(\omega, u) * s_0^*(\omega, -u),$$

is performed. (In theory, there is circular convolution aliasing if $Y_0 \leq L$.) It turns out that the digital signal theory origin of both reasons for zero-padding, that is, $\Delta_{k_y} \leq \pi / Y_0$ and avoiding circular convolution aliasing, is the same.

We should also point out that if $Y_0 > L$ (which means that the synthetic aperture data must be zero-padded), it is better to perform zero-padding *prior* to the sampling rate conversion from Δ_{uc} to Δ_u, which was described in the previous section. It turns out that zero-padding in the slow-time u domain reduces the numerical (slight wraparound) errors in the slow-time compression and decompression processing.

Algorithm The digital reconstruction algorithm (software) for cross-range imaging can be summarized by the following (see Figure 2.10):

Step 1. If $Y_0 > L$, zero-pad the data in the slow-time u domain to achieve the minimum length synthetic aperture extent of $u \in [-L_{\min}, L_{\min}]$, where $L_{\min} = \max(L, Y_0)$. For notational simplicity we also denote the number of resultant samples on the synthetic aperture after zero-padding (if it is necessary) with M_c.

Step 2. Increase the PRF by the slow-time compression, k_u domain zero-padding, and slow-time decompression. The resultant data correspond to unaliased samples of $s(\omega, u)$ at synthetic aperture points separated by Δ_u, that is, $s(\omega, u_i)$, where

$$u_i = \left(i - \frac{M}{2} - 1\right) \Delta_u, \qquad i = 1, 2, \ldots, M,$$

and

$$M = 2\left\lceil \frac{L_{\min}}{\Delta_u} \right\rceil,$$

is the number of the slow-time measurements in the processed (zero-padded) synthetic aperture $[-L_{\min}, L_{\min}]$.

Step 3. In the case of a squint target area, perform baseband conversion in the slow-time domain on the synthetic aperture data. This is achieved by multiplying $s(\omega, u_i)$ with $\exp(-j2k \sin\theta_c u_i)$. This operation is not essential for a broadside target area, since $\theta_c = 0$.

Step 4. Compute the DFT of the slow-time baseband samples of $s(\omega, u_i)$. The result is a set of samples of $S(\omega, k_u)$ centered around the carrier slow-time frequency $2k \sin\theta_c$, within the baseband $\pm\pi/\Delta_u$, and separated by $\Delta_{k_u} = \pi / L_{\min}$, that is,

$$k_{ui} = 2k \sin\theta_c + \left(i - \frac{M}{2} - 1\right) \Delta_{k_u}, \qquad i = 1, 2, \ldots, M.$$

Note that from the DFT equation we have

$$M\Delta_u\Delta_{k_u} = 2\pi.$$

Step 5. Compute the reference signal slow-time Doppler spectrum

$$S_0(\omega, k_u) = \exp\left(-j\sqrt{4k^2 - k_u^2}\, X_c - jk_u Y_c\right),$$

at the above-mentioned k_{ui} sample points.

Step 6. Perform the slow-time matched filtering via

$$F(k_{ui}) = S(\omega, k_{ui}) S_0^*(\omega, k_{ui}).$$

Step 7. Compute the inverse DFT of $F(k_{ui})$ with respect to the slow-time Doppler samples k_{ui} to obtain the baseband target function $f(y)$ at the samples

$$y_i = \left(i - \frac{M}{2} - 1\right)\Delta_u, \qquad i = 1, 2, \ldots, M.$$

Display $|f(y_i)|$ versus y_i.

For target identification purposes, it is useful to form and perhaps store the coherent (complex) version of the reconstructed target function $f(y)$. Thus it is useful to remove any redundancy in the $f(y)$ array before storing it. Next we examine two methods for reducing the size of the reconstructed target function $f(y)$ array.

Slow-time Doppler Domain Subsampling

In some applications (e.g., certain spotlight SAR systems), the size of the aperture $2L$ might be much larger than the size of the target area $2Y_0$. Clearly there is no need for zero-padding the synthetic aperture data $s(\omega, u)$ in the slow-time domain. The resultant target function in the k_y domain has a sample spacing of

$$\Delta_{k_y} = \frac{\pi}{L}.$$

However, the Nyquist sample spacing for the cross-range target function is

$$\Delta_{k_y} \le \frac{\pi}{Y_0}.$$

Since L is much greater than Y_0, we satisfy the Nyquist criterion for $F(k_y)$ in the k_y domain. In fact, the samples $F(k_{ui})$ contain a large amount of redundancy, so we are overdoing it.

To reduce the computational load and array sizes, one might perform slow-time matched filtering only at a subsampled set of k_{ui}'s. Sampling rate reduction can be achieved via interpolation; this, however, increases the computational load. A simple approach is to skip certain samples. For instance, if $L = 3.8Y_0$, then the user could reduce the sampling rate by 3.8 via interpolation. Or, skip every two samples; in this case there is still some redundancy in the data which provides a guard band in the cross-range y domain [which reduces inverse DFT errors in obtaining $f(y)$ from $F(k_y)$]. The user should be aware of the k_{ui} value which also corresponds to the middle sample. Moreover, if the number of subsampled data is an odd number, the $F(k_{ui})$ array should be appended with a zero to make the size of the array even (for inverse DFT processing).

Reducing Bandwidth of Reconstructed Image

We now examine another procedure for reducing the size of the $f(y)$ array; this is encountered in spotlight SAR systems. The reconstructed target function was found to be

$$f(y) = \sum_n \sigma_n i_n(\omega, y - y_n),$$

where

$$i_n(\omega, y) = |\Omega_n| \exp(j\Omega_{nc} y) \operatorname{sinc}\left(\frac{1}{2\pi}|\Omega_n| y\right)$$

and

$$\Omega_{nc} \approx 2k \sin\theta_n(0),$$

$$|\Omega_n| \approx \frac{4kL}{x_n} \cos^2\theta_n(0),$$

are, respectively, the center Doppler frequency and length of the slow-time Doppler band Ω_n. Clearly the slow-time Doppler carrier frequency Ω_{nc} results in unnecessary expansion of the slow-time Doppler bandwidth of $f(y)$. Unfortunately, since this carrier depends on the coordinates of the target (*shift-varying*), we cannot simply multiply $f(y)$ with a *single* carrier to bring it to the baseband.

For instance, for the baseband conversion of the nth target signature, that is,

$$\sigma_n i_n(\omega, y - y_n),$$

the following phase function (demodulating signal) should be used:

$$\begin{aligned}\exp(-j\Omega_{nc} y) &= \exp[-j2k \sin\theta_n(0) y] \\ &= \exp\left(-j2k \frac{y_n}{\sqrt{X_c^2 + y_n^2}} y\right).\end{aligned}$$

The *instantaneous frequency* of the demodulating signal with respect to y is

$$-2k \frac{y_n}{\sqrt{X_c^2 + y_n^2}}.$$

The signature of the nth target in the reconstructed cross-range image is concentrated around $y = y_n$. Thus the above instantaneous frequency within a small neighborhood of y_n can be approximated via

$$-2k\frac{y}{\sqrt{X_c^2 + y^2}}.$$

Note that the above instantaneous frequency is invariant in the cross-range of the target y_n. Moreover this instantaneous frequency is equal to the instantaneous frequency with respect to y of the following phase-modulated signal:

$$\exp\left[-jk\frac{y^2}{\sqrt{X_c^2 + y^2}}\right].$$

Thus we could use this phase on the entire reconstructed image for baseband conversion of all target signatures in the scene:

$$f_c(y) = f(y)\exp\left(-jk\frac{y}{\sqrt{X_c^2 + y^2}}y\right).$$

We call $f_c(y)$ the *bandwidth-compressed* reconstructed target function (Figure P2.10).

A more accurate analysis could be performed to show that the baseband conversion function has the following form (recall the functional form of the spherical PM signal):

$$\exp(-j\Omega_{nc}y) \approx \exp\left(-j2k\sqrt{X_c^2 + y^2}\right).$$

Moreover, with the shift of the origin to the center of the target area in the general squint case, the bandwidth compressed target function becomes

$$f_c(y) = f(y)\exp\left[-j2k\sqrt{X_c^2 + (Y_c + y)^2} + j2k\sin\theta_c y\right].$$

[Compare Figure P2.10, compressed target function spectrum $|F_c(k_y)|$, and Figure P2.8, target function spectrum before compression $|F(k_y)| = |S(\omega, k_u)|$.]

2.9 SYNTHETIC APERTURE-DEPENDENT TARGET REFLECTIVITY

The consequence of finite synthetic aperture measurements of the spherical PM signal on its Fourier spectrum can be restated via the following windowing operations. The measured spherical PM signal for the nth target can be represented via

$$s_n(\omega, u) = \sigma_n a_n(\omega, x_n, y_n - u)\exp\left[-j2k\sqrt{x_n^2 + (y_n - u)^2}\right],$$

where

$$a_n(\omega, x_n, y_n - u) = \begin{cases} 1 & \text{for } |u| \le L, \\ 0 & \text{otherwise,} \end{cases}$$

is a *rectangular window* function in the synthetic aperture u domain. [Note that this window function is not a function of the coordinates of the target (x_n, y_n); the above notation for $a_n(\omega, x_n, y_n - u)$ is the general form of what we will encounter in our future discussion.] The slow-time Fourier transform of this signal can be represented via

$$S_n(\omega, k_u) = A_n(\omega, k_u) \exp\left(-j\sqrt{4k^2 - k_u^2}\, x_n - jk_u y_n\right),$$

where

$$A_n(\omega, k_u) = \begin{cases} 1 & \text{for } k_u \in \Omega_n = [2k \sin\theta_n(L), 2k \sin\theta_n(-L)], \\ 0 & \text{otherwise,} \end{cases}$$

is a *shifted rectangular window* function in the slow-time Doppler k_u domain.

- The window $A_n(\omega, k_u)$ is not the Fourier transform of the window $a_n(\omega, x_n, y_n - u)$ with respect to the slow-time u domain. In fact these two rectangular window functions are related via the following (this will be shown):

  $$a_n(\omega, x_n, y_n - u) = A_n\left[2k \sin\theta_n(u), \omega\right].$$

  The support of both window functions increases as the size of the synthetic aperture, that is, $2L$, increases. This is a property of the PM signals and their Fourier transform.

AM-PM Signal Model

In this section we generalize the above results for an arbitrary window or *amplitude* function which is applied to the spherical PM signal in the synthetic aperture u domain. The origin of such an amplitude function can be traced back to the physical properties of both the radar and the target. In the case of a radar-induced amplitude function, the type, coordinates, and the orientation of the radar determine the amplitude function. The amplitude function which is induced by the target is dictated by the resonance effects and orientation of the target. We will provide a system representation of the radar and target-induced amplitude functions in Chapter 3. At this point we are interested in signal (e.g., Fourier) properties of a spherical PM signal which is amplitude modulated.

Suppose that the measurement of the nth target signature contains a function of (ω, u) that amplitude modulates the spherical PM signal. This function, which we refer to as the target *amplitude pattern* or function, is related to the physical properties of the radar and target, which will be discussed later. The target amplitude pattern is a *complex* function (i.e., it contains both magnitude and phase) that depends on the relative coordinates of the radar and target, that is, (x_n, y_n) and $(0, u)$. We denote the amplitude pattern for a target with coordinates (x_n, y_n) when the radar is at $(0, u)$ by $a_n(\omega, x_n, y_n - u)$; note that the vector $(x_n, y_n - u)$ is the relative coordinates of the target with respect to the radar.

The target amplitude pattern replaces the factor $\sigma_n a_n(\omega, x_n, y_n - u)$ in the system model; that is,

$$\sigma_n a_n(\omega, x_n, y_n - u) \longrightarrow a_n(\omega, x_n, y_n - u).$$

Thus the signature of the nth target in the measured (ω, u) domain becomes

$$s_n(\omega, u) = \underbrace{a_n(\omega, x_n, y_n - u)}_{\text{AM}} \underbrace{\exp\left[-j2k\sqrt{x_n^2 + (y_n - u)^2}\right]}_{\text{PM}},$$

which is an amplitude-modulated–phase-modulated (AM–PM) signal. In most practical radar problems, the fluctuations of both magnitude and phase of the AM component (amplitude pattern) are much smaller than those of the PM component (spherical PM signal). Figure 2.11a shows an example of AM-PM spherical signal $s_n(\omega, u)$.

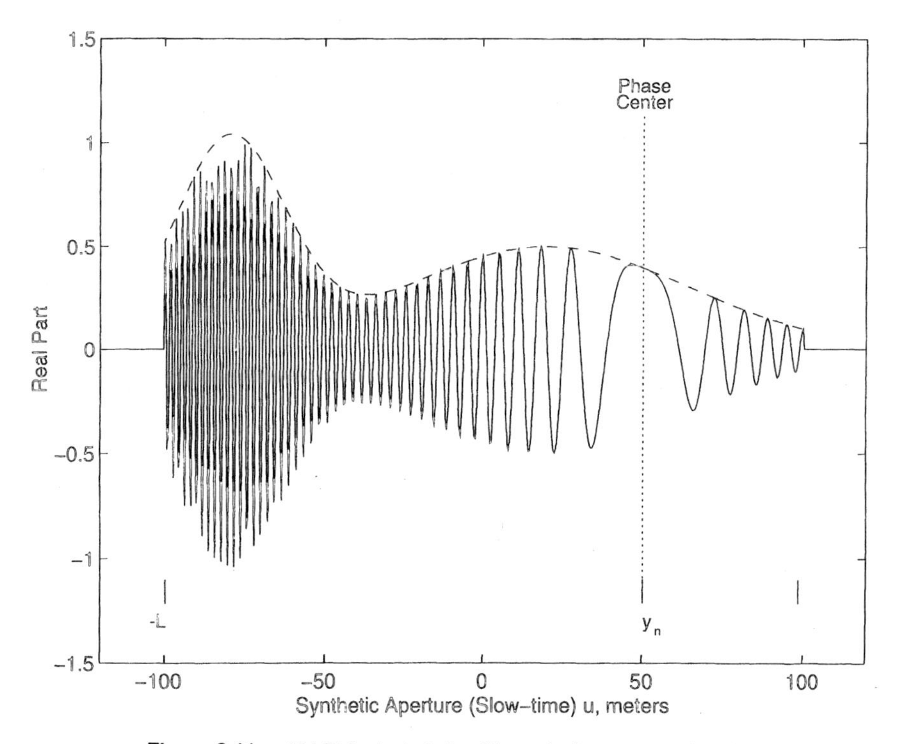

Figure 2.11a. AM-PM spherical signal in synthetic aperture u domain.

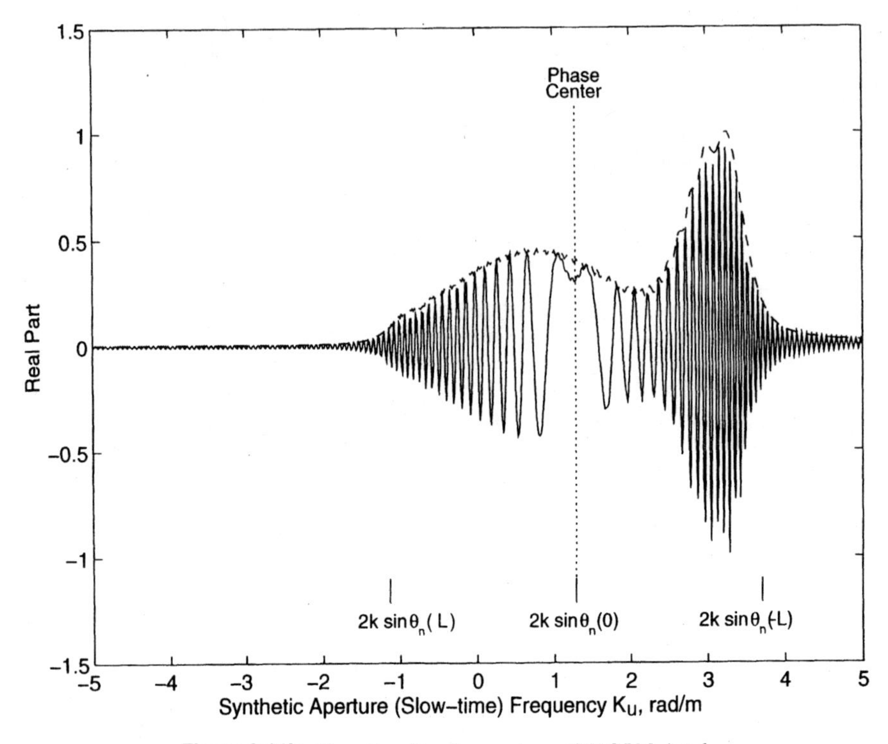

Figure 2.11b. Slow-time Doppler spectrum of AM-PM signal.

Slow-time Fourier Transform of AM-PM Signal

We wish to find the Fourier transform of the nth target signature with respect to the slow-time u domain. For this we construct the Fourier integral:

$$\begin{aligned} S_n(\omega, k_u) &= \int_{-\infty}^{\infty} s_n(\omega, u) \exp(-jk_u u)\, du \\ &= \int_{-\infty}^{\infty} a_n(\omega, x_n, y_n - u) \exp\left[-j2k\sqrt{x_n^2 + (y_n - u)^2} - jk_u u\right] du \end{aligned}$$

The integrand in the above integral is composed of an amplitude function

$$a_n(\omega, x_n, y_n - u),$$

and a phase function

$$\exp\left[-j2k\sqrt{x_n^2 + (y_n - u)^2} - jk_u u\right].$$

In practice, the fluctuations of the amplitude function are much slower than those of the phase function. In this case we can approximate the above integral to be equal to the integrand at the point where the instantaneous frequency of the phase function is zero in the integral u domain, that is, its phase center in the u domain. This is known as the method of *stationary phase* [pap].

For a *fixed* k_u we denote the phase center of the phase function by $\mathbf{u}(k_u)$. This phase center is found by equating the instantaneous frequency of the phase function to zero; that is,

$$\frac{\partial}{\partial u}\left[-2k\sqrt{x_n^2+(y_n-u)^2}-k_u u\right]_{u=\mathbf{u}(k_u)}=0,$$

which yields the following solution for $\mathbf{u}(k_u)$ in terms of k_u:

$$2k\sin\theta_n[\mathbf{u}(k_u)]=k_u.$$

Using this fact, the slow-time Fourier transform of the AM-PM synthetic aperture signal $s_n(\omega, u)$ can be found to be the following:

$$S_n(\omega, k_u)=A_n(\omega, k_u)\exp\left(-j\sqrt{4k^2-k_u^2}\,x_n-jk_u y_n\right),$$

where $A_n(\cdot)$ is an amplitude pattern in the slow-time Doppler domain which is related to the slow-time domain amplitude pattern $a_n(\cdot)$ via the following *scale* transformation:

$$A_n\left[\omega, 2k\sin\theta_n(u)\right]=a_n(\omega, x_n, y_n-u),$$

or

$$A_n\left(\omega, 2k\frac{y}{\sqrt{x^2+y^2}}\right)=a_n(\omega, x, y).$$

Figure 2.11*b* shows the slow-time Doppler spectrum $S_n(\omega, k_u)$ for the AM-PM spherical signal $s_n(\omega, u)$ of Figure 2.11*a*.

In our discussion we do not carry the constant $1/r_n^2$ which only affects the strength of the signature of the target and not its relative phase; the phase is the critical component in coherent SAR imaging.

The above result indicates that the amplitude pattern $a_n(\cdot)$, which puts a u-dependent weight on the spherical PM signal component of $s_n(\omega, u)$, puts a similar weight on its Fourier transform phase function in the slow-time Doppler k_u domain.

- The effective support of $S_n(\omega, k_u)$ in the k_u domain (the slow-time Doppler band) increases as the effective support of $s_n(\omega, u)$ in the u domain increases, since both are PM signals that are amplitude modulated with $a_n(\cdot)$ and its scaled version $A_n(\cdot)$.

In the next two examples, we examine two specific classes of AM-PM SAR signals. The AM component in these cases is induced by the manner in which the synthetic aperture data are collected by the user. These, as we will see in Chapters 5 and 6, define the information (signal) subspace in spotlight and stripmap SAR systems.

Example 1 *Spotlight SAR* In spotlight SAR systems, the radar-induced amplitude function is a *rectangular window* function in the synthetic aperture u domain; that is,

$$a_n(\omega, x_n, y_n - u) = \begin{cases} 1 & \text{for } |u| \le L, \\ 0 & \text{otherwise.} \end{cases}$$

Note that this amplitude function is invariant of the coordinates of the target (x_n, y_n). As we showed in Section 2.4 (*spherical PM signal within a finite aperture*), the slow-time Doppler domain amplitude function for this scenario is

$$A_n(\omega, k_u) = \begin{cases} 1 & \text{for } k_u \in \Omega_n = [2k \sin\theta_n(L), 2k \sin\theta_n(-L)], \\ 0 & \text{otherwise.} \end{cases}$$

which is a shifted rectangular window function in the slow-time Doppler k_u domain that depends on the coordinates of the target.

In Section 2.4 we showed that the slow-time Doppler support band of $S_n(\omega, k_u)$ or, equivalently, $A_n(\omega, k_u)$, that is,

$$\Omega_n = [2k \sin\theta_n(L), 2k \sin\theta_n(-L)],$$

is a bandpass region in the slow-time Doppler domain k_u; the center Doppler frequency of this band is approximately located at

$$\Omega_{nc} \approx 2k \sin\theta_n(0),$$

and the length of the band Ω_n can be approximated via

$$|\Omega_n| \approx \frac{4kL}{x_n} \cos^2\theta_n(0).$$

We now examine two numerical examples of the spotlight SAR signal.

Consider a spotlight SAR (cross-range) imaging system with the following parameters: fast-time frequency 300 MHz; $(X_c, Y_c) = (500, 0)$ m; $L = 100$ m. Figure 2.12*a* and *b*, respectively, show the real parts of the spotlight SAR signal $s(\omega, u)$ and its slow-time Fourier transform $S(\omega, k_u)$ for a target located at $(x_n, y_n) = (500, 0)$ m (broadside target). Figure 2.12*c* and *d* show the same signals for a squint target located at $(x_n, y_n) = (500, 60)$ m. Note that the Fourier spectrum of the spotlight SAR signal with a rectangular window contains ringing within the theoretical support band of $A_n(\omega, k_u)$ in the k_u domain; one can also observe undesirable side lobes outside this support band.

The passband ringing and the side lobes have a degrading effect in the cross-range reconstruction which is obtained via

$$\begin{aligned} F(k_y) &= S(\omega, k_u) S_0^*(\omega, k_u) \\ &= S(\omega, k_u) \exp\left(-j\sqrt{4k^2 - k_u^2}\, X_c - j k_u Y_c\right). \end{aligned}$$

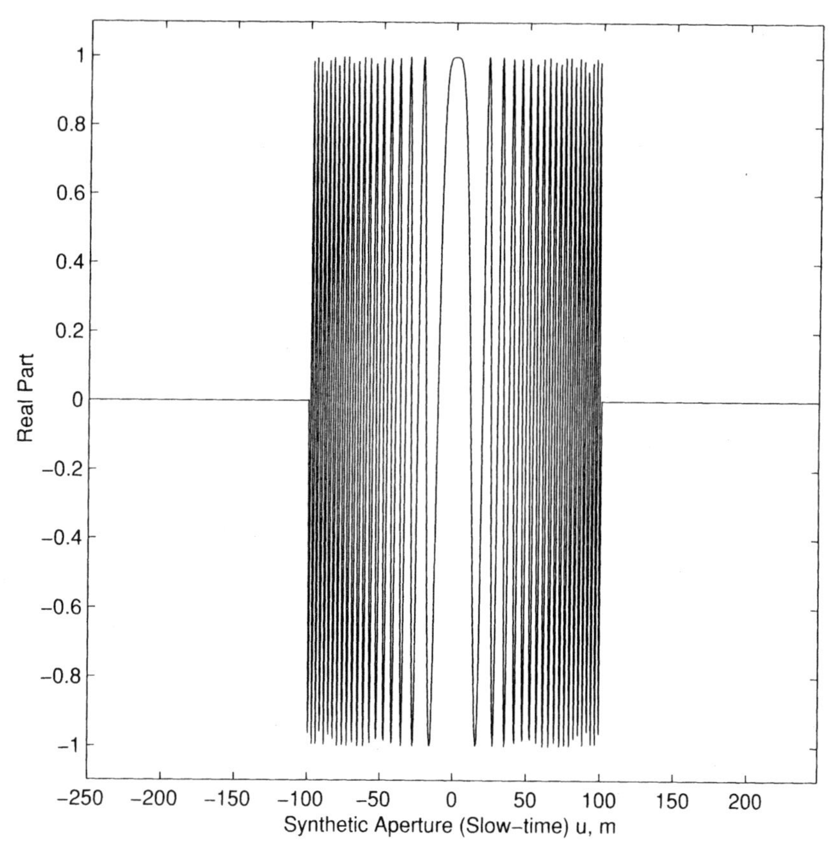

Figure 2.12a. Real part of spotlight SAR signal $s(\omega, u)$ for a target located at $(x_n, y_n) = (500, 0)$ m (broadside target).

(The reconstruction problem and its associated point spread function are discussed at the end of this section.) Figure 2.12*e* shows the reconstructed target function in the cross-range domain when both the broadside and squint targets are present.

One may use windowing techniques in the aperture domain to reduce the undesirable ringing effects of the rectangular aperture. Figure 2.13 shows the same outputs as in Figure 2.12 when a Hamming window is applied to the spotlight SAR signal. Note that the passband ringing and side lobes are significantly reduced in the two slow-time Doppler spectra (Figure 2.13*b* and *d*). The windowing has also reduced the *effective* support band (e.g., the 3-dB bandwidth) of the targets in the slow-time Doppler domain. One can observe the widening of the point spread function in the reconstruction of Figure 2.13*e*. However, as one would anticipate, the side lobes in this reconstruction are weaker than the side lobes in the reconstruction with the rectangular window of Figure 2.12*e*.

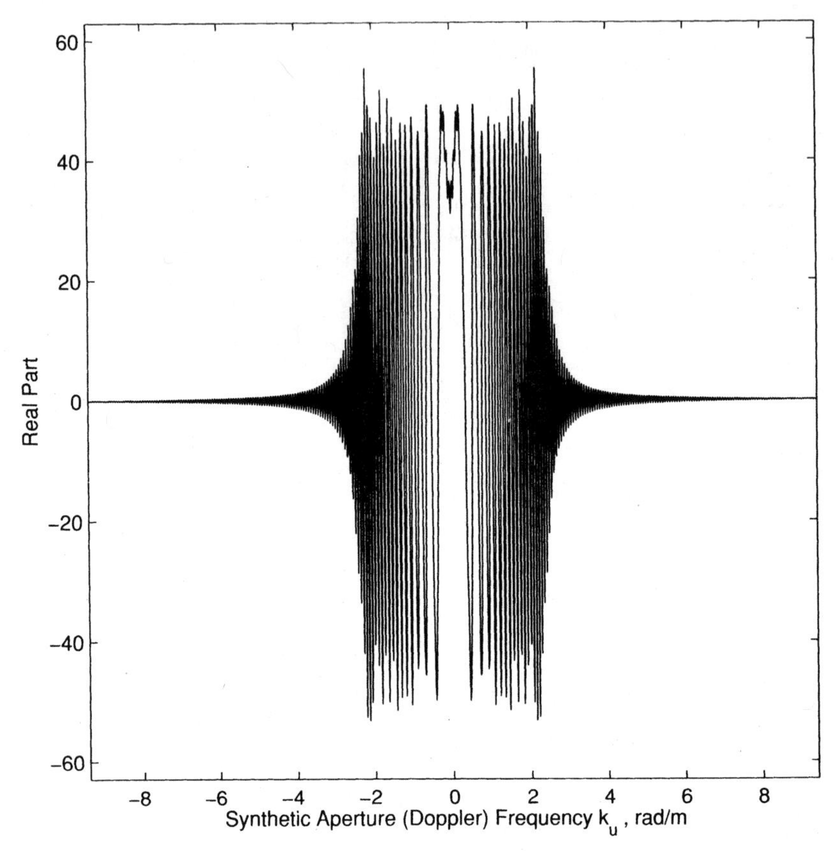

Figure 2.12b. Real part of spotlight SAR signal slow-time Fourier transform $S(\omega, k_u)$ for a target located at $(x_n, y_n) = (500, 0)$ m (broadside target).

One could design a window that provides a smooth tapering near the edges of the synthetic aperture (i.e., $|u| \approx L$) to reduce spectral ringing and side lobes while preserving most of the slow-time Doppler bandwidth to improve cross-range resolution. An example of such a window is

$$\exp(-\eta|u|^{\zeta}),$$

which we refer to as the *power (or exponential) window*. The constants (η, ζ) are chosen by the user based on the desired shape of the power window. One could solve for (η, ζ) by specifying the window value at two distinct points in the synthetic aperture domain. (Note that all the derivatives of the window exist if ζ is an even integer.) The results of using the power window on spotlight SAR signals are shown in Figure 2.14. In this case the window is chosen to be -3 dB and -20 dB at $u = 0.9L$ and $u = L$, respectively. The resultant slow-time Doppler spectra in Figure

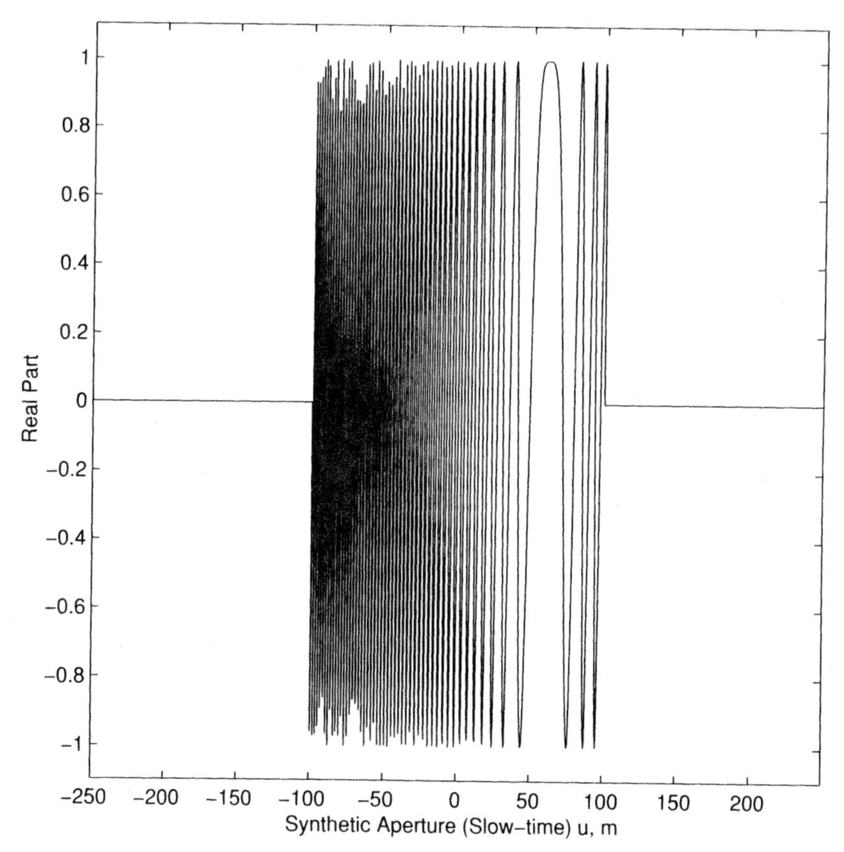

Figure 2.12c. Real part of spotlight SAR signal $s(\omega, u)$ for a target located at $(x_n, y_n) = (500, 60)$ m (squint target).

2.14*b* and *d* contain a relatively weak amount of ringing and side lobes. The main lobes of the cross-range reconstructions in Figure 2.12*e* (rectangular window) and Figure 2.14*e* (power window) have approximately the same width (the same cross-range resolution). However, the side lobes in Figure 2.14*e* are weaker than the side lobes in Figure 2.12*e*.

Example 2 *Stripmap SAR* In the study of the radiation pattern of a planar radar in Chapter 3, we will encounter an amplitude pattern which is the sinc-squared function

$$A_n(\omega, k_u) = \sigma_n \operatorname{sinc}^2\left(\frac{D}{4\pi}k_u\right),$$

where D, which we will show to be the *diameter* of the radar, is a known constant. *Note that this amplitude function is invariant of ω in the k_u domain.* Such a slow-time

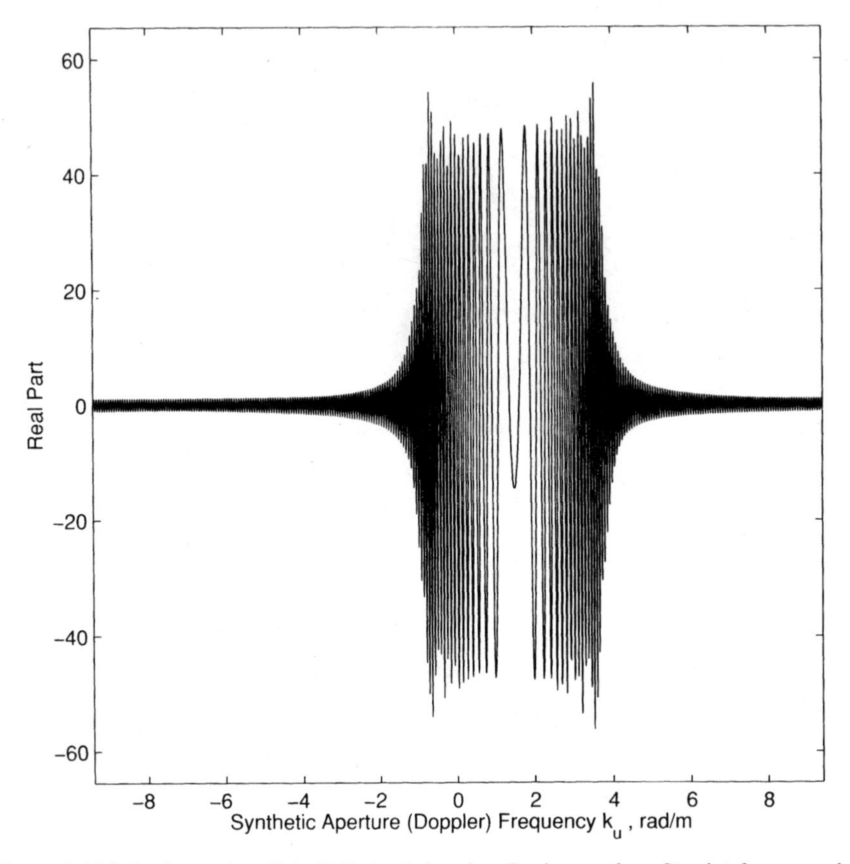

Figure 2.12d. Real part of spotlight SAR signal slow-time Fourier transform $S(\omega, k_u)$ for a target located at $(x_n, y_n) = (500, 60)$ m (squint target).

Doppler domain amplitude function will be encountered in stripmap SAR systems (Chapter 6).

The support of this amplitude function, which is approximated to be the main lobe of the sinc-squared function, is within

$$k_u \in \left[\frac{-4\pi}{D}, \frac{4\pi}{D} \right].$$

Thus the slow-time Doppler support band of the nth target in stripmap SAR is

$$\Omega_n = \left[\frac{-4\pi}{D}, \frac{4\pi}{D} \right].$$

Unlike the spotlight SAR signal slow-time support band, the slow-time Doppler support band of a stripmap SAR signal is *invariant* of the coordinates of the target.

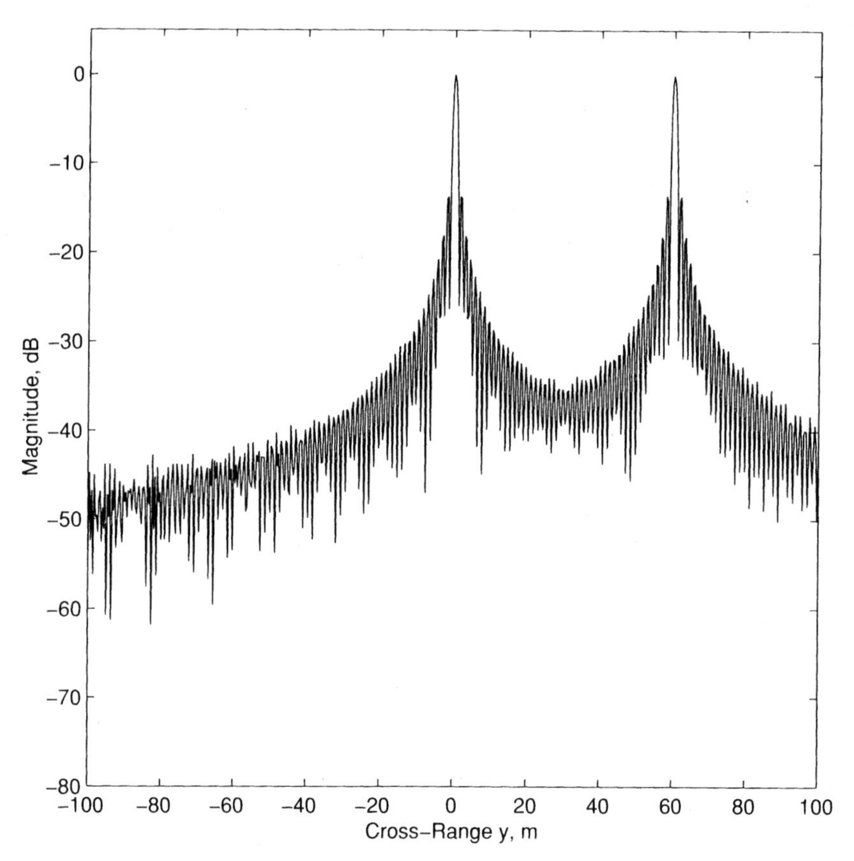

Figure 2.12e. Reconstructed broadside and squint targets in cross-range domain from their spotlight SAR signal.

Moreover the center Doppler frequency of this band is at the zero Doppler frequency, that is,

$$\Omega_{nc} = 0,$$

and the length of the slow-time Doppler band Ω_n is

$$|\Omega_n| = \frac{8\pi}{D}.$$

Using the sinc-squared expression for the amplitude pattern, the individual target signature in the synthetic aperture domain becomes

$$s_n(\omega, u) = \sigma_n \operatorname{sinc}^2 \left[\frac{kD}{2\pi} \sin\theta_n(u) \right] \exp\left[-j2k\sqrt{x_n^2 + (y_n - u)^2} \right].$$

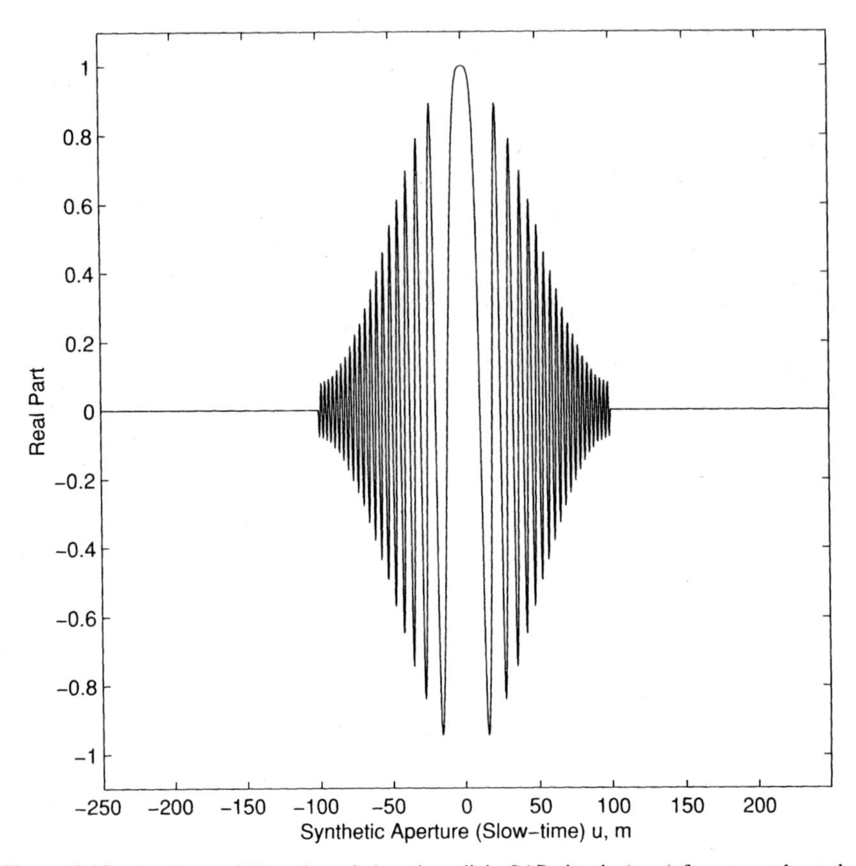

Figure 2.13a. Real part of Hamming-windowed spotlight SAR signal $s(\omega, u)$ for a target located at $(x_n, y_n) = (500, 0)$ m (broadside target).

The main lobe of the AM component of $s_n(\omega, u)$ in the synthetic aperture u domain, that is,

$$a_n(\omega, x_n, y_n - u) = \sigma_n \operatorname{sinc}^2\left[\frac{kD}{2\pi} \sin\theta_n(u)\right]$$

$$= \sigma_n \operatorname{sinc}^2\left[\frac{kD}{2\pi} \frac{y_n - u}{\sqrt{x_n^2 + (y_n - u)^2}}\right]$$

is within

$$u \in [y_n - B, y_n + B],$$

where

$$B = \frac{x_n \lambda}{D},$$

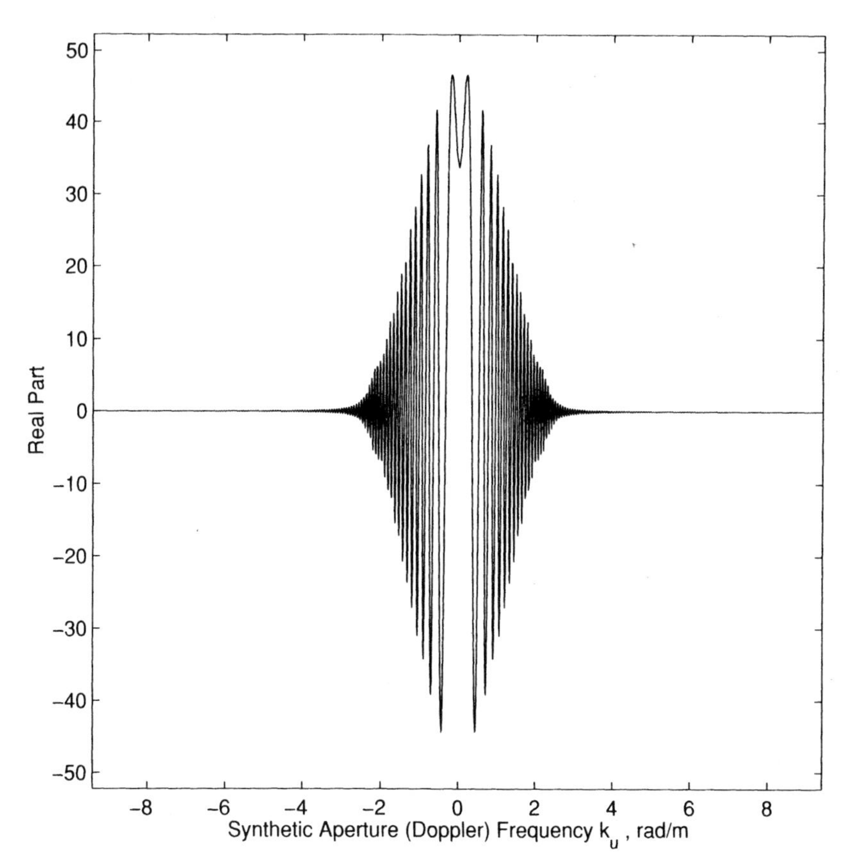

Figure 2.13b. Real part of Hamming-windowed spotlight SAR signal slow-time Fourier transform $S(\omega, k_u)$ for a target located at $(x_n, y_n) = (500, 0)$ m (broadside target).

is called the *half-beamwidth.* The main lobe of the sinc function is called the *beamwidth* of the synthetic aperture signal $s_n(\omega, u)$. For a general $A_n(\cdot)$ the effective support of $s_n(\omega, u)$ is called the beamwidth of the signal.

- For $A_n(\omega, k_u) = \sigma_n \operatorname{sinc}^2(Dk_u/4\pi)$, the support band of $S_n(\omega, k_u)$ in the slow-time Doppler k_u domain is $8\pi/D$, which is *invariant of the radar frequency* ω. However, the support band of $s_n(\omega, u)$ in the slow-time u domain, that is, the beamwidth $2B = 2x_n\lambda/D$ which increases with the wavelength λ, is inversely proportional to the radar frequency ω.
- For $A_n(\omega, k_u) = \sigma_n \operatorname{sinc}^2(Dk_u/4\pi)$, some texts define the effective support band of $s_n(\omega, u)$ to be $2B = x_n\lambda/D$, or

$$B = \frac{x_n\lambda}{2D},$$

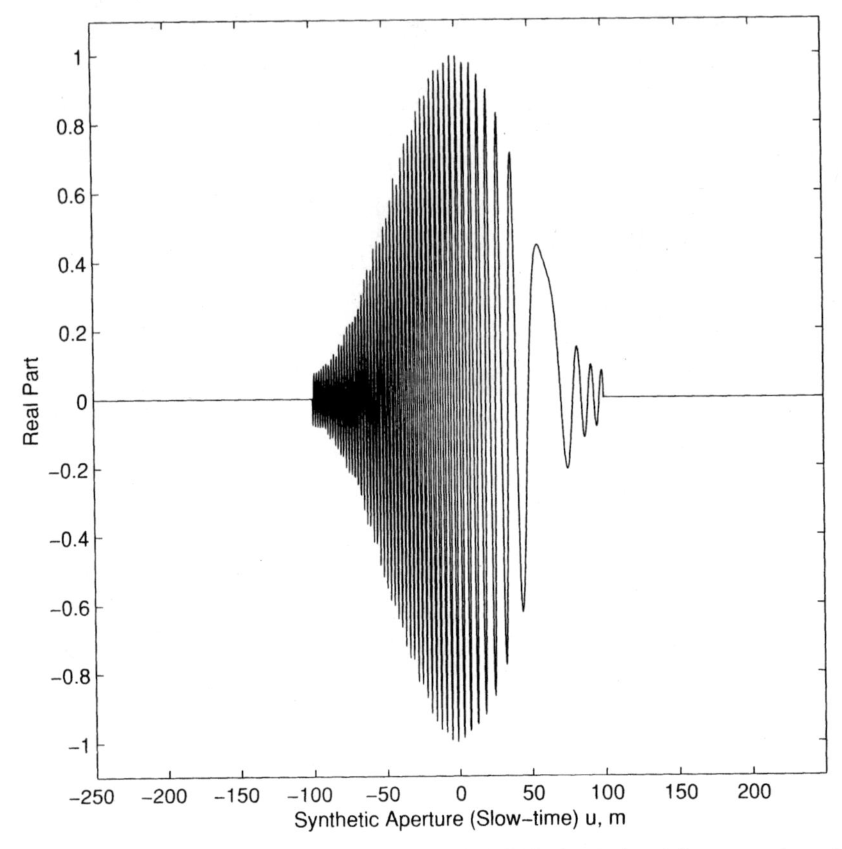

Figure 2.13c. Real part of Hamming-windowed spotlight SAR signal $s(\omega, u)$ for a target located at $(x_n, y_n) = (500, 60)$ m (squint target).

which is one-half of the main lobe of its AM component. It turns out that one must use different definitions for B to identify a *conservative* measure for alias-free processing of the aperture signal, and cross-range resolution. In this text we use

$$\begin{cases} B \geq \dfrac{x_n \lambda}{D}, & \text{to determine the Nyquist sampling rate in the slow-time domain,} \\ B = \dfrac{x_n \lambda}{2D}, & \text{to determine the cross-range resolution.} \end{cases}$$

If we use the beamwidth definition of $B = x_n \lambda / D$ in the expression for the slow-time Doppler support band of the nth target, which is

$$\Omega_n = \left[\frac{-4\pi}{D}, \frac{4\pi}{D} \right],$$

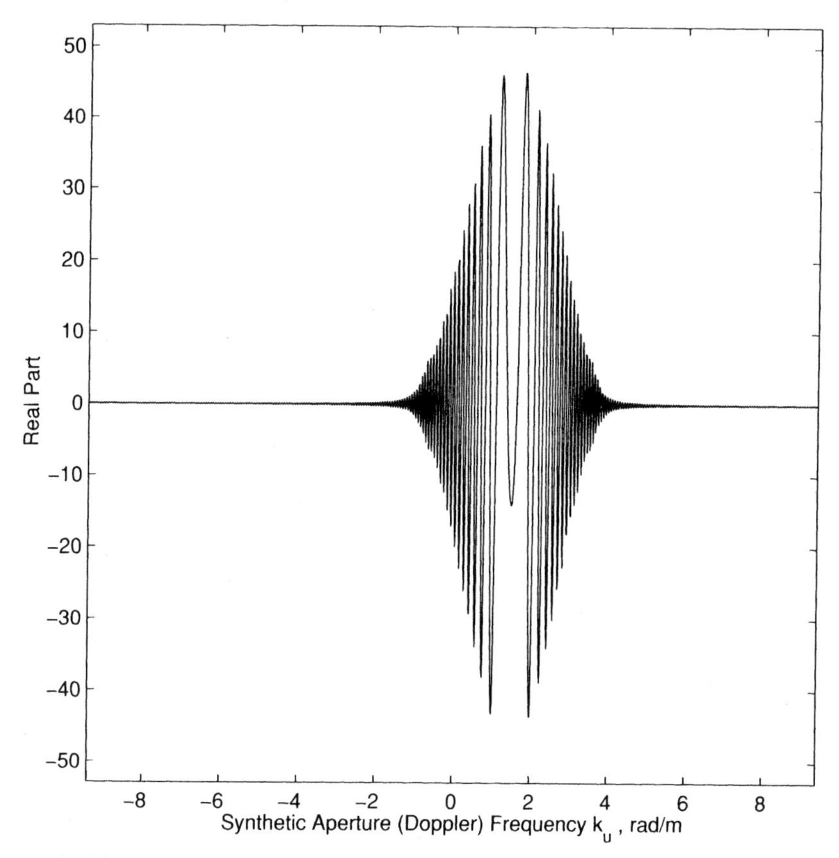

Figure 2.13d. Real part of Hamming-windowed spotlight SAR signal slow-time Fourier transform $S(\omega, k_u)$ for a target located at $(x_n, y_n) = (500, 60)$ m (squint target).

we obtain

$$\Omega_n = \left[\frac{-4\pi B}{x_n \lambda}, \frac{4\pi B}{x_n \lambda}\right]$$
$$= \left[\frac{-2kB}{x_n}, \frac{2kB}{x_n}\right].$$

The above slow-time support band for the stripmap SAR signal (which is invariant of the coordinates of the nth target) is the same as the slow-time support band for the spotlight SAR signal for a broadside target (see Section 2.4), that is,

$$\Omega_n = \left[\frac{-2kL}{x_n}, \frac{2kL}{x_n}\right],$$

with $L = B$. The physical meaning of this phenomenon is that in stripmap SAR a target, irrespective of its cross-range coordinates y_n, possesses a *broadside* spotlight

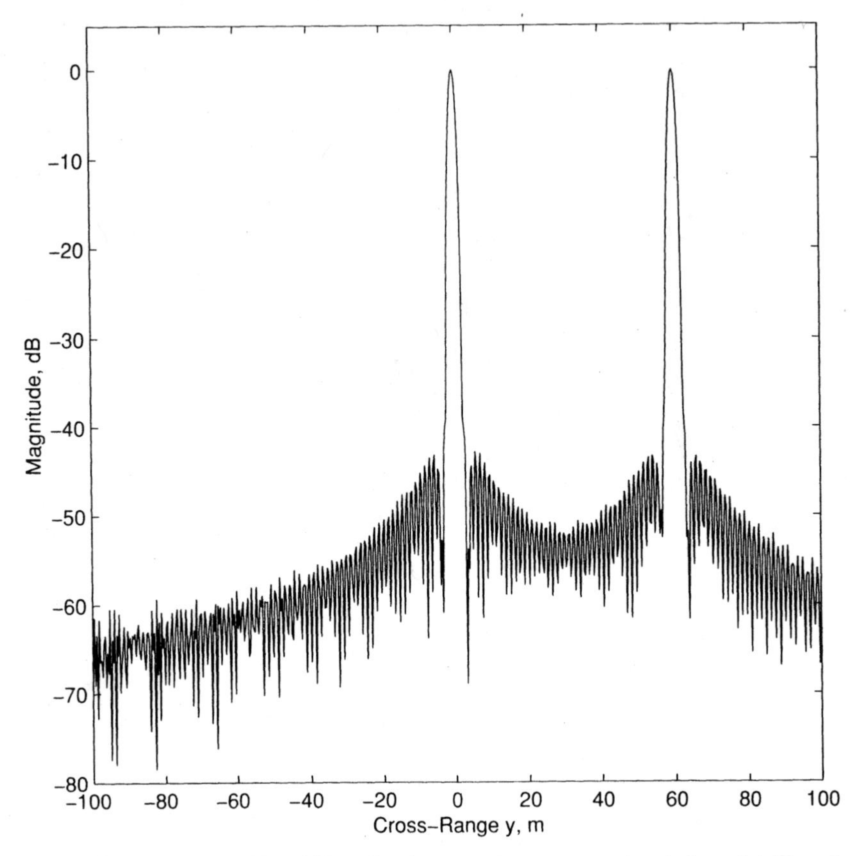

Figure 2.13e. Reconstructed broadside and squint targets in cross-range domain from their Hamming-windowed spotlight SAR signal.

SAR signature within an effective aperture of $[-B, B]$ which is *shifted* to $u = y_n$. We examine numerical examples of the stripmap SAR signal next.

As we mentioned earlier, examples of the sinc-squared amplitude pattern in stripmap SAR systems will be studied in Chapter 3. For our numerical example of a stripmap SAR system, we consider a *cleaner* (i.e., smoother with less side lobes) amplitude pattern which is generated by a hanning window:

$$a_n(\omega, x_n, y_n - u) = 0.5 + 0.5 \cos\left[\frac{\pi}{B}(y_n - u)\right],$$

for $u \in [y_n - B, y_n + B]$, and zero otherwise. For simulation we use the following parameters in this stripmap SAR (cross-range) imaging system: fast-time frequency 300 MHz; $(X_c, Y_c) = (500, 0)$ m; $B = 160$ m. Figure 2.15*a* and *b*, respectively, shows the real parts of the stripmap SAR signal $s(\omega, u)$ and its slow-time Fourier transform $S(\omega, k_u)$ for a target located at $(x_n, y_n) = (500, 0)$ m (broadside target).

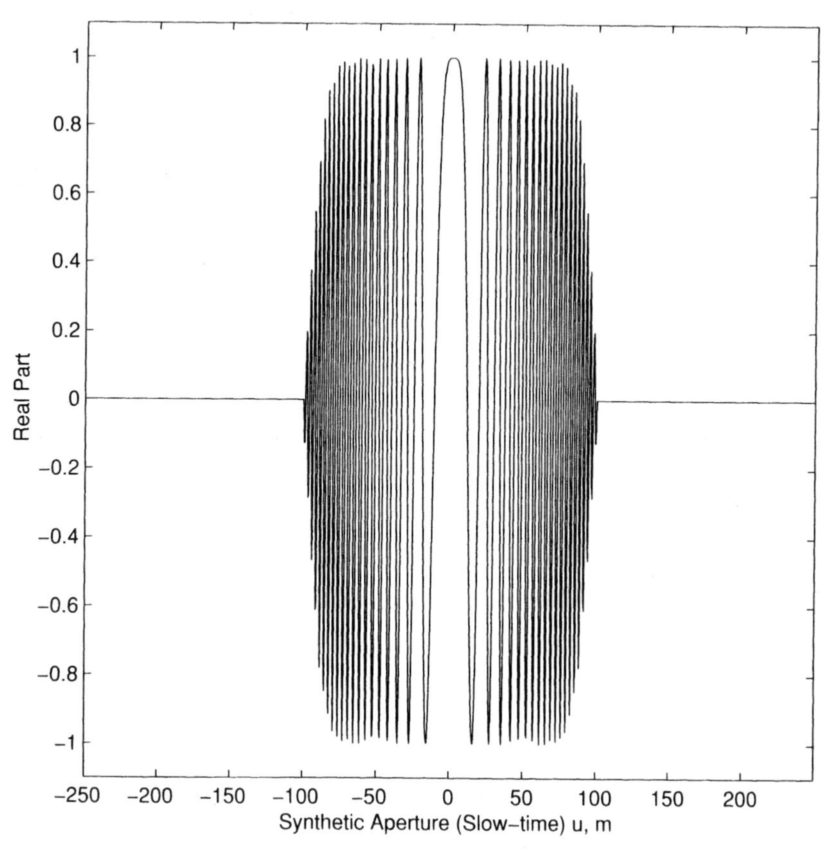

Figure 2.14a. Real part of power-windowed spotlight SAR signal $s(\omega, u)$ for a target located at $(x_n, y_n) = (500, 0)$ m (broadside target).

Figure 2.15*c* and *d* shows the same signals for a squint target located at $(x_n, y_n) = (500, 60)$ m. Note that the slow-time Doppler spectra for the broadside and squint targets, Figure 2.15*b* and *d*, exhibit the same slow-time Doppler support band in the k_u domain; this is not the case in the spotlight spectra of Figure 2.12*b* and *d*. For reconstruction we use the following slow-time matched filtering with the reference signal:

$$F(k_y) = S(\omega, k_u) S_0^*(\omega, k_u)$$
$$= S(\omega, k_u) \exp\left(-j\sqrt{4k^2 - k_u^2}\, X_c - j k_u Y_c\right).$$

Figure 2.15*e* is the reconstructed target function in the cross-range domain when both the broadside and squint targets are present.

In some of the near-range stripmap SAR systems, the beamwidth of the radar $2B$ is larger than the finite aperture $2L$ where the synthetic aperture measurements are

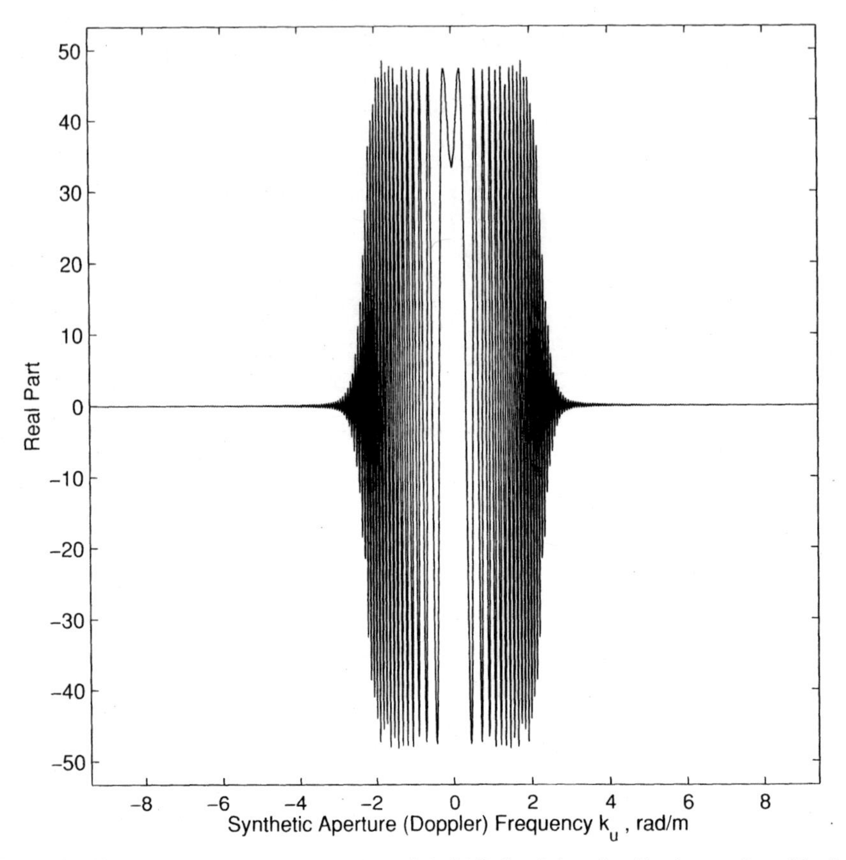

Figure 2.14b. Real part of power-windowed spotlight SAR signal slow-time Fourier transform $S(\omega, k_u)$ for a target located at $(x_n, y_n) = (500, 0)$ m (broadside target).

made; we refer to $u \in [-L, L]$ as the *spotlight window* of the stripmap SAR system. Figure 2.16 shows the results for the above-mentioned stripmap SAR system when a spotlight window of $[-L, L] = [-100, 100]$ m is used. Note that the slow-time spectra of the broadside target (Figure 2.16*b*) and the squint target (Figure 2.16*d*) exhibit the slow-time spectral properties of a spotlight SAR system; see the slow-time spectra of spotlight SAR signals in Figures 2.12*b* and *d*. This class of SAR signals will be further examined in Section 2.10.

We conclude this example with an observation on the point spread function of the broadside and squint targets in Figure 2.16*e*. From this figure one can observe that the PSF signature of the broadside target is *sharper* than the PSF of the squint target; thus the resolution of the targets near the broadside is better than the resolution of the squint targets. Note that the amplitude patterns, that is, $a_n(\omega, x_n, y_n - u)$, of both the broadside and squint targets are *nonzero* in the spotlight window, of $[-L, L] = [-100, 100]$. However, the amplitude pattern of the squint target is tapered off within the spotlight window, while the amplitude pattern of the broadside

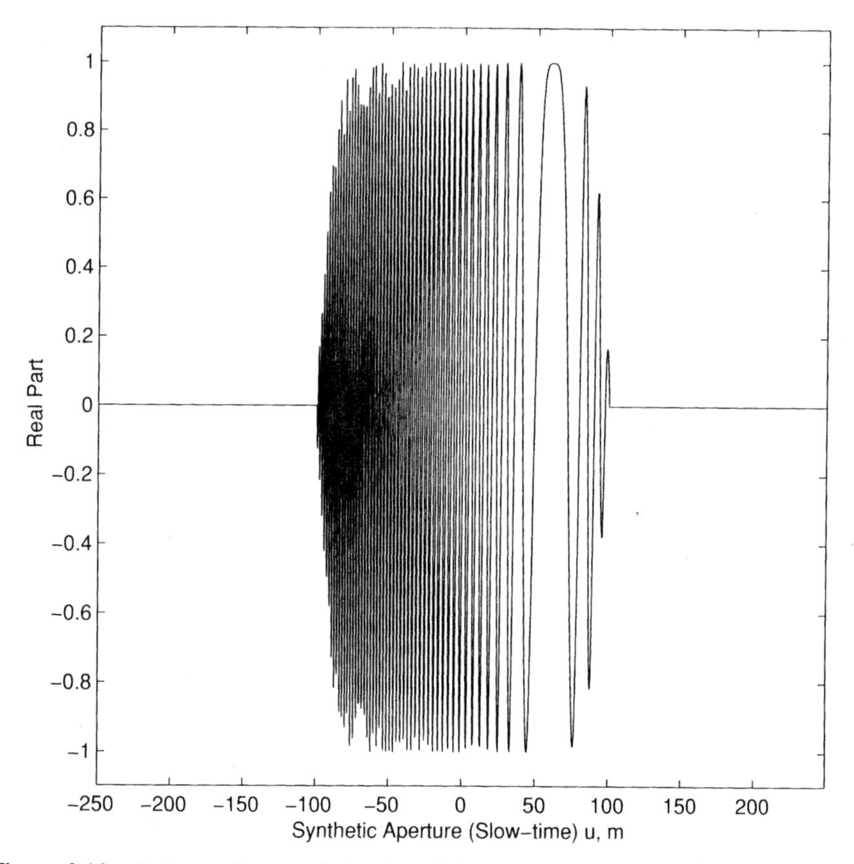

Figure 2.14c. Real part of power-windowed spotlight SAR signal $s(\omega, u)$ for a target located at $(x_n, y_n) = (500, 60)$ m (squint target).

target is approximately flat within the spotlight window. In fact what we refer to as the *effective* beamwidth of a target is a better measure of how *observable* a target is within a synthetic aperture.

There are many measures for defining the effective beamwidth of a target. We use a definition based on what is referred to as the *noise equivalent bandwidth* in communication systems [car]. If the average radar cross section (RCS) of the nth target is σ_n, then its effective beamwidth within a spotlight window $[-L, L]$ is defined via

$$2B_n = \frac{1}{|\sigma_n|^2} \int_{-L}^{L} |a_n(\omega, x_n, y_n - u)|^2 \, du.$$

In this case the cross-range resolution for the nth target is determined via

$$\Delta_{y_n} \approx \frac{x_n \lambda}{4B_n \cos^2 \theta_n(0)},$$

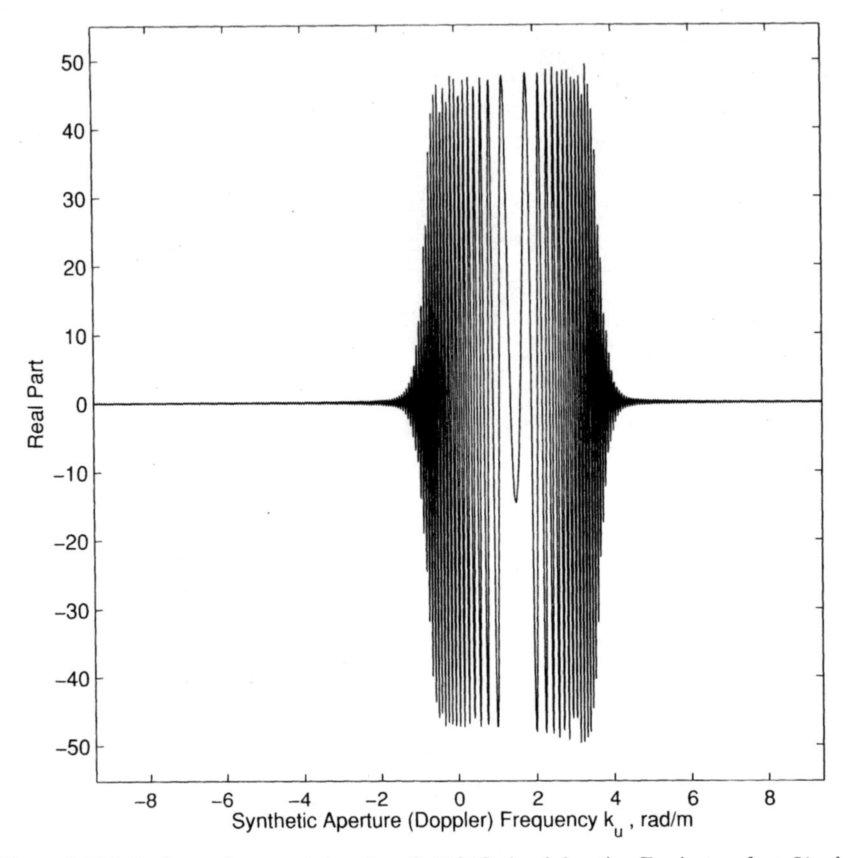

Figure 2.14d. Real part of power-windowed spotlight SAR signal slow-time Fourier transform $S(\omega, k_u)$ for a target located at $(x_n, y_n) = (500, 60)$ m (squint target).

which is the same as the cross-range resolution that we derived in Section 2.6 for a target with a constant amplitude pattern, that is,

$$\Delta_{y_n} = \frac{x_n \lambda}{4L \cos^2 \theta_n(0)},$$

with $L = B_n$. Section 2.10 will provide a discussion on the cross-range resolution for a different class of SAR signals.

Reconstruction

In the previous two examples of spotlight and stripmap SAR systems, we observed that the point spread function of the cross-range imaging systems depends on the amplitude pattern. Our purpose in this section is to obtain an analytical expression for the point spread function in terms of the amplitude pattern. For this, we re-examine

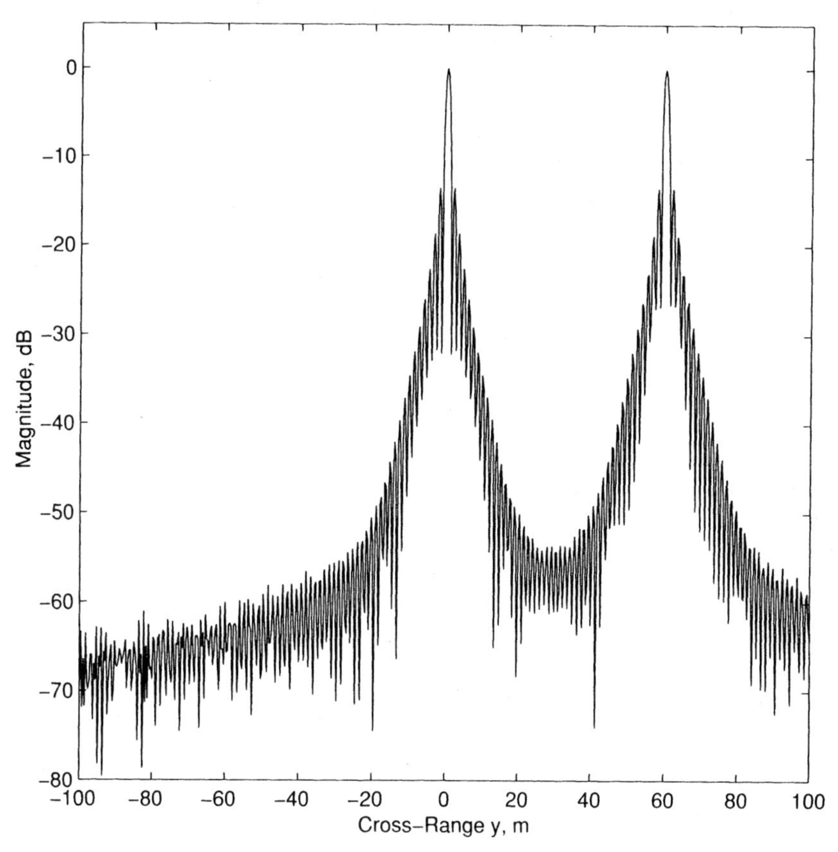

Figure 2.14e. Reconstructed broadside and squint targets in cross-range domain from their power-windowed spotlight SAR signal.

the problem of *one-dimensional* cross-range imaging of a target scene located at a known fixed range $x_n = X_c$ and unknown cross-range points y_n, $n = 1, 2, \ldots$. It is assumed that we make measurements of the AM-PM synthetic aperture signal, that is, $s_n(\omega, u)$, within the finite interval $[-L, L]$ in the synthetic aperture u domain. If we absorb the finite aperture effect window in the amplitude pattern $a_n(\omega, x_n, y_n - u)$, then we obtain the following model for the measured signal (after fast-time baseband conversion):

$$s(\omega, u) = \sum_n a_n(\omega, x_n, y_n - u) \exp\left[-j2k\sqrt{x_n^2 + (y_n - u)^2}\right],$$

and its slow-time Doppler spectrum

$$S(\omega, k_u) = \sum_n A_n(\omega, k_u) \exp\left(-j\sqrt{4k^2 - k_u^2}\, x_n - jk_u y_n\right).$$

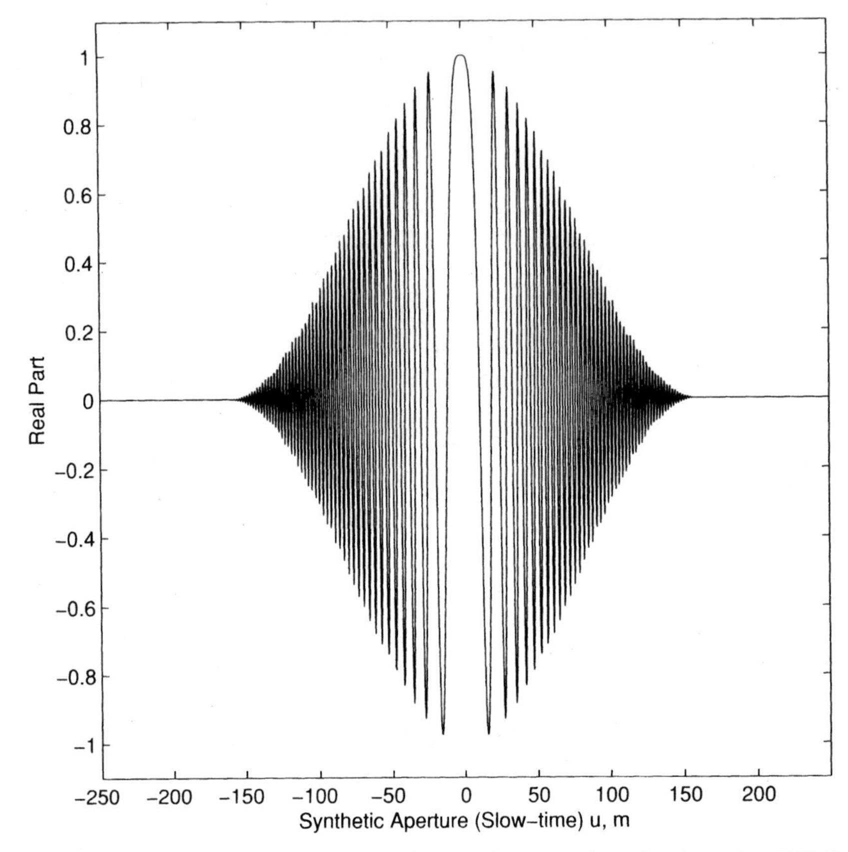

Figure 2.15a. Real part of stripmap SAR signal $s(\omega, u)$ for a target located at $(x_n, y_n) = (500, 0)$ m (broadside target).

Again, we consider the case of a common range for all the targets in the imaging scene; that is, $x_n = X_c$, $n = 1, 2, \ldots$. After slow-time matched filtering with $S_0(\omega, k_u)$, the reconstructed target function in the spatial frequency domain becomes

$$
\begin{aligned}
F(k_y) &= S(\omega, k_u) S_0^*(\omega, k_u) \\
&= \sum_n A_n(\omega, k_u) \exp(-j k_y y_n),
\end{aligned}
$$

where S_0^* is the complex conjugate of S_0, and the cross-range spatial frequency mapping is defined via $k_y = k_u$.

We showed that the slow-time domain AM signal $a_n(\omega, x_n, y_n - u)$ and the slow-time Doppler domain AM signal $A_n(\omega, k_u)$ are related via the following scale transformation:

$$
A_n [\omega, 2k \sin\theta_n(u)] = a_n(\omega, x_n, y_n - u),
$$

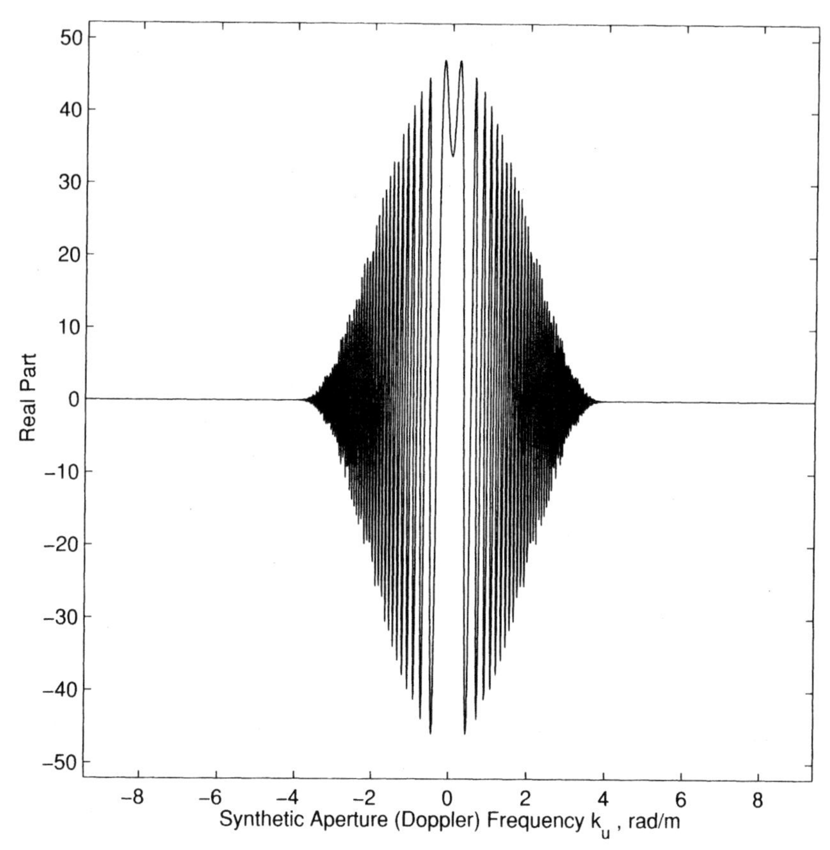

Figure 2.15b. Real part of stripmap SAR signal slow-time Fourier transform $S(\omega, k_u)$ for a target located at $(x_n, y_n) = (500, 0)$ m (broadside target).

where

$$\theta_n(u) = \arctan\left(\frac{y_n - u}{x_n}\right),$$

is the aspect angle of the nth target when the radar is located at $(0, u)$.

Thus the slow-time domain functional information in the AM signal $a_n(\omega, x_n, y_n - u)$ is copied via a one-to-one transformation into the slow-time Doppler k_u domain. Moreover, in the matched-filtered reconstruction process, the functional information in the slow-time Doppler k_u domain is directly mapped into the target spatial frequency k_y domain, since $k_y = k_u$. Hence the slow-time domain AM signal $a_n(\omega, x_n, y_n - u)$ acts as a spatial frequency k_y domain (magnitude and/or phase) filter for the nth target.

One can develop the domain mapping that translates the AM signal in the slow-time u domain into the target spatial frequency k_y domain. For this, we combine the

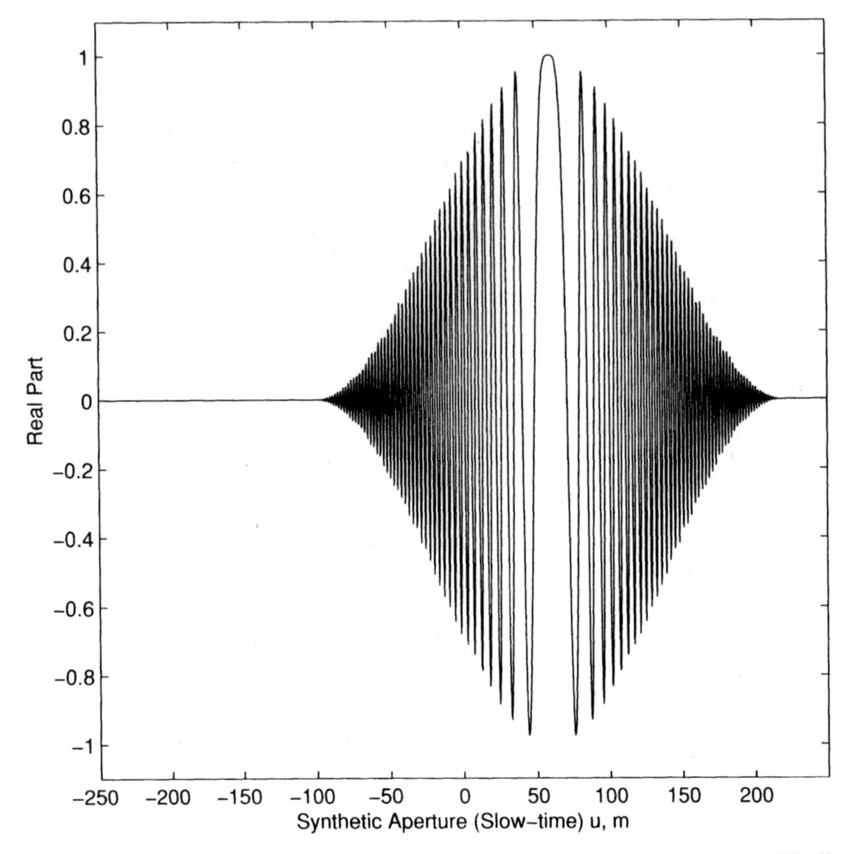

Figure 2.15c. Real part of stripmap SAR signal $s(\omega, u)$ for a target located at $(x_n, y_n) = (500, 60)$ m (squint target).

slow-time to the slow-time Doppler mapping

$$k_u = 2k \sin \theta_n(u),$$

with the reconstruction mapping

$$k_u = k_y.$$

The resultant mapping from the spatial frequency k_y domain to the slow-time u domain is

$$u = y_n - \frac{k_y}{k_x} x_n,$$

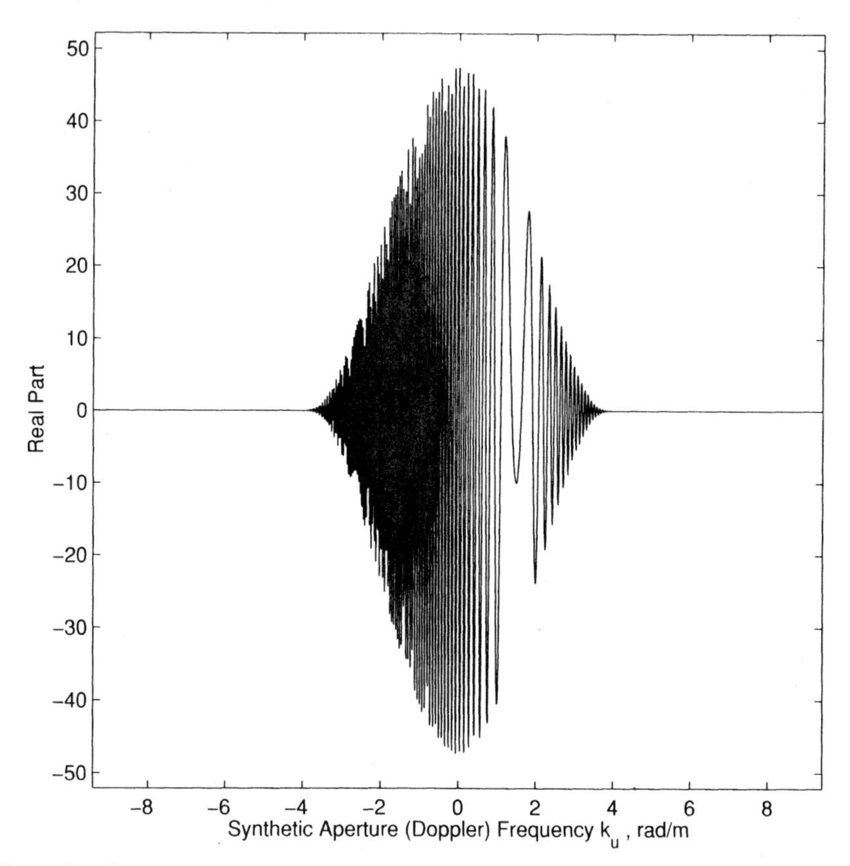

Figure 2.15d. Real part of stripmap SAR signal slow-time Fourier transform $S(\omega, k_u)$ for a target located at $(x_n, y_n) = (500, 60)$ m (squint target).

where $k_x = \sqrt{4k^2 - k_y^2}$. The resultant spatial frequency filter for the nth target becomes

$$\begin{aligned} \text{PSF}_n(k_y) &= A_n(\omega, k_y) \\ &= a_n(\omega, x_n, y_n - u), \end{aligned}$$

where u is related to k_y via the above transformation. Note that the above mapping between the u and k_y domains is only valid for the *amplitude* functions. In fact the matched-filtered reconstruction is based on the Fourier transformation of the spherical PM signal with respect to the slow-time u domain.

To determine the reconstructed target function in the cross-range domain, we obtain the inverse spatial Fourier transform of the matched-filtered signal $F(k_y)$ with respect to k_y; this yields

$$f(y) = \sum_n f_n \text{psf}_n(y - y_n),$$

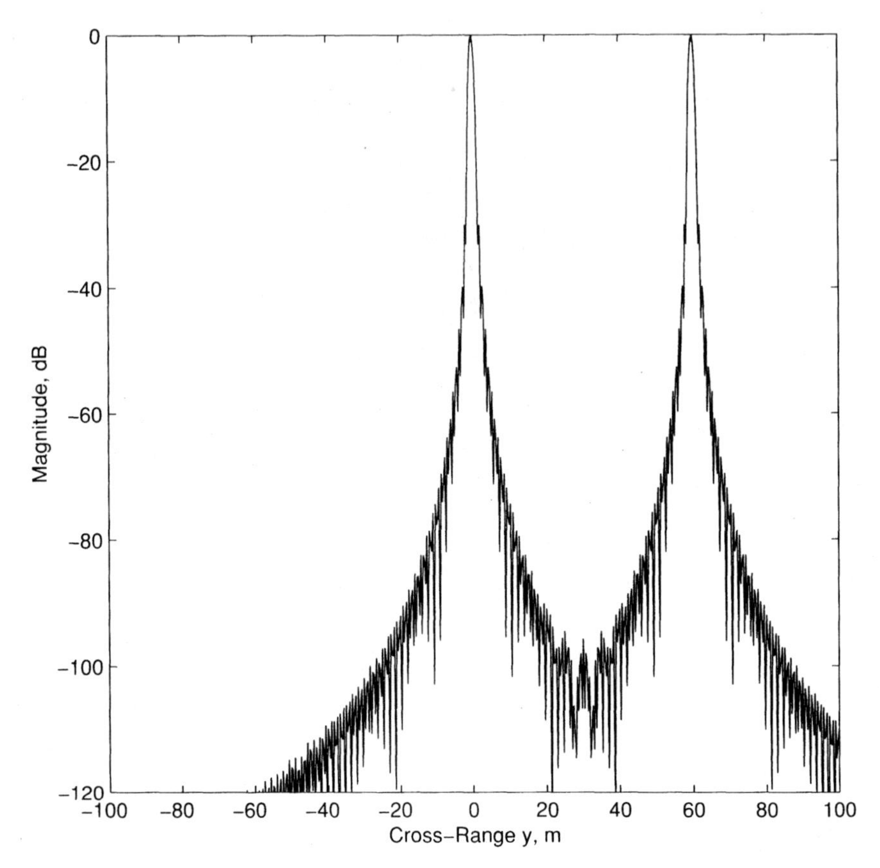

Figure 2.15e. Reconstructed broadside and squint targets in cross-range domain from their stripmap SAR signal.

where the *point spread function* for the nth target is defined via the following:

$$\begin{aligned} \text{psf}_n(y) &= \mathcal{F}^{-1}_{(k_u)}\left[\text{PSF}_n(k_y)\right] \\ &= \mathcal{F}^{-1}_{(k_u)}\left[A_n(\omega, k_y)\right]. \end{aligned}$$

Depending on the nature of the amplitude pattern (resonance in a cavity, surface waves, etc.), and its and its phase and amplitude fluctuations, we observe a *smearing* of the target in the cross-range domain; this phenomenon is similar to what we observed for a fast-time frequency-dependent target reflectivity in range imaging which resulted in range extension. If the amplitude pattern is unknown, the user cannot do much to counter its undesirable cross-range extension effect.

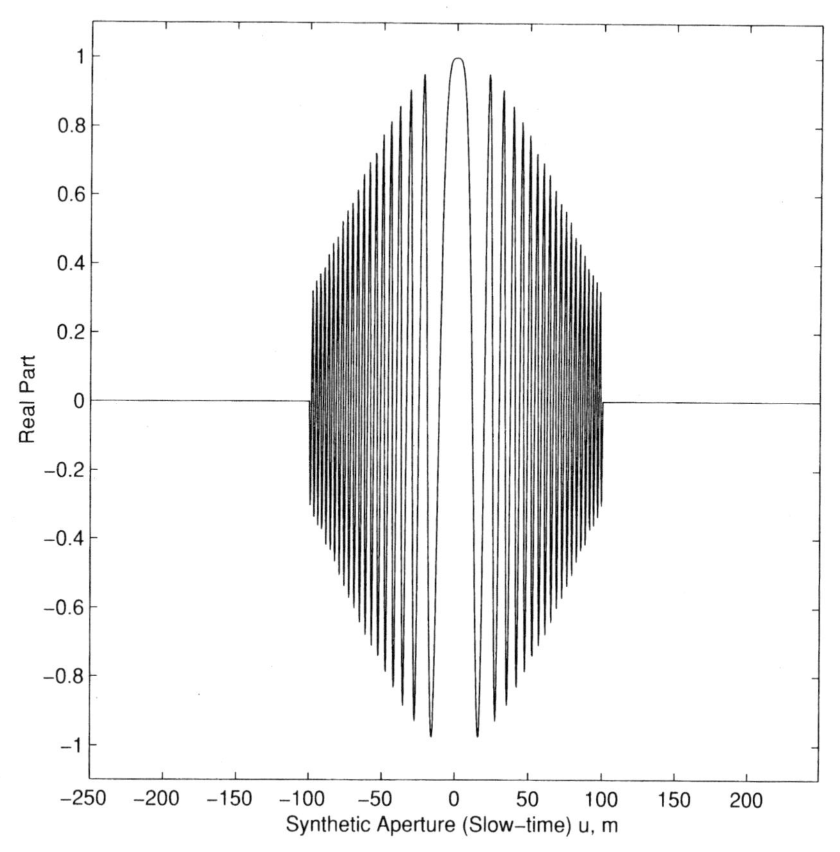

Figure 2.16a. Real part of stripmap SAR signal $s(\omega, u)$ for a target located at $(x_n, y_n) = (500, 0)$ m (broadside target); a spotlight window of $[-L, L] = [-100, 100]$ m is applied to stripmap SAR signal.

Representation in Slow-time Angular Doppler Domain

In Section 2.4 we defined the *angular* slow-time Doppler domain via

$$\phi = \arcsin\left(\frac{k_u}{2k}\right),$$

or

$$k_u = 2k \sin\phi.$$

Recall that the slow-time domain AM signal $a_n(\omega, x_n, y_n - u)$ and the slow-time Doppler domain AM signal $A_n(\omega, k_u)$ are related via the following scale transfor-

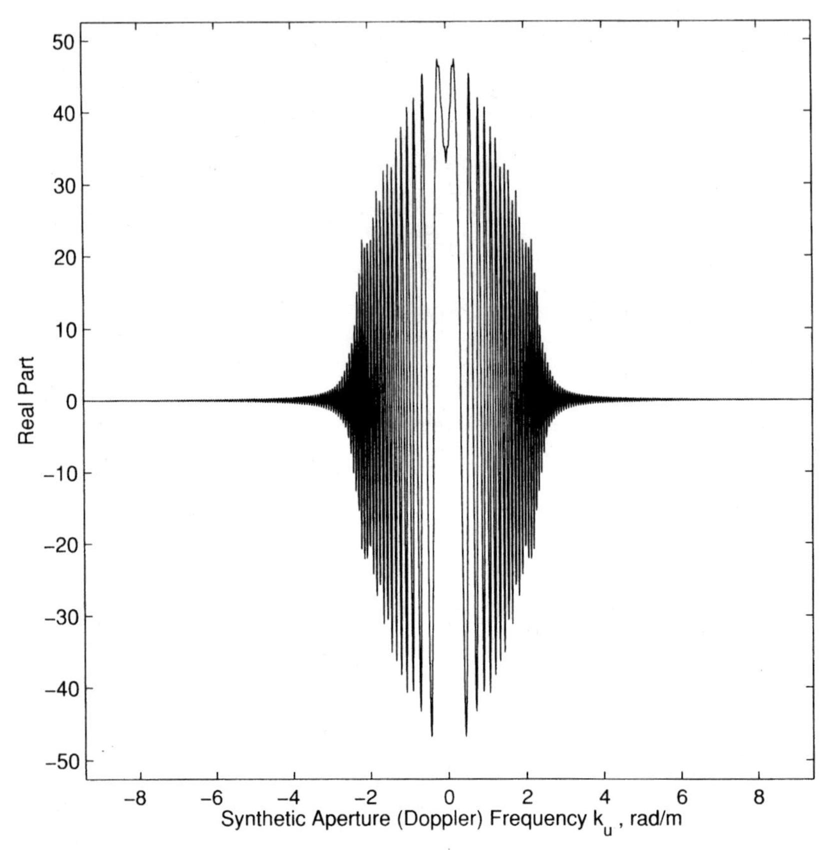

Figure 2.16b. Real part of stripmap SAR signal slow-time Fourier transform $S(\omega, k_u)$ for a target located at $(x_n, y_n) = (500, 0)$ m (broadside target); a spotlight window of $[-L, L] = [-100, 100]$ m is applied to stripmap SAR signal.

mation:

$$A_n(\omega, k_u) = \frac{1}{r_n^2} a_n(\omega, x_n, y_n - u),$$

where the mapping is defined via

$$k_u = 2k \sin \theta_n(u).$$

Thus the scale mapping of the AM signals in the angular slow-time Doppler domain is

$$\phi = \theta_n(u).$$

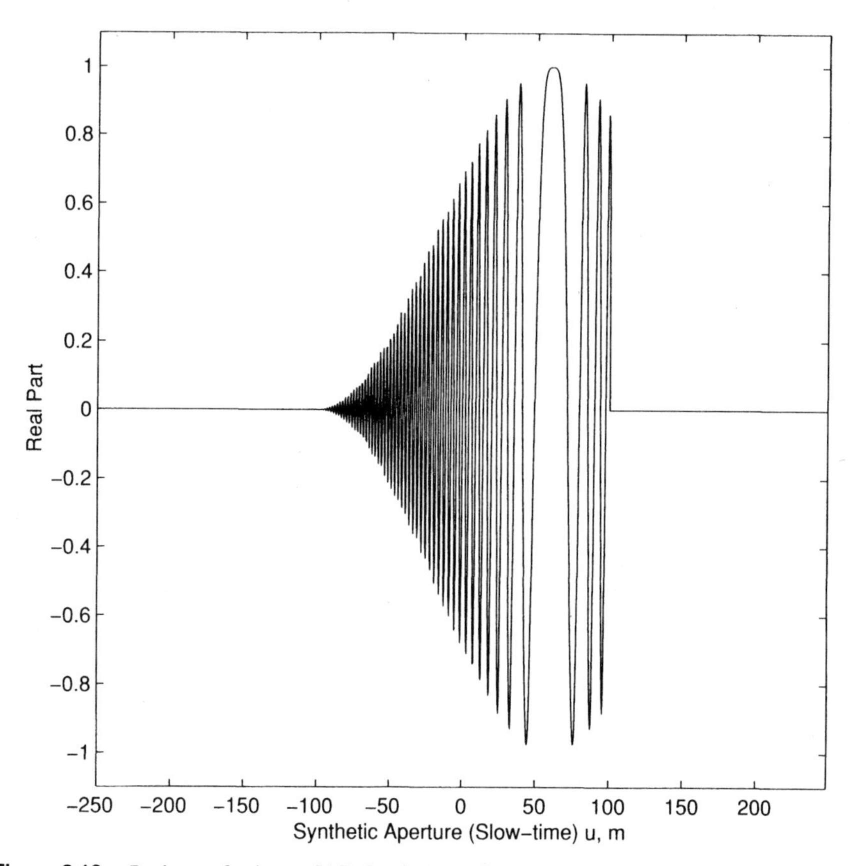

Figure 2.16c. Real part of stripmap SAR signal $s(\omega, u)$ for a target located at $(x_n, y_n) = (500, 60)$ m (squint target); a spotlight window of $[-L, L] = [-100, 100]$ m is applied to stripmap SAR signal.

The above indicates that the AM signature of a target in the aspect angle $\theta_n(u)$ domain, that is, $a_n(\cdot)$, directly maps into the slow-time angular ϕ domain to form the AM component of $S_n(\omega, k_u)$.

This phenomenon is demonstrated in Figure 2.17*a* and *b*. Figure 2.17*a* shows the aspect angles of the nth target at $u = -L$, $u = 0$, $u = L$, and the closest point of approach (CPA) $u = y_n$ which are, respectively, $\theta_n(-L)$, $\theta_n(0)$, $\theta_n(L)$, and $\theta_n(y_n) = 0$. Figure 2.17*b* shows the slow-time Doppler mapping of the AM component in the slow-time angular Doppler ϕ domain. For this purpose the slow-time angular Doppler ϕ domain is shown as the *polar* angle domain of

$$k_x = \sqrt{4k^2 - k_u^2} = 2k\cos\phi,$$

$$k_y = k_u = 2k\sin\phi,$$

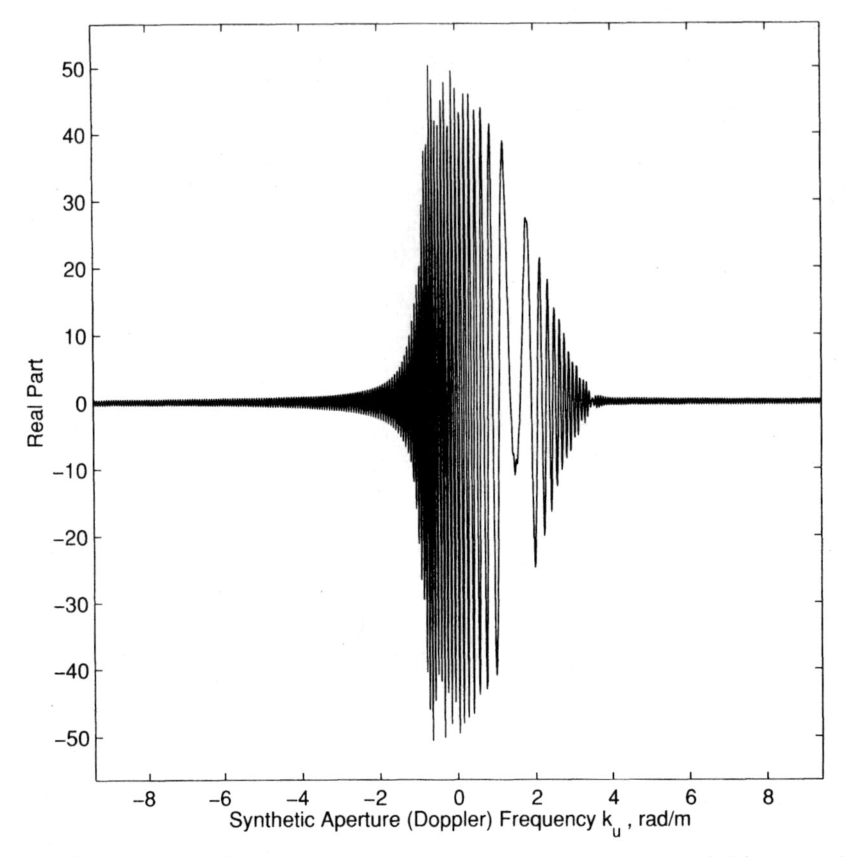

Figure 2.16d. Real part of stripmap SAR signal slow-time Fourier transform $S(\omega, k_u)$ for a target located at $(x_n, y_n) = (500, 60)$ m (squint target); a spotlight window of $[-L, L] = [-100, 100]$ m is applied to stripmap SAR signal.

which we earlier called the range spatial frequency and cross-range spatial frequency domains, respectively.

Before closing this section, we use the slow-time angular Doppler variable ϕ to redefine the slow-time Fourier transform of the AM-PM signal, that is,

$$S_n(\omega, k_u) = A_n(\omega, k_u) \exp\left(-j\sqrt{4k^2 - k_u^2}\, x_n - jk_u y_n\right),$$

via

$$S_n(\omega, 2k\sin\phi) = A_n(\omega, 2k\sin\phi) \exp\left(-j2k\cos\phi x_n - j2k\sin\phi y_n\right).$$

Also we define the AM signal of the nth target in the slow-time angular Doppler domain by

$$A_{pn}(\omega, \phi) = A_n(\omega, 2k\sin\phi).$$

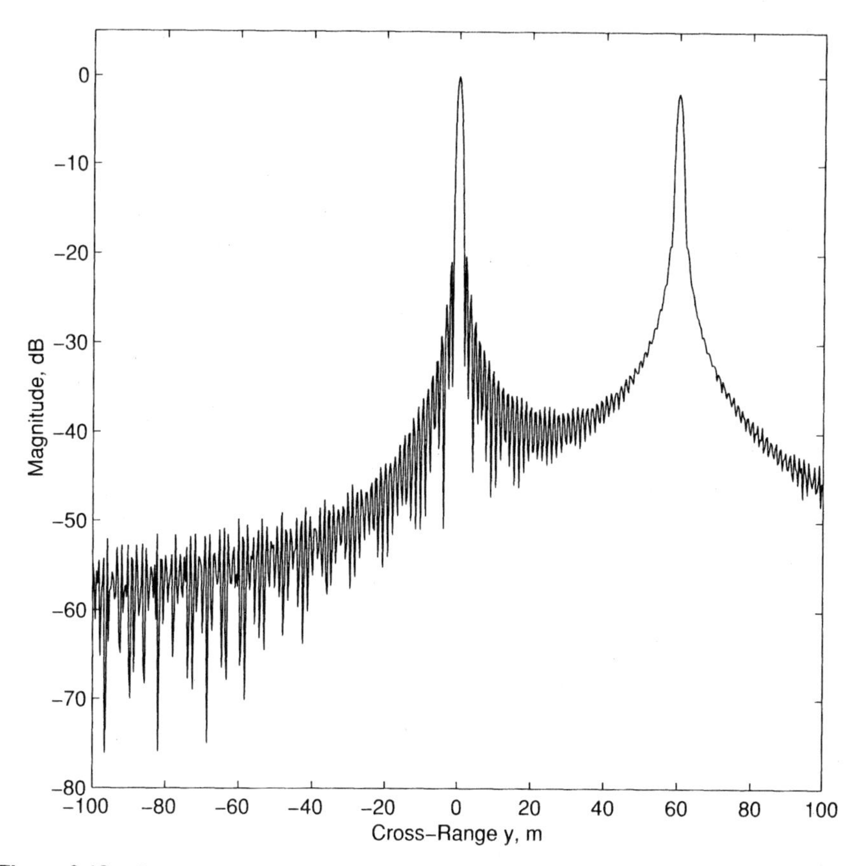

Figure 2.16e. Reconstructed broadside and squint targets in cross-range domain from their stripmap SAR signal; a spotlight window of $[-L, L] = [-100, 100]$ m is applied to stripmap SAR signal.

Thus we may rewrite the AM-PM signal in the angular slow-time Doppler domain by

$$S_n(\omega, 2k \sin \phi) = A_{pn}(\omega, \phi) \exp\left(-jk_x x_n - jk_y y_n\right).$$

The equivalence of the aspect angle $\theta_n(u)$ domain and the slow-time angular Doppler ϕ domain in the mapping of the AM signature of a target would play a key role in characterizing the target SAR image; this is crucial for target *recognition* or *detection* in SAR images.

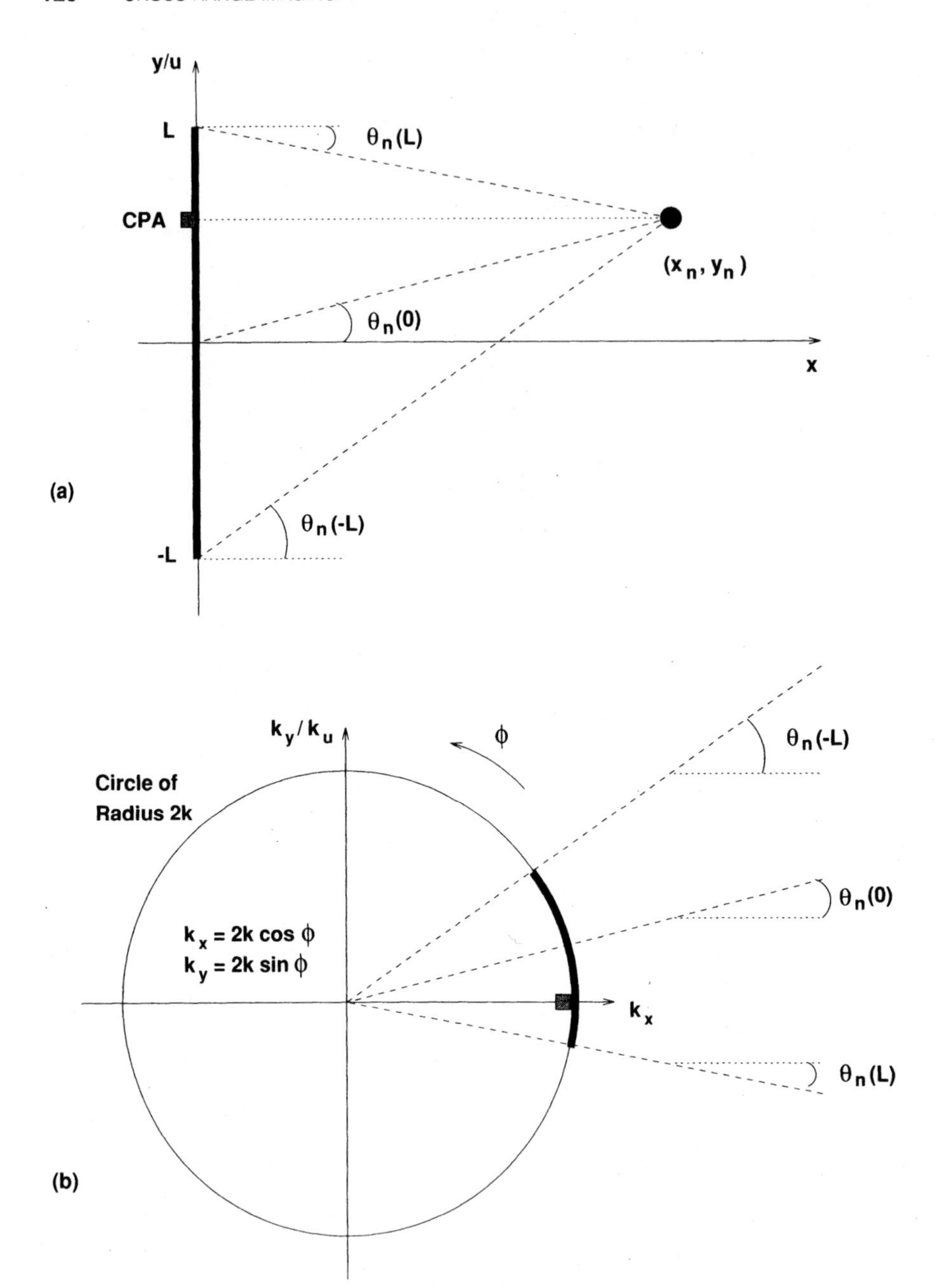

Figure 2.17. (a) Aspect angles of n-th target at $u = -L$, $u = 0$, $u = L$, and closest point of approach (CPA) $u = y_n$; (b) slow-time Doppler mapping of AM component in slow-time angular Doppler ϕ domain.

2.10 RECONSTRUCTION VIA TARGET SIGNATURE SLOW-TIME MATCHED FILTERING

The radar-induced amplitude patterns which we studied in the earlier two examples of spotlight and stripmap SAR systems were *real* functions. This is not generally the case in the high-resolution stripmap SAR systems that possess a relatively large beamwidth. The presence of a *complex* amplitude pattern, whether it is radar-induced and/or target-induced, results in degradations in the reconstructed cross-range image.

- In the stripmap SAR systems, the phase of the amplitude function $A_n(k_u, \omega)$ is negligible near zero slow-time Doppler k_u values while it increases with $|k_u|$. Based on experimentation, some of the users of the wide-beamwidth stripmap SAR systems have noticed that the quality of the reconstructed SAR image improves as the higher slow-time Doppler information is filtered out or windowed; that is, the slow-time Doppler SAR data which contain complex amplitude patterns are discarded or weakened. However, reducing the processed slow-time Doppler band results in a cross-range resolution that is inferior to the resolution anticipated by the full band slow-time Doppler processing. (Recall the effect of Hamming windowing on the spotlight SAR data in the previous section.)

We now examine scenarios in which some knowledge of the functional properties of the amplitude pattern is available, and the manner in which this information can be used to improve cross-range resolution; we refer to this as *target signature slow-time matched filtering*. The approach is similar to what we referred to as *target signature matched filtering* in Section 1.7 (frequency-dependent target reflectivity). We will examine target signature slow-time matched filtering in what we refer to as generalized spotlight and stripmap SAR systems. Based on the relative values of (x_n, y_n), L, and B, we encounter one of the following three scenarios in the SAR imaging problem (see Figure 2.18).

Type 1: Generalization of Spotlight SAR

System Model This case deals with signals that will be seen in our discussion on *spotlight* SAR. For the first scenario shown in Figure 2.18*a*, the AM component of $s_n(\omega, u)$ does not vary significantly for $u \in [-L, L]$. *We refer to these AM-PM signals as Type 1 SAR signals.* (Examples of this class of SAR signals were studied in the previous section, Figures 2.12, 2.13, 2.14, and 2.16.) Much of the analysis that follows for AM-PM spotlight SAR signals is similar to the analyses for the spherical PM signal within a finite aperture in Sections 2.4 to 2.8.

The Fourier transform of the generalized AM-PM spotlight SAR signal with respect to the slow-time u domain is

$$S(\omega, k_u) = \sum_n I_n(\omega, k_u) A_n(\omega, k_u) \exp\left(-j\sqrt{4k^2 - k_u^2}\, x_n - jk_u y_n\right),$$

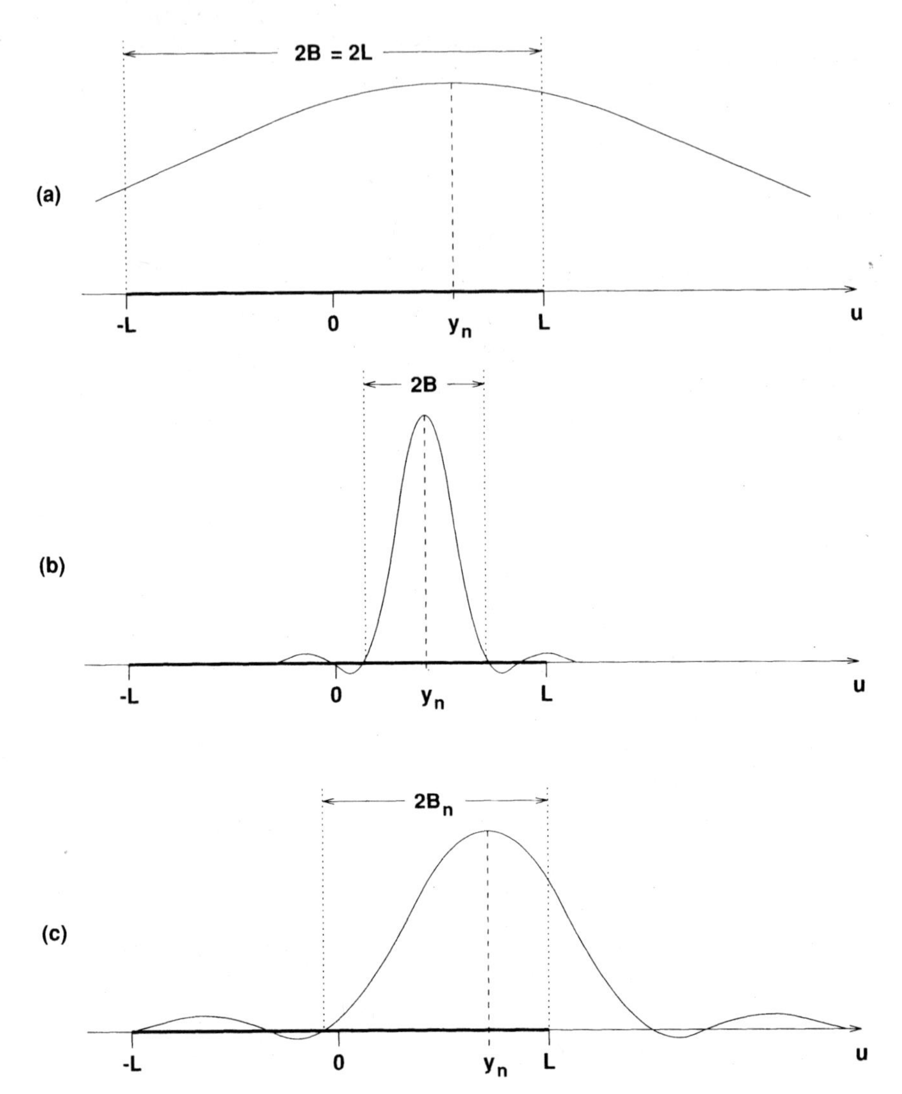

Figure 2.18. Depiction of support of SAR signal in synthetic aperture domain: (a) generalization spotlight SAR; (b) generalization of stripmap SAR; (c) partially observable SAR.

where I_n is a rectangular window which is defined via

$$I_n(\omega, k_u) = \begin{cases} 1 & \text{for } k_u \in \Omega_n = [2k \sin\theta_n(L), 2k \sin\theta_n(-L)], \\ 0 & \text{otherwise.} \end{cases}$$

- The window $I_n(\omega, k_u)$ is the support band of the signature of the nth target in the slow-time Doppler domain. This band depends on the aspect angles of the target with respect to the radar at the two extreme ends of the synthetic aperture, that is, $\theta_n(L)$ and $\theta_n(-L)$.

Slow-time Matched-Filtered Reconstruction Again, consider the case where all of the targets are located at a known range $X_c = x_n$, and the amplitude pattern has the following functional form:

$$A_n(\omega, k_u) = \sigma_n A(\omega, k_u),$$

where $A(\omega, k_u)$ is a *known* amplitude pattern. (The target dependence appears only in the power of the amplitude pattern.) Then the cross-range reconstruction for this problem is also performed via slow-time matched filtering with $A^*(\cdot)$; that is,

$$\begin{aligned} F(k_y) &= S(\omega, k_u) A^*(\omega, k_u) \exp\left(j\sqrt{4k^2 - k_u^2}\, X_c\right) \\ &= |A(\omega, k_u)|^2 \sum_n \sigma_n I_n(\omega, k_u) \exp(-jk_u y_n) \end{aligned}$$

with $k_y = k_u$.

The matched-filtered reconstructed target function in the cross-range domain is

$$f(y) = \sum_n f_n \text{psf}_{I_n}(y - y_n),$$

where the *point spread function* for the nth target is defined via the following:

$$\text{psf}_{I_n}(u) = \mathcal{F}^{-1}_{(k_u)}\left[|A(\omega, k_u)|^2 I_n(\omega, k_u)\right].$$

- The cross-range point spread function of the imaging system varies with the target location in Type 1 (spotlight SAR). The same is not true for Type 2 (stripmap SAR) which will be examined.

Cross-range Resolution For Type 1 the cross-range resolution is approximately the same as the cross-range resolution for the spherical PM signal which does not contain an AM component; that is,

$$\Delta_{y_n} = \frac{x_n \lambda}{4L\cos^2\theta_n(0)}.$$

Synthetic Aperture Sampling The sampling constraint in the slow-time domain is also the same as the spherical PM case; that is,

$$\Delta_u \le \frac{R_c \lambda}{4(Y_0 + L)\cos\theta_c}.$$

For Type 1, the use of slow-time compression does result in bandwidth reduction similar to the spherical PM case. In this case the Nyquist sample spacing for the slow-time compressed signal is

$$\Delta_{uc} \le \frac{R_c \lambda}{4Y_0 \cos\theta_c}.$$

Other signal processing issues which are associated with the generalized AM-PM spotlight SAR signals (reducing PRF via slow-time compression, cross-range gating via slow-time compression, zero-padding in the synthetic aperture domain, slow-time Doppler domain subsampling, reducing bandwidth of reconstructed image, etc.) are the same as the ones for the spherical PM signal in Sections 2.7 and 2.8.

Type 2: Generalization of Stripmap SAR

System Model For this case the entire signature of the target, which resides in $[y_n - B, y_n + B]$, falls in the measurement interval $[-L, L]$; see Figure 2.18*b*. (An example of this class of stripmap SAR signals was studied in the previous section, Figure 2.15.) For instance, if we consider a broadside target area, that is, $y_n \in [-Y_0, Y_0]$, then

$$Y_0 + B \leq L.$$

This case corresponds to the signal encountered in *stripmap* SAR, which will be discussed in the later chapters. *We refer to these AM-PM signals as Type 2 SAR signals.*

We have the following for the measured signature of the nth target:

$$s_n(\omega, u) = a_n(\omega, x_n, y_n - u) \exp\left[-j2k\sqrt{x_n^2 + (y_n - u)^2} \right],$$

which is the AM-PM signal that was analyzed in the previous section. Hence in the slow-time Doppler domain we have

$$S_n(\omega, k_u) = A_n(\omega, k_u) \exp\left(-j\sqrt{4k^2 - k_u^2}\, x_n - jk_u y_n \right).$$

- The entire support band of $A_n(\omega, k_u)$ is *observable* in $S_n(\omega, k_u)$. The extent of this band plays the key role in the resolvability of the target, that is, its cross-range resolution.

Consider the case of multiple targets in an imaging scene, all of which satisfy the requirement of Type 2. The SAR signal model becomes

$$s(\omega, u) = \sum_n a_n(\omega, x_n, y_n - u) \exp\left[-j2k\sqrt{x_n^2 + (y_n - u)^2} \right].$$

In the slow-time Doppler domain, the total signal becomes

$$S(\omega, k_u) = \sum_n A_n(\omega, k_u) \exp\left(-j\sqrt{4k^2 - k_u^2}\, x_n - jk_u y_n \right).$$

Slow-time Matched-Filtered Reconstruction Suppose that the targets are all located at a known range, for example, $x_n = X_c$. The amplitude pattern has the following functional form

$$a_n(\omega, x_n, y_n - u) = \sigma_n a(\omega, x_n, y_n - u),$$

or

$$A_n(\omega, k_u) = \sigma_n A(\omega, k_u),$$

where $a(\omega, x, y)$ or $A(\omega, k_u)$ is a *known* amplitude pattern. (The target dependence appears only in the power of the amplitude pattern.) Then the target function in the spatial frequency domain is reconstructed via the following inverse equation:

$$\begin{aligned} F(k_y) &= \frac{S(\omega, k_u)}{A(\omega, k_u)} \exp\left(j\sqrt{4k^2 - k_u^2}\, X_c\right) \\ &= \sum_n \sigma_n \exp(-jk_u y_n), \end{aligned}$$

where the spatial frequency cross-range mapping is defined via $k_y = k_u$. The above reconstruction involves a deconvolution operation $S(\omega, k_u)/A(\omega, k_u)$ that resembles the source deconvolution operation $S(\omega/P(\omega)$ encountered in one-dimensional range imaging.

The practical method to implement this deconvolution, which is not sensitive to additive noise, is via the following amplitude pattern (as well as phase) *matched-filtering* operation in the slow-time domain

$$\begin{aligned} F(k_y) &= S(\omega, k_u) A^*(\omega, k_u) \exp\left(j\sqrt{4k^2 - k_u^2}\, X_c\right) \\ &= |A(\omega, k_u)|^2 \sum_n \sigma_n \exp(-jk_u y_n), \end{aligned}$$

where A^* is the complex conjugate of A.

The matched-filtered reconstructed target function in the cross-range domain is

$$f(y) = \sum_n \sigma_n \text{psf}(y - y_n),$$

where the *point spread function* for this imaging system is defined by

$$\text{psf}(u) = \mathcal{F}^{-1}_{(k_u)}\left[|A(\omega, k_u)|^2\right].$$

Note that this *point spread function is invariant of the coordinates of the target.*

- If the amplitude function $A(\omega, k_u)$ is a *real* signal or *its phase is very small*, then there is no need to perform the above matched filtering. In this case the point spread function of the imaging system is equal to the inverse Fourier transform of $A(\omega, k_u)$.

The reconstruction for Type 2 may also be derived from the reconstruction for Type 1. For the scenario encountered in Type 1, the support band of $A(\omega, k_u)$ is simply a subset of the support band of $I_n(\omega, k_u)$. Thus we have the following point spread function for Type 2 which is invariant of the target parameters:

$$\begin{aligned} \text{psf}_{I_n}(u) &= \mathcal{F}^{-1}_{(k_u)}\left[|A(\omega, k_u)|^2\right] \\ &= \text{psf}(u). \end{aligned}$$

Clearly the reconstruction for Type 1 (spotlight SAR) is the generalized version of the one shown for Type 2 (stripmap SAR). However, the same is not true for the cross-range resolution and slow-time sampling constraints of the two systems. These are discussed next.

Cross-range Resolution The point spread function dictates the resolution of the imaging system. For instance, when $A(\omega, k_u) = \text{sinc}^2(Dk_u/4\pi)$, the cross-range resolution is approximately equal to (use $B = x_n\lambda/2D$ to determine the cross-range resolution)

$$\begin{aligned} \Delta_{y_n} &= \frac{x_n\lambda}{4B} \\ &= \frac{D}{2}. \end{aligned}$$

For a general amplitude pattern $a(\omega, x, y)$ of a stripmap SAR system, one may also borrow the definition of *noise equivalent bandwidth* in communication systems [car] to identify the *effective* beamwidth; that is,

$$2B = \int_{-\infty}^{\infty} |a(\omega, x_n, y)|^2 \, dy.$$

[The average magnitude of $a(\omega, x_n, y)$ is assumed to be one in the above.]

- When $s_n(\omega, u)$ is an AM-PM signal with a support band $u \in [y_n - B, y_n + B]$, where $B = x_n\lambda/2D$, which is a subset of the synthetic aperture measurement interval $[-L, L]$, the cross-range resolution is invariant of the length of the window in the u domain (measurement interval) $2L$, the coordinates of the target, or the wavelength $\lambda = 2\pi/k$.
- The cross-range resolution depends on the cross-range point spread function of the imaging system which is dictated by the inverse Fourier transform of $|A(\omega, k_u)|^2$. For a planar radar, *the cross-range resolution depends only on the diameter of the radar*. As the radar diameter D is decreased, the beamwidth of the radar (the slow-time interval over which the target is illuminated), that is, $2B$, increases. Moreover the use of a smaller radar improves the cross-range resolution which is $D/2$. However, a smaller radar outputs less power, which affects the target signal-to-noise power ratio.

Synthetic Aperture Sampling As we mentioned earlier, the slow-time Doppler k_u support band of $S_n(\omega, k_u)$ or $S(\omega, k_u)$ is solely dictated by the support band of $A(\omega, k_u)$. Thus the Nyquist sampling rate for $s(\omega, u)$ in the slow-time domain is dictated by the slow-time Doppler support of $A(\omega, k_u)$ and is invariant of the location of the targets and the size of the synthetic aperture.

We mentioned that the beamwidth, that is, the support of the amplitude pattern $a(\omega, x, y)$, in the slow-time domain is $[-B, B]$. Moreover the two amplitude functions are related via the following:

$$a(\omega, x, y) = A\left(2k\frac{y}{\sqrt{x^2+y^2}}, \omega\right).$$

Thus the support band of $A(\omega, k_u)$ in the slow-time Doppler k_u domain can be expressed in terms of the beamwidth to be the following:

$$\begin{aligned}\Omega_A &= \left[-2k\sin\left[\arctan\left(\frac{B}{x_n}\right)\right], 2k\sin\left[\arctan\left(\frac{B}{x_n}\right)\right]\right]\\ &\approx \left[-2k\frac{B}{x_n}, 2k\frac{B}{x_n}\right].\end{aligned}$$

Thus the sample spacing in the slow-time u domain should satisfy

$$\Delta_u \le \frac{x_n\lambda}{4B}.$$

For the case of a planar radar antenna where the amplitude function is $A(\omega, k_u) = \text{sinc}^2(Dk_u/4\pi)$, the half-beamwidth is approximately $B = x_n\lambda/D$ (note that the larger value for B is used here to determine the Nyquist sampling rate in the slow-time domain). In this case the support band of the synthetic aperture signal $S(\omega, k_u)$ in the slow-time Doppler domain is approximately equal to $[-4\pi/D, 4\pi/D]$. Thus the Nyquist sampling rate in the slow-time domain is

$$\Delta_u \le \frac{D}{4},$$

which is equal to one-half of the diameter of the radar and is *independent of the coordinates of the targets*. To reduce aliasing errors, it is advisable to use an even higher value for the half-beamwidth B to determine Δ_u. Note that for the above selection of the half-beamwidth B, the Nyquist sample spacing in the slow-time domain $D/4$ is one-half of the cross-range resolution $\Delta_y = D/2$.

- Reminder: For the spherical PM signal with no AM component, the Nyquist sampling rate in the slow-time domain, that is,

$$\Delta_u \le \frac{x_n\lambda}{4(Y_0+L)},$$

is proportional to the size of the synthetic aperture $2L$, the extent of the target area in the cross-range domain $2Y_0$, and the radar frequency, while it is inversely proportional to the target area range.

Slow-time Compression Consider the slow-time compressed version of the AM-PM SAR signal, that is,

$$s_c(\omega, u) = s(\omega, u) \exp\left(j2k\sqrt{X_c^2 + u^2}\right)$$

$$\approx \sum_n \sigma_n a(\omega, x_n, y_n - u) \exp\left[-j2k(x_n - X_c) + j\frac{2ky_n}{X_c}u\right].$$

The contribution (signature) of the nth target in this slow-time compressed signal is approximately a sinusoid windowed with $a(\cdot)$ within the interval $[y_n - B, y_n + B]$ in the slow-time u domain. The distribution of this signal in the slow-time Doppler k_u domain is

$$S_c(\omega, k_u) \approx \sum_n \sigma_n \exp[-j2k(x_n - X_c) - jk_u y_n]\mathbf{a}\left[B\left(k_u - \frac{2ky_n}{X_c}\right)\right],$$

where $\mathbf{a}(k_u)$ is the Fourier transform of $a(\omega, x, u)$ with respect to u.

Provided that the support band of the Fourier transformed signal $\mathbf{a}(k_u)$ is much smaller than $2kY_0/X_c$, then the support band of the slow-time compressed signal in the slow-time Doppler k_u domain is

$$\Omega_c = \left[-2k\frac{Y_0}{X_c}, 2k\frac{Y_0}{X_c}\right],$$

and the Nyquist sampling rate in the slow-time domain for the slow-time compressed SAR signal becomes

$$\Delta_{uc} \le \frac{X_c \lambda}{4Y_0}.$$

This result is identical to the Nyquist sampling rate which we derived earlier for the slow-time compressed spherical PM signal that contained no AM component.

As we mentioned earlier, the Nyquist sample spacing for the AM-PM signal of Type 2 depends only on the support band of $A(\omega, k_u)$, and it is invariant in the parameters of the targets. For instance, in the case of a planar radar, we have $\Delta_u \le D/4$. Thus if $\Delta_{uc} < \Delta_u$ or, equivalently,

$$\frac{X_c \lambda}{4Y_0} < \frac{D}{4},$$

then the slow-time compression results in the *expansion* of the bandwidth of the AM-PM signal.

However, if the slow-time compression results in the slow-time Doppler bandwidth expansion, we could artificially (digitally) increase the bandwidth of $s(\omega, u)$ via zero-padding its DFT $S(\omega, k_u)$ in the k_u domain to achieve the sample spacing Δ_{uc}; then, we could perform the slow-time compression. It turns out that we will use the slow-time compression as a signal processing tool in certain SAR systems. We will show that the slow-time compressed signal has certain utilities for *digital spotlighting* in stripmap SAR systems.

Type 3: Partial Observability

In the third case, a portion of the signature of the nth target with extent $2B_n$ falls in the measurement domain; see Figure 2.18c. *We refer to these AM-PM signals as Type 3 SAR signals.* For this scenario one may also develop the point spread function and the resultant resolution. Suppose that $\xi_n \leq 1$ is the percentage of the main lobe of the signature of the nth target which falls in the measurement interval $[-L, L]$. Then, in the case of a planar radar, the cross-range resolution for this target is

$$\Delta y_n \approx \frac{D}{2\xi_n}.$$

These type of targets appear near the cross-range edges of a measured SAR database and are not important.

2.11 MATLAB ALGORITHMS

The code for cross-range imaging is identified as *cross-range imaging*. The program simulates the fast-time baseband converted data $s(\omega, u)$. [The relationship between $\mathbf{s}(t, u)$ and its fast-time baseband converted signal $s(\omega, u)$ is trivial.] The algorithm is written for the general squint case for which squint baseband processing in the slow-time Doppler domain is required; the broadside case is automatically taken care of (no change is required). The algorithm sets the slow-time (synthetic aperture) sampling for the compressed signal $s_c(\omega, u)$ which results in aliased $s(\omega, u)$ data. The code performs compression, slow-time upsampling (zero-padding in the slow-time Doppler k_u domain), and decompression to retrieve unaliased (clean) $s(\omega, u)$ data from the aliased data.

The code contains the parameters for two scenarios. In one case, a broadside target area with $Y_0 < L$ is considered (Case 1). In the other case, a squint target area with $Y_0 > L$ is studied (Case 2). This case requires zero-padding in the synthetic aperture domain; this is also taken care of by the algorithm, and no supervision is necessary. The code provides the following outputs for Case 1(a), and Case 2(b):

- P2.1 Real part of aliased measured synthetic aperture signal $s(\omega, u)$ (baseband converted in the slow-time for the general squint case)
- P2.2 Aliased synthetic aperture signal spectrum $|S(\omega, k_u)|$

P2.3 Real part of compressed synthetic aperture signal $s_c(\omega, u)$

P2.4 Compressed synthetic aperture signal spectrum $|S_c(\omega, k_u)|$

P2.5 Compressed synthetic aperture signal spectrum $|S_c(\omega, k_u)|$ after zero-padding

P2.6 Real part of compressed synthetic aperture signal $s_c(\omega, u)$ after upsampling

P2.7 Real part of synthetic aperture signal $s(\omega, u)$ after upsampling

P2.8 Synthetic aperture signal spectrum $|S(\omega, k_u)|$ after upsampling

P2.9 Cross-range reconstruction via slow-time matched filtering $|f(y)|$

P2.10 Compressed target function spectrum $|F_c(k_y)|$

3

SAR RADIATION PATTERN

Introduction

An important issue in an information transmission system, for example, a data communication system over air, is to model and incorporate the effect of the *input/output response* of the transmitter, the medium (air), and the receiver in the analysis of the signals sent via this system. The simplest approach to this problem is to model the input/output response of the transmitter, the medium, and the receiver via a *linear time-invariant* system; the overall *transfer function* of these three is called the *channel* transfer function. More complicated models, which are based on *linear time-varying* systems, are currently being used to accurately model the channel effect.

A SAR transmitter/receiver is also an information processing system in which the user must be aware of the *channel effect*; that is, the effect of the radar transmitter, the propagation medium, and the radar receiver in the echoed signals stored in a computer for imaging. Yet a SAR channel is a *multidimensional* system that is a function of not only time (or radar frequency) but also space.

What we refer to as *SAR radiation pattern* is a measure that identifies the SAR channel effect. SAR radiation pattern is a measure of the relative *phase* as well as *power* (or *magnitude*) of the echoed signal due to a given target recorded in the computer; SAR radiation pattern is a *complex* (phase and magnitude) multidimensional signal. SAR radiation pattern of a target is a function of the following:

1. Spatial coordinates of the target and radar
2. Frequency of the transmitted radar signal
3. Physical properties of the target and radar

In fact, the SAR channel effect or SAR radiation pattern can be represented via a *spatially varying* model. For the targets that exhibit a reflectivity that varies with the radar signal frequency, the model becomes even more complicated. The effect of the transmitting/receiving radar antenna in a SAR radiation pattern can be quantified in most circumstances and can be incorporated into the signal processing and imaging phases of a SAR system. This is true only for *stable* radars; that is, the radars that exhibit the same radiation pattern in different experiments (over time). System modeling of the transmitting/receiving radiation pattern of a stable radar is the main topic addressed in this chapter.

Our analysis, however, does not include two important topics: the effect of radar instability and radar signal propagation through *inhomogeneous* media. (Certain SAR systems have been investigated for imaging problems that involve wave propagation through several layers of the atmosphere; these layers possess a varying index of refraction and, as a result, wave propagation speed. Such a scenario is referred to as wave propagation through inhomogeneous media.)

It turns out that these two effects vary over time; that is, they possess some type of *stochastic* behavior. In this case it is not feasible to incorporate these effects in our *deterministic* solution for SAR imaging. (Some have suggested stochastic-based imaging methods for applications such as positron emission tomography systems of diagnostic medicine which contain *random* data reception. However, the suggested methods are based on a heuristic; there has not been a stochastic solution that shows robustness and significant improvement over a simple common sense filtering approach.) It turns out that these two random SAR channel effects can be viewed as a *multiplicative* noise which can be dealt with [edd; sko; s92a]. There are various methods for dealing with SAR multiplicative noise [carr; s92a]; these are not discussed in this book.

Outline

In the previous chapter we established signal theory principles that are instrumental in analyzing the SAR radiation pattern. The basis of our approach was the synthetic aperture Fourier or slow-time Doppler analysis of synthetic aperture signals. In particular, we examined the effect of an amplitude function in the Fourier properties of the spherical PM signal (AM-PM signal); this was done from an analytical point of view that did not identify the *physical* origin of the amplitude function. In this chapter we show that a SAR channel effect or radiation pattern can be modeled via an AM-PM signal. Moreover the signal properties of an AM-PM signal are used to study the spatial and Fourier domain properties of a radar radiation pattern in the transmit and receive modes and to study its interaction with targets.

Our discussion begins with the modeling of the wave propagation from the radar to the target; this is called the radar transmit mode radiation pattern and is discussed in Section 3.1. In this analysis we show the role of the radar *type* in its transmit mode radiation pattern; examples are provided for planar, parabolic, and circular radars.

Our entire analysis of SAR imaging systems in this book is based on a *two-dimensional* spatial domain. However, the true physical model involves a *three-dimensional* spatial domain. It turns out that the two-dimensional SAR model is a simple *mathematical* representation of the true three-dimensional SAR model; the simpler two-dimensional SAR model reduces the number of variables that we have to carry and lets us concentrate on more important signal properties of a SAR signal. We examine a SAR system in three-dimensional spatial domain, and the manner in which it can be mathematically represented via a two-dimensional SAR system in Section 3.2. After this section we use only the two-dimensional SAR model.

In Section 3.3 we model the wave propagation from the target to the radar; this is called the radar receive mode radiation pattern. We combine the transmit mode and receive mode radiation patterns of a SAR system in a *serial* fashion to form what we refer to as the transmit-receive mode radar radiation pattern. In addition to the round-trip delay for wave propagation from the radar to the target and back to the radar, this channel model contains two multiplicative *amplitude* functions or patterns. One is due to the radar effect in the transmit mode; the other characterizes the radar contribution in the receive mode.

Our analytical development continues with a discussion on including the effect of the physical properties a target in its echoed SAR signal (Section 3.4). The origin and nature of the target effect can be traced back to our discussion on frequency-dependent target reflectivity in Chapter 1, and synthetic aperture-dependent target reflectivity in Chapter 2. The target effect adds another *multiplicative* component (amplitude pattern) to the serial network which represents the SAR radiation pattern of a target. Our discussion in this chapter ends with a brief discussion of the radar polarization problem, and a generalization of the scalar wave theory of Sections 3.1 to 3.4 to the vector wave theory.

The Matlab code of this chapter that is discussed in Section 3.6 and related figures exhibit the transmit mode and transmit-receive mode radiation patterns of two types of radar at two different frequencies. Figures P3.1*a–d* to P3.8*a–d* show these for the following cases:

1. A planar radar antenna at 1 GHz
2. A planar radar antenna at 2 GHz
3. A parabolic radar antenna (without feed) at 1 GHz
4. A parabolic radar antenna (without feed) at 2 GHz

As we pointed out earlier, we will identify and discuss the properties of planar radar antennas and parabolic radar antennas (without feed) in Section 3.1.

Mathematical Notations and Symbols

The following is a list of mathematical symbols used in this chapter, and their definitions:

$a(\omega, x, y)$	Transmit-receive mode radar amplitude pattern
$a_n(\omega, x, y)$	Amplitude pattern for nth target
$a_n^{HH}(\omega, x, y)$	HH mode amplitude pattern for nth target
$a_n^{HV}(\omega, x, y)$	HV mode amplitude pattern for nth target
$a_n^{VH}(\omega, x, y)$	VH mode amplitude pattern for nth target

$a_n^{VV}(\omega, x, y)$	VV mode amplitude pattern for nth target
$a_R(\omega, x, y)$	Receive mode radar amplitude pattern
$a_R^H(\omega, x, y)$	H-receive mode radar amplitude pattern
$a_R^V(\omega, x, y)$	V-receive mode radar amplitude pattern
$a_T(\omega, x, y)$	Transmit mode radar amplitude pattern
$a_T^H(\omega, x, y)$	H-transmit mode radar amplitude pattern
$a_T^V(\omega, x, y)$	V-transmit mode radar amplitude pattern
$A(\omega, k_u)$	Slow-time Doppler transmit-receive mode radar amplitude pattern
$A_n(\omega, k_u)$	Slow-time Doppler amplitude pattern for nth target
$A_R(\omega, k_u)$	Slow-time Doppler receive mode radar amplitude pattern
$A_T(\omega, k_u)$	Slow-time Doppler transmit mode radar amplitude pattern
B	Half-beamwidth of amplitude pattern; beamwidth of amplitude pattern is $2B$
B_y	Half-beamwidth of amplitude pattern in cross-range domain (three-dimensional); beamwidth of amplitude pattern is $2B_y$
B_z	Half-beamwidth of amplitude pattern in altitude domain (three-dimensional); beamwidth of amplitude pattern is $2B_z$
c	Wave propagation speed $c = 3 \times 10^8$ m/s for radar waves $c = 1500$ m/s for acoustic waves in water $c = 340$ m/s for acoustic waves in air
D	Diameter of radar
D_x	Effective diameter of radar in range domain (three-dimensional)
D_y	Diameter of radar in cross-range domain (three-dimensional)
D_z	Diameter of radar in altitude domain (three-dimensional)
$h(\omega, x, y)$	Transmit-receive mode radar radiation pattern
$h_n(\omega, x, y)$	Transmit-receive mode radar-target radiation pattern for nth target
$h_R(\omega, x, y)$	Receive mode radar radiation pattern
$h_T(\omega, x, y)$	Transmit mode radar radiation pattern
$h_{T0}(\omega, x, y)$	Transmit mode radiation pattern of an ideal source; that is, a radar with a zero diameter $D = 0$
$i(\ell)$	Complex amplitude function across radar antenna surface
k	Wavenumber: $k = \omega/c$
k_u	Spatial frequency or wavenumber domain for synthetic aperture u; slow-time Doppler (frequency) domain
k_x	Spatial frequency or wavenumber domain for range x
k_y	Spatial frequency or wavenumber domain for cross-range y
$p(t)$	Transmitted radar signal; radar signal is a single tone with frequency ω in this chapter
r	Radial distance in spatial domain

$\mathbf{S}$	Radar surface contour
$\mathrm{sinc}(\cdot)$	sinc function: $\mathrm{sinc}(a) = \sin(\pi a)/\pi a$
u	Synthetic aperture or slow-time domain
x	Range domain
$x_e(\ell)$	Range coordinate of differential element on radar surface
x_f	Focal range of planar, parabolic, and circular radar antenna at near-field; radius of circular radar antenna
x_{fy}	Focal range of radar antenna due to D_y (three-dimensional)
x_{fz}	Focal range of radar antenna due to D_z (three-dimensional)
x_n	Range of nth target
x_s	Slant-range domain (three-dimensional)
x_t	Transition range of planar radar antenna
X_c	Center point of target area in range domain
X_{s0}	Half-size of illuminated target area in slant-range domain; size of illuminated target area (radar beamwidth) in slant-range domain is $2X_{s0}$
X_0	Half-size of illuminated target area in range domain; size of illuminated target area (radar beamwidth) in range domain is $2X_0$
y	Cross-range domain
$y_e(\ell)$	Cross-range coordinate of differential element on radar surface
y_n	Cross-range of nth target
Y_c	Center point of target area in cross-range domain
Y_0	Half-size of illuminated target area in cross-range domain; size of illuminated target area (radar beamwidth) in cross-range domain is $2Y_0$
Z_c	Center point of target area in altitude (elevation) domain
$\mathcal{F}$	Forward Fourier transform operator
$\mathcal{F}^{-1}$	Inverse Fourier transform operator
ℓ	Radar surface domain
λ	Wavelength: $\lambda = 2\pi c/\omega$
ω	Single frequency of radar signal
ϕ	Slow-time angular spectral (Doppler) domain
ϕ_d	Divergence angle of planar, parabolic, and circular radar antenna
ϕ_{dy}	Divergence angle of radar antenna in cross-range domain (three-dimensional)
ϕ_{dz}	Divergence angle of radar antenna in altitude domain (three-dimensional)
$\theta_n(u)$	Aspect angle of nth target when radar is at $(0, u)$
θ_z	Radar grazing angle or slant angle (three-dimensional)

3.1 TRANSMIT MODE RADAR RADIATION PATTERN

Our purpose in this section is to develop the Fourier properties of a radar transmit mode radiation pattern. *This is not the synthetic aperture radiation pattern.* We are concerned with the spatial and Fourier properties of the radiation pattern of the *physical radar antenna* at a fixed slow-time (or a fixed radar position on the synthetic aperture). We show that this radiation pattern has the signal properties of the AM-PM signals that were discussed in the previous chapter.

We consider a radar antenna that transmits (or absorbs) microwave energy via a radiating *surface*; such a radar is also called a *dish*. There exists another class of radars which are called *horns*. Analyzing the functional properties of the radiation pattern of a radar horn is more complicated than those of a radar dish. However, it can be shown that most radar horns exhibit radiation pattern spectral (Fourier) properties similar to the spectral properties that we will develop for a *planar* radar dish.

We begin with a physical radar antenna (dish) which is located at the origin, that is, the slow-time $u = 0$. We formulate the problem for one of the fast-time frequencies of the radar signal by assuming that the transmitted signal is a single-tone sinusoid of the form $p(t) = \exp(j\omega t)$. The physical radar antenna (dish) is composed of differential elements on its aperture (surface). These differential elements are located at the coordinates $[x_e(\ell), y_e(\ell)]$, $\ell \in \mathbf{S}$, in the spatial (x, y) domain where $\mathbf{S}$ represents the contour that defines the antenna surface.

In the transmit mode the radiation experienced at an arbitrary point (x, y) in the spatial domain due to the radiation from the differential element located at

$$(x, y) = [x_e(\ell), y_e(\ell)], \qquad \ell \in \mathbf{S},$$

is

$$\begin{aligned}
&\frac{1}{r} i(\ell) p\left[t - \frac{\sqrt{[x - x_e(\ell)]^2 + [y - y_e(\ell)]^2}}{c}\right] d\ell \\
&\quad = \frac{1}{r} i(\ell) \exp(j\omega t) \exp\left[-jk\sqrt{[x - x_e(\ell)]^2 + [y - y_e(\ell)]^2}\right] d\ell,
\end{aligned}$$

where $r = \sqrt{x^2 + y^2}$, and $i(\ell)$ is an amplitude function which represents the relative strength of that element. This amplitude function, which depends on the curvature of the radar antenna surface, is identified by the manufacturer of the radar, or can be assumed to be a constant (i.e., invariant in ℓ). We do not carry the term $\exp(j\omega t)$ in our discussion; this term disappears after the baseband conversion of the fast-time domain data.

The total radiation experienced at the spatial point (x, y), that is, the *physical radar antenna transmit mode radiation pattern*, is the sum of the radiations from all the differential elements on the surface of the radar. We identify the radar antenna transmit mode radiation pattern at the spatial point (x, y) and fast-time frequency ω

via

$$h_T(\omega, x, y) = \frac{1}{r} \int_{\ell \in \mathbf{S}} i(\ell) \underbrace{\exp\left[-jk\sqrt{[x - x_e(\ell)]^2 + [y - y_e(\ell)]^2}\right]}_{\text{Spherical PM signal}} d\ell.$$

Based on our discussion on the spectral properties of a spherical PM signal, we can write the following Fourier decomposition for the spherical PM signal on the right side of the above transmit mode radiation pattern:

$$\exp\left[-jk\sqrt{[x - x_e(\ell)]^2 + [y - y_e(\ell)]^2}\right]$$
$$= \int_{-k}^{k} \exp\left[-j\sqrt{k^2 - k_u^2}\,[x - x_e(\ell)] - jk_u[-y_e(\ell)]\right] dk_u.$$

Substituting this Fourier decomposition in the expression for h_T, and after interchanging the order of integration over ℓ and k_u, we obtain

$$h_T(\omega, x, y) = \frac{1}{r} \int_{-k}^{k} \exp\left(-j\sqrt{k^2 - k_u^2}\,x - jk_u y\right)$$
$$\times \underbrace{\int_{\ell \in \mathbf{S}} i(\ell) \exp\left[j\sqrt{k^2 - k_u^2}\,x_e(\ell) + jk_u y_e(\ell)\right] d\ell}_{\text{Amplitude pattern } A_T(\omega, k_u)}\, dk_u.$$

Thus we can rewrite the radar antenna transmit mode radiation pattern via the following Fourier decomposition which is identical to that of an amplitude-modulated spherical PM signal (with $2k$ replaced with k):

$$h_T(\omega, x, y) = \frac{1}{r} \int_{-k}^{k} A_T(\omega, k_u) \exp\left(-j\sqrt{k^2 - k_u^2}\,x - jk_u y\right) dk_u,$$

where the amplitude pattern in the slow-time Doppler domain is

$$A_T(\omega, k_u) = \int_{\ell \in \mathbf{S}} i(\ell) \exp\left[j\sqrt{k^2 - k_u^2}\,x_e(\ell) + jk_u y_e(\ell)\right] d\ell.$$

Clearly this amplitude pattern depends on the shape and size of the radar antenna, that is, **S**, as well as its fast-time frequency ω. Next we examine this amplitude pattern for three types of radar antenna.

- In the above derivation we did not utilize the near-field *Fresnel* approximation or the far-field *Fraunhofer* approximation [chr; goo]. The only approximation used in deriving $A_T(\omega, k_u)$ was that the *evanescent waves were negligible.* The key strength of the above derivation is in expressing the radiation pattern

$h_T(\omega, x, y)$ in terms of a weighted sum of spatial domain plane waves of the form

$$\exp\left(-j\sqrt{k^2 - k_u^2}\,x - jk_u y\right) = \exp\big(-jk\cos\phi x - jk\sin\phi y\big),$$

where $\phi = \arctan(k_u/k)$ (angular spectral or Doppler domain in transmit mode) is the propagation angle of the plane wave. Fresnel and Fraunhofer approximations are based on approximations of this plane wave (phase function); these approximations could result in significant phase errors and degradations in an array imaging problem [s94]. However, the next discussion will indicate that the region of the support for the amplitude pattern $a_T(\omega, x, y)$, or what we refer to as the *beamwidth*, is similar to what is predicted by the Fresnel approximation in the near-field, and the Fraunhofer approximation in the far-field. *Making approximations in the region of the support of a radiation pattern (amplitude function) is not as critical (dangerous) as approximating its phase function.*

Example 1 *Planar Radar Antenna* A planar transmitting radar antenna corresponds to a *flat* two-dimensional aperture in the three-dimensional spatial domain; the differential transmitting elements on this aperture are assumed to be *in-phase* which implies that there is no relative phase difference (delay) in their transmissions. We examine the flat-type antenna in the two-dimensional spatial domain, and for simplicity, we call it a planar radar antenna. We will re-examine this class of radars in the three-dimensional spatial domain later in this section.

For a planar radar antenna the surface of the radar is identified via

$$[x_e(\ell), y_e(\ell)] = (0, \ell) \qquad \text{for } \ell \in \left[-\frac{D}{2}, \frac{D}{2}\right],$$

where D is the diameter of the radar antenna. Uniform illumination along the physical aperture is assumed; that is, $i(\ell) = 1$ for $\ell \in [-D/2, D/2]$, and zero otherwise. Substituting these specifications in the model for the amplitude pattern A_T, we obtain

$$A_T(\omega, k_u) = \int_{-D/2}^{D/2} \exp(jk_u\ell)d\ell$$

$$= D\,\text{sinc}\left(\frac{1}{2\pi}Dk_u\right).$$

- The transmit mode amplitude pattern of a planar radar antenna in the slow-time Doppler domain k_u is a sinc function that depends only on the size of the radar antenna and is *invariant in the radar fast-time frequency* ω.

While the support band of $A_T(\omega, k_u)$, which is the main lobe of the sinc function,

$$k_u \in \left[\frac{-2\pi}{D}, \frac{2\pi}{D}\right],$$

is invariant in the radar frequency, this does not imply that the transmit mode radar beamwidth is invariant in ω too. In fact, from the amplitude mapping property of the AM-PM signals (see Section 2.9), we have

$$a_T(\omega, x, y) = \frac{1}{r} A_T \left[k \sin \left[\arctan \left(\frac{y}{x} \right) \right], \omega \right]$$
$$= \frac{D}{r} \operatorname{sinc} \left(\frac{kDy}{2r} \right).$$

The support band (beamwidth) of a_T in the cross-range domain is

$$y \in [-B, B]$$
$$\approx \left[\frac{-r\lambda}{D}, \frac{r\lambda}{D} \right],$$

which varies with the target range as well as the radar fast-time frequency ω. *Note that the support band of both A_T and a_T is a **decreasing** function of the planar radar antenna diameter D.*

The dependency of the support of the spatial amplitude pattern $a_T(\omega, x, y)$ for a planar radar antenna at a fixed range x and versus λ (wavelength) and cross-range y is shown in Figure 3.1*a* (Figures P3.1*a*–*b*). The dependency of the support of the Doppler amplitude pattern $A_T(\omega, k_u)$ for a planar radar antenna versus ω and cross-range Doppler k_u is shown in Figure 3.1*b* (Figures P3.3*a*–*b*).

We can redefine the beamwidth via the following:

$$B = r \sin \phi_d,$$

where

$$\phi_d = \arcsin \left(\frac{\lambda}{D} \right)$$

is called the *divergence angle* [chr]. A planar radar radiation pattern is confined in a cone-shaped region with an axial angle of ϕ_d. This is demonstrated for two different sizes of planar aperture D in Figure 3.2*a* and *b*. Figure 3.3*a* and *b* shows simulated amplitude patterns of a planar radar $|a_T(\omega, x, y)|$ at 1 GHz (wavelength $\lambda = 0.1$ meter) for $D = 1$ meter and $D = 2$ meters, respectively. (The Matlab code for this chapter was used for the simulation at various range values.)

The amplitude relationship between $a_T(\omega, x, y)$ and $A_T(\omega, k_u)$ is valid at far-field, that is, $r \gg \lambda$. While our future analysis on SAR is concerned with the far-field radiation pattern of a radar antenna, it is worthwhile to point out some of the near-field properties of radars; we do this without showing the derivations. For the radiation patterns in Figures 3.2 and 3.3, we observe that the beam is approximately collimated (nondiverging) from $x = 0$ to

$$x_t = \frac{D}{4\lambda};$$

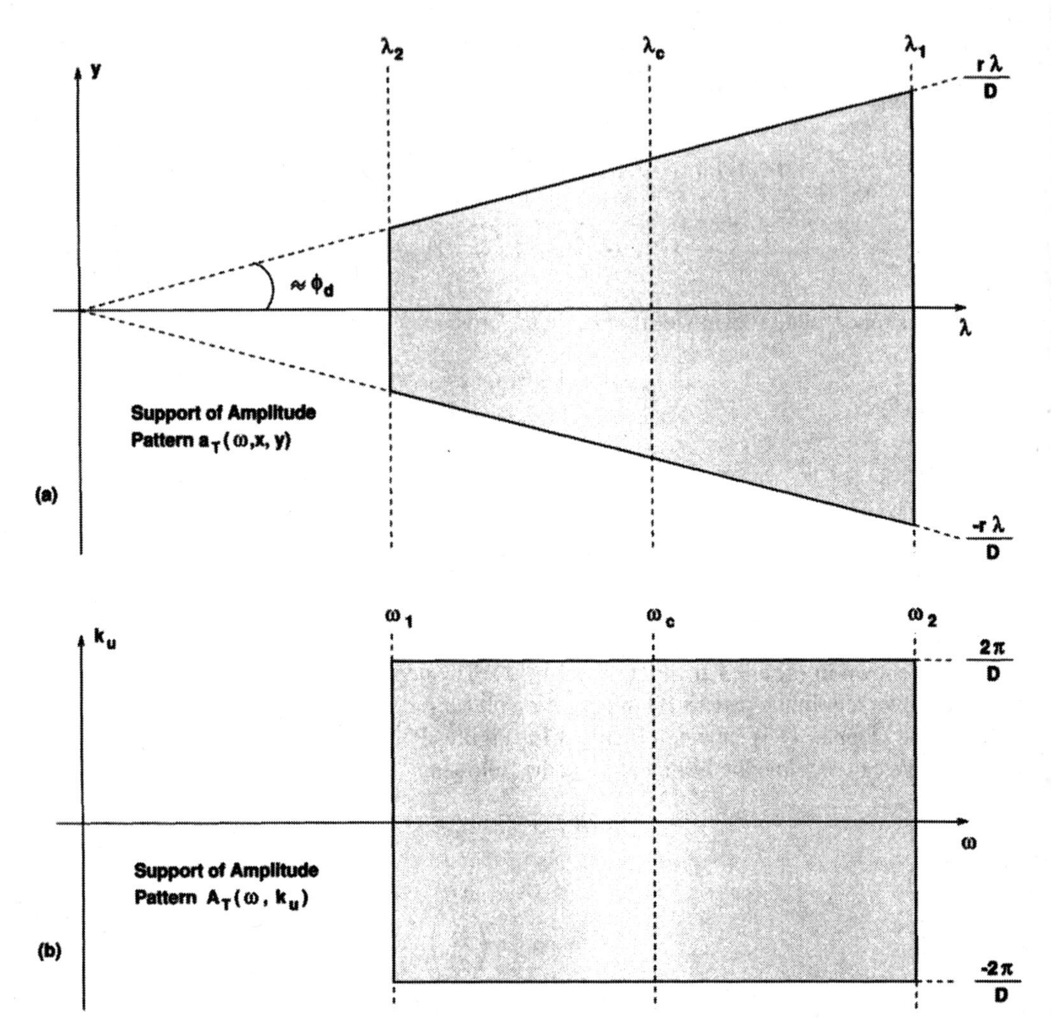

Figure 3.1. Planar radar aperture in transmit mode: (a) support of spatial amplitude pattern $a_T(\omega, x, y)$ and (b) support of Doppler amplitude pattern $A_T(\omega, k_u)$.

x_t is called the transition range of the beam [chr]. In the simulated example of Figure 3.3, $x_t = 2.5$ meters for $D = 1$ meter, and $x_t = 5$ meters for $D = 2$ meters. For $x > x_t$, the radiation pattern starts to exhibit a diverging behavior. For $x \gg x_t$ (far-field), the radiation pattern resembles a cone with an axial angle of ϕ_d; this was the outcome of our earlier discussion on the far-field transmit-mode radiation pattern of a radar.

There exists another class of radar antennas, parabolic antennas with a *feed*, that are effectively the same as a planar antenna. (The feed component makes them a

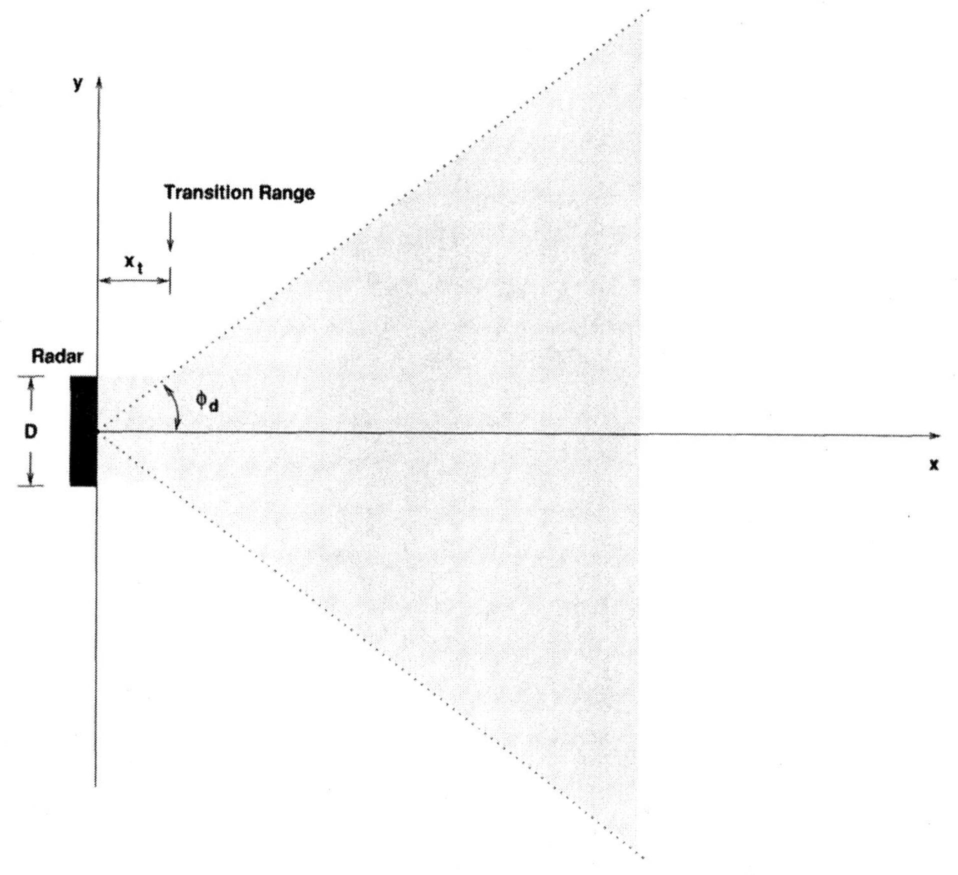

Figure 3.2a. Planar radar aperture in transmit mode: depiction of spatial amplitude pattern $a_T(\omega, x, y)$.

planar antenna.) Next we examine the class of parabolic radar antennas that does not have a feed system.

Example 2 *Parabolic Radar Antenna* Most curved radar antennas have parabolic surfaces that may be identified via

$$[x_e(\ell), y_e(\ell)] = \left(\frac{\ell^2}{2x_f}, \ell \right) \qquad \text{for } \ell \in \left[-\frac{D}{2}, \frac{D}{2} \right],$$

where x_f, called the *focal range*, is a known constant. In this case, with uniform illumination along the physical aperture, the Doppler amplitude pattern becomes

$$A_T(\omega, k_u) = \int_{-D/2}^{D/2} \exp\left(j\sqrt{k^2 - k_u^2}\, \frac{\ell^2}{2x_f} + jk_u\ell \right) d\ell.$$

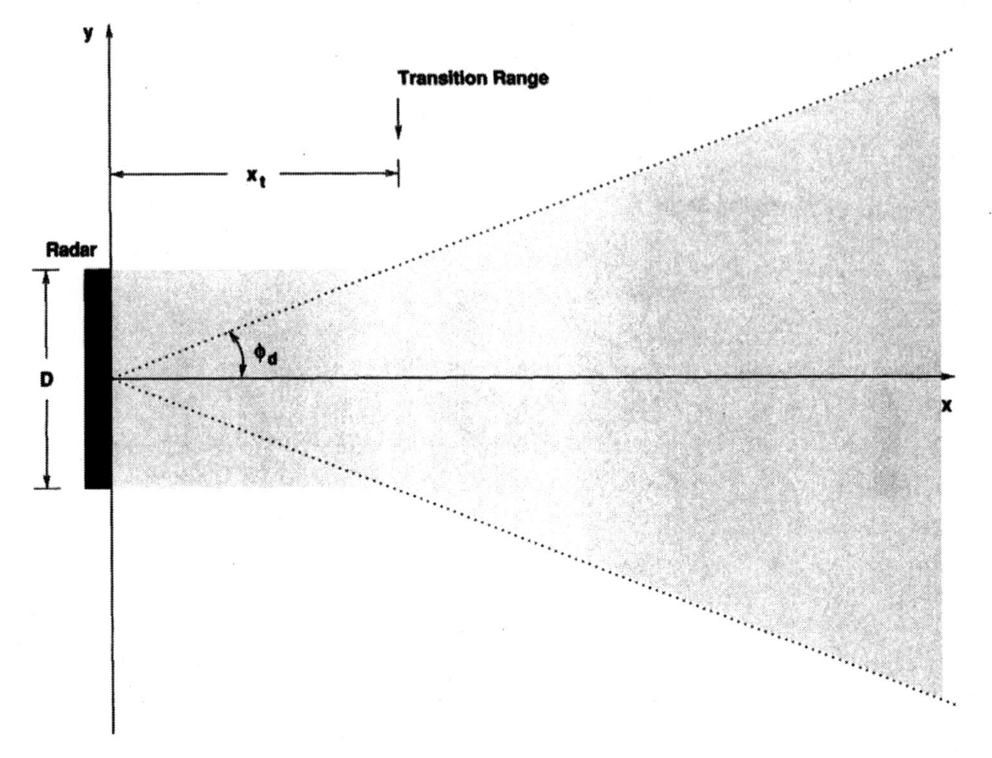

Figure 3.2b. Same as (a) when aperture diameter is doubled.

[The amplitude pattern $i(\ell)$ is related to the Jacobian of the surface integral over **S**, and is suppressed in the above formula.]

We define the divergence angle via

$$\phi_d = \arctan\left(\frac{D}{2x_f}\right).$$

When the condition

$$k \sin \phi_d \gg \frac{2\pi}{D}$$

is satisfied, we can use the properties of AM-PM signals to show that the support band of the amplitude pattern A_T is

$$k_u \in \left[-k \sin \phi_d, k \sin \phi_d\right].$$

This result states that, unlike the case of the planar radar antenna, the slow-time Doppler band increases with the radar fast-time frequency ω, and the cross-range

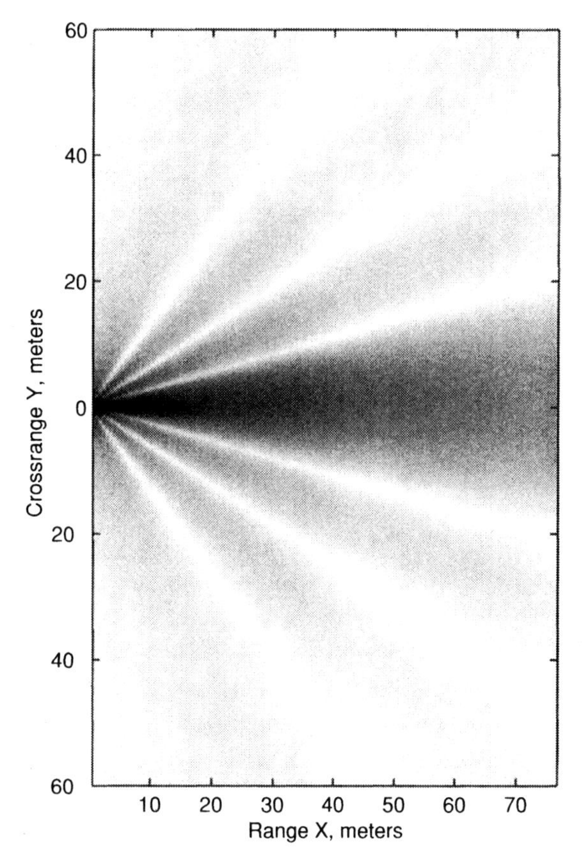

Figure 3.3a. Planar radar aperture in transmit mode: simulated amplitude pattern $|a_T(\omega, x, y)|$ at 1 GHz for $D = 1$ m.

extent of the dish D. The slow-time Doppler band is also an increasing function of the curvature of the parabolic dish $1/2x_f$. Moreover, using the fact that

$$a_T(\omega, x, y) = \frac{1}{r} A_T \left[k \sin \left[\arctan \left(\frac{y}{x} \right) \right], \omega \right],$$

the support band (beamwidth) of a_T in the cross-range domain becomes

$$\begin{aligned} y &\in [-B, B] \\ &= \left[-r \sin \phi_d, r \sin \phi_d \right] \end{aligned}$$

which, unlike the support band of A_T, *is invariant in the radar fast-time frequency* ω. The support band of both A_T and a_T is an *increasing* function of the parabolic radar antenna diameter D. *Reminder: The opposite is true for the planar radar antenna.*

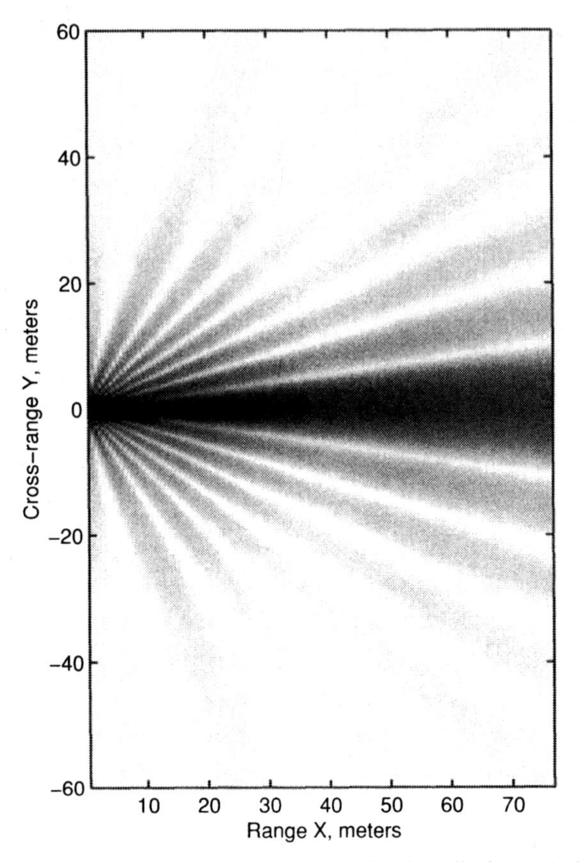

Figure 3.3b. Planar radar aperture in transmit mode: simulated amplitude pattern $|a_T(\omega, x, y)|$ at 1 GHz for $D = 2$ m.

- At the far-field, both planar and parabolic radars exhibit a cone-shaped diverging beam pattern with an axial angle of ϕ_d; that is, the half-beamwidth is governed by

$$B = r \sin \phi_d.$$

However, the divergence angle of the two radar types show different dependency on the radar diameter and frequency:

$$\phi_d = \begin{cases} \arcsin\left(\dfrac{\lambda}{D}\right), & \text{planar radar,} \\ \arctan\left(\dfrac{D}{2x_f}\right), & \text{parabolic radar.} \end{cases}$$

The condition $k \sin \phi_d \gg 2\pi/D$ is satisfied for high-frequency SAR systems, for example, X or C band. However, it might not be valid for certain UHF and VHF systems. Perhaps the best way to identify the slow-time Doppler band of such systems is through simulation of the radiation pattern.

Consider the radar transmit mode radiation pattern at a fixed range x. For this, the dependency of the support of the spatial amplitude pattern $a_T(\omega, x, y)$ for a parabolic radar antenna versus λ (wavelength) and cross-range y is shown in Figure 3.4*a* (Figures P3.1*c*–*d*); the dependency of the support of the Doppler amplitude pattern $A_T(\omega, k_u)$ for a planar radar antenna versus ω and cross-range Doppler k_u is shown in Figure 3.4*b* (Figures P3.3*c*–*d*).

Figure 3.5*a* and *b* show examples of the transmit mode radiation pattern for a parabolic radar antenna with two aperture diameter D values. Figure 3.6*a* and *b* show simulated radiation patterns of a parabolic radar $|a_T(\omega, x, y)|$ at 1 GHz (wavelength $\lambda = 0.1$ meter) with $x_f = 2$ meters for $D = 1$ meter and $D = 2$ meters, respectively. (The Matlab code for this chapter was used for the simulation at various

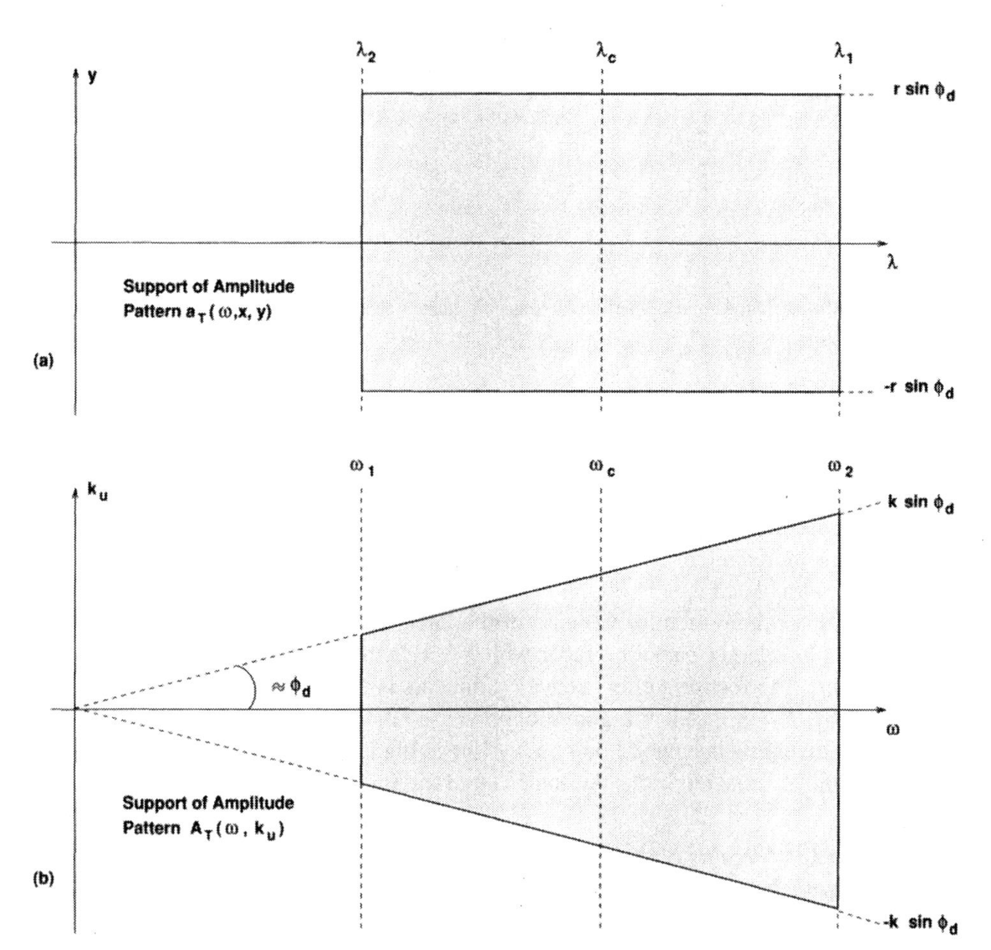

Figure 3.4. Parabolic radar aperture in transmit mode: (a) support of spatial amplitude pattern $a_T(\omega, x, y)$ and (b) support of Doppler amplitude pattern $A_T(\omega, k_u)$.

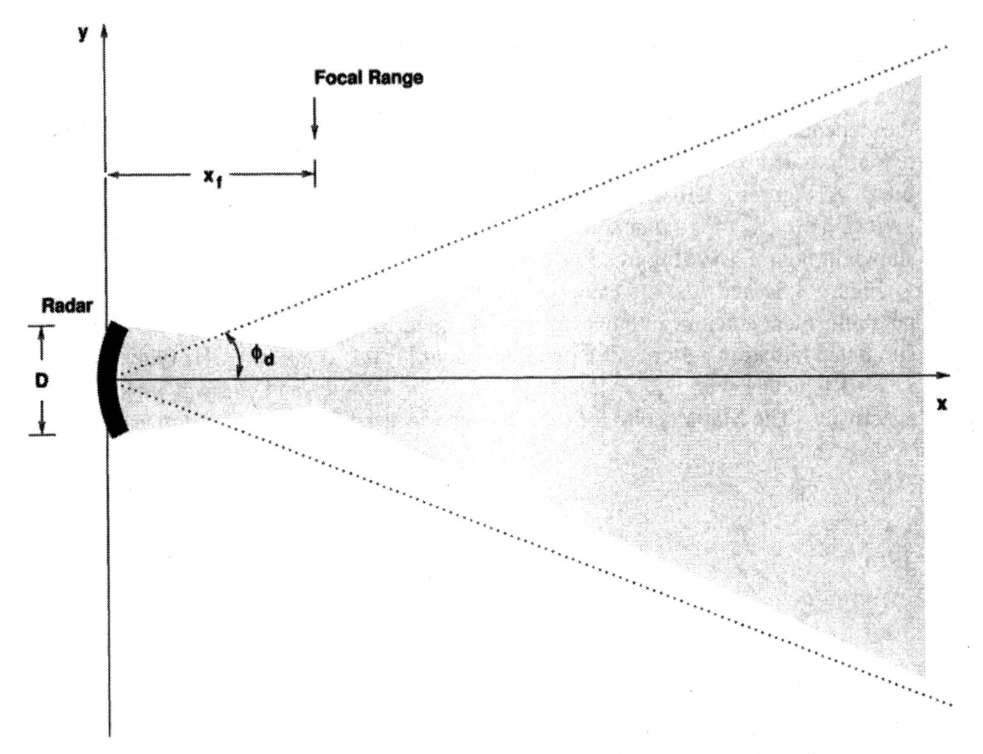

Figure 3.5a. Parabolic radar aperture in transmit mode: depiction of spatial amplitude pattern $a_T(\omega, x, y)$.

range values.) At the near range, these classes of radars exhibit a focusing property at $(x, y) = (x_f, 0)$. The origin of this focusing phenomenon can be found in focusing with a lens [goo; chr]. In this case a *quadratic* phase across a flat aperture (planar radar) is used for focusing. A larger aperture D yields a sharper focal point in the cross-range domain [chr].

For the simulated data in Figure 3.6, the focal range is supposed to be at $x_f = 2$ meters. This can be seen for the case of the larger radar, that is, $D = 2$ meters in Figure 3.6*b*; a larger parabolic radar with $x_f = 2$ meters would exhibit the same focal point. The focusing effect at $x_f = 2$ meters is not as obvious for the smaller radar with $D = 1$ meter in Figure 3.6*a*. For $x > x_f$, the radiation pattern starts to exhibit a diverging behavior. For $x \gg x_f$ (far-field), the radiation pattern resembles a cone with an axial angle ϕ_d; the same is true for a planar radar.

Example 3 *Circular Radar Antenna* The surface of a circular radar antenna has the following pattern:

$$[x_e(\ell), y_e(\ell)] = \left(x_f - \sqrt{x_f^2 - \ell^2},\ \ell\right) \qquad \text{for } \ell \in \left[-\frac{D}{2}, \frac{D}{2}\right],$$

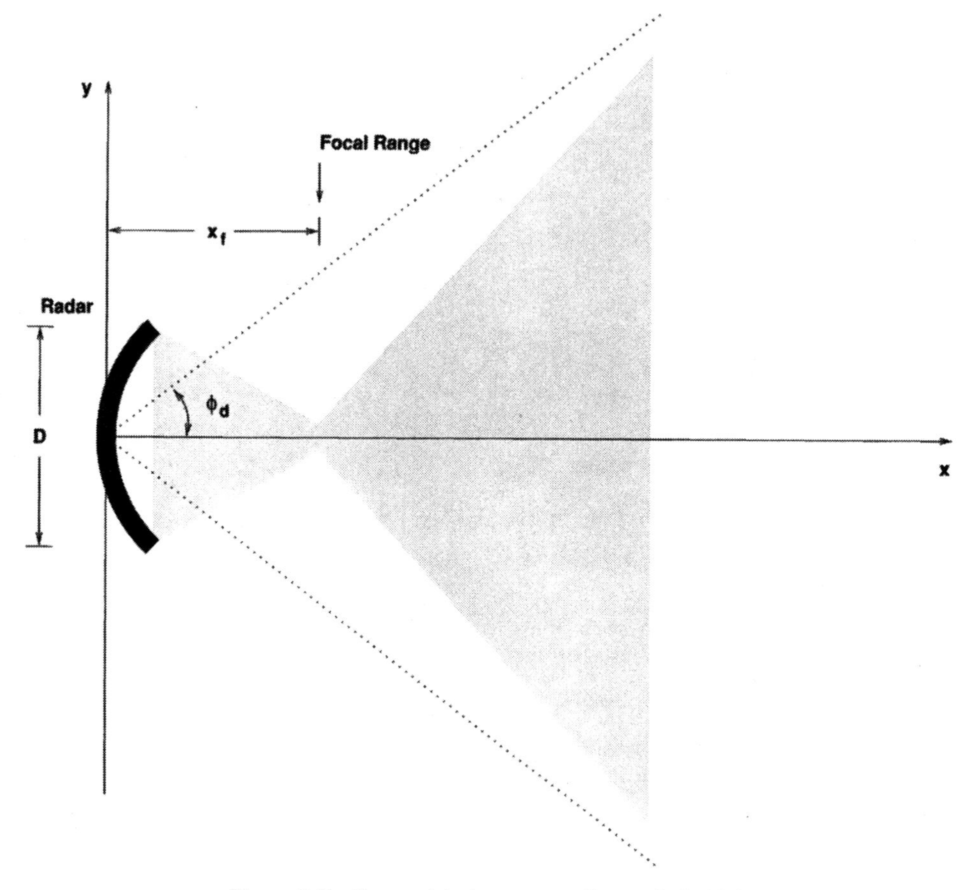

Figure 3.5b. Same as (a) when aperture diameter is doubled.

where x_f is the known radius of the circular aperture. Note that with a uniform illumination (i.e., $i_\ell = 1$ across the aperture), the radiation from all the differential transmitters are added constructively (coherently) at the *focal* point $(x, y) = (x_f, 0)$.

This is similar to the focusing effect for the parabolic radars. In fact, with the first-order Taylor series approximation of

$$\sqrt{x_f^2 - \ell^2} \approx x_f - \frac{\ell^2}{2x_f},$$

a circular radar surface can be approximated via the surface of a parabolic radar, that is,

$$[x_e(\ell), y_e(\ell)] \approx \left(\frac{\ell^2}{2x_f}, \ell \right) \qquad \text{for } \ell \in \left[-\frac{D}{2}, \frac{D}{2} \right].$$

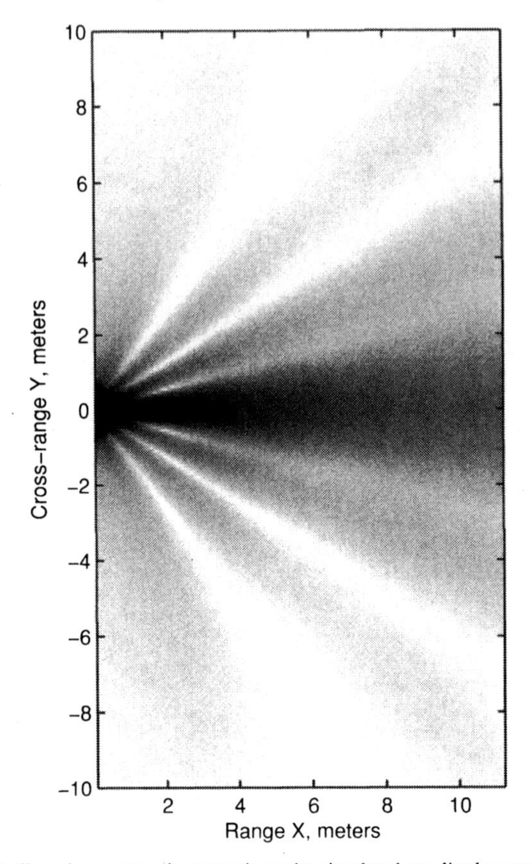

Figure 3.6a. Parabolic radar aperture in transmit mode: simulated amplitude pattern $|a_T(\omega, x, y)|$ at 1 GHz for $D = 1$ m.

Thus, for a circular radar antenna, we expect to derive properties that are similar to those of the planar radar antennas.

For uniform illumination along the physical aperture, the amplitude pattern is

$$A_T(\omega, k_u) = \int_{-D/2}^{D/2} \exp\left[j\sqrt{k^2 - k_u^2}\left(x_f - \sqrt{x_f^2 - \ell^2}\right) + jk_u\ell \right] d\ell.$$

[The amplitude pattern $i(\ell)$ is related to the Jacobian of the surface integral over **S**, and is suppressed in the above equation.]

We define the divergence angle via

$$\phi_d = \arcsin\left(\frac{D}{2x_f}\right),$$

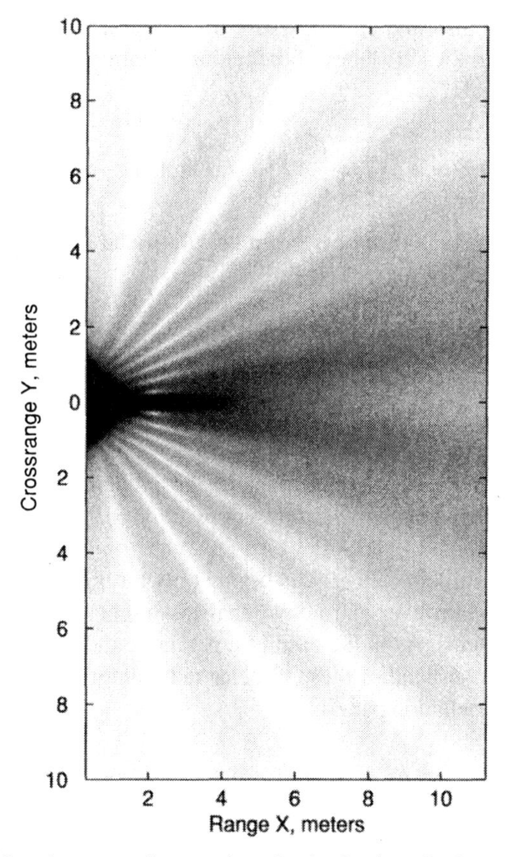

Figure 3.6b. Parabolic radar aperture in transmit mode: simulated amplitude pattern $|a_T(\omega, x, y)|$ at 1 GHz for $D = 2$ m.

which is the same as the divergence angle for a parabolic radar. Under the constraint

$$k \sin \phi_d \gg \frac{2\pi}{D},$$

one can use the properties of AM-PM signals to show that the support band of the amplitude pattern A_T is

$$k_u \in \left[-k \sin \phi_d, k \sin \phi_d\right].$$

This slow-time Doppler band also increases with the radar fast-time frequency ω, and the cross-range extent of the circular dish D. However, it decreases with the radius of the dish x_f. *Note that the curvature of the dish also decreases as x_f increases.* Thus the parabolic radar dish and circular radar dish exhibit the same properties. Their main difference is in the side lobes of the radiation pattern.

Using the fact that

$$a_T(\omega, x, y) = \frac{1}{r} A_T \left[k \sin \left[\arctan \left(\frac{y}{x}\right)\right], \omega\right],$$

the support band (beamwidth) of a_T of a circular radar antenna in the cross-range domain becomes

$$\begin{aligned} y &\in [-B, B] \\ &= \left[\frac{-rD}{2x_f}, \frac{rD}{2x_f}\right] = \left[-r \sin \phi_d, r \sin \phi_d\right]. \end{aligned}$$

Thus the support band of a_T is proportional to the circular radar antenna diameter D. Unlike the support band of A_T, *the support band of a_T is invariant in the radar fast-time frequency ω.*

The condition $k \sin \phi_d \gg 2\pi/D$ can be met only for high-frequency SAR systems in X and C bands. However, the above analysis might not be suitable for certain UHF and VHF systems. As in the case of a parabolic radar antenna, the analysis of the slow-time Doppler band of these circular radar antennas may be performed via simulation of their radiation pattern.

Synthetic Aperture (Slow-time) Dependence

Next we consider the scenario when the physical radar is located at an arbitrary point in the slow-time domain, for example, u on the synthetic aperture. The principles that we developed earlier for $u = 0$ now hold for a shifted version of the cross-range domain; the amount of the shift is u. Thus, when the radar is located at u, its transmit mode radiation pattern at fast-time frequency ω has the following Fourier decomposition:

$$\begin{aligned} h_T(\omega, x, y - u) &= \int_{-k}^{k} A_T(\omega, k_u) \exp\left[-j\sqrt{k^2 - k_u^2}\, x - jk_u(y - u)\right] dk_u \\ &= \mathcal{F}^{-1}_{(k_u)}\left[A_T(\omega, k_u) \exp\left[-j\sqrt{k^2 - k_u^2}\, x - jk_u y\right]\right]. \end{aligned}$$

Moreover this transmit mode radiation pattern can also be expressed in terms of an amplitude pattern and a spherical PM signal as shown in the following:

$$h_T(\omega, x, y - u) = \underbrace{a_T(\omega, x, y - u)}_{\text{Amplitude pattern}} \underbrace{\exp\left[-jk\sqrt{x^2 + (y - u)^2}\right]}_{\text{Spherical PM signal}}$$

The two amplitude patterns $a_T(\cdot)$ and $A_T(\cdot)$ are related via the following:

$$a_T(\omega, x, y-u) = \frac{1}{\sqrt{x^2+(y-u)^2}} A_T\left[\frac{k(y-u)}{\sqrt{x^2+(y-u)^2}}, \omega\right]$$
$$\approx \frac{1}{r} A_T\left[\frac{k(y-u)}{\sqrt{x^2+(y-u)^2}}, \omega\right].$$

When the radar antenna length is zero (an *ideal* transmitting element), that is, $D = 0$, the radiation pattern is *omni-directional*, and the amplitude pattern takes on the form $a_T(\omega, x, y-u) = 1/r$.

Note that we approximated the u-dependent amplitude function $1/\sqrt{x^2+(y-u)^2}$ with $1/r$, which is inavarint in the slow-time u. It has been suggested that the variations of the amplitude function $1/\sqrt{x^2+(y-u)^2}$ with respect to u can be used for high-resolution SAR imaging.* Whether the theoretical foundation of this approach is correct or not, the user should realize that such amplitude variations with respect to the synthetic aperture (slow-time) u are well below the noise (A/D quantization noise, additive noise, motion multiplicative noise, etc.) level of a practical SAR system.

3.2 RADIATION PATTERN IN THREE-DIMENSIONAL SPATIAL DOMAIN

In Chapters 1 and 2 we considered a two-dimensional spatial domain to represent the imaging scene. Thus far in this chapter we have stayed with that representation, and after this section we will still view the imaging scene as a two-dimensional spatial domain. It turns out that the use of a two-dimensional spatial domain instead of a three-dimensional one greatly simplifies our notation and analysis; this is done by making certain variables *transparent* to the reader and, thus, simplifying a notationally complicated imaging problem.

However, we intend to provide a more realistic view of the problem by exposing the reader to the transparent variables of the three-dimensional spatial domain. This helps the reader to understand the limitations on the multidimensional imaging capability of a conventional SAR system. Once again, we should emphasize that we will return to the two-dimensional spatial domain representation after this discussion.

Radar Footprint

A realistic SAR system corresponds to an imaging geometry in the three-dimensional spatial domain. In a typical scenario a radar with dimensions D_y by D_z is positioned

*J. Lee, O. Arikan, and D. Munson, "Formulation of a general imaging algorithm for high-resolution synthetic aperture radar," *Proc. ICSASSP*, Atlanta, May 1996.

at some point in the spatial domain, for example, $(x, y, z) = (0, 0, 0)$ (origin). The radar possesses a *two-dimensional surface*. The radar surface could be planar (flat) or curved, and its outer contour could trace a rectangle, a circle, an ellipse, and so on. For instance, the radar in Figure 3.7*a* and *b* appears to have a rectangular outer contour. It turns out that the shape of the outer contour of a radar plays a role in forming the outer shape of the main lobe and the side lobes of its radiation pattern [goo]. However, the role of the radar outer contour is not as dramatic as the role of its curvature (flat or curved). Recall that the radiation pattern of a flat aperture radar has quite different properties with respect to the radar size and frequency as compared to a curved radar.

First, consider the scenario when the radar surface is on the plane $x = 0$. In this case one could go through an analytical derivation, which is very similar to what we presented earlier for planar, parabolic, and circular radars in the two-dimensional domain and indicates that the resultant beam pattern in the three-dimensional spatial domain is within a three-dimensional cone. The axis of the cone is the line $y = z = 0$ (i.e., the x-axis), and the axial divergence angles of the cone are

$$\phi_{dy} = \begin{cases} \arcsin\left(\dfrac{\lambda}{D_y}\right), & \text{planar radar,} \\ \arctan\left(\dfrac{D_y}{2x_{fy}}\right), & \text{curved radar;} \end{cases}$$

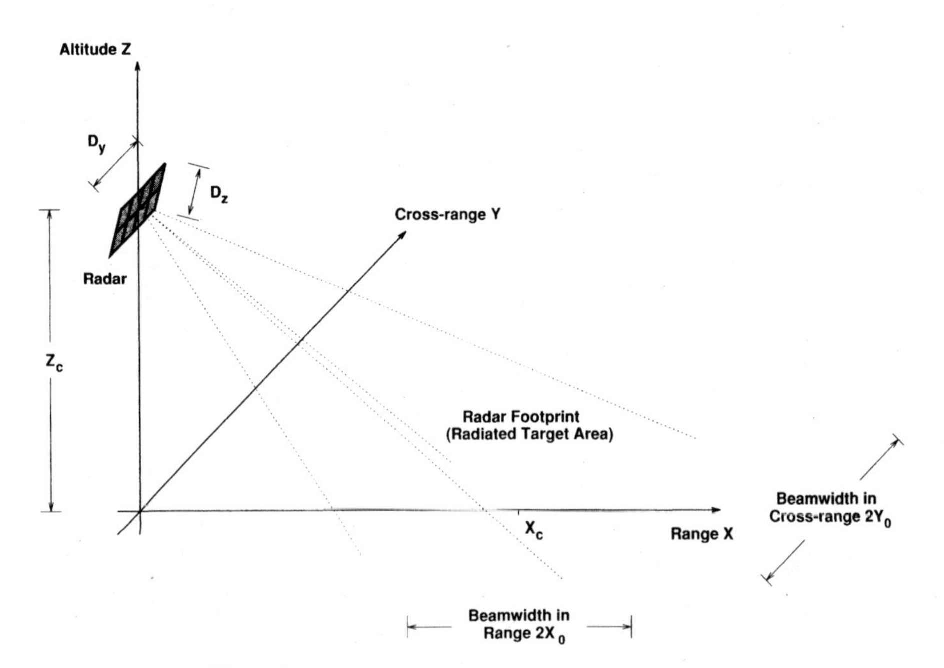

Figure 3.7a. Radar footprint with beam steering at broadside.

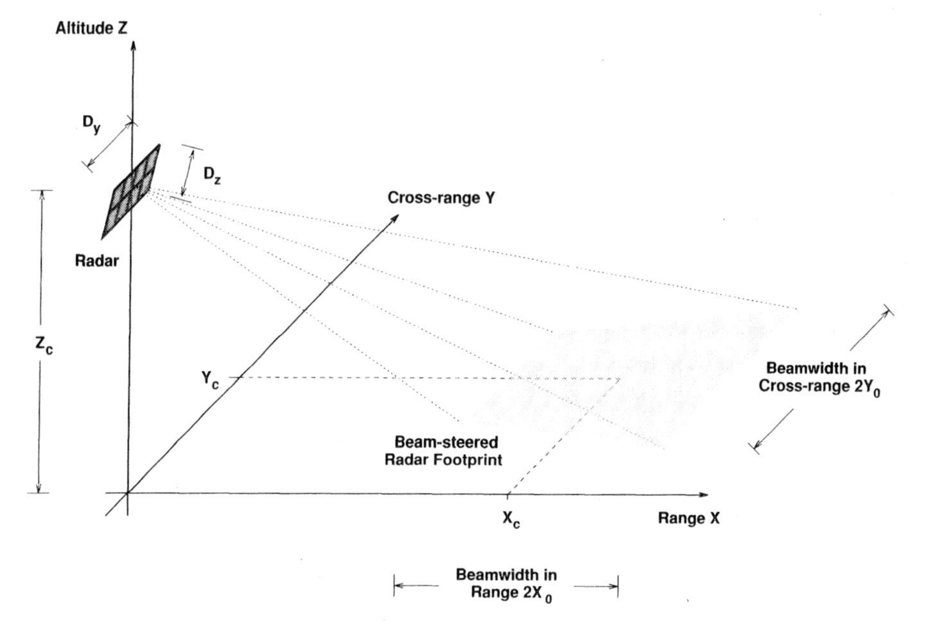

Figure 3.7b. Radar footprint with beam steering at squint.

and

$$\phi_{dz} = \begin{cases} \arcsin\left(\dfrac{\lambda}{D_z}\right), & \text{planar radar,} \\ \arctan\left(\dfrac{D_z}{2x_{fz}}\right), & \text{curved radar.} \end{cases}$$

The outer contour of the cone could be a rectangle (for a radar with a rectangular outer contour) or an ellipse (for a radar with an elliptical outer contour) [goo]. (*If one considers the other errors or ambiguities in a radar system, the exact knowledge of the shape of the beam pattern outer contour becomes an irrelevant issue.*)

We could also define the half-beamwidth in the (y, z) domain for the radar via

$$B_y = r \sin \phi_{dy},$$

$$B_z = r \sin \phi_{dz},$$

where $r = \sqrt{x^2 + y^2 + z^2}$ is the radial distance of an arbitrary point in the spatial domain (x, y, z) from the center of the radar at the origin $(0, 0, 0)$. Thus the support region of the radiation pattern in the spatial domain is $y \in [-B_y, B_y]$ and $z \in [-B_z, B_z]$ at a given radial distance r.

In a practical SAR system, the radar-carrying aircraft has a relative average height (altitude or elevation) with respect to the target area. We denote this mean height by Z_c, and the domain that represents the height variations in the target by the variable

z, the *altitude* domain. The radar-carrying aircraft moves along a linear path in the spatial domain, which we identify with the variable y, the *cross-range* or *azimuth* domain. The third dimension, which is perpendicular to the (y, z) (cross-range and altitude) plane is identified with the variable x, the *range* domain.

In a typical SAR system the radar is *directed* or *steered* at a specific target (ground) area; this is similar to directing a flashlight to *see* a specific target area in the dark. (We will briefly examine radar mechanical and electronic beam-steering in the next chapter.) For instance, the radar can be directed such that the illuminated target area is centered around some mean range value, for example, $x = X_c$, and cross-range $y = 0$; see Figure 3.7*a*. The radar orientation and radiation on the plane $y = 0$ are shown in Figure 3.8.

In this scenario the line that connects the center of the target area at $(x, y, z) = (X_c, 0, Z_c)$ to the center of the radar aperture at $(x, y, z) = (0, 0, 0)$ is *perpendicular to the radar surface* (in the case of a phased array radar, the line is perpendicular to the radiation wavefront); see the dotted line in Figure 3.8. (Recall the flashlight analogy.) This implies that the beamwidth in the cross-range is dictated by D_y with the divergence angle ϕ_{dy} and the beamwidth $B_y = r \sin \phi_{dy}$. Moreover the illuminated

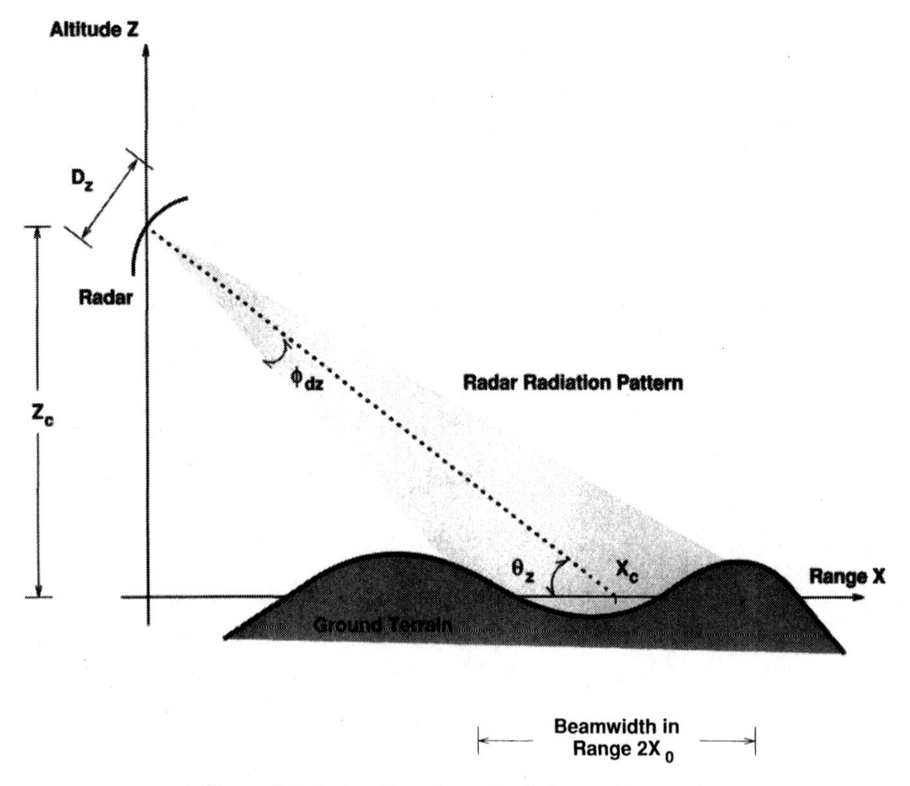

Figure 3.8. Radar orientation and radiation on plane $y = 0$.

target area in the cross-range domain is identified via

$$y \in \left[-Y_0, Y_0 \right],$$

where

$$Y_0 = B_y.$$

Similarly the beamwidth on the (x, z) plane is dictated by D_z with the divergence angle ϕ_{dz} and the beamwidth $B_z = r \sin \phi_{dz}$. However, due to the height of the radar, the illuminated target area in the range domain is *approximately* within (see Figure 3.8)

$$x \in \left[X_c - X_0, X_c + X_0 \right],$$

where

$$\begin{aligned} X_0 &= \frac{B_z}{\sin \theta_z} \\ &= \frac{r \sin \phi_{dz}}{\sin \theta_z} \end{aligned}$$

and

$$\theta_z = \arctan \left(\frac{Z_c}{X_c} \right),$$

is called the radar depression angle, grazing angle, or slant angle.

In general, it is also possible to direct or steer the radar beam around a target area centered around $(x, y) = (X_c, Y_c)$; see Figure 3.7*b*. In this case the illuminated target area, which is also referred to as the *radar footprint*, is approximately within

$$\begin{aligned} x &\in \left[X_c - X_0, X_c + X_0 \right], \\ y &\in \left[Y_c - Y_0, Y_c + Y_0 \right], \end{aligned}$$

where

$$\begin{aligned} X_0 &= \frac{r \sin \phi_{dz}}{\sin \theta_z}, \\ Y_0 &= r \sin \phi_{dy}. \end{aligned}$$

In the above, the radar is assumed to be located at the origin in the spatial domain, and $r = \sqrt{x^2 + y^2 + z^2}$ is the radial distance of an arbitrary point (x, y, z) in the illuminated target area from the radar.

Slant-Range

The discussion in the previous section on *radar footprint* was about the *observability* of a point in space when the radar beam is steered (moved) around; this provides information on the AM component of an AM-PM SAR signal in the three-dimensional spatial domain. In this section we study the PM component of this signal.

When the radar is located at $(x, y, z) = (0, u, 0)$ on the synthetic aperture, the transmit mode radiation pattern at an arbitrary point in space can be expressed via

$$h_T(\omega, x, y-u, z) = \underbrace{a_T(\omega, x, y-u, z)}_{\text{Amplitude pattern}} \underbrace{\exp\left[-jk\sqrt{x^2 + (y-u)^2 + z^2}\right]}_{\text{Spherical PM signal}}.$$

Consider the spherical PM signal in the above transmit mode radiation pattern. We define a new variable, which we refer to as the *slant-range* via

$$x_s = \sqrt{x^2 + z^2}.$$

Using this variable, the spherical PM signal can be rewritten as follows:

$$\exp\left[-jk\sqrt{x^2 + (y-u)^2 + z^2}\right] = \exp\left[-jk\sqrt{x_s^2 + (y-u)^2}\right].$$

The above indicates that *the (x, z) points on the y plane with a fixed slant-range x_s experience the same phase delay in the transmit mode radiation pattern.*

As we will see in our discussion on SAR imaging, the spherical PM signal is the key component which carries useful information for *imaging*; the amplitude function translates into some kind of a window function (observable band) in the spatial frequency domain which dictates *resolution* of a reflector in space. In fact it can be shown that the three-dimensional transmit mode radiation pattern can be expressed in a *mathematical* two-dimensional domain of the slant-range and cross-range (x_s, y) via reasonable approximations on the amplitude pattern:

$$h_T(\omega, x, y-u, z) = \underbrace{a_T(\omega, x_s, y-u)}_{\text{Amplitude pattern}} \underbrace{\exp\left[-jk\sqrt{x_s^2 + (y-u)^2}\right]}_{\text{Spherical PM signal}}$$

The above indicates that the range and altitude (x, z) can be combined to form a one-dimensional slant-range x_s domain; that is, the actual (x, z) variables are transparent to the user. From this point we combine the range and altitude variables to formulate the SAR problem in the two-dimensional (x_s, y) domain. Moreover, for notational simplicity, we call the imaging plane the range and cross-range domains and identify it as (x, y) (i.e., drop the subscript in x_s and the term *slant* in slant-range).

Finally we should also identify the *observable* target region in the slant-range. This can be shown to be approximately within

$$x_s \in \left[-X_{s0}, X_{s0} \right],$$

where

$$X_{s0} = \frac{r \sin \phi_{dz}}{\tan \theta_z}.$$

Once again, for notational simplicity, we replace X_{s0} with X_0, and $\sqrt{X_c^2 + Z_c^2}$ with X_c. We also replace D_z with what we refer to as the *effective* range aperture D_x. D_x is related to D_z and the slant angle θ_z; for a planar radar $D_x \approx D_z \tan \theta_z$, and for a curved radar $D_x \approx D_z / \tan \theta_z$.

3.3 TRANSMIT-RECEIVE MODE RADAR RADIATION PATTERN

The analysis of the *receive* mode radiation pattern of a physical radar is identical to the analysis of its transmit mode radiation pattern. For a unit omni-directional reflector located at (x, y) in the spatial domain and at the fast-time frequency ω, the receive mode radiation pattern of a physical radar located at $(0, u)$ has the following Fourier decomposition:

$$\begin{aligned} h_R(\omega, x, y - u) &= \int_{-k}^{k} A_R(\omega, k_u) \exp\left[-j\sqrt{k^2 - k_u^2}\, x - jk_u(y - u) \right] dk_u \\ &= \mathcal{F}^{-1}_{(k_u)}\left[A_R(\omega, k_u) \exp\left(-j\sqrt{k^2 - k_u^2}\, x - jk_u y \right) \right]. \end{aligned}$$

Similar to the analysis of the transmit mode radiation pattern and the analysis of AM-PM signals in Chapter 2, the receive mode radiation pattern can be expressed in terms of an amplitude pattern and a spherical PM signal; that is,

$$h_R(\omega, x, y - u) = \underbrace{a_R(\omega, x, y - u)}_{\text{Amplitude pattern}} \underbrace{\exp\left[-jk\sqrt{x^2 + (y - u)^2} \right]}_{\text{Spherical PM signal}},$$

where

$$a_R(\omega, x, y - u) = \frac{1}{r} A_R\left[\frac{k(y - u)}{\sqrt{x^2 + (y - u)^2}}, \omega \right].$$

In most radar systems the transmit and receive amplitude patterns are the same; that is,

$$A_R(\omega, k_u) = A_T(\omega, k_u).$$

The transmit-receive mode radiation pattern of an active monostatic physical radar of a SAR system is the product of its radiation patterns at its transmit and receive modes; that is,

$$h(\omega, x, y - u) = h_T(\omega, x, y - u) h_R(\omega, x, y - u).$$

Consider the Fourier transform of the transmit-receive radiation pattern with respect to the slow-time u. The multiplication of h_T and h_R in the slow-time u domain translates into the following convolution integral in the slow-time Doppler k_u domain:

$$\begin{aligned}
&\mathcal{F}_{(u)}[h(\omega, x, y - u)] \\
&\quad = \int_{-2k}^{2k} A_T(\omega, \rho) \exp\left[-j\sqrt{k^2 - \rho^2}\, x - j\rho y\right] \\
&\qquad \times A_R(\omega, k_u - \rho) \exp\left[-j\sqrt{k^2 - (k_u - \rho)^2}\, x - j(k_u - \rho) y\right] d\rho \\
&\quad = \exp(-jk_u y) \int_{-2k}^{2k} A_T(\omega, \rho) A_R(\omega, k_u - \rho) \\
&\qquad \times \exp\left[-j\left[\sqrt{k^2 - \rho^2} + \sqrt{k^2 - (k_u - \rho)^2}\right] x\right] d\rho
\end{aligned}$$

for $k_u \in [-2k, 2k]$.

In the above convolution integral, the amplitude patterns $A_T(\omega, \rho)$ and $A_R(\omega, k_u - \rho)$ are slowly fluctuating functions of ρ as compared to the range-dependent phase-modulated (PM) function

$$\exp\left[-j\left[\sqrt{k^2 - \rho^2} + \sqrt{k^2 - (k_u - \rho)^2}\right] x\right].$$

Using the method of stationary phase, the ρ domain integral of the convolution can be approximated by the value of the integrand when the instantaneous frequency of the PM wave is zero (i.e., at its phase center). This occurs at $\rho = k_u/2$. In this case the convolution becomes

$$\mathcal{F}_{(u)}[h(\omega, x, y - u)] = A(\omega, k_u) \exp\left(-j\sqrt{4k^2 - k_u^2}\, x - jk_u y\right),$$

where

$$A(\omega, k_u) = A_T\left(\omega, \frac{k_u}{2}\right) A_R\left(\omega, \frac{k_u}{2}\right)$$

(Figure P3.7*a–d*). For instance, in the case of a planar radar antenna, the amplitude pattern becomes the sinc-squared function (Figure P3.7*a–b*), that is,

$$A(\omega, k_u) = \text{sinc}^2(Dk_u).$$

The transmit-receive mode radar radiation pattern may be expressed in terms of the following amplitude pattern and a phase function:

$$h(\omega, x, y-u) = \underbrace{a(\omega, x, y-u)}_{\text{Amplitude pattern}} \underbrace{\exp\left[-j2k\sqrt{x^2+(y-u)^2}\right]}_{\text{Spherical PM signal}},$$

with

$$\begin{aligned} a(\omega, x, y-u) &= a_T(\omega, x, y-u) a_R(\omega, x, y-u) \\ &= \frac{1}{r^2} A\left[\frac{2k(y-u)}{\sqrt{x^2+(y-u)^2}}, \omega\right] \end{aligned}$$

(Figure P3.5*a–d*). For notational simplicity, we do not carry the amplitude function $1/r^2$ in our future discussions.

3.4 TRANSMIT-RECEIVE MODE RADAR-TARGET RADIATION PATTERN

In our discussion in the previous section on the radar transmit-receive mode radiation pattern, we assumed that the target is a unit reflector. This is not true for most physical structures. The type as well as shape of a target determines its interaction with the radar illumination. For instance, the characteristics of the SAR signature of a human-made metallic cylinder are quite different from those of a tree with the same size. *This difference is a key feature that can be used to discriminate a human-made structure from foliage.*

Why not use shape information to distinguish targets in SAR images? Because of the radar wave resonance in cavities, surface waves, and the like, for metallic targets, the SAR image of these targets does not resemble their optical image. Thus *shape* (magnitude) information in a SAR image might not be a reliable source for target detection and identification. However, a target *coherent* SAR radiation pattern or coherent SAR image contains distinct features that could be used for target detection and identification.

We now present a study to quantify the coherent SAR signature of targets. A target exhibits an *amplitude pattern*, which contains phase as well as *magnitude*, when illuminated by a radar system such as SAR. *A target amplitude pattern varies with the radar fast-time frequency and aspect angle*. For a radar located at $(0, u)$ and a target located at (x_n, y_n), we denote the target amplitude pattern with $a_n(\omega, x_n, y_n - u)$. Note that

$$\arctan\left(\frac{y_n - u}{x_n}\right)$$

is the target aspect angle with respect to the radar broadside.

The coherent SAR signature of the nth target is its contribution in the total echoed signal from the target scene. When the radar is located at $(0, u)$ and for the fast-time frequency ω, the coherent SAR signature of the nth target, that is, the *transmit-receive mode radar-target radiation pattern* or *SAR radiation pattern* for the nth target, can be expressed via the following:

$$h_n(\omega, x_n, y_n - u) = \underbrace{h_T(\omega, x_n, y_n - u)}_{\text{Radar-to-target}}\, \underbrace{a_n(\omega, x_n, y_n - u)}_{\text{Target}}\, \underbrace{h_R(\omega, x_n, y_n - u)}_{\text{Target-to-radar}};$$

in the above model, $h_T(\cdot)$ is the radar transmit mode radiation pattern, $h_R(\cdot)$ is the radar receive mode radiation pattern.

Substituting for $h_T(\cdot)$ and $h_R(\cdot)$ in terms of their amplitude patterns and spherical PM signal in the nth target SAR radiation pattern, we obtain

$$\begin{aligned} h_n(\omega, x_n, y_n - u) &= a_T(\omega, x_n, y_n - u) \exp\left[-jk\sqrt{x_n^2 + (y_n - u)^2}\right] \\ &\quad \times a_n(\omega, x_n, y_n - u) \\ &\quad \times a_R(\omega, x_n, y_n - u) \exp\left[-jk\sqrt{x_n^2 + (y_n - u)^2}\right]. \end{aligned}$$

A block diagram representation of the generation of the SAR radiation pattern for the nth target is shown in Figure 3.9.

Using the expression for the radar transmit-receive mode radiation pattern

$$\begin{aligned} h(\omega, x, y) &= h_T(\omega, x, y) h_R(\omega, x, y) \\ &= a(\omega, x, y) \exp\left(-j2k\sqrt{x^2 + y^2}\right), \end{aligned}$$

where

$$a(\omega, x, y) = a_T(\omega, x, y) a_R(\omega, x, y),$$

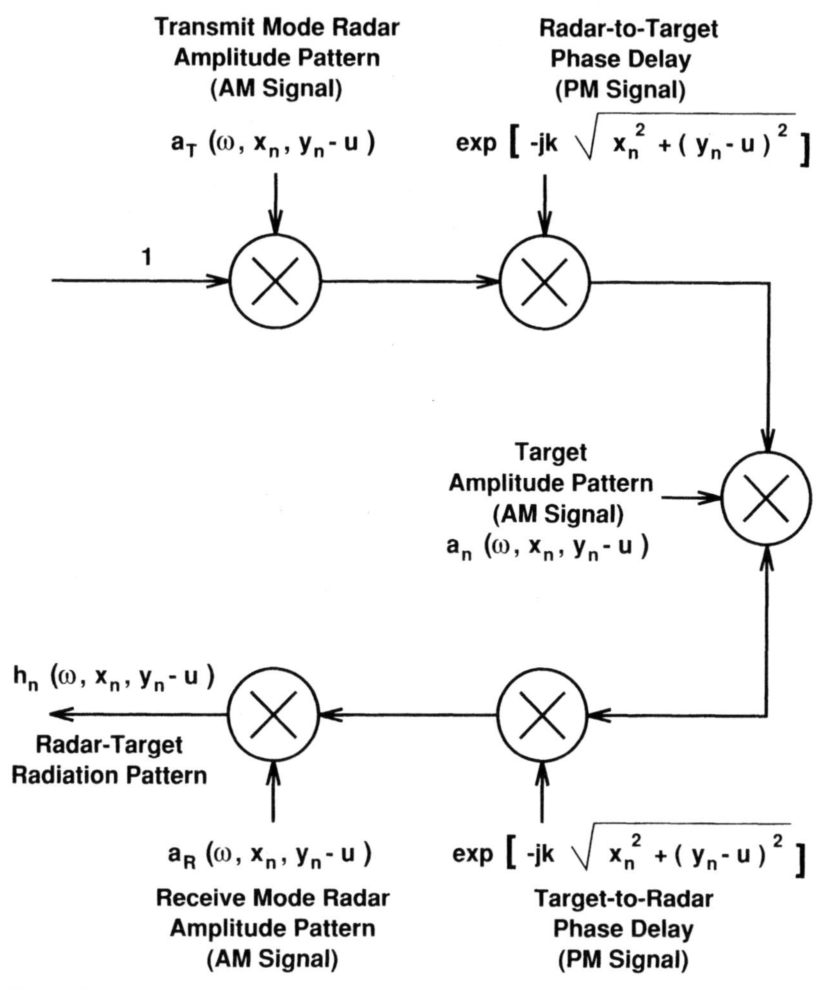

Figure 3.9. Block diagram representation of generation of SAR radiation pattern for nth target.

in the SAR radiation pattern of the nth target, we obtain

$$\begin{aligned}
&h_n(\omega, x_n, y_n - u) \\
&\quad = a_n(\omega, x_n, y_n - u) h(\omega, x_n, y_n - u) \\
&\quad = \underbrace{a_n(\omega, x_n, y_n - u) a(\omega, x_n, y_n - u)}_{\text{Amplitude pattern}} \underbrace{\exp\left[-j2k\sqrt{x_n^2 + (y_n - u)^2}\right]}_{\text{Spherical PM signal}}.
\end{aligned}$$

We showed earlier that the radar amplitude pattern $a(\omega, x_n, y_n - u)$ is related to the physical size and to the type of the radar used. The magnitude of the radar amplitude pattern $|a(\omega, x_n, y_n - u)|$ dictates its *power* at the coordinates of the nth target at the fast-time frequency ω. The spherical PM signal $\exp\left[-j2k\sqrt{x_n^2 + (y_n - u)^2}\right]$ is the round trip phase delay of the echo from the nth target at the fast-time frequency ω.

Based on our discussion of AM-PM signals, we have the following slow-time Fourier transform for the transmit-receive mode radar-target radiation pattern:

$$\mathcal{F}_{(u)}[h_n(\omega, x_n, y_n - u)] = A(\omega, k_u) A_n(\omega, k_u) \exp\left(-j\sqrt{4k^2 - k_u^2}\, x_n - jk_u y_n\right),$$

where the target and radar amplitude patterns in the u and k_u domains are related via the following:

$$\begin{aligned} a_n(\omega, x_n, y_n - u) &= A_n\left[2k \sin\theta_n(u), \omega\right] \\ a(\omega, x_n, y_n - u) &= A\left[2k \sin\theta_n(u), \omega\right], \end{aligned}$$

and

$$\theta_n(u) = \arctan\left(\frac{y_n - u}{x_n}\right)$$

is the nth target *aspect angle* when the radar is located at $(0, u)$.

- Reminder: The amplitude functions $a(\cdot)$ [or $a_n(\cdot)$] and $A(\cdot)$ [or $A_n(\cdot)$] are *scaled* transformations of each other, and not Fourier transform pairs.

In summary, we recognize the following three components in the Fourier transform of the transmit-receive mode radar-target radiation pattern:

1. Active radar amplitude pattern: $A(\omega, k_u)$
2. Target amplitude pattern: $A_n(\omega, k_u)$
3. Target coordinates (phase history): $\exp\left(-j\sqrt{4k^2 - k_u^2}\, x_n - jk_u y_n\right)$.

In Section 2.9 we showed the equivalence of the aspect angle $\theta_n(u)$ domain and the slow-time angular Doppler domain, that is,

$$\phi = \arcsin\left(\frac{k_u}{2k}\right),$$

in the mapping of the amplitude pattern of a target. We now examine this mapping in wide-bandwidth SAR systems via a simulated example shown in Figure 3.10.

The parameters of this simulated SAR problem are radar bandwidth [100, 500] MHz; synthetic aperture $u \in [-L, L] = [-100, 100]$ m; target coordinates

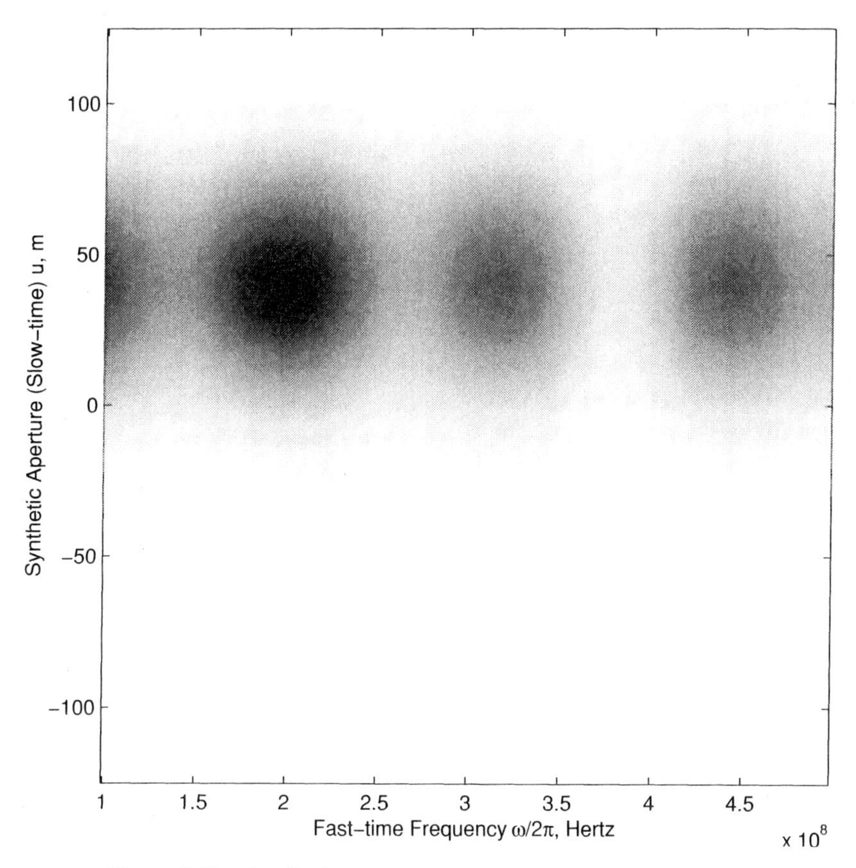

Figure 3.10a. Amplitude pattern in domains of SAR signal: (ω, u) domain.

$(x_n, y_n) = (50, 80)$ m. Figure 3.10*a* is the simulated AM signal

$$a(\omega, x_n, y_n - u)a_n(\omega, x_n, y_n - u),$$

in the fast-time frequency and slow-time (ω, u) domain. This AM signal is chosen to be a Gaussian signal with a standard deviation of 25 m which is centered at $u = 40$ m (peak or *flash* point); the AM signal also possesses some fluctuations in the fast-time frequency ω domain. Note that the peak (flash) point of the AM signal in the synthetic aperture u domain, that is, $u = 40$ m is not the closest point of approach (CPA) for the target which is at $u = y_n = 80$ m. Figure 3.10*b* is the same AM signal in the fast-time frequency and aspect angle $[\omega, \theta_n(u)]$ domain.

Figure 3.10*c* is the slow-time Doppler AM signal

$$A(\omega, k_u)A_n(\omega, k_u),$$

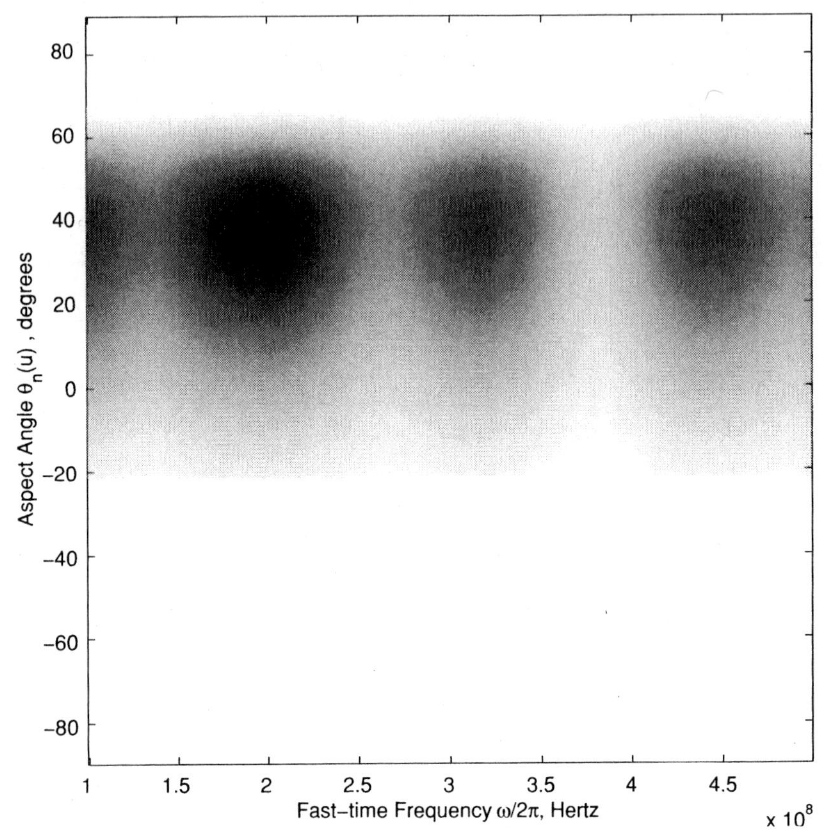

Figure 3.10b. Amplitude pattern in domains of SAR signal: $\left[\omega, \theta_n(u)\right]$ domain.

as a function of the fast-time frequency and slow-time angular Doppler (ω, ϕ). Note that Figure 3.10*b* and *c* are the same. Figure 3.10*d* shows the slow-time Doppler AM signal $A(\omega, k_u)A_n(\omega, k_u)$ in the fast-time frequency and slow-time Doppler (ω, k_u) domain. Finally, Figure 3.10*e* is the transformation of this slow-time Doppler AM signal in the two-dimensional range spatial frequency and cross-range spatial frequency domain which we defined in Section 2.9 via

$$k_x = \sqrt{4k^2 - k_u^2} = 2k \cos \phi,$$

$$k_y = k_u = 2k \sin \phi.$$

As we mentioned in Section 2.9, the equivalence of a target AM signature in the aspect angle $\theta_n(u)$ domain and the slow-time angular Doppler ϕ domain is crucial to understanding how the SAR image of the target is formed, and how it could be exploited in SAR target detection and recognition problems [s95; you].

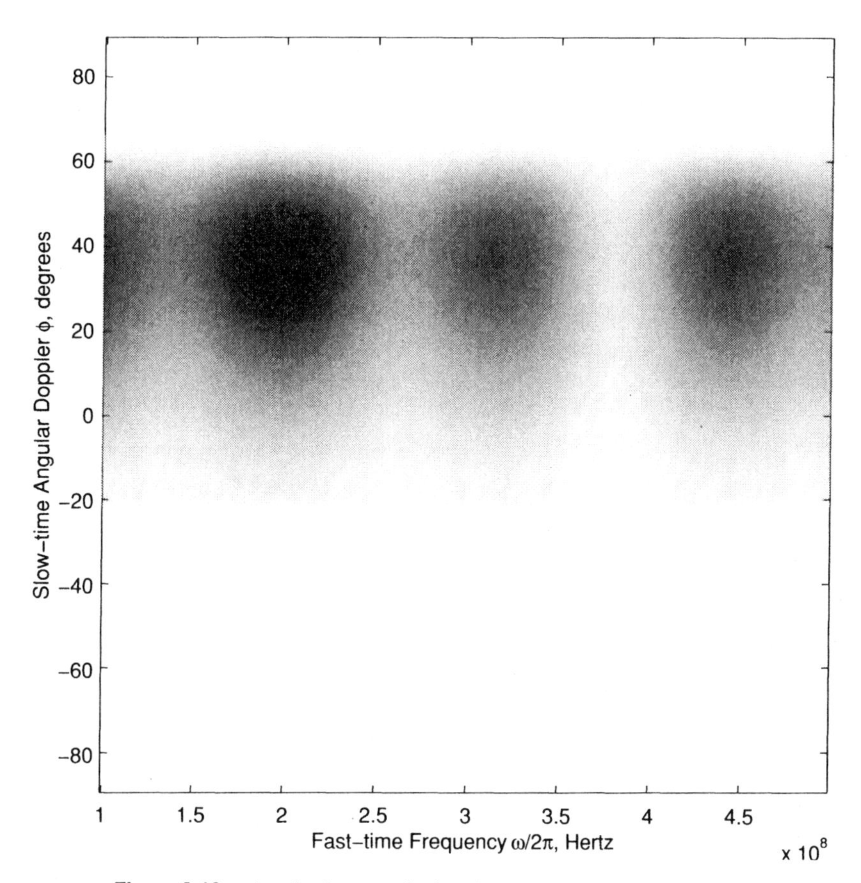

Figure 3.10c. Amplitude pattern in domains of SAR signal: (ω, ϕ) domain.

3.5 POLARIZATION

The analysis in this book is based on *scalar* wave theory. However, when a target is radiated with an electromagnetic or acoustic wave in one *mode*, the resultant scattering may contain other modes as well as the mode of the transmitted radiation. For instance, in acoustic scattering, a longitudinal (compressional) transmitted radiation can produce shear and surface waves as well as longitudinal scattered waves.

In electromagnetic wave theory, one deals with two modes of radiation: horizontally (H) polarized waves, and vertically (V) polarized waves. An H wave impinging on a target could produce V waves as well as H waves; a V wave interacting with a target may also yield H waves as well as V waves.

The scalar wave theory that we formulated in the previous sections may be generalized to incorporate these mode changes of electromagnetic waves, and quantify the resultant scattering. For instance, suppose that a radar transmitter, which is located at the spatial coordinates (0, 0), is radiating a target area with an H wave; the radar

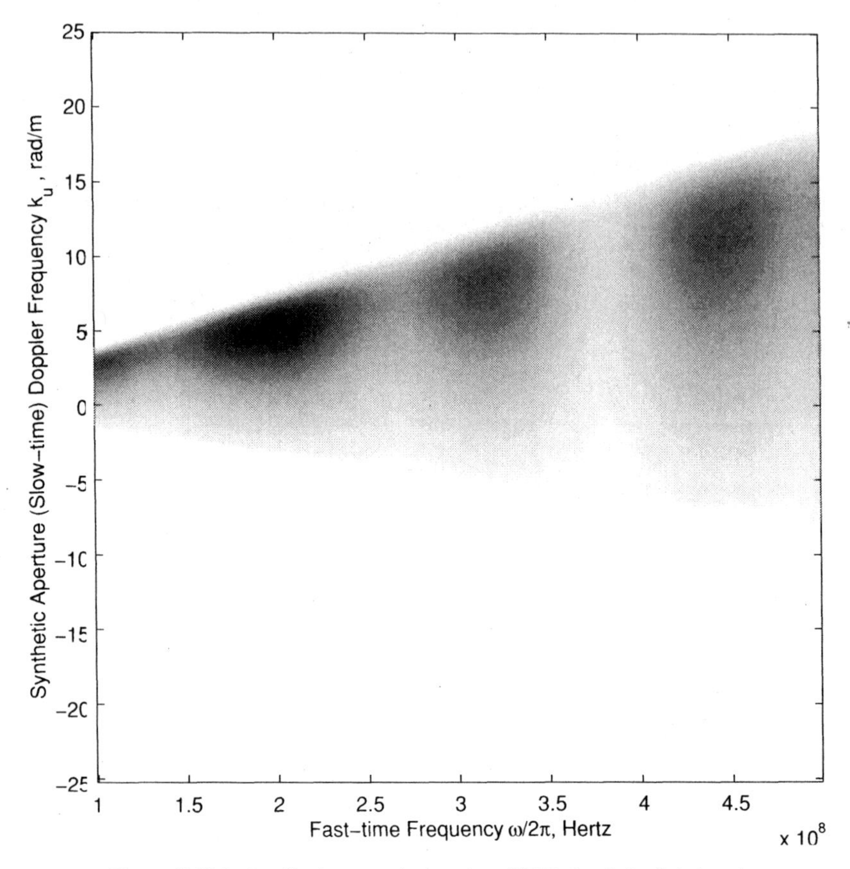

Figure 3.10d. Amplitude pattern in domains of SAR signal: (ω, k_u) domain.

amplitude pattern in H mode, label it

$$a_T^H(\omega, x, y),$$

is a measure that can be used to quantify the magnitude (power) and relative phase distribution of the resultant H wave in the spatial (x, y) domain at the fast-time frequency ω. If the radar transmits a V wave, we may use the notation $a_T^V(\omega, x, y)$ to identify its V wave amplitude pattern.

The nth target which experiences such an H mode or V mode radiation produces H mode and V mode waves the magnitudes and phases of which are related to the physical properties of the target. These can be identify by these four amplitude patterns:

1. H mode scattering for H mode radiation: $a_n^{HH}(\omega, x, y)$
2. V mode scattering for H mode radiation: $a_n^{HV}(\omega, x, y)$

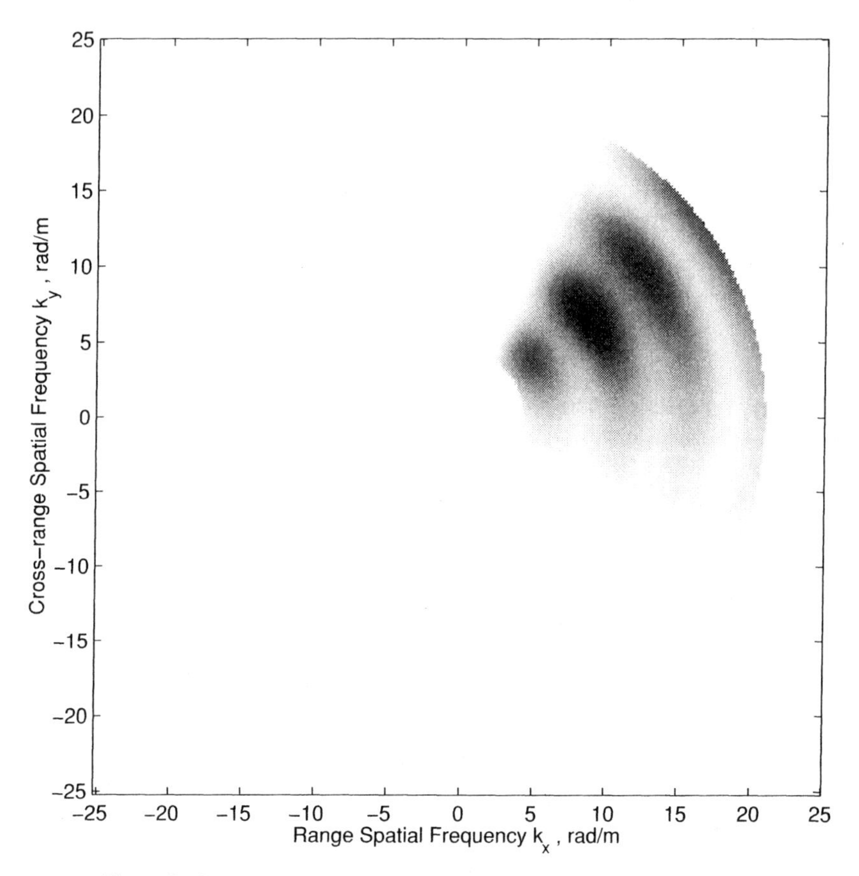

Figure 3.10e. Amplitude pattern in domains of SAR signal: (k_x, k_y) domain.

3. H mode scattering for V mode radiation: $a_n^{VH}(\omega, x, y)$
4. V mode scattering for V mode radiation: $a_n^{VV}(\omega, x, y)$

When the radar is switched to its receive mode, it also exhibits two different amplitude patterns for recording H mode and V mode waves; these, respectively, can be denoted with $a_R^H(\omega, x, y)$ and $a_R^V(\omega, x, y)$.

We can now identify the transmit-receive radar-target radiation pattern via four different amplitude pattern combinations. Suppose that the transmitting/receiving radar is located at $(0, u)$ and that the coordinates of the nth target are (x_n, y_n). In this case the four combined transmit-receive radar-target amplitude patterns are as follows:

1. H mode transmission and H mode reception:

$$a_T^H(\omega, x_n, y_n - u)a_n^{HH}(\omega, x_n, y_n - u)a_R^H(\omega, x_n, y_n - u)$$

2. H mode transmission and V mode reception:

$$a_T^H(\omega, x_n, y_n - u)a_n^{HV}(\omega, x_n, y_n - u)a_R^V(\omega, x_n, y_n - u)$$

3. V mode transmission and H mode reception:

$$a_T^V(\omega, x_n, y_n - u)a_n^{VH}(\omega, x_n, y_n - u)a_R^H(\omega, x_n, y_n - u)$$

4. V mode transmission and V mode reception:

$$a_T^V(\omega, x_n, y_n - u)a_n^{VV}(\omega, x_n, y_n - u)a_R^V(\omega, x_n, y_n - u)$$

The above four modes of the radar, which are referred to as HH, HV, VH, and VV, provide the user with a *vector* information base to characterize the target signature in a SAR system. Such a four-dimensional information base provides a rich source for identifying targets in SAR reconnaissance problems [jam]. (For most targets, the HV and VH modes are observed to yield similar information.)

As we stated earlier, the analysis in this book uses scalar wave theory. While one could use a vector/matrix notation to provide the general analysis for all four modes of a radar, we restrict our formulation to the scalar wave theory to simplify the notation; our results on the scalar wave theory can be easily generalized for the vector wave theory.

3.6 MATLAB ALGORITHMS

The code for examining a radar radiation pattern is identified as *radar radiation pattern*. Four cases are considered: a planar radar antenna at 1 GHz (a) and 2 GHz (b); and a parabolic (without a feed) radar antenna at 1 GHz (c) and 2 GHz (d). The radiation pattern is simulated within $y \in [-4B, 4B]$, where $\pm B$ is the theoretical beamwidth of the radar. The sample spacing in the cross-range y domain is chosen such that the slow-time Doppler band of the DFT of the radiation pattern is within $k_y \in [-4k_{u\max}, 4k_{u\max}]$, where $\pm k_{u\max}$ is the theoretical Doppler support band of $A_T(\omega, k_u)$.

In addition to the radar radiation pattern, its phase deviation from the radiation pattern of an ideal source (i.e., a radar with a zero diameter $D = 0$ [s94])

$$h_{T0}(\omega, x, y) = \exp\left(-jk\sqrt{x^2 + y^2}\right),$$

that is, the phase of

$$h_T(\omega, x, y)h_{T0}^*(\omega, x, y),$$

where h_{T0}^* is the complex conjugate of h_{T0}, is also examined. This phase deviation is also shown in the slow-time Doppler domain. For this, the following slow-time

Fourier transform for the ideal source is used:

$$\exp\left(-j\sqrt{k^2 - k_u^2}\,x\right).$$

The receive-mode radiation pattern is assumed to be equal to the transmit mode radiation pattern.

The following outputs are for a planar radar antenna at 1 GHz (a) and 2 GHz (b), and a parabolic (without a feed) radar antenna at 1 GHz (c) and 2 GHz (d):

P3.1 Real part and magnitude of the radar transmit-mode radiation pattern $h_t(\omega, x, y)$ versus the cross-range y

P3.2 Unwrapped phase deviation of $h_T(\omega, x, y)$ from $h_{T0}(\omega, x, y)$ versus the cross-range y

P3.3 Real part and magnitude of the radar transmit mode radiation pattern spectrum versus the slow-time Doppler k_u

P3.4 Unwrapped phase deviation of the radar transmit mode radiation pattern spectrum from that of the ideal source versus the slow-time Doppler k_u

P3.5 Real part and magnitude of the radar transmit-receive mode radiation pattern $h(\omega, x, y) = h_T^2(\omega, x, y)$ versus the cross-range y

P3.6 Unwrapped phase deviation of $h(\omega, x, y)$ from $h_0(\omega, x, y) = h_{T0}^2(\omega, x, y)$ versus the cross-range y

P3.7 Real part and magnitude of the radar transmit-receive mode radiation pattern spectrum versus the slow-time Doppler k_u

P3.8 Unwrapped phase deviation of the radar transmit-receive mode radiation pattern spectrum from that of the ideal source versus the slow-time Doppler k_u

4

GENERIC SYNTHETIC APERTURE RADAR

Introduction

Synthetic aperture radar (SAR) is an imaging modality that addresses the general problem of forming a target region reflectivity function in the multidimensional spatial domain of range, cross-range, and altitude (x, y, z). Our present discussion is simplified by combining the range and altitude variables to form the slant-range domain, that is, $x_s = \sqrt{x^2 + z^2}$; see the discussion on the radar radiation pattern in the three-dimensional spatial domain in Section 3.2. Thus the imaging problem is formulated in the two-dimensional slant-range and cross-range (x_s, y) domains. For convenience we call the imaging plane range and cross-range, and identify it as (x, y).

SAR image formation is based on a two-dimensional signal theory which heavily benefits from the range imaging and cross-range imaging principles of Chapters 1 and 2. The major difference is that we do not assume that the targets in the imaging scene are at a fixed range or cross-range. The target region is assumed to be composed of *stationary* targets located at unknown coordinates (x_n, y_n), $n = 1, 2, \ldots$.

- While we use a discrete model to represent the target area, our results are also applicable in the case of a continuous target model. The use of a discrete target model is for notational convenience.

The transmitting/receiving radar is mounted on an aircraft and takes on the coordinates $(0, u)$, which we call the synthetic aperture or slow-time domain. For the reflectivity of the nth target, we consider an amplitude pattern $a_n(\omega, x_n, y_n - u)$, which varies with the fast-time frequency, and the target relative range and aspect angle with respect to the radar. This corresponds to a *range extended* and *cross-range extended* target function.

Before dealing with the SAR models which incorporate the radar and target radiation patterns, we provide a brief discussion on a simpler scenario which we refer to as the *Generic SAR* model. In the generic SAR model we use a target signature and radar radiation pattern that do not vary with the aspect angle and frequency. As a result, *without the presence of range extension and cross-range extension effects*, the reader should develop a better feel for the problem and the signals that are involved. We do bring out the role of a target amplitude pattern later in the chapter and an interpretation of its effects; this will assist us in addressing the motion compensation problem in SAR systems.

Outline

The main purpose of this chapter is to expose the reader to the basic and traditional issues of SAR systems. These are SAR system modeling with simple (omnidirectional) radars and targets; SAR reconstruction, and its digital implementation; and motion compensation. The signal processing and imaging problems of spotlight and stripmap systems SAR in Chapters 5 and 6 will be formulated and studied using these basic principles.

We begin by developing a system model that relates the measurement in a generic SAR problem to the unknown target function; this is done in Section 4.1. The measurement domain is the combination of what we refer to as the time (or fast-time) domain in range imaging, and the synthetic aperture (or slow-time) domain in cross-range imaging. The discussion in this section provides a feel for the manner in which the target information in the two-dimensional spatial (x, y) domain is transformed into the two-dimensional slow-time and fast-time (t, u) measurement domain.

In the next two sections (Sections 4.2 and 4.3) the two-dimensional Fourier decomposition of the measurement is constructed. Once again, the analysis is very much dependent on the same Fourier principles and properties that were developed in Chapters 1 and 2 and used for range and cross-range imaging. This two-dimensional Fourier analysis provides a framework for the reconstruction of the two-dimensional target function in the spatial frequency domain; this, which we refer to as SAR *wavefront* reconstruction, is shown in Section 4.4. The next three sections (Sections 4.5–4.7) discuss various methods for the digital (practical) implementation of the reconstruction, and their merits from the point of view of accuracy and speed of computation.

In Section 4.8 we examine the effects of a frequency and synthetic aperture target reflectivity in the SAR signal and reconstructed image. We raise this issue here because we will end up using its multidimensional signal theory in Sections 4.9 and 4.10 for motion compensation in SAR systems.

There exist other SAR imaging methods, for example, polar format processing [wal], range-Doppler imaging [cut], and chirp scaling [ran94] based on approximations that put restrictions on the size of the target area, the radar frequency, and/or the extent of the synthetic aperture. In the next two sections we briefly outline the former two methods, since these two are currently used in many operational SAR systems. However, we do not provide an extensive analysis of these methods, their signal processing issues, and when and why these algorithms fail; these topics are already discussed in the literature, for example, [carr; cur; s92j; s94]. That would divert the reader's attention from our main goal, which is to demonstrate that SAR imaging can be achieved via simple codes (see Matlab algorithms for Chapters 5 and 6) without the need to make approximations in the SAR signal model.

In Section 4.11 plane wave approximation-based polar format processing, and its various versions that use narrow-bandwidth and narrow-beamwidth assumptions, are examined. A method for improving (removing the model-based approximation errors of) the polar format processing is discussed. Conventional inverse SAR (ISAR) modeling and imaging via polar format processing is outlined in Section 4.12. Fresnel approximation-based range-Doppler imaging is examined in Section 4.13.

In Section 4.14 we present the wavefront reconstruction for three-dimensional SAR imaging from azimuthal and elevation (planar) synthetic aperture measurements. In Section 4.15 the chapter closes with a discussion on using pulse diversity for Electronic Countercountermeasure (ECCM) in SAR systems.

Mathematical Notations and Symbols

We will be dealing with multidimensional signals and their Fourier transforms in this chapter. In particular, we identify a target function in the two-dimensional spatial domain via $f(x, y)$. The one-dimensional *marginal* Fourier transform of this signal with respect to the variable x is denoted with [s94, ch. 2]

$$\begin{aligned} F_x(k_x, y) &= \mathcal{F}_{(x)}\left[f(x, y)\right] \\ &= \int_x f(x, y) \exp\big(-jk_x x\big)\, dx. \end{aligned}$$

Also the one-dimensional marginal Fourier transform of $f(x, y)$ with respect to the variable y is defined by

$$\begin{aligned} F_y(x, k_y) &= \mathcal{F}_{(y)}[f(x, y)] \\ &= \int_y f(x, y) \exp\big(-jk_y y\big) dy. \end{aligned}$$

The two-dimensional Fourier transform of the target function $f(x, y)$ with respect to (x, y) is the following two-dimensional signal in the spatial frequency (k_x, k_y) domain:

$$\begin{aligned} F(k_x, k_y) &= \mathcal{F}_{(x,y)}[f(x, y)] \\ &= \int_y \int_x f(x, y) \exp\big(-jk_x x - jk_y y\big)\, dx\, dy. \end{aligned}$$

The two-dimensional Fourier transform of $f(x, y)$ is related to the two marginal Fourier transforms of $f(x, y)$ via

$$F(k_x, k_y) = \mathcal{F}_{(y)}[F_x(k_x, y)] = \int_y F_x(k_x, y) \exp(-jk_y y)\, dy$$

and

$$F(k_x, k_y) = \mathcal{F}_{(x)}[F_y(x, k_y)] = \int_x F_y(x, k_y) \exp(-jk_x x)\, dx.$$

These Fourier relationships for the target function are shown in Figure 4.1*a*.

We will also process a two-dimensional SAR signal, which we identify with the boldface $\mathbf{s}(t, u)$, and its one-dimensional marginal Fourier transform with respect to the variable t as well as its two-dimensional Fourier transform. (We do not need to define the marginal Fourier transform of $\mathbf{s}(t, u)$ with respect to u, though this signal has been used in certain SAR signal processing [s94, ch. 5].) However, we also need to define the SAR signal with subscripts, for example, $\mathbf{s}_0(t, u)$ (reference SAR signal), and $\mathbf{s}_M(t, u)$ (fast-time matched-filtered SAR signal). Thus, to avoid lengthy subscripts, we use the following to identify the marginal Fourier transform of $\mathbf{s}(t, u)$ with respect to the variable t:

$$s(\omega, u) = \mathcal{F}_{(t)}[\mathbf{s}(t, u)] = \int_x \mathbf{s}(t, u) \exp(-j\omega t)\, dt.$$

The two-dimensional Fourier transform of the SAR signal $\mathbf{s}(t, u)$ with respect to (t, u) is denoted with

$$S(\omega, k_u) = \mathcal{F}_{(t,u)}[\mathbf{s}(t, u)] = \int_u \int_t \mathbf{s}(t, u) \exp(-j\omega t - jk_u u)\, dt\, du.$$

The two-dimensional Fourier transform of $\mathbf{s}(t, u)$ is related to the marginal Fourier transform of $\mathbf{s}(t, u)$ with respect to the variable t via

$$S(\omega, k_u) = \mathcal{F}_{(u)}[s(\omega, u)] = \int_u s(\omega, u) \exp(-jk_u u)\, du.$$

These Fourier relationships for the SAR signal are shown in Figure 4.1*b*. Similar notations are used for the Fourier transforms of the three-dimensional target function $f(x, y, z)$, and the three-dimensional SAR signal $\mathbf{s}(t, u, v)$.

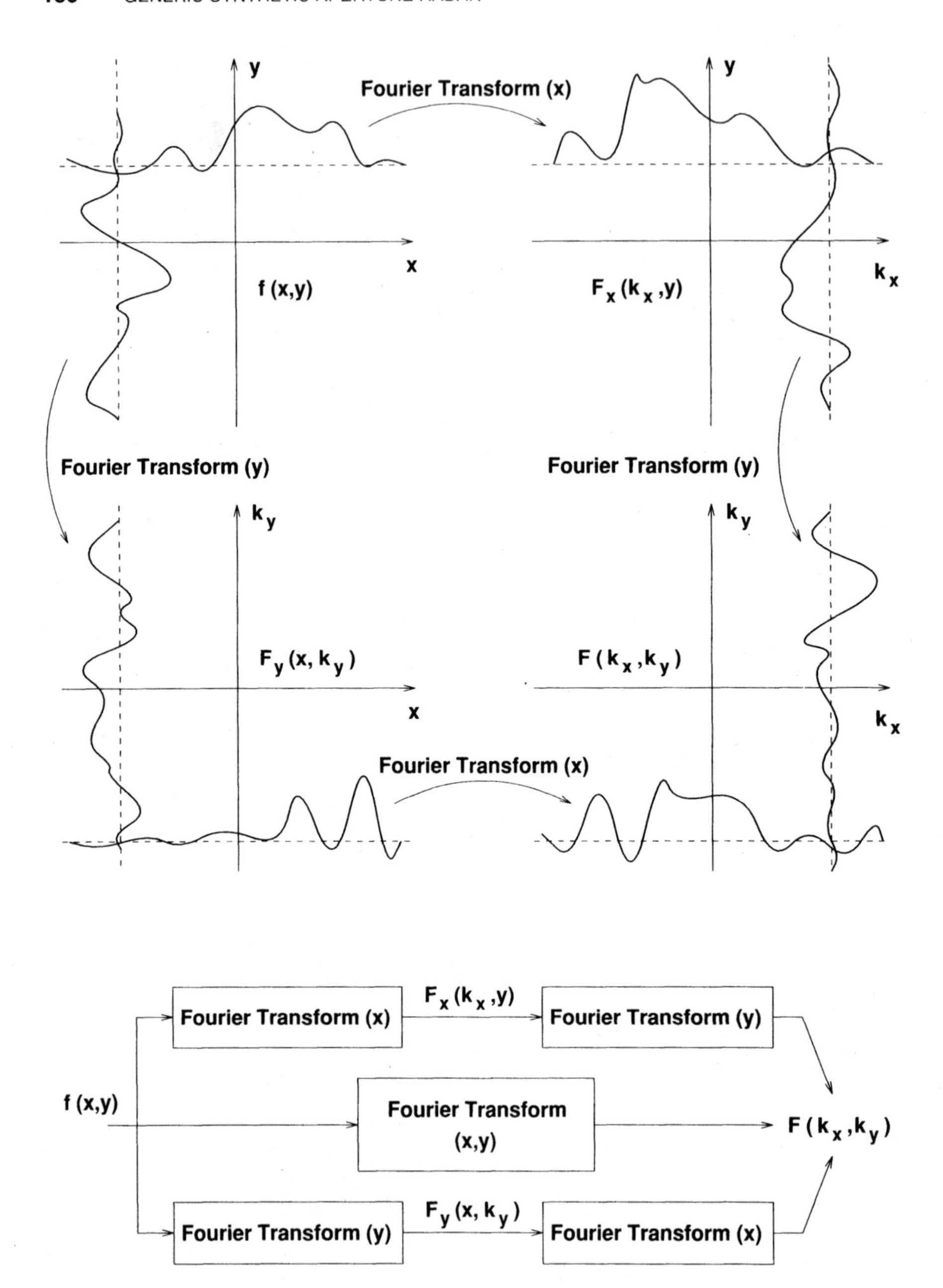

Figure 4.1a. Marginal Fourier transforms of two-dimensional signals: target function $f(x, y)$.

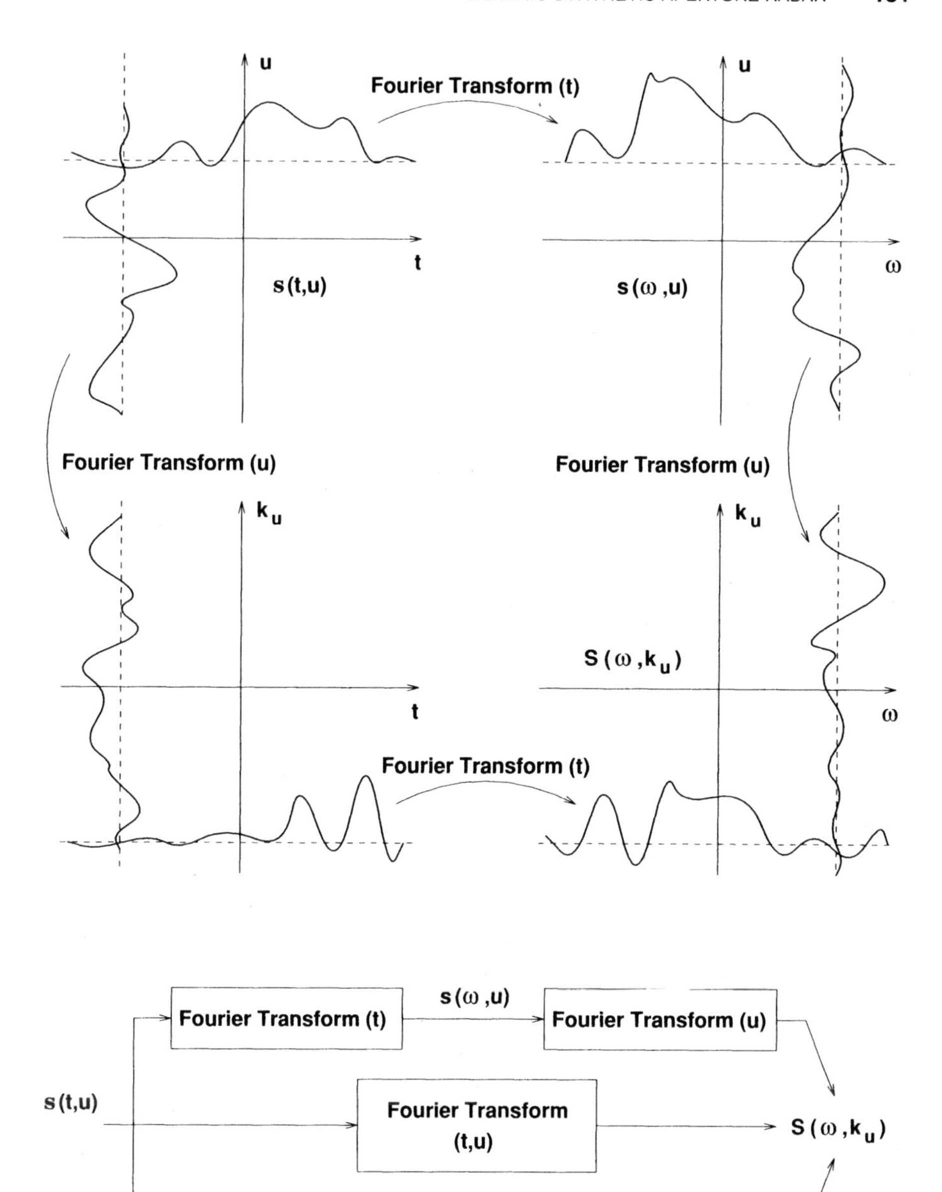

Figure 4.1b. Marginal Fourier transforms of two-dimensional signals: SAR signal $s(t, u)$.

The following is a list of mathematical symbols used in this chapter, and their definitions:

Symbol	Definition
$a_{en}(\omega, x, y)$	Motion phase error function of nth target
$a_n(\omega, x, y)$	Amplitude pattern for n-target (general case)
$A_n(\omega, k_u)$	Slow-time Doppler amplitude pattern for n-target (general case)
a_t	Acceleration of moving target in ISAR
c	Wave propagation speed $c = 3 \times 10^8$ m/s for radar waves $c = 1500$ m/s for acoustic waves in water $c = 340$ m/s for acoustic waves in air
$f(x, y)$	Target function in spatial domain
$F(k_x, k_y)$	Two-dimensional Fourier transform of target function
$f(x, y, z)$	Target function in three-dimensional spatial domain
$F(k_x, k_y, k_z)$	Three-dimensional Fourier transform of target function
$f_b(x, y)$	Baseband target function in spatial domain; origin in spatial domain is at center of baseband target function
$F_b(k_x, k_y)$	Two-dimensional Fourier transform of baseband target function
$F_b(k_x, k_y, k_z)$	Three-dimensional Fourier transform of baseband target function
$f_0(x, y)$	Ideal target function in spatial domain
$F_0(k_x, k_y)$	Two-dimensional Fourier transform of ideal target function
$f_0(x, y, z)$	Ideal target function in three-dimensional spatial domain
$F_0(k_x, k_y, k_z)$	Three-dimensional Fourier transform of ideal target function
$F_x(k_x, y)$	One-dimensional Fourier transform of target function with respect to range x
$F_y(x, k_y)$	One-dimensional Fourier transform of target function with respect to cross-range y
$F_{0x}(k_x, y)$	One-dimensional Fourier transform of ideal target function with respect to range x
$g_m(\omega)$	Nonlinear transformation from ω to k_x for a discrete k_{um}
$h(k_x)$	Sinc interpolating kernel
$h_w(k_x)$	Windowed sinc interpolating kernel
$H_{en}(k_x, k_y)$	Mapping of nth target motion phase error function in spatial frequency domain
$H^*_{exy}(k_x, k_y)$	Spatially varying motion compensation filter
$H_n(k_x, k_y)$	Mapping of nth target amplitude pattern in spatial frequency domain
j_t	Jerk of moving target in ISAR
$J(\omega, k_u)$	Jacobian of transformation from (k_x, k_y) to (ω, k_u)
$J_m(\omega)$	Jacobian (derivative) of transformation $g_m(\omega)$

k	Wavenumber: $k = \omega/c$
k_c	Wavenumber at carrier frequency: $k_c = \omega_c/c$
k_n	Wavenumber sampled points for SAR signal
k_u	Spatial frequency or wavenumber domain for azimuthal synthetic aperture u; slow-time Doppler (frequency) domain
k_{um}	Slow-time Doppler sampled points for SAR signal
k_{un}	Slow-time Doppler location of nth target in polar format processing
k_v	Spatial frequency or wavenumber domain for elevation synthetic aperture v
k_x	Spatial frequency or wavenumber domain for range x
k_{xc}	Center (carrier) range spatial frequency of reconstructed SAR image
k_{xmn}	Available range spatial frequency sampled points
k_y	Spatial frequency or wavenumber domain for cross-range y
k_z	Spatial frequency or wavenumber domain for elevation z
k_{yc}	Center (carrier) cross-range spatial frequency of reconstructed SAR image
k_{ymn}	Available cross-range spatial frequency sampled points
k_0	Baseband wavenumber: $k_0 = \omega_0/c$
M	Number of slow-time samples
n	Index representing a specific target [used with (σ_n, x_n, y_n)]
N	Number of fast-time frequency samples
N_s	Half-number of sinc side lobes which are used for interpolation
N_x	Number of range samples in target reconstruction
N_y	Number of cross-range samples in target reconstruction
$p(t)$	Transmitted radar signal
$p(t, u)$	Transmitted radar signal with pulse diversity
$P(\omega)$	Fourier transform of transmitted radar signal
$P(\omega, u)$	Fast-time Fourier transform of transmitted radar signal with pulse diversity
$\text{psf}_n(x, y)$	Point spread function in spatial domain for nth target
$\text{psf}_t(t)$	Point spread function in fast-time domain for matched filtering
PRF	Radar pulse repetition frequency in slow-time
PRI	Radar pulse repetition interval in slow-time
$r_{en}(u)$	Radial motion error of nth target
$r_{exy}(u)$	Radial motion error of the target at (x, y)
$r_{exyz}(u)$	Radial motion error of the target at (x, y, z)
R_c	Radial distance of center of target from radar at $u = 0$
R_i	Radial distance of target at (x, y) from radar at $(0, u_i)$

$R_0(\tau)$	Radial distance of center of target from radar
$\mathbf{s}(t, u)$	Echoed SAR signal from target area
$\mathbf{s}(t, u, v)$	Echoed SAR signal from target area (3D SAR)
$s(\omega, u)$	One-dimensional Fourier transform of SAR signal with respect to fast-time
$s(\omega, u, v)$	One-dimensional Fourier transform of SAR signal with respect to fast-time in 3D SAR system
$S(\omega, k_u)$	Two-dimensional Fourier transform of SAR signal with respect to fast-time and slow-time
$S(\omega, k_u, k_v)$	Three-dimensional Fourier transform of SAR signal with respect to fast-time and two slow-times
$s_c(\omega, u)$	Slow-time compressed SAR signal
$\mathbf{s}_M(t, u)$	Fast-time matched-filtered echoed signal
$s_M(\omega, u)$	One-dimensional Fourier transform of matched-filtered SAR signal with respect to fast-time
$s_n(\omega, u)$	One-dimensional Fourier transform of nth target SAR signal with respect to fast-time
$S_n(\omega, k_u)$	Two-dimensional Fourier transform of nth target SAR signal with respect to fast-time and slow-time
$\mathbf{s}_0(t, u)$	Reference SAR signal
$s_0(\omega, u)$	One-dimensional Fourier transform of reference SAR signal with respect to fast-time
$S_0(\omega, k_u)$	Two-dimensional Fourier transform of reference SAR signal with respect to fast-time and slow-time
$\mathbf{s}_{0i}(t, u)$	Reference SAR signal at range bin x_i for range stacking reconstruction algorithm
$s_{0i}(\omega, u)$	One-dimensional Fourier transform of reference SAR signal at range bin x_i with respect to fast-time
$\text{sinc}(\cdot)$	sinc function: $\text{sinc}(a) = \sin(\pi a)/\pi a$
t	Fast-time domain
t_i	Round-trip delay of echoed signal for target at (x, y) when radar is at $(0, u_i)$
$t_{ij}(u)$	Round-trip delay of echoed signal for target at (x_i, y_j) when radar is at $(0, u)$
t_n	Fast-time location of nth target in polar format processing
u	Synthetic aperture or slow-time domain (azimuthal)
u_i	Synthetic aperture sampled points for synthetic aperture signal
v	Synthetic aperture or slow-time domain (elevation)
v_r	Speed of radar-carrying aircraft in SAR
v_t	Speed of moving target in ISAR
$w(k_x)$	Window for interpolating kernel

x	Range domain
$x_e(u)$	Range motion error
x_i	Range bins for range stacking reconstruction algorithm
x_m	Range of in-scene target 2
x_n	Range of nth target
x_ℓ	Range of in-scene target 1
x_{θ_c}	Range domain when spatial domain is rotated by squint angle θ_c
X_c	Center point of target area in range domain
X_0	Half-size of target area in range domain. Size of target area in range domain is $2X_0$.
y	Cross-range or azimuth domain
$y_e(u)$	Cross-range motion error
y_m	Cross-range of in-scene target 2
y_n	Cross-range of nth target
y_ℓ	Cross-range of in-scene target 1
y_{θ_c}	Cross-range domain when spatial domain is rotated by squint angle θ_c
Y_c	Center point of target area in cross-range domain; squint cross-range
z	Elevation or altitude domain
$z_e(u)$	Elevation motion error
z_n	Elevation of nth target
Z_c	Center point of target area in altitude (elevation) domain
$\delta(\cdot)$	Delta or impulse function
Δ_k	Sample spacing of SAR signal in wavenumber k domain
Δ_{k_u}	Sample spacing of SAR signal in slow-time Doppler domain
Δ_{k_x}	Sample spacing of reconstructed target function in range spatial frequency domain
Δ_{k_y}	Sample spacing of reconstructed target function in cross-range spatial frequency domain
Δ_u	Sample spacing of SAR signal in synthetic aperture domain
Δ_x	Sample spacing of reconstructed target function in range domain
Δ_y	Sample spacing of reconstructed target function in cross-range domain
Δ_ω	Sample spacing of SAR signal in fast-time frequency domain
Δ_τ	Sample spacing of SAR signal in slow-time (in seconds) domain
$\mathcal{F}$	Forward Fourier transform operator
$\mathcal{F}^{-1}$	Inverse Fourier transform operator
λ	Wavelength: $\lambda = 2\pi c/\omega$
ω	Temporal frequency domain for fast-time t

ω_c	Radar signal carrier or center frequency
ω_n	Fast-time frequency sampled points for SAR signal
ω_0	Radar signal half-bandwidth in radians; radar signal baseband bandwidth is $2\omega_0$
σ_n	Reflectivity of nth target
τ	Slow-time domain in seconds
θ_c	Average squint angle of target area; squint angle of center of target area
$\theta_n(u)$	Aspect angle of nth target when radar is at $(0, u)$
$\theta_0(u)$	Aspect angle of center of target area when radar is at $(0, u)$
$\theta_0(\tau)$	Aspect angle of center of target at slow-time τ
θ_z	Average depression (elevation/grazing) angle of target area

4.1 SYSTEM MODEL

The simplified generic SAR system model can be developed as follows. Consider a stationary target region composed of a set of point reflectors with reflectivity σ_n located at the coordinates (x_n, y_n) $(n = 1, 2, \ldots)$ in the spatial (x, y) domain; see Figure 4.2*a* (broadside target area) and Figure 4.2*b* (squint target area). The variable x is used for the range domain (also slant range or down range), and the variable y identifies the cross-range (also called azimuth or along the track). A radar located at $(0, u)$ in the spatial domain illuminates the target area with a multifrequency (large-bandwidth) signal $p(t)$. The radar radiation pattern is assumed to be *omni-directional*. The measured echoed signal is

$$\mathbf{s}(t, u) = \sum_n \sigma_n p\left[t - \frac{2\sqrt{x_n^2 + (y_n - u)^2}}{c}\right],$$

where

$$\frac{2\sqrt{x_n^2 + (y_n - u)^2}}{c}$$

is the round-trip delay from the radar to the nth target. An amplitude function

$$\frac{1}{\sqrt{x_n^2 + (y_n - u)^2}} \approx \frac{1}{\sqrt{x_n^2 + y_n^2}},$$

which is related to the wave divergence [s94], is absorbed in σ_n on the right side of the above model.

The aircraft coordinates $(0, u)$ are changed; that is, the position in the synthetic aperture u domain is varied to form a two-dimensional measured signal $\mathbf{s}(t, u)$. For the generic SAR model, we assume that the synthetic aperture measurements are

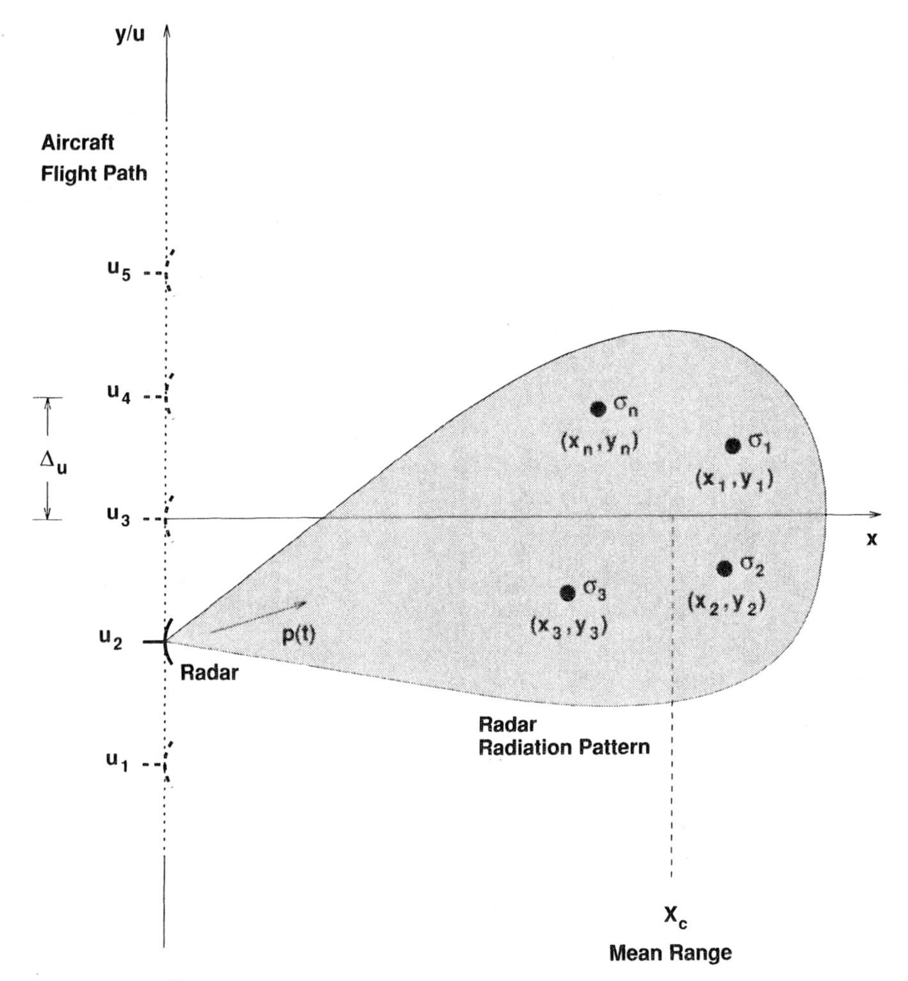

Figure 4.2a. SAR imaging system geometry: broadside target area.

made within an infinite synthetic aperture $u \in (-\infty, \infty)$. The point $u = y_n$ on the synthetic aperture is called the closest point of approach (CPA) of the radar for the nth target.

As we pointed out earlier, the time domain t is called the *fast-time* domain. This is due to the fact that it corresponds to the time delays associated with the propagation of the echoed signals that propagate at the relatively fast speed of light, that is, $c = 3 \times 10^8$ m/s. The synthetic aperture domain u is also referred to as the *slow-time* domain. This term originates from the fact that the synthetic aperture is generated via the motion of a vehicle (e.g., an aircraft) that carries the radar. The motion of the vehicle is much slower than the motion (propagation) of the echoed signals, since the speed of the vehicle is much slower than the speed of light.

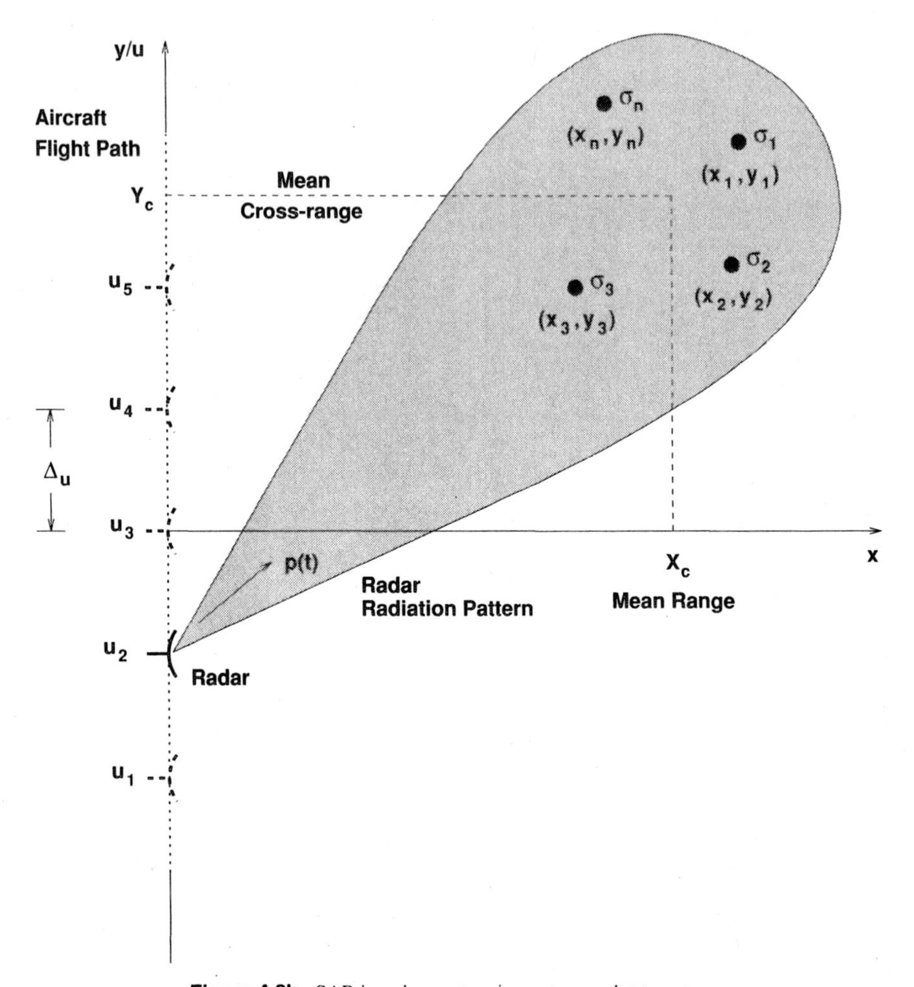

Figure 4.2b. SAR imaging system geometry: squint target area.

In the case of a synthetic aperture u which is generated by the motion of a radar-carrying aircraft with a constant speed v_r, the slow-time domain in the units of time (i.e., seconds) is defined via

$$\tau = \frac{u}{v_r};$$

that is, the slow-time τ domain is a *scaled* version of the synthetic aperture u domain. Much of the original SAR terminology for the Fourier processing of synthetic aperture data was in terms of the frequency domain of the slow-time variable τ, which is called the slow-time Doppler domain. We will use this terminology in our analysis too, though this analysis is based on more concrete sampled aperture (array) processing principles which were outlined in Chapter 2. In the discussion that follows, we refer to u as the synthetic aperture or the slow-time domain.

Figure 4.2*a* and *b* shows that the measurement of the SAR signal $\mathbf{s}(t, u)$ in the synthetic aperture u domain is made at evenly spaced points in that domain. We denote the sample spacing in the synthetic aperture domain with Δ_u. At each of the evenly spaced points in the synthetic aperture domain (i.e., u_i's), the radar radiates the target area with its signal (burst) $p(t)$. The number of radar bursts per second is called the radar pulse repetition frequency (PRF). In this case the slow-time interval between two consecutive bursts (transmissions) of the radar, which is called the pulse repetition interval (PRI), is

$$\text{PRI} = \frac{1}{\text{PRF}}.$$

Thus the sample spacing of the SAR signal in the slow-time τ domain is

$$\begin{aligned}\Delta_\tau &= \text{PRI}\\ &= \frac{1}{\text{PRF}}.\end{aligned}$$

The sample spacing of the SAR signal $\mathbf{s}(t, u)$ in the synthetic aperture domain can be related to the radar-carrying aircraft speed and the sample spacing in the slow-time domain (or the radar PRF) via

$$\begin{aligned}\Delta_u &= v_t \Delta_\tau\\ &= \frac{v_t}{\text{PRF}}.\end{aligned}$$

A synthetic aperture may also be generated by the motion of the target relative to a stationary radar. These systems are called inverse synthetic aperture radar (ISAR). If the target moves with a constant speed v_t, the slow-time domain is related to the synthetic aperture via

$$\tau = \frac{u}{v_t}.$$

(A discussion on ISAR systems and its geometry will be provided in Section 4.12 and Figure 4.16.) We will examine the implications of nonlinear motion in SAR and ISAR systems, and methods for compensating for them later in this chapter.

Figure 4.3*a* and *b* provide a graphical representation of how a target SAR signature is formed in the measurement (t, u) domain. Figure 4.3*a* represents the target spatial domain. A unit reflector is located at (x, y). Figure 4.3*a* also shows the flight path of the radar-carrying aircraft with a vertical dotted line. On this dotted line, which represents the synthetic aperture u domain, three positions of the radar (aircraft) are identified: u_1, u_2, and u_3. The values of u_1 and u_3 are chosen arbitrarily. However, the location u_2 on the synthetic aperture corresponds to the closest point of

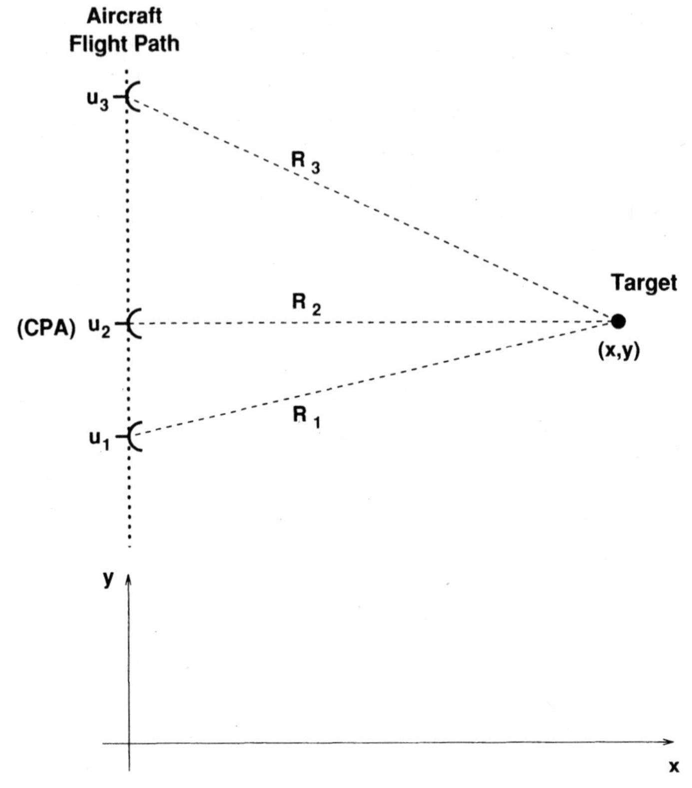

Figure 4.3a. Graphical representation of how a target SAR signature is formed in measurement (t, u) domain: radar and target in spatial domain.

approach (CPA) of the radar to the target at (x, y). Note the CPAs of the targets in an imaging scene are not all the same; the CPA of a target depends on its coordinates.

In Figure 4.3*a*, the distance of the target from the three radar coordinates are identified via

$$R_i = \sqrt{x^2 + (y - u_i)^2} \qquad \text{for } i = 1, 2, 3.$$

The round-trip fast-time delay for the wave propagation from the radar to the target and back to the radar is

$$\begin{aligned} t_i &= \frac{2R_i}{c} \\ &= \frac{2\sqrt{x^2 + (y - u_i)^2}}{c} \end{aligned}$$

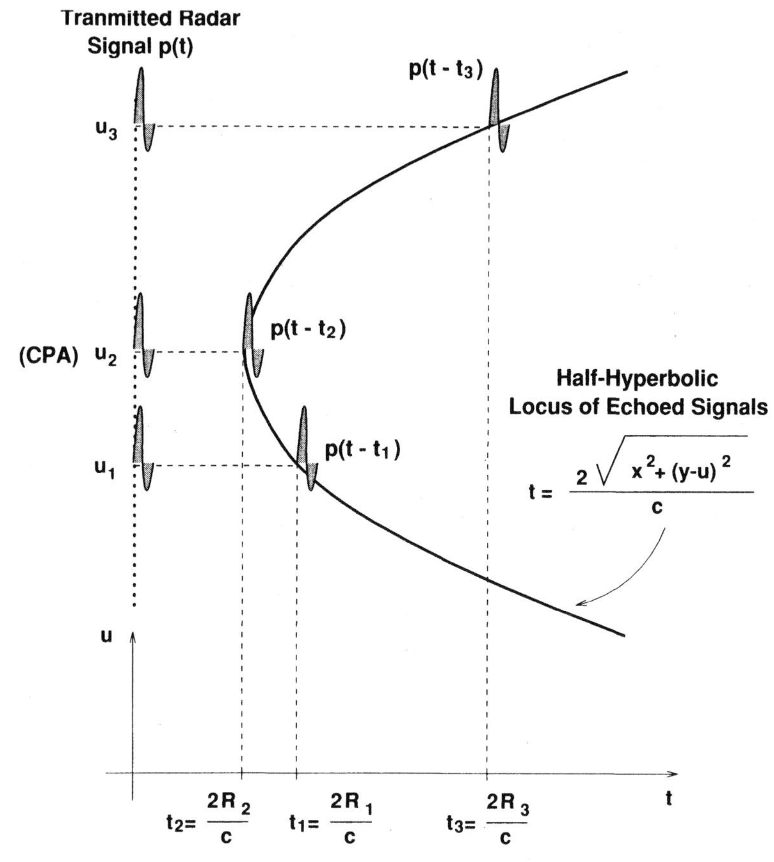

Figure 4.3b. Graphical representation of how a target SAR signature is formed in measurement (t, u) domain: half-hyperbolic target signature in SAR measurement domain.

for $i = 1, 2, 3$. For a general value of the synthetic aperture location of the radar u, the round-trip fast-time delay is

$$t = \frac{2\sqrt{x^2 + (y-u)^2}}{c}.$$

The above is the equation of the following half-hyperbola in the (t, u) domain:

$$\frac{c^2}{4}t^2 - (u-y)^2 = x^2 \qquad \text{for } t \geq 0.$$

This half-hyperbola is shown in Figure 4.3*b* which is a representation of the measurement (t, u) domain. The three shaded blips on the vertical dotted line ($t = 0$ line) at u_i's represent the transmitted radar signal $p(t)$ at the fast-time $t = 0$. The three shaded blips at t_i's are the echoed signals $p(t - t_i)$ of the target for the three synthetic aperture locations of the radar.

If $p(t)$ is not a *sharp* blip (similar to an impulse function) in the fast-time domain, and possesses a relatively long duration in the fast-time domain (e.g., a chirp signal), then it is difficult to *visualize* the (t, u) domain hyperbolic SAR signature of a target. The scene in fact contains many targets, and the (t, u) domain hyperbolic SAR signatures of the targets may overlap with each other. An example is shown in Figure 4.4*a*. This figure shows the (t, u) domain *spotlight* SAR signature of a set of simulated reflectors; the transmitted radar signal is a chirp. (The Matlab algorithm for the next chapter is used for this simulation.)

One way to improve the visualization of SAR (t, u) domain data is to fast-time matched filtered $\mathbf{s}(t, u)$ with the transmitted radar signal $p(t)$; that is,

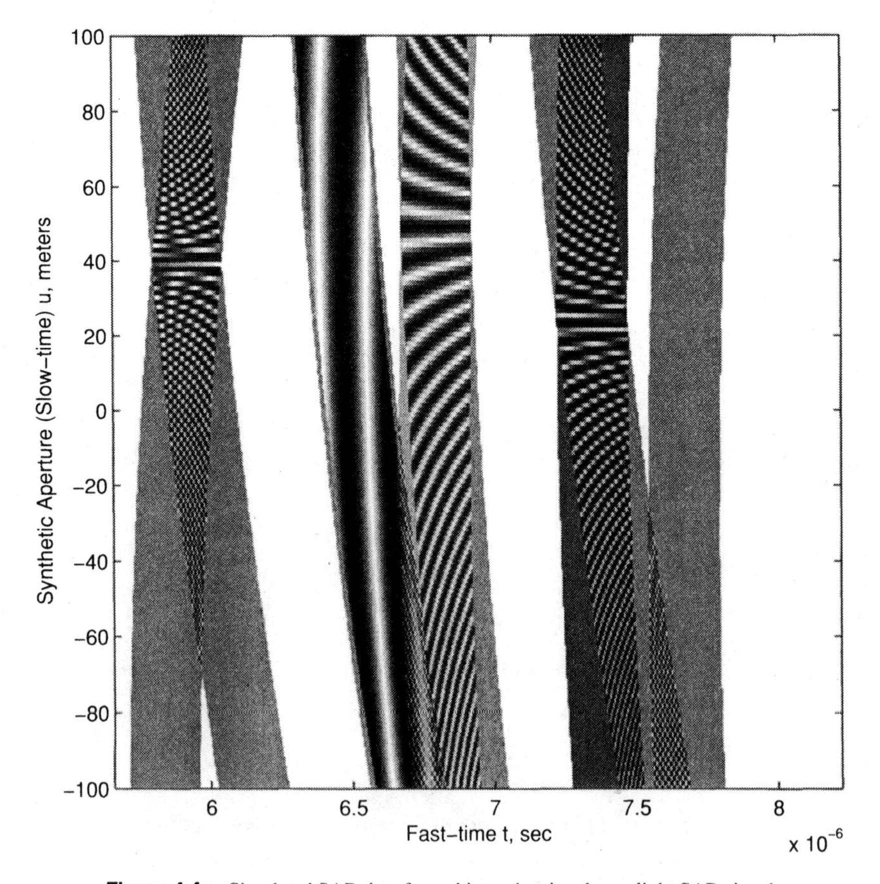

Figure 4.4a. Simulated SAR data for a chirp radar signal: spotlight SAR signal.

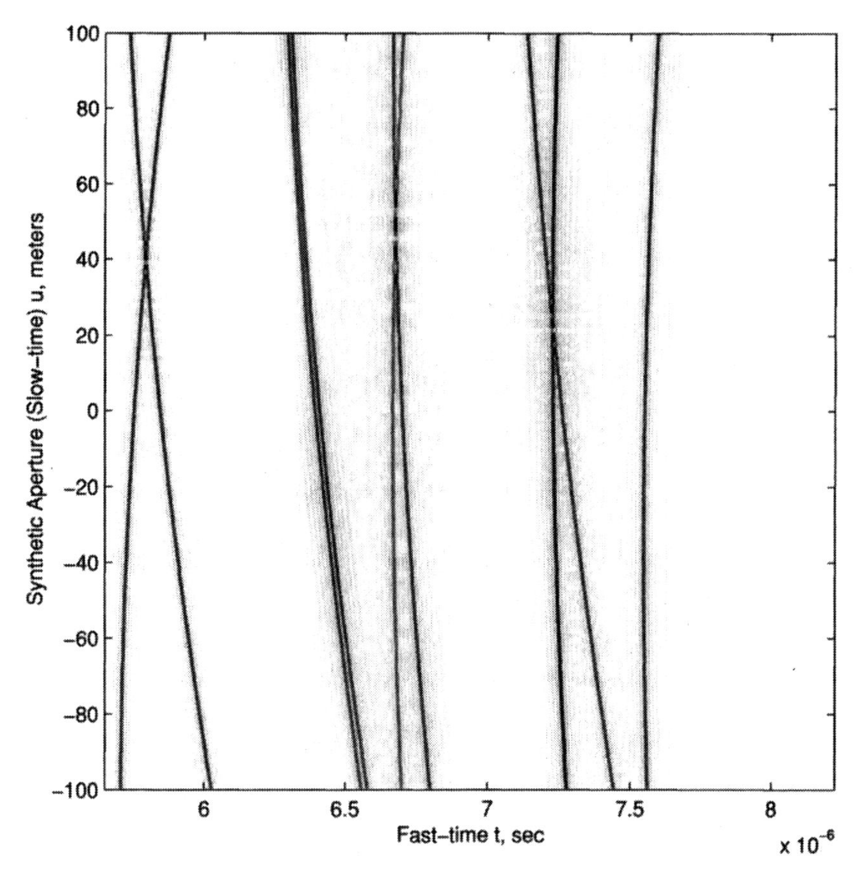

Figure 4.4b. Simulated SAR data for a chirp radar signal: fast-time matched-filtered spotlight SAR signal.

$$\begin{aligned} \mathbf{s}_M(t,u) &= \mathbf{s}(t,u) * p^*(-t) \\ &= \sum_n \sigma_n \, \mathrm{psf}_t \left[t - \frac{2\sqrt{x_n^2 + (y_n - u)^2}}{c} \right] \end{aligned}$$

where, as we showed in *range imaging* (Chapter 1), the point spread function is

$$\mathrm{psf}_t(t) = \mathcal{F}^{-1}_{(\omega)}\left[|P(\omega)|^2 \right];$$

ω represents the fast-time frequency domain.

In the case of FM radar signals, such as chirp signals, the fast-time duration of the point spread function $\mathrm{psf}_t(t)$ is shorter than that of the transmitted radar signal $p(t)$. Thus the (t, u) domain hyperbolic SAR signatures of the targets in the imaging scene are more visible in the fast-time matched filtered signal $\mathbf{s}_M(t, u)$. Figure 4.4b

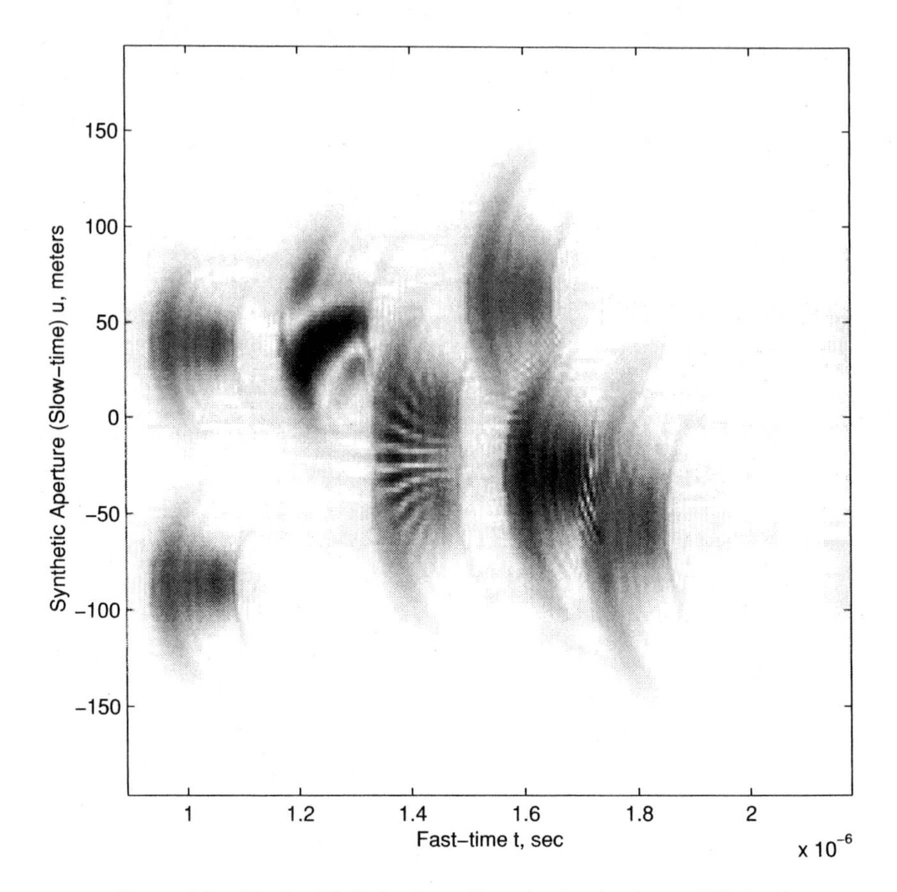

Figure 4.4c. Simulated SAR data for a chirp radar signal: stripmap SAR signal.

is the fast-time matched filtered version of the spotlight SAR data in Figure 4.4*a*. The half-hyperbolic (t, u) domain SAR signature of the targets are more visible in this figure.

We will discuss another SAR modality in Chapter 6 which is called a *stripmap* or *side-looking* SAR. Figure 4.4*c* shows a simulated stripmap SAR $\mathbf{s}(t, u)$ where the transmitted radar signal is a chirp; the Matlab code for Chapter 6 is used to generate this database. Figure 4.4*d* is the fast-time matched-filtered signal $\mathbf{s}_M(t, u)$ for the stripmap SAR data of Figure 4.4*d*. Note that the hyperbolic targets signatures appear *truncated* in Figure 4.4*c*–*d*. This phenomenon, which is related to the radiation pattern of a side-looking SAR, will be discussed in Chapter 6.

We should point out that visualizing the (t, u) domain SAR signature of targets does not have any impact on developing the *reconstruction* problem in SAR, which is achieved in the two-dimensional frequency domain of the measured signal. The above discussion was presented to provide an understanding of the nature of the (t, u) domain SAR signal.

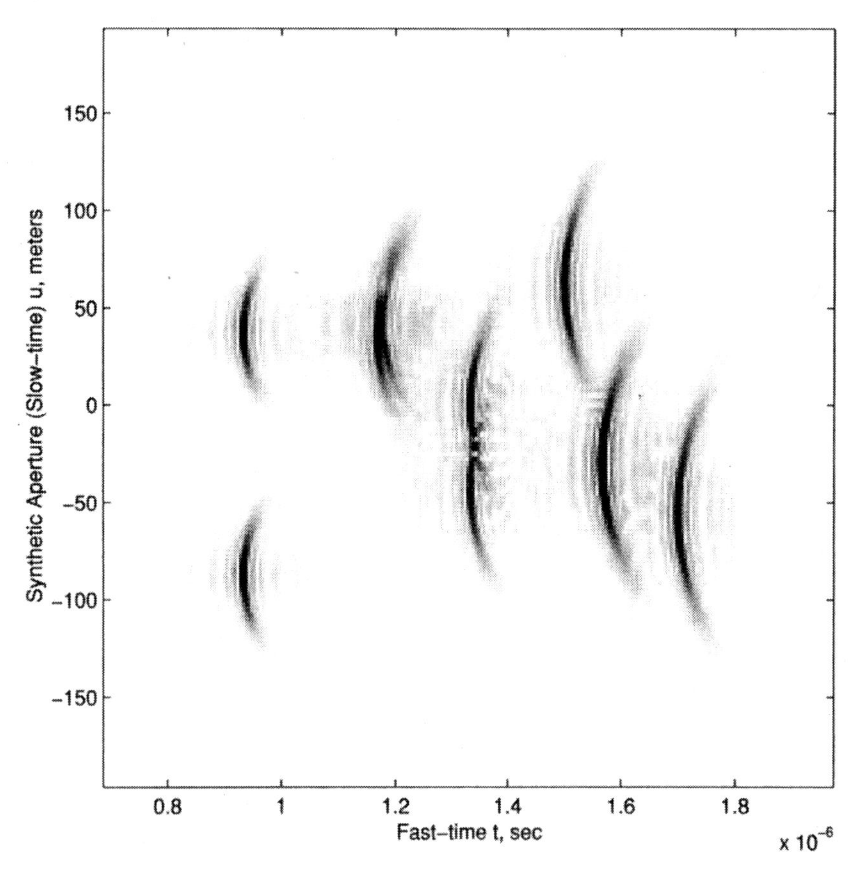

Figure 4.4d. Simulated SAR data for a chirp radar signal: fast-time matched-filtered stripmap SAR signal.

4.2 FAST-TIME FOURIER TRANSFORM

The Fourier transform of the generic SAR signal $\mathbf{s}(t, u)$ with respect to the fast-time t is

$$s(\omega, u) = P(\omega) \sum_n \sigma_n \underbrace{\exp\left[-j2k\sqrt{x_n^2 + (y_n - u)^2}\right]}_{\text{Spherical PM signal}},$$

where $k = \omega/c$ is the wavenumber. As one can see in the above model, the SAR signal in the (ω, u) domain is composed of a linear combination of the spherical PM signals, that is,

$$\exp\left[-j2k\sqrt{x_n^2 + (y_n - u)^2}\right].$$

In Chapters 2 and 3, we examined the Fourier properties of these signals, which were the key for cross-range imaging. Similar principles will be the foundation of image formation in spotlight and stripmap SAR systems.

4.3 SLOW-TIME FOURIER TRANSFORM

Our next operation involves a Fourier transformation with respect to the slow-time or the synthetic aperture u domain. For the generic SAR model, we assume the measurements are made for $u \in (-\infty, \infty)$. The Fourier transform of the spherical PM signal with respect to the slow-time u is (constants and amplitude functions are suppressed in the following)

$$\mathcal{F}_{(u)}\left[\exp\left[-j2k\sqrt{x_n^2 + (y_n - u)^2}\right]\right] = \exp\left(-j\sqrt{4k^2 - k_u^2}\, x_n - jk_u y_n\right)$$

for $k_u \in [-2k, 2k]$. We refer to k_u as the synthetic aperture frequency domain, or slow-time frequency domain. However, to give a meaning to this variable in terms of the classical SAR imaging system in the 1950s, we call the variable k_u the *slow-time Doppler* domain.

Using the slow-time Fourier property of the spherical PM signal, the Fourier transform of $s(\omega, u)$ with respect to the slow-time u is

$$S(\omega, k_u) = P(\omega) \sum_n \sigma_n \underbrace{\exp\left(-j\sqrt{4k^2 - k_u^2}\, x_n - jk_u y_n\right)}_{\text{Linear phase function of } (x_n, y_n)}.$$

The fact that the phase function

$$\exp\left(-j\sqrt{4k^2 - k_u^2}\, x_n - jk_u y_n\right)$$

is a *linear* phase function of (x_n, y_n) will play a key role in formulating a practical reconstruction algorithm in SAR imaging systems.

4.4 RECONSTRUCTION

We can rewrite the SAR signal in the (ω, k_u) in the following form by defining two new *functions* of (ω, k_u):

$$S(\omega, k_u) = P(\omega) \sum_n \sigma_n \exp\left[-jk_x(\omega, k_u)x_n - jk_y(\omega, k_u)y_n\right],$$

where the two new functions are defined to be

$$k_x(\omega, k_u) = \sqrt{4k^2 - k_u^2},$$
$$k_y(\omega, k_u) = k_u.$$

We call these two functions the *SAR spatial frequency mapping* or *transformation.*

We define the *ideal* target function in the spatial domain via

$$f_0(x, y) = \sum_n \sigma_n \delta(x - x_n, y - y_n),$$

which has the following two-dimensional spatial Fourier transform:

$$F_0(k_x, k_y) = \underbrace{\sum_n \sigma_n \underbrace{\exp\big(- jk_x x_n - jk_y y_n \big)}_{\text{Linear phase function}}}_{\text{Linear combination}}.$$

Note that $F_0(k_x, k_y)$ is also composed of a linear combination of a linear phase function of (x_n, y_n), $n = 1, 2, \ldots,$ which is due to the Fourier shift property.

Next we use the expression for $F_0(k_x, k_y)$ in the SAR signal $S(\omega, k_u)$; this yields

$$S(\omega, k_u) = P(\omega) F_0\big[k_x(\omega, k_u), k_y(\omega, k_u)\big]$$

where $\big[k_x(\omega, k_u), k_y(\omega, k_u)\big]$ are governed by the SAR spatial frequency mapping. For the reconstruction, that is, imaging $f_0(x, y)$ or $F_0(k_x, k_y)$ from the Fourier transform of the measured signal $S(\omega, k_u)$, we have

$$F_0\big[k_x(\omega, k_u), k_y(\omega, k_u)\big] = \frac{S(\omega, k_u)}{P(\omega)}.$$

As we pointed out in *range imaging* (Chapter 1), the above is a theoretical reconstruction, since $p(t)$ is a bandlimited signal (or $P(\omega)$ has a finite support in the ω domain). Moreover $S(\omega, k_u)$ is zero for $|k_u| > 2k$.

The practical reconstruction is via fast-time matched filtering; that is,

$$\begin{aligned} F\big[k_x(\omega, k_u), k_y(\omega, k_u)\big] &= P^*(\omega) S(\omega, k_u) \\ &= |P(\omega)|^2 \sum_n \sigma_n \exp\big(- jk_x x_n - jk_y y_n \big) \end{aligned}$$

for $k_u \in [-2k, 2k]$ and $\omega \in [\omega_c - \omega_0, \omega_c + \omega_0]$.

As we mentioned in Chapter 2, the origin of the above-mentioned approach for image formation from aperture data can be found in the wave equation inversion theory [gab], which is also known as *wavefront reconstruction* or *holography* [goo]. In the early 1960s SAR researchers recognized the application of this theory for image formation from synthetic aperture data [lei62; lei64]. The principal limitation that they encountered was the implementation issue. The early SAR image formation techniques utilize an approximation (Fresnel) to reconstruct images via *analog* means (a lens); see the discussion on range-Doppler SAR imaging which appears later in this chapter.

With the introduction of powerful digital computers and digital signal processing algorithms, there is no need for using the Fresnel approximation and analog image formation in SAR. Various SAR digital image formation methods that utilize the wavefront reconstruction theory are outlined in the next sections. We refer to these as SAR *wavefront* reconstruction algorithms.

As we pointed out earlier in this chapter, the *generic SAR* with an infinite synthetic aperture is not a realistic imaging system. We are using the generic SAR in this chapter to familiarize the reader with certain basic properties of more realistic spotlight and stripmap SAR systems which will be discussed in the next chapter. We could examine the resolution, two-dimensional point spread function, and data acquisition and signal processing issues for a generic SAR system. However, this would have no impact on what we will be facing in the realistic spotlight and stripmap systems.

Yet there exists a critical signal processing issue with a generic SAR system which will also be encountered in the realistic spotlight and stripmap SAR systems. This issue is related to the practical discrete (computer-based) implementation of SAR reconstruction which involves *interpolation*. This is discussed in the next section.

4.5 DIGITAL RECONSTRUCTION VIA SPATIAL FREQUENCY INTERPOLATION

The desired information in SAR imaging is the distribution of the target area reflectivity function in the spatial domain, that is, $f(x, y)$. The generic SAR reconstruction, which was outlined in the previous section, provides a mapping of the acquired data in the (ω, k_u) domain into the (k_x, k_y) domain. In practice, the measurement SAR system provides the user with evenly spaced samples of $S(\omega, k_u)$ on a rectangular grid in the (ω, k_u) domain. However, *due to the* "nonlinear" *nature of the two-dimensional mapping from the* (ω, k_u) *domain into the* (k_x, k_y) *domain, the resultant database for* $F(k_x, k_y)$ *is* unevenly spaced; an example is shown in Figure 4.5. Moreover we require the knowledge of $F(k_x, k_y)$ on a (uniform) rectangular grid to retrieve a sampled version of $f(x, y)$ via two-dimensional FFTs.

Suppose that there are M synthetic aperture samples of the SAR signal $s(\omega, u)$ with sample spacing Δ_u in the u domain. (Note that we are practically assuming a finite synthetic aperture.) Thus, from the FFT equations [s94, ch. 1], after performing FFT with respect to the discrete samples of $s(\omega, u)$ in the u domain, we obtain M samples of $S(\omega, k_u)$ in the slow-time Doppler k_u domain with the following sample spacing:

$$\Delta_{k_u} = \frac{2\pi}{M\Delta_u}.$$

The SAR mapping into the k_y domain is evenly spaced, since $k_y = k_u$, and we have the values of $S(\omega, k_u)$ at evenly spaced values of k_u. The same is not true for

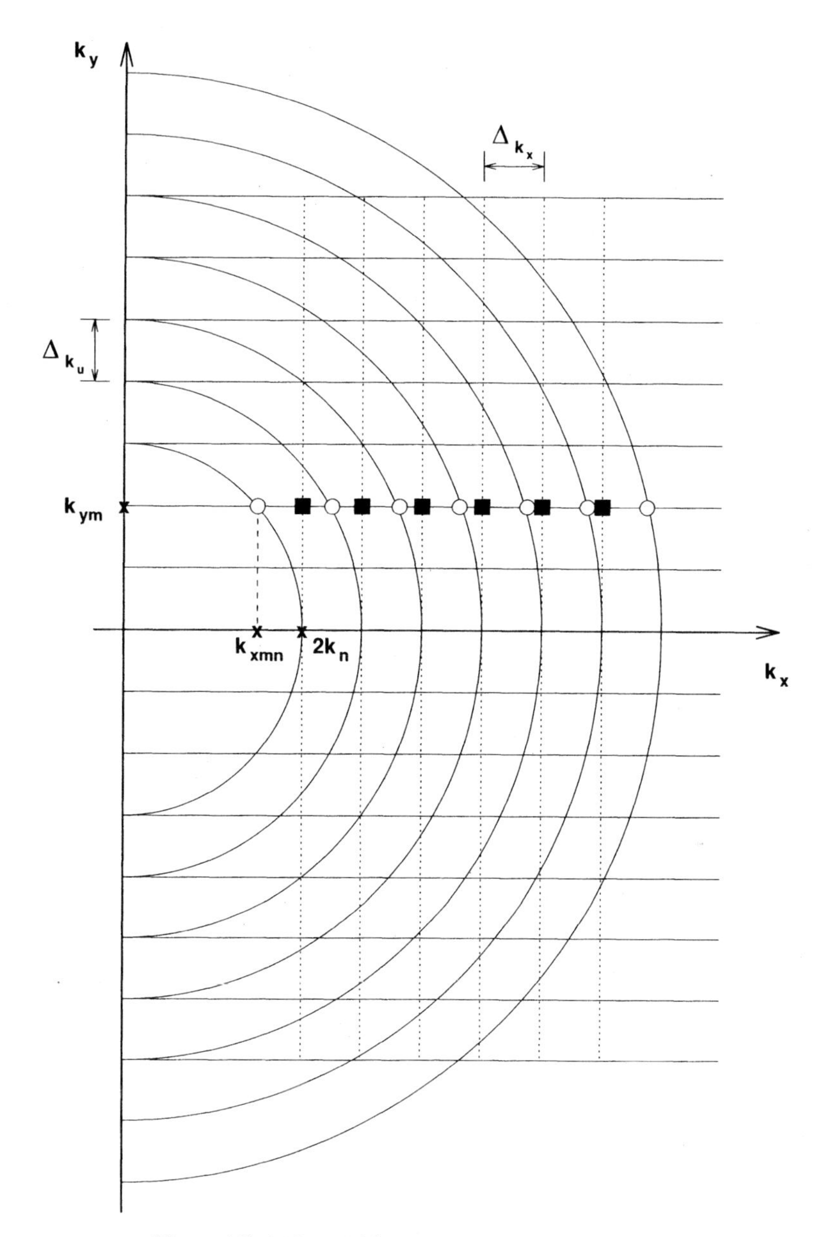

Figure 4.5. SAR spatial frequency mapping for discrete data.

$$k_x = \sqrt{4k^2 - k_u^2},$$

that is, a nonlinear function of (ω, k_u). Thus, at a fixed discrete value of

$$k_{ymn} = k_{um} = m\Delta_{k_u},$$

our problem is a *one-dimensional* interpolation in the k_x domain from the unevenly spaced values of k_x at

$$k_{xmn} = \sqrt{4k_n^2 - k_{um}^2},$$

where

$$k_n = n\Delta_k = n\frac{\Delta_\omega}{c}$$

are evenly spaced points in the $k = \omega/c$ domain and Δ_ω is the sample spacing of the measured SAR data in the fast-time frequency ω domain. (Fast-time frequency sampled points for the SAR signal are $\omega_n = ck_n$.)

In the long-range imaging applications of high-frequency (e.g., X or C band) spotlight or stripmap SAR, the range of available slow-time Doppler k_u values is much smaller than the wavenumber, that is, $|k_u| \ll k$. Moreover the carrier frequency of the radar signal, ω_c, is much larger than the bandwidth of the radar signal, $2\omega_0$. The same is not true in the reconnaissance with stripmap SAR at UHF-VHF frequencies.

- There are several factors that make the condition $|k_u| \ll k$ to be true in the practical high-frequency spotlight and stripmap SAR systems. For instance, in stripmap SAR, the condition is true due to the fact that the physical radar antenna beamwidth is very small. In spotlight SAR, $|k_u| \ll k$ is valid, since the radar swath mean range, X_c, is much greater than the synthetic aperture length, $2L$, and the radar footprint in cross-range. These will be examined in our future analysis of spotlight and stripmap SAR systems.

With the conditions $|k_u| \ll k$ and $2\omega_0 \ll \omega_c$ satisfied, one may avoid the interpolation problem by using the approximation

$$k_x \approx 2k$$

for *only* the *assignment of the* (ω, k_u) *domain data in the* (k_x, k_y) *spatial frequency domain*. This implies that we also have evenly spaced data coverage in the spatial frequency domain of k_x.

- The use of the approximation $k_x \approx 2k$ in the spatial frequency data assignment causes slight degradations in the reconstructed image for these high-frequency SAR systems. However, if the approximation $k_x \approx 2k$ is used in the phase function

$$\exp\left[-jk_x(\omega, k_u)x_n\right],$$

the resultant phase errors could severely degrade the reconstructed image. The above phase function should not be approximated in any SAR system.

Due to the large beamwidth used in, for example, reconnaissance with UHF-VHF SAR, the condition $|k_u| \ll k$ cannot be met in these SAR imaging systems. Moreover, in these SAR systems, the radar signal bandwidth is comparable to its carrier frequency, for example, the carrier frequency is 300 MHz and the bandwidth is 200 MHz. In these problems the interpolation in the k_x domain is an *essential* (but not *very critical*) step in the reconstruction algorithm.

Baseband Conversion of Target Area

Broadside Mode SAR Prior to the interpolation we must perform one more step. The illuminated target area in the range domain is determined by the radar range swath. Suppose that the radar range swath is identified via the region $x \in [X_c - X_0, X_c + X_0]$, where X_c is its mean range and $2X_0$ is the size of the radar footprint in the range domain. Thus the signal $F(k_x, k_y)$ is a *bandpass* signal in the k_x domain, since its inverse Fourier transform is centered at $x = X_c$. (The inverse Fourier transform of a *lowpass* signal in the k_x domain has its spectrum centered at $x = 0$.)

For the interpolation step, we need to transform this signal to the lowpass. For this purpose we perform the following baseband conversion:

$$F_b(k_x, k_y) = F(k_x, k_y)\exp(jk_x X_c).$$

Note that the origin in the spatial (x, y) domain is at the center of the baseband target function $f_b(x, y)$. The reconstruction equation for the lowpass target function becomes

$$\begin{aligned} F_b\big[k_x(\omega, k_u), k_y(\omega, k_u)\big] &= P^*(\omega)S(\omega, k_u)\exp\big[jk_x(\omega, k_u)X_c\big] \\ &= P^*(\omega)\exp\left(j\sqrt{4k^2 - k_u^2}\,X_c\right)S(\omega, k_u). \end{aligned}$$

With the addition of the phase function

$$\exp\left(j\sqrt{4k^2 - k_u^2}\,X_c\right),$$

the above baseband version of the SAR reconstruction has some similarities to what we derived for *slow-time matched-filtered* reconstruction in cross-range imaging.

This fact becomes obvious in the following. Consider the generic SAR signature of a unit reflector located at $(x, y) = (X_c, 0)$ in the spatial domain; that is,

$$s_0(t, u) = p\left[t - \frac{2\sqrt{X_c^2 + u^2}}{c}\right]$$

for $u \in (-\infty, \infty)$. We call $s_0(t, u)$ the *reference* generic SAR signal. Using the steps that we used to derive $S(\omega, k_u)$ from $s(t, u)$, one can obtain the following for the two-dimensional (fast-time and slow-time) Fourier transform of the generic reference signal with respect to (t, u):

$$S_0(\omega, k_u) = P(\omega) \exp\left(-j\sqrt{4k^2 - k_u^2}\, X_c\right).$$

Using the expression for $S_0(\omega, k_u)$ in the SAR baseband reconstruction, we obtain

$$F_b\left[k_x(\omega, k_u), k_y(\omega, k_u)\right] = S(\omega, k_u) S_0^*(\omega, k_u).$$

This indicates that the baseband target function in the spatial frequency domain is reconstructed via a *two-dimensional matched filtering* of the measured SAR signal with the generic reference SAR signal. This is a conceptually appealing result, since it conforms to the matched-filtering principles for reconstruction in *range imaging* (Chapter 1) and *cross-range imaging* (Chapter 2).

Squint Mode SAR In certain SAR systems the target area of interest is not located at (or facing) the center of the synthetic aperture. These systems are called *squint* SAR. For a squint SAR system the distance of the center of the target area in the cross-range domain from the center of the synthetic aperture, that is, $u = 0$, is a known nonzero value which we denote with Y_c; this parameter is called the *mean* or *squint cross-range*. The case of $Y_c = 0$ is referred to as a *broadside* SAR system.

For squint SAR systems the baseband conversion of the target area is performed via

$$F_b(k_x, k_y) = F(k_x, k_y) \exp(jk_x X_c + k_y Y_c).$$

Thus the reconstruction equation for the lowpass target function is

$$\begin{aligned} F_b\left[k_x(\omega, k_u), k_y(\omega, k_u)\right] &= P^*(\omega) S(\omega, k_u) \exp\left[jk_x(\omega, k_u) X_c + jk_y(\omega, k_u) Y_c\right] \\ &= P^*(\omega) \exp\left(j\sqrt{4k^2 - k_u^2}\, X_c + jk_u Y_c\right) S(\omega, k_u). \end{aligned}$$

The squint case may also be formulated via two-dimensional matched filtering. Consider the generic SAR signature of a unit reflector located at the center of the target area $(x, y) = (X_c, Y_c)$ in the spatial domain; that is,

$$s_0(t, u) = p\left[t - \frac{2\sqrt{X_c^2 + (Y_c - u)^2}}{c}\right],$$

for $u \in (-\infty, \infty)$, which is *reference* generic SAR signal for squint mode. The two-dimensional (fast-time and slow-time) Fourier transform of this reference signal with

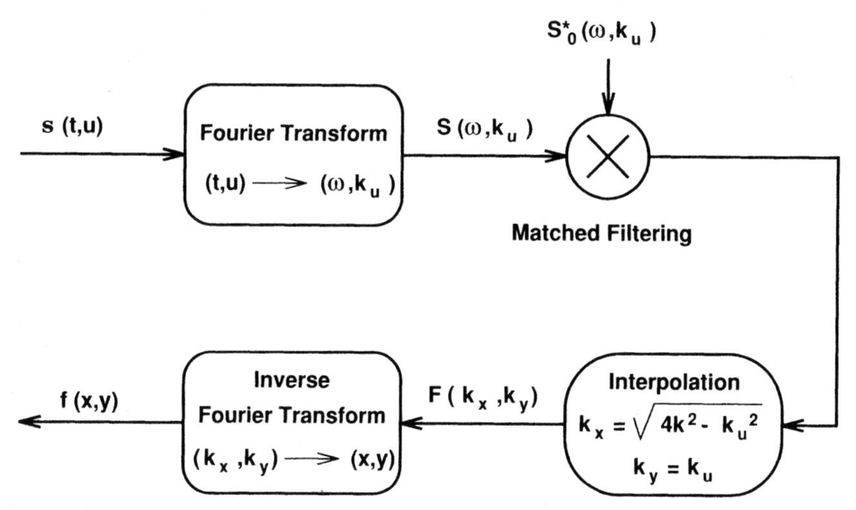

Figure 4.6. Block diagram of generic SAR digital reconstruction algorithm via spatial frequency domain interpolation.

respect to (t, u) is

$$S_0(\omega, k_u) = P(\omega) \exp\left(-j\sqrt{4k^2 - k_u^2}\, X_c - jk_u Y_c\right).$$

Thus the SAR baseband reconstruction can be formed via two-dimensional matched filtering in both fast-time and slow-time domains:

$$F_b\big[k_x(\omega, k_u), k_y(\omega, k_u)\big] = S(\omega, k_u) S_0^*(\omega, k_u).$$

For notational simplicity we denote the baseband target function by $F(k_x, k_y)$ instead of $F_b(k_x, k_y)$ in our future discussions. Generic SAR digital reconstruction via spatial frequency interpolation is summarized in Figure 4.6. We are now ready for the interpolation phase of the reconstruction algorithm.

Interpolation from Evenly Spaced Data

We first review interpolation from evenly spaced data, that is, $k_{xn} = n\Delta_{k_x}$. (Note that the Nyquist criterion in the range spatial frequency domain is $\Delta_{k_x} \le \pi / X_0$, since the baseband size of the target area in the range domain is $2X_0$.) The interpolation equation for evenly spaced data is

$$F(k_x, k_{ymn}) = \sum_n F(n\Delta_{k_x}, k_{ymn}) h(k_x - n\Delta_{k_x}),$$

where

$$h(k_x) = \text{sinc}\left(\frac{k_x}{\Delta_{k_x}}\right),$$

is the sinc interpolating function.

To reduce the computational cost of the above interpolation, one may use a *tapered* or *windowed* version of $h(k_x)$, labeled $h_w(k_x)$, in the above interpolation; that is,

$$F(k_x, k_{ymn}) \approx \sum_{|k_x - n\Delta_{k_x}| \leq N_s \Delta_{k_x}} F(k_{xn}, k_{ymn}) h_w(k_x - n\Delta_{k_x}),$$

where

$$h_w(k_x) = \begin{cases} h(k_x)w(k_x) & \text{for } |k_x| \leq N_s \Delta_{k_x}, \\ 0 & \text{otherwise.} \end{cases}$$

The parameter N_s, which denotes the half-number of the sinc side lobes used for interpolation, is a constant chosen by the user that determines the support region (size) of the window, that is, $[-N_s\Delta_{k_x}, N_s\Delta_{k_x}]$. The signal $w(k_x)$ can be chosen to be any window function. In our applications we select $w(k_x)$ to be the *Hamming* window; that is,

$$w(k_x) = \begin{cases} 0.54 + 0.46\cos\left(\dfrac{\pi k_x}{N_s \Delta_{k_x}}\right) & \text{for } |k_x| \leq N_s \Delta_{k_x}; \\ 0 & \text{otherwise.} \end{cases}$$

Clearly, the support of $h_w(k_x)$ or $w(k_x)$ in the k_x domain is

$$k_x \in [-N_s\Delta_{k_x}, N_s\Delta_{k_x}].$$

Interpolation from Unevenly Spaced Data

Consider the SAR interpolation problem where, for a fixed k_{um}, the k_x domain is a nonlinear transformation, labeled $g_m(\cdot)$, of the wavenumber (radar signal frequency ω) domain; that is,

$$\begin{aligned} k_x &= g_m(\omega) \\ &= \sqrt{4k^2 - k_{um}^2}. \end{aligned}$$

The samples of $F(k_x, k_{ymn})$ are available at unevenly spaced points in the k_x domain which are identified by k_{xmn}'s; these are mappings of evenly spaced points in the fast-time frequency ω or the wavenumber $k = \omega/c$ domain, that is,

$$k_{xmn} = g_m(n\Delta_\omega),$$

where Δ_ω is the evenly spaced sample spacing in the ω domain. The unevenly spaced sampled points k_{xmn} are shown with white circles in Figure 4.5.

For this problem, the interpolation equation becomes [s88]

$$F(k_x, k_{ymn}) \approx \sum_{|k_x - k_{xmn}| \le N_s \Delta_{k_x}} J_m(n\Delta_\omega) F(k_{xmn}, k_{ymn}) h_w(k_x - k_{xmn}),$$

where

$$\begin{aligned} J_m(\omega) &= \frac{d g_m(\omega)}{d\omega} \\ &= \frac{4k}{c\sqrt{4k^2 - k_{um}^2}} \end{aligned}$$

is the Jacobian (derivative) of the transformation from k to k_x. *Note that the interpolation equation for unevenly spaced data contains an additional term, that is, the magnitude of the Jacobian function at the sampling point.* (The denominator of this Jacobian function cancels the amplitude function $\sqrt{4k^2 - k_u^2}$ which should have been incorporated with the slow-time Fourier transform of the spherical PM signal [s92ja; s94].)

The values of $F(k_x, k_{ymn})$ are interpolated from $F(k_{xmn}, k_{ymn})$ at a set of evenly spaced points in the k_x domain; these evenly spaced points are identified by black squares in Figure 4.5. The sample spacing of the evenly spaced points in the k_x domain, Δ_{k_x}, is chosen by the user based on the radar swath length, that is, $2X_0$; this sample spacing should satisfy the following Nyquist criterion

$$\Delta_{k_x} \le \frac{\pi}{X_0}.$$

We mentioned that the transmitted radar signal $p(t)$ is a bandpass signal in the fast-time frequency band $[\omega_{\min}, \omega_{\max}] = [\omega_c - \omega_0, \omega_c + \omega_0]$. In this case, for the narrowband broadside-mode SAR systems, the coverage of the SAR data in the spatial frequency k_x domain is approximately a bandpass region in the interval (this will be shown in Chapters 5 and 6)

$$k_x \in \left[2(k_c - k_0), 2(k_c + k_0)\right],$$

where $k_c = \omega_c/c$, and $k_0 = \omega_0/c$. For the inverse discrete Fourier transform (DFT) from the k_x domain to the x domain, this bandpass database should be *shifted* by $2k_c$. As a result the shifted database is within the *lowpass* band (i.e., centered around $k_x = 0$)

$$k_x \in [-2k_0, 2k_0],$$

which is suitable for DFT processing. (A squint SAR system has a similar problem; these will be examined in Chapters 5 and 6.)

The same is not true for wideband SAR systems. In that case the range of available k_x values is dictated by several parameters of the SAR system. This can be incorporated in the interpolation. The spotlight and stripmap Matlab algorithms for Chapters 5 and 6 are designed to deal with the general interpolation problem of a wideband SAR system.

4.6 DIGITAL RECONSTRUCTION VIA RANGE STACKING

Consider the inverse two-dimensional Fourier integral equation for the target function:

$$f(x, y) = \int_{k_y} \int_{k_x} F(k_x, k_y) \exp\big(jk_x x + jk_y y\big) dk_x dk_y,$$

where $x \in [-X_0, X_0]$ is the baseband range domain. We wish to rewrite the right-hand side of the above double integral in terms of the measurement domain and the measured SAR signal.

For this, we use the SAR mapping which relates the target spatial frequency domain (k_x, k_y) to the measurement Fourier domain (ω, k_u), and the reconstruction equation

$$F\big[k_x(\omega, k_u), k_y(\omega, k_u)\big] = P^*(\omega) \exp\left(j\sqrt{4k^2 - k_u^2}\, X_c + jk_u Y_c\right) S(\omega, k_u).$$

Using these, we make two-dimensional transformations from (k_x, k_y) to (ω, k_u) in the inverse Fourier integral for $f(x, y)$, and after substitution for $F(\cdot)$, we obtain

$$f(x, y) = \int_{k_u} \int_{\omega} P^*(\omega) \exp\left(j\sqrt{4k^2 - k_u^2}\, X_c + jk_u Y_c\right) S(\omega, k_u),$$
$$\times \exp\left(j\sqrt{4k^2 - k_u^2}\, x + jk_u y\right) J(\omega, k_u)\, d\omega\, dk_u,$$

where

$$J(\omega, k_u) = \frac{4k}{c\sqrt{4k^2 - k_u^2}}$$

is the Jacobian of the transformation from (k_x, k_y) to (ω, k_u). The Jacobian function is a slowly fluctuating amplitude function, and as a result its contribution in the reconstruction is negligible. In fact we have neglected an amplitude function $\sqrt{4k^2 - k_u^2}$ in the slow-time Fourier transform of the spherical PM signal which would be canceled with the denominator of the Jacobian function.

In the following discussion we assume that the Jacobian function, or any other amplitude function, is absorbed in the SAR signal spectrum $S(\omega, k_u)$; that is,

$$f(x, y) = \int_{k_u} \int_{\omega} P^*(\omega) \exp\left(j\sqrt{4k^2 - k_u^2}\, X_c + jk_u Y_c\right) S(\omega, k_u)$$
$$\times \exp\left(j\sqrt{4k^2 - k_u^2}\, x + jk_u y\right) d\omega\, dk_u.$$

Suppose that we are interested in forming the SAR image at the range bins x_i's, where $x_i \in [-X_0, X_0]$. Consider the above reconstruction for $f(x, y)$ at a fixed range $x = x_i$:

$$f(x_i, y) = \int_{k_u} \int_{\omega} P^*(\omega) \exp\left(j\sqrt{4k^2 - k_u^2}\, X_c + jk_u Y_c\right) S(\omega, k_u)$$
$$\times \exp\left(j\sqrt{4k^2 - k_u^2}\, x_i + jk_u y\right) d\omega\, dk_u.$$

Next we develop two algorithms for implementing the above reconstruction at the desired range bins.

Algorithm 1: Fast-time Slow-time Matched Filtering Consider the SAR signal from a reflector located at $(x, y) = (X_c + x_i, Y_c)$:

$$\mathbf{s}_{0i}(t, u) = p\left[t - \frac{2\sqrt{(X_c + x_i)^2 + (Y_c - u)^2}}{c}\right],$$

for $u \in (-\infty, \infty)$. We call this signal the *reference* SAR signal at the reference range bin $x = x_i$. This reference range and its signal are depicted for a broadside and squint SAR systems in Figure 4.7*a* and *b*, respectively. (Note that the range distance of the baseband target at $x = x_i$ from the radar flight path $x = 0$ is $X_c + x_i$.)

The fast-time Fourier transform of this reference SAR signal is

$$s_{0i}(\omega, u) = P(\omega) \exp\left[-j2k\sqrt{(X_c + x_i)^2 + (Y_c - u)^2}\right].$$

Also, based on our discussion on the slow-time Doppler properties of the SAR spherical PM signal, the two-dimensional Fourier transform of the reference SAR signal $\mathbf{s}_{0i}(t, u)$ with respect to (t, u) can be found to be

$$S_{0i}(\omega, k_u) = P(\omega) \exp\left[-j\sqrt{4k^2 - k_u^2}\,(X_c + x_i) - jk_u Y_c\right].$$

After rearrangement the target function reconstruction at (x_i, y) can be expressed by

$$f(x_i, y) = \int_{k_u} \int_{\omega} \underbrace{P^*(\omega) \exp\left[j\sqrt{4k^2 - k_u^2}\,(X_c + x_i) + jk_u Y_c\right]}_{S_{0i}^*(\omega, k_u)}$$

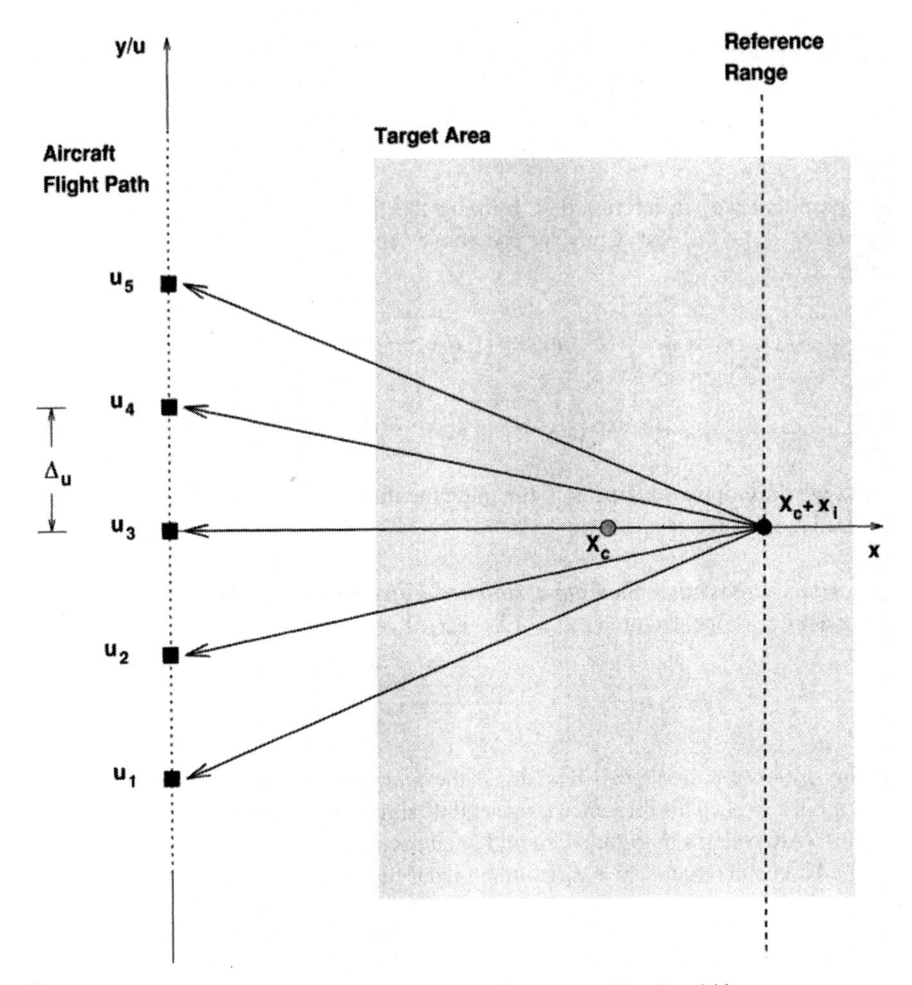

Figure 4.7a. Reference SAR signal at range x_i for range stacking: broadside target area.

$$
\begin{aligned}
&\times S(\omega, k_u) \exp\big(jk_u y\big) d\omega\, dk_u \\
&= \int_{k_u} \left[\int_{\omega} \underbrace{S_{0i}^*(\omega, k_u) S(\omega, k_u)}_{\text{Matched filtering at range } x_i} \; d\omega \right] \exp\big(jk_u y\big)\, dk_u,
\end{aligned}
$$

where $S_{0i}^*(\omega, k_u)$ is the complex conjugate of $S_{0i}(\omega, k_u)$. Note that the one-dimensional Fourier transform of the target function at range x_i with respect to the cross-range y is

$$
\begin{aligned}
F_y(x_i, k_y) &= \mathcal{F}_{(y)}[f(x_i, y)] \\
&= \int_y f(x_i, y) \exp\big(-jk_y y\big) dy.
\end{aligned}
$$

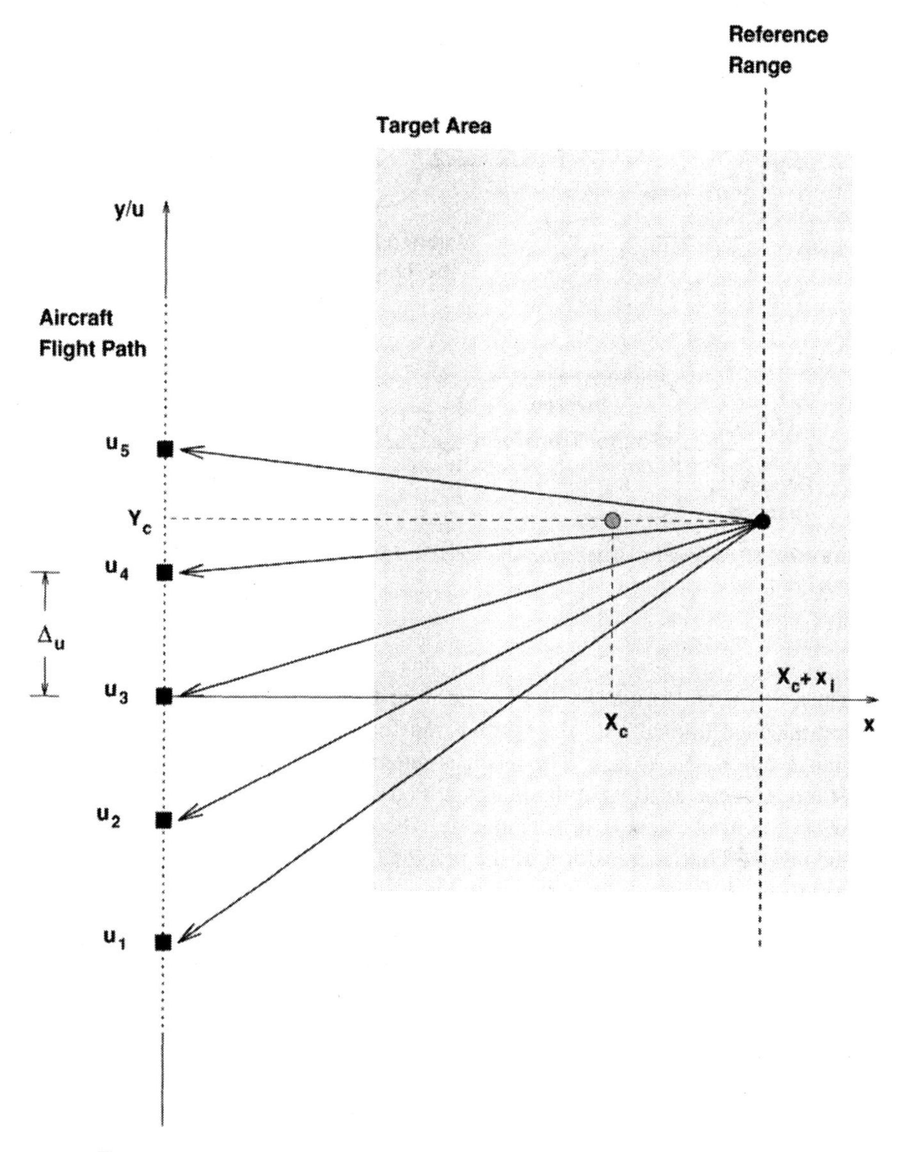

Figure 4.7b. Reference SAR signal at range x_i for range stacking: squint target area.

[$F_y(x, k_y)$ is the marginal Fourier transform of $f(x, y)$ with respect to the variable y.] Thus the reconstruction for $F_y(x_i, y)$ takes the form

$$F_y(x_i, k_y) = \int_\omega S_{0i}^*(\omega, k_u) S(\omega, k_u)\, d\omega,$$

where

$$k_y = k_u.$$

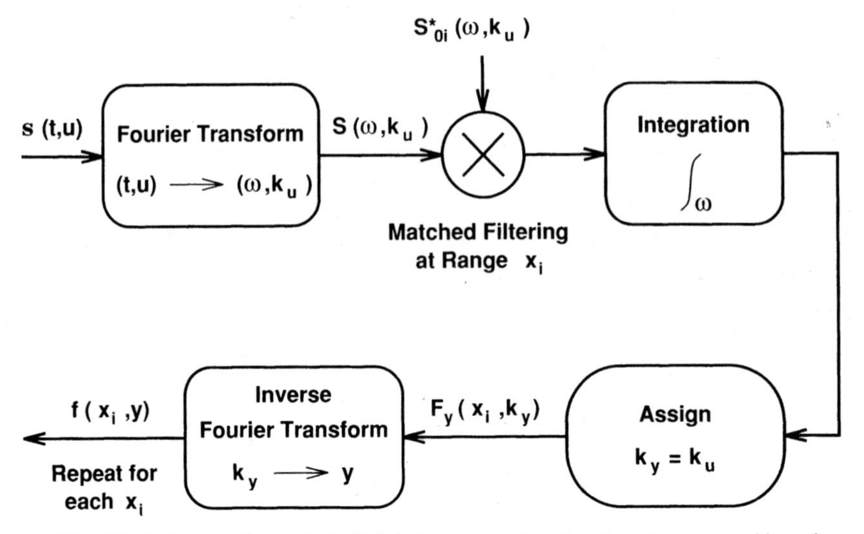

Figure 4.7c. Block diagram of generic SAR digital reconstruction algorithm via range stacking: algorithm 1.

Thus, for image formation, the Fourier transform of the SAR signal, $S(\omega, k_u)$, is first matched filtered with the Fourier transform of the reference SAR signal at the range bin x_i, $S_{0i}(\omega, k_u)$. The result is integrated (summed) over the available fast-time frequencies to yield the marginal Fourier transform of the target function $F_y(x_i, k_y)$, with $k_y = k_u$, at the range x_i. The target function $f(x_i, y)$ is formed by the inverse Fourier transform of the marginal Fourier transform $F_y(x_i, k_y)$ with respect to k_y. This algorithm is shown in Figure 4.7*c*.

Algorithm 2: Slow-time Fast-time Matched Filtering By changing the order of the integrals in the reconstruction equation

$$f(x_i, y) = \int_{k_u} \left[\int_{\omega} S_{0i}^*(\omega, k_u) S(\omega, k_u) d\omega \right] \exp\left(j k_u y\right) dk_u,$$

we obtain

$$f(x_i, y) = \int_{\omega} \left[\int_{k_u} S_{0i}^*(\omega, k_u) S(\omega, k_u) \exp\left(j k_u y\right) dk_u \right] d\omega.$$

The above integral in the k_u domain is an inverse Fourier integral. Thus we can rewrite it via

$$f(x_i, y) = \int_{\omega} \left[s(\omega, u) * s_{0i}^*(\omega, -u) \right] d\omega,$$

with

$$y = u,$$

where $*$ denotes convolution in the synthetic aperture (slow-time) u domain.

Thus, for image formation at range x_i, the SAR signal $s(\omega, u)$ is matched filtered with the reference SAR signal $s_{0i}(\omega, u)$ in the slow-time domain. The result is integrated (summed) over the available fast-time frequencies ω to reconstruct the target function $f(x_i, y)$, with the mapping $y = u$. This algorithm is summarized in Figure 4.7*d*.

Note that both of the above algorithms form the target function at the individual range bins x_i's, and then put the outcomes together to form a two-dimensional image of the target function. We call these two methods the *range stacking* digital reconstruction for SAR. In the case of their Matlab codes, Algorithm 1 is less computationally intensive than Algorithm 2. This is due to the fact that Algorithm 1 requires only one inverse DFT from the k_y to the y domain; Algorithm 2 requires an inverse DFT from the k_y to the y domain for each available fast-time frequency ω. The code for the range stacking Algorithm 1 is included in the spotlight and stripmap SAR Matlab algorithms for Chapters 5 and 6.

The range stacking algorithm does not require interpolation, and thus it does not suffer from the truncation errors. Moreover it does not contain DFT wraparound errors in the k_x domain. (The reader could observe this by using the codes in Chapters 5 and 6.) Yet the Matlab implementation of the range stacking method is more time-consuming than the interpolation-based method of Section 4.5. This is not the case in the *parallel* implementation of the range stacking algorithm.

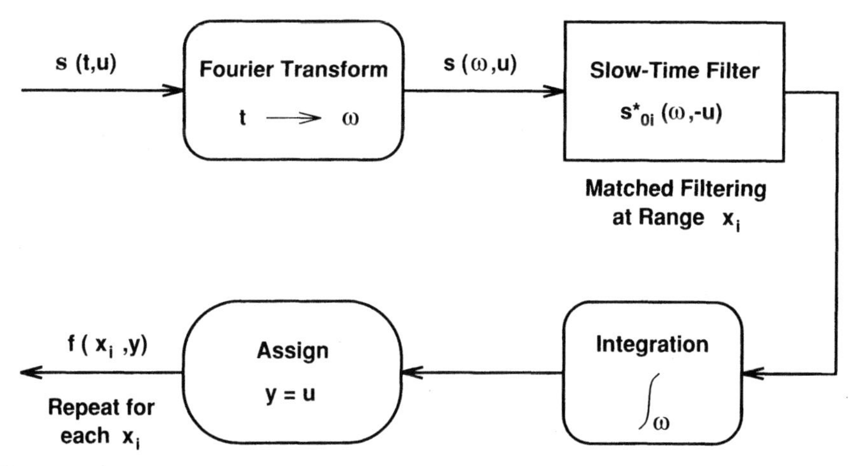

Figure 4.7d. Block diagram of generic SAR digital reconstruction algorithm via range stacking: algorithm 2.

One final note on the *complex* coherent data in $f(x, y)$ which is formed via the range stacking method. The support region for the target function in the (k_x, k_y) domain is a bandpass region centered around a carrier spatial frequency, for example,

$$(k_x, k_y) = (k_{xc}, k_{yc}).$$

This is due to the fact that the radar signal is a bandpass signal; also, in squint mode SAR systems, one deals with another form of shift in the spatial frequency domain that makes the data to be within a bandpass region in both k_x and k_y domains.

For a given SAR system, the value of the carrier spatial frequency (k_{xc}, k_{yc}) depends on the type of the SAR system (spotlight, stripmap, broadside, squint, narrowband, wideband, etc.), and its parameters (size of the target area, size of the synthetic aperture, etc.). The user can identify this carrier based on the SAR digital signal analysis in Chapters 5 and 6.

Once the carrier spatial frequency (k_{xc}, k_{yc}) is determined, the user should multiply the range stack reconstruction of $f(x, y)$ with the range domain carrier, that is,

$$f(x, y) \exp\big(-jk_{xc}x\big),$$

to obtain the *unaliased* coherent baseband data in the (k_x, k_y) domain. [It is not necessary to perform baseband conversion in the cross-range y domain with the carrier frequency k_{yc}. This baseband conversion is performed in the digital signal processing of $s(\omega, u)$; see the discussion on the *squint* case in Chapter 5.] The above multiplication with the carrier is not essential if the user is only interested in the magnitude of the reconstruction $|f(x, y)|$.

4.7 DIGITAL RECONSTRUCTION VIA TIME DOMAIN CORRELATION AND BACKPROJECTION

As we mentioned earlier, SAR image formation is performed via the fast-time and slow-time matched filtering of the SAR signal with the reference SAR signal; that is, the two-dimensional convolution of $\mathbf{s}(t, u)$ with $\mathbf{s}_0^*(-t, -u)$ which is followed by mapping the (ω, k_u) domain onto the (k_x, k_y) domain. SAR imaging may also be formulated via convolving the SAR signal with a *shift-varying* filter. In this section we present two SAR digital reconstruction algorithms that utilize this principle.

Time Domain Correlation Algorithm

The basic principle behind the TDC imaging method is simply *correlation* implementation of the matched filtering [ba]. Suppose that we are interested in forming the target function at a set of two-dimensional sampled points (x_i, y_j)'s in the spatial domain. The TDC processor correlates the SAR signature at a given grid point

(x_i, y_j) $(x_i \in [X_c - X_0, X_c + X_0])$, which is

$$p\left[t - \frac{2\sqrt{x_i^2 + (y_j - u)^2}}{c}\right],$$

with the measured SAR data in the fast-time and slow-time (t, u) domain, that is, $\mathbf{s}(t, u)$. The result is a measure of reflectivity at that grid point. The imaging equation for TDC is

$$\begin{aligned} f(x_i, y_j) &= \int_u \int_t \mathbf{s}(t, u) p^*\left[t - \frac{2\sqrt{x_i^2 + (y_j - u)^2}}{c}\right] dt\, du \\ &= \int_u \int_t \mathbf{s}(t, u) p^*\left[t - t_{ij}(u)\right] dt\, du, \end{aligned}$$

where $p^*(t)$ is the complex conjugate of $p(t)$, and

$$t_{ij}(u) = \frac{2\sqrt{x_i^2 + (y_j - u)^2}}{c}.$$

In practice, the two-dimensional integral in the fast-time and slow-time domain (t, u) is converted into a double sum over the available discrete values of (t, u), that is, the domain of the measured SAR data. The reconstruction is performed for discrete values of the spatial (x, y) domain on a uniform grid, that is, (x_i, y_j)'s. To reduce the numerical errors, the measured signal $\mathbf{s}(t, u)$ is *upsampled* (interpolated) in both the fast-time and slow-time domains.

Similar to the range stacking, the target function formed via the TDC method has to be multiplied with a phase function for baseband conversion in the range spatial frequency k_x domain. This is achieved via

$$f(x, y) \exp\left(-jk_{xc}x\right).$$

The main drawback of the TDC method is its high computational cost which is due to the two-dimensional discrete sum that is performed for correlation [and, perhaps, upsampling in the (t, u) domain]. The Fourier domain imaging method performs the same operation via fast DFTs and multiplication in the frequency domain. The range stacking method also utilizes DFTs.

The correlation for the TDC method may also be performed in the (ω, u) domain via utilizing the following identity which can be obtained from the general Parseval's theorem:

$$\begin{aligned} &\int_t \mathbf{s}(t, u) p^*\left[t - t_{ij}(u)\right] dt \\ &\quad = \int_\omega s(\omega, u) P^*(\omega) \exp\left[j\omega t_{ij}(u)\right] d\omega. \end{aligned}$$

Substituting this identity in the TDC imaging equation, one obtains the following:

$$f(x_i, y_j) = \int_u \int_\omega s(\omega, u) P^*(\omega) \exp\big[j\omega t_{ij}(u)\big]\, d\omega\, du.$$

It is not difficult to see that the (ω, u) domain correlation method to reconstruct the target function can be converted into the convolution-based reference signal matched-filtering of the range stack reconstruction method.

Backprojection Algorithm

We denote the fast-time matched-filtered SAR signal via

$$\mathbf{s}_M(t, u) = \mathbf{s}(t, u) * p^*(-t),$$

where $*$ denotes convolution in the fast-time domain. Using this in the TDC equation, we obtain

$$\begin{aligned} f(x_i, y_j) &= \int_u \mathbf{s}_M \left[\frac{2\sqrt{x_i^2 + (y_j - u)^2}}{c}, u \right] du \\ &= \int_u \mathbf{s}_M\big[t_{ij}(u), u\big]\, du \end{aligned}$$

where

$$t_{ij}(u) = \frac{2\sqrt{x_i^2 + (y_j - u)^2}}{c}$$

is the round-trip delay of the echoed signal for the target at (x_i, y_j) when the radar is at $(0, u)$. Thus, to form the target function at a given grid point (x_i, y_j) in the spatial domain, one could coherently add the data at the fast-time bins that corresponds to the location of that point for all synthetic aperture locations u.

This algorithm is known as the *backprojection* SAR reconstruction. This is due to the fact that for a given synthetic aperture location u, the fast-time data of $\mathbf{s}_M(t, u)$ are traced back in the fast-time domain (backprojected) to isolate the return of the reflector at (x_i, y_j). A block diagram for the backprojection algorithm is shown in Figure 4.8.

To implement this method in practice, the available discrete fast-time samples of $\mathbf{s}_M(t, u)$ must be interpolated to recover

$$\mathbf{s}_M\big[t_{ij}(u), u\big].$$

If a sufficiently accurate interpolator was not used, this would result in the loss of high-resolution information. Note that SAR reconstruction via spatial frequency domain in Section 4.5 is also based on some form of interpolation. Yet the spatial fre-

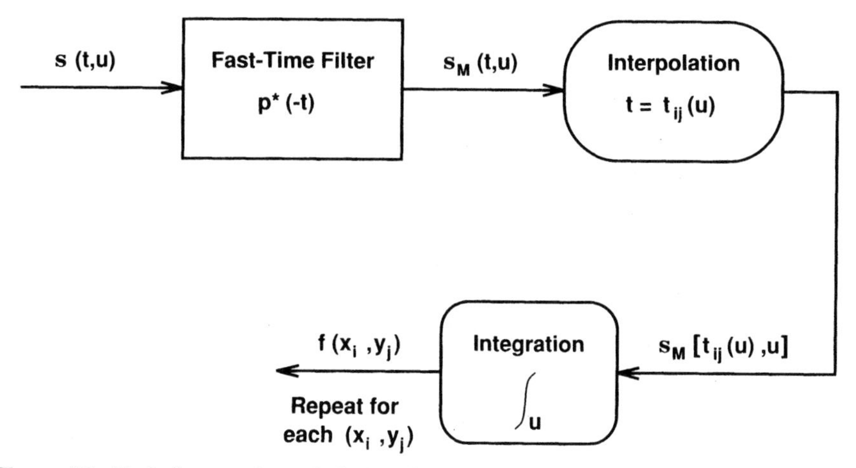

Figure 4.8. Block diagram of generic SAR digital reconstruction algorithm via backprojection algorithm.

quency (k_x, k_y) interpolation results in degradation in the edges of the reconstructed image in the spatial (x, y) domain.

The target function formed via the backprojection method should also be multiplied with a phase function for baseband conversion in the range spatial frequency k_x domain, that is,

$$f(x, y) \exp\big(-jk_{xc}x\big).$$

4.8 FREQUENCY AND SYNTHETIC APERTURE DEPENDENT TARGET REFLECTIVITY

In this section we examine the effect of a fast-time frequency and synthetic aperture-dependent target reflectivity in the reconstructed SAR image in the spatial frequency and spatial domains. This analysis not only is useful for understanding imaging information in a spotlight or stripmap SAR system but also will become valuable in the next section to quantify and identify the effects of motion errors in a SAR system.

Our approach is based on slow-time Fourier properties of the SAR AM-PM signal which was established in Chapter 2 with the help of the method of stationary phase. Earlier we identified the fast-time frequency and synthetic aperture-dependent amplitude pattern for the nth target via $a_n(\omega, x_n, y_n - u)$. We also introduced the SAR AM-PM signature of the nth target in the measured (ω, u) domain to be

$$s_n(\omega, u) = \underbrace{a_n(\omega, x_n, y_n - u)}_{\text{AM}} \underbrace{\exp\left[-j2k\sqrt{x_n^2 + (y_n - u)^2}\,\right]}_{\text{PM}}.$$

Provided that the fluctuations of the AM signal are small as compared with the fluctuations of the PM signal, the SAR AM-PM signal has the following slow-time Fourier transform:

$$S_n(\omega, k_u) = \underbrace{A_n(\omega, k_u)}_{\text{AM}} \underbrace{\exp\left(-j\sqrt{4k^2 - k_u^2}\, x_n - jk_u y_n\right)}_{\text{PM}}$$

which is also an AM-PM signal. Moreover the slow-time domain and slow-time Doppler domain AM signals are related to each other via the following scale transformation:

$$A_n\big[\omega, 2k \sin\theta_n(u)\big] = a_n(\omega, x_n, y_n - u),$$

where

$$\theta_n(u) = \arctan\left(\frac{y_n - u}{x_n}\right),$$

is the aspect angle of the nth target when the radar is located at $(0, u)$.

The above result indicates that the information in the AM signal $a_n(\omega, x_n, y_n - u)$ directly maps into the (ω, k_u) domain. Moreover, for the reconstruction, the (ω, k_u) domain information is directly mapped onto the target spatial frequency domain, that is, (k_x, k_y) domain. Thus the AM signal $a_n(\omega, x_n, y_n - u)$ can be viewed as a spatial frequency (k_x, k_y) domain (magnitude and/or phase) filter for the nth target.

To identify the transformation from the (ω, u) domain to the (k_x, k_y) domain and the resultant filter, we combine the SAR mapping

$$k_x(\omega, k_u) = \sqrt{4k^2 - k_u^2},$$

$$k_y(\omega, k_u) = k_u,$$

and the slow-time to the slow-time Doppler mapping

$$k_u = 2k \sin\theta_n(u).$$

This yields the following mapping from the (k_x, k_y) domain to the (ω, u) domain (note that $k = \omega/c$):

$$2k = \sqrt{k_x^2 + k_y^2},$$

$$u = y_n - \frac{k_y}{k_x} x_n.$$

In this case the resultant filter for the nth target becomes

$$H_n(k_x, k_y) = a_n(\omega, x_n, y_n - u),$$

where (ω, u) are related to (k_x, k_y) via the above transformation.

The total measured SAR signal in the (ω, u) domain is

$$s(\omega, u) = P(\omega) \sum_n s_n(\omega, u)$$
$$= P(\omega) \sum_n a_n(\omega, x_n, y_n - u) \exp\left[-j2k\sqrt{x_n^2 + (y_n - u)^2}\right]$$

and its slow-time Doppler spectrum is

$$S(\omega, k_u) = P(\omega) \sum_n A_n(\omega, k_u) \exp\left(-j\sqrt{4k^2 - k_u^2}\, x_n - jk_u y_n\right).$$

With the inclusion of the target amplitude pattern, the matched-filtered reconstruction becomes

$$F\big[k_x(\omega, k_u), k_y(\omega, k_u)\big] = P^*(\omega) S(\omega, k_u)$$
$$= |P(\omega)|^2 \sum_n A_n(\omega, k_u) \exp\big(-jk_x x_n - jk_y y_n\big).$$

Thus the point spread function for the nth target in the spatial domain is

$$\text{psf}_n(x, y) = \mathcal{F}^{-1}_{((k_x, k_y))}\Big[|P(\omega)|^2 A_n(\omega, k_u)\Big],$$

where (k_x, k_y) are related to (ω, k_u) via the SAR spatial frequency mapping.

4.9 MOTION COMPENSATION USING GLOBAL POSITIONING SYSTEM

A practical issue which is associated with SAR systems involves compensation for nonlinear motion components (motion errors) of the radar-carrying vehicle. This problem arises due to the fact that the vehicle cannot maintain a constant speed on a linear course during the SAR data acquisition. These vehicles are commonly equipped with an instrument, called global positioning system (GPS), that provides an estimate of the coordinates of the vehicle during the data acquisition (i.e., at each pulse transmission or PRI). The GPS data are not very accurate. Yet they provide a fairly good database to remove the *gross* nonlinear motion components that are very destructive.

Once a GPS-based motion compensated SAR image is formed, one can use a measure of the sharpness of the targets in the imaging scene for enhanced motion compensation [carr]; this is known as *in-scene target*-based motion compensation. This section provides methods for GPS-based motion compensation for SAR problems; the next section treats the problem of motion compensation using in-scene targets. A signal theory tool which is crucial for both is the spatial frequency modeling of motion errors; this is described next.

Spatial Frequency Modeling of Motion Errors

We now exploit the slow-time Fourier properties of the SAR AM-PM signal and its interpretation as a filter in the target spatial frequency domain, which were discussed in the previous section, to model the effect of motion errors in the reconstructed SAR image. We denote the vehicle trajectory in the spatial domain as a function of the slow-time via

$$\left[x_e(u), u + y_e(u)\right],$$

where $x_e(u)$ and $y_e(u)$, respectively, are the motion errors in the range and cross-range domains. (We will discuss the three-dimensional motion errors later.)

Suppose that a unit omni-directional reflector is located at (x_n, y_n). The measured SAR signature of this target in the (ω, u) domain is

$$s_n(\omega, u) = \exp\left[-j2k\sqrt{\left[x_n - x_e(u)\right]^2 + \left[y_n - u - y_e(u)\right]^2}\right].$$

We can rewrite the above signature via

$$s_n(\omega, u) = a_{en}(\omega, x_n, y_n - u)\exp\left[-j2k\sqrt{x_n^2 + (y_n - u)^2}\right],$$

where

$$a_{en}(\omega, x_n, y_n - u) = \exp\left[j2kr_{en}(u)\right],$$

and

$$r_{en}(u) = -\sqrt{\left[x_n - x_e(u)\right]^2 + \left[y_n - u - y_e(u)\right]^2} + \sqrt{x_n^2 + (y_n - u)^2}.$$

We call $a_{en}(\omega, x_n, y_n - u)$ and $r_{en}(u)$, respectively, to be the motion phase error function and radial error for the nth target.

Thus, provided that the fluctuations of $a_{en}(\omega, x_n, y_n - u)$ are smaller than the fluctuations of the motion error-free SAR PM signal, the motion phase error function can be modeled as a filter in the spatial frequency (k_x, k_y) domain of the nth target. Note that this is a *spatially varying* filter function; that is, the filter varies with the coordinates of the target. In fact we showed earlier that the spatially varying filter is

$$\begin{aligned} H_{en}(k_x, k_y) &= a_{en}(\omega, x_n, y_n - u) \\ &= \exp\left[2kr_{en}(u)\right] \end{aligned}$$

where

$$2k = \sqrt{k_x^2 + k_y^2},$$

$$u = y_n - \frac{k_y}{k_x}x_n.$$

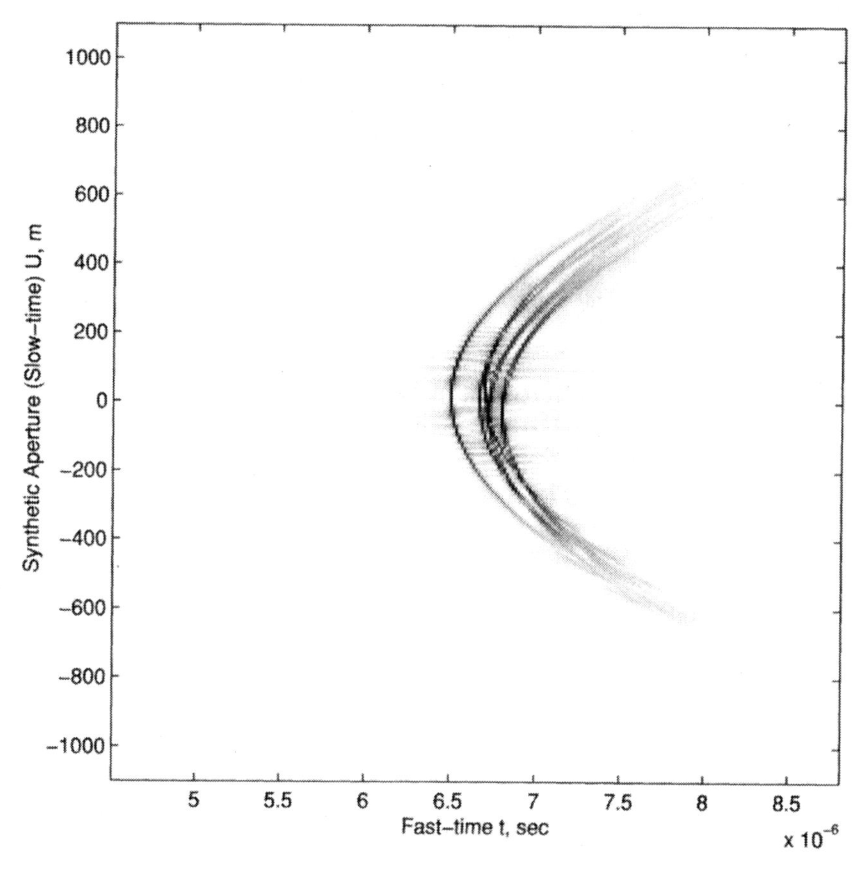

Figure 4.9a. SAR signal $\mathbf{s}(t, u)$ with no motion errors.

To demonstrate the effect of motion errors and the motion compensation algorithms, we consider a wide-bandwidth ([30, 90] MHz) and wide-beamwidth VHF stripmap SAR system, which will be examined in Chapter 6. Figure 4.9*a* shows the SAR signal $\mathbf{s}(t, u)$, with no motion errors, of seven unit reflectors; Figure 4.9*b* is the reconstructed SAR image of these targets. Figure 4.10*a* to *c*, respectively, shows simulated range motion error $x_e(u)$, cross-range motion error $y_e(u)$, and the resultant radial motion error $r_{en}(u)$ for the target at $(x_n, y_n) = (0, 0)$. Figure 4.10*d* is the SAR signature $\mathbf{s}(t, u)$ of the targets in the presence of the motion errors of Figures 4.10*a* and *b*. The targets appear to be completely washed out in the reconstructed SAR image from this database which appears in Figure 4.10*e*.

Narrow-Beamwidth Motion Compensation

As we mentioned earlier, a GPS system could be used to obtain an estimate of the vehicle trajectory and, as a result, the motion error $[x_e(u), y_e(u)]$. This information

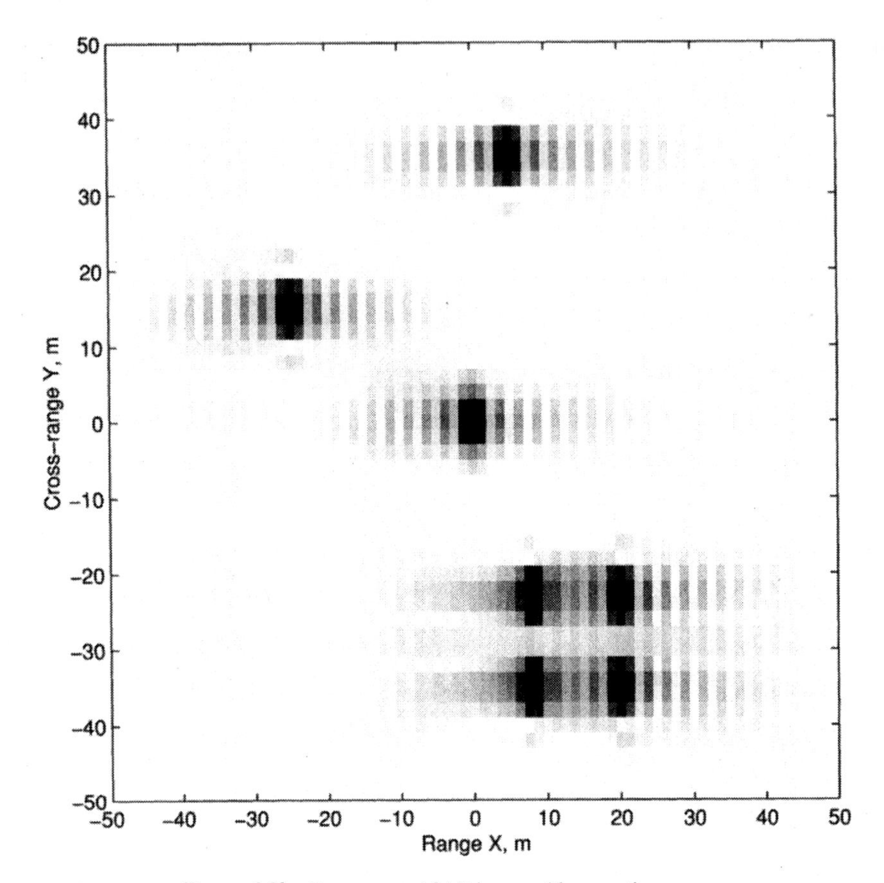

Figure 4.9b. Reconstructed SAR image with no motion errors.

can be used for motion compensation. In this section we show a GPS-based motion compensation that is useful when the target radial distance is much larger than the slow-time interval (synthetic aperture length) over which it is radiated with the radar signal; this is known as a narrow-beamwidth (look angle) for synthetic aperture. (We assumed an infinite aperture or slow-time for a generic SAR system. This is not the case in the practical spotlight and stripmap SAR systems.) In the next section we treat the general case of wide-beamwidth SAR systems.

In the *broadside* narrow-beamwidth SAR systems, the target range x_n is much larger than its cross-range y_n, the synthetic aperture u, and the motion errors. In this case the radial motion error for the nth target can be approximated via

$$r_{en}(u) \approx x_e(u).$$

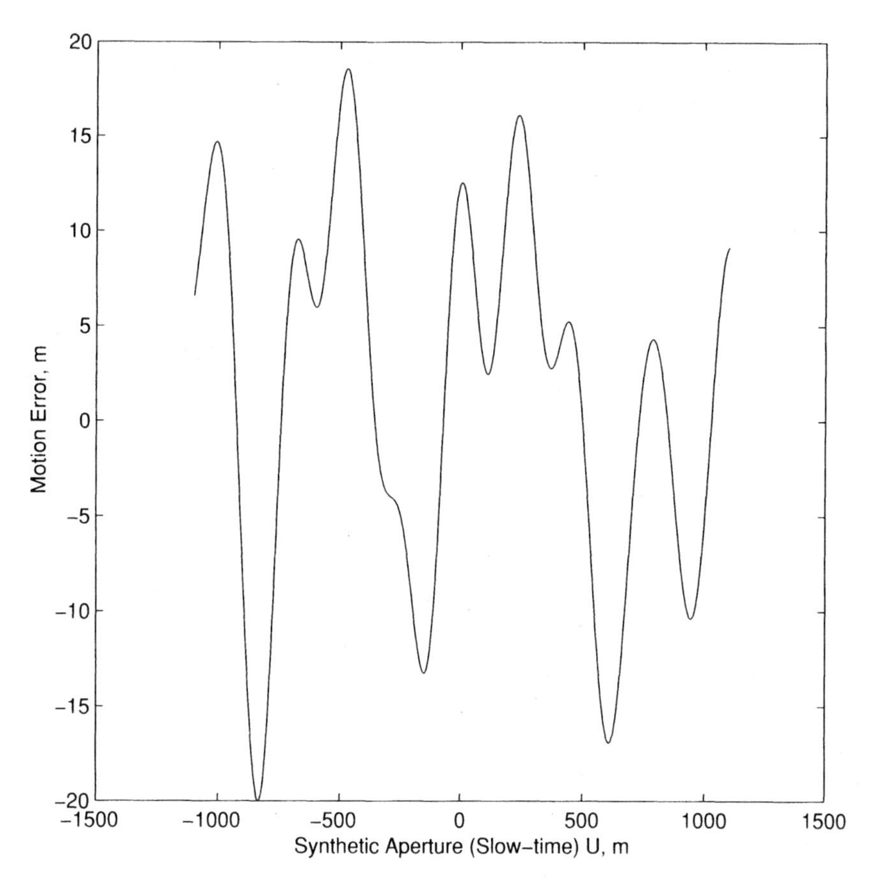

Figure 4.10a. Simulated range motion error $x_e(u)$.

The motion error phase function and filter function for the nth target become

$$\begin{aligned} H_{en}(k_x, k_y) &= a_{en}(\omega, x_n, y_n - u) \\ &\approx \exp\big[j2kx_e(u)\big] \end{aligned}$$

which is *invariant* in the coordinates of the target. Thus the user could perform motion compensation on the measured SAR signal in the (ω, u) domain via

$$s(\omega, u)\exp\big[-j2kx_e(u)\big].$$

We refer to the above operation as the narrow-beamwidth motion compensation.

For the general *squint* narrow-beamwidth SAR systems, the radial motion error function is approximated via

$$r_{en}(u) \approx x_e(u)\cos\theta_c + y_e(u)\sin\theta_c,$$

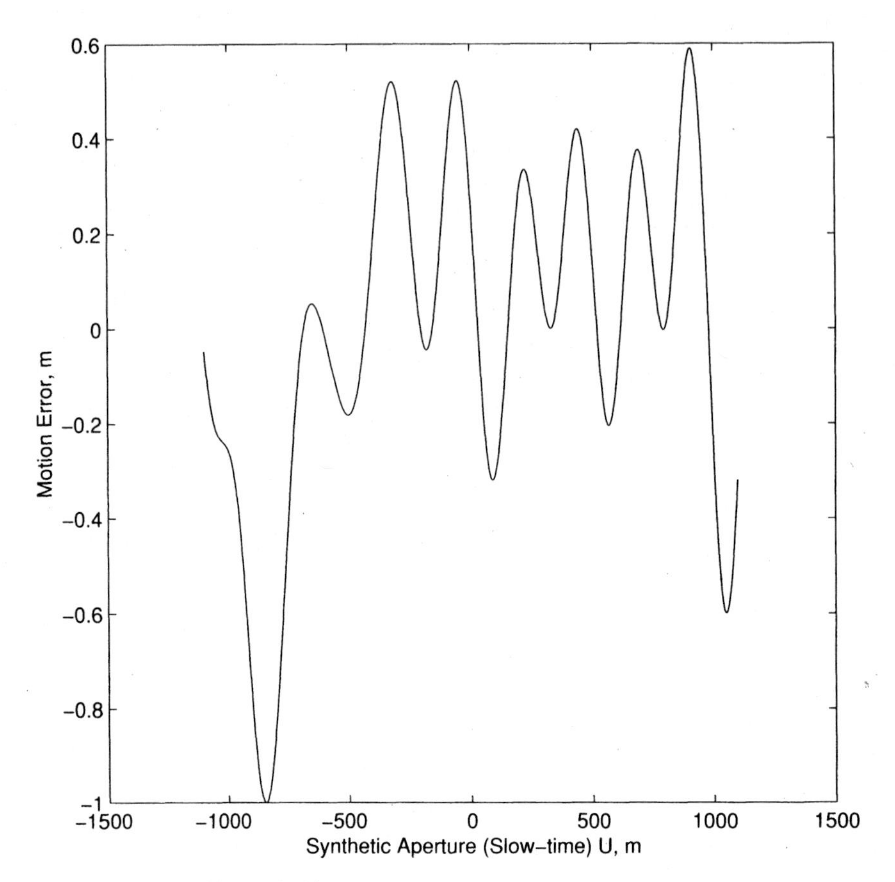

Figure 4.10b. Simulated cross-range motion error $y_e(u)$.

where

$$\theta_c = \arctan\left(\frac{Y_c}{X_c}\right)$$

is the average squint angle of the target area. In this case the narrow-beamwidth motion compensation is performed via

$$s(\omega, u) \exp\left[- j2kx_e(u) \cos\theta_c - j2ky_e(u) \sin\theta_c \right].$$

Consider the simulated SAR problem with motion errors of Figure 4.10*a* and *b*. Figure 4.11*a* shows the radial motion error $r_{en}(u)$ for the target at $(x_n, y_n) = (0, 0)$ after the narrow-beamwidth motion compensation. Figure 4.11*b* is the SAR signal $\mathbf{s}(t, u)$ of the targets after the narrow-beamwidth motion compensation. The reconstructed SAR image from this database is shown in Figure 4.11*c* which is a signifi-

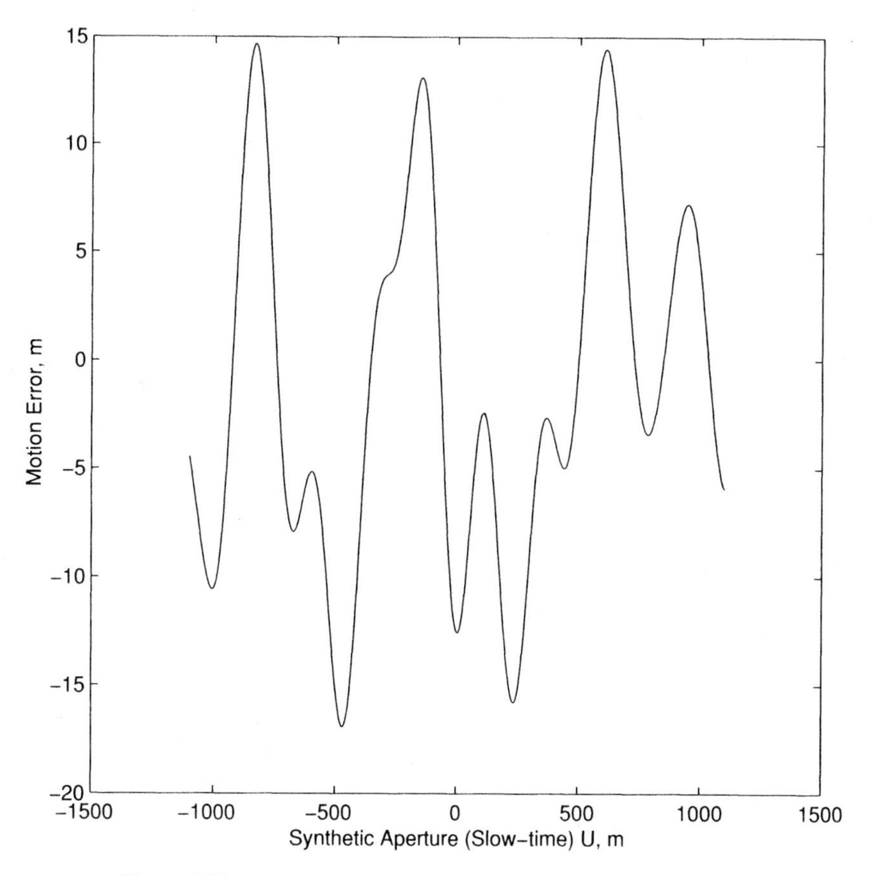

Figure 4.10c. Radial motion error $r_{en}(u)$ for a target at $(x_n, y_n) = (0, 0)$.

cant improvement over the SAR reconstruction without any motion compensation in Figure 4.10*e*.

Wide-Beamwidth Motion Compensation

Next we examine the motion compensation for the wide-beamwidth SAR systems. In most practical wide-beamwidth SAR systems, the motion phase error function

$$a_{en}(\omega, x_n, y_n - u)$$

has fluctuations comparable to that of the SAR PM signal. In this case we cannot use the slow-time Fourier property of the AM-PM signals by which we developed an expression for the filter function $H_{en}(k_x, k_y)$. Moreover the relatively high fluctuations of the motion error phase function often result in an aliased database in the slow-time domain.

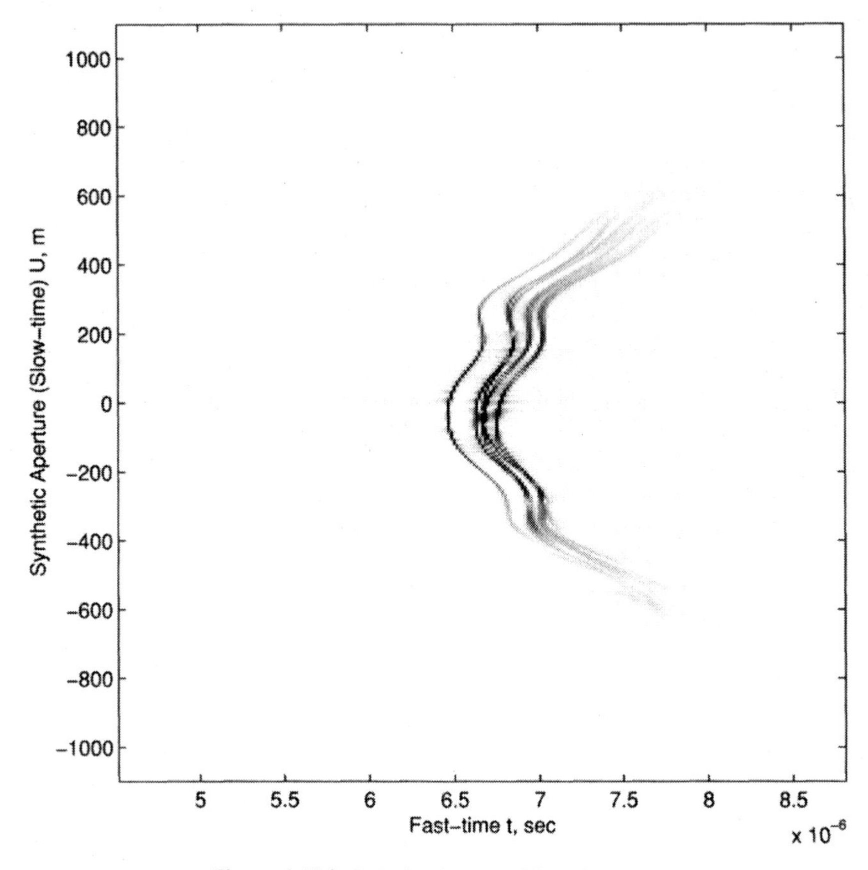

Figure 4.10d. SAR signal $s(t, u)$ with motion errors.

One way to address these two issues is to perform narrow-beamwidth motion compensation on the measured SAR signal. In this case the radial motion error function becomes

$$r_{en}(u) = -\sqrt{\left[x_n - x_e(u)\right]^2 + \left[y_n - u - y_e(u)\right]^2} + \sqrt{x_n^2 + (y_n - u)^2} - x_e(u)\cos\theta_c - y_e(u)\sin\theta_c.$$

In practical SAR systems this modified radial motion error has a smaller dynamic range than the radial motion error function without narrow-beamwidth motion compensation. This reduces the fluctuations and the bandwidth of the motion phase error function

$$a_{en}(\omega, x_n, y_n - u) = \exp\left[j2kr_{en}(u)\right].$$

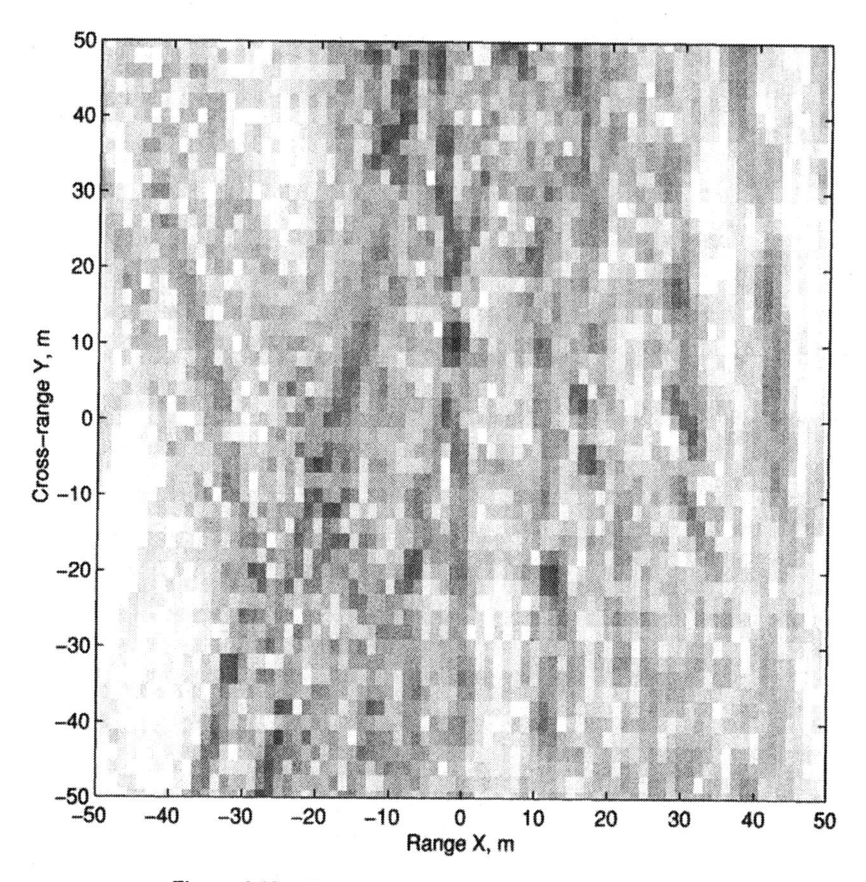

Figure 4.10e. Reconstructed SAR image with motion errors.

For wide-beamwidth motion compensation, the user first performs narrow-beamwidth motion compensation on the (ω, u) domain SAR signal. A SAR image is reconstructed using the narrow-beamwidth motion compensated SAR data. Then this SAR image is passed through the shift-varying filter

$$H^*_{exy}(k_x, k_y) = \exp\big[-j2kr_{exy}(u)\big],$$

where

$$2k = \sqrt{k_x^2 + k_y^2},$$

$$u = y - \frac{k_y}{k_x}x,$$

and r_{exy} is the radial motion error for the target at (x, y) after the narrow-beamwidth motion compensation.

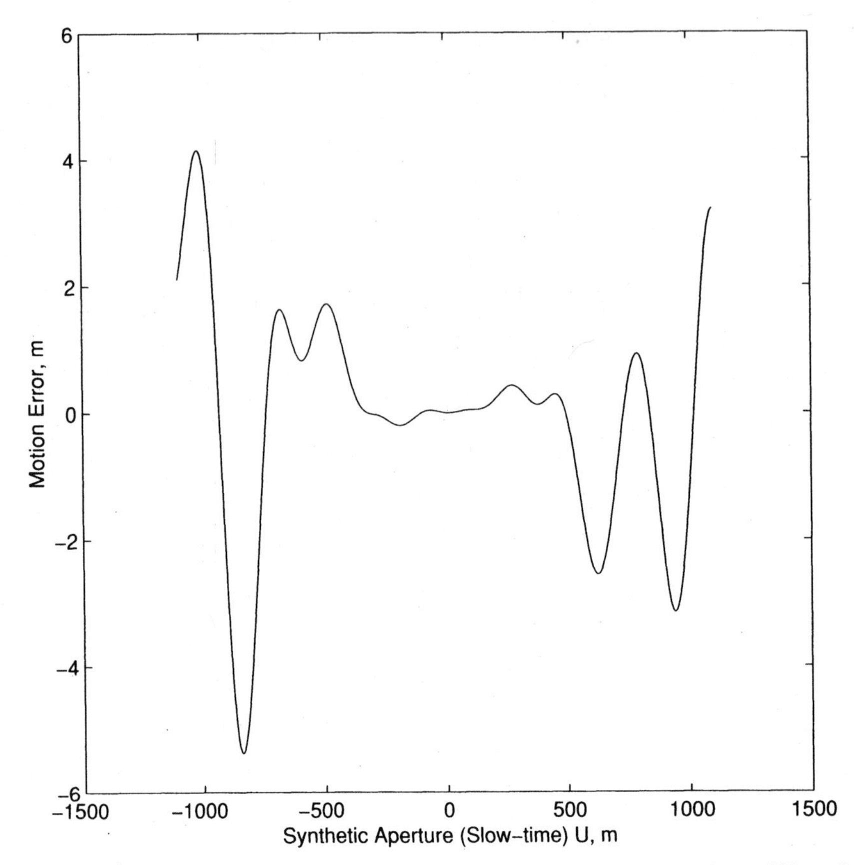

Figure 4.11a. Radial motion error for a target at $(x_n, y_n) = (0, 0)$ after narrow-beamwidth motion compensation.

Consider the simulated SAR problem with motion errors of Figures 4.10*a* and *b*. Figure 4.12 is the SAR reconstruction using the above-mentioned wide-beamwidth motion compensation algorithm. Note that the wide-beamwidth motion compensation filtering is performed on the image in Figure 4.11*c* which is the SAR reconstruction after the narrow-beamwidth motion compensation.

Three-Dimensional Wide-Beamwidth Motion Compensation

As we mentioned in Chapter 3, the two-dimensional analytical SAR model is a mathematical representation of the three-dimensional spatial domain. In practical SAR systems, one deals with motion errors in the third dimension, that is, elevation (or height), as well as the range and cross-range domains. The GPS-based motion compensation, which was formulated in the previous section, can also be used for motion compensation using three-dimensional GPS data.

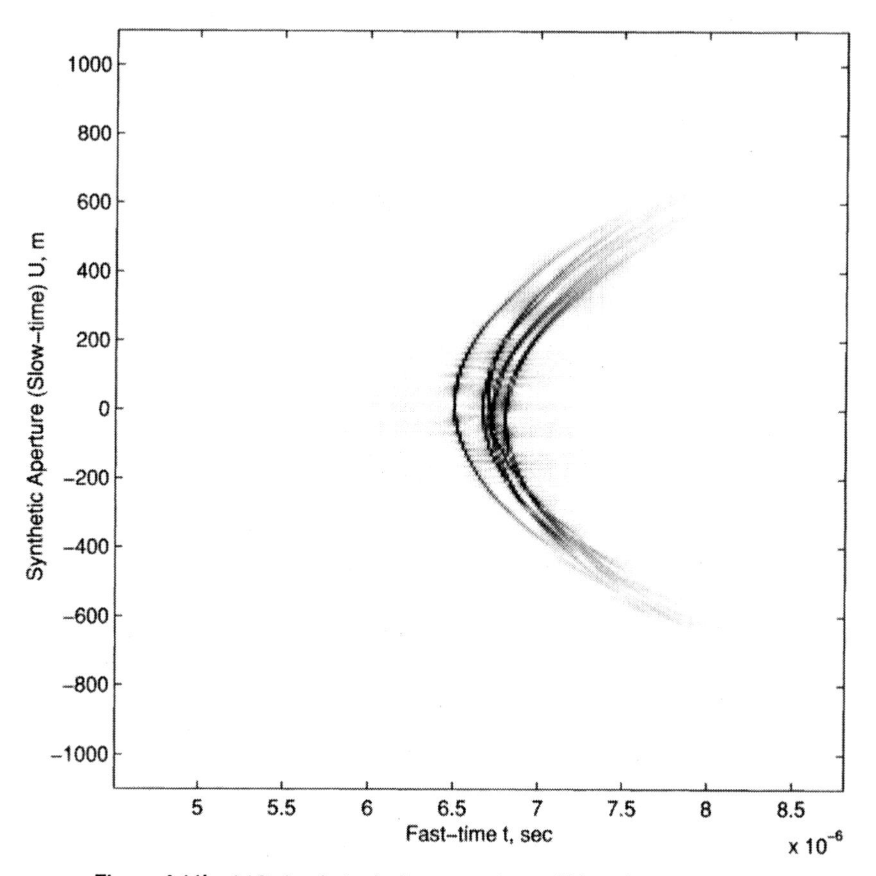

Figure 4.11b. SAR signal $s(t, u)$ after narrow-beamwidth motion compensation.

To show this, we first consider the motion error model in the three-dimensional geometries. In this case the vehicle trajectory can be expressed via

$$\left[x_e(u), u + y_e(u), z_e(u)\right],$$

where $x_e(u)$, $y_e(u)$, and $z_e(u)$, respectively, are the motion errors in the range, cross-range, and elevation domains. The distance of a target located at (x_n, y_n, z_n) from the radar at the slow-time u is

$$\sqrt{\left[x_n - x_e(u)\right]^2 + \left[y_n - u - y_e(u)\right]^2 + \left[z_n - z_e(u)\right]^2}.$$

Thus the radial motion error function becomes

$$r_{en}(u) = -\sqrt{\left[x_n - x_e(u)\right]^2 + \left[y_n - u - y_e(u)\right]^2 + \left[z_n - z_e(u)\right]^2} + \sqrt{x_n^2 + (y_n - u)^2 + z_n^2}.$$

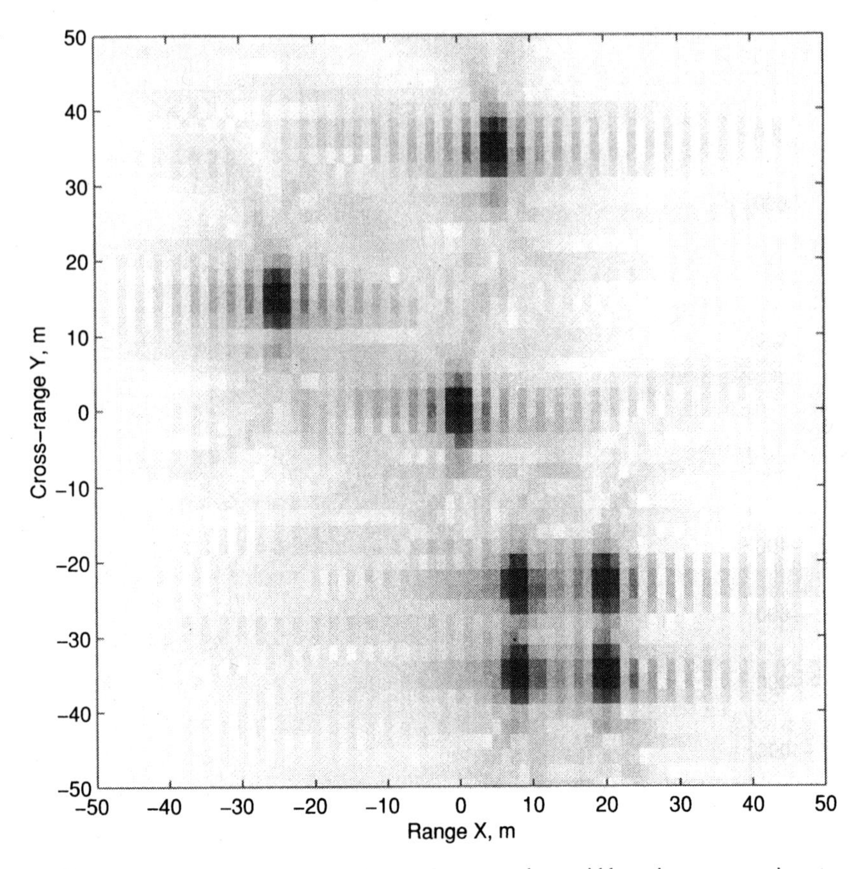

Figure 4.11c. Reconstructed SAR image after narrow-beamwidth motion compensation.

For the narrow-beamwidth motion compensation, the radial motion error function is approximated via

$$r_{en}(u) \approx x_e(u)\cos\theta_z\cos\theta_c + y_e(u)\sin\theta_c + z_e(u)\sin\theta_z\cos\theta_c,$$

where

$$\theta_z = \arctan\left(\frac{Z_c}{X_c}\right)$$

is the average depression (elevation) angle of the target area; Z_c is the mean elevation of the target area. In this case the modified radial motion error function for wide-beamwidth motion compensation is

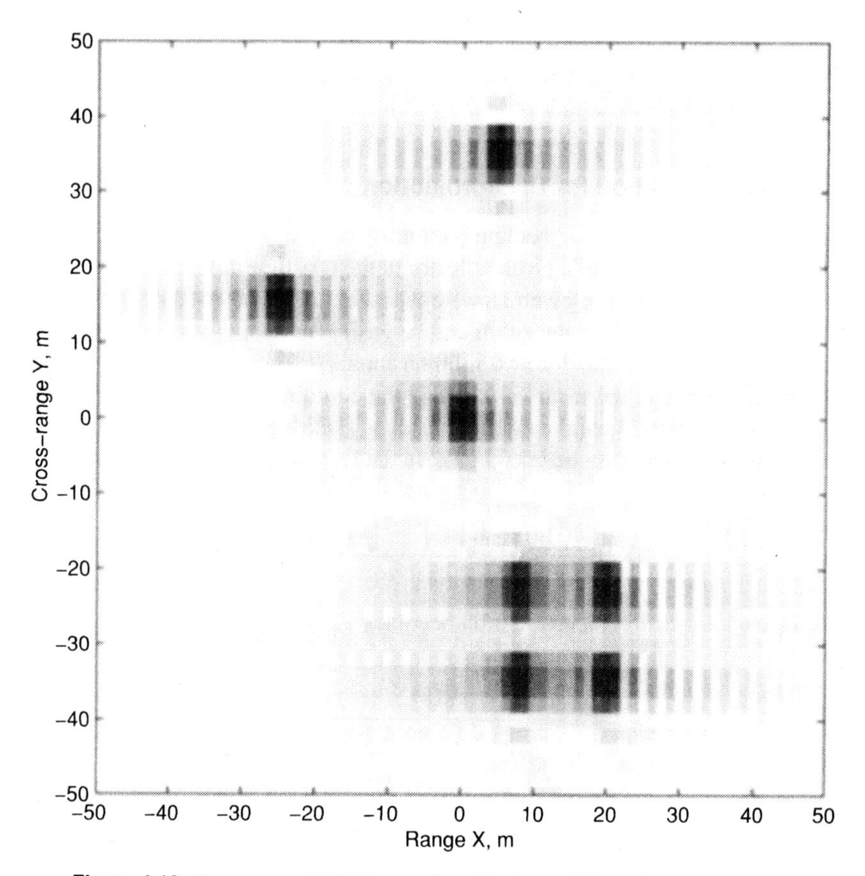

Figure 4.12. Reconstructed SAR image after wide-beamwidth motion compensation.

$$
\begin{aligned}
r_{en}(u) = & -\sqrt{\left[x_n - x_e(u)\right]^2 + \left[y_n - u - y_e(u)\right]^2 + \left[z_n - z_e(u)\right]^2} \\
& + \sqrt{x_n^2 + (y_n - u)^2 + z_n^2} \\
& - x_e(u)\cos\theta_z\cos\theta_c - y_e(u)\sin\theta_c - z_e(u)\sin\theta_z\cos\theta_c.
\end{aligned}
$$

Moreover the spatially varying filter function is found via

$$
H^*_{exyz}(k_x, k_y) = \exp\left[-j2kr_{exyz}(u)\right],
$$

where

$$
\begin{aligned}
2k &= \sqrt{k_x^2 + k_y^2}, \\
u &= y - \frac{k_y}{k_x}\sqrt{x^2 + z^2},
\end{aligned}
$$

and r_{exyz} is the radial motion error for the target at (x, y, z) after the narrow-beamwidth motion compensation. (In the above, k_x represents the spatial frequency domain for the slant-range, that is, $\sqrt{x^2 + z^2}$.)

Motion Compensation for Backprojection

As we mentioned earlier, the backprojection reconstruction algorithm is based on tracing back the signature of a given reflector in the fast-time domain of the matched-filtered signal $s_M(t, u)$ at a given slow-time u and coherently adding the results at the available u values. The algorithm can be easily modified to incorporate known motion errors. We show this for a two-dimensional SAR problem.

At the slow-time u, the radar is located at the coordinates $[x_e(u), u + y_e(u)]$. For a given reflector at the spatial coordinates (x_i, y_j), the signature of this target in the matched-filtered signal can be traced back to the fast-time:

$$t_{ij}(u) = \frac{2\sqrt{\left[x_i - x_e(u)\right]^2 + \left[y_j - u - y_e(u)\right]^2}}{c}.$$

Thus the backprojection reconstruction becomes

$$f(x_i, y_j) = \int_u s_M \left[\frac{2\sqrt{\left[x_i - x_e(u)\right]^2 + \left[y_j - u - y_e(u)\right]^2}}{c}, u \right] du.$$

The algorithm is much simpler to implement than the Fourier-based motion compensation which was described earlier. Its main disadvantage is its high computational cost. In fact the Fourier reconstruction followed by the spatially varying filtering for motion compensation requires significantly less computer time than the backprojection algorithm. Moreover, as we will see in the next section, the spatial frequency interpretation of the motion errors provides a multidimensional signal processing framework for motion compensation using in-scene targets.

4.10 MOTION COMPENSATION USING IN-SCENE TARGETS

Next we examine the problem of removing residual motion errors that are due to inaccuracy of the GPS data. This analysis is also based on the spatial frequency interpretation of motion errors. The objective here is to exploit the *signature* of the targets that appear in the SAR imaging scene for motion compensation.

Narrow-Beamwidth Motion Compensation

Suppose that there exists a distinct and perfectly omni-directional reflector in the imaging scene; we call this target the *in-scene target*. We identify the coordinates of the in-scene target by (x_ℓ, y_ℓ). We also require that the SAR image of the in-scene target be fairly localized in the GPS-based motion compensated image. In this case we can extract the SAR image signature of the in-scene target, that is, $h_{e\ell}(x, y)$ which is shifted by (x_ℓ, y_ℓ).

The two-dimensional Fourier transform of this signature is $H_{e\ell}(k_x, k_y)$ which is the motion error filter function for the in-scene target in the spatial frequency domain. Our objective is to remove the residual motion errors in the entire SAR image from the knowledge of the motion error filter function of the in-scene target, that is, $H_{e\ell}(k_x, k_y)$.

In the case of the narrow-beamwidth SAR systems, the radial motion error function for the ℓth target can be approximated by

$$r_{e\ell}(u) \approx x_e(u)\cos\theta_c + y_e(u)\sin\theta_c,$$

where θ_c is the average target area squint angle. (For most stripmap systems, $\theta_c = 0$.) This approximation states that the radial motion error function is invariant in the target coordinates.

In this case the inverse spatial Fourier transform of the ℓth target signature in the reconstructed SAR image is

$$H_{e\ell}(k_x, k_y) \approx \exp\left[j2kx_e(u)\cos\theta_c + j2ky_e(u)\sin\theta_c\right],$$

which is also invariant in the coordinates of the ℓth target. In other words, for any target in the scene, for example, the nth target which is located at the coordinates (x_n, y_n), the spatial frequency filter function that corresponds to the motion errors is

$$H_{en}(k_x, k_y) \approx \exp\left[j2kx_e(u)\cos\theta_c + j2ky_e(u)\sin\theta_c\right].$$

Thus, for narrow-beamwidth motion compensation, the Fourier spectrum of the reconstructed target function is multiplied with the complex conjugate of $H_{e\ell}(k_x, k_y)$; that is,

$$F(k_x, k_y)H^*_{e\ell}(k_x, k_y).$$

Wide-Beamwidth Motion Compensation

Next we address the problem of wide-beamwidth motion compensation in a two-dimensional SAR scene. It is assumed that there are two distinct and perfectly omnidirectional reflectors in the imaging scene; we call these targets the two *in-scene targets*. We identify the coordinates of the two in-scene targets by (x_ℓ, y_ℓ) and (x_m, y_m). Moreover we require that the SAR images of the two in-scene targets are fairly localized in the GPS-based wide-beamwidth motion compensated image. In this case we can extract each of these two signatures, which are $h_{e\ell}(x, y)$ shifted by (x_ℓ, y_ℓ) and $h_{em}(x, y)$ shifted by (x_m, y_m).

The two-dimensional Fourier transform of these two signatures are $H_{e\ell}(k_x, k_y)$ and $H_{em}(k_x, k_y)$ which are the motion error filter functions in the spatial frequency domain. Our objective is to remove the residual motion errors in the entire SAR image from the knowledge of the motion error filter functions of the two in-scene targets, that is, $H_{e\ell}(k_x, k_y)$ and $H_{em}(k_x, k_y)$.

In practice, the residual range and cross-range motion errors, labeled $[x_e(u), y_e(u)]$, are much smaller than the coordinates of the two in-scene targets. In this case the radial motion errors for the two in-scene targets can be approximated via

$$r_{e\ell}(u) \approx x_e(u)\cos\theta_\ell(u) + y_e(u)\sin\theta_\ell(u),$$

$$r_{em}(u) \approx x_e(u)\cos\theta_m(u) + y_e(u)\sin\theta_m(u).$$

Thus the motion error filter functions of the two in-scene targets can be approximated by

$$H_\ell(k_x, k_y) \approx \exp\left[j2kx_e(u)\cos\theta_\ell(u) + j2ky_e(u)\sin\theta_\ell(u)\right],$$

with

$$2k = \sqrt{k_x^2 + k_y^2},$$

$$u = y_\ell - \frac{k_y}{k_x}x_\ell,$$

and

$$H_m(k_x, k_y) \approx \exp\left[j2kx_e(u)\cos\theta_m(u) + j2ky_e(u)\sin\theta_m(u)\right],$$

with

$$2k = \sqrt{k_x^2 + k_y^2},$$

$$u = y_m - \frac{k_y}{k_x}x_m.$$

(Note that the functional mapping for u varies with the coordinates of the two in-scene targets.)

Using the above mappings from the (k_x, k_y) domain to the (ω, u) domain, one can transform the motion error filter functions into the motion error phase functions:

$$a_\ell(\omega, x_\ell, y_\ell - u) \approx \exp\left[j2kx_e(u)\cos\theta_\ell(u) + j2ky_e(u)\sin\theta_\ell(u)\right],$$

$$a_m(\omega, x_m, y_m - u) \approx \exp\left[j2kx_e(u)\cos\theta_m(u) + j2ky_e(u)\sin\theta_m(u)\right].$$

Since (x_ℓ, y_ℓ) and (x_m, y_m) are known, $\theta_\ell(u)$ and $\theta_m(u)$ are also known. Thus the residual motion errors can be found at a fixed u by extracting the phase of $a_\ell(\omega, \cdot)$ and $a_m(\omega, \cdot)$ at any available fast-time frequency ω, and solving the resultant system of two equations with two unknowns. For this the phase functions of $a_\ell(\omega, \cdot)$ and $a_m(\omega, \cdot)$ should be *unwrapped* in the slow-time u domain. The resultant solutions for $[x_e(u), y_e(u)]$ at the available fast-time frequencies could be averaged to improve accuracy.

Due to the difficulties associated with implementing phase-unwrapping algorithms (and certain ambiguities), the above-mentioned procedure might not be a viable option to solve for the residual motion errors. There exists a direct way to determine the spatially varying motion error filter function at a given point in the spatial domain (SAR image).

Suppose that we wish to construct the motion error filter function $H_n(k_x, k_y)$ for a reflector at the coordinates (x_n, y_n). Consider a given spatial frequency point (k_x, k_y) which maps into the following values of the fast-time frequency ω and the slow-time u:

$$2k = \sqrt{k_x^2 + k_y^2},$$

$$u = y_n - \frac{k_y}{k_x} x_n.$$

The radial motion error function for the nth target is

$$r_{en}(u) \approx x_e(u) \cos\theta_n(u) + y_e(u) \sin\theta_n(u).$$

Our claim is that the motion phase error function for the nth target, that is,

$$\exp\left[j2kx_e(u)\cos\theta_n(u) + j2ky_e(u)\sin\theta_n\right],$$

can be constructed by the direct or conjugate product of the motion phase errors functions of the two in-scene targets at two *different* fast-time frequencies, that is, the direct or conjugate product of [note that $u = y_n - (k_y x_n / k_x)$ is fixed]

$$\exp\left[j2k_\ell x_e(u)\cos\theta_\ell(u) + j2k_\ell y_e(u)\sin\theta_\ell\right]$$

and

$$\exp\left[j2k_m x_e(u)\cos\theta_m(u) + j2k_m y_e(u)\sin\theta_m\right],$$

where (k_ℓ, k_m) or (ω_ℓ, ω_m) are unknown which are to be determined.

By equating the motion phase error of the nth target at (ω, u) with the product of the motion phase error of the ℓth target at (u, ω_ℓ) and the mth target at (u, ω_m), we obtain

$$\left[k_\ell \cos\theta_\ell(u) + k_m \cos\theta_m(u)\right] x_e(u) + \left[k_\ell \sin\theta_\ell(u) + k_m \sin\theta_m(u)\right] y_e(u)$$
$$= k\cos\theta_n(u) x_e(u) + k\sin\theta_n(u) y_e(u).$$

By setting the coefficients of $x_e(u)$ and $y_e(u)$ on the two sides of the above equation equal to each other, we obtain

$$k_\ell \cos\theta_\ell(u) + k_m \cos\theta_m(u) = k\cos\theta_n(u),$$

$$k_\ell \sin\theta_\ell(u) + k_m \sin\theta_m(u) = k\sin\theta_n(u).$$

Solving for (k_ℓ, k_m) from these two equations yields

$$k_\ell = \frac{\sin\left[\theta_m(u) - \theta_n(u)\right]}{\sin\left[\theta_m(u) - \theta_\ell(u)\right]} k,$$

$$k_m = \frac{\sin\left[\theta_n(u) - \theta_\ell(u)\right]}{\sin\left[\theta_m(u) - \theta_\ell(u)\right]} k.$$

If the solution for k_ℓ and/or k_m is a negative number, then the complex conjugate of the ℓth and/or mth in-scene target motion phase error is used to construct the motion phase error for the nth target.

If the solution for k_ℓ or k_m corresponds to a fast-time frequency which is outside the radar signal bandwidth, then an acceptable solution does not exist. Moreover, as we will see in our discussion on stripmap SAR systems in Chapter 6, a target is not *observable* by the radar at all aspect angles (i.e., u values). In this case the user could only perform motion compensation at the slow-time intervals (u values) over which *both* of the in-scene targets are observable by (or fall in the radiation pattern of) the radar.

Three-Dimensional Wide-Beamwidth Motion Compensation

For the three-dimensional motion compensation problems, three in-scene targets are needed to remove the effects of the unknown residual motion errors $[x_e(u), y_e(u), z_e(u)]$. In this case a procedure similar to the one for the two-dimensional wide-beamwidth problems yields three equations for three unknown fast-time frequencies for the three in-scene targets' motion phase errors. A similar solution then follows.

4.11 POLAR FORMAT PROCESSING

The SAR reconstruction method in Section 4.5 is a relatively new scheme which was developed in the late 1980s. Prior to the introduction of this method, SAR image formation was done using two approximation-based imaging algorithms: polar format processing which is outlined in this section (and its application in ISAR in Section 4.12) and range-Doppler imaging which will be examined in Section 4.13.

Reconstruction via polar format processing, and GPS-based motion compensation for this algorithm, are described next.

Plane Wave Approximation-Based Reconstruction

Polar format processing is based on an approximation of the SAR system model which was developed earlier in Sections 4.1 and 4.2. To show this, consider the Fourier transform of the generic SAR signal with respect to the fast-time t (see Section 4.2); that is,

$$s(\omega, u) = P(\omega) \sum_n \sigma_n \exp\left[-j2k\sqrt{x_n^2 + (y_n - u)^2}\right].$$

In the above model, the origin in the spatial domain is located at the synthetic aperture with $u = 0$. To facilitate our analysis, we move the origin to the center of the target area. Suppose that the center of the target area is at (X_c, Y_c). The SAR system model in the resultant shifted spatial domain (i.e., after the coordinate transformation) is

$$s(\omega, u) = P(\omega) \sum_n \sigma_n \exp\left[-j2k\sqrt{(X_c + x_n)^2 + (Y_c + y_n - u)^2}\right].$$

Consider the SAR signal for a unit reflector which is located at the center of the target scene; that is,

$$s_0(\omega, u) = P(\omega) \exp\left[-j2k\sqrt{X_c^2 + (Y_c - u)^2}\right].$$

Earlier in this chapter and Chapter 2, we called this signal the reference SAR signal. Moreover, in Chapter 2, we also defined the slow-time compressed SAR signal via

$$\begin{aligned} s_c(\omega, u) &= s(\omega, u) s_0^*(\omega, u) \\ &= |P(\omega)|^2 \sum_n \sigma_n \exp\left[-j2k\sqrt{(X_c + x_n)^2 + (Y_c + y_n - u)^2}\right] \\ &\quad \times \exp\left[j2k\sqrt{X_c^2 + (Y_c - u)^2}\right]. \end{aligned}$$

Consider the following expansion and approximation of the target distance to the radar:

$$\begin{aligned} \sqrt{(X_c + x_n)^2 + (Y_c + y_n - u)^2} &= \sqrt{X_c^2 + (Y_c - u)^2 + 2X_c x_n + 2Y_c y_n + \cdots} \\ &= \sqrt{X_c^2 + (Y_c - u)^2} \\ &\qquad + \cos\theta_0(u) x_n + \sin\theta_0(u) y_n + \cdots \\ &\approx \sqrt{X_c^2 + (Y_c - u)^2} + \cos\theta_0(u) x_n + \sin\theta_0(u) y_n, \end{aligned}$$

where

$$\theta_0(u) = \arctan\left(\frac{Y_c - u}{X_c}\right)$$

is the aspect angle of the radar with respect to the center of the target area at the slow-time (synthetic aperture position) u.

Using the above approximation in the expression for the slow-time compressed SAR signal, we obtain

$$s_c(\omega, u) \approx |P(\omega)|^2 \sum_n \sigma_n \exp\left[-jk_x(\omega, u)x_n - jk_y(\omega, u)y_n\right],$$

where

$$k_x(\omega, u) = 2k \cos\theta_0(u),$$

$$k_y(\omega, u) = 2k \sin\theta_0(u).$$

Note that the pair $[2k, \theta_0(u)]$ is the polar function mapping of the (k_x, k_y) domain. The target function reconstruction in the spatial frequency domain is formed via

$$\begin{aligned} F\left[k_x(\omega, u), k_y(\omega, u)\right] &= s_c(\omega, u) \\ &\approx |P(\omega)|^2 \sum_n \sigma_n \exp\left[-jk_x(\omega, u)x_n - jk_y(\omega, u)y_n\right]. \end{aligned}$$

The above is known as the polar format processing or reconstruction [aus; wal]; the polar format processing is summarized in Figure 4.13. The name polar format processing stems from the fact that the samples of the SAR signal in the (ω, u) domain are mapped into the polar samples of the target function in the spatial frequency (k_x, k_y) domain. [This is not quite right, since $\theta_0(u)$ is not a linear function of u; see the discussion on the narrow-beamwidth approximation which follows.] Examples of such samples in the (k_x, k_y) domain that are generated by the mapping of evenly spaced samples of the (ω, u) domain are shown in Figure 4.14*a* (broadside) and *b* (squint); the samples are identified by the $*$ symbol.

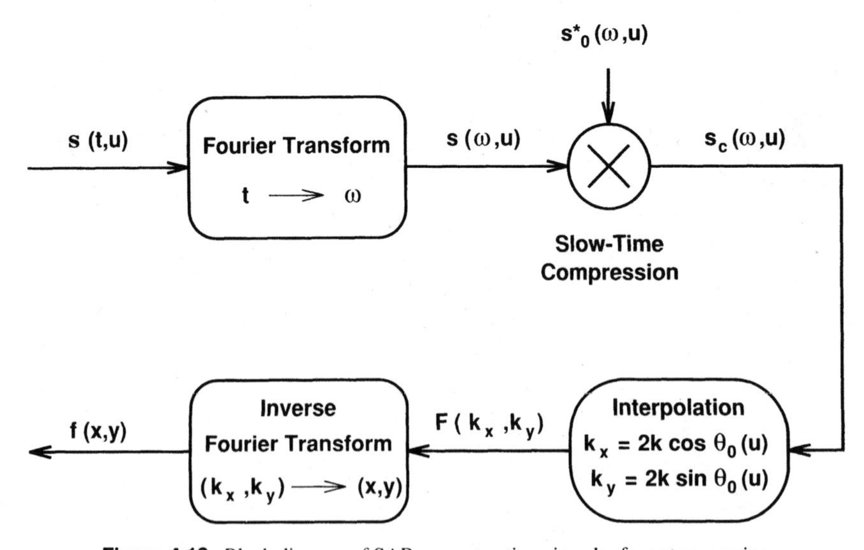

Figure 4.13. Block diagram of SAR reconstruction via polar format processing.

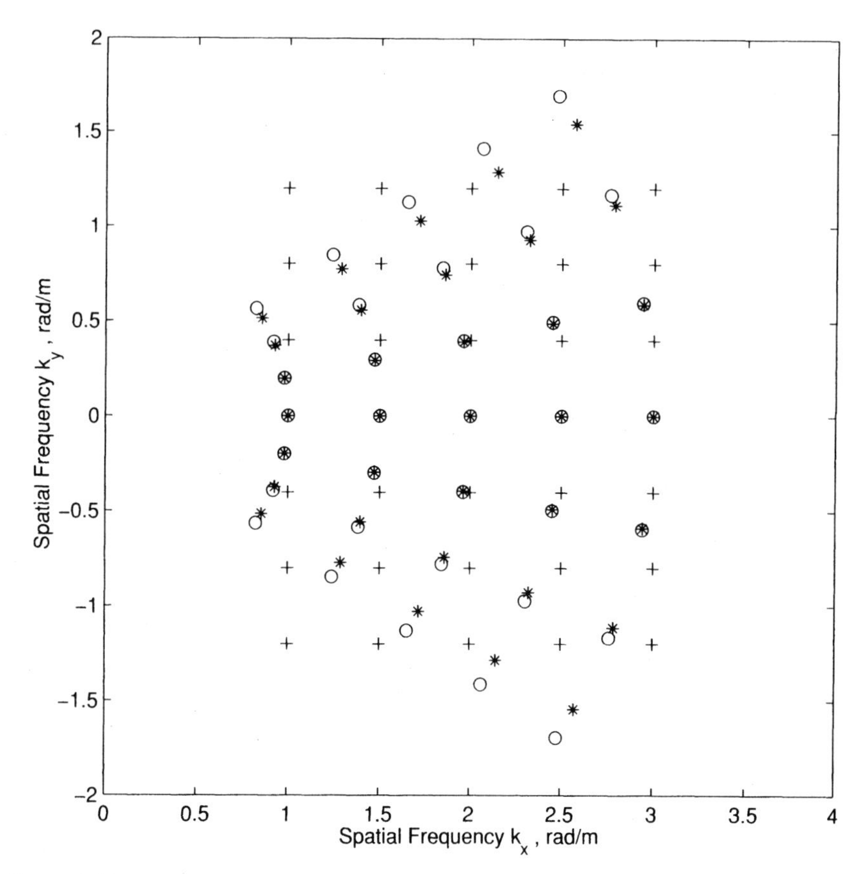

Figure 4.14a. Polar format samples $*$, with narrow-beamwidth approximation $\circ$, and with narrow-bandwidth and narrow-beamwidth approximations $+$: broadside target area.

For target reconstruction in the spatial domain, the samples of the two-dimensional function $F(k_x, k_y)$ on a rectilinear grid are interpolated from its available polar samples; the spatial domain image $f(x, y)$ is obtained via the two-dimensional FFT of the interpolated rectilinear samples.

The approximation used in the polar format processing is based on neglecting the wavefront curvature in the spherical PM signal; this is known as the plane wave approximation of a spherical wave function [s92j; s94]. For the validity of this approximation, certain conditions must be met in terms of the size of the target area, the radar frequency, and the extent of the synthetic aperture u [s94]. Later in this section we will examine a procedure to compensate for the wavefront curvature errors of the polar format processing; the approach is similar to the motion compensation method which was examined earlier in Section 4.9. Next we outline the narrow-beamwidth approximation commonly used in conjunction with the polar format processing.

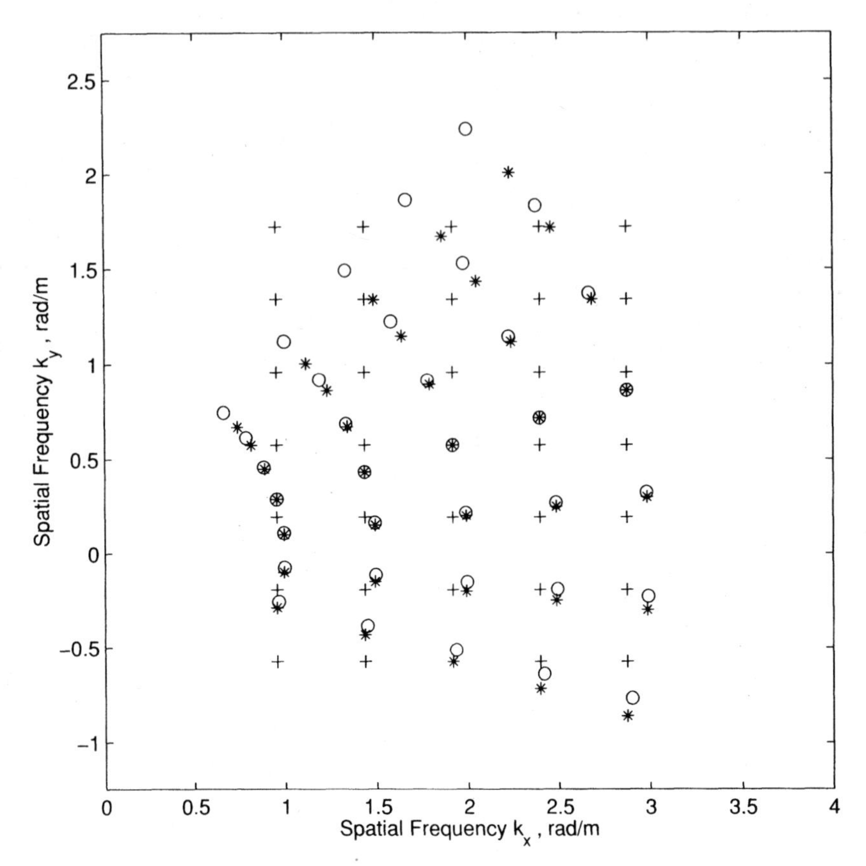

Figure 4.14b. Polar format samples $*$, with narrow-beamwidth approximation $\circ$, and with narrow-bandwidth and narrow-beamwidth approximations $+$: squint target area.

Narrow-Beamwidth Approximation

As we mentioned earlier, the mapping of the evenly spaced samples in the measurement domain (ω, u) to the target spatial frequency (k_x, k_y) domain, that is,

$$k_x(\omega, u) = 2k \cos \theta_0(u),$$

$$k_y(\omega, u) = 2k \sin \theta_0(u),$$

is not polar, since $\theta_0(u)$ is a nonlinear function of u. Most polar format processors utilize the following approximation to linearize this mapping; the approach is based on assuming a narrow-beamwidth for the transmitting/receiving radar of the SAR system.

- We should emphasize that the narrow-beamwidth approximation can be avoided in plane wave approximation-based reconstruction. One could simply develop a two-dimensional interpolation scheme to convert the available unevenly spaced samples of $F(k_x, k_y)$, which are the result of the mapping of the evenly spaced samples of $s_c(\omega, u)$ in the (ω, u) domain, into its evenly spaced samples in the (k_x, k_y) domain [s94, ch. 2]. This interpolation is not any simpler or more complicated than the polar interpolation.

We consider the general squint case where the center of the target area is located at (X_c, Y_c) when the radar is at $u = 0$. We denote the polar coordinates for the center of the target area by (θ_c, R_c); that is,

$$\theta_c = \arctan\left(\frac{Y_c}{X_c}\right),$$

$$R_c = \sqrt{X_c^2 + Y_c^2}.$$

Provided that $u \ll R_c$ (narrow-beamwidth approximation), the aspect angle of the center of the target area when the radar is at $(0, u)$, that is,

$$\theta_0(u) = \arctan\left(\frac{Y_c - u}{X_c}\right),$$

can be approximated by the following linear function of the synthetic aperture u domain:

$$\theta_0(u) \approx \theta_c - \frac{\cos\theta_c}{R_c}u.$$

With the above approximation the mapping from the measurement (ω, u) domain into the target spatial frequency (k_x, k_y) domain, that is,

$$k_x(\omega, u) \approx 2k\cos\left(\theta_c - \frac{\cos\theta_c}{R_c}u\right)$$

$$k_y(\omega, u) \approx 2k\sin\left(\theta_c - \frac{\cos\theta_c}{R_c}u\right)$$

becomes polar. These samples are identified by the symbol "o" in the examples of Figure 4.14*a* and *b*.

We should point out that the examples which are shown in Figure 4.14*a* and *b* correspond to a wide-bandwidth VHF SAR system. The main source of error in polar format processing in the high frequency (e.g., X band) SAR systems is the plane wave approximation [s94].

Narrow-Bandwidth and Narrow-Beamwidth Approximation

Another approximation that is used for near-broadside SAR image formation via polar format processing imposes constraints on the radar bandwidth as well as the radar beamwidth in order to come up with a *quick* target reconstruction that does not require any interpolation. From the mathematical point of view, all the nonlinear functions of (ω, u) in the mapping onto the (k_x, k_y) domain are neglected. The following provides a physical interpretation of this approximation.

In this approach the radar beamwidth is approximated to be narrow enough such that the circular curvature in the mapping of the range spatial frequency domain can be ignored; that is,

$$\begin{aligned} k_x &= 2k \cos \theta_0(u) \\ &\approx 2k \cos \theta_c, \end{aligned}$$

which is invariant in the synthetic aperture u domain. In addition to a small radar beamwidth, the radar bandwidth is assumed to be much smaller than its carrier frequency (i.e., $|k - k_c| \ll k_c$ for the available wavenumbers k) such that we have

$$\begin{aligned} k_y &= 2k \sin \theta_0(u) \\ &\approx 2k \left(\sin \theta_c - \frac{\cos^2 \theta_c}{R_c} u \right) \\ &\approx 2k \sin \theta_c - 2k_c \frac{\cos^2 \theta_c}{R_c} u, \end{aligned}$$

where k_c is the wavenumber at the carrier frequency ($k_c = \omega_c / c$). Note that k_y is a linear function of both u and k (or ω) in the above model.

The combination of the two approximations for (k_x, k_y) yields

$$\begin{aligned} k_x(\omega) &\approx 2k \cos \theta_c, \\ k_y(\omega, u) &\approx 2k \sin \theta_c - 2k_c \frac{\cos^2 \theta_c}{R_c} u, \end{aligned}$$

which corresponds to a *hexagonal* sampling in the target spatial frequency domain. For target reconstruction, one still must convert the hexagonal samples into rectilinear samples in the (k_x, k_y) domain via a one-dimensional interpolation in the k_y domain.

Provided that the squint angle is very small, that is, $\theta_c \ll 1$ (near-broadside case), we can further approximate k_y via

$$k_y \approx 2k_c \sin \theta_c - 2k_c \frac{\cos^2 \theta_c}{R_c} u.$$

Note that in the above approximation k_y is a linear function of the synthetic aperture u, and is invariant in the radar frequency ω.

To summarize the narrow bandwidth and narrow-beamwidth approximations for the near-broadside case, we can write

$$\begin{aligned} k_x(\omega) &\approx 2k \cos \theta_c, \\ k_y(u) &\approx 2k_c \sin \theta_c - 2k_c \frac{\cos^2 \theta_c}{R_c} u. \end{aligned}$$

The resultant samples for the examples of Figure 4.14a and b are identified by the "+" symbol.

Based on the above approximations, the evenly-spaced samples of the slow-time compressed signal $s_c(\omega, u)$ map into the evenly spaced samples of the target function two-dimensional Fourier transform $F(k_x, k_y)$ (within a bandpass region). Thus, to form the spatial domain target reconstruction $f(x, y)$, one could simply obtain the two-dimensional FFT of the sampled $s_c(\omega, u)$. This algorithm is shown in Figure 4.15 for the broadside case, $\theta_c = 0$.

- Polar format processing based on the narrow-bandwidth and narrow-beamwidth approximations at near-broadside results in a separable two-dimensional processing of the slow-time compressed signal $s_c(\omega, u)$ for image formation; the FFT processing in the slow-time u and fast-time ω domains, respectively, yield the target function distribution in the azimuth and range domains.

To determine the sample spacing in the resultant reconstruction, one could use the FFT equation [s94, ch. 1]. Suppose that there are (N, M) samples of $s_c(\omega, u)$ with sample spacing $(\Delta_u, \Delta_\omega)$. (Clearly we are assuming a finite synthetic aperture here.) Then the sample spacings of the mapped data from $s_c(\omega, u)$ to $F(k_x, k_y)$, via the above-mentioned narrow-bandwidth and narrow-beamwidth approximations, in the (k_x, k_y) domain are

$$\Delta_{k_x} = \frac{2\cos\theta_c}{c}\Delta_\omega,$$

$$\Delta_{k_y} = \frac{2k_c\cos^2\theta_c}{R_c}\Delta_u.$$

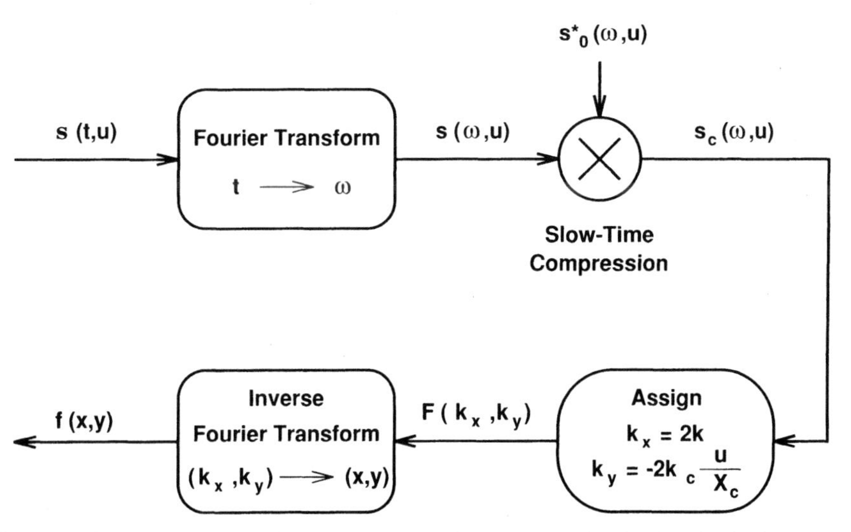

Figure 4.15. Block diagram of SAR reconstruction via plane approximation-based polar format processing with narrow-bandwidth and narrow-beamwidth approximations.

The number of samples in the (k_x, k_y) domain are

$$(N_x, N_y) = (N, M).$$

Thus, from the FFT equations, we have

$$\Delta_x = \frac{2\pi}{N_x \Delta_{k_x}} = \frac{\pi c}{N_\omega \cos\theta_c \Delta_\omega},$$

$$\Delta_y = \frac{2\pi}{N_y \Delta_{k_y}} = \frac{\pi R_c}{M k_c \cos^2\theta_c \Delta_u}.$$

Note that for the broadside case, that is, $Y_c = \theta_c = 0$ and $R_c = X_c$, the spatial sample spacing becomes

$$\Delta_x = \frac{\pi c}{N_\omega \Delta_\omega},$$

$$\Delta_y = \frac{\pi X_c}{M k_c \Delta_u}.$$

Finally some authors have noticed that some of the geometric distortions, in particular, the spatially varying shifting, of the targets in a polar format processing image can be compensated for via a transformation of the (polar format processed) SAR image. In fact a more accurate analysis of the polar format processing would indicate that the signature of the nth target, which is located at the polar spatial coordinates $[\theta_n(0), r_n]$, appears at

$$t_n \approx \frac{2r_n}{c},$$

$$k_{un} \approx 2k_c \sin\left[\theta_n(0) - \theta_c\right],$$

in the polar format processed image, which is the following two-dimensional Fourier transform:

$$\mathcal{F}_{(\omega,u)}[s_c(\omega, u)].$$

(Note that the Fourier transform with respect to the variable ω is an inverse one.)

Thus one can use this property to remove some of the geometric distortions in the SAR image obtained via polar format processing. The resultant SAR image still contains spatial-varying geometric distortions as well as blurring (smearing). Yet this polar format processed image and the slow-time compressed SAR signal have utility in *digital spotlighting* SAR data; this will be discussed in Chapters 5 and 6.

Wavefront Curvature Compensation

The plane wave approximation of polar format processing results in a *spatially varying* smearing and shifting of the targets in the imaging SAR scene [s92j; s94]. Similar to the motion errors in Section 4.9, we can identify this spatially varying smearing and shifting function and try to compensate for it. For this purpose we first identify the radial wavefront curvature error which is introduced by the plane wave approximation for the nth; that is,

$$r_{en}(u) = -\sqrt{(X_c + x_n)^2 + (Y_c + y_n - u)^2} + \sqrt{X_c^2 + (Y_c - u)^2}$$
$$+ \cos\theta_0(u)x_n + \sin\theta_0(u)y_n.$$

Polar format processing maps the slow-time compressed signal in the (ω, u) domain onto the target spatial frequency (k_x, k_y) domain. Thus the radial wavefront curvature error directly contaminates the target function Fourier transform. For the nth target, the contaminating filter function is

$$H_{en}(k_x, k_y) = \exp\left[2kr_{en}(u)\right],$$

where

$$2k = \sqrt{k_x^2 + k_y^2},$$
$$u = Y_c - \frac{k_y}{k_x}X_c.$$

Note that the transformation from (k_x, k_y) to the (ω, u) domain is invariant in the coordinates of the target.

Thus, to compensate for wavefront curvature errors of the plane wave approximation, the SAR image which is formed by the polar format processing SAR image is passed through the following shift-varying filter:

$$H_{exy}^*(k_x, k_y) = \exp\left[-j2kr_{exy}(u)\right],$$

and r_{exy} is the radial wavefront curvature error for the target at (x, y). The size of the filter and the signal processing issues for the implementation of this spatially varying filter depend on the type of the SAR system (spotlight or stripmap).

Motion Compensation Using Global Positioning System

In addition to wavefront curvature compensation, the user could also modify the filter to compensate for motion errors from the GPS data. Suppose that the GPS motion error signals are $[x_e(u), y_e(u)]$. Then the radial error function which contains both the radial wavefront curvature error and the radial motion error is

$$r_{en}(u) = -\sqrt{\left[X_c + x_n - x_e(u)\right]^2 + \left[Y_c + y_n - u - y_e(u)\right]^2} + \sqrt{X_c^2 + (Y_c - u)^2}$$
$$+ \cos\theta_0(u)x_n + \sin\theta_0(u)y_n.$$

The procedure for design of the spatially varying filter to remove the effect of the above radial error function is the same as the one described in the previous section.

4.12 CONVENTIONAL ISAR MODELING AND IMAGING

We mentioned earlier that inverse synthetic aperture radar (ISAR) is an imaging modality that utilizes the relative motion of a target (e.g., an aircraft or a missile) with respect to a stationary transmitting/receiving radar to image that target. The imaging system geometry for an ISAR system is shown in Figure 4.16. The trans-

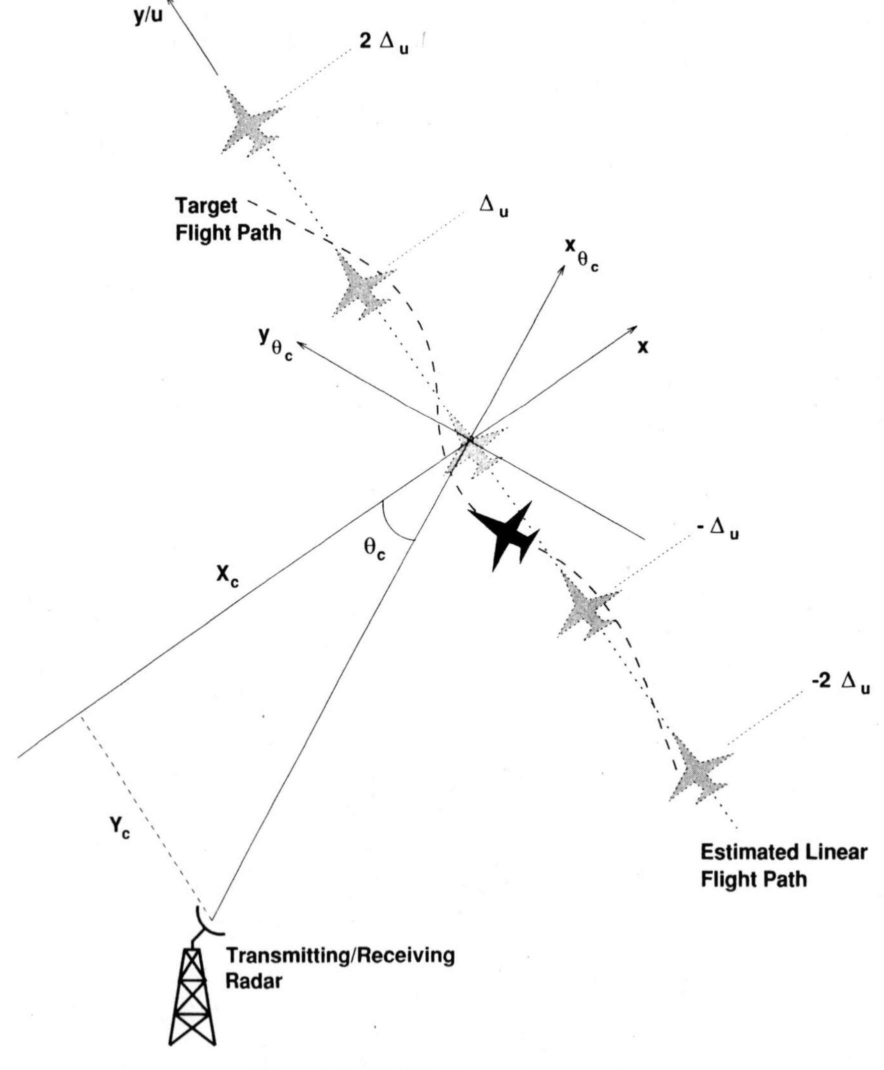

Figure 4.16. ISAR imaging system geometry.

mitting/receiving radar is stationary. The target (e.g., an aircraft) is moving along a trajectory shown as a dashed line in the figure. To formulate the ISAR imaging problem in terms of the SAR principles, the target should possess a constant linear motion. This is not the case in practice. In fact the user encounters *nonlinear* components in the target motion; these are analogous to the motion errors $[x_e(u), y_e(u)]$ in SAR.

One can also use motion compensation methods in ISAR to remove the nonlinear motion terms. The outcome would be the ISAR imaging of the target when it is moving along a line with a constant speed v_t. This line is shown as a dotted line (estimated linear flight path) in Figure 4.16. In this idealized model the radar captures ISAR signature of the moving target when it is located at the points $\ldots, -2\Delta_u, -\Delta_u, 0, \Delta_u, 2\Delta_u, \ldots$ along the linear flight path; this is shown in Figure 4.16. Note that the continuous slow-time domain τ and the continuous synthetic aperture domain u are related by

$$u = v_t \tau.$$

The parameter Δ_u, the sample spacing in the synthetic aperture domain, is related to the target constant speed and the sample spacing in the slow-time domain via

$$\Delta_u = v_t \Delta_\tau.$$

The sample spacing in the synthetic aperture domain is the slow-time interval between two consecutive bursts (transmissions) of the radar. This parameter is also called the pulse repetition interval (PRI):

$$\text{PRI} = \Delta_\tau.$$

The number of radar bursts per second, which is the inverse of PRI, is called the pulse repetition frequency (PRF):

$$\text{PRF} = \frac{1}{\text{PRI}}.$$

The conventional methods for modeling an ISAR signal rely on approximations that represent the slow-time varying distance of the moving target from the radar via a polynomial function of the slow-time τ. For instance, the distance of the center of mass of the target from the radar as a function of the slow-time τ is modeled via

$$R_0(\tau) = R_c + v_t \tau + \tfrac{1}{2} a_t \tau^2 + \tfrac{1}{6} j_t \tau^3,$$

where the parameters (v_t, a_t, j_t) are, respectively, speed, acceleration, and jerk of the target. The following discussion provides an analytical study of ISAR signature of a moving target via these approximations, and the polar format-based reconstruction methods.

ISAR Model

Consider a moving target that possesses both linear and nonlinear motion functions of the slow-time τ. We select the origin in the spatial domain to be the center of this target. The distance of the center of the target to the radar as a function of the slow-time τ is modeled by

$$R_0(\tau) = \sqrt{[X_c - x_e(\tau)]^2 + [Y_c - y_e(\tau) - v_t\tau]^2},$$

where $(0, v_t\tau)$ and $[x_e(\tau), y_e(\tau)]$, respectively, correspond to the linear and nonlinear components of the target motion, and (X_c, Y_c) are the target squint parameters; see Figure 4.16.

- The slow-time dependent radial distance of the center of the target area from the radar is generally a high-order polynomial function of the slow-time τ.

Let $p(t)$ be the transmitted radar signal. The echoed signal from a reflector at (x_n, y_n) on the moving target at the slow-time τ is

$$\mathbf{s}_n(t, \tau) = p\left[t - \frac{2\sqrt{[X_c + x_n - x_e(\tau)]^2 + [Y_c + y_n - y_e(\tau) - v_t\tau]^2}}{c}\right].$$

The Fourier transform of this echoed signal with respect to the fast-time t is

$$s_n(\omega, \tau) = P(\omega)\exp\left[-j2k\sqrt{[X_c + x_n - x_e(\tau)]^2 + [Y_c + y_n - y_e(\tau) - v_t\tau]^2}\right],$$

where $k = \omega/c$ is the wavenumber. For notational simplicity we suppress $P(\omega)$ in the following discussion. [The effect of the radar signal spectrum is simply an amplitude function after fast-time matched-filtering with $P^*(\omega)$.]

Slow-time Compression or Motion Compensation

The *reference* ISAR signature of the center of the target in the (ω, τ) domain is

$$\begin{aligned} s_0(\omega, \tau) &= \exp\left[-j2k\sqrt{[X_c - x_e(\tau)]^2 + [Y_c - y_e(\tau) - v_t\tau]^2}\right] \\ &= \exp\left[-j2kR_0(\tau)\right]. \end{aligned}$$

Multiplying the ISAR signature of the target at (x_n, y_n) with the complex conjugate of the above reference phase function, we obtain

$$s_{nc}(\omega,\tau) = s_n(\omega,\tau)s_0^*(\omega,\tau)$$
$$= \exp\left[-j2k\sqrt{[X_c + x_n - x_e(\tau)]^2 + [Y_c + y_n - y_e(\tau) - v_t\tau]^2} + 2jkR_0(\tau)\right].$$

We called the above signal the slow-time compressed SAR (ISAR) signal. In the ISAR literature this signal is also referred to as the *motion-compensated* ISAR signal.

Polar Format Processing

The basic idea in the conventional ISAR imaging is as follows: Provided that the slow-time radial motion history of an in-scene target [in this case, the radial motion of the reflector at the center of the target area, $R_0(\tau)$], and the nonlinear motion components $x_e(t)$ and $y_e(t)$ are much smaller than $R_c = \sqrt{X_c^2 + Y_c^2}$, then one can use the plane wave approximation of the polar format processing to obtain

$$s_{nc}(\omega,\tau) \approx \exp\left[-j\underbrace{2k\frac{X_c}{R_0(\tau)}}_{k_x}x_n - j\underbrace{2k\frac{Y_c - v_t\tau}{R_0(\tau)}}_{k_y}y_n\right]$$
$$= \exp\left[-j\underbrace{2k\cos\theta_0(\tau)}_{k_x}x_n - j\underbrace{2k\sin\theta_0(\tau)}_{k_y}y_n\right]$$

where

$$\theta_0(\tau) = \arctan\left(\frac{Y_c - v_t\tau}{X_c}\right)$$

is the slow-time varying radar aspect angle with respect to the center of the target with $x_e(\tau) = y_e(\tau) = 0$.

After considering the contribution from all the reflectors on the target, it can be seen that the motion compensated (slow-time compressed) ISAR signal in the (ω,τ) domain yields the samples of the two-dimensional Fourier transform of the target reflectivity function at

$$k_x = 2k\cos\theta_0(\tau),$$
$$k_y = 2k\sin\theta_0(\tau).$$

Variations of (ω,τ) yield a cone-shaped (approximately polar) coverage for the reconstruction of the target function in the spatial frequency domain. This reconstruction algorithm is identical to polar format SAR reconstruction which was discussed earlier.

Most conventional ISAR processors are based on the separable two-dimensional polar format processing which is based on the narrow-bandwidth and narrow-beamwidth approximations. With the narrow-beamwidth approximation, we obtain

$$s_{nc}(\omega, \tau) \approx \exp\left(-j\, \underbrace{2k\frac{X_c}{R_c}}_{k_x} x_n - j\, \underbrace{2k\frac{Y_c - v_t\tau}{R_c}}_{k_y} y_n \right).$$

After considering the contribution from all the reflectors on the target, it can be seen that the motion-compensated ISAR signal in the (ω, τ) domain yields the samples of the two-dimensional Fourier transform of the target reflectivity function at

$$k_x = 2k\frac{X_c}{R_c},$$

$$k_y = 2k\frac{Y_c - v_t\tau}{R_c};$$

this transformation from the measurement (ω, τ) domain into the target spatial frequency (k_x, k_y) domain also yields a cone-shaped coverage.

The above reconstruction is similar to the one we described earlier for the narrow-beamwidth SAR polar format processing. Now, with the narrow-bandwidth approximation, we obtain

$$k_x = 2k\frac{X_c}{R_c},$$

$$k_y = 2k_c\frac{Y_c - v_t\tau}{R_c},$$

which yields a two-dimensional separable processing for image formation.

The resultant spatial domain reconstruction may contain a fair amount of information in its *magnitude* about the location of various reflectors on the target. However, the phase information in the reconstructed image is degraded due to the (plane wave) approximation used for modeling the target motion. The amount of phase degradations depends on the values of (x_n, y_n): at $(x_n, y_n) = (0, 0)$, there is no phase degradation; the phase error increases as (x_n, y_n) increase.

This type of ISAR processing has been suggested for aircraft imaging, and identification from coherent ISAR images. Unfortunately, the above-mentioned approximation alters the phase signature of a target ISAR image. It has also been suggested that the phase errors can be reduced with *submillimeter* motion compensation. However, the best the ISAR motion compensation algorithm can do is to make sure that the phase error (e.g., due to nonlinear motion) is zero at $(x_n, y_n) = (0, 0)$, that is, the center of the target.

One of the worst misconceptions in the ISAR literature is that most authors are not aware of and/or don't appreciate the significance of the target squint angle θ_c. In fact the narrow-bandwidth and narrow-beamwidth SAR polar format processing at broadside ($Y_c = \theta_c = 0$ and $X_c = R_c$) are used by most authors to model the ISAR spatial frequency coverage; that is,

$$k_x = 2k,$$

$$k_y = \frac{-2k_c v_t \tau}{X_c}.$$

This approximation results in spatial smearing and warping of the target function, as well as loss of coherent information, in the reconstructed ISAR image.

If the squint angle is ignored, the resultant ISAR image can be shown to be an approximation of the target distribution in $f(x, y)$ when the spatial domain is rotated by the squint angle θ_c. This rotated or squint spatial coordinates are identified by $(x_{\theta_c}, y_{\theta_c})$ in Figure 4.16, where

$$\begin{bmatrix} x_{\theta_c} \\ y_{\theta_c} \end{bmatrix} = \begin{bmatrix} \cos\theta_c & \sin\theta_c \\ -\sin\theta_c & \cos\theta_c \end{bmatrix} \begin{bmatrix} x \\ y \end{bmatrix}.$$

To demonstrate this, we consider a realistic narrowband ISAR system which is studied in Chapter 5 with the following parameters: The radar carrier frequency is 9.25 GHz; the baseband bandwidth of the radar signal is 0.5 GHz (note that $0.5 \ll 9.25$, i.e., the narrow bandwidth assumption is valid); $(X_c, Y_c) = (5.23, 2.08)$ km, or $R_c = 5.63$ km and $\theta_c = 21.65$ degrees. We will examine the ISAR signals and images for this system using the real ISAR data from a commercial aircraft in Chapter 5.

In this section, we show the results for a simulated aircraft which is composed of 15 unit reflectors. Figure 4.17 is the ISAR reconstruction of the aircraft using the wavefront SAR algorithm of Sections 4.4 and 4.5. Note that the reflectors appear as *slanted* cross-shaped structures. This is a typical point spread function for a SAR/ISAR system with a nonzero squint angle; this will be shown in Chapter 5. Figure 4.18*a* is the wavefront ISAR reconstruction when the spatial domain is rotated by the squint angle $\theta_c = 21.65$ degrees; Figure 4.18*b* is a close-up of two of the reflectors in this image. Note that the cross-shaped point spread functions appear to be straight in Figure 4.18*a* and *b*.

Figure 4.19*a* is the conventional ISAR reconstruction using the narrow-bandwidth and narrow-beamwidth approximations in polar format processing; Figure 4.19*b* is the close-up of the same targets that appear in Figure 4.18*b*. Note the shift and smearing of the two targets in the close-up image. We used simulated ISAR data to exhibit the squint rotation which is inherent in this type of polar format processing. One should not be misled by the straight shape of the point spread function in the polar format reconstruction of Figure 4.19*a*.

4.13 RANGE-DOPPLER IMAGING

We mentioned earlier that the early SAR systems used an analog-based method that had a lens for image formation. This analog-based SAR image formation system depended on approximation to the SAR signal model which is known as the Fresnel

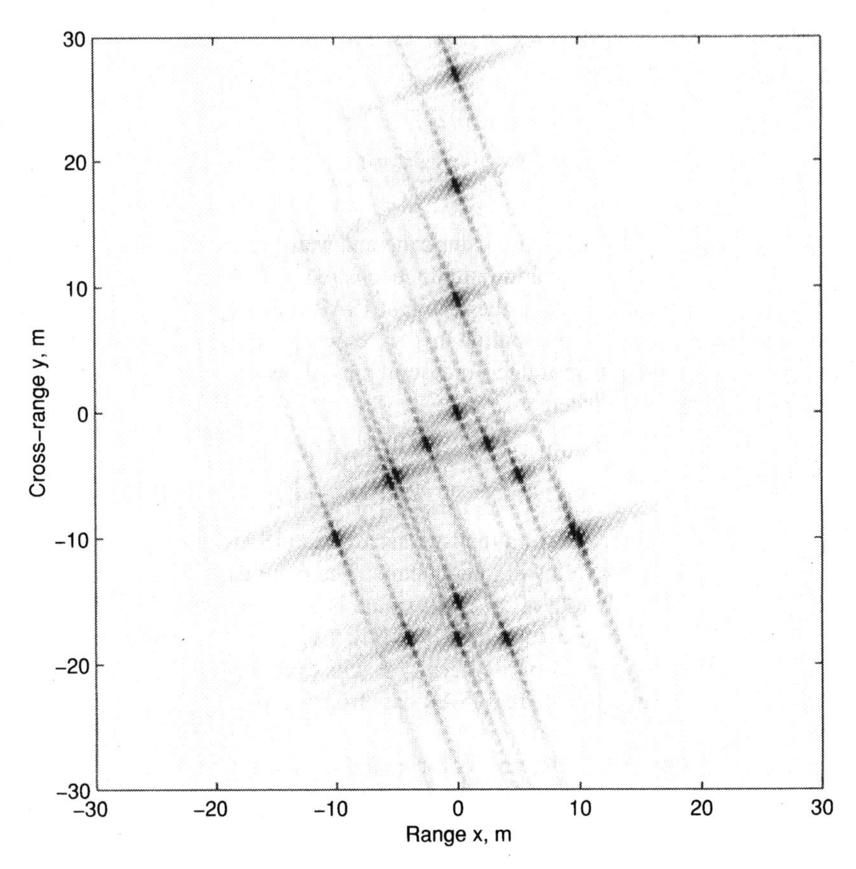

Figure 4.17. ISAR reconstruction using wavefront SAR algorithm.

approximation. The Fresnel approximation-based SAR imaging assumes a narrow-beamwidth for the SAR system. The result is a separable two-dimensional processing of the fast-time and slow-time SAR signal $\mathbf{s}(t, u)$ which, respectively, yield the target function distribution in the range and azimuth domains.

Note that this is not equivalent to the polar format processing based on the narrow-bandwidth and narrow-beamwidth approximations at near-broadside, in which the FFT processing of the slow-time compressed signal $s_c(\omega, u)$ in the slow-time u and fast-time ω domains, respectively, yield the target function distribution in the azimuth and range domains. Both of these *separable* two-dimensional SAR imaging methods are attractive to the users who cannot or do not wish to develop sophisticated SAR digital signal processing algorithms which, for example, require interpolation. In fact some of the *low-resolution* SAR systems are based on these methods.

In the following we briefly outline the Fresnel approximation-based SAR imaging for the broadside case which was introduced in the early 1960s. One may also modify this scheme for the squint mode SAR systems.

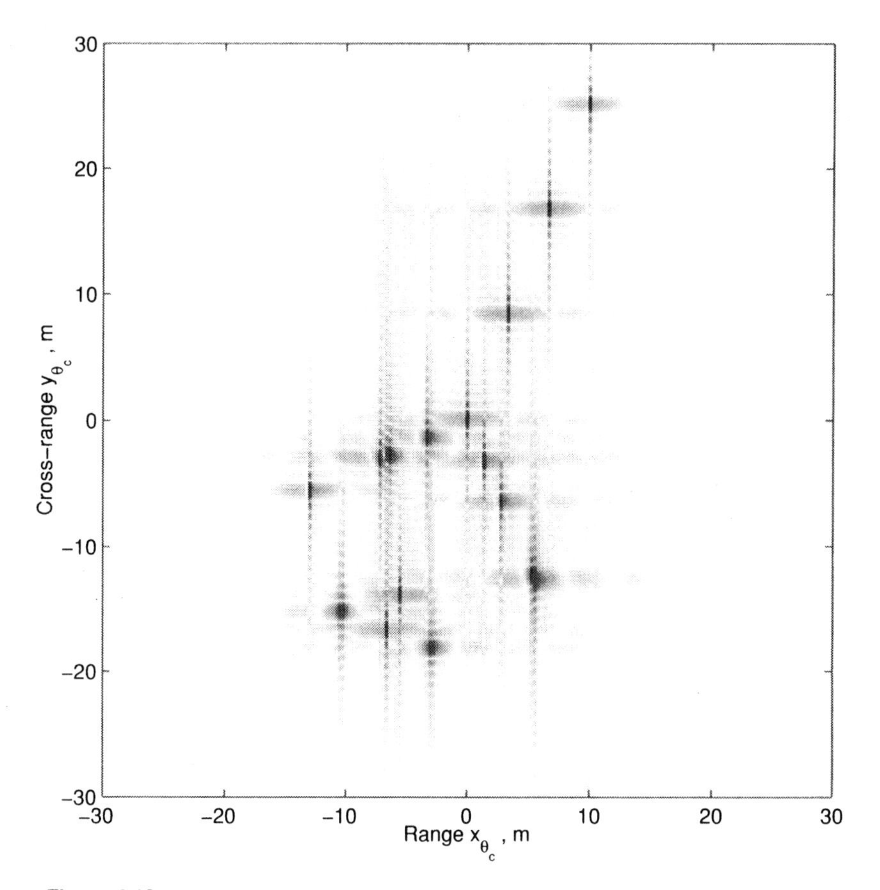

Figure 4.18a. ISAR reconstruction in squint spatial domain using wavefront SAR algorithm.

Fresnel Approximation-Based Reconstruction

Consider the Fourier transform of the generic SAR signal with respect to the fast-time t (see Section 4.2) when the origin is shifted to the center of the target area at $(X_c, 0)$; that is,

$$s(\omega, u) = P(\omega) \sum_n \sigma_n \exp\left[-j2k\sqrt{(X_c + x_n)^2 + (y_n - u)^2}\right].$$

The Taylor series expansion of the radar distance from the nth target at the slow-time u around X_c is

$$\sqrt{(X_c + x_n)^2 + (y_n - u)^2} = X_c + x_n + \frac{(y_n - u)^2}{2X_c} + \cdots.$$

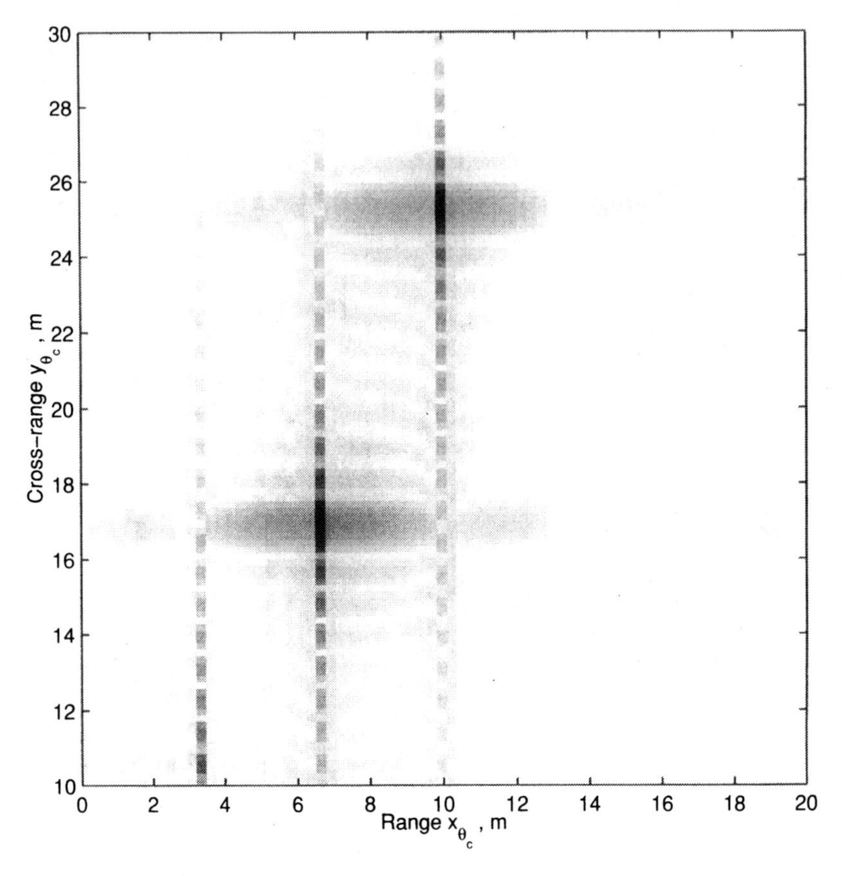

Figure 4.18b. A close-up of two reflectors.

By neglecting the higher-order terms in the above Taylor series, the SAR signal model becomes

$$s(\omega, u) \approx P(\omega) \sum_n \sigma_n \exp\left[-j2k(X_c + x_n) - jk\frac{(y_n - u)^2}{X_c}\right]$$

$$= P(\omega) \exp\big(-j2kX_c\big) \sum_n \sigma_n \exp\left[-j\underbrace{2k}_{k_x} x_n - j\frac{k(y_n - u)^2}{X_c}\right].$$

The above expression is the Fresnel approximation for the SAR signal model [cut; goo]. Note that we have identified $2k$ as the range spatial frequency domain, that is, $k_x = 2k$. The reason for this assignment will become obvious in the following discussion.

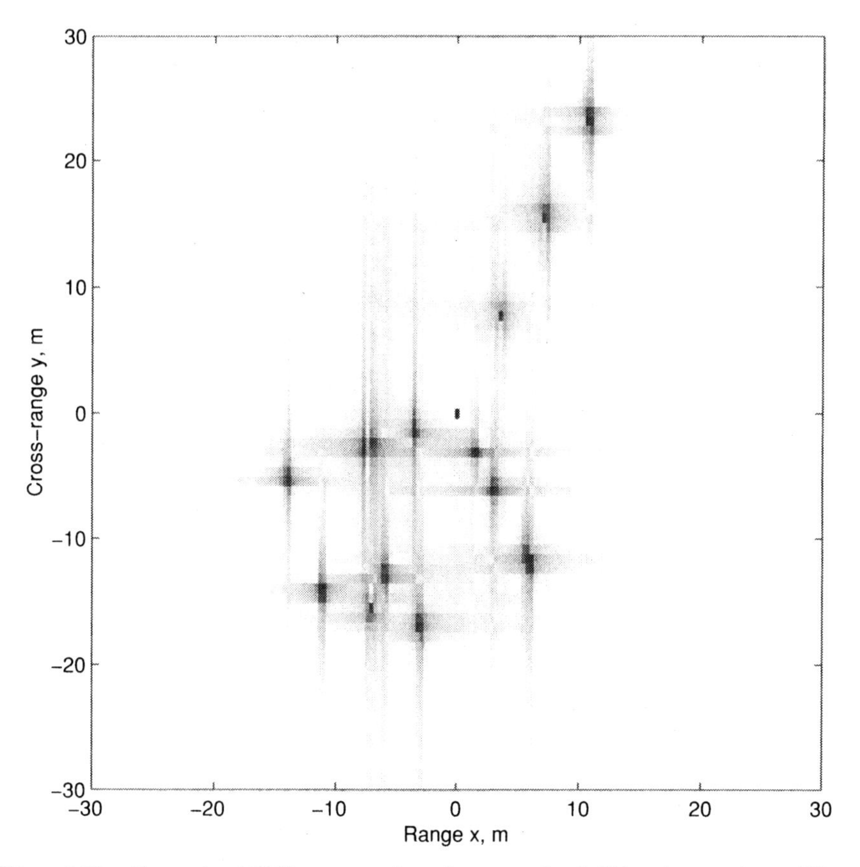

Figure 4.19a. Conventional ISAR reconstruction using narrow-bandwidth and narrow-beamwidth assumptions in polar format processing.

Consider the ideal target function in the spatial domain; that is,

$$f_0(x, y) = \sum_n \sigma_n \delta(x - x_n, y - y_n).$$

The one-dimensional Fourier transform of the ideal target function with respect to the range x domain is

$$\begin{aligned} F_{0x}(k_x, y) &\approx \mathcal{F}_{(x)}[f_0(x, y)] \\ &= \sum_n \sigma_n \exp(-jk_x x_n)\delta(y - y_n). \end{aligned}$$

Using this notation, the Fresnel approximation-based SAR signal model can be rewritten via the following:

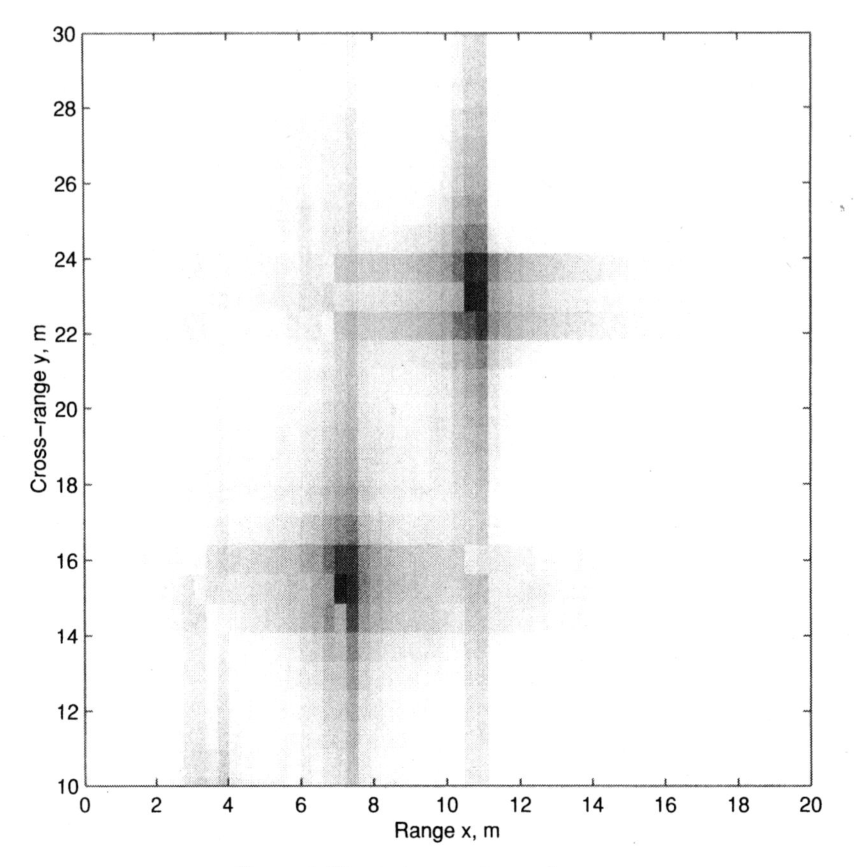

Figure 4.19b. A close-up of two reflectors.

$$s(\omega, u) \approx P(\omega) F_{0x}(2k, u) * \exp\left(-j\frac{ku^2}{X_c}\right),$$

where $*$ denotes convolution in the slow-time u domain.

The SAR imaging problem, that is, recovering the target function $f(x, y)$ or its one-dimensional Fourier transform $F_{0x}(k_x, y)$, can now be viewed as the deconvolution of the known chirp signal

$$\exp\left(-j\frac{ku^2}{X_c}\right),$$

from the SAR signal $s(\omega, u)$. In the matched-filtered form, this imaging method is [cut; s94, sec. 4.7]

$$F_x(k_x, y) \approx P^*(\omega)s(\omega, u) * \exp\left(j\frac{ku^2}{X_c}\right),$$

where

$$F_x(k_x, y) = \mathcal{F}_{(x)}[f(x, y)],$$

is the one-dimensional Fourier transform of the target function with respect to the range x domain, and

$$k_x = 2k,$$

$$y = u.$$

The spatial domain target function $f(x, y)$ is constructed by the inverse Fourier transform of the reconstructed $F_x(k_x, y)$ with respect to k_x.

Narrow-Bandwidth and Narrow-Beamwidth Approximation

The above Fresnel approximation-based SAR image formation can be further approximated and simplified. Suppose that the radar bandwidth is much smaller than its carrier frequency; it is a narrow-bandwidth radar signal where $|k - k_c| \ll k_c$ for the available wavenumbers k. Moreover the range of the available synthetic aperture values u is much smaller than the target range, that is, narrow-beamwidth approximation. In this case the slow-time chirp signal in the Fresnel approximated-based SAR signal model can be approximated via

$$\exp\left(-j\frac{ku^2}{X_c}\right) \approx \exp\left(-j\frac{k_c u^2}{X_c}\right).$$

Now the Fresnel approximation-based SAR imaging becomes

$$F_x(k_x, y) \approx P^*(\omega)s(\omega, u) * \exp\left(j\frac{k_c u^2}{X_c}\right),$$

where

$$k_x = 2k = \frac{2\omega}{c},$$

$$y = u.$$

The above imaging equation can be implemented by a separable two-dimensional filtering operation on the SAR signal in the fast-time and slow-time domains: The fast-time domain filter impulse function is $p^*(-t)$ (the fast-time matched filter), and

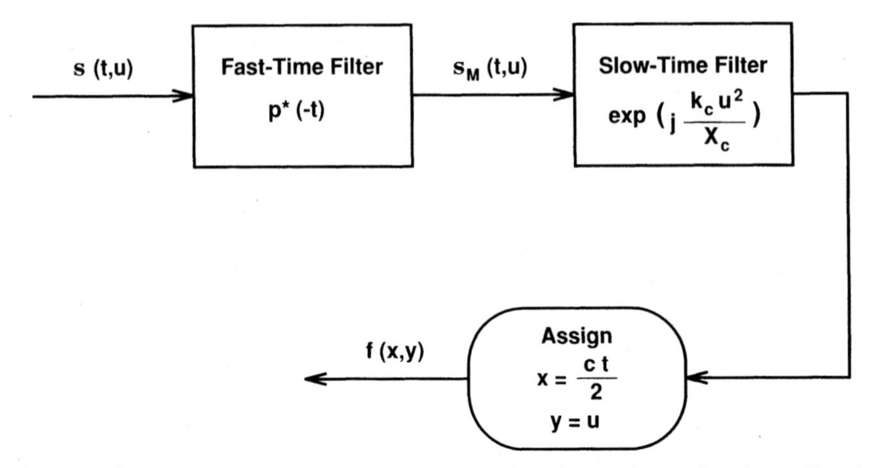

Figure 4.20. Block diagram of SAR reconstruction via Fresnel approximation-based range-Doppler imaging using narrow-bandwidth and narrow-beamwidth approximation.

the slow-time domain filter impulse function is $\exp\left(jk_c u^2/X_c\right)$; that is,

$$f(x, y) \approx \underbrace{\left[\underbrace{s(t, u) * p^*(-t)}_{\text{Convolution in } t} \right] * \exp\left(j\frac{k_c u^2}{X_c}\right)}_{\text{Convolution in } u}$$

$$= s_M(t, u) * \exp\left(j\frac{k_c u^2}{X_c}\right),$$

where the range and azimuth are scaled mappings of the fast-time and slow-time domains

$$x = \frac{ct}{2},$$
$$y = u.$$

Note that there is no need for interpolation in the above *approximation-based* reconstruction. Range-Doppler imaging using narrow-bandwidth and narrow-beamwidth approximation is summarized in the block diagram of Figure 4.20.

4.14 THREE-DIMENSIONAL IMAGING WITH TWO-DIMENSIONAL AZIMUTH AND ELEVATION SYNTHETIC APERTURES

In this section we examine three-dimensional imaging of a target scene via two-dimensional *planar* synthetic aperture measurements made by a wide-bandwidth transmitting/receiving radar. A prototype planar SAR system for three-dimensional imaging has been developed and tested for interior imaging of buildings. The system is composed of a radar moved on a rail to form an azimuthal synthetic aperture u. Also a lift is used to position the rail at various elevation synthetic aperture points, which we denote with v. We develop the system model and reconstruction for this three-dimensional SAR imaging modality via wavefront theory principles which are similar to the ones presented in Sections 4.1 to 4.4 for two-dimensional SAR imaging.

System Model

Consider a stationary target region composed of a set of point reflectors with reflectivity σ_n that are located at the coordinates (x_n, y_n, z_n) $(n = 1, 2, \ldots)$ in the three-dimensional spatial (x, y, z) domain; x is the range domain, y is the azimuth domain, and z is used to identify the elevation domain. A radar located at $(0, u, v)$ in the spatial domain illuminates the target area with a wide-bandwidth signal $p(t)$; the radar radiation pattern is assumed to be omni-directional.

The parameters (u, v) are the azimuthal and elevation coordinates of the radar. These are varied to form a two-dimensional planar synthetic aperture. The resultant SAR signal can be expressed via

$$\mathbf{s}(t, u, v) = \sum_n \sigma_n p\left[t - \frac{2\sqrt{x_n^2 + (y_n - u)^2 + (z_n - v)^2}}{c}\right],$$

where

$$\frac{2\sqrt{x_n^2 + (y_n - u)^2 + (Z - n - v)^2}}{c}$$

is the round-trip delay from the radar to the nth target when the radar is located at $(0, u, v)$. Note that in this problem we have two different slow-time variables: the azimuthal slow-time u and the elevation slow-time v.

The Fourier transform of the SAR signal $\mathbf{s}(t, u, v)$ with respect to the fast-time t is

$$s(\omega, u, v) = P(\omega) \sum_n \sigma_n \underbrace{\exp\left[-j2k\sqrt{x_n^2 + (y_n - u)^2 + (z_n - v)^2}\right]}_{\text{Two-dimensional spherical PM signal}}.$$

At a fixed fast-time frequency ω, the SAR signal is composed of a linear combination of two-dimensional PM modulated signals, that is,

$$\exp\left[-j2k\sqrt{x_n^2 + (y_n - u)^2 + (z_n - v)^2}\right],$$

which we refer to as the two-dimensional *spherical* PM signals.

Reconstruction

One can use the Fourier decomposition of the three-dimensional free space Green's function [mor] or the method of stationary phase [pap] to show that the two-dimensional Fourier transform of the two-dimensional spherical PM signal with respect to (u, v), when $u \in (-\infty, \infty)$ and $v \in (-\infty, \infty)$, is

$$\begin{aligned}\mathcal{F}_{(u,v)}&\left[\exp\left[-j2k\sqrt{x_n^2+(y_n-u)^2+(z_n-v)^2}\right]\right]\\ &=\exp\left(-j\sqrt{4k^2-k_u^2-k_v^2}\,x_n-jk_u y_n-jk_v z_n\right)\end{aligned}$$

for $\sqrt{k_u^2+k_v^2} \le 2k$, and zero otherwise; k_u is the azimuthal synthetic aperture (Doppler) frequency domain for u, and k_v is the elevation synthetic aperture (Doppler) frequency domain for v.

Using this Fourier decomposition, the two-dimensional Fourier transform of $s(\omega, u, v)$ with respect to (u, v) can be shown to be

$$S(\omega, k_u, k_v) = P(\omega)\sum_n \sigma_n \underbrace{\exp\left(-j\sqrt{4k^2-k_u^2-k_v^2}\,x_n - jk_u y_n - jk_v z_n\right)}_{\text{Linear phase function of }(x_n, y_n, z_n)}.$$

We can redefine this Fourier transform via

$$\begin{aligned}&S(\omega, k_u, k_v)\\ &\quad= P(\omega)\sum_n \sigma_n \exp\left[-jk_x(\omega,k_u,k_v)x_n - jk_y(\omega,k_u,k_v)y_n - jk_z(\omega,k_u,k_v)z_n\right],\end{aligned}$$

where

$$\begin{aligned}k_x(\omega, k_u, k_v) &= \sqrt{4k^2-k_u^2-k_v^2},\\ k_y(\omega, k_u, k_v) &= k_u,\\ k_z(\omega, k_u, k_v) &= k_v,\end{aligned}$$

represent the three-dimensional SAR spatial frequency mapping.

We define the three-dimensional ideal target function in the spatial domain via

$$f_0(x, y, z) = \sum_n \sigma_n \delta(x - x_n, y - y_n, z - z_n),$$

and its three-dimensional spatial Fourier transform

$$F_0(k_x, k_y, k_z) = \sum_n \sigma_n \exp(-jk_x x_n - jk_y y_n - jk_z z_n).$$

Thus the SAR signal $S(\omega, k_u, k_v)$ can be rewritten via

$$S(\omega, k_u, k_v) = P(\omega) F_0\big[k_x(k_u, k_v, \omega), k_y(\omega, k_u, k_v), k_z(\omega, k_u, k_v)\big]$$

where $\big[k_x(\omega, k_u, k_v), k_y(\omega, k_u, k_v), k_z(\omega, k_u, k_v)\big]$ are governed by the SAR spatial frequency mapping.

Based on this model for the SAR signal, the practical target reconstruction via fast-time matched filtering is obtained via

$$\begin{aligned} F\big[k_x(\omega, k_u, k_v), k_y(\omega, k_u, k_v), k_z(\omega, k_u, k_v)\big] &= P^*(\omega) S(\omega, k_u, k_v) \\ &= |P(\omega)|^2 \sum_n \sigma_n \exp\big(- jk_x x_n - jk_y y_n - jk_z z_n\big) \end{aligned}$$

for $\sqrt{k_u^2 + k_v^2} \le 2k$ and $\omega \in [\omega_c - \omega_0, \omega_c + \omega_0]$.

In practice, the available discrete evenly-spaced samples of (t, u, v) transform into evenly spaced samples of (k_y, k_z) and unevenly spaced samples of the range spatial frequency k_x. Thus one has to transform these unevenly spaced samples into evenly spaced ones in the k_x domain. The procedure is similar to the one described in Section 4.5 for two-dimensional SAR imaging. First, the *baseband* target function is formed; then the interpolation is done in the k_x domain.

If the target area is centered at (X_c, Y_c, Z_c) in the spatial domain, then the baseband target function is

$$F_b(k_x, k_y, k_z) = F(k_x, k_y, k_z) \exp(jk_x X_c + k_y Y_c + jk_z Z_c).$$

The reconstruction equation for the baseband target function is

$$\begin{aligned} &F_b\big[k_x(\omega, k_u, k_v), k_y(\omega, k_u, k_v), k_z(\omega, k_u, k_v)\big] \\ &\quad = P^*(\omega) \exp(jk_x X_c + k_y Y_c + jk_z Z_c) S(\omega, k_u, k_v). \end{aligned}$$

One can also develop range stack and backprojection digital reconstruction methods for three-dimensional SAR imaging with a planar synthetic aperture.

4.15 ELECTRONIC COUNTER-COUNTERMEASURE VIA PULSE DIVERSITY

As we pointed out in Chapter 1, in military reconnaissance applications of SAR, the enemy may use electronic counter measures (ECM) to confuse the SAR processor and, as a result, produce erroneous information in the reconstructed SAR image. One of the ECM schemes used to jam a SAR system relies on estimating (copying) the radar signal $p(t)$ and transmitting a replica of the copy at random or regular time intervals; this is called a repeat ECM jamming signal.

Since the user must rely on multiple transmissions of the radar signal in SAR [at each slow-time u, a $p(t)$ is transmitted], the enemy could exploit these multiple transmissions to train its signal processor to adaptively obtain a good estimate of the radar signal $p(t)$. To counter this problem, the user could *vary* $p(t)$ at each synthetic aperture position to confuse the enemy's receiving radar and adaptive signal processor (electronic counter-countermeasure, ECCM).

Suppose that the transmitted radar signal at slow-time u is identified by $p(t, u)$ and that the resultant echoed SAR signal is $\mathbf{s}(t, u)$. Thus the SAR signal is

$$\mathbf{s}(t, u) = \sum_n \sigma_n p \left[u, t - \frac{2\sqrt{x_n^2 + (y_n - u)^2}}{c} \right].$$

The fast-time matched-filtered SAR signal is formed via

$$\mathbf{s}_M(t, u) = \mathbf{s}(t, u) * p^*(-t, u),$$

where $*$ denotes convolution in the fast-time t domain and $p^*(-t, u)$ is the complex conjugate of $p(-t, u)$.

After performing Fourier transform with respect to the fast-time, we obtain the following for the SAR signal and its fast-time matched-filtered form:

$$s(\omega, u) = P(\omega, u) \sum_n \sigma_n \exp\left[-j2k\sqrt{x_n^2 + (y_n - u)^2} \right],$$

and

$$s_M(\omega, u) = |P(\omega, u)|^2 \sum_n \sigma_n \exp\left[-j2k\sqrt{x_n^2 + (y_n - u)^2} \right],$$

where $P(\omega, u)$ is the one-dimensional Fourier transform of $p(t, u)$ with respect to the fast-time t.

It can be seen that the fast-time matched-filtered SAR signal contains the familiar spherical PM signal which carries information on the slow-time Doppler spread of each target. The only difference from the SAR model, which we dealt with in Sections 4.1 and 4.2, is the replacement of the magnitude signal $|P(\omega)|^2$, which is invariant in u, with $|P(\omega, u)|^2$ which depends on the slow-time u. Thus, provided that the fast-time bandwidth of $p(t, u)$ does not vary with u, that is, $P(\omega, u) \neq 0$ for $\omega \in [\omega_c - \omega_0, \omega_c + \omega_0]$, then one can perform some kind of amplitude equalization in $S_M(\omega, u)$ to counter the variations of $|P(\omega, u)|^2$.

In theory, this can be achieved via

$$\frac{s_M(\omega, u)}{|P(\omega, u)|^2}$$

within the radar bandwidth. However, this might boost the additive noise with respect to the signal power. In this case the user should design radar signals $p(t, u)$ whose magnitude spectrum in the fast-time frequency ω domain does not vary significantly with u. A simple solution for this problem is to use a continuous pseudo random phase (continuous phase modulation, CPM [pro]) at each u to amplitude modulate the carrier frequency; this can be achieved via a digitally programmable radar transmitter.

If the user is interested in using a chirp signal (e.g., to reduce the A/D converter sampling rate via deramping; see Chapter 1), an amplitude modulation of the chirp signal with a pseudorandom CPM could also be used for ECCM to confuse the enemy's adaptive signal processor. This issue was examined in Chapter 1, and a method for converting the deramped signal into the matched-filtered signal was outlined in that chapter.

5

SPOTLIGHT SYNTHETIC APERTURE RADAR

Introduction

We now turn our discussion to a particular type of SAR system known as *spotlight synthetic aperture radar*. This SAR imaging modality utilizes *mechanical* or *electronic* beam steering of a physical radar. The purpose of the beam steering is to irradiate a *finite* target area centered around the point (X_c, Y_c) in the spatial domain as the radar is moved along a straight line. The mechanical or electronic beam steering, which maintains or *focuses* the physical radar radiation pattern on the same target area while the radar-carrying aircraft moves along the synthetic aperture, is called *analog spotlighting*.

Spotlight SAR is more "modern" than *stripmap SAR*, which is discussed in Chapter 6. However, the reader should not be misled by the word *modern* in assuming that spotlight SAR is preferable to stripmap SAR. Spotlight SAR and stripmap SAR are two different SAR imaging *modalities*. In fact, in SAR reconnaissance problems, the user might first run a stripmap SAR data collection to obtain an image of a relatively large target area, for example, a battlefield. Then, by analyzing this image and identifying smaller target areas that appear to be of interest, a spotlight SAR data collection for each of the smaller target areas is run to obtain *high-resolution* imaging information. (See also the introductory discussion on stripmap SAR in Chapter 6.)

Beyond the application differences of the two SAR systems, the classical *range-Doppler* imaging, which is based on the Fresnel approximation (see Section 4.13), was originally introduced for the stripmap SAR systems. However, when the spotlight SAR system was envisioned in the 1970s, the Fresnel approximation turned out to fail rapidly in this SAR imaging modality. What is referred to as *polar format processing*, which is based on the plane wave approximation (see Section 4.11), was introduced at that time for imaging from spotlight SAR data.

We should emphasize that both range-Doppler imaging and polar format processing are approximation-based imaging methods in SAR. The two methods have different modes of validity that vary with the parameters of a spotlight or stripmap SAR system. For instance, polar format processing is preferable to range-Doppler imaging in the X band ISAR system which is cited later in this chapter. However, when polar format processing is used on the UHF band SAR system which is examined in this and the next chapters, the outcome is far worse than the image that is formed by range-Doppler imaging [s95].

SAR *wavefront* reconstruction, which is the basis of our work, does not utilize the approximations that are used in either range-Doppler or polar format processing. Our analysis of spotlight and stripmap SAR systems in Chapters 5 and 6 will exhibit this fact. As far as we are concerned, spotlight SAR and stripmap SAR are two distinct SAR imaging modalities that have different utilities. Our main objective in Chapters 5 and 6 is to develop signal properties of the data collected in these two SAR systems. Then, using these theoretical principles, we present digital signal and image processing methods for transforming spotlight and stripmap SAR data into imaging information.

What are the distinct features of a spotlight SAR imaging system? A spotlight SAR system with a synthetic aperture length of $2L$ is our best imitation of a *physical* aperture with the same length. In fact the data of a spotlight SAR system exhibit most of the functional properties of the data of an echo-mode sampled physical aperture (array) of equal length with *omni-directional* (zero-size) elements. Thus a spotlight SAR exhibits the resolution properties of a physical aperture with a diameter of $2L$; this, as we will see in Chapter 6, is not true in a stripmap SAR system. Note that in a practical spotlight SAR, a physical radar with a nonzero diameter, and not an omni-directional radar, is used for data collection.

Another factor that separates spotlight SAR and stripmap SAR is the variation of their corresponding point spread function in the spatial domain. At a fixed *range*, a spotlight SAR system exhibits a point spread function that varies with the *cross-range* coordinate of the targets in the imaging scene. We will show in Chapter 6 that this is not the case in the stripmap SAR systems.

There are other differences between the two SAR systems; some result in or are related to the above two factors (resolution and shift-varying point spread function) and are apparent to the user, such as the two-dimensional bandwidth of the SAR signal. These issues, which are crucial in appropriate (e.g., alias-free) processing of SAR data, will become evident in our discussion in Chapters 5 and 6.

Outline

Our study begins with two different methods for analog spotlighting in spotlight SAR systems; these are *mechanical beam steering* (Section 5.1) and *electronic beam steering* (Section 5.2). We study these analog spotlighting mechanisms and their implications in forming the multidimensional spotlight SAR signal subspace. This will help us to relate the target function to the acquired SAR data for imaging purposes.

Knowledge of the issues and tools presented in Sections 5.1 and 5.2 is more crucial than understanding the mathematical details that appear in these two sections. The discussion on analog spotlighting and its implications would be more valuable in more advanced SAR applications, for example, a displaced phase center array (DPCA) SAR system which is being investigated for moving target detection. The analyses in Sections 5.1 and 5.2 are provided for the sake of completeness and future work; the reader may skim through these sections.

Section 5.3 examines the spectral properties of the spotlight SAR signal; this has implications for the selection of the parameters of a spotlight SAR system for acquiring alias-free data and the resolution (a measure of performance) of a spotlight SAR imaging system. The spotlight SAR spectral properties are studied for planar and curved physical radar apertures. An important conclusion of this section is that the slow-time Doppler bandwidth of the spotlight SAR signal is an increasing function of not only the beamwidth angle of the physical radar aperture but also the beamwidth angle of the *synthetic aperture*. This principle, which is established by the SAR wavefront reconstruction theory, was not predicted in either range-Doppler imaging or polar format processing.

One of the outcomes of the two-dimensional spectral analysis of the spotlight SAR signal is the range of *observability* of an individual target in the imaging scene by the radar in the aspect angle domain, and how this is translated to the spatial frequency bandwidth information for that target. This concept is used in Section 5.4 to obtain an analytical expression for the point spread function of the target in the reconstructed SAR image. This analysis indicates that the point spread function varies with the cross-range as well as range coordinates of the target. This phenomenon, which is only predicted by the wavefront reconstruction theory, also results in a shift-varying resolution.

The results of Section 5.3 on the spotlight SAR signal spectral properties are also used to obtain constraints on the selection of the parameters of a spotlight SAR system, for example, pulse repetition frequency (PRF); this is done in Section 5.5. In Chapter 2 we discussed a procedure for reducing PRF via the slow-time compression of the synthetic aperture signal. In Section 5.5 we also use this procedure to reduce the required PRF to acquire spotlight SAR data which can be *converted* to alias-free spotlight SAR data.

In fact the spotlight SAR data acquired via this procedure are *aliased* in the slow-time domain. This is due to the fact that the PRF is selected based on the beamwidth angle of the *physical* radar, though the slow-time Doppler bandwidth is related to the sum of the beamwidth angles of the synthetic aperture as well as the physical radar. We show a procedure to recover alias-free slow-time Doppler data from the acquired aliased spotlight SAR data; this procedure was referred to as the upsampling of synthetic aperture data via slow-time compression in Chapter 2.

Section 5.5 also contains an important discussion on reducing the side lobe leakage effects of the physical radar. Although we use an analog scheme (beam steering) to spotlight a specific target area, the radiation pattern of a realistic radar still cannot be perfectly focused. This imperfection, which is due to the *side lobes* of the radar radiation pattern, produces slow-time Doppler aliasing. To suppress some of the side lobe aliasing effects, we introduce a procedure which we refer to as *digital spotlighting*. The idea behind digital spotlighting is to develop a digital signal processing method, similar to the conventional array beam-forming methods, to deal with side lobes Doppler aliasing. Full aperture and subaperture methods for implementing the digital spotlighting algorithm are discussed, and their merits are examined.

Section 5.6 deals with the problem of digital reconstruction from processed spotlight SAR data. Digital reconstruction via spatial frequency interpolation, range stacking, time domain correlation, and backprojection are studied for the spotlight SAR systems. As we mentioned earlier, the slow-time Doppler bandwidth of the spotlight SAR signal is proportional to the sum of the beamwidth angles of the synthetic aperture and the physical radar. We will show that the spatial frequency bandwidth of the reconstructed spotlight SAR image in the *cross-range* domain is also proportional to the sum of the beamwidth angles of the synthetic aperture and the physical radar. However, we will outline a procedure, which we refer to as spotlight SAR image compression, to reduce this bandwidth. The result is a SAR image whose spatial frequency bandwidth in the cross-range domain is only proportional to the beamwidth angle of the synthetic aperture.

The analytical principles in Sections 5.5 and 5.6 are accompanied with the realistic X band ISAR and UHF band SAR data whose parameters were identified in the introductory chapter of this book. The chapter ends with a discussion on the Matlab code for digital signal processing (upsampling and digital spotlighting) and image reconstruction in the spotlight SAR systems (Section 5.7).

Mathematical Notations and Symbols

The following is a list of mathematical symbols used in this chapter, and their definitions:

$a_{eT}(\omega, x, y)$	Transmit mode amplitude pattern of phased array element
$a_n(\omega, x, y)$	Amplitude pattern for n-target (general case)
$a_s(\omega, x, y)$	Transmit-receive amplitude pattern of phased array
$a_s(\omega, u; x, y)$	Mechanically beam-steered amplitude pattern of radar
$A_n(\omega, k_u)$	Slow-time Doppler amplitude pattern for n-target
$A_s(\omega, k_u)$	Slow-time Doppler amplitude pattern of phased array
$A_s(\omega, k_u; x, y)$	Mechanically beam-steered slow-time Doppler amplitude pattern of radar
B_{x_n}	Support of nth target signature in k_{X_n} domain
B_{y_n}	Support of nth target signature in k_{Y_n} domain
c	Wave propagation speed $c = 3 \times 10^8$ m/s for radar waves $c = 1500$ m/s for acoustic waves in water $c = 340$ m/s for acoustic waves in air
D_x	Diameter of radar in range domain
D_y	Diameter of radar in cross-range domain
$f(x, y)$	Target function in spatial domain

$F(k_x, k_y)$	Two-dimensional Fourier transform of target function
$f_c(x, y)$	Compressed image of target area
$F_c(k_x, k_y)$	Spectrum of compressed image of target area
$f_d(t, k_u)$	Digitally spotlighted polar format processed image
$f_n(x, y)$	Point spread function of nth target in spatial domain
$F_n(k_x, k_y)$	Two-dimensional Fourier transform of point spread function of nth target
$f_{\theta_c}(x_{\theta_c}, y_{\theta_c})$	Squint target function
$F_{\theta_c}(k_{x\theta_c}, k_{y\theta_c})$	Squint target function spectrum
$f_{\theta_c c}(x_{\theta_c}, y_{\theta_c})$	Compressed squint target function
$F_{\theta_c c}(k_{x\theta_c}, k_{y\theta_c})$	Compressed squint target function spectrum
$h_{eT}(\omega, x, y)$	Transmit mode radiation pattern of phased array element
$h_s(\omega, x, y)$	Transmit-receive radiation pattern of phased array
$h_{sR}(\omega, x, y)$	Receive mode radiation pattern of phased array
$h_{sT}(\omega, x, y)$	Transmit mode radiation pattern of phased array
$h_s(\omega, u; x, y)$	Mechanically beam-steered radiation pattern of radar
$J_m(\omega)$	Jacobian of SAR transformation at slow-time Doppler k_{um}
$J_{\max}$	Maximum value of $\lvert J_m(\omega) \rvert$
k	Wavenumber: $k = \omega/c$
k_c	Wavenumber at carrier frequency: $k_c = \omega_c/c$
k_n	Wavenumber sampled points for SAR signal
k_u	Spatial frequency or wavenumber domain for azimuthal synthetic aperture u; slow-time Doppler (frequency) domain
k_{um}	Slow-time Doppler sampled points for SAR signal
$k_{u\min}$	Minimum slow-time Doppler frequency of SAR signal
$k_{u\max}$	Maximum slow-time Doppler frequency of SAR signal
k_{un}	Slow-time Doppler location of nth target in polar format processing
k_x	Spatial frequency or wavenumber domain for range x
k_{xc}	Center (carrier) range spatial frequency of reconstructed SAR image
k_{xmn}	Available range spatial frequency sampled points
$k_{x\max}$	Maximum of available range spatial frequency sampled points k_{xmn}'s
$k_{x\min}$	Minimum of available range spatial frequency sampled points k_{xmn}'s
k_{X_n}	Spatial frequency domain for X_n
k_y	Spatial frequency or wavenumber domain for cross-range y
k_{ymn}	Available cross-range spatial frequency sampled points

$k_{y\max}$	Maximum of available cross-range spatial frequency sampled points k_{ymn}'s
$k_{y\min}$	Minimum of available cross-range spatial frequency sampled points k_{ymn}'s
k_{Y_n}	Spatial frequency domain for Y_n
k_0	Baseband wavenumber: $k_0 = \omega_0/c$
L	Half-size of synthetic aperture
$L_{\min}$	Minimum half-length of processed (zero-padded) synthetic aperture; minimum length of processed (zero-padded) synthetic aperture is $2L_{\min}$
L_s	Half-size of subaperture
M	Number of samples of SAR signal in slow-time u domain
M_c	Number of samples of slow-time compressed SAR signal in slow-time u domain
n	Index representing a specific target
N	Number of fast-time samples
N_s	Number of subapertures for digital spotlighting
N_x	Number of range samples in target reconstruction
N_y	Number of cross-range samples in target reconstruction
$p(t)$	Transmitted radar signal
$p_0(t)$	Transmitted radar signal shifted by reference fast-time T_c
$P(\omega)$	Fourier transform of transmitted radar signal
PRF	Radar pulse repetition frequency in slow-time
PRI	Radar pulse repetition interval in slow-time
r	Radial distance in spatial domain
$r_{\max}$	Farthest radial distance of radar from spotlighted target area
$r_{\min}$	Closest radial distance of radar from spotlighted target area
R_c	Radial distance of center of target from radar at $u = 0$
$\mathbf{s}(t, u)$	Echoed SAR signal from target area
$s(\omega, u)$	One-dimensional Fourier transform of SAR signal with respect to fast-time
$s_b(\omega, u)$	SAR signal after baseband conversion in slow-time domain
$S(\omega, k_u)$	Two-dimensional Fourier transform of SAR signal with respect to fast-time and slow-time
$s_c(\omega, u)$	Slow-time compressed SAR signal
$S_c(\omega, k_u)$	Two-dimensional Fourier transform of slow-time compressed SAR signal with respect to fast-time and slow-time
$s_{cd}(\omega, u)$	Digitally spotlighted slow-time compressed SAR signal
$S_{cd}(\omega, k_u)$	Two-dimensional Fourier transform of digitally spotlighted slow-time compressed SAR signal with respect to fast-time and slow-time

$s_d(\omega, u)$	Digitally spotlighted SAR signal
$S_d(\omega, k_u)$	Two-dimensional Fourier transform of digitally spotlighted SAR signal with respect to fast-time and slow-time
$\mathbf{s}_M(t, u)$	Fast-time matched-filtered echoed signal
$s_M(\omega, u)$	One-dimensional Fourier transform of matched-filtered SAR signal with respect to fast-time
$\mathbf{s}_n(t, u)$	Echoed SAR signal of nth target
$s_n(\omega, u)$	One-dimensional Fourier transform of nth target SAR signal with respect to fast-time
$S_n(\omega, k_u)$	Two-dimensional Fourier transform of nth target SAR signal with respect to fast-time and slow-time
$s_0(\omega, u)$	One-dimensional Fourier transform of reference SAR signal with respect to fast-time
$\text{sinc}(\cdot)$	sinc function: $\text{sinc}(a) = \sin(\pi a)/\pi a$
t	Fast-time domain
$t_{ij}(u)$	Round trip delay of echoed signal for target at (x_i, y_j) when radar is at $(0, u)$
t_n	Fast-time location of nth target in polar format processing
t_{ni}	Fast-time location of nth target in ith subaperture polar format processing
T	Length of fast-time processing interval
T_c	Reference fast-time point
T_f	Ending point of fast-time sampling for SAR signal
$T_{\min}$	Minimum length of fast-time processing interval
T_p	Radar signal pulse duration
T_s	Starting point of fast-time sampling for SAR signal
u	Synthetic aperture or slow-time domain (azimuthal)
v_r	Speed of radar-carrying aircraft in SAR
v_x	Range domain speed of moving target in SAR scene
v_y	Cross-range domain speed of moving target in SAR scene
$W_d(t, k_u)$	Digital spotlight filter for full aperture
$W_{di}(t, k_u)$	Digital spotlight filter for ith subaperture
x	Range domain
x_i	Range bins for range stacking, TDC and backprojection reconstruction algorithms
x_n	Range of nth target
X_n	Range domain when spatial domain is rotated by $\theta_n(0)$
x_{θ_c}	Range domain when spatial domain is rotated by squint angle θ_c
x_{θ_u}	Range domain when spatial domain is rotated by θ_u
X_c	Center point of target area in range domain

X_{cc}	Modified mean range (for squint case)
$X_0(\omega)$	Half-size of spotlighted target area in range domain; size of spotlighted target area in range domain is $2X_0$
y	Cross-range or azimuth domain
y_j	Cross-range bins for TDC and backprojection reconstruction algorithms
y_n	Cross-range of nth target
Y_n	Cross-range domain when spatial domain is rotated by $\theta_n(0)$
y_{θ_c}	Cross-range domain when spatial domain is rotated by squint angle θ_c
y_{θ_u}	Cross-range domain when spatial domain is rotated by θ_u
Y_c	Center point of target area in cross-range domain; squint cross-range
Y_i	Offset of ith subaperture in cross-range domain
$Y_0(\omega)$	Half-size of target area in cross-range domain; size of target area in cross-range domain is $2Y_0(\omega)$
$Y_{0\max}$	Half-size of spotlighted target area in cross-range domain at lowest frequency of planar radar
$Y_{0\min}$	Half-size of spotlighted target area in cross-range domain at highest frequency of planar radar
Δ_k	Sample spacing of SAR signal in wavenumber k domain
Δ_{k_u}	Sample spacing of SAR signal in slow-time Doppler domain
Δ_{k_x}	Sample spacing of reconstructed target function in range spatial frequency domain
Δ_{k_y}	Sample spacing of reconstructed target function in cross-range spatial frequency domain
Δ_t	Sample spacing of spotlight SAR signal in fast-time domain
Δ_u	Sample spacing of SAR signal in synthetic aperture domain
Δ_{uc}	Sample spacing of slow-time compressed SAR signal in synthetic aperture domain
Δ_x	Sample spacing and resolution of reconstructed target function in range domain
Δ_{X_n}	nth target resolution in X_n domain
Δ_y	Sample spacing and resolution of reconstructed target function in cross-range domain
Δ_{Y_n}	nth target resolution in Y_n domain
Δ_ω	Sample spacing of SAR signal in fast-time frequency domain
$\mathcal{F}$	Forward Fourier transform operator
$\mathcal{F}^{-1}$	Inverse Fourier transform operator
λ	Wavelength: $\lambda = 2\pi c/\omega$
λ_c	Wavelength at carrier fast-time frequency: $2\pi c/\omega_c$

$\lambda_{\max}$	Maximum wavelength: $2\pi c/\omega_{\min}$
$\lambda_{\min}$	Minimum wavelength: $2\pi c/\omega_{\max}$
$\lambda_{n\min}$	Minimum wavelength of radiation experienced by nth target
ω	Temporal frequency domain for fast-time t
ω_c	Radar signal carrier or center frequency
ω_n	Fast-time frequency sampled points for SAR signal
ω_0	Radar signal half bandwidth in radians; radar signal baseband bandwidth is $\pm\omega_0$
$\omega_{\max}$	Maximum fast-time frequency of radar
$\omega_{\min}$	Minimum fast-time frequency of radar
$\omega_{n\min}$	Maximum fast-time frequency of radiation experienced by nth target
Ω_n	Slow-time Doppler support band of nth target signature
$\|\Omega_n\|/2$	Baseband Doppler bandwidth of nth target
Ω_{nc}	Center Doppler frequency of nth target
Ω_s	Slow-time Doppler support band of SAR signal
σ_n	Reflectivity of nth target
θ_c	Average squint angle of target area; squint angle of center of target area
θ_{ax}	Largest absolute aspect angle of spotlighted target area with respect to radar synthetic aperture coordinates
$\theta_{\max}$	Largest aspect angle of spotlighted target area with respect to radar synthetic aperture coordinates
$\theta_{\min}$	Smallest aspect angle of spotlighted target area with respect to radar synthetic aperture coordinates
θ_u	Average squint angle of target area when the radar is at $(0, u)$
θ_{ci}	Average squint angle of target area for ith subaperture
$\theta_n(u)$	Aspect angle of nth target when radar is at $(0, u)$
$\theta_0(u)$	Aspect angle of center of target area when radar is at $(0, u)$

5.1 MECHANICALLY BEAM-STEERED SPOTLIGHT SAR

Mechanical Beam Steering

We begin with analysis of radiating a specific target area (analog spotlighting) via mechanical beam steering. The radar antenna used in this scheme possesses a fixed radiation pattern. This radiation pattern attains its maximum energy at broadside, and it gradually tapers off for the off-broadside points in the spatial domain. Without any beam steering (mechanical redirection of the physical antenna) the finite broadside irradiated target area, which is also referred to as the *footprint* of the radar, is within a $\pm X_0$ area in the range domain (range swath or beamwidth in range) and is within a $\pm Y_0$ area in the cross-range domain (beamwidth in cross-range); X_0 is the half-beamwidth in range, and Y_0 is the half-beamwidth in cross-range.

As we showed in Chapter 3, the range and cross-range half-beamwidths (X_0, Y_0) vary with several factors, including the target area range, the type and dimension of the radar used, and perhaps the radar frequency. In some spotlight SAR systems the range and cross-range beamwidths are approximately the same, that is, $X_0 = Y_0$. In this case the radar footprint is a disk of radius X_0 in the spatial (x, y) domain. In what follows we try to present a general analysis by using different symbols for the half-beamwidth in range (X_0), and the half-beamwidth in cross-range (Y_0).

- For a planar radar antenna the range and cross-range half-beamwidths vary with the radar frequency:

$$X_0(\omega) = \frac{R_c \lambda}{D_x},$$

$$Y_0(\omega) = \frac{R_c \lambda}{D_y},$$

where R_c is the radial distance of the radar from the center of the target area and (D_x, D_y) identify the physical diameters of the radar antenna. In this case the radar footprint varies with the radar signal frequency. For high-frequency SAR systems where analog spotlighting is used, the carrier frequency is much larger than the bandwidth of the radar signal. In that case we can approximate the radar footprint at all the available frequencies of the radar signal via

$$X_0(\omega) \approx \frac{R_c \lambda_c}{D_x},$$

$$Y_0(\omega) \approx \frac{R_c \lambda_c}{D_y},$$

where $\lambda_c = 2\pi c/\omega_c$ is the wavenumber at the carrier frequency.

How does the radar maintain its beam on a fixed target area? As the aircraft moves along the synthetic aperture, the radar beam is steered (redirected) via *mechanically rotating* the radar antenna such that the point (X_c, Y_c) (the center of the spotlighted target area) is on its *broadside* line for all slow-time values of u. This is demonstrated in Figure 5.1. We define the following angle:

$$\theta_u = \arctan\left(\frac{Y_c - u}{X_c}\right),$$

which is the boresight angle of the radar at $(0, u)$ with respect to the center of the target area. Note that the boresight angle of the radar θ_u is a variable, while the squint angle

$$\theta_c = \arctan\left(\frac{Y_c}{X_c}\right)$$

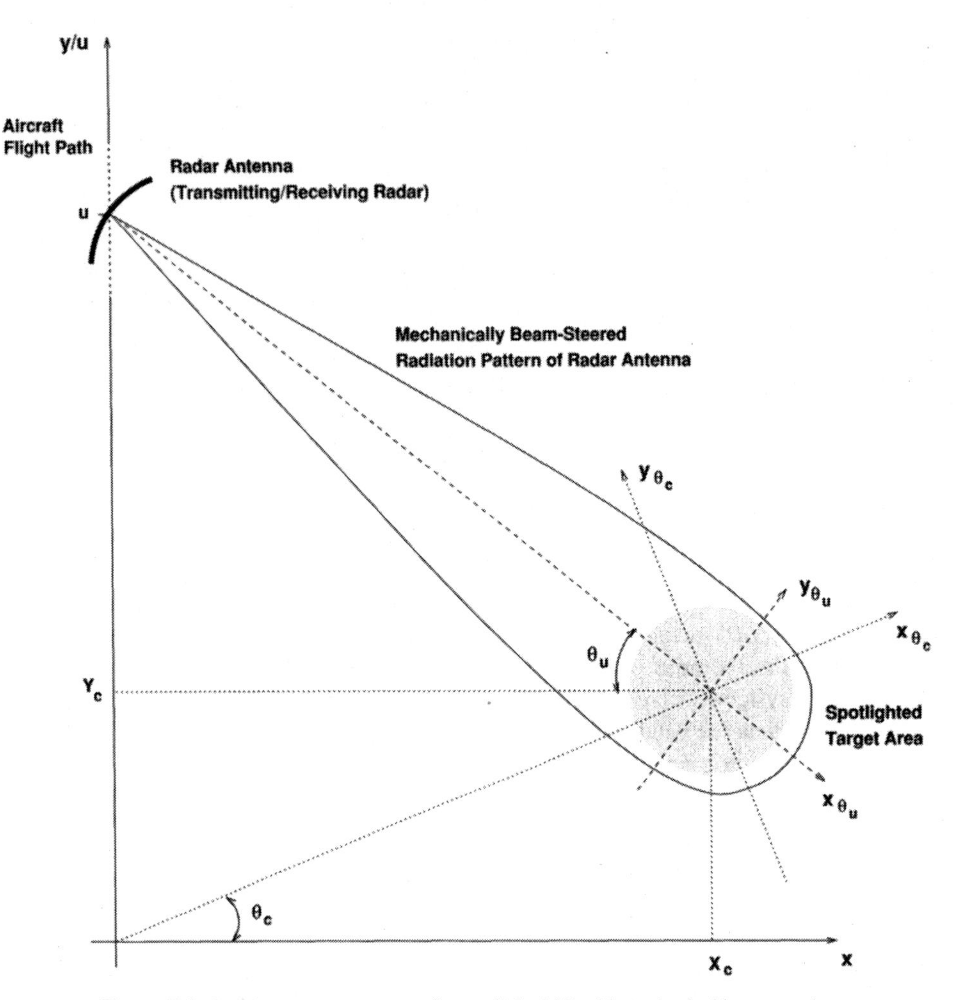

Figure 5.1. Imaging system geometry for spotlight SAR with mechanical beam steering.

is a constant in a given spotlight SAR problem; see Figure 5.1. For $u = 0$ the two angles are the same.

We define the rotational transformation of the $(x - X_c, y - Y_c)$ axis by θ_u via the following:

$$\begin{bmatrix} x_{\theta_u} \\ y_{\theta_u} \end{bmatrix} = \begin{bmatrix} \cos\theta_u & \sin\theta_u \\ -\sin\theta_u & \cos\theta_u \end{bmatrix} \begin{bmatrix} x - X_c \\ y - Y_c \end{bmatrix}.$$

We should emphasize that the spatial coordinates $(x_{\theta_u}, y_{\theta_u})$ vary with the synthetic aperture u, while the target area squint spatial coordinates, which were defined in

Chapter 4 to be

$$\begin{bmatrix} x_{\theta_c} \\ y_{\theta_c} \end{bmatrix} = \begin{bmatrix} \cos\theta_c & \sin\theta_c \\ -\sin\theta_c & \cos\theta_c \end{bmatrix} \begin{bmatrix} x - X_c \\ y - Y_c \end{bmatrix},$$

are fixed in a given SAR problem; see Figure 5.1.

- The distance of the radar from the center of the target area at the slow-time u is $\sqrt{X_c^2 + (Y_c - u)^2}$ which varies with the radar coordinates. This implies that the radar footprint (X_0, Y_0) varies with the slow-time u. Moreover, due to the variations of the radar aspect (look) angle θ_u with respect to u, the radar footprint corresponds to the radiated area in the $(x_{\theta_u}, y_{\theta_u})$ domain and not the (x, y) domain. Despite these two factors, in practical spotlight SAR systems, the synthetic aperture size is much smaller than the mean radial distance, $|u| \ll R_c$, such that the radar footprint variations with respect to u are negligible.

We begin our analysis with a scenario when the radar beam is *not* mechanically rotated. We mentioned that the transmit-receive mode amplitude pattern of the radar without beam steering is

$$\begin{aligned} a(\omega, x, y - u) &= A\left[\omega, \frac{2k(y-u)}{\sqrt{x^2 + (y-u)^2}}\right] \\ &\approx A\left[\omega, \frac{2k(y-u)}{r}\right], \end{aligned}$$

where $u = 0$ corresponds to the *broadside* illumination of the target area.

To develop the beam-steered transmit-receive radiation pattern, we realize the fact that the radar is at the *broadside* in the $(x_{\theta_u}, y_{\theta_u})$ domain (see Figure 5.1). Thus the beam-steered transmit-receive amplitude pattern as a function of (ω, u) at the spatial point (x, y) is

$$\begin{aligned} a_s(\omega, u; x, y) &\approx A\left[\omega, \frac{2k(y_{\theta_u} - 0)}{r}\right] \\ &= A\left[\omega, \frac{2k[(x - X_c)\sin\theta_u + (y - Y_c)\cos\theta_u]}{r}\right]. \end{aligned}$$

Note that we did not use the notation $a_s(\omega, x, y - u)$ to identify the radar transmit-receive amplitude pattern, since that would be mathematically incorrect.

The spherical phase function, which carries information on the relative coordinates of a target with respect to the radar (round trip delay information), is unchanged by beam steering. Thus the beam-steered transmit-receive radiation pattern is

$$h_s(\omega, u; x, y) = \underbrace{a_s(\omega, u; x, y)}_{\text{Amplitude function}} \underbrace{\exp\left[-j2k\sqrt{x^2 + (y-u)^2}\right]}_{\text{Spherical PM signal}}.$$

Recall that the transmit-receive mode amplitude pattern of the radar antenna without beam steering in the (ω, k_u) domain is $A(\omega, k_u)$, which is invariant in the coordinates of the target (x, y). However, the beam-steered transmit-receive amplitude pattern, that is, $a_s(\omega, u; x, y)$, results in an amplitude function in the (ω, k_u) domain which *varies with the coordinates of the target*; that is, the beam-steered amplitude function has the form $A_s(\omega, k_u; x, y)$.

System Model

We consider a spotlighted target region composed of a set of stationary targets located at the coordinates (x_n, y_n) $(n = 1, 2, \ldots;)$ the illuminated target area is centered around (X_c, Y_c). The radar illuminates the target area with the signal $p(t)$. The radar signal is bandlimited within the fast-time frequency region

$$\begin{aligned} \omega &\in [\omega_c - \omega_0, \omega_c + \omega_0] \\ &= [\omega_{\min}, \omega_{\max}], \end{aligned}$$

where ω_c and $\pm\omega_0$ are, respectively, the carrier frequency and the baseband frequency of the radar signal; $(\omega_{\min}, \omega_{\max})$ are the minimum and maximum fast-time frequencies of the radar signal.

The synthetic aperture u domain is formed via the motion of the radar-carrying aircraft along a line parallel to the y axis. The synthetic aperture measurements are made at discrete values of u which are separated by

$$\Delta_u = \frac{v_r}{\text{PRF}},$$

where v_r is the speed of the aircraft and PRF is the pulse repetition frequency (i.e., the number of transmitted pulses per second) of the radar. The synthetic aperture measurements are made within a finite aperture $u \in [-L, L]$, where $2L$ is the length of the synthetic aperture.

For notational simplicity we begin our analysis by considering the acquired SAR signal at a fixed fast-time frequency ω and slow-time u on the synthetic aperture, that is, the SAR signal in the (ω, u) domain. This signal can be expressed as the sum of the SAR signatures of the individual targets in the spotlighted area; that is,

$$\begin{aligned} s(\omega, u) &= P(\omega) \sum_n a_n(\omega, x_n, y_n - u) h_s(\omega, u; x_n, y_n) \\ &= P(\omega) \sum_n a_n(\omega, x_n, y_n - u) a_s(\omega, u; x_n, y_n) \exp\left[-j2k\sqrt{x_n^2 + (y_n - u)^2}\right], \end{aligned}$$

where $a_n(\cdot)$ is the nth target amplitude pattern.

Reconstruction

Based on our discussion of SAR radiation pattern and Fourier properties of AM-PM signals of Type 1 (see Sections 2.9–2.10), the Fourier transform of the above SAR signal with respect to the slow-time u is

$$\begin{aligned} S(\omega, k_u) &= P(\omega) \sum_n A_n(\omega, k_u) A_s(\omega, k_u; x_n, y_n) I_n(\omega, k_u) \\ &\quad \times \exp\left(-j\sqrt{4k^2 - k_u^2} x_n - jk_u y_n \right) \\ &= P(\omega) \sum_n A_n(\omega, k_u) A_s(\omega, k_u; x_n, y_n) I_n(\omega, k_u) \\ &\quad \times \exp\left[-jk_x(\omega, k_u) x_n - jk_y(\omega, k_u) y_n \right], \end{aligned}$$

where

$$\begin{aligned} k_x(\omega, k_u) &= \sqrt{4k^2 - k_u^2}, \\ k_y(\omega, k_u) &= k_u. \end{aligned}$$

We mentioned in Chapter 2 that the signal

$$I_n(\omega, k_u) = \begin{cases} 1 & \text{for } k_u \in \Omega_n = [2k \sin\theta_n(-L), 2k \sin\theta_n(L)], \\ 0 & \text{otherwise,} \end{cases}$$

is the rectangular window that identifies the slow-time Doppler band over which the nth target SAR signature is *observable*. The angular interval

$$\left[\theta_n(L), \theta_n(-L)\right]$$

is the radar aspect interval over which the nth target signature is recorded by the radar; see Figure 5.2.

We define the target function in the spatial frequency domain via the following fast-time matched filtered *reconstruction* for the above SAR model in the (ω, k_u) domain

$$\begin{aligned} F\left[k_x(\omega, k_u), k_y(\omega, k_u)\right] &= P^*(\omega) S(\omega, k_u) \\ &= |P(\omega)|^2 \sum_n A_n(\omega, k_u) A_s(\omega, k_u; x_n, y_n) I_n(\omega, k_u) \\ &\quad \times \exp\left[-jk_x(\omega, k_u) x_n - jk_y(\omega, k_u) y_n \right]. \end{aligned}$$

We use the SAR interpolation from the ω to the k_x domain to obtain a database that is suitable for inverse discrete Fourier transformation from the (k_x, k_y) domain into the spatial (x, y) domain; see Section 4.5. The resultant target function in the spatial domain is

$$f(x, y) = \sum_n f_n(x - x_n, y - y_n),$$

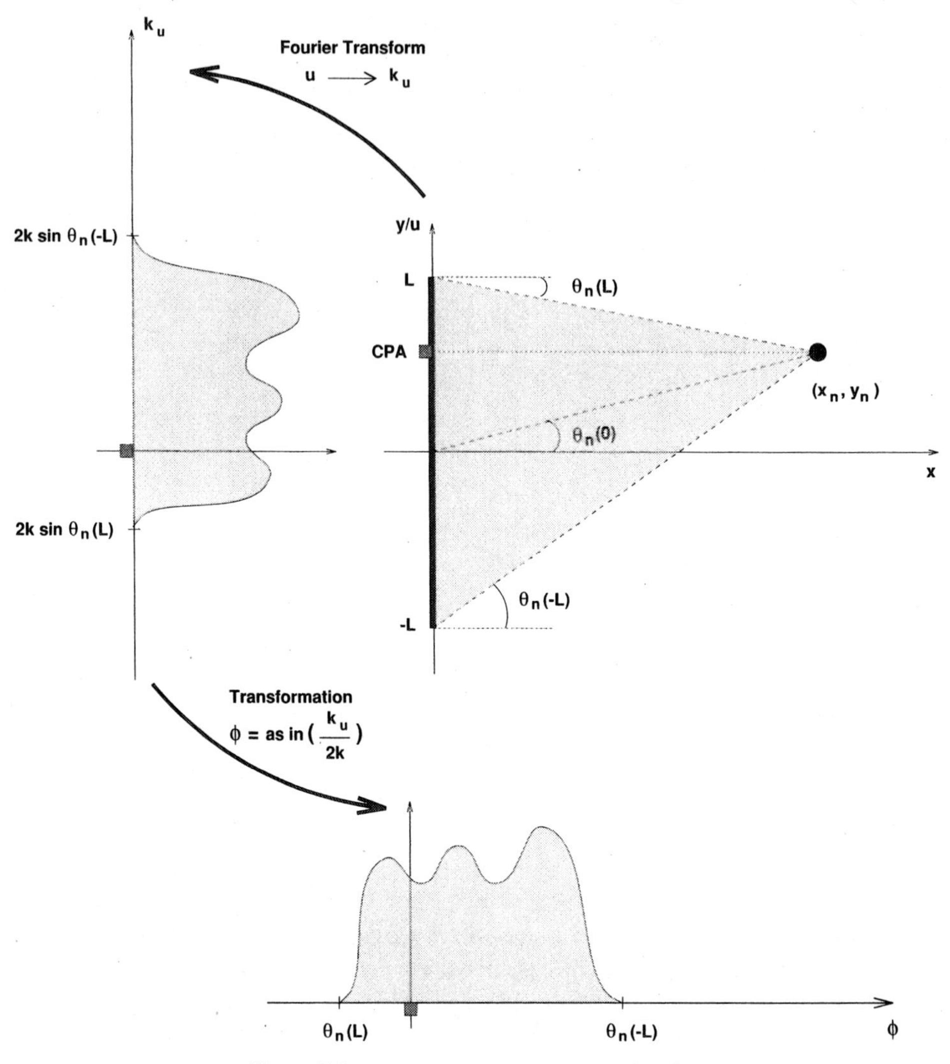

Figure 5.2. Aspect angles of a target in spotlight SAR.

where

$$f_n(x, y) = \mathcal{F}^{-1}_{(k_x, k_y)}[F_n(k_x, k_y)]$$

and

$$F_n\big[k_x(\omega, k_u), k_y(\omega, k_u)\big] = |P(\omega)|^2 A_n(\omega, k_u) A_s(\omega, k_u; x_n, y_n) I_n(\omega, k_u).$$

$F_n(\cdot)$ is the mapping of the product of $A_n(\cdot)$ and $A_s(\cdot)$ from the (ω, k_u) domain to the (k_x, k_y) domain. The function $f_n(x, y)$ is the *point spread function* of the SAR imaging system for the nth target.

Note that the spreading (smearing) of the target function $f_n(x, y)$ increases as the phase of the radar amplitude pattern $A_s(\omega, k_u; x_n, y_n)$ gets larger. To handle this problem, one could utilize a *shift-varying* [i.e., (x, y)-dependent] matched filter $A_s^*(\omega, k_u; x, y)$ in the reconstructed image $f(x, y)$ to remove the phase variations in (or the smearing which is due to) the amplitude pattern of the radar. This operation is similar to the method for motion compensation using GPS data in Section 4.9. As we mentioned earlier in our discussion of slow-time matched filtered cross-range imaging in Section 2.10, there is no need for this type of matched filtering if the amplitude function $A_s(\omega, k_u; x, y)$ is a real signal or its phase is very small.

For the SAR imaging systems where the radar antenna footprint and synthetic aperture length are much smaller than the mean radial distance, that is, $(X_0, Y_0) \ll R_c$ and $L \ll R_c$, we have the following approximation:

$$a_s(\omega, u; x, y) \approx A\left[\omega, \frac{2k[(x - X_c)\sin\theta_c + (y - Y_c)\cos\theta_c]}{r}\right]$$

and

$$A_s(\omega, k_u; x, y) \approx A\left[\omega, \frac{2k[(x - X_c)\sin\theta_c + (y - Y_c)\cos\theta_c]}{r}\right].$$

In this case the radar spotlighted transmit-receive amplitude pattern is invariant in the slow-time u domain. For the narrow-bandwidth spotlight SAR systems where $A(\cdot)$ does not vary much with the available k (or ω), this results in the following weight function:

$$A\left[\omega, \frac{2k_c[(x - X_c)\sin\theta_c + (y - Y_c)\cos\theta_c]}{r}\right]$$

for the target located at (x, y) in the reconstructed image. (k_c is the wavenumber at the radar carrier frequency.)

5.2 ELECTRONICALLY BEAM-STEERED SPOTLIGHT SAR

Electronic Beam Steering

We have discussed the spotlight SAR system which uses mechanical beam steering. Now we turn our attention to another SAR system that utilizes a linear phased array to spotlight the desired target area via *electronic* phasing of its elements. Suppose that the physical radar is an array with diameter D composed of elements with diameter D_e; see Figure 5.3. We denote the transmit mode and receive mode radiation pattern of an element located at the origin by $h_{eT}(\omega, x, y)$ and $h_{eR}(\omega, x, y)$, respectively. When the radar is located at $(0, u)$ on the aircraft flight path, the elements of the array are positioned at $(0, u + v)$, $v \in [-D/2, D/2]$; v is the location of an element with respect to the *center* of the array. Thus the transmit mode radiation of the element located at $(0, u + v)$ is $h_{eT}(\omega, x, y - u - v)$.

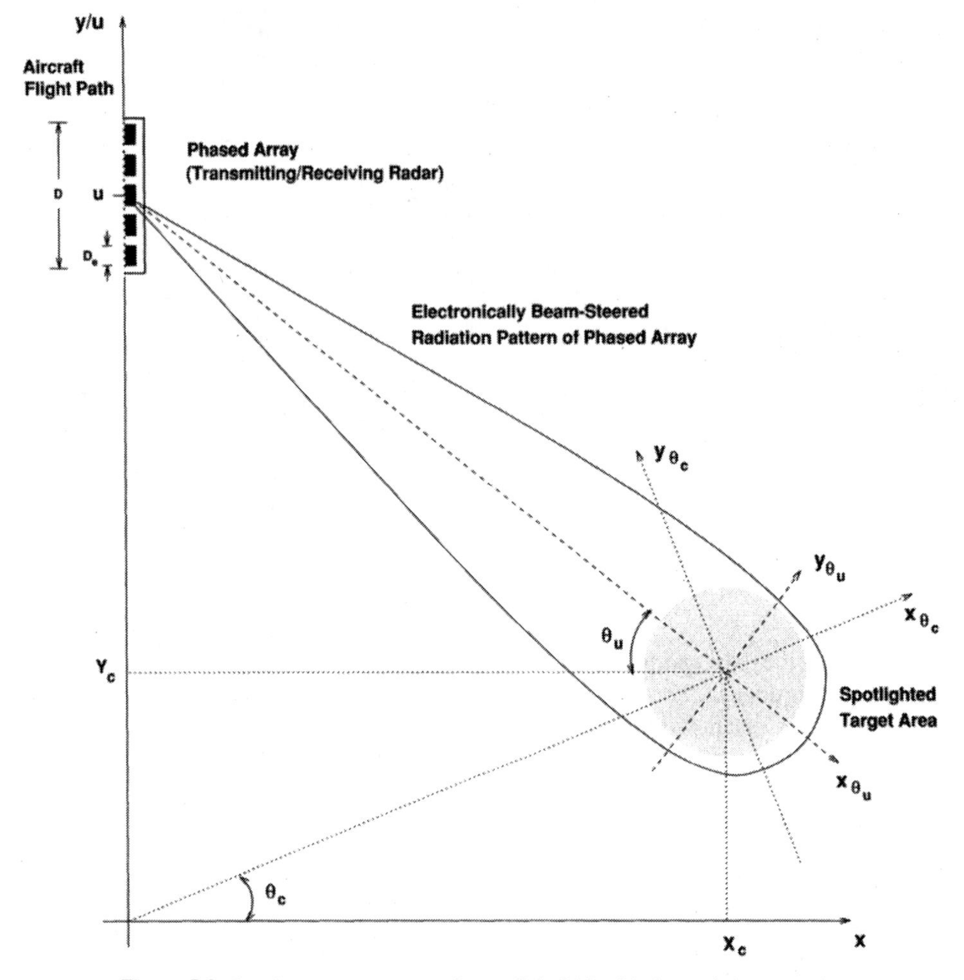

Figure 5.3. Imaging system geometry for spotlight SAR with electronic beam steering.

To spotlight the desired target region when the radar is located at $(0, u)$, we associate a *linear* phase of the form

$$\exp(-jk\sin\theta_u v)$$

to the element located at $(0, u + v)$. The wave experienced at a reflector at (x, y) is the sum of the radiation patterns of the array elements, that is,

$$h_{sT}(\omega, x, y - u) = \int_{-D/2}^{D/2} h_{eT}(\omega, x, y - u - v)\exp(-jk\sin\theta_u v)dv.$$

In practice, the array is composed of finite discrete elements. For notational simplicity we have expressed the above transmit mode radiation pattern as an *integral* (instead of a sum) in the v domain. Reference [s94, ch. 3] provides a discussion on the implications of having a discrete aperture (array).

Consider the amplitude-phase representation of the radiation pattern for a single element; that is,

$$\begin{aligned} h_{eT}(\omega, x, y-u-v) \\ = a_{eT}(\omega, x, y-u-v) \exp\left[-jk\sqrt{x^2+(y-u-v)^2}\right]. \end{aligned}$$

Using this representation in the transmit mode radiation pattern of the phased array, one obtains the following:

$$\begin{aligned} h_{sT}(\omega, x, y-u) = \int_{-D/2}^{D/2} & a_{eT}(\omega, x, y-u-v) \\ & \times \exp\left[-jk\sqrt{x^2+(y-u-v)^2}\right] \exp(-jk\sin\theta_u v)\, dv. \end{aligned}$$

Since the amplitude function is a slowly fluctuating signal, and $D \ll X_c$ and L, we can write the following:

$$a_{eT}(\omega, x, y-u-v) \approx a_{eT}(\omega, x, y-u).$$

Moreover, due to the fact that $D \ll X_c$ and L, we can approximate the phase function via the following:

$$\begin{aligned} &\exp\left[-jk\sqrt{x^2+(y-u-v)^2}\right] \\ &\approx \exp\left[-jk\sqrt{x^2+(y-u)^2} + jk\frac{(y-u)}{\sqrt{X_c^2+u^2}}v\right] \\ &\approx \exp\left[-jk\sqrt{x^2+(y-u)^2}\right] \exp\left(jk\sin\theta_u v - j\frac{ky}{X_c}v\right). \end{aligned}$$

Using these approximations in the transmit mode radiation pattern of the phased array, we obtain the following:

$$\begin{aligned} h_{sT}(\omega, x, y-u) \approx \int_{-D/2}^{D/2} & a_{eT}(\omega, x, y-u) \exp\left[-jk\sqrt{x^2+(y-u)^2}\right] \\ & \times \exp\left(jk\sin\theta_u v - j\frac{ky}{X_c}v\right) \exp(-jk\sin\theta_u v) dv. \end{aligned}$$

After some rearrangements the above can be rewritten as follows:

$$h_{sT}(\omega, x, y-u) = \underbrace{a_{eT}(\omega, x, y-u)\exp\left[-jk\sqrt{x^2+(y-u)^2}\right]}_{h_{eT}(\omega,x,y-u)} \int_{-D/2}^{D/2}$$

$$\times \exp\left(-j\frac{ky}{X_c}v\right)dv = h_{eT}(\omega, x, y-u)D\text{sinc}\left(\frac{kDy}{2\pi X_c}\right).$$

- Thus the transmit mode radiation pattern of the phased array, that is,

$$h_{sT}(\omega, x, y-u) = h_{eT}(\omega, x, y-u)D\ \text{sinc}\left(\frac{kDy}{2X_c}\right),$$

  is the product of the radiation pattern of a single element located at $(0, u)$, that is, $h_{eT}(\omega, x, y-u)$, and an amplitude pattern, that is, $D\ \text{sinc}(kDy/2\pi X_c)$. This amplitude pattern dictates the support of the spotlighted target area, which is approximately $|y| \le Y_0 = X_c\lambda/D$.

With the above result the transmit-receive radiation pattern of the phased array at (x, y) becomes

$$\begin{aligned} h_s(\omega, x, y-u) &= h_{sT}(\omega, x, y-u)h_{sR}(\omega, x, y-u) \\ &= D^2\ \text{sinc}^2\left(\frac{kDy}{2\pi X_c}\right)h_{eT}(\omega, x, y-u)h_{eR}(\omega, x, y-u) \\ &= D^2\ \text{sinc}^2\left(\frac{kDy}{2\pi X_c}\right)h_e(\omega, x, y-u), \end{aligned}$$

where

$$h_e(\omega, x, y-u) = a_e(\omega, x, y-u)\exp\left[-j2k\sqrt{x^2+(y-u)^2}\right]$$

is the transmit-receive mode radiation pattern of an element. As we mentioned earlier, the term $D^2\ \text{sinc}^2(kDy/2\pi X_c)$ is an amplitude function that dictates the support of the spotlighted target area.

System Model

Suppose that the spotlighted target region is composed of a set of stationary targets which are located at the coordinates (x_n, y_n) $(n = 1, 2, \ldots)$ which are centered around (X_c, Y_c) in the spatial domain. The radar signal used to irradiate the target area is $p(t)$ which is bandlimited within the fast-time frequency region $\omega \in$

$[\omega_c - \omega_0, \omega_c + \omega_0]$. The synthetic aperture measurements are made at discrete values of u, which are separated by Δ_u, within a finite aperture $u \in [-L, L]$; $2L$ is the length of the synthetic aperture.

We perform our analysis on the acquired SAR signal at a fixed fast-time frequency ω and slow-time u on the synthetic aperture, that is, the SAR signal in the (ω, u) domain. This signal, which is the sum of the SAR signatures of the individual targets in the spotlighted area, can be written as follows:

$$\begin{aligned} s(\omega, u) &= P(\omega) \sum_n a_n(\omega, x_n, y_n - u) h_s(\omega, x_n, y_n - u) \\ &= P(\omega) \sum_n a_n(\omega, x_n, y_n - u) a_e(\omega, x_n, y_n - u) \\ &\quad \times D^2 \operatorname{sinc}^2\left(\frac{kDy_n}{2\pi X_c}\right) \exp\left[-j2k\sqrt{x_n^2 + (y_n - u)^2}\right], \end{aligned}$$

where $a_n(\cdot)$ is the nth target amplitude pattern. For notational simplicity we do not carry the amplitude function $D^2 \operatorname{sinc}^2(kDy_n/2\pi X_c)$ in the following discussion.

Reconstruction

Taking the Fourier transform of the above SAR signal with respect to the slow-time u yields

$$\begin{aligned} S(\omega, k_u) &= P(\omega) \sum_n A_n(\omega, k_u) A_e(\omega, k_u) I_n(\omega, k_u) \exp\left(-j\sqrt{4k^2 - k_u^2} x_n - jk_u y_n\right) \\ &= P(\omega) A_e(\omega, k_u) \sum_n A_n(\omega, k_u) I_n(\omega, k_u) \\ &\quad \times \exp\left[-jk_x(\omega, k_u) x_n - jk_y(\omega, k_u) y_n\right], \end{aligned}$$

where

$$\begin{aligned} k_x(\omega, k_u) &= \sqrt{4k^2 - k_u^2}, \\ k_y(\omega, k_u) &= k_u, \end{aligned}$$

and

$$I_n(\omega, k_u) = \begin{cases} 1 & \text{for } k_u \in \Omega_n = [2k \sin\theta_n(-L), 2k \sin\theta_n(L)], \\ 0 & \text{otherwise,} \end{cases}$$

is the rectangular window that identifies the slow-time Doppler band over which the nth target SAR signature is *observable*.

The target function in the spatial frequency domain is reconstructed via the following fast-time and slow-time matched filtered inversion for the above SAR model in the (ω, k_u) domain

$$\begin{aligned} F\big[k_x(\omega, k_u), k_y(\omega, k_u)\big] &= P^*(\omega)A_e^*(\omega, k_u)S(\omega, k_u) \\ &= |P(\omega)A_e(\omega, k_u)|^2 \sum_n A_n(\omega, k_u)I_n(\omega, k_u) \\ &\quad \times \exp\big[-jk_x(\omega, k_u)x_n - jk_y(\omega, k_u)y_n\big]. \end{aligned}$$

The slow-time matched filter, that is, $A_e^*(\omega, k_u)$, is most likely to be unnecessary. This is due to the fact that the size of an element, D_e, is very small as compared with D and L. In this case $A_e(\cdot)$ does not contain much phase and amplitude fluctuation in the target SAR signature support band, that is, Ω_n.

Next the SAR interpolation from the ω to the k_x domain is used to obtain a database that is suitable for inverse discrete Fourier transformation from the (k_x, k_y) domain into the spatial (x, y) domain. The resultant target function in the spatial domain is

$$f(x, y) = \sum_n f_n(x - x_n, y - y_n),$$

where

$$f_n(x, y) = \mathcal{F}^{-1}_{(k_x, k_y)}[F_n(k_x, k_y)]$$

and

$$F_n\big[k_x(\omega, k_u), k_y(\omega, k_u)\big] = |P(\omega)A_e(\omega, k_u)|^2 A_n(\omega, k_u)I_n(\omega, k_u).$$

The signal $F_n(\cdot)$ is the mapping of $A_n(\cdot)$ which is windowed by

$$|P(\omega)A_e(\omega, k_u)|^2$$

from the (ω, k_u) domain to the (k_x, k_y) domain. The function $f_n(x, y)$ is the *point spread function* of the SAR imaging system for the nth target.

5.3 BANDWIDTH OF SPOTLIGHT SAR SIGNAL

In this section we examine the spectral properties and two-dimensional bandwidth of the spotlight SAR signal. This analysis provides us with principles that are crucial for determining the Nyquist sampling rate and alias-free processing of the spotlight SAR signal. This spectral analysis is also useful for understanding the nature of the *shift-varying* point spread function of spotlight SAR systems which is a measure of the resolution of the spotlight SAR system.

Since we are only concerned with the spectral support of the spotlight SAR signal, we can assume that the radar and target amplitude patterns are constants. As a result the analysis is applicable in both mechanically beam-steered and electronically beam-steered spotlight SAR systems.

We begin with the spectral analysis of SAR signature of a single target in the spotlighted area. Then we consider the spectral support of the measured SAR signal which is the union of the SAR spectral supports of all individual targets in the spotlighted (target) area. Our discussion includes two separate analyses of the spectral support of the spotlight SAR signal: one for the case of a curved radar aperture; the other for the case of a planar radar aperture. These separate analyses are particularly useful for wide-bandwidth spotlight SAR systems. For the narrow-bandwidth spotlight SAR systems, the analysis which will be given for a curved radar aperture is also sufficient for a planar radar aperture.

Single Target

Curved Radar Aperture Consider the case when the radar possesses a curved aperture. Thus the spotlighted target area in the cross-range domain, that is, $[Y_c - Y_0, Y_c + Y_0]$, is the same for all the fast-time frequencies of the radar. We begin with the analysis of the spotlight SAR signal of a single target, for example, the nth target which is located at (x_n, y_n) within the illuminated (spotlighted) target area.

Our objective is to determine the bandwidth of the SAR signature of the nth target, that is,

$$\mathbf{s}_n(t, u) = \sigma_n p \left[t - \frac{2\sqrt{x_n^2 + (y_n - u)^2}}{c} \right],$$

for $u \in [-L, L]$, where σ_n is the nth target reflectivity. (The target and radar amplitude patterns can be assumed to be constants in the analysis of the bandwidth of the spotlight SAR.) Note that as long as the nth target falls in the spotlighted area, then this target is *observable* to the radar at all the available fast-time frequencies ω.

The spotlight SAR signature of the nth target in the fast-time frequency and slow-time frequency (ω, k_u) domain is

$$S_n(\omega, k_u) = \sigma_n P(\omega) I_n(\omega, k_u) \exp\left(-j\sqrt{4k^2 - k_u^2}x_n - jk_u y_n \right)$$

where

$$I_n(\omega, k_u) = \begin{cases} 1 & \text{for } k_u \in \Omega_n = [2k \sin\theta_n(-L), 2k \sin\theta_n(L)], \\ 0 & \text{otherwise.} \end{cases}$$

In Section 2.10 we identified this class of SAR signals to belong to Type 1.

For this class of SAR signals, at a given fast-time frequency ω, the slow-time Doppler band of a single reflector at (x_n, y_n) in spotlight SAR is dictated by the support band of the slow-time Doppler window function $I_n(\omega, k_u)$ in the k_u domain which is

$$\Omega_n = \left[2k \sin\theta_n(L), 2k \sin\theta_n(-L)\right];$$

this slow-time support band linearly increases with the fast-time frequency ω. Consider the target and the synthetic aperture geometry in Figure 5.2. Figure 5.4 shows the support band of the spotlight SAR signal $S_n(\omega, k_u)$ or $I_n(\omega, k_u)$ for this SAR geometry.

Two cases are identified in Figure 5.4. In one case the nth target is a *broadside* one, that is, $y_n = 0$. In this scenario the support band of the SAR signal $S_n(\omega, k_u)$ is centered around $k_u = 0$, that is, the zero slow-time Doppler frequency for all the available fast-time frequency ω values. In the other case, which is labeled as a *squint* target where $y_n \neq 0$, the support band of the spotlight SAR signal is centered around

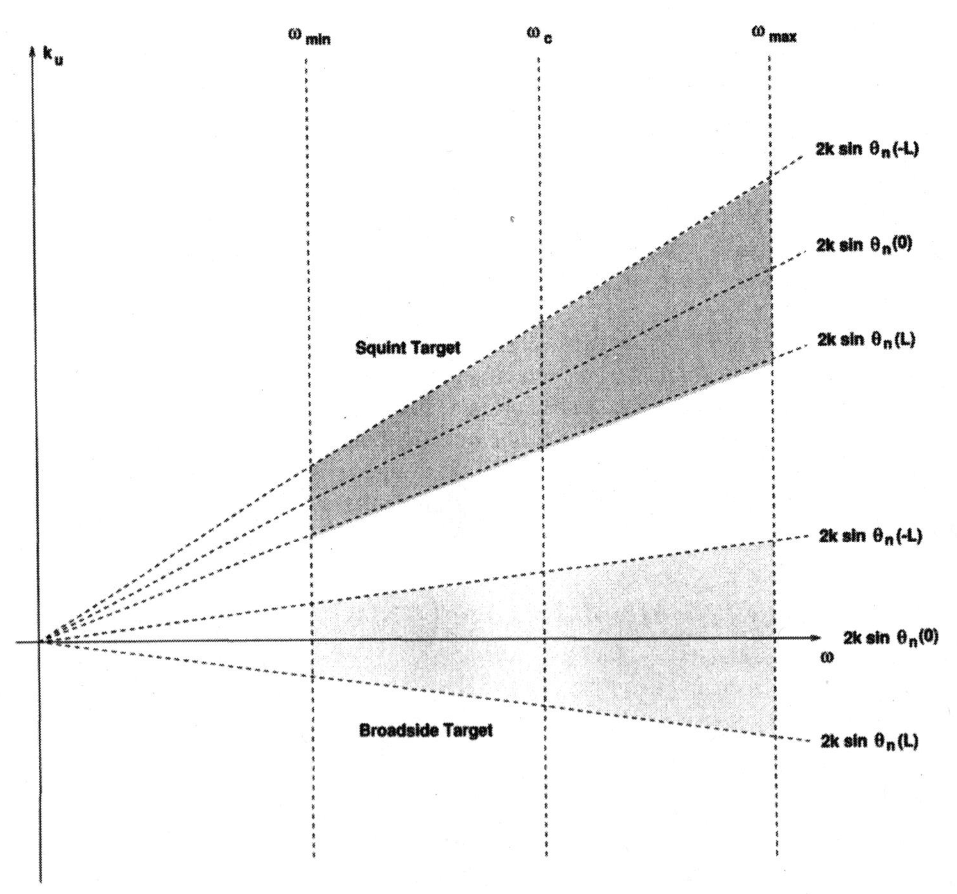

Figure 5.4. Two-dimensional spectral support band of nth target spotlight SAR signal $S_n(\omega, k_u)$ for broadside and squint targets.

the line

$$k_u = 2k \sin \theta_n(0),$$

in the (ω, k_u) domain, where

$$\theta_n(0) = \arctan\left(\frac{y_n}{x_n}\right).$$

The broadside target case is a special case of the squint target scenario with $y_n = 0$ or $\theta_n(0) = 0$.

Planar Radar Aperture In the case of a planar radar, the radar half-beamwidth in the cross-range domain depends on the radar fast-time frequency via

$$Y_0(\omega) = \frac{R_c \lambda}{D_y},$$

where D_y is the radar diameter in the cross-range domain. In this case a target area within

$$\left[Y_c - Y_{0\min}, Y_c + Y_{0\min}\right],$$

where

$$Y_{0\min} = \frac{R_c \lambda_{\min}}{D_y}$$

and $\lambda_{\min} = 2\pi / k_{\max}$, is observable at all the radar fast-time frequencies. However, a target which falls between this region and

$$\left[Y_c - Y_{0\max}, Y_c + Y_{0\max}\right],$$

where

$$Y_{0\max} = \frac{R_c \lambda_{\max}}{D_y}$$

and $\lambda_{\max} = 2\pi / k_{\min}$, is observable within a portion of the radar fast-time band $[\omega_{\min}, \omega_{\max}]$.

For instance, if the nth target falls within this region, then the target is radar observable within the wavelengths $[\lambda_{n\min}, \lambda_{\max}]$, where

$$\lambda_{n\min} = \frac{y_n D_y}{R_c}.$$

In other words, the target is not observable to the radar fast-time frequencies whose wavelengths fall within $[\lambda_{\min}, \lambda_{n\min}]$. The spectral support of the SAR signature of this target resembles the one shown in Figure 5.4; the only difference is that the spectral support band for this target in the fast-time frequency domain is limited to

$$\omega \in [\omega_{\min}, \omega_{n\max}],$$

where $\omega_{n\max} = 2\pi c/\lambda_{n\min}$.

Note that for a narrow-bandwidth spotlight SAR system, we have

$$Y_0(\omega) \approx \frac{R_c \lambda_c}{D_y},$$

which does not vary with the radar fast-time frequency. Thus the results that we obtained for a curved aperture can be used for a planar radar aperture of a narrow-bandwidth spotlight SAR system also.

Target Area

Next we use the results of the previous discussion on the spectral support of a single-target SAR signature to develop principles that govern the SAR spectral support of the echoed signal from all individual targets in the spotlighted scene, that is, the measured SAR signal.

Curved Radar Aperture The echoed SAR signal from the target area $\mathbf{s}(t, u)$ is the *sum* of the echoed SAR signals from the individual targets in the imaging scene; that is,

$$\mathbf{s}(t, u) = \sum_n \mathbf{s}_n(t, u).$$

Suppose that the illuminated target area in the cross-range domain is within

$$y_n \in \left[Y_c - Y_0, Y_c + Y_0\right],$$

where Y_0 is the half-beamwidth of the radar in the cross-range domain. *Note that in the case of a curved radar aperture, Y_0 does not vary with the radar fast-time frequency ω.* Thus, at a given fast-time frequency ω, the Doppler support band of the total echoed SAR signal $s(\omega, u)$ in the k_u domain, labeled Ω_s, is the *union* of the Doppler support bands of all $s_n(\omega, u)$'s in the slow-time Doppler frequency domain; that is (for notational simplicity the squint cross-range Y_c is assumed to be a non-negative value in the following),

$$\begin{aligned}\Omega_s &= \bigcup_{x_n \in [X_c - X_0, X_c + X_0]} \ \bigcup_{y_n \in [Y_c - Y_0, Y_c + Y_0]} \Omega_n \\ &= \left[2k \sin\theta_{\min}, 2k \sin\theta_{\max}\right]\end{aligned}$$

where

$$\theta_{\min} = \begin{cases} \arctan\left(\frac{Y_c - Y_0 - L}{X_c - X_0}\right) & \text{if } Y_c - Y_0 - L \le 0, \\ \arctan\left(\frac{Y_c - Y_0 - L}{X_c + X_0}\right) & \text{otherwise,} \end{cases}$$

$$\theta_{\max} = \arctan\left(\frac{Y_c + Y_0 + L}{X_c - X_0}\right),$$

are, respectively, the smallest and largest aspect angle of the reflectors within the spotlighted target area with respect to the radar for $u \in [-L, L]$.

Provided that $L \ll R_c$ and $(X_0, Y_0) \ll R_c$, then the spectral support of the SAR signal can be approximated via

$$\begin{aligned} \Omega_s &\approx \left[2k \sin\theta_c - 2k\frac{Y_0 + L}{X_c}\cos^2\theta_c, 2k \sin\theta_c + 2k\frac{Y_0 + L}{X_c}\cos^2\theta_c\right] \\ &= \left[2k \sin\theta_c - 2k\frac{Y_0 + L}{X_{cc}}, 2k \sin\theta_c + 2k\frac{Y_0 + L}{X_{cc}}\right], \end{aligned}$$

where

$$X_{cc} = \frac{X_c}{\cos^2\theta_c}.$$

Note that the SAR signal spectral support Ω_s is linearly increasing with the fast-time frequency ω; the slope of the linear function is proportional to the spotlighted target area in the cross-range, $2Y_0$, plus the length of the synthetic aperture, $2L$. The SAR signal spectral support is inversely proportional to the squint radial range R_c, and the squint angle θ_c.

- The beamwidth (divergence) angle of the physical curved radar is $\pm Y_0/X_{cc}$ (see Section 3.1), and the beamwidth look angle of the synthetic aperture is $\pm L/X_{cc}$. Thus, for a curved physical aperture, the slow-time Doppler bandwidth of the spotlight SAR signal is proportional to the *sum* of the beamwidth angle of the physical radar and beamwidth look angle of the synthetic aperture.

Figure 5.5*a* shows the two-dimensional spectral support of the total echoed SAR signal when the radar half-beamwidth in the cross-range domain Y_0 is invariant in the fast-time frequency (a constant). In this case a fixed target region in the cross-range domain is observable to the radar at all the available fast-time frequencies ω. As we mentioned in Chapter 3, this scenario is encountered with the curved radar apertures. Figure 5.5*a* shows the SAR signal spectral support for a broadside ($Y_c = 0$) target area and a squint ($Y_c \neq 0$) target area.

Planar Radar Aperture As we pointed out earlier in our discussion on the observability of a target by a planar radar in the fast-time frequency domain, the targets

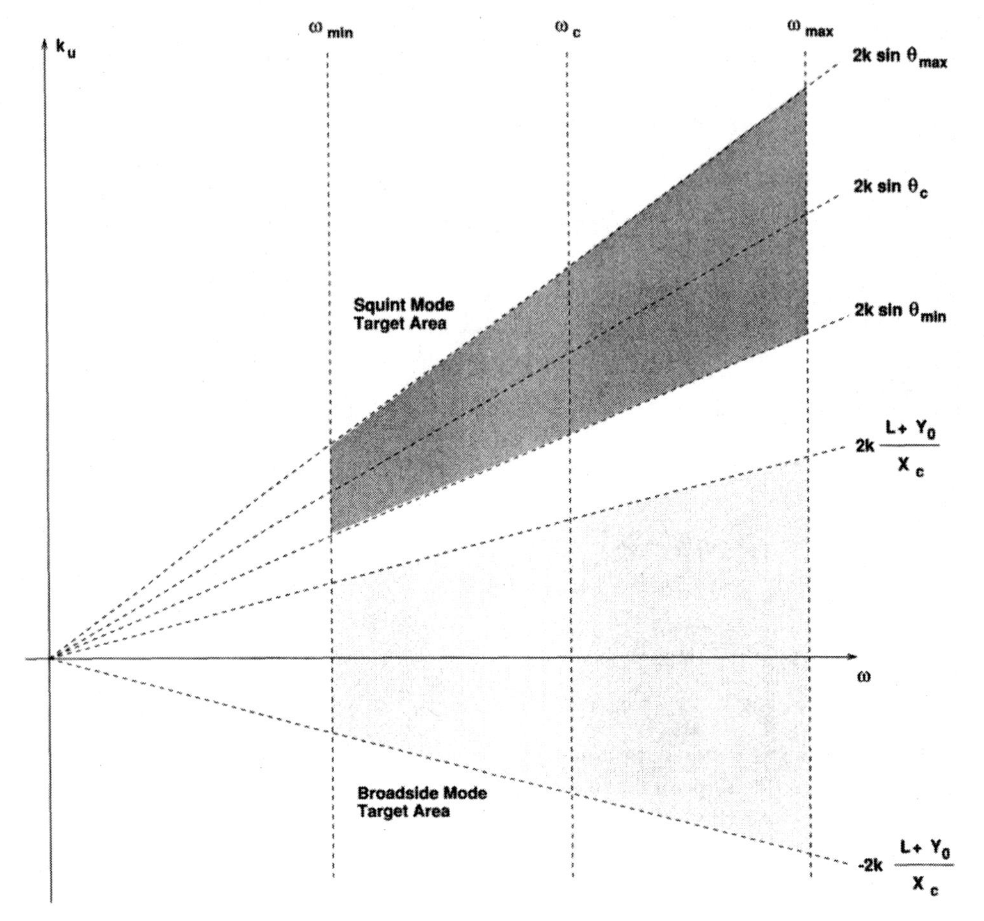

Figure 5.5a. Two-dimensional spectral support band of SAR signal for broadside and squint target areas: curved radar aperture.

that fall within the region

$$[Y_c - Y_{0\min}, Y_c + Y_{0\min}],$$

where $Y_{0\min} = R_c \lambda_{\min} / D_y$, are observable at all the available fast-time frequencies of the radar. However, the targets that fall between above region and

$$[Y_c - Y_{0\max}, Y_c + Y_{0\max}],$$

where $Y_{0\max} = R_c \lambda_{\max} / D_y$, are only partially observable in the fast-time frequency ω domain.

This has an interesting implication in the spectral support band of the spotlight SAR signal with a planar radar aperture. Substituting the fast-time dependent $Y_0(\omega) = R_c \lambda / D_y$ in the expression for the two-dimensional spectral support of the

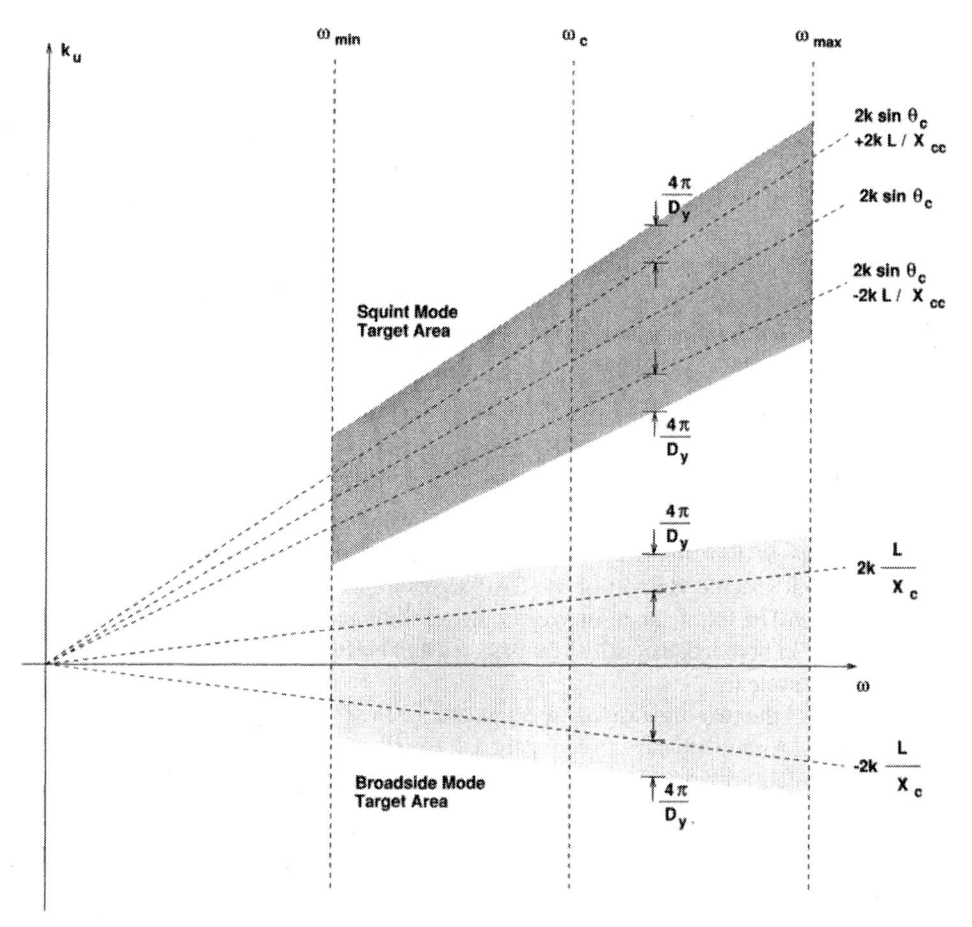

Figure 5.5b. Two-dimensional spectral support band of SAR signal for broadside and squint target areas: planar radar aperture.

spotlight SAR signal with a curved radar aperture, we obtain

$$
\begin{aligned}
\Omega_s &\approx \left[2k\sin\theta_c - 2k\frac{L}{X_{cc}} - \frac{4\pi}{D_y},\, 2k\sin\theta_c + 2k\frac{L}{X_{cc}} + \frac{4\pi}{D_y}\right] \\
&= \left[2k\sin\theta_c - 2k\left(\frac{L}{X_{cc}} + \frac{\lambda}{D_y}\right),\, 2k\sin\theta_c + 2k\left(\frac{L}{X_{cc}} + \frac{\lambda}{D_y}\right)\right].
\end{aligned}
$$

This spectral support is shown in Figure 5.5*b* for a broadside target area, and a squint target area. *Reminder: In Section 3.1 we showed that the divergence angle of a planar aperture radar is* λ/D_y.

- The beamwidth (divergence) angle of the physical planar radar is $\pm\lambda/D_y$ (see Section 3.1), and the beamwidth look angle of the synthetic aperture is

$\pm L/X_{cc}$. Thus, for a planar physical aperture, the slow-time Doppler bandwidth of the spotlight SAR signal is proportional to the *sum* of the beamwidth angle of the physical radar and the beamwidth look angle of the synthetic aperture.

Note that the SAR signal spectral support Ω_s in Figure 5.5*b* is also linearly increasing with the fast-time frequency ω; however, the slope of the linear function is only proportional to the length of the synthetic aperture, $2L$. The effect of the radar in this spectral support is a slow-time Doppler spread of $\pm 4\pi/D_y$ around the center Doppler frequency $2k \sin\theta_c$; this slow-time Doppler spread is *invariant* in the fast-time frequency ω.

5.4 RESOLUTION AND POINT SPREAD FUNCTION

The principles that govern two-dimensional resolution and point spread function of spotlight SAR systems are discussed next. The analysis is based on the two-dimensional spectral support of the SAR signature of a single target discussed in Section 5.3. The following results regarding resolution and point spread function are applicable in both mechanically beam-steered and electronically beam-steered spotlight SAR systems.

Consider the two-dimensional spectral, that is, the (ω, k_u) domain support of the SAR signal for the nth target; see Figure 5.4. Based on the SAR reconstruction principles, the data within this spectral region are mapped onto the target function spatial frequency (spectral) (k_x, k_y) domain via

$$k_x(\omega, k_u) = \sqrt{4k^2 - k_u^2},$$
$$k_y(\omega, k_u) = k_u.$$

The spectral support mapping of the SAR signal $S_n(\omega, k_u)$ of Figure 5.4 onto the target function $F_n(k_x, k_y)$ spectral domain is shown in Figure 5.6. (In the case of a planar radar and a target that is partially observable in the radar fast-time frequency band, $\omega_{\max}$ should be replaced with $\omega_{n\max}$.)

Note that Figure 5.6 only represents the *support* region for $F_n(k_x, k_y)$ or the *bandwidth* of the spatial domain target function $f_n(x, y)$; there exist all kinds of phase and amplitude fluctuations within this band which, for example, represent (see the discussion on spotlight SAR reconstruction in Sections 5.1 and 5.2) the following:

Target location	$\exp\big(-jk_x x_n - jk_y y_n\big)$
Target amplitude pattern	$A_n(\omega, k_u)$
Magnitude-squared of the radar signal spectrum and radar amplitude pattern	$\lvert P(\omega)\rvert^2$

Yet the (k_x, k_y) domain spectral support in Figure 5.6 provides a starting point for the user to comprehend the nature of target point spread functions which appear as

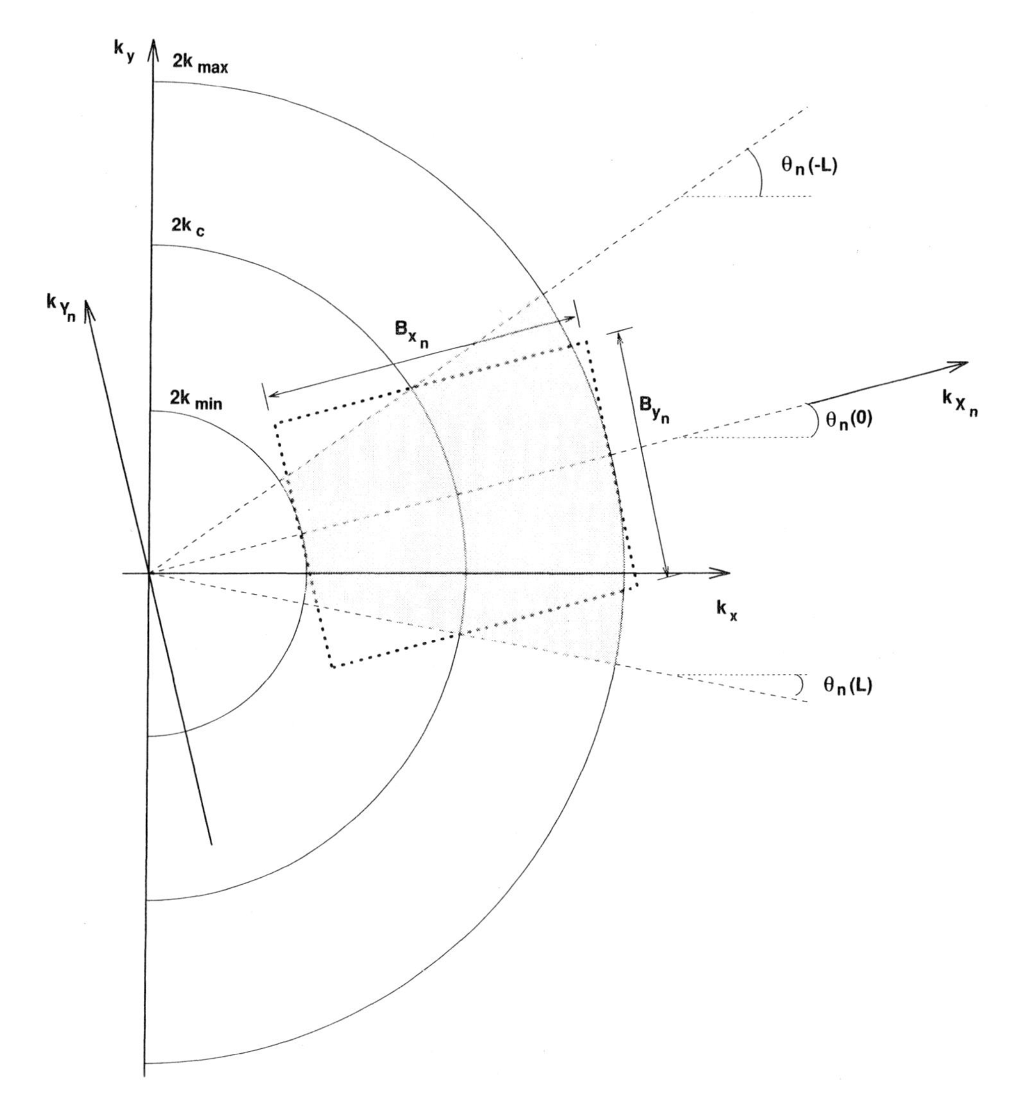

Figure 5.6. Spectral support band of nth target reconstruction $F_n(k_x, k_y)$ in spatial frequency domain.

spotlight SAR images. Consider the nth target spectral support in the domain which is generated by rotating the (k_x, k_y) domain by $\theta_n(0)$; that is (see Figure 5.6),

$$\begin{bmatrix} k_{X_n} \\ k_{Y_n} \end{bmatrix} = \begin{bmatrix} \cos\theta_n(0) & \sin\theta_n(0) \\ -\sin\theta_n(0) & \cos\theta_n(0) \end{bmatrix} \begin{bmatrix} k_x \\ k_y \end{bmatrix}.$$

In this rotated spectral domain, the target support region is *approximately* within

$$B_{x_n} = 2(k_{\max} - k_{\min}) \qquad \text{in the } k_{X_n} \text{ domain,}$$
$$B_{y_n} = 2k_c\big[\sin\theta_n(-L) - \sin\theta_n(L)\big] \quad \text{in the } k_{Y_n} \text{ domain,}$$

which is a rectangle in the (k_{X_n}, k_{Y_n}) domain.

We also identify the following transformed spatial coordinates (see Figure 5.7):

$$\begin{bmatrix} X_n \\ Y_n \end{bmatrix} = \begin{bmatrix} \cos\theta_n(0) & \sin\theta_n(0) \\ -\sin\theta_n(0) & \cos\theta_n(0) \end{bmatrix} \begin{bmatrix} x - x_n \\ y - y_n \end{bmatrix}$$

which are generated by shifting the origin by (x_n, y_n) and rotating the resultant coordinates by $\theta_n(0)$. Note that the spectral domain (k_{X_n}, k_{Y_n}) in Figure 5.6 is the spatial frequency domain for (X_n, Y_n) in Figure 5.7 [s94, ch. 2].

The inverse spatial Fourier transform of the rectangle in Figure 5.6 with respect to (k_{X_n}, k_{Y_n}) is composed of the following separable two-dimensional sinc functions in the (X_n, Y_n) domain:

$$\operatorname{sinc}\left(\frac{B_{x_n} X_n}{2\pi}\right) \operatorname{sinc}\left(\frac{B_{y_n} Y_n}{2\pi}\right).$$

This two-dimensional sinc pattern is shown as a cross-shaped structure in Figure 5.7.

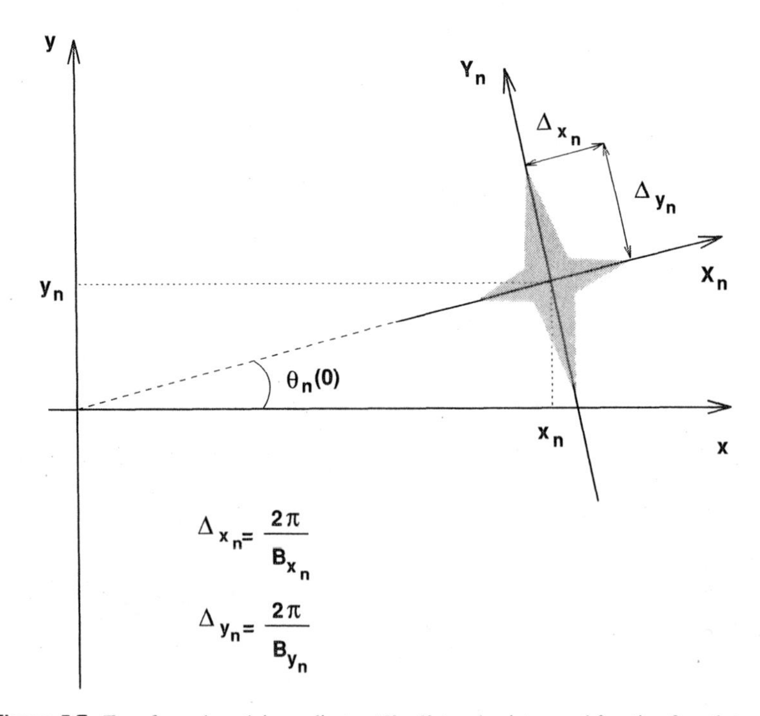

Figure 5.7. Transformed spatial coordinates (X_n, Y_n) and point spread function for nth target.

This two-dimensional sinc pattern represents the point spread function of the spotlight SAR systems for an *ideal* reflector (i.e., the target and radar amplitude patterns are constants) which is located at (x_n, y_n). The orientation (rotation angle) of the sinc pattern in the original spatial (x, y) domain, which is $\theta_n(0)$, varies with the nth target coordinates. Moreover B_{y_n} (and possibly B_{x_n} in the case of a planar radar) also varies with the coordinates of the nth target. Thus the ideal point spread function of the spotlight SAR systems is *shift-varying*.

The main lobes of the two sinc functions in the X_n and Y_n domains are, respectively, within $\pm\Delta_{x_n}$ and $\pm\Delta_{y_n}$, where

$$\Delta_{x_n} = \frac{2\pi}{B_{x_n}},$$

$$\Delta_{y_n} = \frac{2\pi}{B_{y_n}},$$

which after substituting for (B_{x_n}, B_{y_n}) become

$$\Delta_{x_n} = \frac{\pi}{(k_{\max} - k_{\min})} = \frac{\pi c}{2\omega_0},$$

$$\Delta_{y_n} = \frac{\pi}{k_c\big[\sin\theta_n(-L) - \sin\theta_n(L)\big]} \approx \frac{r_n \lambda_c}{4L\cos\theta_n(0)}.$$

(Recall the range resolution and cross-range resolution principles of Chapters 1 and 2.) The pair $(\Delta_{x_n}, \Delta_{y_n})$ can be viewed as the resolution of the nth target in the (X_n, Y_n) domain. Note that the resolution pair are also dependent on the coordinates of the target, that is, spatially varying.

We should point out that the above analysis, and the results and conclusions that followed, were based on approximating the target spectral support in Figure 5.6 via a rectangular region. This is a good approximation for narrow-bandwidth spotlight SAR systems where the radar carrier frequency is much larger than its baseband frequency ($\omega_0 \ll \omega_c$), for example, an X band SAR.

In the case of the wideband spotlight SAR systems, the shift-varying point spread function cannot be approximated by a two-dimensional sinc pattern, and it does not resemble a cross-shaped structure. For instance, for wideband UHF SAR systems, the point spread function looks like a *funnel*. Yet the point spread function can be analytically constructed by working on a more tedious algebra. Or, the user could simulate the point spread function numerically via, for example, the Matlab code for the spotlight SAR systems for this chapter.

5.5 DATA ACQUISITION AND SIGNAL PROCESSING

Next we consider the sampling constraints and processing issues associated with the digital signal processing of the spotlight SAR signal. Much of this analysis is based on the tools and principles that we established in Chapters 1 and 2 for digital signal processing of echoed and synthetic aperture signals. The current discussion of spotlight SAR signals utilizes those tools in conjunction with the two-dimensional spectral analysis of the spotlight SAR signal which appeared in Section 5.3.

Fast-time Domain Sampling and Processing

Fast-time Interval of Sampling For the general squint case, the target area is within the range region $[X_c - X_0, X_c + X_0]$ and the cross-range region $[Y_c - Y_0, Y_c + Y_0]$. (For notational simplicity, we assume that $Y_c \geq 0$ in the following discussion.) Thus for $u \in [-L, L]$, the closest radial distance of the radar to any reflector in the target area is

$$r_{\min} = \begin{cases} X_c - X_0, & \text{if } Y_c - Y_0 - L \leq 0, \\ \sqrt{(X_c - X_0)^2 + (Y_c - Y_0 - L)^2} & \text{otherwise.} \end{cases}$$

The first echoed signal arrives at the receiving radar at the fast-time

$$T_s = \frac{2r_{\min}}{c}.$$

Moreover the farthest radial distance of the radar, for $u \in [-L, L]$, to any reflector in the target area is

$$r_{\max} = \sqrt{(X_c + X_0)^2 + (Y_c + Y_0 + L)^2}.$$

The echoed signal from this radial distance lasts until the fast-time

$$T_f = \frac{2r_{\max}}{c} + T_p,$$

where T_p is the duration of the transmitted pulsed radar signal $p(t)$.

Thus, to capture the echoed signals from all of the reflectors in the spotlighted target area, we must acquire the time samples of the echoed SAR signal $\mathbf{s}(t, u)$ in the following time interval:

$$t \in [T_s, T_f].$$

The above fast-time interval for sampling the SAR signal, which corresponds to the echoed signals from the points within the radial range $[r_{\min}, r_{\max}]$, is also called the *range gate* of the SAR system A/D fast-time sampler.

It is possible to select a minimum fast-time (range) gate for each position u of the radar on the synthetic aperture. This would reduce the number of the fast-time samples that should be acquired. However, for the subsequent digital signal processing, all the fast-time Fourier (DFT) processing for the available u values should be performed on a *common* fast-time observation window that contains all the echoed signals, for example, the above-mentioned $[T_s, T_f]$. For this, the user should zero-pad the minimum fast-time gate for each u accordingly.

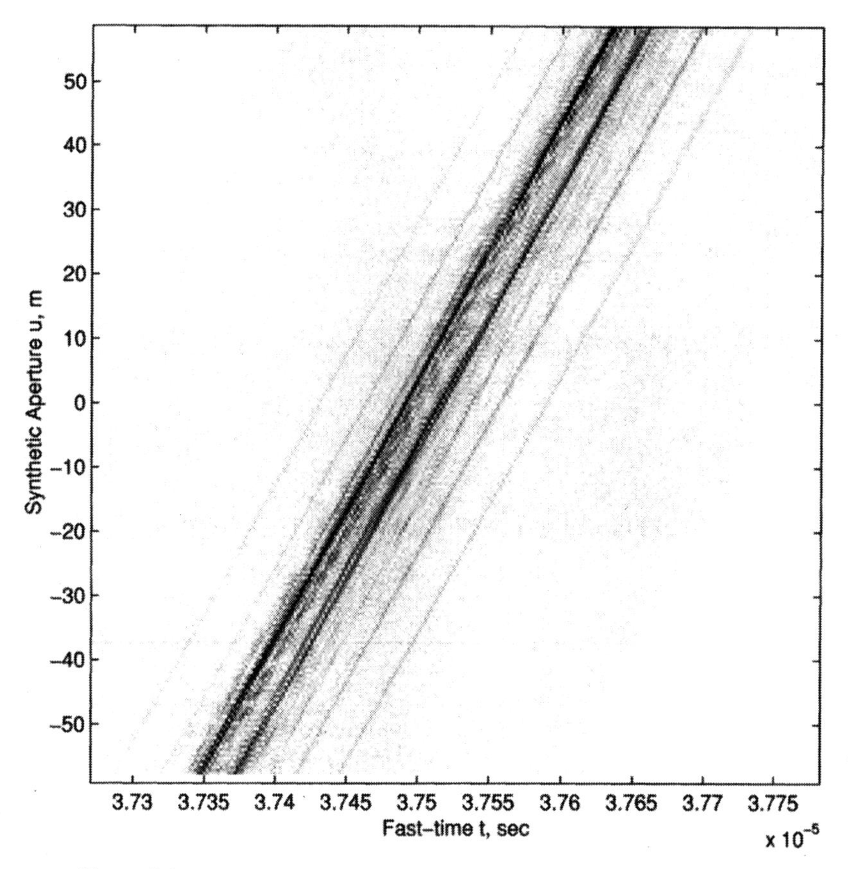

Figure 5.8a. Measured SAR (t, u) domain data: X band ISAR data of an aircraft.

An example of such a zero-padded database for the tracking ISAR system is shown in Figure 5.8*a* (Figures P5.1 and P5.2). Figure 5.8*b* is the (t, u) domain data for the FOPEN stripmap SAR system within a relatively small synthetic aperture; as we mentioned before, this SAR database belongs to the class of Type 1 SAR signals and exhibits the signal properties of the spotlight SAR signals. The (t, u) domain ISAR and SAR data in Figure 5.8*a* and *b* are fast-time matched filtered and digitally spotlighted (via the algorithm which will be discussed later) to provide a clearer picture of the hyperbolic target signatures.

Fast-time Sample Spacing Since the baseband bandwidth of the radar signal is $\pm\omega_0$, the A/D fast-time sample spacing should satisfy the following Nyquist criterion:

$$\Delta_t \leq \frac{\pi}{\omega_0}.$$

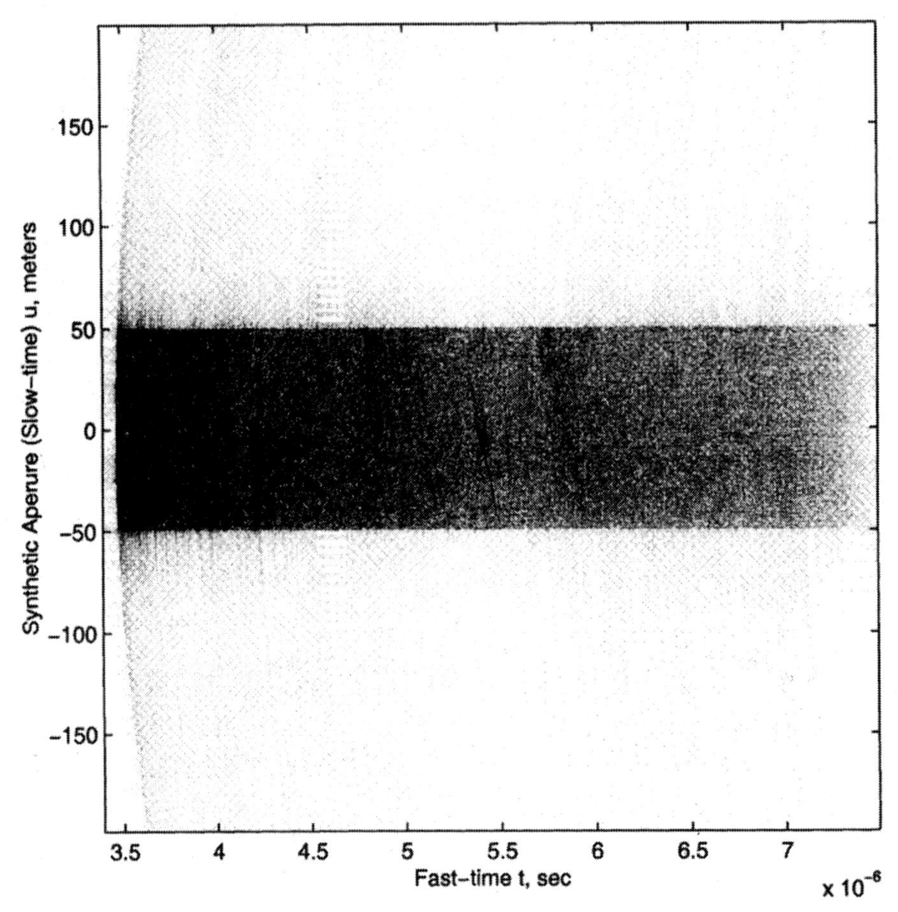

Figure 5.8b. Measured SAR (t, u) domain data: UHF band SAR data of foliage area (after digital spotlighting).

As we mentioned in Section 1.6, in the case of a chirp radar signal, the Nyquist constraint for fast-time domain sampling could be less restrictive for the deramped echoed signal. In such scenarios the upsampling scheme, which was described in Section 1.6, could be utilized to recover the alias-free echoed signal.

Finally the required number of fast-time samples within the fast-time gate $t \in [T_s, T_f]$ is

$$N = 2\left\lceil \frac{T_f - T_s}{2\Delta_t} \right\rceil,$$

where $\lceil a \rceil$ denotes the smallest integer which is larger than a. Since the above operation may increase the duration of the fast-time gate, one should fix the starting point of the time gate T_s and redefine the ending point of the time gate via

$$T_f = T_s + (N - 1)\Delta_t.$$

The other option is to fix the ending point of the time gate T_f and redefine the starting point of the time gate via

$$T_s = T_f - (N - 1)\Delta_t.$$

Fast-time Reference Point for Matched Filtering In our discussion of matched filtering for range imaging in Chapter 1, we pointed out that the matched filter is usually selected based on the echoed signal from the center of the target area. In SAR imaging the fast-time matched filter is commonly chosen to be the echoed signal from the center of the target area [i.e., $(x, y) = (X_c, Y_c)$] when the radar is at $u = 0$. Since the distance of the radar at $(0, u) = (0, 0)$ from the target at (X_c, Y_c) is R_c, this echoed signal is

$$p_0(t) = p\big(t - T_c\big),$$

where

$$T_c = \frac{2R_c}{c}$$

is called the reference fast-time point. In this case the fast-time matched-filtered SAR signal is formed via (Figure P5.2)

$$\mathbf{s}_M(t, u) = \mathbf{s}(t, u) * p_0^*(-t),$$

where $*$ denotes convolution in the fast-time domain, and $p_0^*(-t)$ is the complex conjugate of $p_0(-t)$.

The user, however, may select other values for the reference fast-time point T_c; the choice usually depends on the relative values of the range swath and the pulse duration. For instance, the user could choose the reference fast-time point to be the midpoint of the fast-time gate; that is,

$$T_c = \frac{T_f - T_s}{2}.$$

After choosing a reference fast-time point T_c and performing the above fast-time matched filtering with $p_0^*(-t)$, the fast-time sample number $(N/2) + 1$ (i.e., the sample in the middle of the fast-time array) in the matched-filtered signal $\mathbf{s}_M(t, u)$ corresponds to the fast-time point $t = T_c$. We will examine the implication of this in the spotlight SAR digital reconstruction algorithms.

Slow-time Domain Sampling and Processing

Baseband Slow-time Sample Spacing Consider the two-dimensional spectral support of the spotlight SAR signal in Figure 5.5*a* for a curved radar and Figure 5.5*b* for a planar radar. The spotlight SAR signal belongs to the class of AM-PM signals that was identified as Type 1 in Section 2.10. For this class of SAR signals, at a given fast-time frequency ω, the *baseband* Nyquist sample spacing in the u domain for recording $s(\omega, u)$ is

$$\begin{aligned}\Delta_u &\leq \frac{\pi}{k(\sin\theta_{\max} - \sin\theta_{\min})}\\ &\approx \frac{X_{cc}\lambda}{4(Y_0 + L)}.\end{aligned}$$

To select a slow-time sample spacing suitable at all the available fast-time frequencies $\omega \in [\omega_{\min}, \omega_{\max}]$, the above constraint must be satisfied for the worst scenario, that is, the smallest wavelength λ (see Figure 5.5*a* and *b*); this yields

$$\Delta_u \leq \frac{X_{cc}\lambda_{\min}}{4(Y_0 + L)},$$

where $\lambda_{\min} = 2\pi c/\omega_{\max}$. The above constraint is valid in the case of both a curved radar where Y_0 is invariant in the fast-time frequency ω (see Figure 5.5*a*) and a planar radar with $Y_{0\max} = R_c\lambda_{\max}/D_y$ (see Figure 5.5*b*).

- If the radar range swath $2X_0$ is comparable to the mean range X_c, then X_{cc} should be replaced with the minimum radial distance of the radar from the target area, that is, $r_{\min}$, in the Nyquist constraint for the slow-time domain sample spacing:

$$\Delta_u \leq \frac{r_{\min}\lambda_{\min}}{4(Y_0 + L)}.$$

As we mentioned earlier, the radar *pulse repetition interval* (PRI) is governed by

$$\text{PRI} = \frac{\Delta_u}{v_r},$$

where v_r is the speed of the radar-carrying aircraft. The number of the transmitted radar pulses (i.e., the slow-time samples) within the synthetic aperture $u \in [-L, L]$ is

$$M = 2\left\lceil \frac{L}{\Delta_u} \right\rceil.$$

Slow-time Baseband Conversion The above constraint for the slow-time sample spacing Δ_u is for the *baseband* spotlight SAR signal in the synthetic aperture u domain. As we showed in our discussion of Type 1 SAR signals in Section 2.9, for the general squint target area, the spotlight SAR signal is a *bandpass* signal.

The slow-time Doppler band Ω_s for the spotlight SAR signal at a given fast-time frequency ω is approximately centered around (see Sections 2.4 and 2.9)

$$k_u \approx 2k \sin\theta_c,$$

where θ_c is the squint angle. Thus the baseband conversion of the spotlight SAR signal in the slow-time domain can be achieved by

$$s_b(\omega, u) = s(\omega, u) \exp\big(- j2k \sin\theta_c u \big).$$

Figure 5.9*a* is the two-dimensional fast-time and slow-time spectrum of this baseband signal for the ISAR signal of Figure 5.8*a*. This signal appears to be fairly alias-free. Thus the PRF of the ISAR system was sufficiently high for the recovery of this baseband ISAR signal. For the SAR database of Figure 5.8*b*, the squint angle is zero; thus there is no need for the slow-time baseband conversion of this SAR database. Figure 5.9*b* shows the two-dimensional spectrum $S(\omega, k_u)$ for the SAR

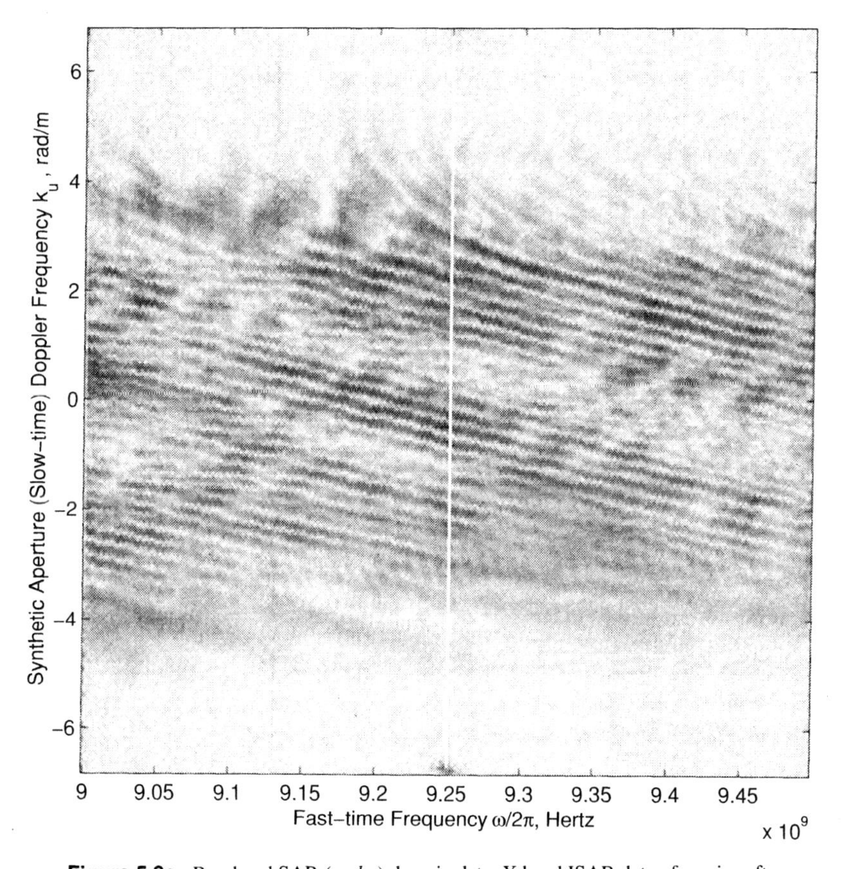

Figure 5.9a. Baseband SAR (ω, k_u) domain data: X band ISAR data of an aircraft.

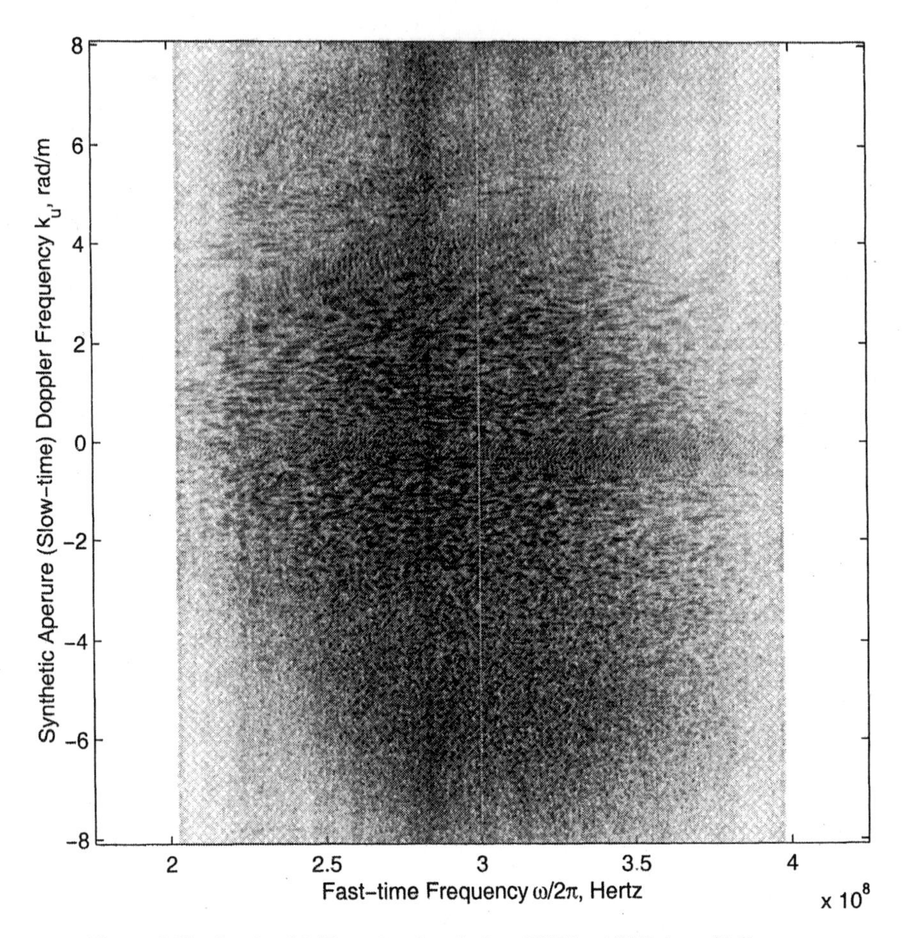

Figure 5.9b. Baseband SAR (ω, k_u) domain data: UHF band SAR data of foliage area.

data of Figure 5.8*b*. This SAR spectrum appears to contain some aliasing in the slow-time Doppler k_u domain.

Note that the slow-time Doppler carrier frequency used for the slow-time baseband conversion, that is, $2k \sin\theta_c$, varies with the fast-time frequency ω. However, it turns out that using a slow-time Doppler carrier that is invariant in the fast-time frequencies could simplify the digital reconstruction of the spotlight SAR; this issue will be discussed in the next section.

For instance, for the narrowband spotlight SAR systems, one may select the common slow-time Doppler carrier frequency $2k_c \sin\theta_c$ that is invariant in the fast-time frequency ω; that is, the slow-time baseband conversion is achieved via

$$s(\omega, u) \exp\big(-j2k_c \sin\theta_c u\big).$$

This operation simply shifts the spectrum of the SAR signal $S(\omega, k_u)$ in the slow-time Doppler k_u domain by $2k_c \sin\theta_c$; the amount of the slow-time Doppler shift is the same at all of the available fast-time frequencies.

Figure 5.10 is the two-dimensional fast-time and slow-time spectrum of the resultant signal for the ISAR signal of Figure 5.8*a* (Figure P5.3). Note that this signal exhibits slow-time Doppler domain *wrap around* aliasing at the lowest and highest fast-time frequencies.

In fact the slow-time Doppler bandwidth of the SAR signal, which is baseband converted with a common slow-time Doppler carrier, is generally larger than that of the baseband signal $s_b(\omega, u)$. To use a common slow-time Doppler carrier for slow-time baseband conversion, the synthetic aperture sample spacing should satisfy (see Figure 5.5*a*)

$$\Delta_u \leq \frac{2\pi}{k_{u\max} - k_{u\min}},$$

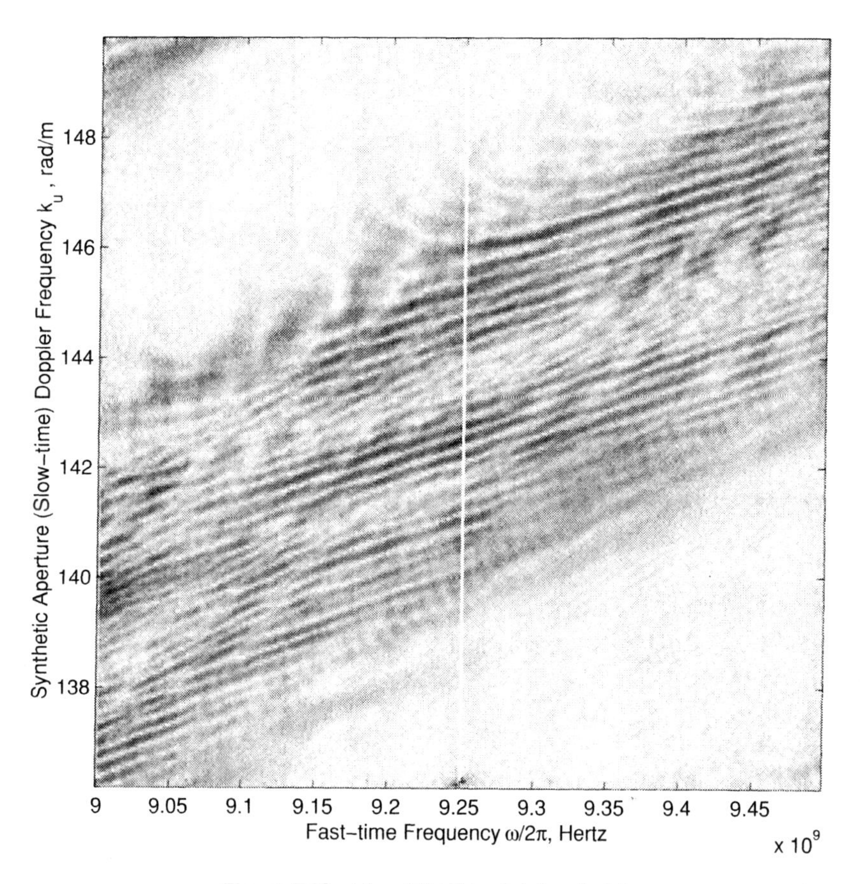

Figure 5.10. Aliased ISAR (ω, k_u) domain data.

where (for notational simplicity it is assumed that $Y_c \geq 0$)

$$k_{u\min} = \min\left[2k_{\min} \sin\theta_{\min}, 2k_{\max} \sin\theta_{\min}\right],$$

$$k_{u\max} = 2k_{\max} \sin\theta_{\max}.$$

We should point out that as long as the baseband slow-time sampling constraint, that is,

$$\Delta_u \leq \frac{X_{cc}\lambda_{\min}}{4(Y_0 + L)},$$

is satisfied, the baseband signal with a common slow-time Doppler carrier would suffer from wraparound aliasing and not an *overlapping* aliasing. In this case this signal can be recovered from the alias-free samples of $s_b(\omega, u)$ in the synthetic aperture domain. For this, one must zero-pad the samples of $S_b(\omega, k_u)$ in the slow-time Doppler k_u domain such that the inverse DFT of the resultant samples, that is, the samples of $s_b(\omega, u)$ in the u domain, satisfy the above constraint on Δ_u. Note that this produces an *upsampled* version of $s_b(\omega, u)$. These samples can then be converted into the slow-time Doppler domain baseband data via the common slow-time Doppler frequency $2k_c \sin\theta_c$.

Figure 5.11 shows the resultant signal for the ISAR example. For this, the baseband ISAR signal $S_b(\omega, k_u)$ in Figure 5.9*a*, which is composed of 256 slow-time Doppler samples, is padded with 128 zeros in both the negative and positive axes of k_u (256 zeros); this results in the doubling of the slow-time sampling rate for $s_b(\omega, u)$ in the synthetic aperture u domain. The upsampled database is converted to the samples of the ISAR signal via

$$s(\omega, u) = s_b(\omega, u) \exp\left(j2k \sin\theta_c u\right).$$

The ISAR signal in Figure 5.11 is the DFT of the samples of

$$s(\omega, u) \exp\left(-j2k_c \sin\theta_c u\right),$$

with respect to the synthetic aperture u domain.

Reducing PRF via Slow-time Compression

In Section 2.7 we showed that the slow-time compressed version of the spotlight SAR signal, that is,

$$s_c(\omega, u) = s(\omega, u) s_0^*(\omega, u),$$

where

$$s_0(\omega, u) = \exp\left[-j2k\sqrt{X_c^2 + (Y_c - u)^2}\right],$$

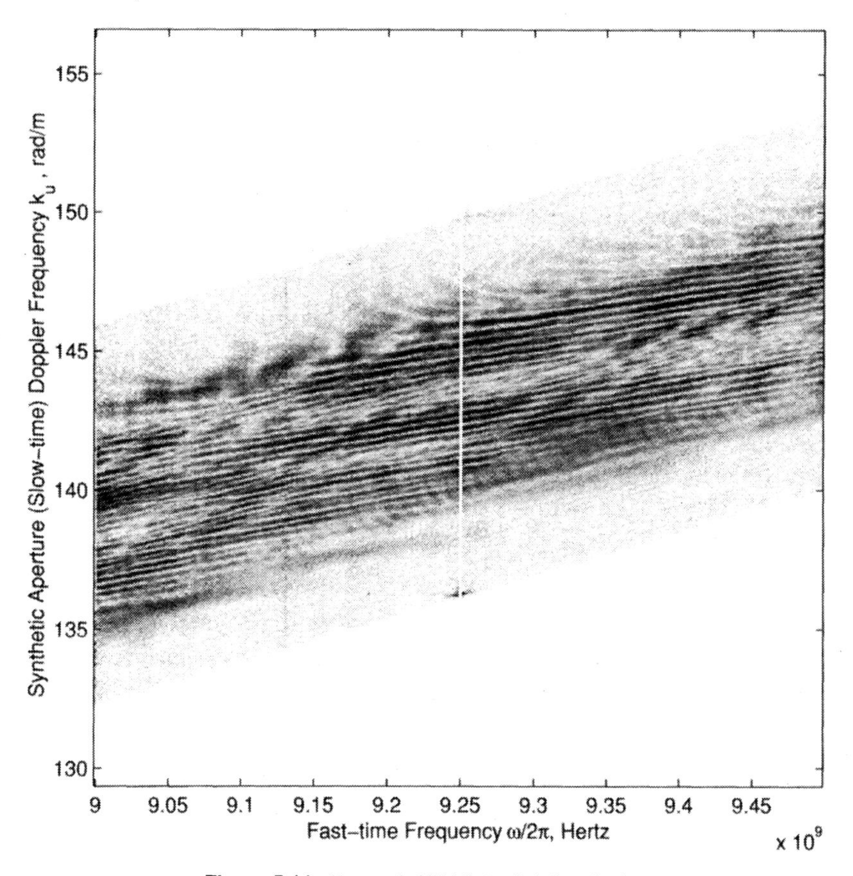

Figure 5.11. Upsampled ISAR (ω, k_u) domain data.

is the reference SAR signal in the (ω, u) domain, which requires a less restrictive sampling constraint in the synthetic aperture domain. For the general squint case, at a given fast-time frequency ω, the Nyquist rate for the slow-time compressed spotlight SAR signal is

$$\begin{aligned}\Delta_{uc} &\leq \frac{R_c \lambda}{4Y_0 \cos\theta_c} \\ &= \frac{X_{cc}\lambda}{4Y_0},\end{aligned}$$

which is invariant in the size of the synthetic aperture $2L$.

Within the radar signal fast-time frequency band $\omega \in [\omega_{\min}, \omega_{\max}]$, the most restrictive slow-time sample spacing for the slow-time compressed SAR signal is

$$\Delta_{uc} = \frac{X_{cc}\lambda_{\min}}{4Y_0},$$

which is less restrictive than the constraint on the slow-time baseband sampling Δ_u for the SAR signal $s(\omega, u)$. Based on the slow-time sample spacing of the slow-time compressed SAR signal $s_c(\omega, u)$, the number of the transmitted radar pulses (i.e., the slow-time samples) within the synthetic aperture $u \in [-L, L]$ is

$$M_c = 2\left\lceil \frac{L}{\Delta_{uc}} \right\rceil.$$

Suppose that the original SAR measurements are made at the Nyquist sampling rate of the slow-time compressed SAR signal, that is, Δ_{uc}. Thus the *pulse repetition frequency* (PRF) of the radar is v_r/Δ_{uc}, where v_r is the aircraft speed. However, the SAR signal requires a higher PRF to be measured without aliasing since $\Delta_u < \Delta_{uc}$. Thus the measured SAR data with slow-time sample spacing Δ_{uc}, that is, $s(\omega, ic\Delta_{uc})$ for $ic = 1, 2, M_c$, are aliased. These aliased measured SAR data are compressed in the slow-time domain via

$$s_c(\omega, ic\Delta_{uc}) = s(\omega, ic\Delta_{uc}) \exp\left[j2k\sqrt{X_c^2 + (Y_c - ic\Delta_{uc})^2} \right].$$

The resultant slow-time compressed spotlight SAR signal is *alias-free* since the measurements sampling rate in the slow-time domain is Δ_{uc}. This signal is upsampled by the ratio of

$$\frac{\Delta_{uc}}{\Delta_u} = \frac{Y_0 + L}{Y_0},$$

such that the resultant signal, $s_c(\omega, i\Delta_u)$, $i = 1, 2, M$, satisfies the Nyquist rate for the SAR signal, Δ_u. This upsampling can be performed via interpolation of the compressed SAR signal $s_c(\omega, u)$ in the slow-time u domain, or zero-padding its slow-time Fourier transform $S_c(\omega, k_u)$ in the slow-time Doppler k_u domain. The resultant upsampled $s_c(\omega, i\Delta_u)$ is then decompressed via

$$s(\omega, i\Delta_u) = s_c(\omega, i\Delta_u) \exp\left[-j2k\sqrt{X_c^2 + (Y_c - i\Delta_u)^2} \right],$$

to yield the alias-free SAR signal.

- The above digital signal processing-based compression, upsampling, and decompression effectively increases the radar PRF which is now v_r/Δ_u. This operation heavily exploits the functional properties of the SAR (spherical PM) signal as some form of *a priori information* or *redundancy* to increase the slow-time sampling rate of the measured SAR data. This operation cannot be performed on an arbitrary signal.

This method has been used to retrieve alias-free ISAR data of an airborne DC-9 aircraft in [s94, sec. 5.8]. In the example cited there the radar PRF was relatively low, which caused severe slow-time Doppler domain aliasing. A *step fast-time frequency* data collection for the DC-9 ISAR signal made the *effective* PRF to be relatively low.

The same is not true for the ISAR data in Figure 5.8*a* which were collected with a chirp *pulsed* radar; see the two-dimensional spectrum of this database in Figure 5.9*a*. Thus there is no need to upsample the ISAR database in the slow-time domain via the above-mentioned scheme. The spectrum of the slow-time domain compressed signal $S_c(\omega, k_u)$ for the ISAR data is shown in Figure 5.12*a* (Figure P5.4).

Figure 5.12*b* is the spectrum of the slow-time domain compressed signal $S_c(\omega, k_u)$ for the SAR data of Figure 5.8*b*. We showed that the two-dimensional spectrum of the SAR data $S(k_u, \omega)$ in Figure 5.9*b* are aliased in the slow-time Doppler k_u domain. Thus the SAR database could benefit from the above-mentioned upsampling which is achieved via zero-padding the spectrum of its slow-time compressed version of Figure 5.12*b*.

In the next section we examine the utility of the slow-time domain compressed signal in retrieving the SAR/ISAR signature of the desired target area, a process that we call *digital spotlighting*. For the SAR database of Figure 5.8*a*, the digital

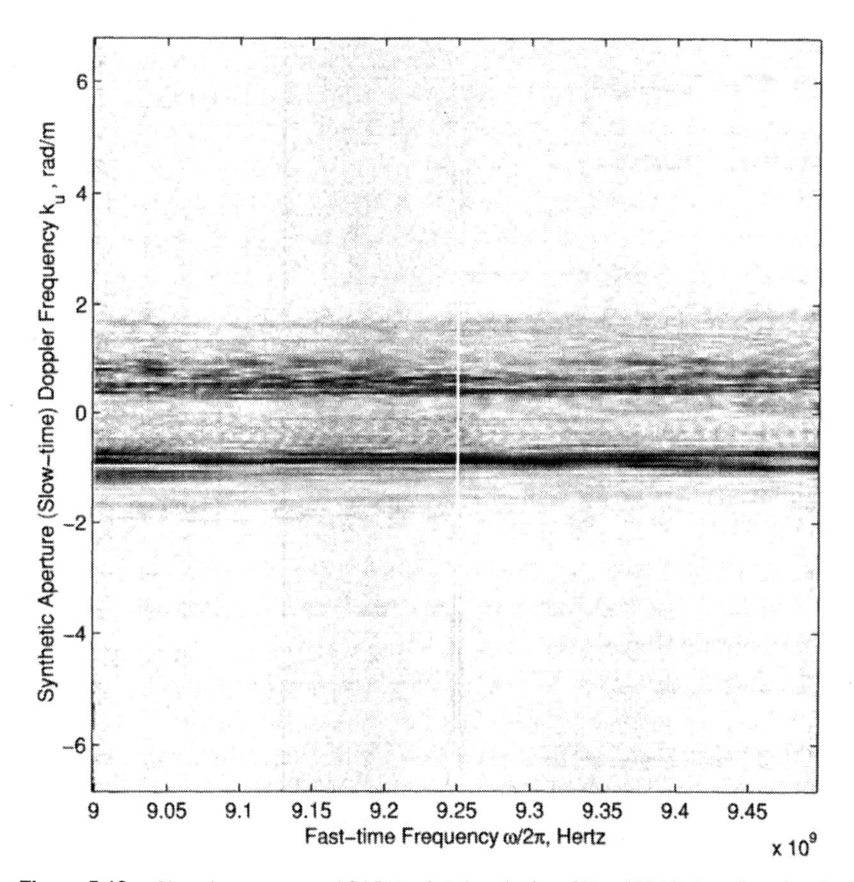

Figure 5.12a. Slow-time compressed SAR (ω, k_u) domain data: X band ISAR data of an aircraft.

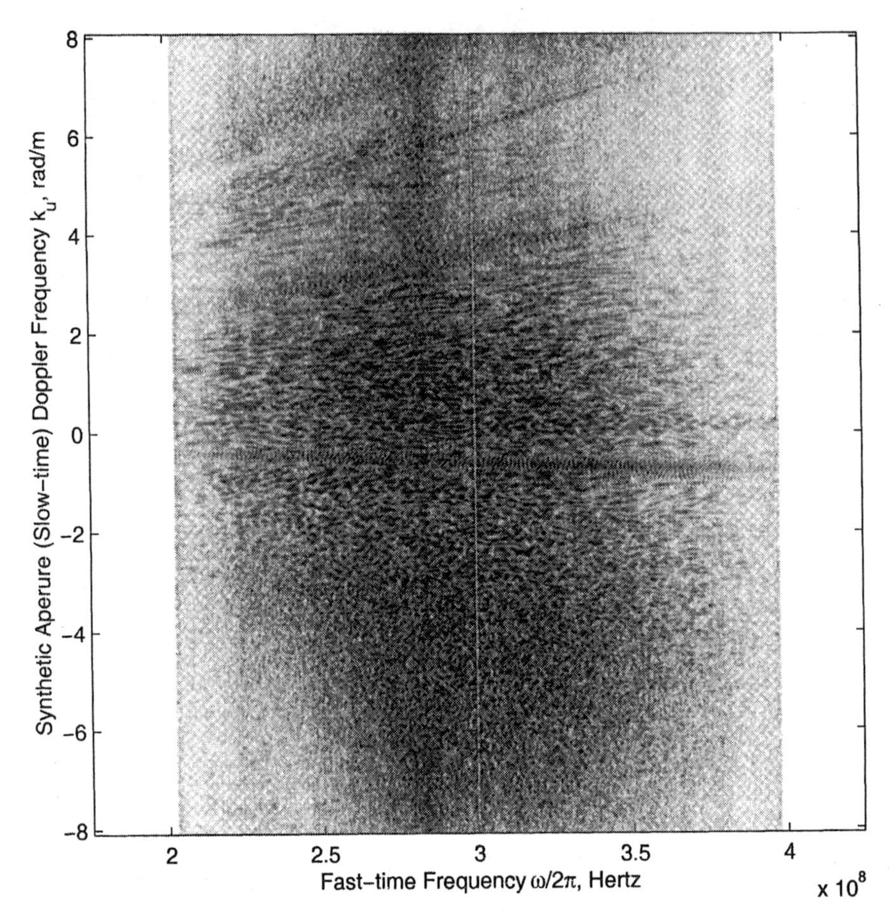

Figure 5.12b. Slow-time compressed SAR (ω, k_u) domain data: UHF band SAR data of foliage area and human-made targets.

spotlighting also results in the removal of most of the slow-time Doppler domain aliasing.

Digital Spotlighting

In Section 2.7 we showed that one can perform cross-range gating on the SAR signal via filtering the slow-time compressed SAR signal in the slow-time Doppler k_u domain. The same scheme can be used on the wideband spotlight SAR signal to extract the SAR signature of a specific target area. We refer to this as *digital spotlighting*, since we rely on digital signal processing of the measured SAR data, as opposed to the *analog spotlighting* methods which were described in Sections 5.1 and 5.2, to retrieve the SAR signature of a desired target region.

- In theory, the analog spotlighting provides the SAR signature of a finite (desired) target area. However, the radar radiation pattern is not perfectly focused and contains side lobes. In fact what we refer to as the spotlighted target area is the main lobe of the radar radiation pattern. Thus it is desirable to suppress the effects of the side lobes of the radar radiation pattern which translate into the SAR signature of the targets that reside outside the desired target area.
- In Section 5.5 we showed that the baseband Nyquist rate for the synthetic aperture sample spacing is

$$\Delta_u \leq \frac{X_{cc}\lambda_{\min}}{4(Y_0 + L)};$$

that is, a larger illuminated target area ($2Y_0$) requires a finer slow-time sample spacing. Thus, if Y_0 is chosen based on the size of the main lobe of the radar radiation pattern, then the echoed signals due to the side lobes of the radar radiation pattern (i.e., the SAR signature of the targets with the cross-range values that fall outside the region $[Y_c - Y_0, Y_c + Y_0]$) will be *aliased*. Such an aliased component usually appears as background clutter in the reconstructed SAR image. The presence of this aliasing clutter might not be apparent to the user. However, the aliasing does cause the quality of the SAR image to deteriorate and, in some circumstances, corrupt the coherent (phase) information in the SAR image.

Cross-range Gating We showed in Section 2.7 that for the general squint target area, the compressed SAR signal in the slow-time Doppler k_u domain is

$$S_c(\omega, k_u) \approx \sum_n \sigma_n \exp[-j2k(r_n - R_c)] \operatorname{sinc}\left[\frac{L}{\pi}\left(k_u - \frac{2k\cos\theta_c(y_n - Y_c)}{r_n}\right)\right].$$

Thus the signature of the nth target appears as a sinc-like blip at

$$k_u = \frac{2k\cos\theta_c(y_n - Y_c)}{r_n}.$$

Provided that (narrow-swath and narrow-beamwidth assumption)

$$\frac{1}{r_n} \approx \frac{1}{R_c},$$

the location of the sinc-like blip in the slow-time Doppler domain can be approximated via

$$\begin{aligned} k_u &\approx \frac{2k\cos\theta_c(y_n - Y_c)}{R_c}, \\ &= \frac{2k(y_n - Y_c)}{X_{cc}}. \end{aligned}$$

Note that the location of this sinc-like blip in the slow-time Doppler k_u domain is *linearly* increasing (if $y_n > 0$) or decreasing (if $y_n < 0$) with the fast-time frequency ω.

For digital spotlighting, that is, to suppress the SAR signature of the targets that reside outside the desired target area $y \in [Y_c - Y_0, Y_c + Y_0]$, the following slow-time Doppler components of the slow-time compressed signal are filtered:

$$S_c(\omega, k_u) = 0 \qquad \text{for } |k_u| > \frac{2kY_0}{X_{cc}}.$$

The boundaries of this slow-time filter in the (ω, k_u) domain are identified via two straight lines in Figure 5.12*a* and *b*.

Range and Cross-range Gating The approximation $1/r_n \approx 1/R_c$, which was used to develop the above (ω, k_u) domain digital spotlight filter, works well for the SAR/ISAR systems where the radar swath and radar beamwidth are much smaller than the mean radial distance of the target area R_c. This assumption is not valid in the UHF SAR example of Figure 5.8*b*. In this case the above digital spotlight filter rejects the signature of the targets with

$$\frac{|y_n|}{r_n} > \frac{Y_0}{X_{cc}},$$

or

$$|\sin\theta_n(0)| > \frac{Y_0}{X_{cc}},$$

where $\sin\theta_n(0) = \arctan(y_n/x_n)$ is the aspect of the nth target when the radar is at $(0, u) = (0, 0)$. Thus the above digital spotlight filter is effectively an *angular* filter in the spatial (x, y) domain. In fact, in the case of the squint SAR/ISAR problems, this digital spotlight filter may be approximated by a rectangle in the $(x_{\theta_c}, y_{\theta_c})$ domain (see Figures 5.1 and 5.3); this is not the case in the (x, y) domain.

The problem we encounter in using this filter for the broadside SAR data of Figure 5.8*b* is that neither the radar swath nor the radar beamwidth is narrow in this system. It turns out that an analysis of aperture data in the (t, k_u) domain can address this issue [s91]. Perhaps the best way to view this approach is to remember that a *crude* SAR image of the target area can be formed from the slow-time compressed SAR data. This was the basis of the *polar format processing* which was described in Section 4.11. The resultant image suffers from smearing, shifting, and geometric distortion of the target area. While these are not desirable in the final reconstructed image, it is the case that this crude image can be used for the purpose of digital spotlighting.

To formulate this concept, consider the target function which is formed via narrow-bandwidth and narrow-beamwidth polar format processing at near broadside; that is (see Figure 4.15),

$$F(k_x, k_y) = s_c(\omega, u),$$

where

$$k_x = 2k \cos\theta_c,$$

$$k_y = 2k_c \sin\theta_c - 2k_c \frac{\cos^2\theta_c}{R_c} u.$$

Thus the target function $f(x, y)$ is simply the two-dimensional Fourier transform of the slow-time compressed signal, that is, the slow-time compressed SAR signal in the (t, k_u) domain.

In Section 4.11 we pointed out that a more accurate analysis of the SAR image which is formed via polar format processing would indicate that the signature of the nth target, which is located at the polar spatial coordinates $[\theta_n(0), r_n]$, appears at

$$t_n \approx \frac{2r_n}{c},$$

$$k_{un} \approx 2k_c \sin\left[\theta_n(0) - \theta_c\right],$$

in the polar format processed SAR image, that is, the two-dimensional Fourier transform

$$\mathcal{F}_{(\omega,u)}[s_c(\omega, u)],$$

which we refer to as the polar format processed reconstruction. (The Fourier transform with respect to the variable ω is an inverse one.)

Note that the above transformation from (r_n, θ_n) to (t_n, k_{un}) can be translated into the following rectilinear coordinates in the spatial domain:

$$x_n = \frac{ct_n}{2}\cos(\phi_n + \theta_c),$$

$$y_n = \frac{ct_n}{2}\sin(\phi_n + \theta_c),$$

where

$$\phi_n = \arcsin\left(\frac{k_{un}}{2k_c}\right).$$

Thus, for digital spotlighting the desired target area, that is,

$$x \in [X_c - X_0, X_c + X_0] \quad \text{and} \quad y \in [Y_c - Y_0, Y_c + Y_0],$$

the polar format processed image in the (t, k_u) domain should be passed through a two-dimensional filter in the (t, k_u) domain with the following passband:

$$\left|\frac{ct}{2}\cos(\phi + \theta_c) - X_c\right| < X_0,$$

$$\left| \frac{ct}{2} \sin(\phi + \theta_c) - Y_c \right| < Y_0,$$

where

$$\phi = \arcsin\left(\frac{k_u}{2k_c}\right)$$

is the *angular* slow-time Doppler domain at the carrier frequency.

We should point out that this definition of the angular slow-time Doppler ϕ domain is suitable for the *narrow-bandwidth* SAR systems. One could improve the results of digital spotlight filtering by converting the SAR signal $S_c(\omega, k_u)$ into the (ω, ϕ) domain where

$$\phi = \arcsin\left(\frac{k_u}{2k}\right)$$

is the angular slow-time Doppler domain which varies with the fast-time frequency. For this purpose, at a fixed ω or k, one must interpolate $S_c(\omega, k_u)$ from the k_u domain to the ϕ domain. Then this database is converted into the (t, ϕ) domain via an inverse Fourier transform with respect to ω. The digital spotlight window is then applied to this (t, ϕ) domain database. The digital-spotlighted data are then brought back to the (ω, ϕ) domain via a forward Fourier transform with respect to t. The resultant (ω, ϕ) domain data are interpolated to recover the (ω, k_u) domain data via the wide-bandwidth mapping $\phi = \arcsin\left(k_u/2k\right)$, or

$$k_u = 2k \sin\phi.$$

Thus, for the wide-bandwidth SAR systems, one needs to implement one-dimensional interpolation from the k_u domain to the ϕ domain, and then back from the ϕ domain to the k_u domain. It turns out that using the narrow-bandwidth approximation, that is, $\phi = \arcsin\left(k_u/2k_c\right)$, for digital spotlighting results in slight degradations near the edges of the reconstructed SAR image; these are negligible effects especially when the purpose is to image a large target area. (One may reduce and/or completely remove these degradations by using a larger digital spotlight filter.) In the discussion and the results that follow, we use the narrow-bandwidth approximation for digital spotlighting.

The procedure for digital spotlighting is shown in Figure 5.13 and can be summarized as follows. The reference SAR signal is defined via

$$s_0(\omega, u) = \exp\left[-j2k\sqrt{X_c^2 + (Y_c - u)^2} + j2kR_c\right];$$

the addition of the phase term $2kR_c$ to the reference signal ensures that the reference fast-time point T_c is unchanged. The digital spotlight filter in the (t, k_u) domain is

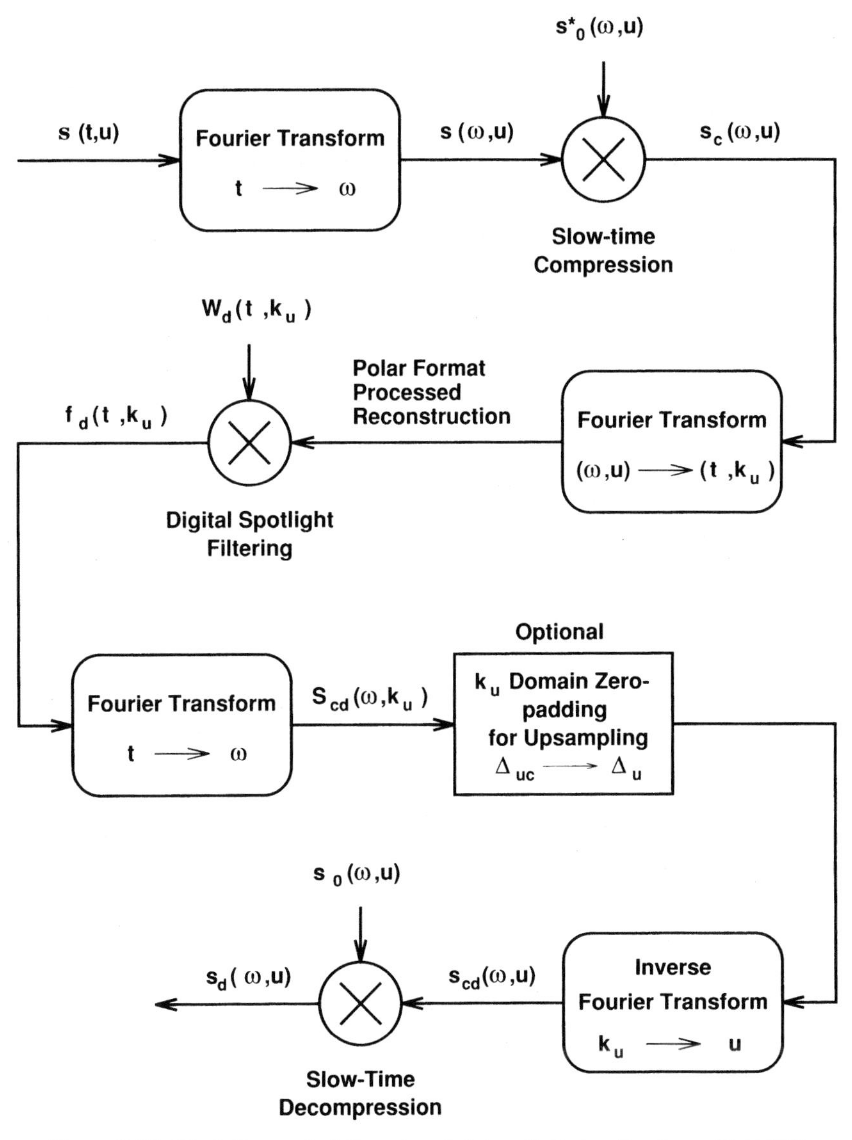

Figure 5.13. Block diagram for full aperture digital spotlight algorithm for spotlight SAR.

defined via

$$W_d(t, k_u) = \begin{cases} 1 & \text{for } |\frac{ct}{2}\cos(\phi + \theta_c) - X_c| < X_0 \text{ and } |\frac{ct}{2}\sin(\phi + \theta_c) - Y_c| < Y_0, \\ 0 & \text{otherwise.} \end{cases}$$

Polar format processed reconstruction with the digital spotlight filter is (Figure P5.5)

$$f_d(t, k_u) = W_d(t, k_u)\mathcal{F}_{(\omega,u)}[s_c(\omega, u)].$$

Then the digital-spotlighted slow-time compressed SAR signal in the (ω, k_u) domain is obtained by

$$S_{cd}(\omega, k_u) = \mathcal{F}_{(t)}[f_d(t, k_u)].$$

At this point, if the user wishes to upsample the SAR data in the synthetic aperture u domain (i.e., sample spacing conversion from Δ_{uc} to Δ_u), an appropriate number of zeros should be added to the samples of $S_{cd}(\omega, k_u)$ in the slow-time Doppler k_u domain; see Figure 2.10. After the optional zero-padding in the k_u domain, the digital-spotlighted slow-time compressed signal in the (ω, u) domain is formed via

$$s_{cd}(\omega, u) = \mathcal{F}^{(-1)}_{(k_u)}[S_{cd}(\omega, k_u)].$$

Finally the digital-spotlighted SAR signal in the (ω, u) domain is constructed from the following slow-time decompression (mixing with the reference SAR signal):

$$s_d(\omega, u) = s_{cd}(\omega, u)s_0(\omega, u).$$

Figure 5.14*a* shows the polar format processed image after digital spotlight filtering for the ISAR database of Figure 5.8*a* (Figure P5.5); $(X_0, Y_0) = (30, 30)$ m is used for the digital spotlight filter. Note that the digital spotlight filter in Figure 5.14*a*, which resembles a parallelogram, is for a *squint* ISAR system where $(X_0, Y_0) \ll R_c$. This digital spotlight filtering does not significantly alter the baseband spectrum of the ISAR data which appears in Figure 5.9*a*.

Figure 5.14*b* is the polar format processed image and the digital spotlight filter for the SAR database of Figure 5.8*b*; $(X_0, Y_0) = (250, 400)$ m are used for the digital spotlight filter. In this figure the digital spotlight filter (the shaded area) is superimposed on the polar format reconstruction. The digital spotlight filter in Figure 5.14*b* is for a *broadside* SAR system with a relatively *large-range swath.* The sample spacing in the synthetic aperture u domain of the measured SAR data is $\Delta_u = 0.39$ m. This translates into the following slow-time Doppler support band for the measured SAR data:

$$k_u \in \left[-\frac{\pi}{\Delta_u}, \frac{\pi}{\Delta_u}\right] = [-8, 8] \text{ rad/m}.$$

We zero-pad the digitally-spotlighted SAR data in Figure 5.14*b* with sufficient zeros to double the slow-time Doppler k_u domain bandwidth; that is, the zero-padded slow-time Doppler support band for the SAR data becomes

$$k_u \in [-16, 16] \text{ rad/m}.$$

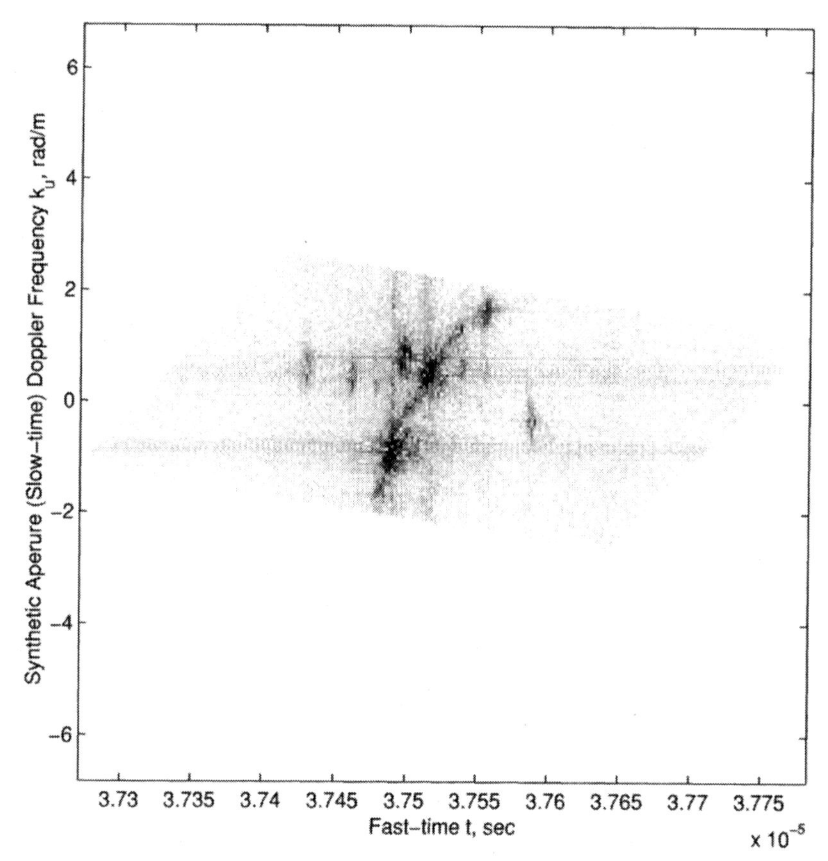

Figure 5.14a. Polar format processed (t, k_u) domain data with digital spotlight filter (shaded area) superimposed: X band ISAR data of an aircraft.

This slow-time Doppler band corresponds to a new synthetic aperture sample spacing of $\Delta_u = 0.39/2 = 0.195$ m; that is, the synthetic aperture u domain data are upsampled by a factor of two. [Due to the processing that follows, it is more computationally efficient to perform the k_u domain zero-padding after Fourier transforming the (t, k_u) domain data into the (ω, k_u) domain data.]

After Fourier transforming the zero-padded (t, k_u) domain data into the (ω, u) domain data, we obtain the digitally spotlighted and upsampled version of $s_{cd}(\omega, u)$, that is, the slow-time compressed SAR signal. We then transform this database to the SAR signal $s_d(\omega, u)$ via the *decompression* operation in the slow-time domain [i.e., multiplication with the reference signal $s_0(\omega, u)$.] The Fourier transform of this signal with respect to the slow-time u is the two-dimensional spectrum of the digitally spotlighted (filtered) signal $S_d(\omega, k_u)$ for the SAR data; this spectrum is shown in Figure 5.15 (Figure P5.6).

Note that the two-dimensional SAR spectrum in Figure 5.15 does not suffer from the slow-time Doppler aliasing of the original SAR spectrum in Figure 5.9*b*. How-

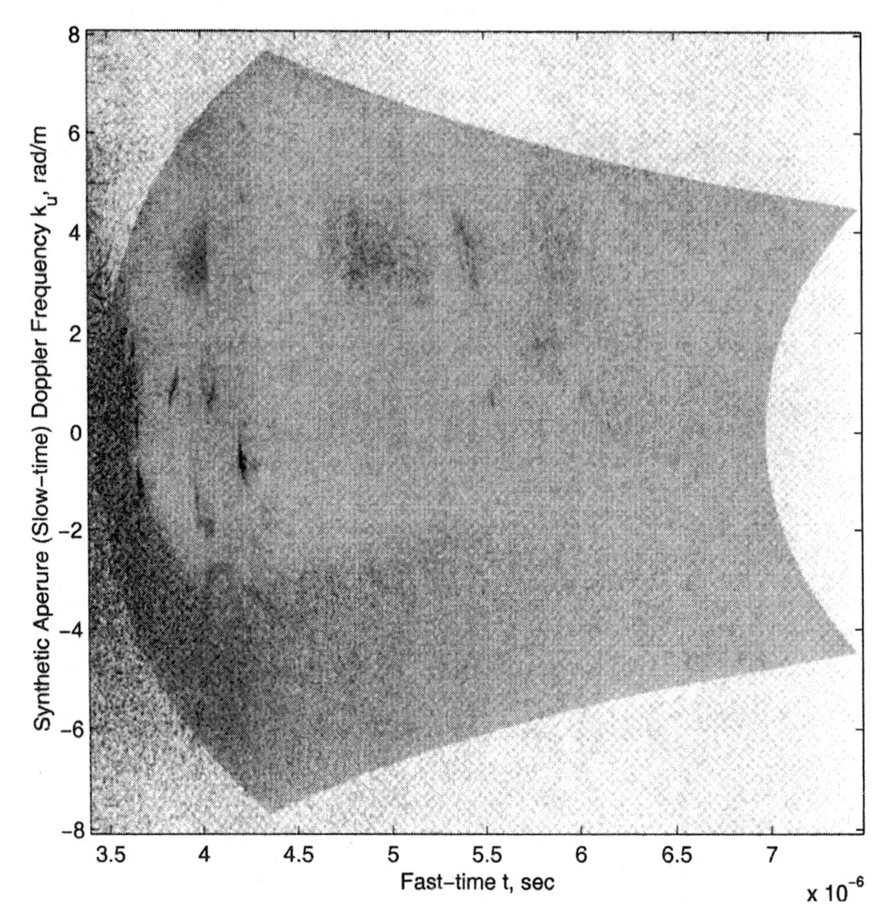

Figure 5.14b. Polar format processed (t, k_u) domain data with digital spotlight filter (shaded area) superimposed: UHF band SAR data of foliage area and human-made targets.

ever, its slow-time Doppler support band is approximately within $k_u \in [-8, 8]$ rad/m which is comparable to the original measured SAR data spectral support band. This implies that there was no need for upsampling the digitally spotlighted data; the digital spotlight filter effectively removed most of the slow-time Doppler domain aliasing.

We should emphasize that this operation works only for the SAR data with the synthetic aperture of $[-L, L] = [-50, 50]$ m. As we will see in Chapter 6, when the synthetic aperture is increased to $[-L, L] = [-200, 200]$, we cannot rely on only the digital spotlighting to remove the slow-time Doppler domain aliasing. In that case we will use the upsampling of the slow-time compressed SAR signal to handle the slow-time Doppler domain aliasing.

As we stated earlier, for the digitally spotlighted SAR data of Figure 5.14b, the desired target area in the cross-range domain was chosen to be $[-Y_0, Y_0] =$

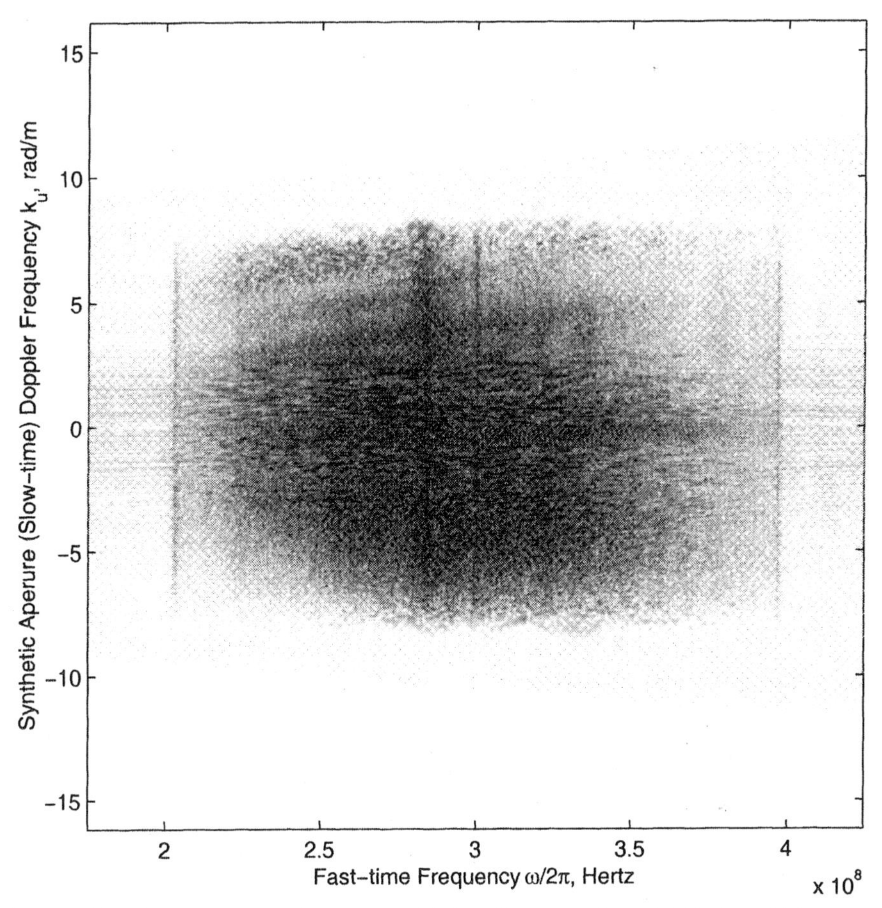

Figure 5.15. SAR (ω, k_u) domain data after digital spotlighting: UHF band SAR data of foliage area and human-made targets.

$[-400, 400]$ m. With the synthetic aperture of $[-L, L] = [-50, 50]$ m and the closest target located at $r_{\min} = 510$ m, the radar aspect (look) angles are within $[-\theta_{\max}, \theta_{\max}]$, where

$$\theta_{\max} = \arctan\left(\frac{Y_0 + L}{r_{\min}}\right) = 41.4°.$$

(Note that $r_{\min}$ is used instead of X_c or X_{cc} for this SAR system where the range swath is comparable to the mean range swath.) Thus the Nyquist sample spacing in the synthetic aperture u domain at the carrier frequency 300 Mhz and range $r_{\min} = 510$ m (the closest range) is

$$\Delta_u \leq \frac{2\pi}{4k_c \sin\theta_{\max}} = 0.38 \text{ m},$$

which is approximately equal to the synthetic aperture sample spacing of the measured SAR data, that is, 0.39 m.

The Nyquist sample spacing in the synthetic aperture u domain at the highest radar fast-time frequency 400 Mhz and range $r_{\min} = 510$ m (the closest range) is

$$\Delta_u \leq \frac{2\pi}{4k_{\max} \sin \theta_{\max}} = 0.28 \text{ m};$$

thus this Nyquist criterion is violated with the synthetic aperture sample spacing of the measured SAR data, that is, 0.39 m. However, the digitally spotlighted SAR spectrum in Figure 5.15 does not appear to contain much aliasing at the higher fast-time frequencies. There are two reasons for this. First, the radar energy drops off rapidly at the higher fast-time frequencies which makes their aliased data weak and difficult to observe.

The second reason for not having much aliasing at the highest fast-time frequencies in Figure 5.15 is the digital spotlighting operation. Since we used the narrow-bandwidth approximation for digital spotlight filtering, the highest slow-time Doppler frequency passed through the digital spotlight filter is

$$k_{u\max} = 2k_c \sin\left[\arctan\left(\frac{Y_0}{r_{\min}}\right)\right] = 7.8 \text{ rad/m}.$$

This translates into the following half-width cross-range at the highest fast-time frequency:

$$Y_0(\omega_{\max}) = r_{\min} \tan\left[\arcsin\left(\frac{k_{u\max}}{2k_{\max}}\right)\right] = 266 \text{ m}.$$

Thus the digitally spotlighted cross-range area at the nearest range and the highest fast-time frequency is $y \in [-Y_0(\omega_{\max}), Y_0(\omega_{\max})]$. For the synthetic aperture of $[-L, L] = [-50, 50]$ m, this yields the following Nyquist criterion for the synthetic aperture sample spacing:

$$\Delta_u \leq \frac{2\pi}{4k_{\max} \sin \arctan[\{(Y_0(\omega_{\max}) + L)/r_{\min}\}]} = 0.36 \text{ m},$$

which is fairly close to the synthetic aperture sample spacing of the measured SAR data (0.39 m).

One of the properties of the narrowband digital spotlight filter is to generate a SAR database whose two-dimensional spectrum belongs to the class of spectra which was shown in Figure 5.5*b*; in that scenario the extent of the illuminated (spotlighted) target area in the cross-range domain varied with the fast-time frequency. For the digitally spotlighted SAR data of Figure 5.15, the half-width cross-range of the digitally spotlighted target area at the lowest, center, and highest fast-time frequencies are

$$Y_0(\omega) = \begin{cases} r_{\min} \tan\left[\arcsin\left(\dfrac{k_{u\max}}{2k_{\min}}\right)\right] = r_{\min} \tan 68^\circ = 1248 \text{ m} & \text{at 200 MHz;} \\ r_{\min} \tan\left[\arcsin\left(\dfrac{k_{u\max}}{2k_c}\right)\right] = r_{\min} \tan 38^\circ = 400 \text{ m} & \text{at 300 MHz;} \\ r_{\min} \tan\left[\arcsin\left(\dfrac{k_{u\max}}{2k_{\max}}\right)\right] = r_{\min} \tan 28^\circ = 266 \text{ m} & \text{at 400 MHz.} \end{cases}$$

Note that with $Y_0(\omega_{\min}) = 1248$ m at the lowest frequency, the maximum radar aspect angle with $L = 50$ m is

$$\arctan\left[\frac{Y_0(\omega_{\min}) + L}{r_{\min}}\right] = 68.6^\circ.$$

Yet the beamwidth angle of the *physical* radar [i.e., the support (main lobe) of the radar transmit/receive amplitude pattern in the cross-range domain] at 200 MHz is smaller than $\pm 68.6^\circ$. We will show this in our discussion of this *stripmap* SAR database with a larger synthetic aperture in Chapter 6 (stripmap SAR systems). Our study will indicate that the beamwidth angle of the physical radar is approximately ± 45 (i.e., the total beamwidth angle of 90°).

Finally we should point out that the digital spotlighting provides a digital signal processing approach for removing *some* of the aliasing due to the side lobes of the radar radiation pattern. The resultant digitally spotlighted SAR data still contain some *leakage* from the targets which fall outside the digital spotlight filter. Some of this leakage is actually aliased in the (t, k_u) domain.

The portion of the leakage that is not aliased can be dealt with by using a larger Y_0 than the one chosen for the digital spotlight filter in the subsequent digital signal processing. This implies that the digitally spotlighted data $f_d(t, k_u)$ should be zero-padded in the k_u domain *beyond the theoretical Nyquist slow-time Doppler k_u band* (i.e., upsampling in the synthetic aperture u domain). Moreover, for the digital SAR reconstruction methods that form the image in the (k_x, k_y) domain (frequency domain interpolation-based method) or the (x, k_y) domain (range stacking method), a finer sample spacing in the k_y domain (which corresponds to a larger Y_0) should be used. (See the digital signal processing issues associated with the digital reconstruction algorithms in Section 5.6.)

Subaperture Digital Spotlighting

The digital spotlighting algorithm described in the previous section was based on the approximation that the nth target signature in the polar format processed SAR image appears at the following fast-time:

$$t_n \approx \frac{2r_n}{c},$$

where r_n is the nth target radial distance from the radar when the radar-carrying aircraft is at the center of the synthetic aperture, that is, $u = 0$. In spotlight SAR systems where the size of the synthetic aperture $2L$ is relatively large, the variations of the fast-time of arrival of the echoed signal from the nth target with respect to the slow-time u becomes large. In this case t_n cannot be used as a common fast-time of arrival for the nth target signature at all $u \in [-L, L]$.

A simple remedy for this problem is to divide the synthetic aperture $[-L, L]$ into smaller subapertures and perform digital spotlighting on the slow-time compressed SAR signal $s_c(\omega, u)$ within these subapertures. For instance, suppose that there are N_s subapertures with length $2L_s$ within $u \in [-L, L]$; these subapertures are

$$[Y_i - L_s, Y_i + L_s], \qquad i = 1, \ldots, N_s,$$

where

$$Y_i = (2i - 1)L_s - L.$$

Thus the SAR data in the ith subaperture can be viewed as the SAR data from a spotlight SAR system with squint range and cross-range parameters

$$(X_c, Y_c + Y_i),$$

and the squint angle

$$\theta_{ci} = \arctan\left(\frac{Y_c + Y_i}{X_c}\right).$$

The fast-time of arrival of the nth target signature in the ith subaperture can now be approximated via

$$t_{ni} \approx \frac{2\sqrt{x_n^2 + (y_n + Y_i)^2}}{c}.$$

The polar format processed image of the ith subaperture is formed by

$$\mathcal{F}_{(\omega,u)}\Big[s_c(\omega, u);\, u \in [Y_i - L_s, Y_i + L_s]\Big].$$

To develop the (t, k_u) domain digital spotlight filter for the ith subaperture, one could follow the steps that were used in the previous section with the squint cross-range $Y_c + Y_i$ and the squint angle θ_{ci}. For digital spotlighting within the ith subaperture, the resultant digital spotlight filter used on the polar format processed image of the ith subaperture becomes

$$W_{di}(t, k_u) = \begin{cases} 1 & \text{for } |\frac{ct}{2}\cos(\phi + \theta_{ci}) - X_c| < X_0 \text{ and} \\ & |\frac{ct}{2}\sin(\phi + \theta_{ci}) - Y_c - Y_i| < Y_0, \\ 0 & \text{otherwise.} \end{cases}$$

After slow-time upsampling via zero-padding in the k_u domain (if it is essential), the digitally spotlighted and upsampled SAR data of the N_s subapertures are appended to each other to form the SAR data that will be used for image formation.

Subaperture Size It is desirable to use the size of the subaperture $2L_s$ to be as small as possible to improve the estimate of the fast-time arrival of the targets in the imaging scene. However, if L_s is chosen to be too small, then the rectangular window spreading of the subaperture in the k_u domain, that is, its main lobe $\pm\pi/L_s$, becomes more dominant than the theoretical bandwidth of the slow-time compressed SAR signal in the k_u domain, which is

$$\left[\frac{-2kY_0}{X_{cc}}, \frac{2kY_0}{X_{cc}}\right] \approx \left[\frac{-4\pi}{D_y}, \frac{4\pi}{D_y}\right].$$

(The right-hand side of the above is for a planar radar.)

A simple rule of thumb is to choose L_s such that the rectangular window spreading of the subapertures is an order of magnitude smaller than (1/100 of) the theoretical slow-time Doppler band of $s_c(\omega, u)$; that is,

$$\frac{\pi}{L_s} \approx \frac{2k_{\min} Y_0}{100 X_{cc}},$$

which yields

$$\begin{aligned} L_s &\approx \frac{25 X_{cc} \lambda_{\max}}{Y_0} \\ &\approx 25 D_y. \end{aligned}$$

In most practical SAR systems, this corresponds to approximately a few hundred slow-time samples.

The ISAR and SAR examples examined in this chapter have a relatively smaller synthetic aperture and thus do not require subaperture digital spotlighting. We will deal with a stripmap SAR problem in the next chapter which will benefit from subaperture digital spotlighting.

5.6 RECONSTRUCTION ALGORITHMS AND SAR IMAGE PROCESSING

Next we examine the digital reconstruction algorithms for the spotlight SAR systems and the image processing issues associated with the formation of a spotlight SAR image. This analysis is based on the principles that were outlined in Sections 4.5 to 4.12 where we introduced four SAR digital reconstruction methods: spatial frequency-based interpolation, range stacking, time domain correlation, and backprojection. These methods are now specialized for the spotlight SAR systems; the Matlab codes for these algorithms are discussed at the end of this chapter.

Digital Reconstruction via Spatial Frequency Interpolation

Two-dimensional Frequency Domain Matched Filtering Digital reconstruction via spatial frequency interpolation for generic SAR systems was discussed in Section 4.5; the block diagram for this algorithm was shown in Figure 4.5. The basic principle for SAR image formation via this approach is the two-dimensional frequency domain matched filtering of the measured SAR signal with the reference signal, that is,

$$S(\omega, k_u) S_0^*(\omega, k_u),$$

where the two-dimensional frequency domain matched filter

$$S_0^*(\omega, k_u) = \underbrace{P^*(\omega)}_{\text{Fast time matched filter}} \quad \underbrace{\exp\left(j\sqrt{4k^2 - k_u^2}X_c + jk_uY_c\right)}_{\text{Target area baseband conversion}}$$

is composed of two components: $P^*(\omega)$ which is the fast-time domain matched filter and the phase function $\exp\left(j\sqrt{4k^2 - k_u^2}X_c + jk_uY_c\right)$ which brings the origin to the center of the target area in the spatial domain.

As we pointed out in our discussion of reconstruction for the two spotlight SAR systems of Sections 5.1 and 5.2, the physical radar beam-steered radiation pattern introduces additional amplitude functions in the two-dimensional frequency (ω, k_u) domain; these have implications in the *spatially varying* point spread function $f_n(x, y)$ of the spotlight SAR system. For the electronically beam-steered spotlight SAR, an additional matched filtering with the amplitude pattern of the phased array element, that is, $A_e^*(\omega, k_u)$ should be performed. (This amplitude pattern is typically very wideband in the slow-time Doppler k_u domain and contains negligible amounts of phase; its effects can be ignored.)

Zero-padding in Fast-time Domain In our discussion of SAR spatial frequency mapping and interpolation in Section 4.5., we identified the Jacobian function

$$J_m(\omega) = \frac{4k}{c\sqrt{4k^2 - k_{um}^2}},$$

which is the Jacobian of the transformation from ω to k_x at a given discrete slow-time Doppler k_{um}. The maximum value of $|J_m(k)|$ in the region of the available data and *within the two-dimensional bandwidth of the spotlight SAR signal* in the (k_u, ω) domain is

$$J_{\max} = \frac{2}{\cos\theta_{\text{ax}}},$$

where

$$\theta_{\text{ax}} = \max\left[|\theta_{\min}|, |\theta_{\max}|\right]$$

is the maximum *absolute* aspect angle of the spotlighted target area with respect to the radar synthetic aperture coordinates. For $Y_c \geq 0$ we have

$$\begin{aligned}\theta_{ax} &= \theta_{max} \\ &= \arctan\left(\frac{Y_c + Y_0 + L}{X_c - X_0}\right)\end{aligned}$$

Note the aspect angle θ_{ax} corresponds to the *largest slow-time Doppler frequency* in the spotlight SAR signal.

Then the *common* (worst-case scenario) unevenly spaced Nyquist constraint over the entire region of available data (within the two-dimensional bandwidth of the spotlight SAR signal) becomes [s88]

$$\Delta_k J_{max} \leq \Delta_{k_x}.$$

Moreover the Nyquist constraint in the range spatial frequency domain k_x is

$$\Delta_{k_x} \leq \frac{\pi}{X_0},$$

where $2X_0$ is the size of the spotlighted target area in the range domain. Combining the above two inequalities produces

$$\begin{aligned}\Delta_k &\leq \frac{\pi}{J_{max} X_0} \\ &= \frac{\pi \cos\theta_{ax}}{2X_0}.\end{aligned}$$

The sample spacing in the ω domain is related to Δ_k via

$$\Delta_\omega = \Delta_k c.$$

Moreover, based on DFT equations [s94, ch. 1], the length of the SAR data in the fast-time t domain is

$$\begin{aligned}T &= \frac{2\pi}{\Delta_\omega} \\ &= \frac{2\pi}{\Delta_k c}.\end{aligned}$$

Based on the constraint on Δ_k, the fast-time interval of processing should satisfy

$$T \geq \frac{4X_0}{c\cos\theta_{ax}}.$$

Note that the fast-time interval for measuring the SAR data is $T_f - T_s$. Thus the minimum fast-time interval for processing the SAR data is

$$T_{min} = \max\left[T_f - T_s, \frac{4X_0}{c\cos\theta_{ax}}\right].$$

This could be met by *zero-padding* the measured SAR data in the fast-time domain prior to its Fourier transformation into the ω domain.

Zero-padding in Synthetic Aperture Domain This issue was discussed earlier in Section 2.8, on digital reconstruction in cross-range imaging. The sample spacing of the DFT of the measured SAR signal with respect to the synthetic aperture u in the slow-time Doppler k_u domain is

$$\Delta_{k_u} = \frac{\pi}{L},$$

where L is the half-width of the synthetic aperture. Based on the SAR mapping from the measurement spectral (ω, k_u) domain into the target spectral (k_x, k_y) domain, we have

$$k_y = k_u.$$

Thus the available samples of $F(k_x, k_y)$ in the k_y domain (i.e., $k_{ymn} = k_{um}$ values) are also separated by

$$\Delta_{k_y} = \frac{\pi}{L}.$$

However, the extent of the target area in the cross-range domain is $2Y_0$. Thus, to avoid aliasing in the cross-range y domain, the sample spacing in the k_y domain should satisfy the following Nyquist rate:

$$\Delta_{k_y} \leq \frac{\pi}{Y_0}.$$

If the size of the target area is greater than the length of the synthetic aperture, that is, $Y_0 > L$, then the available samples of $F(k_x, k_y)$ in the k_y domain with sample spacing π/L are aliased. As we pointed out in Section 2.8, this problem can be resolved by zero-padding the SAR signal in the synthetic aperture u domain to create an effective aperture of $[-Y_0, Y_0]$ prior to the DFT processing with respect to the discrete samples of u. In Section 2.8 we also provided another reason (i.e., avoiding circular convolution aliasing) for this zero-padding. In the general case ($Y_0 > L$ or $Y_0 \leq L$), the minimum half-length of the processed (zero-padded) synthetic aperture is chosen to be

$$L_{\min} = \max(L, Y_0).$$

Finally we should point out that the above constraints on the minimum intervals for processing the SAR data in the fast-time and slow-time domains, that is, $(T_{\min}, L_{\min})$, are suited for the *one-dimensional* interpolation from the ω to the k_x domain. If one treats the problem as a (more complicated) *two-dimensional* interpolation problem, then different values for $(T_{\min}, L_{\min})$ should be chosen [s94, p. 285].

Algorithm The implementation of the reconstruction algorithm based on the spatial frequency interpolation of the above-mentioned matched-filtered database also requires the knowledge of the support of the spotlight SAR signal in various domains which were described in Sections 5.3 to 5.5. The reconstruction algorithm is summarized in the following.

Step 1. Perform the (discrete) fast-time matched filtering

$$\mathbf{s}_M(t, u) = \mathbf{s}(t, u) * p_0^*(-t),$$

where

$$p_0(t) = p\big(t - T_c\big),$$

and T_c is the user-prescribed reference fast-time point (e.g., $T_c = 2R_c/c$).

Step 2. Perform the fast-time and slow-time processing (slow-time baseband conversion for the squint case; digital spotlighting; if it is required, upsampling of the data from Δ_{uc} to Δ_u in the slow-time domain after slow-time compression; etc.) which was described in Section 5.5.

Step 3. If $T_f - T_s$ does not provide the sufficient sampling density in the k_x domain, zero-pad the SAR data in the fast-time domain according to the constraint on $T_{\min}$. Also, if $Y_0 > L$, zero-pad the data in the slow-time u domain to achieve the minimum length synthetic aperture extent of $u \in [-L_{\min}, L_{\min}]$, where $L_{\min} = \max(L, Y_0)$.

Step 4. Obtain the two-dimensional DFT of the digitally spotlighted, upsampled, and zero-padded SAR signal; the outcome is the samples $S_d(\omega_n, k_{um})$.

Step 5. If $Y_0 < L$, that is, the size of the target area in the cross-range domain is greater than the length of the synthetic aperture, perform slow-time Doppler domain subsampling on the available $S_d(\omega_n, k_{um})$ data; this option will be discussed later.

Step 6. The DFT algorithm assumes that the fast-time sample number $(N/2)+1$ (i.e., the sample in the middle of the fast-time array) is at $t = 0$; however, this sample corresponds to the actual fast-time value of $t - T_c$. Thus the two-dimensional DFT of the upsampled and digitally spotlighted SAR signal should be modified by the following linear phase function (discrete) multiplication:

$$S_d(\omega, k_u) \exp\big(-j\omega T_c\big),$$

which moves the fast-time origin back to the actual fast-time zero, that is, $t = 0$. The above multiplication with the linear phase function of ω is crucial for all the SAR digital reconstruction algorithms that involve processing of the measured data in the fast-time frequency domain.

Step 7. Perform (discrete) baseband conversion of target area via the addition of the phase function of (X_c, Y_c) to form the target function

$$F(k_x, k_y) = S_d(\omega, k_u) \exp\left(- j\omega T_c\right) \exp\left(j\sqrt{4k^2 - k_u^2} X_c + jk_u Y_c \right),$$

where (Figure P5.7)

$$k_x = \sqrt{4k^2 - k_u^2},$$
$$k_y = k_u.$$

For the available discrete samples of SAR data in the (ω, k_u) *domain, labeled* (ω_n, k_{um})*, this results in a set of unevenly spaced samples of the target function in the spatial frequency* (k_x, k_y) *domain at (see Section 4.5)*

$$k_{xmn} = \sqrt{4k_n^2 - k_{um}^2},$$
$$k_{ymn} = k_{um}.$$

Step 8. Identify the support of the coverage of SAR data in the spatial frequency (k_x, k_y) *domain, labeled*

$$k_x \in \left[k_{x\min}, k_{x\max}\right] \quad \text{and} \quad k_y \in \left[k_{y\min}, k_{y\max}\right],$$

where (see the notes regarding these two parameters in the Matlab code in Section 5.7)

$$k_{x\min} = \min\left[k_{xmn}\right],$$
$$k_{x\max} = \max\left[k_{xmn}\right],$$

and

$$k_{y\min} = \min\left[k_{ymn}\right],$$
$$k_{y\max} = \max\left[k_{ymn}\right].$$

Identify a grid in this region with the following sample spacings:

$$\Delta_{k_x} = \frac{\pi}{X_0},$$
$$\Delta_{k_y} = \frac{\pi}{Y_0}.$$

The resultant number of samples in the grid are

$$N_x = 2\left\lceil \frac{k_{x\max} - k_{x\min}}{2\Delta_{k_x}} \right\rceil \quad \text{in the } k_x \text{ domain,}$$

$$N_y = 2\left\lceil \frac{k_{y\max} - k_{y\min}}{2\Delta_{k_y}} \right\rceil \quad \text{in the } k_y \text{ domain.}$$

Step 9. Interpolate the samples of the target function spectrum $F(k_x, k_y)$ on the unfirm grid from its available unevenly spaced data $F(k_{xmn}, k_{ymn})$.

Step 10. Obtain the two-dimensional inverse DFT of the evenly spaced samples of the target function spectrum $F(k_x, k_y)$ on the uniform grid. This yields (N_x, N_y) evenly spaced samples of the spatial domain target function $f(x, y)$ on a uniform grid with the following sample spacings:

$$\Delta_x = \frac{2\pi}{N_x \Delta_{k_x}} = \frac{2X_0}{N_x},$$

$$\Delta_y = \frac{2\pi}{N_y \Delta_{k_y}} = \frac{2Y_0}{N_y}.$$

Step 11. Reduce the bandwidth of the reconstructed image and remove some of the clutter effects by SAR image compression; this option will be discussed later.

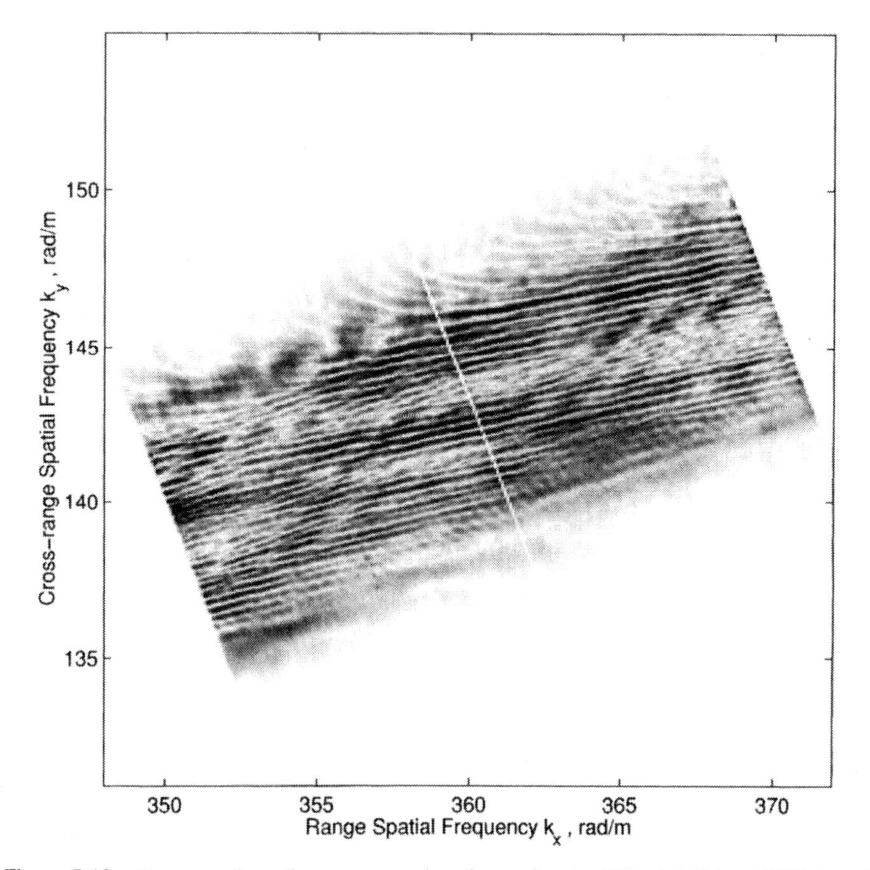

Figure 5.16a. Spectrum of wavefront reconstruction of target function $F(k_x, k_y)$: X band ISAR data of an aircraft.

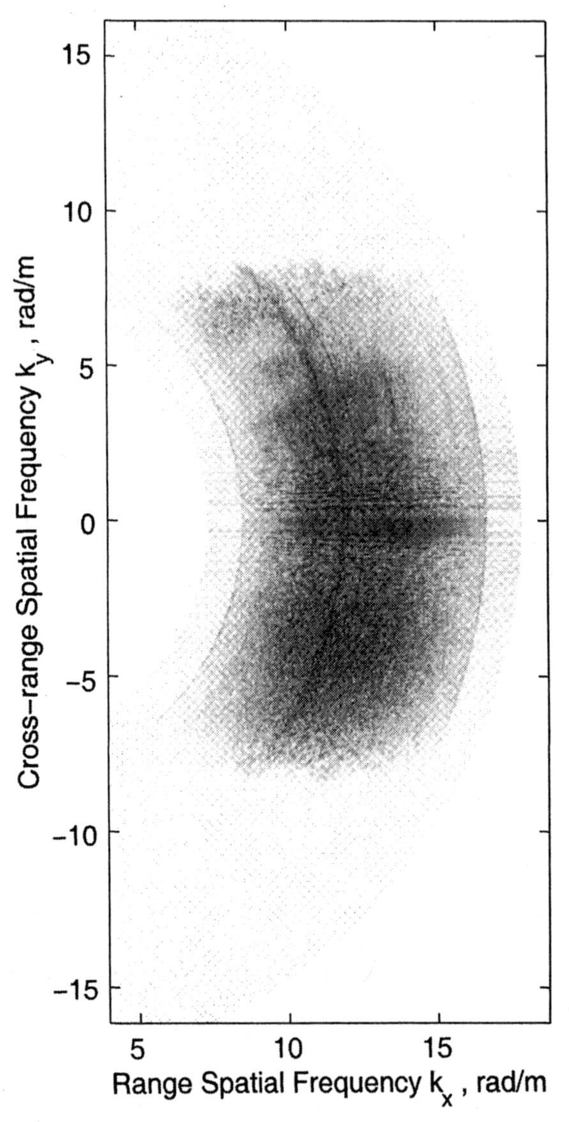

Figure 5.16b. Spectrum of wavefront reconstruction of target function $F(k_x, k_y)$: UHF band SAR data of foliage area and human-made targets.

Figure 5.16*a* and *b*, respectively, shows the reconstruction of the target function spectrum $F(k_x, k_y)$ for the ISAR and SAR examples of Figure 5.8*a* and *b*; this is the output of the algorithm after Step (9) (Figure P5.8). Figure 5.17*a* is the target function reconstruction in the spatial domain, the outcome of Step (10) of the algorithm (Figure P5.9), for the ISAR example. Note that the reflectors on the aircraft appear as cross-shaped streaks as were predicted in our discussion on the spotlight SAR point spread function in Section 5.4.

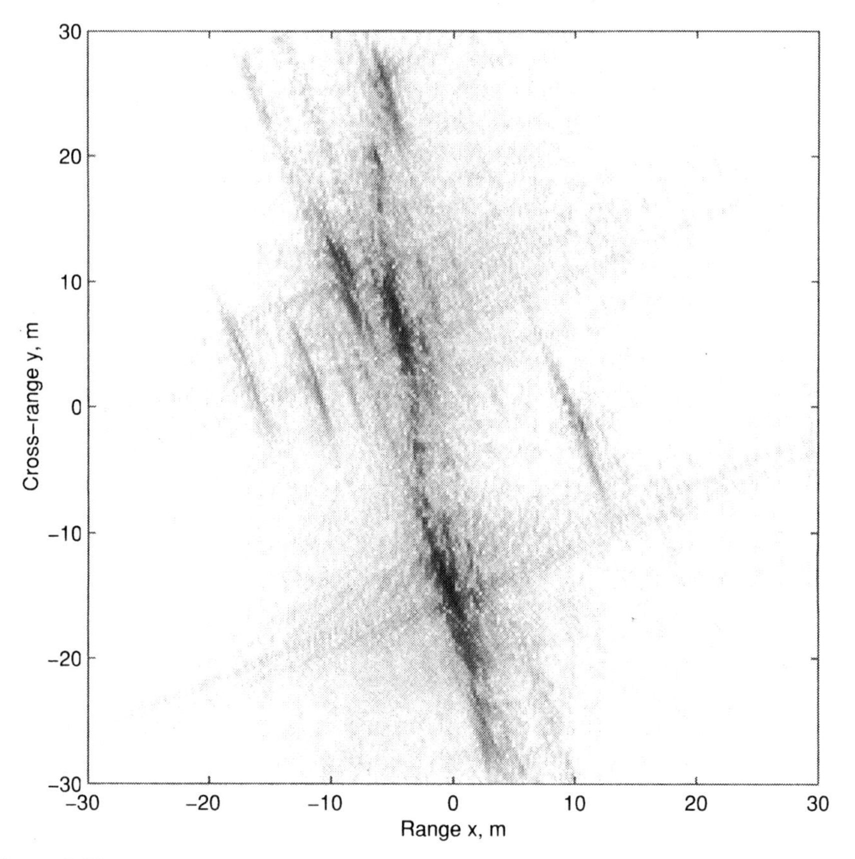

Figure 5.17a. Wavefront reconstruction of target function $f(x, y)$ from X band ISAR data within synthetic apertures: $u \in [-L, L]$ (full aperture).

However, the resolution of these cross-shaped streaks does not follow the theoretical resolution outlined in Section 5.4; in fact the resolution seen in Figure 5.17*a* is much poorer than the theoretical resolution. The reason for this phenomenon is that at this stage of its flight, the aircraft possesses a significant amount of *nonlinear* and *out of slant-plane* motion components (yaw, pitch, and roll) which are not included in our theoretical SAR/ISAR signal model.

To observe this, one can compare the ISAR reconstructions of the aircraft which are formed with the lower half of the synthetic aperture, that is, $u \in [-L, 0]$, and the upper half of the synthetic aperture, which is $u \in [0, L]$; these are shown in Figure 5.17*b* and *c*. Provided that the aircraft motion model had been the same as our theoretical ISAR model, then two ISAR images would have possessed slightly different point spread functions. In that scenario, when the two images are coherently added, the reflector should appear even sharper; this is due to processing a larger effective synthetic aperture of $[-L, L]$.

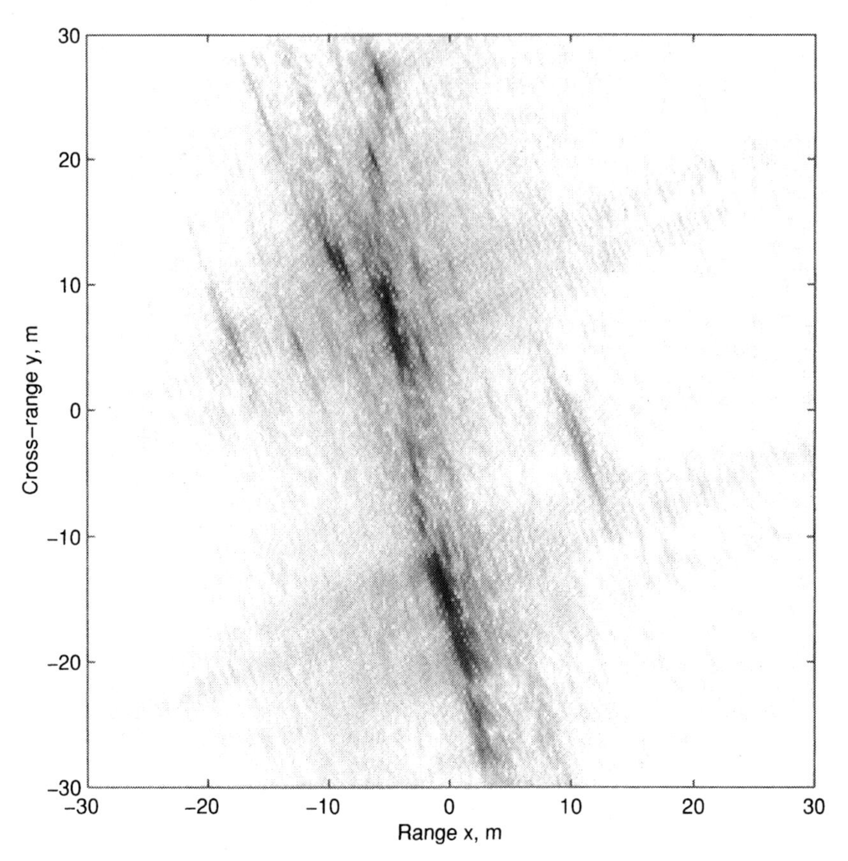

Figure 5.17b. Wavefront reconstruction of target function $f(x, y)$ from X band ISAR data within synthetic apertures: $u \in [-L, 0]$ (lower-half aperture).

This is clearly not the case in Figure 5.17*a*. Note that the coordinates of the wing reflectors vary in these two ISAR images from the lower and upper half synthetic apertures. In fact the *inward* change of the coordinates of the wing reflectors from Figure 5.17*b* to *c* signifies that the aircraft was rolling during the slow-time (synthetic aperture) data acquisition.

Figure 5.18*a* shows the reconstruction for the UHF SAR example of Figure 5.8*b*. The foliage area can be seen at near range on the left side of the image ($x < 600$ m), the lower side of the image (cross-range values $y < -200$ m), and the far range at the upper right side of the image (relatively weak due to the far range). Figure 5.18*b* and *c* shows close-ups of this SAR reconstruction. Figure 5.18*b* contains a portion of the near-range and lower cross-range foliage area, and includes trucks and corner reflectors. Figure 5.18*c* shows a few corner reflectors (note the one at $y = 300$ m at the upper left corner of the image) and a farm area in its upper right corner. (There are several moving targets at or near the farm area which appear as smeared and streaking targets.) We will re-examine this UHF SAR database in Chapter 6.

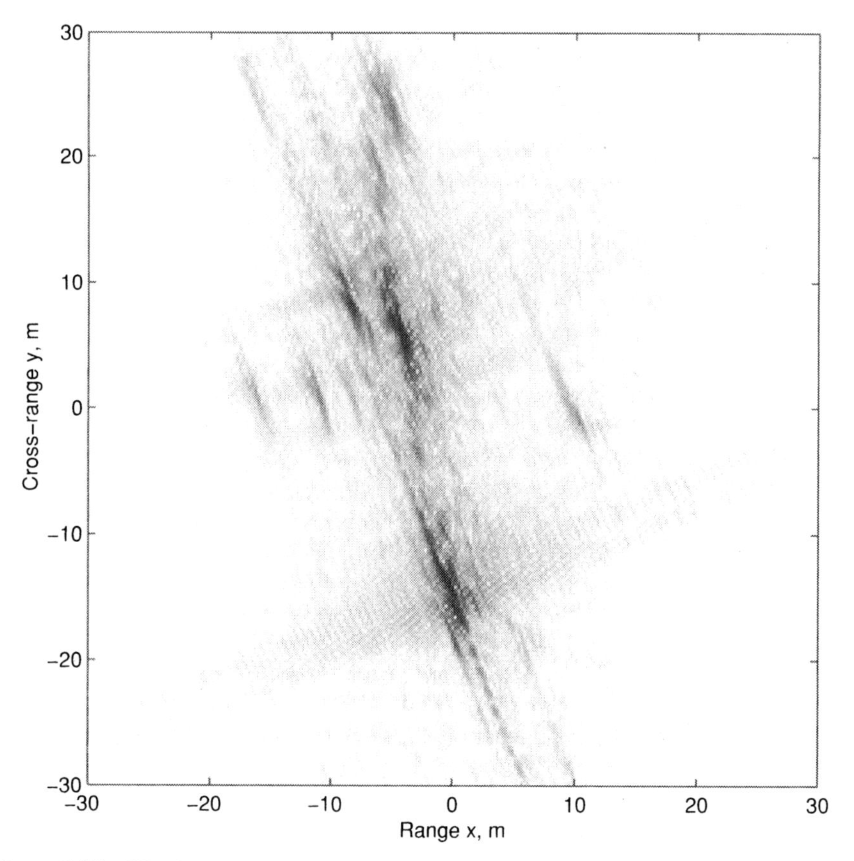

Figure 5.17c. Wavefront reconstruction of target function $f(x, y)$ from X band ISAR data within synthetic apertures: $u \in [0, L]$ (upper-half aperture).

Reconstruction in Squint Spatial Coordinates

In Chapter 4 and earlier in this chapter, we identified the squint spatial coordinates of a SAR system via (see Figures 5.1 and 5.3)

$$\begin{bmatrix} x_{\theta_c} \\ y_{\theta_c} \end{bmatrix} = \begin{bmatrix} \cos\theta_c & \sin\theta_c \\ -\sin\theta_c & \cos\theta_c \end{bmatrix} \begin{bmatrix} x - X_c \\ y - Y_c \end{bmatrix}.$$

When the origin in the spatial (x, y) domain is moved to the center of the target area, that is, (X_c, Y_c), the squint spatial coordinates become

$$\begin{bmatrix} x_{\theta_c} \\ y_{\theta_c} \end{bmatrix} = \begin{bmatrix} \cos\theta_c & \sin\theta_c \\ -\sin\theta_c & \cos\theta_c \end{bmatrix} \begin{bmatrix} x \\ y \end{bmatrix},$$

which is simply the θ_c rotated version of the spatial (x, y) domain.

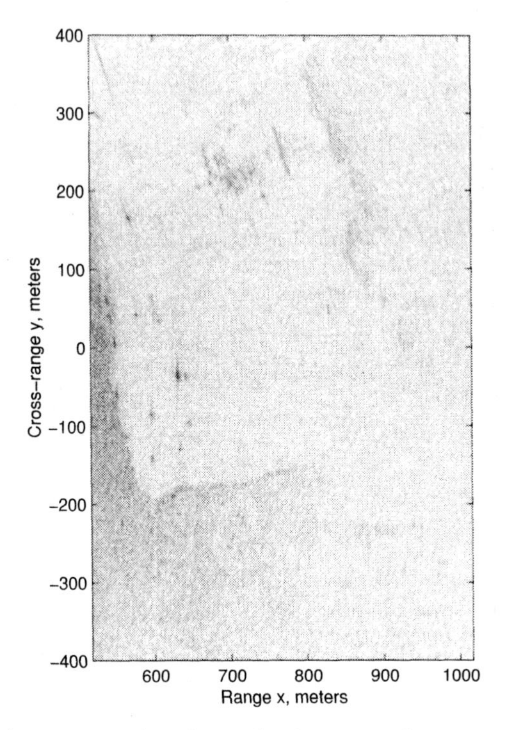

Figure 5.18a. Wavefront reconstruction of target function $f(x, y)$ from UHF band SAR data: target area.

We denote the target function in the squint spatial coordinates by

$$f_{\theta_c}(x_{\theta_c}, y_{\theta_c}) = f(x, y).$$

Thus the two-dimensional spectra of the target function $f(x, y)$ and its rotated version $f_{\theta_c}(x_{\theta_c}, y_{\theta_c})$ are related via [s94, ch. 2]

$$F_{\theta_c}(k_{x\theta_c}, k_{y\theta_c}) = F(k_x, k_y),$$

where

$$\begin{bmatrix} k_{x\theta_c} \\ k_{y\theta_c} \end{bmatrix} = \begin{bmatrix} \cos\theta_c & \sin\theta_c \\ -\sin\theta_c & \cos\theta_c \end{bmatrix} \begin{bmatrix} k_x \\ k_y \end{bmatrix},$$

which we refer to as the squint spatial frequency domain. This spatial frequency rotational transformation is shown in Figure 5.19.

Based on the two-dimensional spectral (ω, k_u) domain properties of the squint spotlight SAR signal (see Section 5.4) and the SAR mapping from the (ω, k_u) to the (k_x, k_y) domain, it can be seen that the SAR data coverage in the squint spatial

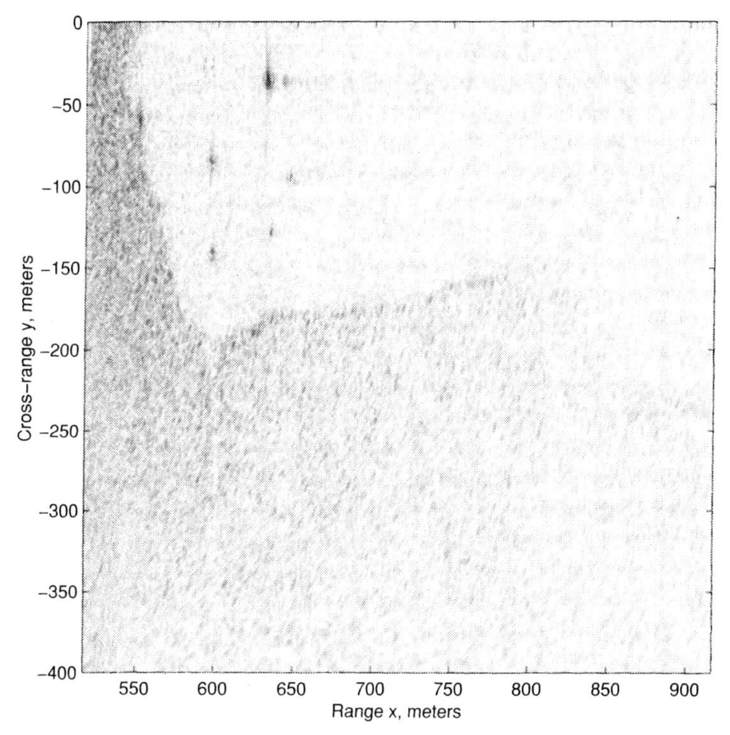

Figure 5.18b. Wavefront reconstruction of target function $f(x, y)$ from UHF band SAR data: close-up of lower target area.

frequency domain is approximately centered at

$$k_{y\theta_c} = 0;$$

that is, it is a lowpass region in the $k_{y\theta_c}$ domain. Two sets of SAR data coverages for a general squint case and the special broadside case (i.e., $\theta_c = 0$) are shown in Figure 5.19.

Due to the lowpass properties of the SAR data coverage in the $k_{y\theta_c}$ domain, the size of the array to represent the samples of $F_{\theta_c}\left(k_{x\theta_c}, k_{y\theta_c}\right)$ is smaller than the size of the array for the representation of $F(k_x, k_y)$. Moreover, to form the reconstruction in the squint spatial frequency domain, the user could directly process the baseband ISAR data (see Figure 5.9*a*), since the slow-time Doppler carrier

$$k_u = 2k \sin \theta_c$$

(which varies with the fast-time frequency) maps onto the line

$$k_{y\theta_c} = 0$$

for all the available fast-time frequency ω values.

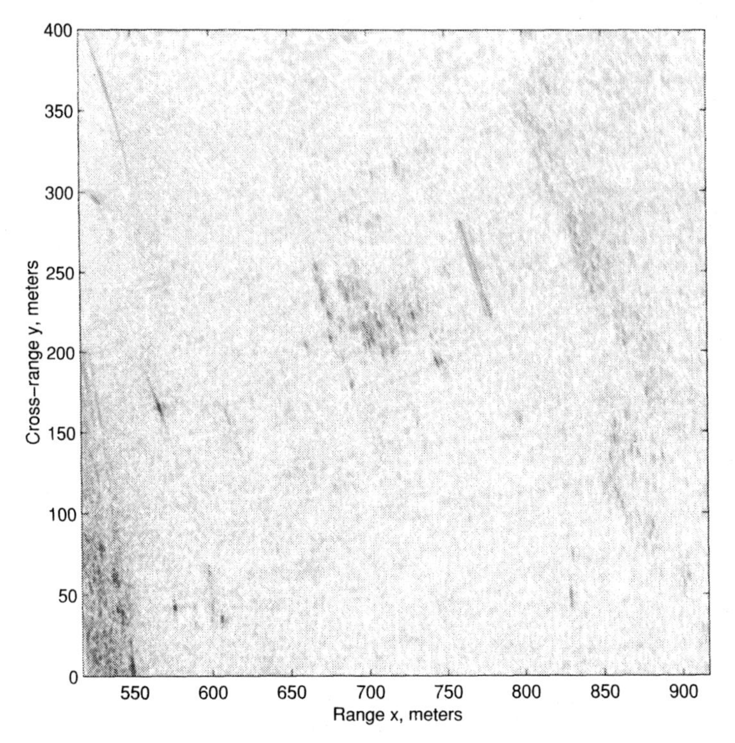

Figure 5.18c. Wavefront reconstruction of target function $f(x, y)$ from UHF band SAR data: close-up of upper target area.

The only drawback of the target function reconstruction in the squint spatial frequency domain is the additional interpolation. In fact, as we showed in Section 4.5, at a given discrete slow-time Doppler frequency k_{um}, we only need to perform interpolation in the k_x domain (see Figure 4.5), since the discrete values of

$$k_{ymn} = k_{um}$$

are evenly spaced in the k_y domain. The same is not true in the squint spatial frequency domain. In fact, to form the squint target function spectrum $F_{\theta_c}(k_{x\theta_c}, k_{y\theta_c})$, the user needs to perform two-dimensional interpolation in Step 9 of the SAR digital reconstruction algorithm which was described in the previous section.

One of the ways to visualize the SAR data coverage and the interpolation problem is to approximate the SAR mapping via [s94, sec. 4.4]

$$k_x \approx \tan\theta_c \left(\frac{2k}{\sin\theta_c} - k_u \right),$$

$$k_y = k_u.$$

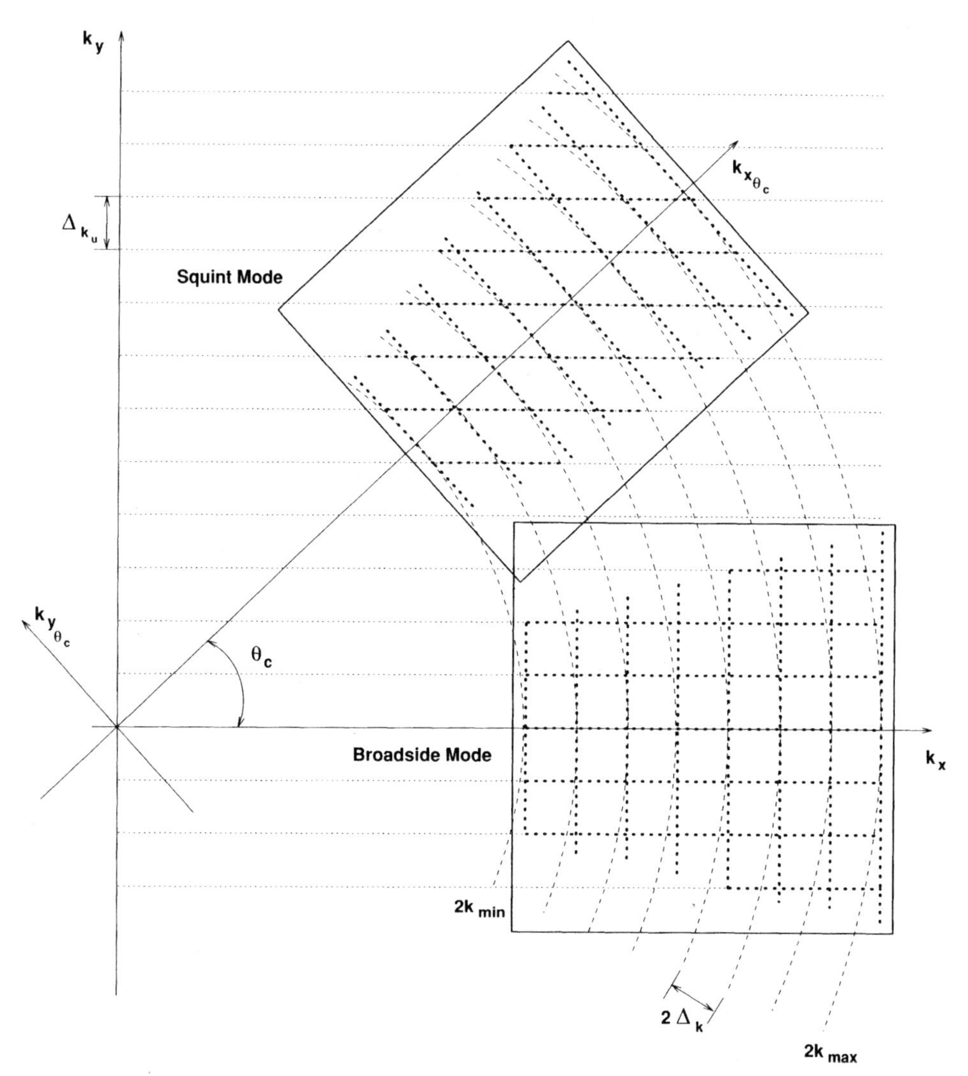

Figure 5.19. Depiction of target function spectral support in squint spatial frequency domain for broadside and squint target areas.

Using these in the expression for the squint spatial frequency coordinates, we obtain

$$k_{x\theta_c} \approx 2k,$$

$$k_{y\theta_c} \approx \frac{k_u}{\cos\theta_c} - 2k\tan\theta_c.$$

In this case $(k_{x\theta_c}, k_{y\theta_c})$ is a *linear* transformation of (ω, k_u). Thus the evenly spaced samples in the (ω, k_u) translate into *hexagonal* samples in the $(k_{x\theta_c}, k_{y\theta_c})$ domain; this is demonstrated in Figure 5.19. Two-dimensional interpolation should be performed to convert these hexagonal samples into rectilinear samples in the squint spatial frequency domain.

Note that the squint database in Figure 5.19 corresponds to a *common* slow-time Doppler carrier frequency ($2k_c \cos\theta_c$) for the slow-time baseband conversion. If one uses the slow-time Doppler carrier frequency $2k \sin\theta_c$, which varies with the fast-time frequency, the depiction of the resultant samples becomes far more complicated. Moreover these samples are not evenly spaced in the $k_{y_{\theta_c}}$. Thus one still needs to perform two-dimensional interpolation to recover rectilinear samples of $F\left(k_{x_{\theta_c}}, k_{y_{\theta_c}}\right)$ from the available SAR data in the (ω, k_u) domain.

Figure 5.20*a* shows the target function spectrum reconstruction in the squint spatial frequency domain for the ISAR example of Figure 5.8*a*. Figure 5.20*b* is the ISAR target function reconstruction $f_{\theta_c}\left(x_{\theta_c}, y_{\theta_c}\right)$ in the squint spatial domain. Note that the point spread functions of the reflectors on the aircraft do not appear *slanted* in this ISAR image; the same was not true for the ISAR reconstruction $f(x, y)$ in Figure 5.17*a*. This phenomenon was predicted in our discussion of the spotlight SAR systems point spread function in Section 5.4.

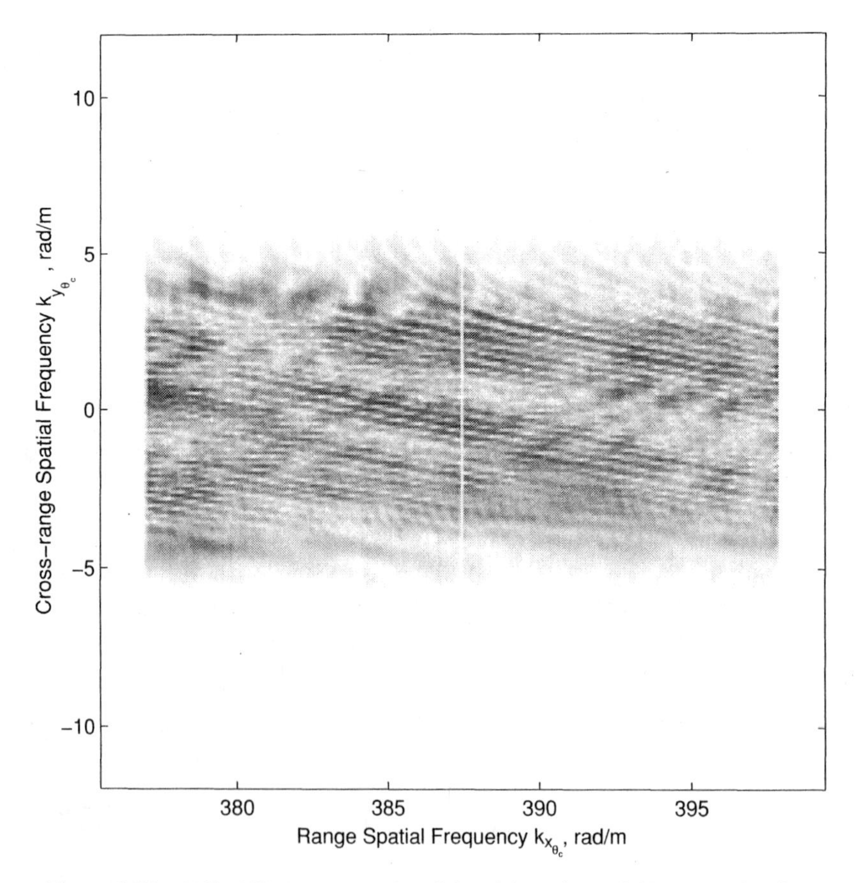

Figure 5.20a. X band ISAR reconstruction of aircraft in squint spatial frequency domain.

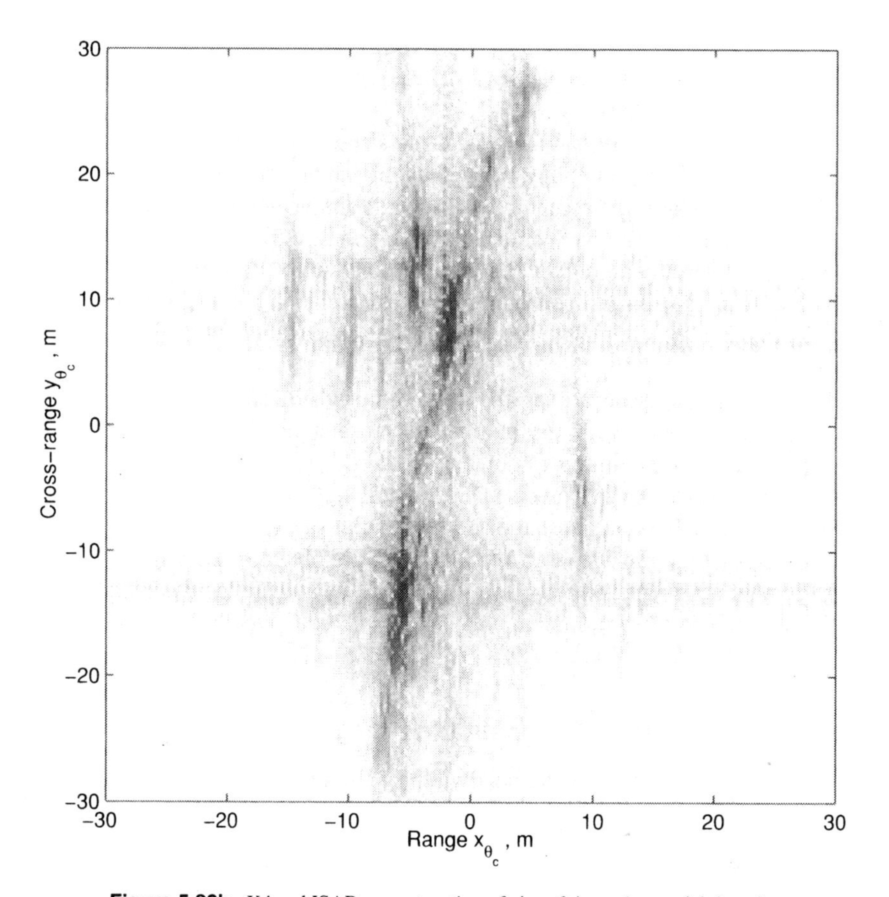

Figure 5.20b. X band ISAR reconstruction of aircraft in squint spatial domain.

Slow-time Doppler Domain Subsampling

This issue was also brought up in Section 2.8 in our discussion of cross-range imaging. In certain spotlight SAR systems, the size of the aperture $2L$ might be larger than the size of the target area in the cross-range domain $2Y_0$. After performing the above-mentioned SAR reconstruction in the spatial frequency domain, the resultant target function in the k_y domain has a sample spacing of

$$\Delta_{k_y} = \frac{\pi}{L}.$$

However, the Nyquist sample spacing for the cross-range target function is

$$\Delta_{k_y} \leq \frac{\pi}{Y_0}.$$

Since L is greater than Y_0, we satisfy the Nyquist criterion for $F(k_y)$ in the k_y domain. In fact the available samples $F(k_{xmn}, k_{ymn})$ contain redundancy (even before interpolation in the k_x domain).

To reduce the computational load and array sizes, one could subsample $F(k_{xmn}, k_{ymn})$ in the k_y domain. Sampling rate reduction can be achieved via interpolation; this, however, increases the computational load. A simple approach is to skip certain samples. The example used in Section 2.8 was as follows: If $L = 3.8Y_0$, then the user could reduce the sampling rate by 3.8 via interpolation; or, skip every two samples. In this case there is still some redundancy in the data which provides a guard band in the cross-range y domain [that reduces inverse DFT errors in obtaining $f(x, y)$ from $F(k_x, k_y)$].

The user should be aware of the k_{um} value which also corresponds to the middle sample. Moreover, if the number of subsampled data is an odd number, the $F(k_{xmn}, k_{ymn})$ array in the k_y domain should be appended with a zero to make the size of the array even (for inverse DFT processing).

Finally we should point out that once the samples of the digital-spotlighted SAR spectrum $S_d(\omega_n, k_{um})$ in Step 4 of the algorithm are obtained, then subsampling in the k_u domain can be performed on this database. Thus there is no need to perform two-dimensional frequency domain matched on the redundant data to form the samples $F(k_{xmn}, k_{ymn})$.

Reducing Bandwidth of Reconstructed Image

In Chapter 2 we showed that the bandwidth of the one-dimensional reconstructed cross-range image can be reduced by multiplying it with a phase-modulated signal in the cross-range y domain. In this section we generalize that approach to reduce (compress) the bandwidth of the reconstructed two-dimensional spotlight SAR image.

As we pointed out in Sections 2.4 and 5.3, the support band of the nth target signature in the slow-time Doppler k_u domain for the class of spotlight SAR signals can be approximated via

$$k_u \in \left[2k \sin\theta_n(0) - \frac{2kL}{r_n}\cos\theta_n(0),\, 2k\sin\theta_n(0) + \frac{2kL}{r_n}\cos\theta_n(0)\right].$$

This approximation indicates that the spotlight SAR signature of the nth target in the slow-time Doppler k_u domain is a bandpass function centered at

$$\begin{aligned}\Omega_{nc} &= 2k\sin\theta_n(0),\\ &= 2k\frac{y_n}{\sqrt{x_n^2 + y_n^2}},\end{aligned}$$

and its baseband bandwidth is

$$\pm\frac{|\Omega_n|}{2} = \pm\frac{2kL}{r_n}\cos\theta_n(0).$$

This baseband bandwidth dictates the cross-range resolution of the target which is (see Section 5.4)

$$\Delta_{y_n} \approx \frac{r_n \lambda_c}{4L \cos\theta_n(0)} \approx \frac{X_{cc}\lambda_c}{4L}.$$

However, the target function which is reconstructed in the spatial frequency (k_x, k_y) domain has the following support band in the $k_y = k_u$ domain:

$$k_u \in \left[\frac{-2k(Y_0 + L)}{X_{cc}}, \frac{2k(Y_0 + L)}{X_{cc}}\right],$$

which is the union of the slow-time Doppler support band of all the targets in the spotlighted scene. This yields the following cross-range sample spacing for the reconstructed target function in the spatial (x, y) domain:

$$\Delta_u \approx \frac{X_{cc}\lambda_c}{4(Y_0 + L)},$$

which is a *finer* sample spacing than the one required by the cross-range resolution Δ_y. This implies that we have redundant information. The percentage of the redundant data is

$$\frac{Y_0}{Y_0 + L}\%.$$

The reason for encountering this redundancy is that the slow-time Doppler support of an off-broadside target, that is, $y_n \neq 0$, is in a *bandpass* region. One may use the following procedure to remove the redundancy. (A similar scheme was used in Section 2.8 to reduce the bandwidth of the reconstructed cross-range image.) The procedure, which we refer to as *reconstructed image compression*, is based on a *spatially varying* demodulating phase function that shifts the slow-time Doppler support band of the off-broadside targets into the lowpass region.

The reconstruction of the nth target, that is, $f_n(x - x_n, y - y_n)$, is a bandpass signal in the cross-range y domain; its carrier (center) frequency in the spatial frequency k_y domain is

$$\begin{aligned} k_y &= 2k \sin\theta - n(0) \\ &= 2k \frac{y_n}{\sqrt{x_n^2 + y_n^2}}. \end{aligned}$$

Thus, one may convert this signal to a lowpass function in the cross-range y domain via the following baseband conversion (demodulation):

$$f_{nc}(x - x_n, y - y_n) = f_n(x - x_n, y - y_n) \underbrace{\exp\left(-j2k_c \frac{y_n}{\sqrt{x_n^2 + y_n^2}} y\right)}_{\text{Demodulating signal}}.$$

- Note that we used a *fixed* carrier which is based on the center frequency of the radar signal, that is, $k = k_c = \omega_c/c$, for the baseband conversion. We cannot use a variable wavenumber k for this operation.

The compressed signal $f_{nc}(x - x_n, y - y_n)$ is *approximately* a lowpass signal with the following support band in the slow-time Doppler k_y domain:

$$k_y \in \left[2(k - k_c)\sin\theta_n(0) - \frac{2kL}{r_n}\cos\theta_n(0), 2(k - k_c)\sin\theta_n(0) + \frac{2kL}{r_n}\cos\theta_n(0)\right].$$

Figure 5.21 shows the two-dimensional spectral support of the nth target signature before the image compression $F_n(k_x, k_y)$, and after the image compression $F_{nc}(k_x, k_y)$.

There is one problem with the above-mentioned compression scheme: the demodulating signal

$$\exp\left(-j2k_c \frac{y_n}{\sqrt{x_n^2 + y_n^2}} y\right)$$

is a function of the coordinates of the target (x_n, y_n). However, we need a demodulating signal that works at any range and cross-range (x, y) point in the reconstructed SAR image. To construct such a *two-dimensional* demodulating signal in the spatial (x, y) domain, we first recognize that the *instantaneous frequency* of the demodulating signal with respect to y is

$$-2k_c \frac{y_n}{\sqrt{x_n^2 + y_n^2}}.$$

Since the signature (contribution) of the nth target in the reconstructed function $f_n(x - x_n, y - y_n)$ is concentrated around $(x, y) = (x_n, y_n)$, this instantaneous frequency within a small neighborhood of (x_n, y_n) can be approximated via

$$-2k_c \frac{y}{\sqrt{x^2 + y^2}}.$$

Note that the above instantaneous frequency is invariant in the coordinates of the target (x_n, y_n). Moreover this (x, y)-dependent instantaneous frequency is equal to the instantaneous frequency with respect to y of the following two-dimensional phase-

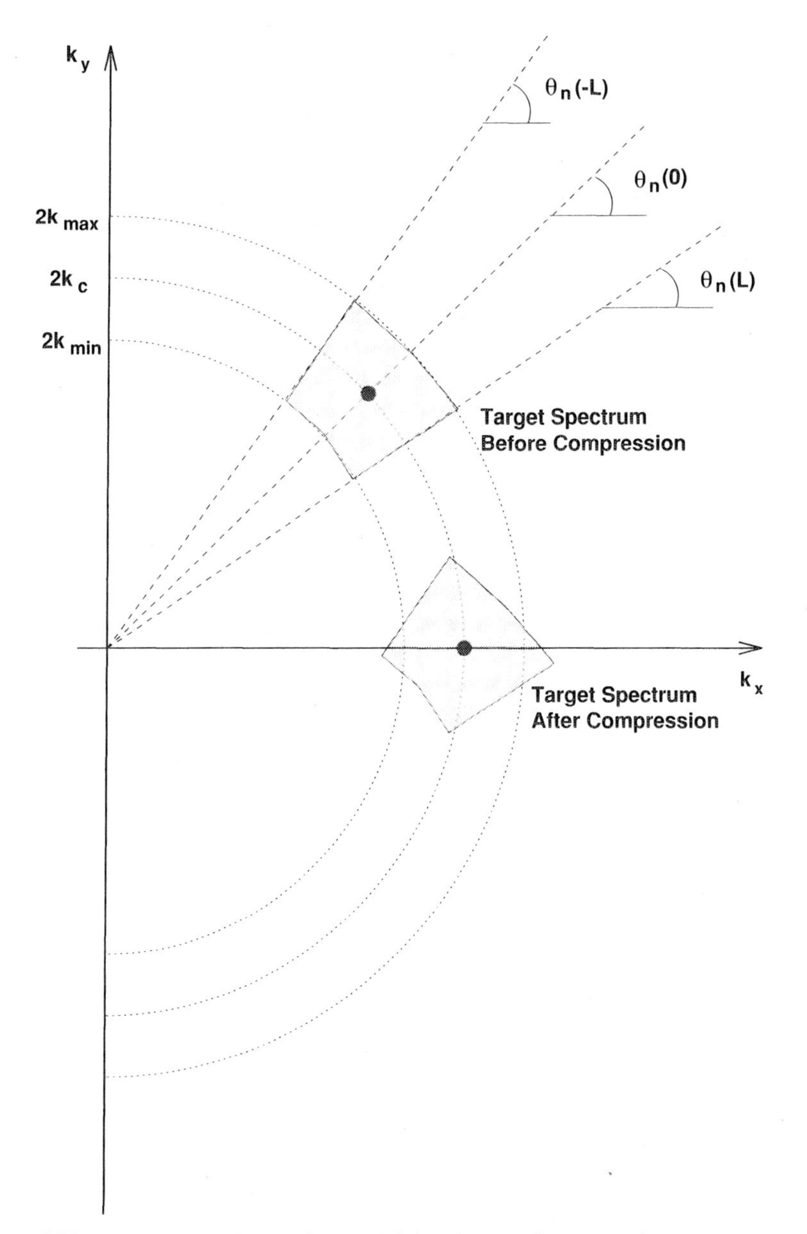

Figure 5.21. Two-dimensional spectral support of the nth target signature: before image compression $F_n(k_x, k_y)$ and after image compression $F_{nc}(k_x, k_y)$.

modulated signal:

$$\exp\left[-jk_c\frac{y^2}{\sqrt{x^2+y^2}}\right].$$

Thus one may perform baseband conversion of the nth target signature via the following:

$$f_{nc}(x-x_n, y-y_n) = f_n(x-x_n, y-y_n)\underbrace{\exp\left(j2k_c\frac{y^2}{\sqrt{x^2+y^2}}\right)}_{\text{Demodulating signal}}.$$

Note that the phase function used for demodulating signal in the above expression is invariant in the target coordinates (x_n, y_n). Thus we may use this demodulating signal for all of the targets in the imaging scene.

The reconstructed target area is the sum of all the signatures of the targets; that is,

$$f(x, y) = \sum_n f_n(x-x_n, y-y_n).$$

Thus the demodulation operation for the target area reconstruction is

$$\begin{aligned} f_c(x, y) &= \sum_n f_{nc}(x-x_n, y-y_n) \\ &= \sum_n f_n(x-x_n, y-y_n)\exp\left(-jk_c\frac{y^2}{\sqrt{x^2+y^2}}\right) \\ &= f(x, y)\underbrace{\exp\left(jk_c\frac{y^2}{\sqrt{x^2+y^2}}\right)}_{\text{Demodulating signal}}. \end{aligned}$$

We refer to this signal as the compressed SAR image.

In the broadside mode spotlight SAR systems with a planar radar [that is, when $Y_0(\omega) = R_c\lambda/D_y$ varies with the fast-time frequency], the compressed image is approximately a lowpass signal with the following support band in the cross-range spatial frequency k_y domain:

$$k_y \in \left[-\frac{2k_{\max}L}{r_{\min}}, \frac{2k_{\max}L}{r_{\min}}\right].$$

In the case of a curved radar aperture where Y_0 is a constant, the above k_y band for the compressed image of a squint spotlight SAR system is increased by

$$\frac{2(k_{\max} - k_c)Y_0}{r_{\min}}$$

in both the negative and positive directions in the k_y domain.

For the general squint case with a planar radar, the spatial frequency domain support band of the compressed spotlight SAR image is within a rectangle region in the squint spatial frequency $(k_{x\theta_c}, k_{y\theta_c})$ domain; the lengths of this rectangular area are

$$2(k_{\max} - k_{\min}) \qquad \text{in the } k_{x\theta_c} \text{ domain,}$$

$$\frac{4k_{\max}L}{r_{\min}} \qquad \text{in the } k_{y\theta_c} \text{ domain.}$$

In the case of a curved radar aperture where Y_0 is a constant, the above $k_{y\theta_c}$ band is increased by $2(k_{\max} - k_c)Y_0/r_{\min}$ in both the negative and positive directions in the $k_{y\theta_c}$ domain.

Note that the demodulating (compression) signal is approximately the following *chirp* signal in the cross-range domain:

$$\exp\left(-jk_c\frac{y^2}{\sqrt{x^2+y^2}}\right) \approx \exp\left(-jk_c\frac{y^2}{X_{cc}}\right).$$

In fact the origin of this signal can be found in the SAR *spherical* signal. A more accurate analysis would show that the reconstructed image compression should be performed via the following demodulation with a spherical signal:

$$f_c(x, y) = f(x, y)\exp\left(j2k_cx - j2k_c\sqrt{x^2+y^2}\right).$$

The compression may be incorporated into the interpolation phase of the reconstruction algorithm. This has some utility but involves a lengthy discussion, so it is not presented here.

With the shift of origin to the center of the target area (X_c, Y_c), the SAR image compression is performed via

$$f_c(x, y) = f(x, y) \times \exp\left[j2k_c(X_c + x) + j2k_c\sin\theta_c y - j2k_c\sqrt{(X_c+x)^2 + (Y_c+y)^2}\right].$$

Using the rotational properties of the two-dimensional Fourier transform [s94, ch. 2], the spotlight SAR image compression in the squint spatial domain $(x_{\theta_c}, y_{\theta_c})$ becomes

$$f_{\theta_c c}(x_{\theta_c}, y_{\theta_c}) = f_{\theta_c}(x_{\theta_c}, y_{\theta_c}) \exp\left[j2k_c(R_c + x_{\theta_c}) - j2k_c\sqrt{(R_c + x_{\theta_c})^2 + y_{\theta_c}^2} \right].$$

As we will see in the numerical examples, there are advantages in forming the compressed SAR image in the squint spatial domain.

Figure 5.22*a* is the two-dimensional compressed ISAR image (Figure P5.10), that is, $F_c(k_x, k_y)$, for the reconstructed ISAR image of Figure 5.17*a*. Note that the spectral support of the compressed image in Figure 5.22*a* is smaller than the support of the spectrum of the reconstruction ISAR image, $F(k_x, k_y)$, which was shown in Figure 5.16*a*. Figure 5.22*b* is the two-dimensional compressed ISAR image in the squint spatial frequency domain, $F_{\theta_c c}(k_{x\theta_c}, k_{y\theta_c})$. (This image is practically the compressed ISAR spectrum when it is rotated by θ_c.) Comparing this image with $F_{\theta_c}(k_{x\theta_c}, k_{y\theta_c})$ in Figure 5.20*a*, one can observe the bandwidth compression effect.

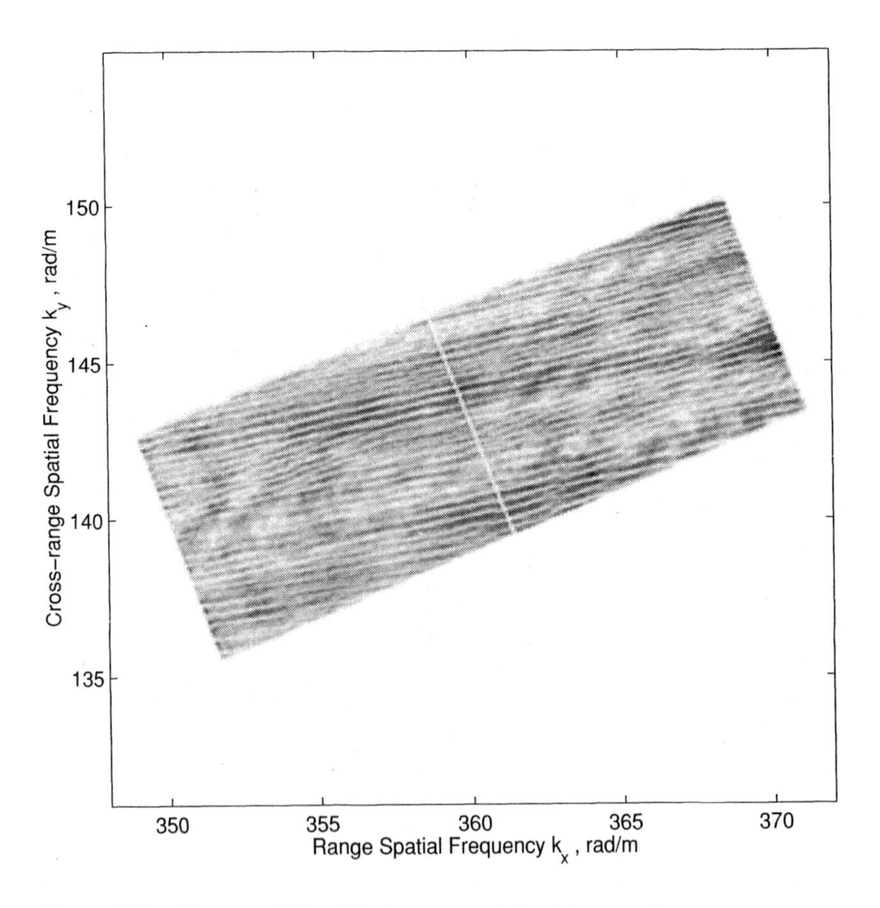

Figure 5.22a. Compressed X band ISAR spectrum of aircraft in spatial frequency domain.

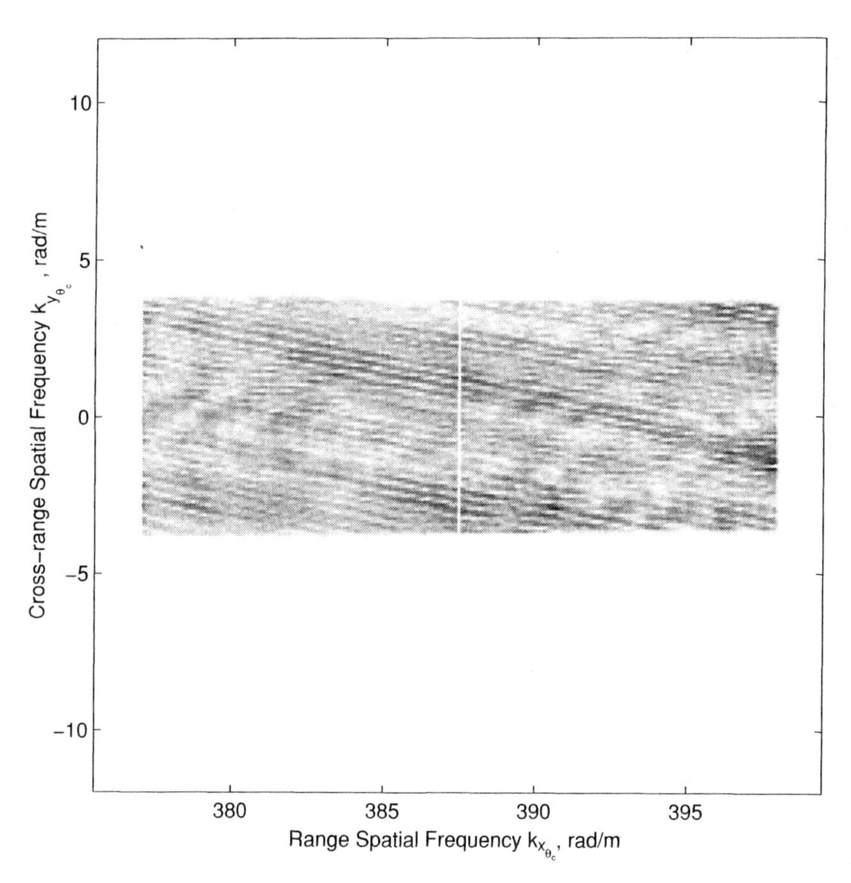

Figure 5.22b. Compressed X band ISAR spectrum of aircraft in squint spatial frequency domain.

Figure 5.23 shows the two-dimensional compressed SAR image, $F_c(k_x, k_y)$, for the reconstructed SAR image of Figure 5.18*a*. By comparing this spectral support, that is, $F_c(k_x, k_y)$, with the one for the reconstructed image without compression $F(k_x, k_y)$ in Figure 5.16*b*, one can observe that SAR image compression brings the signature of squint as well as broadside targets to a common baseband region.

This can also be verified by examining the spectral support of a specific region in the reconstructed image. Figure 5.24*a* is the reconstructed image within a 16 m by 16 m foliage area centered around $(x_n, y_n) = (526, 0)$ m; this corresponds to a *broadside* region in Figure 5.18*a*. Figure 5.24*b* and *c* shows the spectral supports of this broadside target area before and after SAR image compression. Note that for this broadside region, the two spectral supports in Figure 5.24*b* and *c* resides in the same region in the spatial frequency (k_x, k_y) domain.

The equation of the continuous line which is superimposed on the target spectrum in Figure 5.24*b* is

$$k_y = \tan\theta_n(0)k_x,$$

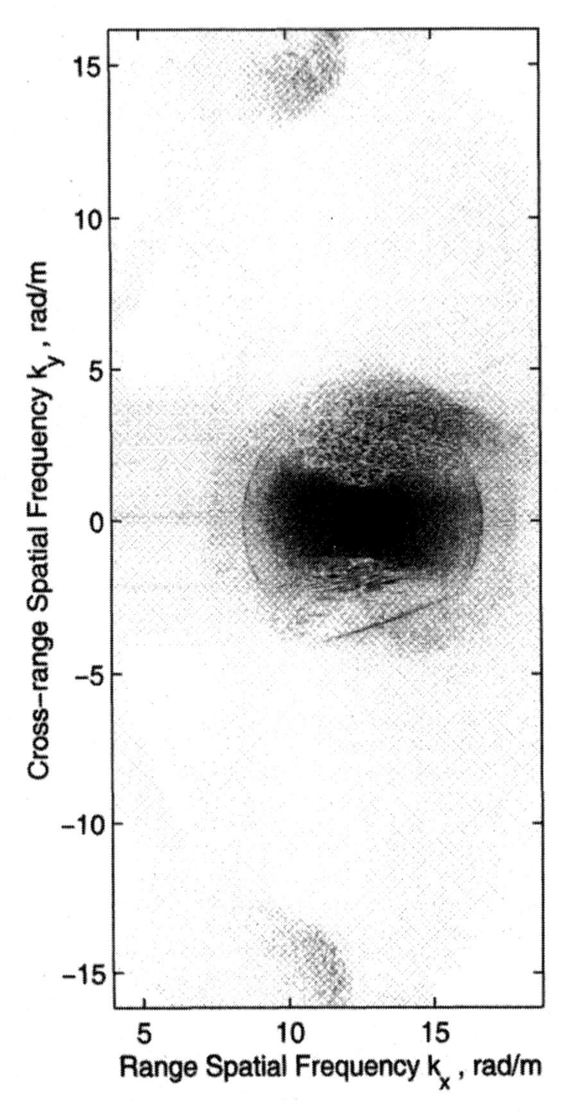

Figure 5.23. Compressed UHF band SAR spectrum of foliage area and human-made targets.

where $\theta_n(0)$ is the aspect of the target area (assumed to be the nth target) when the radar is at $u = 0$; $\theta_n(0) = 0$ for the broadside target region. Also the angular interval over which this region is observed by the radar is

$$\left[\arctan\left(\frac{y_n - L - 8}{x_n - 8}\right), \arctan\left(\frac{y_n + L + 8}{x_n - 8}\right)\right] = [-6.4°, 6.4°].$$

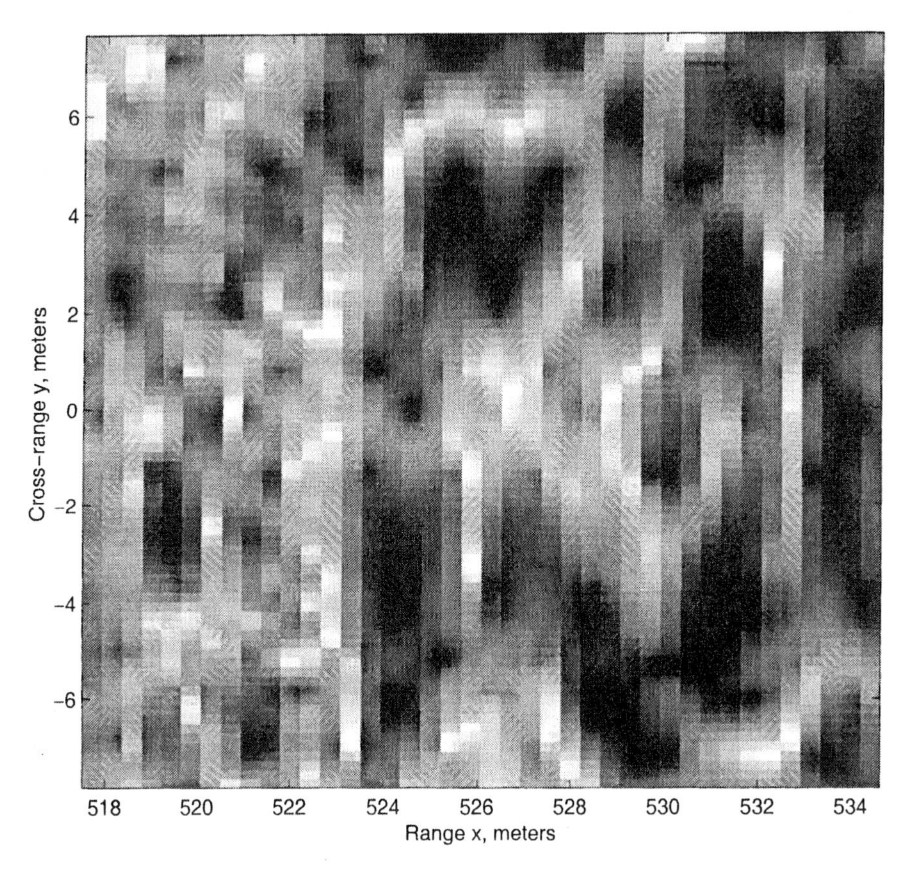

Figure 5.24a. Wavefront reconstruction of broadside target area: spatial domain reconstruction.

In Figure 5.24*b* the lines

$$k_y = \pm \tan 6.4^\circ k_x$$

are shown via dashed lines superimposed on the target spectrum; these two lines correspond to the theoretical spectral support of this broadside foliage area.

Figure 5.25*a* is the reconstructed image within a 16 m by 16 m foliage area centered around $(x_n, y_n) = (526, -388)$ m; this corresponds to a *squint* region in Figure 5.18*a* with $\theta_n(0) = -36.4^\circ$. Figure 5.25*b* and *c* shows the spectral supports of this squint target area before and after SAR image compression; note the location of the line $k_y = \tan\theta_n(0)k_x$ for this squint target area. For the squint target area the shift of the target spectrum from a bandpass region to a lowpass region is evident in Figure 5.25*b* and *c*. The dashed lines identify the theoretical spectral support of this

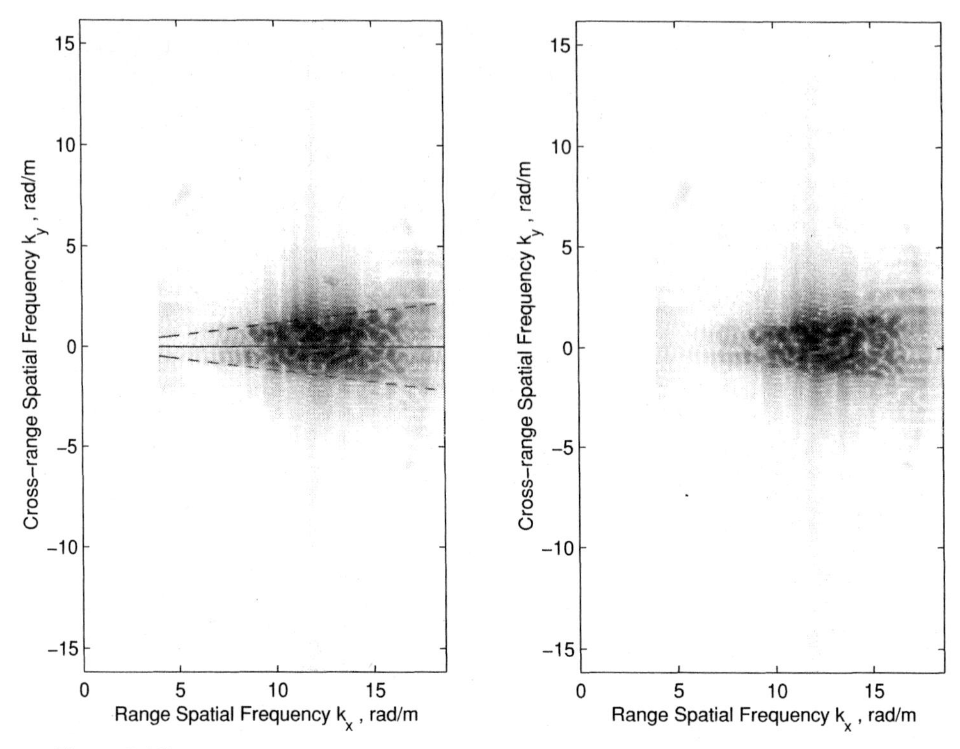

Figure 5.24b-c. Wavefront reconstruction of broadside target area: (b) original spectrum; (c) compressed spectrum.

squint foliage area which are dictated by the aspect angle interval

$$\left[\arctan\left(\frac{y_n - L - 8}{x_n - 8}\right), \arctan\left(\frac{y_n + L + 8}{x_n + 8}\right)\right] = [-40.7°, 31.7°].$$

Note that the spectrum of this squint target area at $(x_n, y_n) = (526, -388)$ m, which is close to the boundary of the digital spotlight filter at $Y_0 = -400$ m, appears filtered out for

$$\sqrt{k_x^2 + k_y^2} > 2k_c.$$

This is the result of using the *narrow-bandwidth* approximation in digital spotlight filtering. As we noted earlier, this causes filtering of the signature of the near-boundary targets (that is, $y_n \approx Y_0$) at the fast-time frequencies which are greater than the carrier frequency.

Figures 5.26 to 5.29, respectively, show similar results for a HEMTT truck, a broadside corner reflector, a slanted corner reflector, and a corner reflector (CR 3) at a relatively large squint angle. Note that these human-made targets exhibit spectral properties that correspond to not only the spotlight bandwidth support $I_n(\omega, k_u)$

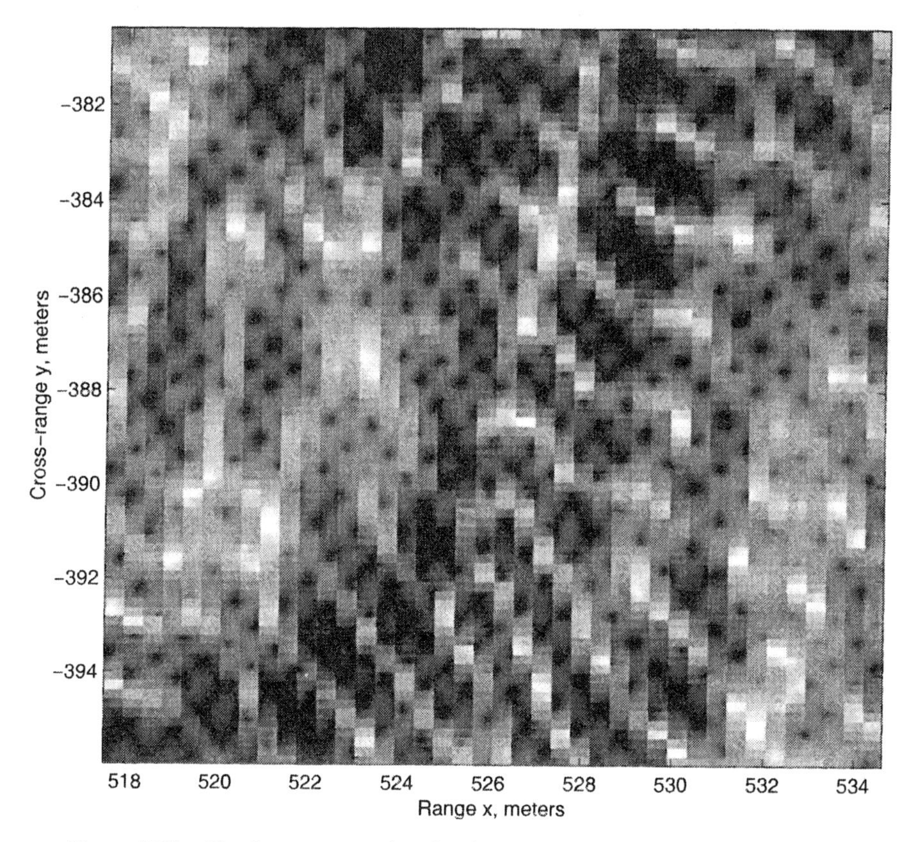

Figure 5.25a. Wavefront reconstruction of squint target area: spatial domain reconstruction.

(that varies with the coordinates of the target) but also their own amplitude pattern $A_n(\omega, k_u)$. The effect of the target amplitude pattern will become more evident in the results of the FOPEN SAR database with a larger synthetic aperture in Chapter 6.

As we mentioned earlier, the SAR image of Figure 5.18*a* is formed from the digitally spotlighted data that possess a fast-time varying cross-range gate (i.e., $Y_0 =$ 400 m at 300 MHz, $Y_0 = 266$ m at 400 MHz, etc.). Thus the resultant SAR database should exhibit the functional properties of a spotlight SAR with a planar radar. The support band of the compressed SAR image in the cross-range spatial frequency k_y domain is

$$k_y \in \left[-\frac{2k_{\max} L}{r_{\min}}, \frac{2k_{\max} L}{r_{\min}} \right] \approx [-2, 2] \text{ rad/m}.$$

Figure 5.23 shows this support band. The signature that appears near $|k_y| \approx 15$ rad/m is likely due to the side lobe leakage which we pointed out in our discussion on digital spotlighting. Figure 5.23 also shows a cluster of relatively weak signatures just outside the band $k_y \in [-2, 2]$ rad/m; these are mainly due to the moving targets in the imaging scene. SAR image compression does shift the spectral support of the

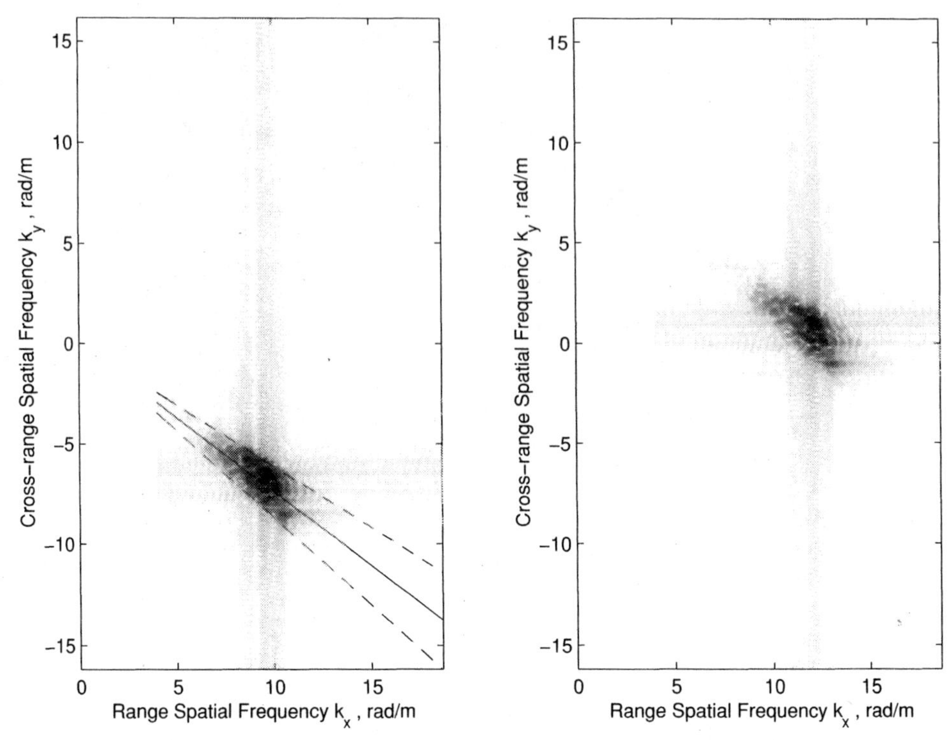

Figure 5.25b-c. Wavefront reconstruction of squint target area: (b) original spectrum; (c) compressed spectrum.

moving target's SAR signature closer to the $k_y = 0$ line. However, the resultant spectrum is still a bandpass signal in the k_y domain. (This is a consequence of the functional properties of the SAR signature of moving targets [s94, ch. 5; yan].)

One could observe the SAR image of *some* of the moving targets by filtering out the components of the compressed SAR image spectrum of Figure 5.23 in the band $k_y \in [-2, 2]$, and displaying the inverse two-dimensional Fourier transform of the resultant filtered spectrum in the spatial (x, y) domain. Such an image shows the streaklike signatures which are typical of moving targets. The moving targets that appear in the filtered image are the ones with a sufficiently large speed in the range domain, which partially or completely shifts the SAR signature of the moving target outside the band $k_y \in [-2, 2]$ [che; s94, ch. 5; yan].

Figure 5.30*a* shows a target area that we refer to as a *moving target*. The spectrum and compressed spectrum of this target area are shown in Figure 5.30*b* and *c*, respectively. When we re-examine the SAR data with a larger synthetic aperture in the next chapter, we will show that the signature in Figure 5.30*a* belongs to a moving target. The properties of moving targets will become evident in the analytical and numerical results of the next chapter, where we will introduce a statistic which we refer to as *SAR ambiguity function* for moving target detection (MTD).

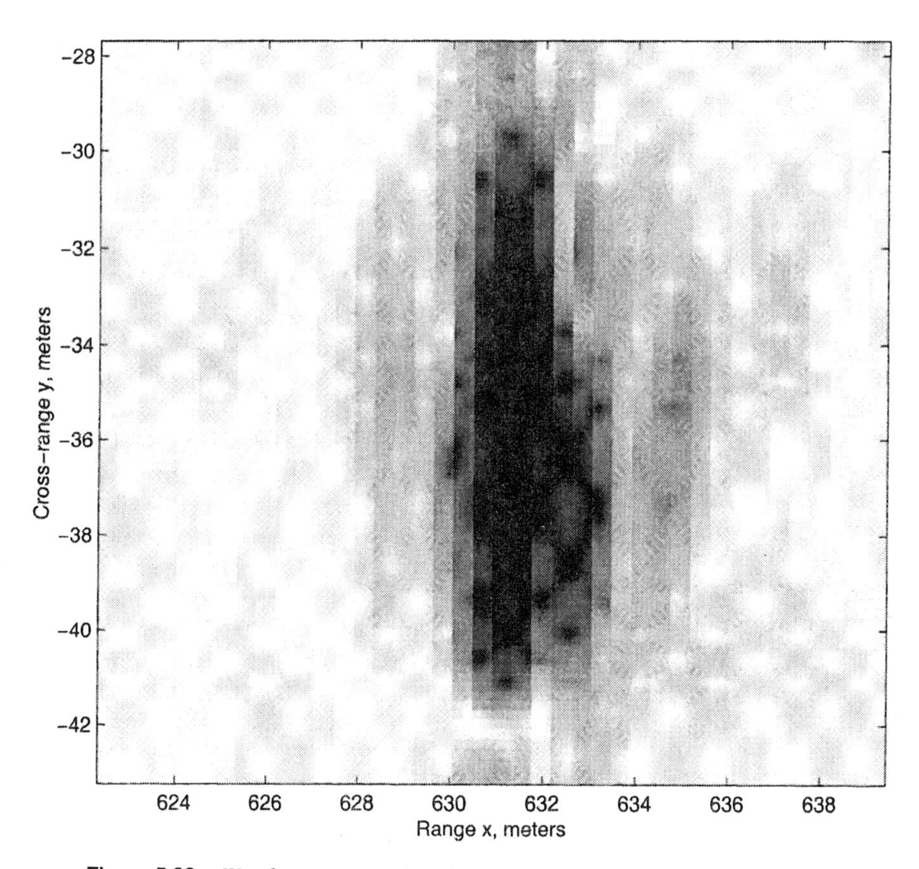

Figure 5.26a. Wavefront reconstruction of HEMTT truck: spatial domain reconstruction.

Digital Reconstruction via Range Stacking

Two digital reconstruction via range-stacking algorithms for a generic SAR system were described in Section 4.6, and their block diagrams were provided in Figure 4.7*c* and *d*. Similar to the SAR digital reconstruction via spatial frequency interpolation, the SAR range-stacking algorithms are based on processing the SAR data in the two-dimensional frequency domain. Thus the additional processing required to implement the range-stacking algorithms for the spotlight SAR systems is similar to the one described in the previous section.

Algorithm The range stack reconstruction algorithm of Figure 4.7*c* for the spotlight SAR systems is summarized in the following: the Matlab code that is discussed in Section 5.7 contains this algorithm (Figures P5.11 and P5.12). (The other range stack algorithm in Figure 4.7*d* can be implemented via similar steps.)

Step 1. Perform Step 1, fast-time domain matched filtering, of the digital reconstruction via spatial frequency.

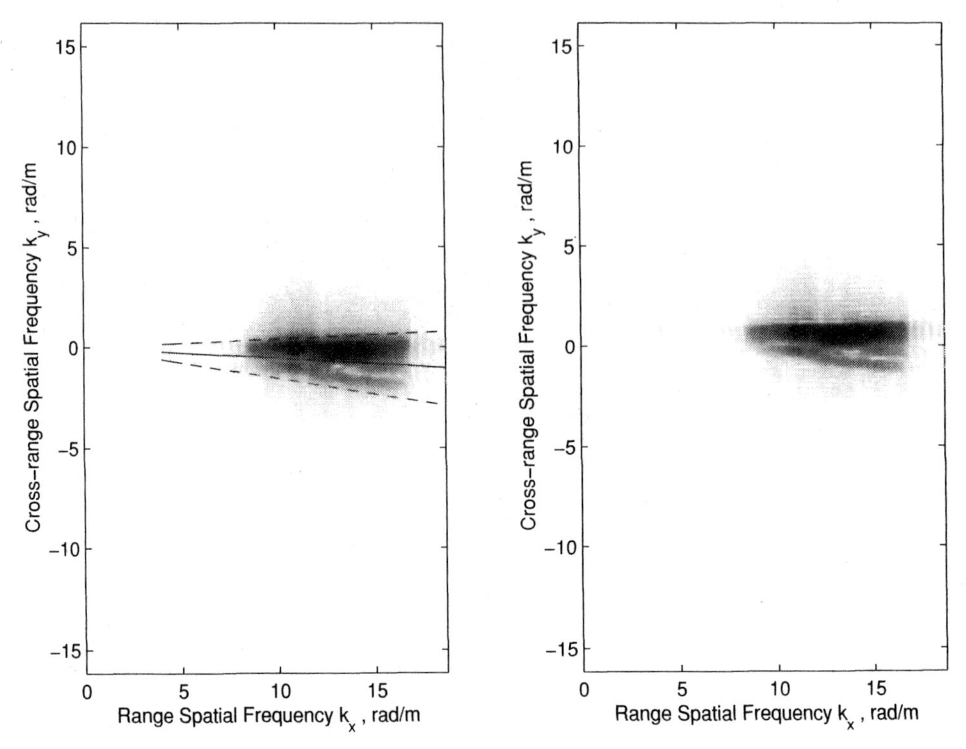

Figure 5.26b-c. Wavefront reconstruction of HEMTT truck: (b) original spectrum; (c) compressed spectrum.

Step 2. Perform Step 2 of the digital reconstruction via spatial frequency interpolation to obtain the two-dimensional DFT of the digitally spotlighted and upsampled SAR signal.

Step 3. If $Y_0 > L$, *zero-pad the data in the slow-time* u *domain to achieve the minimum length synthetic aperture extent of* $u \in [-L_{\min}, L_{\min}]$, *where* $L_{\min} = \max(L, Y_0)$. *(Zero-padding in the fast-time domain is not required, since there is no need for interpolation in the* k_x *domain.)*

Step 4. Obtain the two-dimensional DFT of the digitally spotlighted, upsampled and zero-padded SAR signal; the results are the samples $S_d(\omega_n, k_{um})$.

Step 5. If $Y_0 < L$, *perform slow-time Doppler domain subsampling on the available samples of* $S_d(\omega_n, k_{um})$.

Step 6. Perform Step 6 of the digital reconstruction via spatial frequency interpolation, that is,

$$S_d(\omega, k_u) \exp\big(-j\omega T_c\big),$$

which moves the fast-time origin back to the actual fast-time zero, that is, $t = 0$.

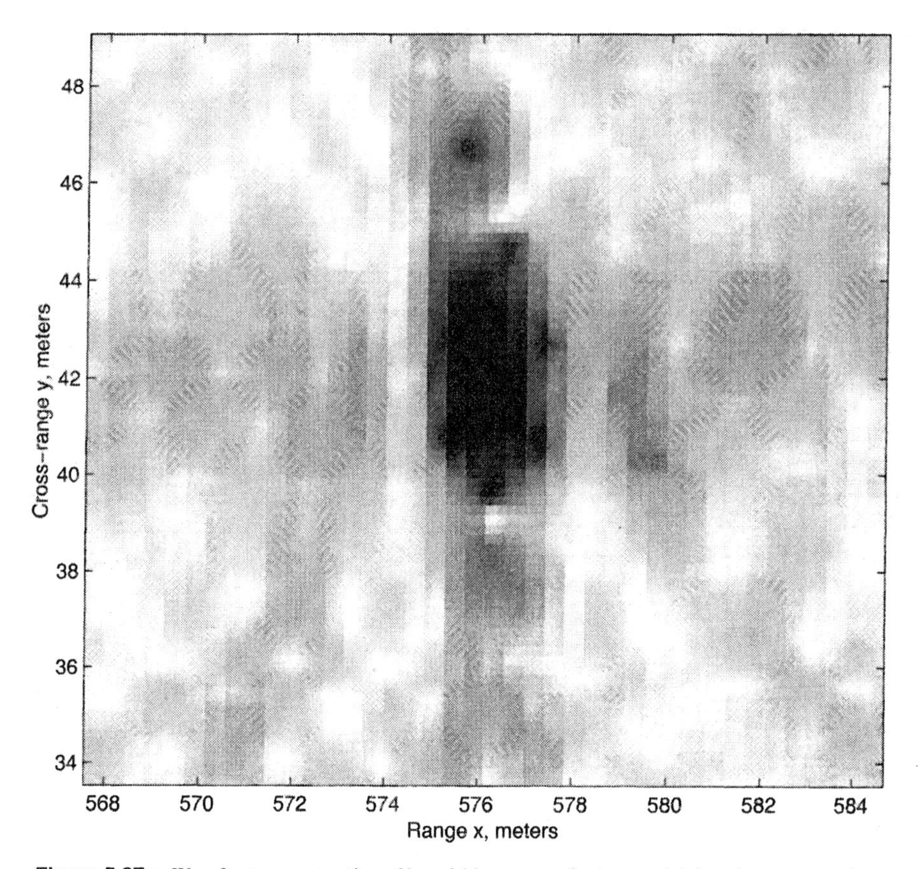

Figure 5.27a. Wavefront reconstruction of broadside corner reflector: spatial domain reconstruction.

Step 7. Identify the spatial domain and spatial frequency domain grids for the target function reconstruction; see Steps (5) and (8) in the digital reconstruction via spatial frequency interpolation. This yields the following sample spacings for the spatial frequency grid:

$$\Delta_{k_x} = \frac{\pi}{X_0},$$

$$\Delta_{k_y} = \frac{\pi}{Y_0}.$$

The number of samples on the grid are

$$N_x = 2\left\lceil \frac{k_{x\max} - k_{x\min}}{2\Delta_{k_x}} \right\rceil \qquad \textit{in the } k_x \textit{ domain,}$$

$$N_y = 2\left\lceil \frac{k_{y\max} - k_{y\min}}{2\Delta_{k_y}} \right\rceil \qquad \textit{in the } k_y \textit{ domain.}$$

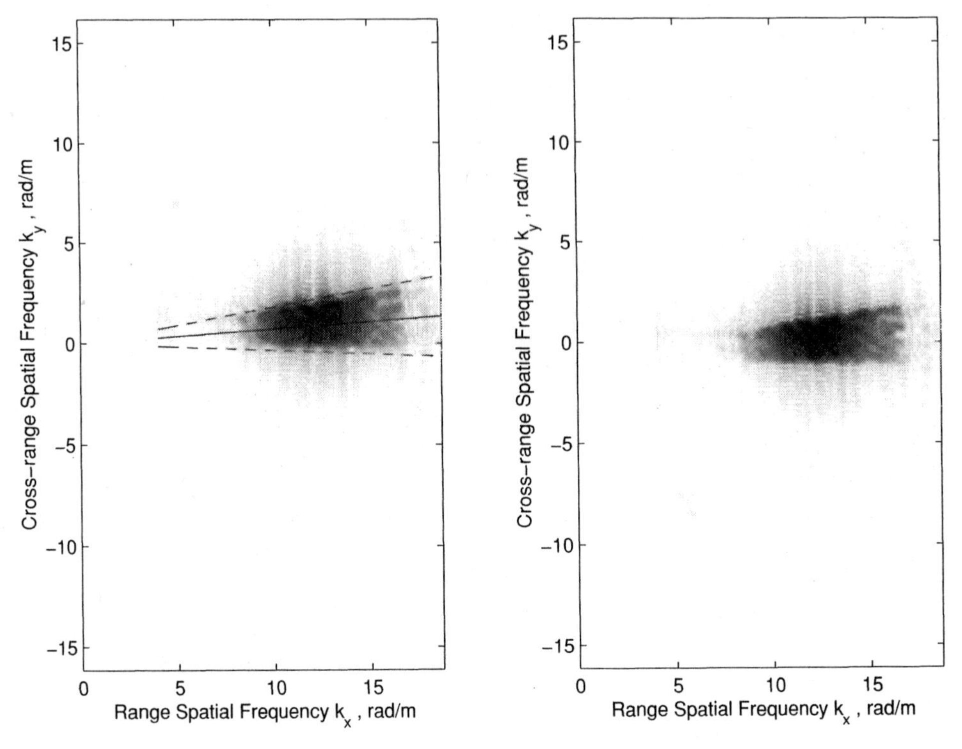

Figure 5.27b-c. Wavefront reconstruction of broadside corner reflector: (b) original spectrum; (c) compressed spectrum.

There are also (N_x, N_y) *evenly spaced samples on the spatial domain* (x, y) *grid with the following sample spacings:*

$$\Delta_x = \frac{2\pi}{N_x \Delta_{k_x}} = \frac{2X_0}{N_x},$$

$$\Delta_y = \frac{2\pi}{N_y \Delta_{k_y}} = \frac{2Y_0}{N_y}.$$

The range bins x_i*'s for range stacking correspond to the above evenly spaced samples in the range* x *domain, that is,*

$$x_i = X_c + \left(i - \frac{N_x}{2} - 1\right) \Delta_x, \qquad i = 1, 2, \ldots, N_x.$$

Step 8. For a given range bin x_i*, perform (discrete) baseband conversion of the target area via the addition of the phase function of* (x_i, Y_c)

$$S_d(\omega, k_u) \exp\left(-j\omega T_c\right) \exp\left(j\sqrt{4k^2 - k_u^2}x_i + jk_u Y_c\right).$$

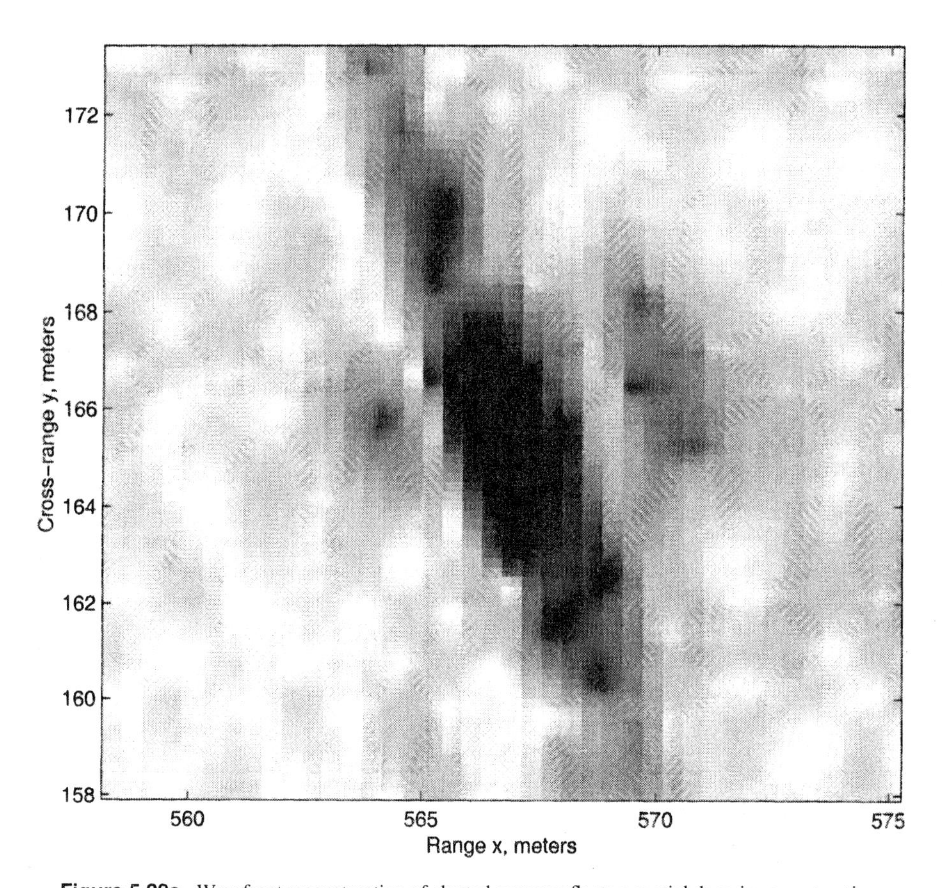

Figure 5.28a. Wavefront reconstruction of slanted corner reflector: spatial domain reconstruction.

Step 9. Integrate (discrete sum) the result over the available fast-time frequencies

$$\int_{\omega} S_d(\omega, k_u) \exp\left(-j\omega T_c\right) \exp\left(j\sqrt{4k^2 - k_u^2}x_i + jk_u Y_c\right) d\omega.$$

Step 10. Obtain the one-dimensional inverse DFT of the outcome with respect to the available discrete values of k_u. The results are the samples of the spatial domain target function $f(x_i, y)$ at the discrete (DFT) samples of $y = u$.

Step 11. Repeat Steps 8 to 10 for all the available range bins x_i's.

Step 12. Perform coherent baseband conversion on the range stack reconstruction via

$$f(x_i, y) \exp\left(-jk_{xc}x_i\right),$$

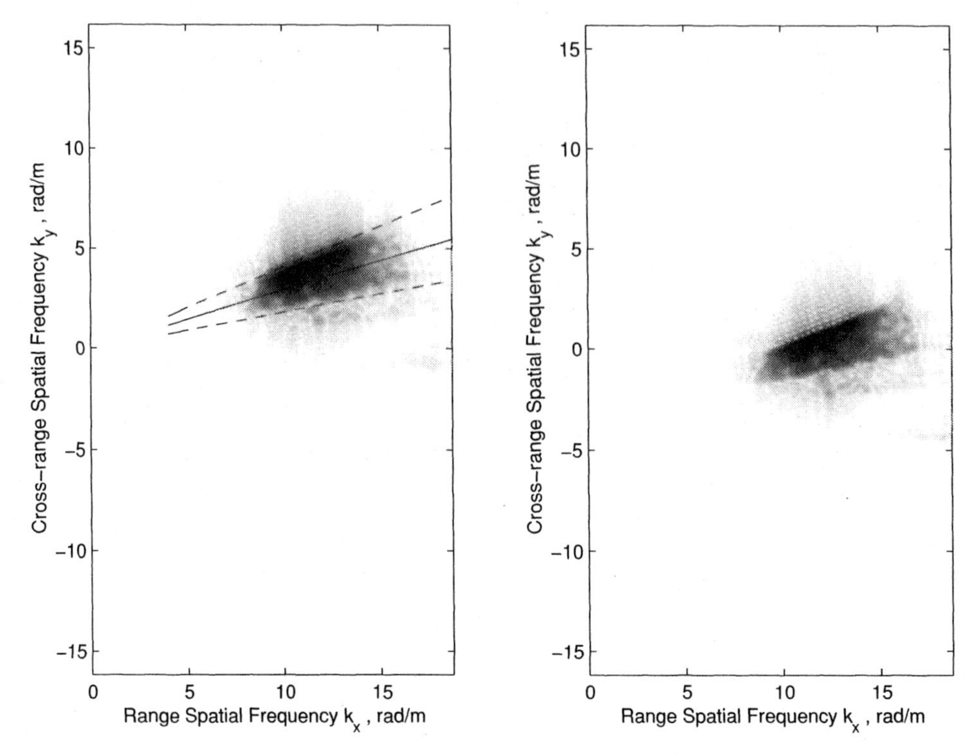

Figure 5.28b-c. Wavefront reconstruction of slanted corner reflector: (b) original spectrum; (c) compressed spectrum.

where

$$k_{xc} = \frac{k_{x\max} + k_{x\min}}{2}.$$

This operation brings the two-dimensional spectrum of the reconstructed target function to the baseband (k_x, k_y) *domain (see Section 4.6).*

Step 13. Reduce the bandwidth of the reconstructed image and remove some of the clutter effects by SAR image compression.

Digital Reconstruction via Time Domain Correlation and Backprojection

Time domain correlation (TDC) and backprojection algorithms, which were described in Section 4.7, can be used for image formation in spotlight SAR systems. The implementation of these algorithms in the spotlight SAR system is described next.

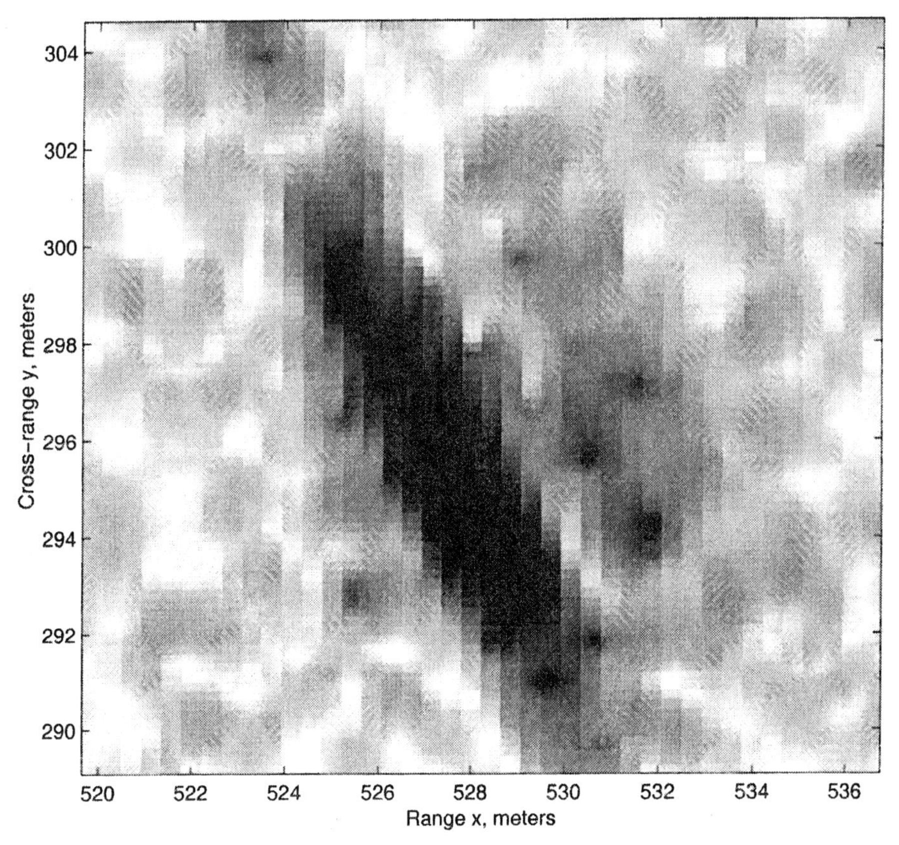

Figure 5.29a. Wavefront reconstruction of corner reflector 3 (wide squint angle): spatial domain reconstruction.

TDC Algorithm For a given pixel point (x_i, y_j), the TDC method performs target reconstruction via (see Section 4.7)

$$f(x_i, y_j) = \int_{-L}^{L} \int_{T_s}^{T_f} \mathbf{s}(t, u) p^*\big[t - t_{ij}(u)\big]\, dt\, du,$$

where $p^*(t)$ is the complex conjugate of the transmitted radar signal $p(t)$, and

$$t_{ij}(u) = \frac{2\sqrt{x_i^2 + (y_j - u)^2}}{c}.$$

Using the Parseval's theorem, the above integral over the fast-time t can be converted to an integral over the fast-time frequency ω; this yields

$$f(x_i, y_j) = \int_{-L}^{L} \int_{\omega_{\min}}^{\omega_{\max}} s(\omega, u) P^*(\omega) \exp\big[j\omega t_{ij}(u)\big]\, d\omega\, du$$

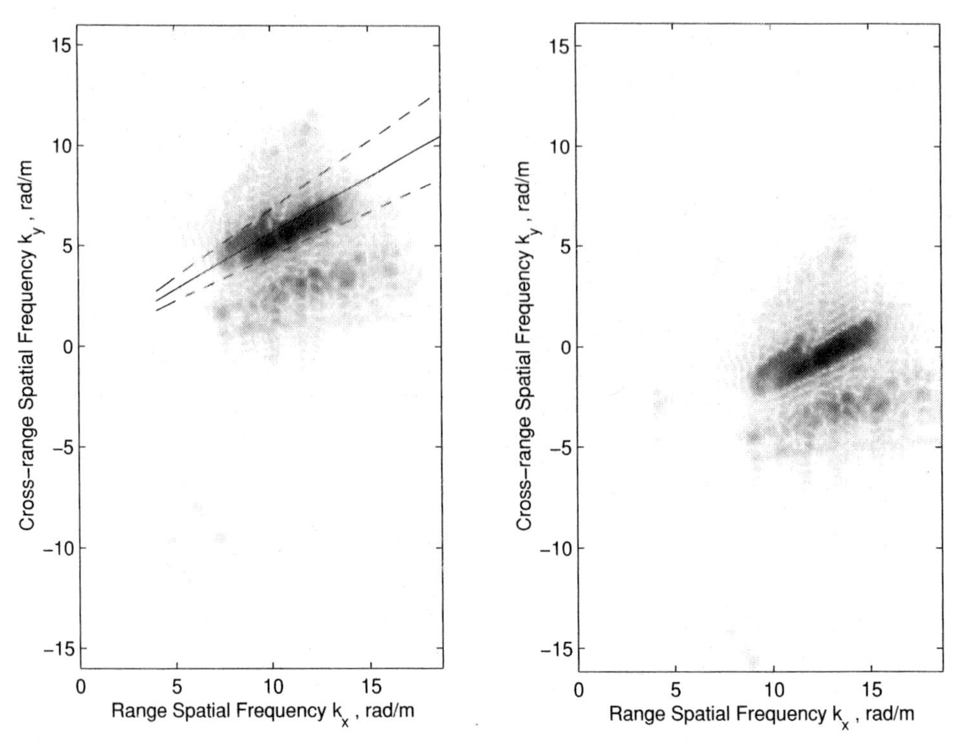

Figure 5.29b-c. Wavefront reconstruction of corner reflector 3 (wide squint angle): (b) original spectrum; (c) compressed spectrum.

for $T_s \leq t_{ij}(u) \leq T_f$; otherwise, the integrand is zero. The following steps are used to implement the above reconstruction equation. The Matlab code for this chapter contains this method (Figures P5.13 and P5.14).

Step 1. Obtain the discrete Fourier transform of the measured SAR signal with respect to the fast-time samples; this yields the two-dimensional samples of $s(\omega, u)$.

Step 2. Bring the reference fast-time point to $t = T_c$ via the addition of the following phase function:

$$s(\omega, u) \exp\big(- j\omega T_c \big).$$

Step 3. For a given pixel point (x_i, y_j), construct the TDC integral in the (ω, u) domain for the u values with $T_s \leq t_{ij}(u) \leq T_f$; this yields the value of the TDC reconstruction for $f(x_i, y_j)$.

Step 4. Repeat Step 3 at all the desired pixel points (x_i, y_j) in the spatial domain.

Step 5. Perform coherent baseband conversion on the TDC reconstruction via

$$f(x_i, y_j) \exp\big(- jk_{xc} x_i \big),$$

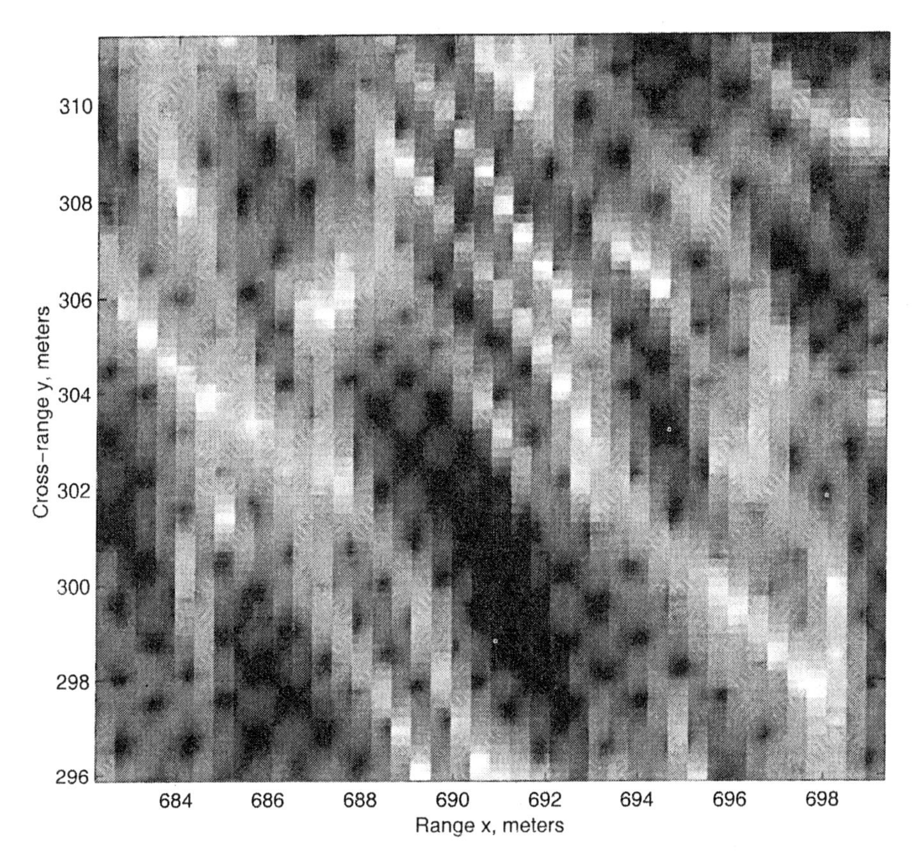

Figure 5.30a. Wavefront reconstruction of moving target: spatial domain reconstruction.

where

$$k_{xc} = \frac{k_{x\max} + k_{x\min}}{2}.$$

This operation brings the two-dimensional spectrum of the reconstructed target function to the baseband (k_x, k_y) domain.

Backprojection Algorithm The backprojection reconstruction at a pixel point (x_i, y_j) is obtained by

$$\begin{aligned} f(x_i, y_j) &= \int_{-L}^{L} \mathbf{s}_M \left[\frac{2\sqrt{x_i^2 + (y_j - u)^2}}{c}, u \right] du \\ &= \int_u \mathbf{s}_M \big[t_{ij}(u), u \big] du \end{aligned}$$

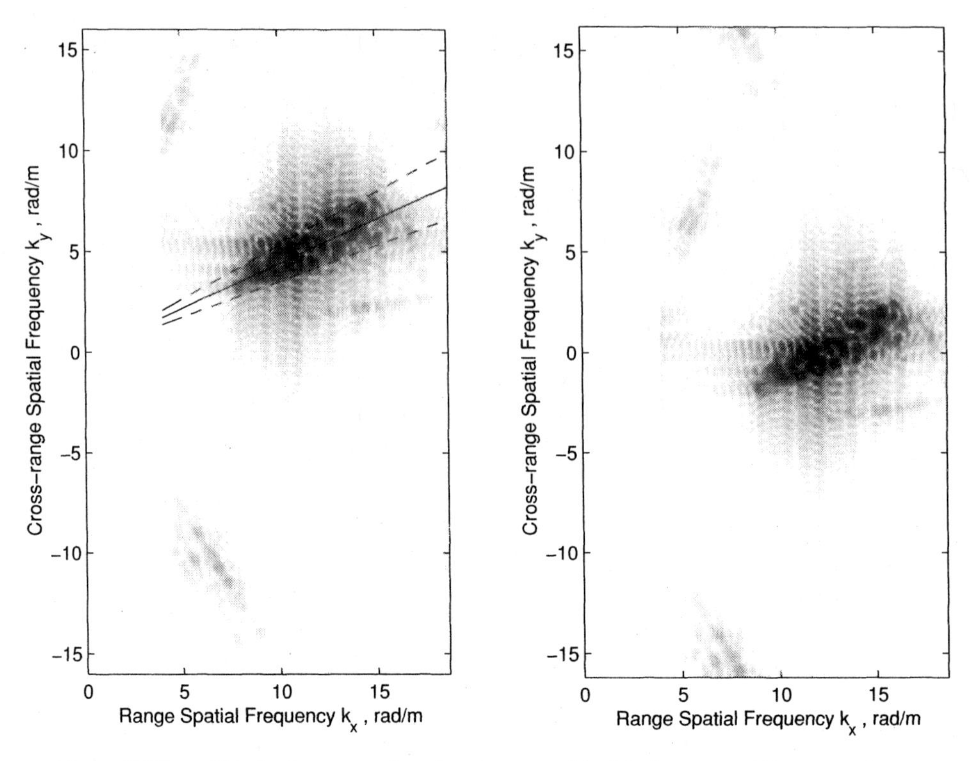

Figure 5.30b-c. Wavefront reconstruction of moving target: (b) original spectrum; (c) compressed spectrum.

where

$$t_{ij}(u) = \frac{2\sqrt{x_i^2 + (y_j - u)^2}}{c}$$

is the round-trip delay of the echoed signal for the target at (x_i, y_j) when the radar is at $(0, u)$. A digital implementation of the above reconstruction, which appears in the Matlab code for this chapter, is described in the following (Figure P5.15 and P5.16).

Step 1. Perform the (discrete) fast-time matched filtering

$$\mathbf{s}_M(t, u) = \mathbf{s}(t, u) * p_0^*(-t),$$

where

$$p_0(t) = p\big(t - T_c\big)$$

and T_c is the user-prescribed reference fast-time point.

Step 2. *Obtain finer samples of the matched-filtered SAR signal in the fast-time domain via zero-padding $s_m(\omega, u)$ in the fast-time frequency ω domain. A fast-time upsampling rate of higher than 100 is typically required.*

Step 3. *Perform the inverse of the fast-time baseband conversion (which was originally done by the hardware I/Q processor), that is,*

$$\mathbf{s}_M(t, u) \exp\left(j\omega_c t\right),$$

on the upsampled data (using the upsampled fast-time t array). This brings back the SAR signal to the bandpass fast-time frequency $\omega = \omega_c$.

Step 4. *Construct the spatial domain target function array $f(x_i, y_j)$, and fill it with zeros.*

Step 5. *For a given pixel point (x_i, y_j) on the reconstruction grid and a given radar position $(0, u)$ on the synthetic aperture, find the nearest upsampled fast-time point to*

$$t_{ij}(u) = \frac{2\sqrt{x_i^2 + (y_j - u)^2}}{c}.$$

Add the value of the matched-filtered SAR signal $\mathbf{s}_M(t, u)$ at that upsampled fast-time point to the array component $f(x_i, y_j)$. Repeat this for all the available discrete values of the synthetic aperture position; this yields the reconstructed value of $f(x_i, y_j)$.

Step 6. *Repeat Step 5 for the desired spatial pixel points (x_i, y_j).*

Step 7. *Perform coherent baseband conversion on the backprojection reconstruction via*

$$f(x_i, y_j) \exp\left(-jk_{xc}x_i\right),$$

where

$$k_{xc} = \frac{k_{x\max} + k_{x\min}}{2}.$$

This operation brings the two-dimensional spectrum of the reconstructed target function to the baseband (k_x, k_y) domain.

Digital spotlighting and upsampling in the synthetic aperture domain could be used to improve the quality of the TDC and backprojection SAR images also. Provided that these algorithms are properly implemented on alias-free SAR data to form images on a grid with appropriate sample spacing (Δ_x, Δ_y), then the resultant coherent SAR image should be very similar to the SAR image that is formed via the spatial frequency interpolation-based or range stacking method.

Unfortunately, the spectral shape and properties of the spotlight SAR signal are *transparent* to the users of the TDC and backprojection SAR image formation methods. Due to this fact these methods have provided a simplistic shortcut for those who do not fully understand SAR signal theory, its Nyquist sampling rates in the measurement (t, u) domain and the reconstruction (x, y) domain, and the preprocessing of the measured SAR data which could enhance the quality of the reconstructed SAR images.

For instance, a user of the backprojection believed that to form an alias-free coherent SAR image, the number of the cross-range samples in the y domain is sufficient to be the same as the number of the synthetic aperture samples in the u domain. He used this presumption to form a backprojection image in a SAR system where the size of the target area in the cross-range was ten times the length of the synthetic aperture (i.e, $Y_0 = 10L$.) This results in the cross-range sample spacing of $\Delta_y = 10\Delta_u$, which produces a severely aliased coherent SAR image.

Next we examine the backprojection reconstruction using the first-run data of Figure 5.8*b*. To prevent any adverse effects which our Fourier digital spotlighting and upsampling might have on the backprojection method, we use the original measured

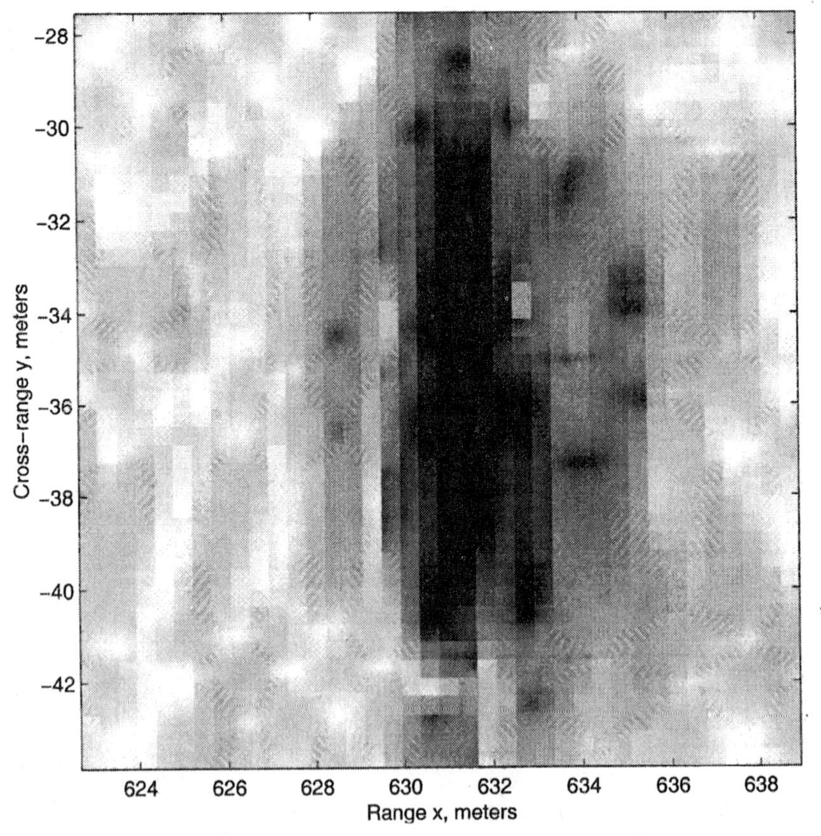

Figure 5.31a. Backprojection reconstruction of HEMTT truck (first run; without digital spotlighting and upsampling): spatial domain reconstruction.

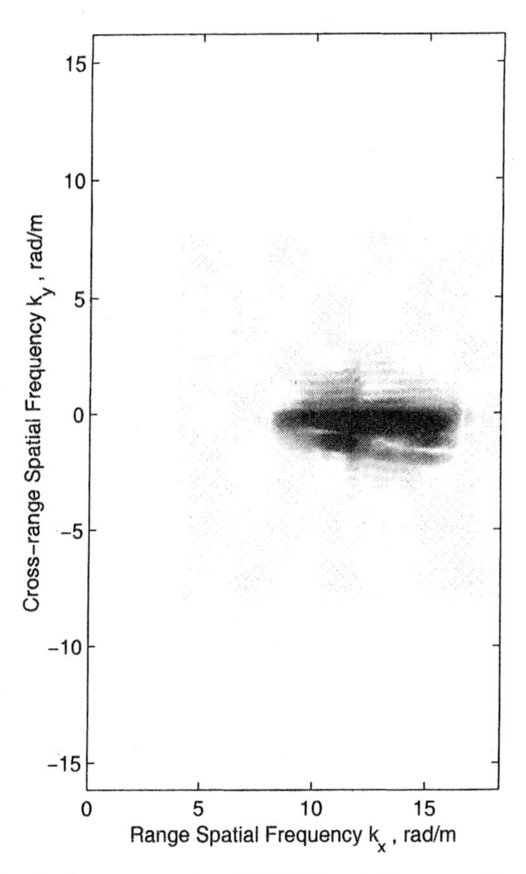

Figure 5.31b. Backprojection reconstruction of HEMTT truck (first run; without digital spotlighting and upsampling): spectrum.

SAR data (i.e, the data not digitally spotlighted and upsampled). Figure 5.31*a* and *b* shows the backprojection reconstruction and its spectrum for HEMTT 4 (as a reference, see Figures 5.18*a*–*b* and 5.26*a*–*c*).

The truck backprojection image in Figure 5.31*a* appears to contain more clutter than the wavefront reconstruction of the same target in Figure 5.26*a*. However, the truck backprojection spectrum looks as *clean* as its wavefront spectrum in Figure 5.26*b*; in fact both spectra are within the theoretical spectral support band of the truck shown in Figure 5.26*b* with dashed lines. So, *where is the spectrum of the additional clutter that appears in the backprojection reconstruction of Figure 5.31a?*

It turns out that both TDC and backprojection algorithms provide an automatic spectral filter in the (ω, k_u) domain based on the coordinates of the target. This filter translates into the theoretical spectral (k_x, k_y) band which is within the dashed lines in Figure 5.26*a*; this explains why the backprojection spectrum in Figure 5.31*b* appears to be relatively clean. However, the backprojection algorithm picks up any other signal within this theoretical band; this includes aliasing and the other types of clutter effects. In fact, if the digitally spotlighted and upsampled data were used, the backprojection reconstruction of the truck would look as clean as its wavefront reconstruction in Figure 5.26*a*.

Figure 5.32*a* and *b* show similar results for corner reflector; the wavefront reconstructions and spectra of this target appeared in Figure 5.29*a* and *b*. The corner reflector backprojection reconstruction shows clutter (aliasing) effects that could be taken care of by preprocessing the SAR data via digital spotlighting and upsampling.

Effect of Slow-time Doppler Filtering

Another commonly seen problem with the implementation of the TDC or backprojection method is the premature slow-time Doppler filtering of the SAR data. There are two reasons for this type of slow-time Doppler filtering, which results in the loss of the higher spatial frequency information and thus resolution. The first reason for filtering out the higher slow-time Doppler data is that the measured SAR data are

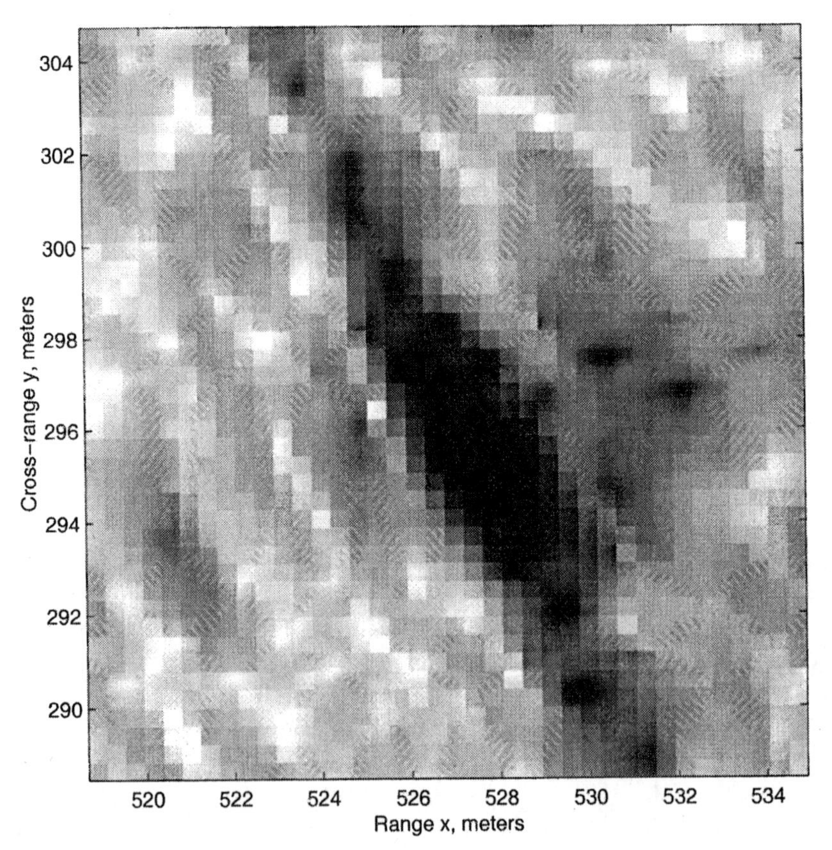

Figure 5.32a. Backprojection reconstruction of corner reflector 3 (first run; without digital spotlighting and upsampling): spatial domain reconstruction.

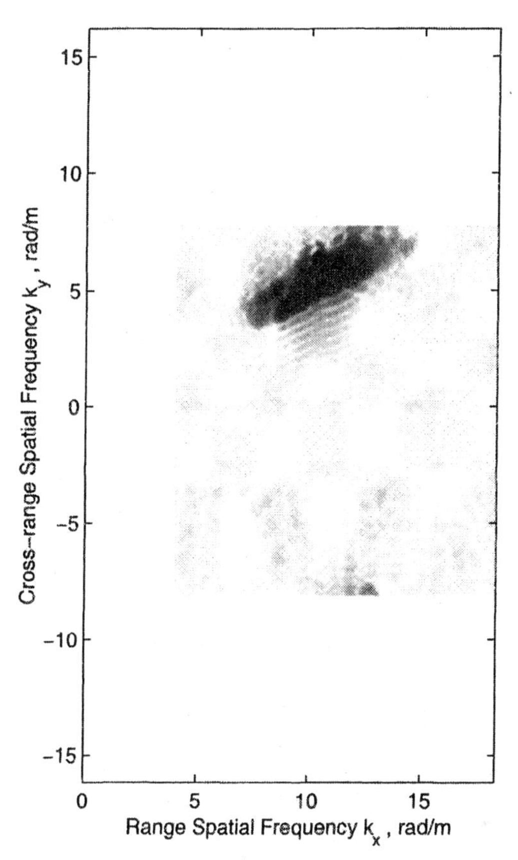

Figure 5.32b. Backprojection reconstruction of corner reflector 3 (first run; without digital spotlighting and upsampling): spectrum.

often aliased in the higher $|k_u|$ values due to either the side lobes of the radar radiation pattern (which were mentioned in our discussion on digital spotlighting) or an insufficient slow-time sampling rate (PRF).

As we showed earlier in our discussion on the two-dimensional spectrum of the spotlight SAR signal and its digital signal processing in the slow-time domain (Sections 5.3 and 5.5), the Nyquist sampling constraint for the spotlight SAR signal in the slow-time domain is

$$\Delta_u \leq \frac{X_{cc}\lambda_{\min}}{4(Y_0 + L)}.$$

Unfortunately, most operational spotlight SAR systems are based on the approximation-based polar format processing which requires a less restrictive Nyquist constraint in the slow-time domain, that is,

$$\Delta_u \leq \frac{X_{cc}\lambda_{\min}}{4Y_0}.$$

[Note that the polar format constraint for the slow-time sample spacing of the SAR signal $s(\omega, u)$ is the same as the Nyquist criterion for the slow-time sample spacing of the slow-time compressed SAR signal $s_c(\omega, u)$, that is, Δ_{u_c}.] Thus these spotlight SAR databases are aliased.

- As we mentioned earlier, digital spotlighting in the (t, k_u) domain of the slow-time compressed SAR data followed by upsampling (zero-padding in the k_u domain) could be used to counter radar radiation pattern side lobe effects, and insufficient PRF of polar format-based data acquisition.

The second reason for discarding the higher slow-time Doppler frequency data by the users of TDC and backprojection methods is simply to reduce the heavy computational burden of these reconstruction algorithms. For instance, for the wide-beamwidth UHF SAR example cited in this chapter, the radar beamwidth angle (at the carrier frequency) is in excess of 80° (i.e., ±40°). Processing of such a database via TDC or backprojection might require days or even weeks.

However, a commonly used and incorrect assumption about the wide-beamwidth UHF SAR systems is that the total beamwidth angle does not exceed 30° (that is, ±15°). Thus, at a given fast-time frequency ω, the SAR data in the slow-time Doppler frequencies

$$|k_u| > 2k \sin 15^\circ,$$

or the slow-time Doppler angles

$$|\phi| = \left|\arcsin\left(\frac{k_u}{2k}\right)\right| > 15^\circ,$$

could be discarded. This slow-time Doppler filtering significantly reduces the computational load of the TDC and backprojection SAR reconstruction methods in addition to degrading the quality of the reconstructed image and loss of information on *moving* targets; this is shown in the following example.

Figure 5.33 shows the two-dimensional spectrum of the reconstructed image which is formed with the backprojection method for the cited wide-beamwidth SAR system when the beamwidth angle is limited to the interval [−15, 15] degrees. Figure 5.34*a* is the reconstructed SAR image in the spatial domain; Figure 5.34*b* and *c* shows the close-ups of this reconstruction. By comparing the slow-time Doppler filtered images in Figure 5.34*a* to *c* with the ones formed with all the available slow-time Doppler data (see Figure 5.18*a*–*c*), we can observe the following.

1. *Degradations caused by the increase of the power of the side lobes.* The signature of the foliage (tree trunks) appear more smeared and spread in the slow-time Doppler filtered images. This increases the likelihood of the contamination of the coherent (complex) signature of human-made targets (e.g., trucks) hidden in the foliage by the signature of the surrounding foliage.

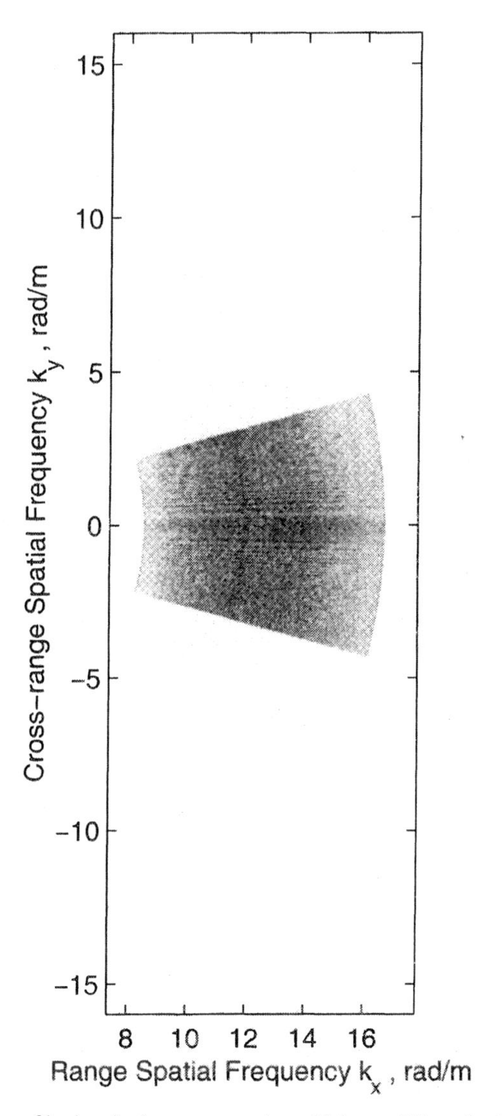

Figure 5.33. Spectrum of backprojection reconstruction with beamwidth angle (Doppler) of $[-15, 15]$ degrees (first run).

2. *Loss of squint target regions.* In the spotlight SAR systems the aspect angles over which a target is observable to the radar depends on the coordinates of the target. For instance, for the nth target, the interval of observability in the aspect angle domain by the radar is (see Figure 5.2)

$$\left[\theta_n(L), \theta_n(-L)\right],$$

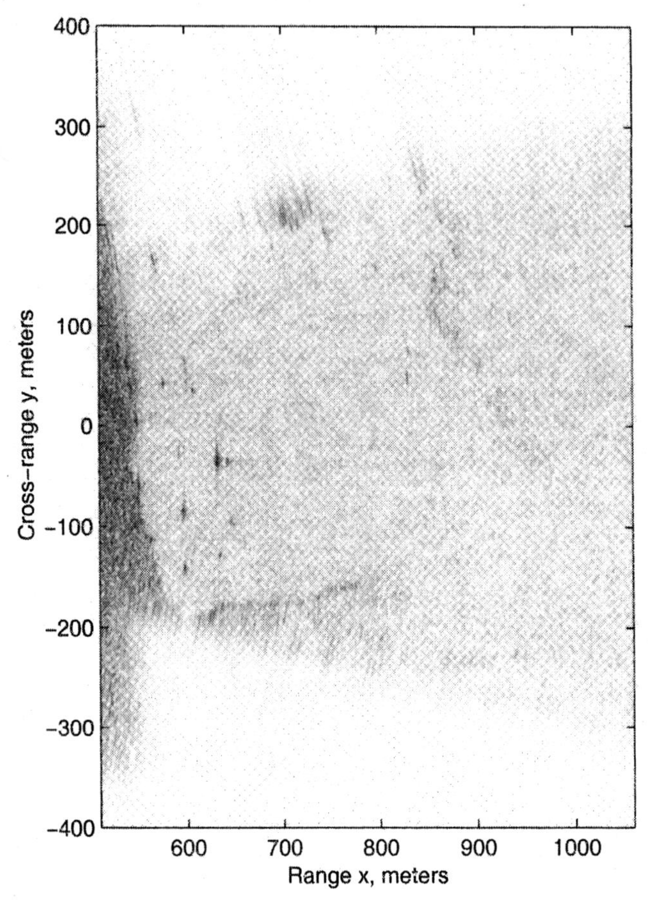

Figure 5.34a. Backprojection reconstruction with beamwidth angle (Doppler) of [−15, 15] degrees (first run): target area.

where

$$\theta_n(u) = \arctan\left(\frac{y_n - u}{x_n}\right).$$

As we showed earlier, the interval of observability of the nth target in the angular slow-time Doppler domain

$$\phi = \arcsin\left(\frac{k_u}{2k}\right)$$

is also dictated by the above aspect angle interval; that is,

$$\phi \in \left[\theta_n(L), \theta_n(-L)\right]$$

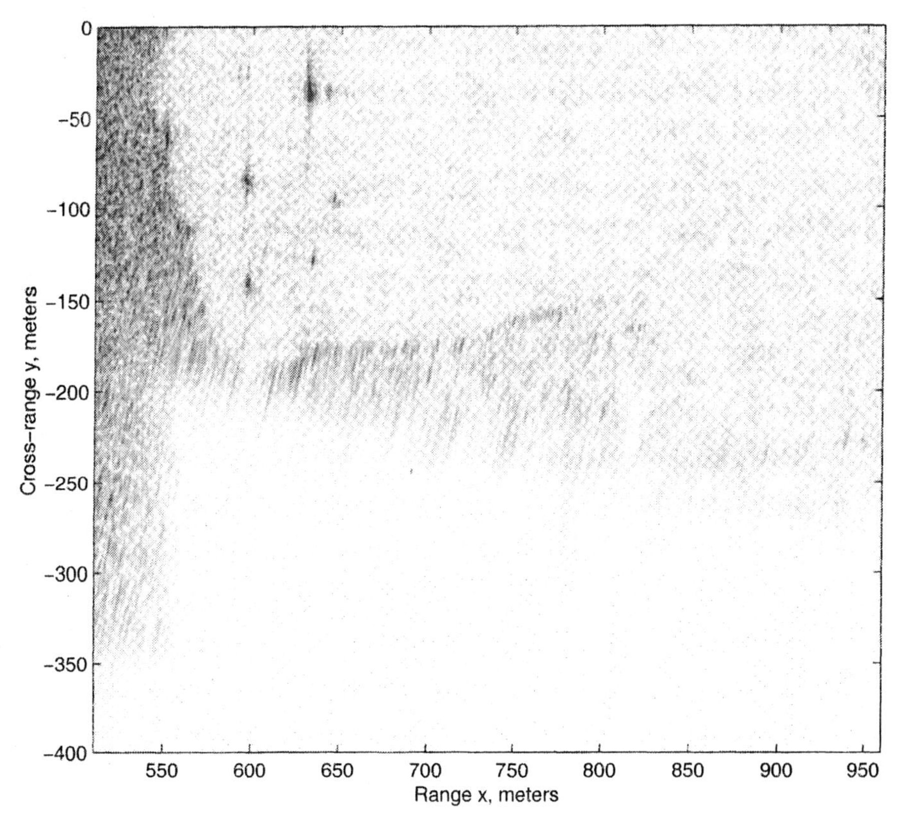

Figure 5.34b. Backprojection reconstruction with beamwidth angle (Doppler) of [−15, 15] degrees (first run): close-up of lower target area.

is the angular slow-time Doppler domain interval over which the nth target signature is measured. Thus, by restricting the processing to the angular slow-time Doppler interval of $\phi \in [-15, 15]$, a portion of the squint target areas are filtered out.

3. *Loss of slanted targets.* The observability of a target, for example, the nth target in the slow-time domain, is dictated not only by the radar spotlighted radiation pattern but also by the target amplitude pattern $a_n(x_n, y_n - u, \omega)$. Depending on the type and orientation of the nth target, its region of observability in its radar aspect angle domain, that is, $\theta_n(u) = \arctan[(y_n - u)/x_n]$, varies. For instance, a *broadside* truck exhibits a strong signature (also called *flash*) for $|\theta_n(u)| \leq 5°$. However, when this truck is rotated by 20 degrees (i.e., *slanted*), its strong SAR signature (flash) appears for the aspect angles that satisfy $|\theta_n(u) - 20°| \leq 5°$. We also showed that the target observability (amplitude pattern) in the aspect angle $\theta_n(u)$ domain directly maps into the angular slow-time Doppler ϕ domain. Thus, the signature of the slanted truck appears at the slow-time

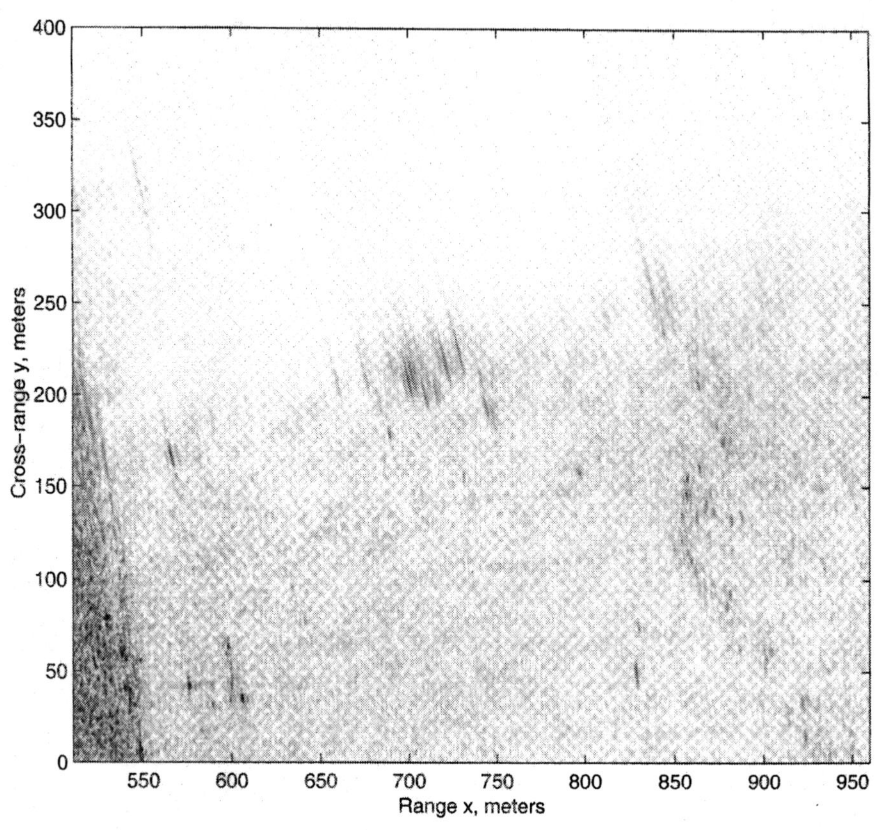

Figure 5.34c. Backprojection reconstruction with beamwidth angle (Doppler) of [−15, 15] degrees (first run): close-up of upper target area.

Doppler angles which satisfy $15^\circ \leq \phi \leq 25^\circ$. In this case, if the SAR data are passed through a lowpass slow-time Doppler filter which rejects the data in the region $|\phi| > 15^\circ$, then the slanted truck signature will be discarded by this slow-time Doppler filtering, and the truck will not appear in the reconstructed SAR image.

4. *Loss of moving targets.* It can be shown that the SAR signature of a moving target in the slow-time Doppler domain is shifted with respect to the SAR signature of the stationary targets [s94, ch. 5; yan]; the amount of the shift in the slow-time Doppler k_u domain at the fast-time frequency ω is

$$2k\frac{v_x}{v_r},$$

where v_r is the along-track speed of the radar-carrying aircraft (in the cross-range y domain), and v_x is the speed of the moving target in the range x domain. This slow-time Doppler k_u domain translates into the following shift in the angular slow-time Doppler ϕ domain (at all the available fast-time frequencies):

$$\arcsin\left(\frac{v_x}{v_r}\right).$$

This shift indicates that a moving target SAR signature exhibits a behavior similar to that of a squint or slanted target in the slow-time Doppler domain [s94, ch. 5; yan]. Clearly the slow-time Doppler filtering of the measured SAR data with a lowpass window of $|\phi| \leq 15°$ could partially or completely wipe out the SAR signature of a moving target.

Finally suppose that we form the spotlight SAR image $f(x, y)$ on a uniform grid that conforms to the Nyquist sample spacing of the compressed SAR image $f_c(x, y)$. While this image of $f(x, y)$ is aliased, it is possible that an *alias-free* version of the compressed SAR image $f_c(x, y)$ can be obtained from the aliased $f(x, y)$ via

$$f_c(x, y) = f(x, y) \exp\left[j2k_c(X_c + x) + j2k_c \sin\theta_c y - j2k_c\sqrt{(X_c + x)^2 + (Y_c + y)^2}\right].$$

This image of $f_c(x, y)$ is upsampled for example, by zero-padding its DFT spectrum $F_c(k_x, k_y)$ such that the resultant grid conforms to the Nyquist sample spacing of $f(x, y)$. Then alias-free samples of $f(x, y)$ can be obtained via the inverse compression (decompression) of these upsampled $f_c(x, y)$ data.

There exists one practical problem with the above-mentioned procedure. As we mentioned earlier, the backprojection method is based on the interpolation of the samples of the matched-filtered SAR signal $\mathbf{s}_M(t, u)$ in the fast-time domain. This interpolation produces errors in the reconstructed SAR image even when a very accurate interpolation method is used. These interpolation errors commonly result in shifted replicas of the target spectrum in the spatial frequency (k_x, k_y) domain.

As a result one cannot assume that the compressed SAR image formed by the backprojection method possesses its theoretical bandwidth. With the above-mentioned compression, upsampling, and decompression of the reconstructed image, the interpolation errors of the backprojection method appear as aliasing, which eventually increase the background clutter (noise) level of the reconstructed image.

Effect of Motion Errors in Slow-time Doppler Spectrum

We examined the motion error model in the fast-time and slow-time (t, u) domain of a generic SAR system in Section 4.9. We also showed that these motion errors translate into phase error functions in the (ω, u) domain of SAR data. Such motion phase errors are also called *multiplicative* noise; similar noise sources are also encountered in other coherent imaging systems. The motion phase error or multiplicative noise in the (ω, u) domain of SAR data becomes a *convolutional* error function in the slow-time Doppler k_u domain; this means that the SAR data with motion errors have a *larger* slow-time Doppler k_u support band than the SAR data without motion errors [s92a].

This phenomenon is particularly evident in the spotlight SAR systems with a relatively large synthetic aperture (slow-time interval of coherent integration). This slow-time Doppler bandwidth expansion commonly results in wraparound slow-time Doppler aliasing which changes the spectral appearance of the SAR signal; see this discussion in Chapter 6 which is accompanied by numerical results. Unfortunately, this is not recognized and/or identified by some users, which could lead to misunderstandings, misleading and incorrect observations, and improper processing of the SAR data.

In order to have a better understanding of spotlight SAR data which are corrupted with motion errors, the user should first use the GPS-based *narrow-beamwidth* motion compensation method which was described in Section 4.9. The resultant data still contain some residual motion errors. However, these residual motion errors do not dramatically alter the nature of the SAR signal in the same manner that the original motion errors do; the residual motion errors result in *localized* smearing and degradation of the SAR data.

The narrow-beamwidth motion compensated spotlight data can be passed through the algorithms that we introduced for preprocessing of the spotlight SAR data, for example, digital spotlighting. After forming the SAR image from this database, the shift-varying filter for the wide-beamwidth motion compensation, which was discussed in Section 4.9, is used for the final refinement of the reconstructed image.

5.7 MATLAB ALGORITHMS

The code for spotlight SAR imaging is identified as *pulsed spotlight SAR simulation and reconstruction*. The simulation is based on the slow-time Nyquist rate for the compressed SAR signal $s_c(\omega, u)$, that is, Δ_{uc}. The Doppler-aliased measured data are compressed, digitally spotlighted, and upsampled to obtain alias-free spotlight SAR data with the slow-time sample spacing of Δ_u.

The Matlab code provides algorithms for digital reconstruction using the spatial frequency interpolation, range stacking, time domain correlation, and backprojection. The code also performs slow-time Doppler subsampling when $Y_0 < L$, and spotlight SAR image compression. Two scenarios are considered: $Y_0 > L$, which requires zero-padding of the SAR signal in the synthetic aperture domain, and $Y_0 < L$, which benefits from slow-time Doppler subsampling of the SAR signal spectrum that reduces the computational load.

To provide a relatively simple and compact code to demonstrate the main properties of a spotlight SAR system, certain procedures are used that restrict the scope of the code and/or increase its computational load. The user could modify the code to remove these restrictions and increase its speed. These issues are listed in the following:

1. The fast-time sampling is based on the bandwidth of the chirp radar signal, and not the deramped signal. In the case of $T_p > T_x$, use the more efficient fast-time sampling rate of the deramped signal, and convert the resultant samples to alias-free matched-filtered echoed data via the procedure which was described in Chapter 1.
2. The program simulates a squint SAR spotlight SAR system with $Y_c \geq 0$. The user could modify the code for $Y_c < 0$.
3. The radar radiation pattern is uniform and invariant of the fast-time frequency in the spotlighted target area. The user could add the radiation pattern by simulating the echoed signal for each individual target $\mathbf{s}_n(t, u)$, convert it to $s_n(\omega, u)$ via DFT with respect to the fast-time samples, apply the appropriate amplitude pattern (which varies with the coordinates of the target), and then add up the contributions of all the targets. Such a procedure is shown in the stripmap SAR Matlab code for Chapter 6.
4. The length of the synthetic aperture is small enough such that full aperture digital spotlighting is used. Subaperture digital spotlighting is provided for the stripmap SAR Matlab code for Chapter 6.
5. Narrow-bandwidth approximation is used for polar format processing and digital spotlighting. The code could be modified to remove this assumption.
6. As we showed in the reconstruction with the realistic ISAR data, it is more economical (from the point of view of size of arrays) to form the target function in the squint spatial domain. The modification of the code requires two-dimensional interpolation in the spatial frequency domain. This is not as simple as the one that appears in our code; however, it is not too difficult.
7. The code selects the minimum and maximum range spatial frequency pair $(k_{x\min}, k_{x\max})$ based on all the available discrete k_{xmn} values. However, these two parameters should be chosen from the k_{xmn} values that fall within the two-dimensional bandwidth of the spotlight SAR signal; that is, the union of

$$k_x \in \left[2k_{\min} \cos \theta_{\min}, 2k_{\min} \cos \theta_{\max}\right]$$

   and

$$k_x \in \left[2k_{\max} \cos \theta_{\min}, 2k_{\max} \cos \theta_{\max}\right].$$

   In the wide-bandwidth spotlight SAR systems, this could reduce the size of the array in the k_x domain. However, the user should include "if" clauses in the interpolation loop to make sure negative array indexes are not addressed.

The outputs of the code are as follows:

P5.1 Measured spotlight SAR signal $\mathbf{s}(t, u)$

P5.2 SAR signal after fast-time matched-filtering $\mathbf{s}_M(t, u)$

P5.3 Aliased spotlight SAR signal spectrum $S(\omega, k_u)$

P5.4 Compressed spotlight SAR signal spectrum $S_c(\omega, k_u)$

P5.5 Polar format SAR reconstruction $f_p(t, k_u)$ with digital spotlight filter. The (t, k_u) is transformed into the (x, y) domain based on the narrow-bandwidth and narrow-beamwidth polar format processing

P5.6 Spotlight SAR signal spectrum after digital spotlighting and upsampling $S_d(\omega, k_u)$

P5.7 Spotlight SAR spatial frequency data coverage

P5.8 Wavefront spotlight SAR reconstruction spectrum

P5.9 Wavefront spotlight SAR reconstruction

P5.10 Compressed spotlight SAR reconstruction spectrum

P5.11 Range stack spotlight SAR reconstruction

P5.12 Range stack spotlight SAR reconstruction spectrum

P5.13 Time domain correlation spotlight SAR reconstruction

P5.14 Time domain correlation spotlight SAR reconstruction spectrum

P5.15 Backprojection spotlight SAR reconstruction

P5.16 Backprojection spotlight SAR reconstruction spectrum

6

STRIPMAP SYNTHETIC APERTURE RADAR

Introduction

This chapter addresses SAR imaging based on processing the echoed data obtained from an airborne radar that does not spotlight a specific target area via beam steering. In this SAR system the radar maintains the same broadside radiation pattern throughout the data acquisition period on a *fixed strip in the (slant) range domain*. For such a data collection scheme, the illuminated cross-range area is varied from one pulse transmission to the next.

This form of SAR or inverse SAR (ISAR) database is mainly encountered in reconnaissance or surveillance problems. Such a SAR imaging system, which provides a map of a terrain within a fixed strip in the range domain, is also known as *stripmap* SAR or *side-looking* SAR.

The stripmap SAR form of target area radiation is analogous to a scenario in which someone is trying to view *all* objects in a dark room with a flashlight. With the flashlight in his right hand (the radar on the aircraft), that person would move his right arm to *scan* the room with the light of the flashlight. This *preliminary* phase of the search is to provide the individual with a general feel for what the room contains. In a similar way, stripmap SAR systems provide imaging information on the general condition and contents of, for example, a terrain area.

Consider again the individual with the flashlight in a dark room. Once he locates an object of interest after scanning the beam of the flashlight throughout the room, he might go around that object with the flashlight to gather more information about it. This phase of search provides that individual with specific information about an object of interest. As we mentioned in Chapter 5, a similar task is performed by spotlight SAR systems to obtain detailed information on targets within a relatively small terrain area.

Stripmap SAR is the original SAR imaging modality which was introduced in the 1950s. However, as we pointed out in Section 5.0 and also earlier in this section, one should view stripmap SAR and spotlight SAR as two different radar imaging modalities that have distinct utilities; the choice of the two SAR systems depends on the application.

The classical image reconstruction for stripmap SAR is based on the Fresnel approximation which utilizes deramping or chirp deconvolution in the synthetic aperture (slow-time) domain. The use of this approximation prohibits the processing of the SAR data over a large synthetic aperture to improve cross-range resolution. The plane wave approximation-based polar format processing performs well over large synthetic apertures (slow-time coherent integration or processing) in certain *spotlight* SAR scenarios. However, one cannot conclude that the polar format processing is also suitable for wide-beamwidth stripmap SAR systems; in fact, the opposite is shown in the analysis of a stripmap SAR system in [s95].

Our approach in this chapter in formulating stripmap SAR processing and imaging is based on the SAR wavefront reconstruction theory, and analysis of the SAR signal in the slow-time domain via the spherical wave Fourier decomposition of the radar radiation pattern which was developed in Chapters 2 and 3. This Fourier-based SAR signal analysis provides not only an accurate digital reconstruction from stripmap SAR data but also an insight into the spectral properties of the stripmap SAR signal; these are crucial for quantifying the performance (resolution and point spread function) of a stripmap SAR system in terms of its parameters, and alias-free data acquisition and processing of stripmap SAR data.

Outline

We begin with developing a signal model for the measured signal in stripmap SAR systems in Section 6.1. This analysis uses the spectral properties of the radiation pattern of a side-looking radar which was outlined in Sections 3.3 and 3.4. We discuss the effect of the radar type on the beamwidth of the radar radiation pattern and identify the role of the radar divergence angle in the stripmap SAR systems. Using these principles, we obtain an expression for the stripmap SAR signal as a function of the fast-time frequency and synthetic aperture coordinates (slow-time) of the radar.

An important feature of the stripmap SAR systems becomes evident in this analysis. In Chapter 5 we showed that the *aspect angle* interval over which a target is observable to a spotlight SAR depends on the coordinates of the target. In the stripmap SAR systems, the interval of observability of a target in the aspect angle domain of the radar depends only on the divergence angle of the radar, and is invariant in the coordinates of the target. This fact plays a key role in the manner in which the two-dimensional Fourier transform (spectrum) of the stripmap SAR signal is formed.

In Section 6.2 we relate the two-dimensional Fourier transform of the stripmap SAR to the two-dimensional Fourier transform of the target area reflectivity function. This becomes the basis of the wavefront reconstruction in the stripmap SAR systems. The general approach is the same one that we developed in Chapter 4 for a generic

SAR system. The main distinction is the role of the observability of a target in the aspect angle domain in forming its slow-time Doppler spectrum.

This also has consequences in the properties and extent of the bandwidth of a stripmap SAR signal, which is discussed in Section 6.3. We begin with a general analysis of the two-dimensional bandwidth of a stripmap SAR signal. We then specialize this analysis for the case of planar and curved aperture radars. The principles developed in Section 6.3 will guide our work in the next two sections.

Range and cross-range resolution for stripmap SAR systems are examined in Section 6.4. The approach is to identify the amplitude functions as well as the support of the spectral signature of a target in the spatial frequency domain of the reconstructed stripmap SAR image. This helps us to provide an analytical expression for the manner in which the target appears in the reconstructed image in the spatial domain, that is, its point spread function; this is also used as a basis to identify the range and cross-range resolution. We will see that unlike the spotlight SAR systems, the point spread function of a target in a stripmap SAR scene is invariant in the coordinates of the target; this property can be traced to the observability of the target in the aspect angle domain or the slow-time Doppler domain which does not depend on the coordinates of the target.

Data acquisition and digital signal processing issues associated with the stripmap SAR systems are outlined in Section 6.5. This analysis provides tools and techniques for alias-free processing of stripmap SAR data. Constraints on measuring the stripmap SAR signal in the fast-time and slow-time domains are examined. Similar to our analysis in Section 5.5 for the spotlight SAR systems, we outline a subaperture digital spotlighting method for the stripmap SAR systems. This scheme provides a *numerical* (digital signal processing) tool to extract the stripmap SAR signature of a *finite* target area (analogous to a finite target area which is radiated by a spotlight SAR) with a minimal effect of side lobes leakage aliasing.

Having access to the stripmap SAR data of a finite target area is also crucial for the success (alias-free implementation) of the digital reconstruction SAR methods that form the target function in its spectral domain, that is, digital reconstruction via spatial frequency interpolation and range stacking. These methods as well as time domain correlation (TDC) imaging or backprojection are discussed in Section 6.6. Image processing issues that are useful for accurate and computationally efficient implementation of these methods are also discussed.

Section 6.7 provides an analytical study on the SAR signature of a moving target. This study provides a basis for understanding how a moving target could appear in the SAR image of a stationary target area (e.g., foliage) which is being investigated by an airborne radar. The problem here is whether one can develop tools for *detecting* and even *imaging* such a moving target. A statistic, which is identified as *SAR ambiguity function*, is developed that provides a measure for moving target detection.

To exhibit the utility and merits of the digital and image processing algorithms of Sections 6.5 and 6.6, two sets of realistic UHF band stripmap SAR data are analyzed in conjunction with our theoretical work; the parameters of these stripmap SAR systems are shown in the introductory chapter of this book. The Matlab code for simulating the data of a stripmap SAR system, its processing through a subaperture digital spotlight algorithm, and reconstruction from the resultant database is discussed in Section 6.8.

Mathematical Notations and Symbols

The following is a list of mathematical symbols used in this chapter, and their definitions:

$a(\omega, x, y)$	Transmit-receive mode radar amplitude pattern
$a_n(\omega, x, y)$	Amplitude pattern for n-target (general case)
$A(\omega, k_u)$	Slow-time Doppler transmit-receive mode radar amplitude pattern
$A_n(\omega, k_u)$	Slow-time Doppler amplitude pattern for n-target
B	Half of radar beamwidth in cross-range domain at range x and fast-time frequency ω; total beamwidth size in cross-range domain is $2B$
$B_{\max}$	Maximum half-beamwidth in cross range domain
B_n	Half of radar beamwidth in cross-range domain at range x_n and fast-time frequency ω; total beamwidth size in cross-range domain is $2B_n$
B_x	Support of nth target signature in k_x domain
B_y	Support of nth target signature in k_y domain
c	Wave propagation speed $c = 3 \times 10^8$ m/s for radar waves $c = 1500$ m/s for acoustic waves in water $c = 340$ m/s for acoustic waves in air
D_x	Diameter of radar in range domain
D_y	Diameter of radar in cross-range domain
$f(x, y)$	Target function in spatial domain
$F(k_x, k_y)$	Two-dimensional Fourier transform of target function
$f_c(x, y)$	Compressed image of target area
$F_c(k_x, k_y)$	Spectrum of compressed image of target area
$f_d(t, k_u)$	Digitally spotlighted polar format processed image
$f_n(x, y)$	Point spread function of nth target in spatial domain
$\mathbf{f}(\mathbf{X}, \mathbf{Y}, \alpha)$	Motion-transformed target function
$F_n(k_x, k_y)$	Two-dimensional Fourier transform of point spread function of nth target
$\mathbf{F}(k_{\mathbf{X}}, k_{\mathbf{Y}}, \alpha)$	Motion-transformed target function spectrum
$h(\omega, x, y)$	Transmit-receive mode radar radiation pattern
k	Wavenumber: $k = \omega/c$
k_c	Wavenumber at carrier frequency: $k_c = \omega_c/c$
k_n	Wavenumber sampled points for SAR signal

k_u	Spatial frequency or wavenumber domain for azimuthal synthetic aperture u; slow-time Doppler (frequency) domain
k_{um}	Slow-time Doppler sampled points for SAR signal
k_x	Spatial frequency or wavenumber domain for range x
k_{xc}	Center (carrier) range spatial frequency of reconstructed SAR image
k_{xmn}	Available range spatial frequency sampled points
$k_{x\max}$	Maximum of available range spatial frequency sampled points k_{xmn}'s
$k_{x\min}$	Minimum of available range spatial frequency sampled points k_{xmn}'s
k_{X_n}	Spatial frequency domain for X_n
$k_{\mathbf{X}}$	Range spatial frequency for motion-transformed domain
k_y	Spatial frequency or wavenumber domain for cross-range y
k_{ymn}	Available cross-range spatial frequency sampled points
$k_{y\max}$	Maximum of available cross-range spatial frequency sampled points k_{ymn}'s
$k_{y\min}$	Minimum of available cross-range spatial frequency sampled points k_{ymn}'s
$k_{\mathbf{Y}}$	Cross-range spatial frequency for motion-transformed domain
k_0	Baseband wavenumber: $k_0 = \omega_0/c$
L	Half-size of synthetic aperture
$L_{\min}$	Minimum half-length of processed (zero-padded) synthetic aperture; minimum length of processed (zero-padded) synthetic aperture is $2L_{\min}$
L_s	Half-size of subaperture
M	Number of samples of SAR signal in slow-time u domain
M_c	Number of samples of slow-time compressed SAR signal in slow-time u domain
n	Index representing a specific target
N	Number of fast-time samples
N_s	Number of subapertures for digital spotlighting
N_x	Number of range samples in target reconstruction
N_y	Number of cross-range samples in target reconstruction
$p(t)$	Transmitted radar signal
$p_0(t)$	Transmitted radar signal shifted by reference fast-time T_c
$P(\omega)$	Fourier transform of transmitted radar signal
PRF	Radar pulse repetition frequency in slow-time
PRI	Radar pulse repetition interval in slow-time
r	Radial distance in spatial domain
r_n	Target radial distance from center of synthetic aperture
$r_{\max}$	Farthest radial distance of radar from irradiated target area
$r_{\min}$	Closest radial distance of radar from irradiated target area
$\mathbf{R}_n$	Digitally spotlighted region around $(\mathbf{X}_{nc}, \mathbf{Y}_{nc})$

$\mathbf{s}(t, u)$	Echoed SAR signal from target area
$s(\omega, u)$	One-dimensional Fourier transform of SAR signal with respect to fast-time
R_c	Radial distance of center of target from radar at $u = 0$
$\mathbf{s}(t, u)$	Echoed SAR signal from target area
$s(\omega, u)$	One-dimensional Fourier transform of SAR signal with respect to fast-time
$s_b(\omega, u)$	SAR signal after baseband conversion in slow-time domain
$S(\omega, k_u)$	Two-dimensional Fourier transform of SAR signal with respect to fast-time and slow-time
$s_c(\omega, u)$	Slow-time compressed SAR signal
$S_c(\omega, k_u)$	Two-dimensional Fourier transform of slow-time compressed SAR signal with respect to fast-time and slow-time
$s_{cd}(\omega, u)$	Digitally spotlighted slow-time compressed SAR signal
$S_{cd}(\omega, k_u)$	Two-dimensional Fourier transform of digitally spotlighted slow-time compressed SAR signal with respect to fast-time and slow-time
$s_d(\omega, u)$	Digitally spotlighted SAR signal
$S_d(\omega, k_u)$	Two-dimensional Fourier transform of digitally spotlighted SAR signal with respect to fast-time and slow-time
$\mathbf{s}_M(t, u)$	Fast-time matched-filtered echoed signal
$s_M(\omega, u)$	One-dimensional Fourier transform of matched-filtered SAR signal with respect to fast-time
$\mathbf{s}_n(t, u)$	Echoed SAR signal of nth target
$s_n(\omega, u)$	One-dimensional Fourier transform of nth target SAR signal with respect to fast-time
$S_n(\omega, k_u)$	Two-dimensional Fourier transform of nth target SAR signal with respect to fast-time and slow-time
$s_0(\omega, u)$	One-dimensional Fourier transform of reference SAR signal with respect to fast-time
$\text{sinc}(\cdot)$	sinc function: $\text{sinc}(a) = \sin(\pi a)/\pi a$
t	Fast-time domain
$t_{ij}(u)$	Round trip delay of echoed signal for target at (x_i, y_j) when radar is at $(0, u)$
t_{ni}	Fast-time location of nth target in ith subaperture polar format processing
T	Length of fast-time processing interval
T_c	Reference fast-time point
T_f	Ending point of fast-time sampling for SAR signal
$T_{\min}$	Minimum length of fast-time processing interval
T_p	Radar signal pulse duration

T_s	Starting point of fast-time sampling for SAR signal
u	Synthetic aperture or slow-time domain (azimuthal)
u_{nf}	Synthetic aperture position at which echoed signal of nth target arrives at fast-time T_f
(u_{nf1}, u_{nf2})	Two solutions for u_{nf}
v_r	Speed of radar-carrying aircraft in SAR
v_x	Range domain speed of moving target in SAR scene
v_{xn}	Range domain speed of nth target
v_y	Cross-range domain speed of moving target in SAR scene
v_{yn}	Cross-range domain speed of nth target
$W_d(t, k_u)$	Digital spotlight filter
$W_{di}(t, k_u)$	Digital spotlight filter for ith subaperture
x	Range domain
x_f	Focal range of radar near-field
x_i	Range bins for range stacking, TDC and backprojection reconstruction algorithms
x_n	Range of nth target
X_c	Midpoint of range swath; center point of target area in range domain
X_0	Half-size of desired target area in range domain; size of desired target area in range domain is $2X_0$
$X_{0\max}$	Maximum half-size of range swath; $2X_{0\max}$ is maximum size of range swath
$\mathbf{X}$	Range in motion-transformed spatial domain
$\mathbf{X}_i$	Range bins in a subpatch for moving target detection
$\mathbf{X}_n$	Range of nth target in motion-transformed spatial domain
$\mathbf{X}_{nc}$	Center range of subpatch for moving target detection
y	Cross-range or azimuth domain
y_j	Cross-range bins for TDC and backprojection reconstruction algorithms
y_n	Cross-range of nth target
Y_e	Half-width of effective radiated target area in cross-range domain; size of effective radiated target area in cross-range domain is $2Y_e$
Y_i	Offset of ith subaperture in cross-range domain
Y_0	Half-size of desired target area in cross-range domain; size of desired target area in cross-range domain is $2Y_0(\omega)$
$\mathbf{Y}$	Cross-range in motion-transformed spatial domain
$\mathbf{Y}_j$	Cross-range bins in a subpatch for moving target detection
$\mathbf{Y}_n$	Cross-range of nth target in motion-transformed spatial domain
$\mathbf{Y}_{nc}$	Center cross-range of subpatch for moving target detection

α	Relative speed of target in SAR scene with respect to radar
α_n	Relative speed of nth target with respect to radar
Δ_k	Sample spacing of SAR signal in wavenumber k domain
Δ_{k_u}	Sample spacing of SAR signal in slow-time Doppler domain
Δ_{k_x}	Sample spacing of reconstructed target function in range spatial frequency domain
Δ_{k_y}	Sample spacing of reconstructed target function in cross-range spatial frequency domain
Δ_t	Sample spacing of stripmap SAR signal in fast-time domain
Δ_u	Sample spacing of SAR signal in synthetic aperture domain
Δ_{uc}	Sample spacing of slow-time compressed SAR signal in synthetic aperture domain
Δ_x	Sample spacing and resolution of reconstructed target function in range domain
Δ_y	Sample spacing and resolution of reconstructed target function in cross-range domain
Δ_ω	Sample spacing of SAR signal in fast-time frequency domain
$\mathcal{F}$	Forward Fourier transform operator
$\mathcal{F}^{-1}$	Inverse Fourier transform operator
$\gamma_n(\omega, u, \alpha)$	Slow-time compressed signal of subpatch $\mathbf{R}_n$
$\Gamma_n(\omega, k_u, \alpha)$	SAR ambiguity function of subpatch $\mathbf{R}_n$
λ	Wavelength: $\lambda = 2\pi c/\omega$
λ_c	Wavelength at carrier fast-time frequency: $2\pi c/\omega_c$
$\lambda_{\max}$	Maximum wavelength: $2\pi c/\omega_{\min}$
$\lambda_{\min}$	Minimum wavelength: $2\pi c/\omega_{\max}$
ω	Temporal frequency domain for fast-time t
ω_c	Radar signal carrier or center frequency
ω_n	Fast-time frequency sampled points for SAR signal
ω_0	Radar signal half bandwidth in radians; radar signal baseband bandwidth is $\pm\omega_0$
$\omega_{\max}$	Maximum fast-time frequency of radar
$\omega_{\min}$	Minimum fast-time frequency of radar
ϕ_d	Divergence angle of radar
σ_n	Reflectivity of nth target
τ	Slow-time domain in seconds
θ_{ax}	Largest absolute aspect angle of target area with respect to radar synthetic aperture coordinates; largest divergence angle of radar
θ_{ci}	Average squint angle of target area for ith subaperture
$\theta_n(u)$	Aspect angle of nth target when radar is at $(0, u)$
$\theta_{0\max}$	Maximum aspect angle of desired target area when radar is at $u = 0$

6.1 SYSTEM MODEL

A stripmap SAR imaging system geometry is shown in Figure 6.1. The airborne radar moves along a line that is called the synthetic aperture domain or the *slow-time* domain. We denote the synthetic aperture domain with u and its frequency domain with k_u. For a fixed synthetic aperture position of the radar, the radar transmits a wide-bandwidth pulsed signal $p(t)$, where t is the *fast-time* domain, and it records the resultant echoed signals from the illuminated area. We denote the fast-time frequency domain with ω.

We examine the imaging problem in the two-dimensional spatial domain of slant range (also called down range) and cross-range (also called azimuth or along-track); this domain is identified via (x, y). The radar height determines the ground swath in the x domain which remains unchanged (not shifted) as the radar is moved (see Figures 3.7*a–b* in Chapter 3); this is due to the fact that the radar does not change its

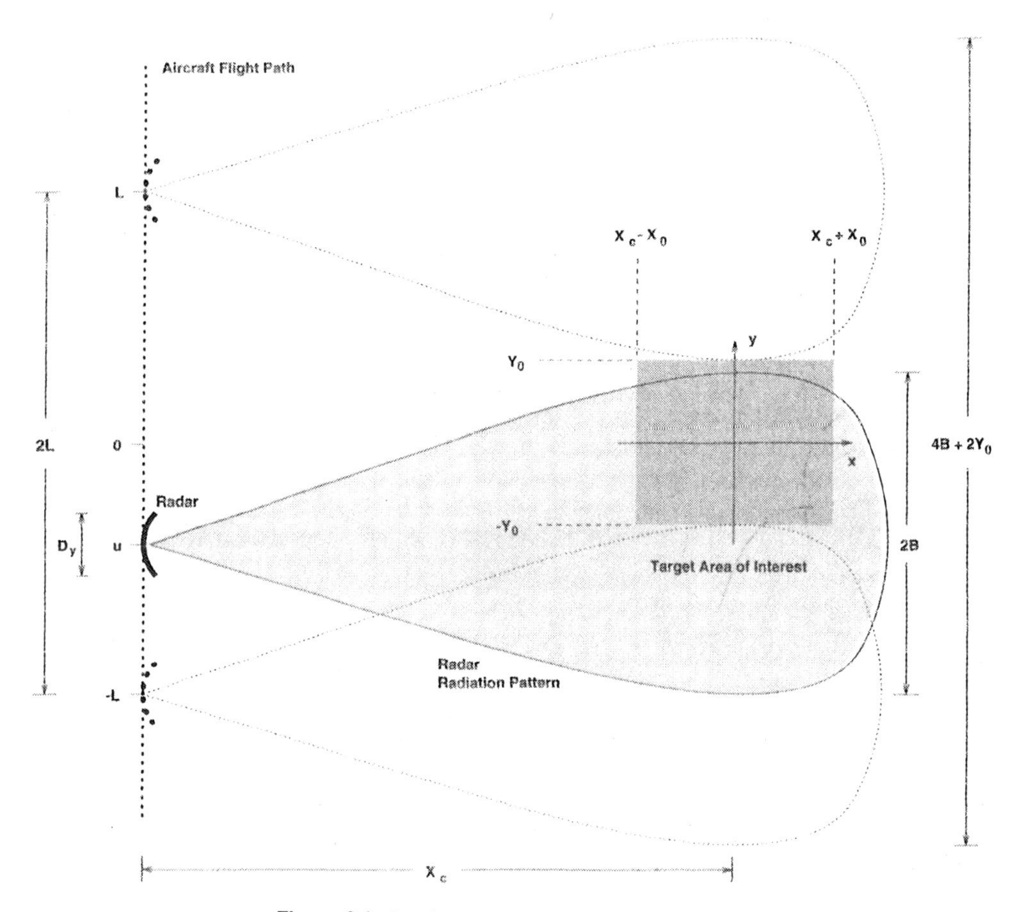

Figure 6.1. Imaging system geometry in stripmap SAR.

coordinates in the x domain during the data acquisition. The radar diameter (length) along the flight path determines its footprint in the y domain that *shifts* as the radar is moved; this is due to the radar motion in the y domain (see Figure 6.1).

We are concerned with the variations of the radar footprint along the track. Clearly, unlike the spotlight SAR systems, the main lobe of the radar radiation pattern in a stripmap SAR system is not focused on a specific target region for all the available slow-time (synthetic aperture) u values. We will outline a digital signal processing of the stripmap SAR data, which we refer to as *digital spotlighting*, to image a specific target area in the along track (cross-range) as well as the range domain.

Radar Radiation Pattern

Unlike the beam-steering radar system used in the spotlight SAR systems, the airborne radar of a stripmap SAR system maintains a fixed radiation pattern for its transmit-receive modes as it is moved along the synthetic aperture. To provide a mathematical representation for this SAR data collection scheme, we consider the radar radiation pattern in the spatial domain (x, y) at a given fast-time frequency, for example, ω, of the radar signal.

The radar radiation pattern when the radar is located at $u = 0$ is denoted by

$$h(\omega, x, y).$$

Then the radar radiation pattern, when it is moved to an arbitrary synthetic aperture location u, is

$$h(\omega, x, y - u),$$

which is a shifted version of $h(\omega, x, y)$ in the cross-range domain. The radar radiation pattern at a fixed value of u is shown in Figure 6.1. The radar radiation pattern depends on the type of radar used and its physical dimensions. We are particularly interested in the effect of the radar type and its diameter along the track, D_y, in its beamwidth in the cross-range domain as the radar-carrying aircraft moves; that is, u is varied.

In Chapter 3 we showed that the transmit-receive radiation pattern of a radar has the following Fourier decomposition:

$$\begin{aligned} h(\omega, x, y - u) &= \int_{-2k}^{2k} A(\omega, k_u) \exp\left[-j\sqrt{4k^2 - k_u^2}\, x - jk_u(y - u)\right] dk_u \\ &= \mathcal{F}_{(k_u)}^{-1}\left[A(\omega, k_u) \exp\left(-j\sqrt{4k^2 - k_u^2}\, x - jk_u y\right)\right], \end{aligned}$$

where $k = \omega/c$ is the wavenumber and $A(\omega, k_u)$ is the Fourier amplitude function of the radiation pattern. The significance of the above decomposition is in representing the radar radiation pattern in terms of a linear combination of plane waves (*linear* phase functions) in the spatial (x, y) domain, that is,

$$\exp\left(-j\sqrt{4k^2 - k_u^2}\,x - jk_u y\right)$$

for $k_u \in [-2k, 2k]$. The plane waves, which are linear phase functions of the spatial domain (x, y) and the synthetic aperture domain u, facilitate the analysis of the stripmap SAR signal for the purpose of imaging; this will be shown later in this chapter.

Our analysis of the SAR radiation pattern in Chapter 3 showed that the radar transmit-receive mode radiation pattern may also be expressed in terms of an amplitude function and a phase function, as shown in the following:

$$h(\omega, x, y-u) = \underbrace{a(\omega, x, y-u)}_{\text{Amplitude function}} \underbrace{\exp\left[-j2k\sqrt{x^2 + (y-u)^2}\right]}_{\text{Spherical PM signal}}.$$

When the radar antenna length is zero (an *ideal* transmitting/receiving element), that is, $D_y = 0$, the radiation pattern is *omni-directional*, and the amplitude function takes on the form

$$a(\omega, x, y) = \frac{1}{r^2},$$

where $r = \sqrt{x^2 + y^2}$. We also showed the following relationship between the amplitude functions in the slow-time domain and slow-time Doppler domain:

$$a(\omega, x, y) \approx \frac{1}{r^2} A\left(\omega, \frac{2ky}{r}\right).$$

As we stated in Chapter 3, for notational simplicity, we do not carry the amplitude function $1/r^2$ in our analysis.

The radar beamwidth is approximated by the main lobe of the radar transmit-receive amplitude pattern $a(x, y, \omega)$ in the cross-range y domain. We examined the beamwidth of the radar as a function of the radar parameters (fast-time frequency, diameter, radar type, etc.) in Chapter 3. Our study indicated that both planar and parabolic radars exhibit a cone-shaped diverging beam pattern with an axial angle of $\pm\phi_d$; the resultant half-beamwidth is governed by

$$\begin{aligned} B &= r \sin \phi_d \\ &= x \tan \phi_d. \end{aligned}$$

However, the divergence angle of the two radar types show different dependency on the radar diameter and frequency:

$$\phi_d = \begin{cases} \arcsin\left(\dfrac{\lambda}{D_y}\right), & \text{planar radar,} \\ \arctan\left(\dfrac{D_y}{2x_f}\right), & \text{curved radar.} \end{cases}$$

(In the above, x_f is the focal range of the curved radar.) For the planar radar when $\lambda \ll D_y$, the half-beamwidth can be approximated via

$$B \approx \frac{r\lambda}{D_y},$$

which is linearly increasing with the range and the wavelength. For a curved radar, the beamwidth is *invariant* in the radar frequency.

A given reflector at (x, y) is *observable* to the radar in the following interval in the synthetic aperture domain:

$$u \in [y - B, y + B].$$

Suppose that we are interested in imaging a target area within the cross-range gate

$$y \in [-Y_0, Y_0],$$

where Y_0 is a chosen constant. Then the synthetic aperture interval over which the stripmap SAR data contains contributions from this cross-range gate is

$$u \in [-L, L],$$

where

$$L = B + Y_0.$$

(Note that B and, thus, L vary with the target range x and the divergence angle ϕ_d of the radar.) This is demonstrated in Figure 6.1.

We denote the *effective* radiated target area in the cross-range domain by $[-Y_e, Y_e]$, where (see Figure 6.1)

$$\begin{aligned} Y_e &= L + Y_0 \\ &= 2B + Y_0; \end{aligned}$$

Y_e is also an increasing function of x and ϕ_d. The synthetic aperture interval $[-L, L]$ also contains contributions from the targets located in the cross-range region

$$Y_0 < |y| \leq Y_e,$$

that reside outside the target area of interest. These, as we will see, have interesting implications on the functional properties of the stripmap SAR signal and its digital signal processing.

Stripmap SAR Signal Model

We denote the recorded stripmap SAR signal with $\mathbf{s}(t, u)$, and its one-dimensional Fourier transform with respect to the fast-time t by $s(\omega, u)$. We identify the target area with an infinite or finite set of stationary targets which are located at the coordinates (x_n, y_n) $(n = 1, 2, \ldots)$. The SAR signal at the synthetic aperture position (slow-time) u and the fast-time frequency ω is

$$
\begin{aligned}
s(\omega, u) &= P(\omega) \sum_n a_n(\omega, x_n, y_n - u) h(\omega, x_n, y_n - u) \\
&= P(\omega) \sum_n a_n(\omega, x_n, y_n - u) a(\omega, x_n, y_n - u) \\
&\quad \times \exp\left[-j2k\sqrt{x_n^2 + (y_n - u)^2} \right],
\end{aligned}
$$

where $a_n(\cdot)$ is the nth target amplitude pattern.

We refer to the above equation as the stripmap SAR signal system model. In the above model the reference zero range, that is, the line $x = 0$, is chosen to be the aircraft flight path. As we mentioned earlier, the observability of the nth target by the radar depends on the radar amplitude pattern $a(\omega, x_n, y_n - u)$ [as well as the nth target amplitude pattern $a_n(\omega, x_n, y_n - u)$]. As is shown in Figure 6.2, if the nth target is an omni-directional reflector [i.e., $a_n(\omega, x_n, y_n - u) = \sigma_n$, a constant], then the nth target is observable to the radar within the following synthetic aperture interval at a given fast-time frequency ω:

$$
u \in [y_n - B_n, y_n + B_n]
$$

where

$$
B_n = x_n \tan \phi_d,
$$

is the radar half-beamwidth at the range x_n and the fast-time frequency ω, and ϕ_d is the divergence angle of the radar. (See the discussion on the radar radiation pattern in the previous section.)

The nth target aspect angle with respect to the radar as a function of the slow-time u is defined via

$$
\theta_n(u) = \arctan\left(\frac{y_n - u}{x_n} \right).
$$

Thus the aspect angle interval over which the nth target is observable to the radar is

$$
\begin{aligned}
[\theta_n(y_n + B_n), \theta_n(y_n - B_n)] &= \left[-\arctan\left(\frac{B_n}{x_n} \right), \arctan\left(\frac{B_n}{x_n} \right) \right] \\
&= [-\phi_d, \phi_d],
\end{aligned}
$$

which is as expected.

- In stripmap SAR, the interval of the synthetic aperture over which a target is observable to the radar (e.g., $u \in [y_n - B_n, y_n + B_n]$ for the nth target) varies

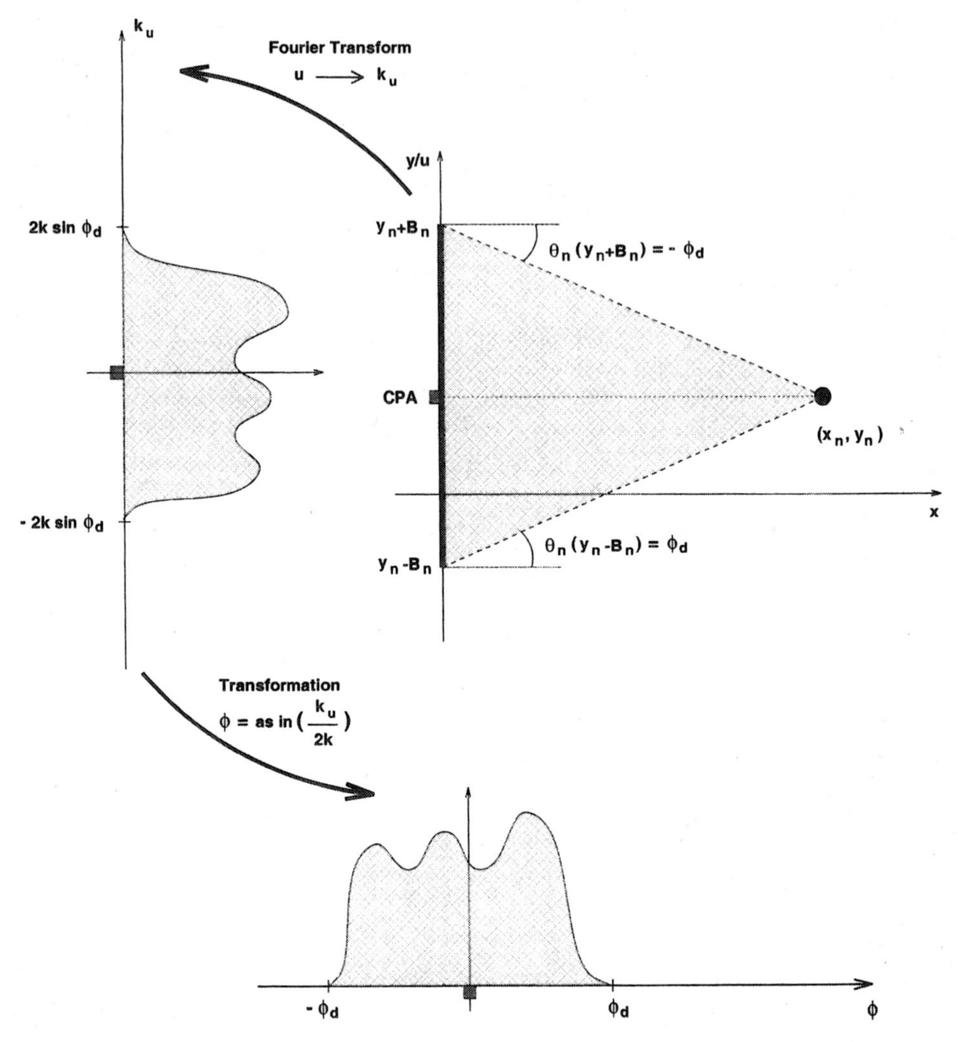

Figure 6.2. Aspect angles of a target in stripmap SAR.

with the coordinates of the target as well as the radar parameters. However, all targets in the imaging scene are observable to the radar within the aspect angle interval $[-\phi_d, \phi_d]$ which depends only on the radar parameters and is invariant in the target coordinates.

The above conclusions exhibit features of the stripmap SAR system that are not observed in the spotlight SAR systems; see Chapter 5.

6.2 RECONSTRUCTION

The stripmap SAR signal model belongs to the class of SAR signals which we referred to as Type 2 in Section 2.9. Using the principles of SAR radiation pattern and Fourier properties of this class of AM-PM signals, the Fourier transform of the above stripmap SAR signal with respect to the slow-time u is

$$\begin{aligned} S(\omega, k_u) &= P(\omega) \sum_n A_n(\omega, k_u) A(\omega, k_u) \exp\left(-j\sqrt{4k^2 - k_u^2}\, x_n - jk_u y_n\right) \\ &= P(\omega) \sum_n A_n(\omega, k_u) A(\omega, k_u) \exp[-jk_x(\omega, k_u)x_n - jk_y(\omega, k_u)y_n], \end{aligned}$$

where

$$\begin{aligned} k_x(\omega, k_u) &= \sqrt{4k^2 - k_u^2}, \\ k_y(\omega, k_u) &= k_u. \end{aligned}$$

Note that there is no rectangular window $I_n(\omega, k_u)$ in the above slow-time and fast-time Fourier domain SAR signature of the nth target. As we mentioned in Chapter 5, the signal $I_n(\cdot)$, which depended on the *target coordinates with respect to the radar*, was the rectangular window that identified the slow-time Doppler band over which the nth target spotlight SAR signature was observable. For the stripmap SAR, the support band of the nth target SAR signature in the frequency domain is dictated by the radar transmit-receive mode amplitude pattern $A(\cdot)$ and the target amplitude pattern $A_n(\cdot)$. Thus the *target coordinates do not affect its stripmap SAR Doppler band.*

For the reconstruction we identify the target function in the spatial frequency domain via the following fast-time and slow-time matched-filtered inversion for the above stripmap SAR model in the (ω, k_u) domain:

$$\begin{aligned} F[k_x(\omega, k_u), k_y(\omega, k_u)] &= P^*(\omega) A^*(\omega, k_u) S(\omega, k_u) \\ &= |P(\omega) A(\omega, k_u)|^2 \sum_n A_n(\omega, k_u) \\ &\quad \times \exp[-jk_x(\omega, k_u)x_n - jk_y(\omega, k_u)y_n]. \end{aligned}$$

The reconstructed target function in the spatial domain is

$$f(x, y) = \sum_n f_n(x - x_n, y - y_n),$$

where

$$f_n(x, y) = \mathcal{F}^{-1}_{(k_x, k_y)}\left[F_n(k_x, k_y)\right]$$

and

$$F_n\left[k_x(\omega, k_u), k_y(\omega, k_u)\right] = |P(\omega)A(\omega, k_u)|^2 A_n(\omega, k_u).$$

The nth target function in the spatial frequency domain, that is, $F_n(\cdot)$, is the mapping of $A_n(\cdot)$ which is windowed by $|P(\cdot)A(\cdot)|^2$ from the (ω, k_u) domain to the (k_x, k_y) domain. The nth target function $f_n(x, y)$ is the *point spread function* of the stripmap SAR imaging system for the nth target.

6.3 BANDWIDTH OF STRIPMAP SAR SIGNAL

Next we study the spectral properties and two-dimensional bandwidth of the stripmap SAR signal. This study reveals the roles of the radar parameters (its type, fast-time band, physical dimensions, etc.) and the coordinates of a target in forming the signal subspace of the target stripmap SAR signature. The results will be used in the future sections to determine the Nyquist sampling rate for alias-free processing of the stripmap SAR signal. Moreover the manner in which a target appears in a stripmap SAR image (i.e., its point spread function) depends on the two-dimensional spectral support of the stripmap SAR signature of the target; this is also a measure of the resolution of the stripmap SAR system.

Since we are only concerned with the spectral support of the stripmap SAR signal, we can assume that the target amplitude pattern is invariant in the fast-time frequency and aspect angle domain, that is, for the nth target, $a_n(\omega, x_n, y_n - u) = \sigma_n$, where σ_n is a constant. Our study includes two separate analyses for the spectral support of the stripmap SAR signal: one for the case of a curved radar aperture; the other for the case of a planar radar aperture. These separate analyses are particularly useful for wide-bandwidth stripmap SAR systems.

Our approach is based on the observation we made in the previous section on the observability of a target by a stripmap radar:

- In stripmap SAR, the interval of the synthetic aperture over which a target is observable to the radar (e.g., $u \in [y_n - B_n, y_n + B_n]$ for the nth target) varies with the coordinates of the target as well as the radar parameters. However, all targets in the imaging scene are observable to the radar within the aspect angle interval $[-\phi_d, \phi_d]$ which depends only on the radar parameters and is invariant in the target coordinates.

This implies that at a fixed fast-time frequency ω, the slow-time Doppler support of the nth target signature in the *angular* slow-time Doppler $\phi = \arcsin(k_u/2k)$ domain is

$$\phi \in [-\phi_d, \phi_d];$$

this is demonstrated in Figure 6.2. This phenomenon can also be observed in our slow-time Fourier transform of the nth target signature. The stripmap SAR signature of the nth target in the (ω, u) domain is

$$\begin{aligned} s_n(\omega, u) &= P(\omega) a_n(\omega, x_n, y_n - u) a(\omega, x_n, y_n - u) \exp\left[-j2k\sqrt{x_n^2 + (y_n - u)^2}\right] \\ &= P(\omega) \sigma_n a(\omega, x_n, y_n - u) \exp\left[-j2k\sqrt{x_n^2 + (y_n - u)^2}\right]. \end{aligned}$$

The Fourier transform of this signal with respect to the slow-time is

$$\begin{aligned} S_n(\omega, k_u) &= P(\omega) A_n(\omega, k_u) A(\omega, k_u) \exp\left(-j\sqrt{4k^2 - k_u^2}\, x_n - jk_u y_n\right) \\ &= P(\omega) \sigma_n A(\omega, k_u) \exp\left(-j\sqrt{4k^2 - k_u^2}\, x_n - jk_u y_n\right) \end{aligned}$$

Thus the spectral support of $S_n(\omega, k_u)$ in the slow-time Doppler k_u domain is dictated by the support band of $A(\omega, k_u)$ in the k_u domain, which is (see Figure 6.2)

$$k_u \in [-2k \sin\phi_d, 2k \sin\phi_d].$$

This slow-time support band does not vary with the target parameters (coordinates). The other limiting factor on the spectral support of the stripmap SAR signal is the limited fast-time bandwidth of the radar signal, which is

$$\begin{aligned} \omega &\in [\omega_{\min}, \omega_{\max}] \\ &\in [\omega_c - \omega_0, \omega_c + \omega_0]. \end{aligned}$$

We are now ready to examine the two-dimensional spectral support of the stripmap SAR signal for the planar and curved radars.

Planar Radar Aperture

For a radar with a planar aperture, the divergence angle, which varies with the radar fast-time frequency ω, is

$$\phi_d = \arcsin\left(\frac{\lambda}{D_y}\right),$$

where $\lambda = 2\pi c/\omega$ is the wavelength. The support of the radar amplitude pattern in the spatial domain $a(\omega, x, y)$ as a function of the wavelength and the cross-range y is shown in Figure 6.3*a*. Note that the support of the amplitude pattern in the cross-range y domain (i.e., the beamwidth) is (approximately) a linear function of the wavelength.

Figure 6.3*b* shows the slow-time Doppler amplitude pattern of a planar radar $A(\omega, k_u)$ as a function of the fast-time frequency ω and the slow-time Doppler k_u

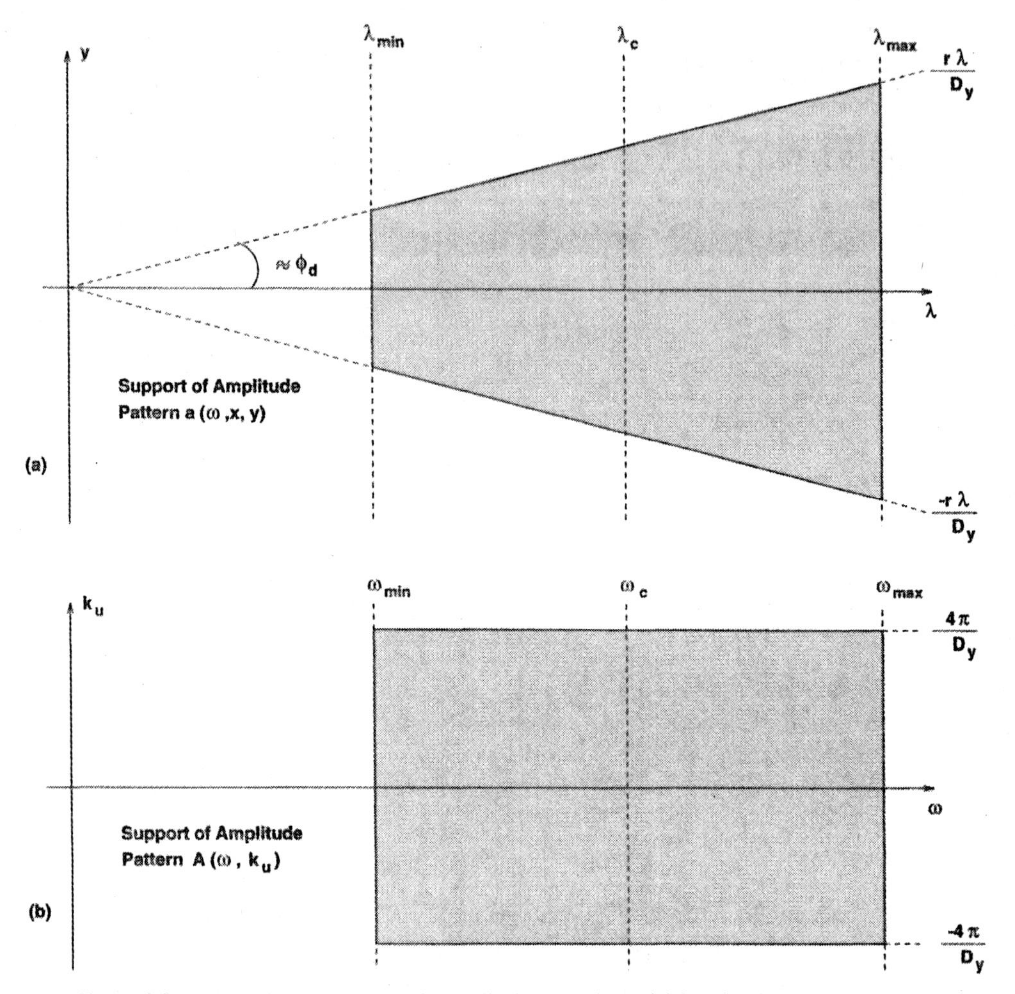

Figure 6.3. Planar radar aperture: (a) radar amplitude pattern in spatial domain $a(\omega, x, y)$ versus wavelength λ and cross-range y; (b) slow-time Doppler amplitude pattern $A(\omega, k_u)$ versus fast-time frequency ω and slow-time Doppler k_u.

domain. In Chapter 3 we showed that this amplitude pattern is a sinc-squared pattern; that is,

$$A(\omega, k_u) = \text{sinc}^2(D_y k_u).$$

Moreover the support of this amplitude pattern in the slow-time Doppler domain can be approximated by the main lobe of the sinc pattern, which is

$$k_u \in \left[\frac{-4\pi}{D_y}, \frac{4\pi}{D_y} \right].$$

Note that this slow-time Doppler support band is invariant in the radar fast-time frequency ω.

Figure 6.3*b* represents the two-dimensional spectral support of a stripmap SAR signal for an arbitrarily-located omni-directional reflector in the imaging scene for a planar radar antenna. The spectral support in Figure 6.3*b* is also the two-dimensional bandwidth of the stripmap SAR signal $\mathbf{s}(t, u)$ for a general target area when a planar radar antenna is used for data collection.

Curved Radar Aperture

In the case of a curved radar aperture, the divergence angle becomes

$$\phi_d = \arctan\left(\frac{D_y}{2x_f}\right),$$

where x_f is the focal range of the radar. This divergence angle does not vary (at least explicitly) with the radar fast-time frequency. Thus the support of the radar amplitude pattern in the spatial domain $a(\omega, x, y)$, which is shown in Figure 6.4*a* as a function of the wavelength and the cross-range y, is invariant in the wavelength (or the radar fast-time frequency ω).

The same is not true for the slow-time Doppler amplitude pattern of a curved radar $A(\omega, k_u)$. Figure 6.4*b* shows this amplitude pattern as a function of the fast-time frequency ω and the slow-time Doppler k_u domain. The slow-time Doppler support band of this amplitude pattern is

$$\begin{aligned} k_u &\in [-2k \sin\phi_d, 2k \sin\phi_d] \\ &\in \left[-2k\frac{D_y}{\sqrt{D_y^2 + 4x_f^2}}, 2k\frac{D_y}{\sqrt{D_y^2 + 4x_f^2}}\right] \end{aligned}$$

This support band is linearly increasing with the wavenumber k (or the radar fast-time frequency ω).

Figure 6.4*b* is the two-dimensional spectral support of a stripmap SAR signal for an arbitrarily located omni-directional reflector in the imaging scene for a curved aperture radar antenna; it is also the two-dimensional bandwidth of the stripmap SAR signal $\mathbf{s}(t, u)$ for a general target area when a curved radar antenna is used for data collection.

6.4 RESOLUTION AND POINT SPREAD FUNCTION

Next we use the two-dimensional spectral analysis of the stripmap SAR signal, which was discussed in the previous section, to develop the resolution and point spread function for the stripmap SAR systems.

Based on the results of Section 6.3, the two-dimensional spectral support of the SAR signature of an omni-directional reflector in the imaging scene is governed by

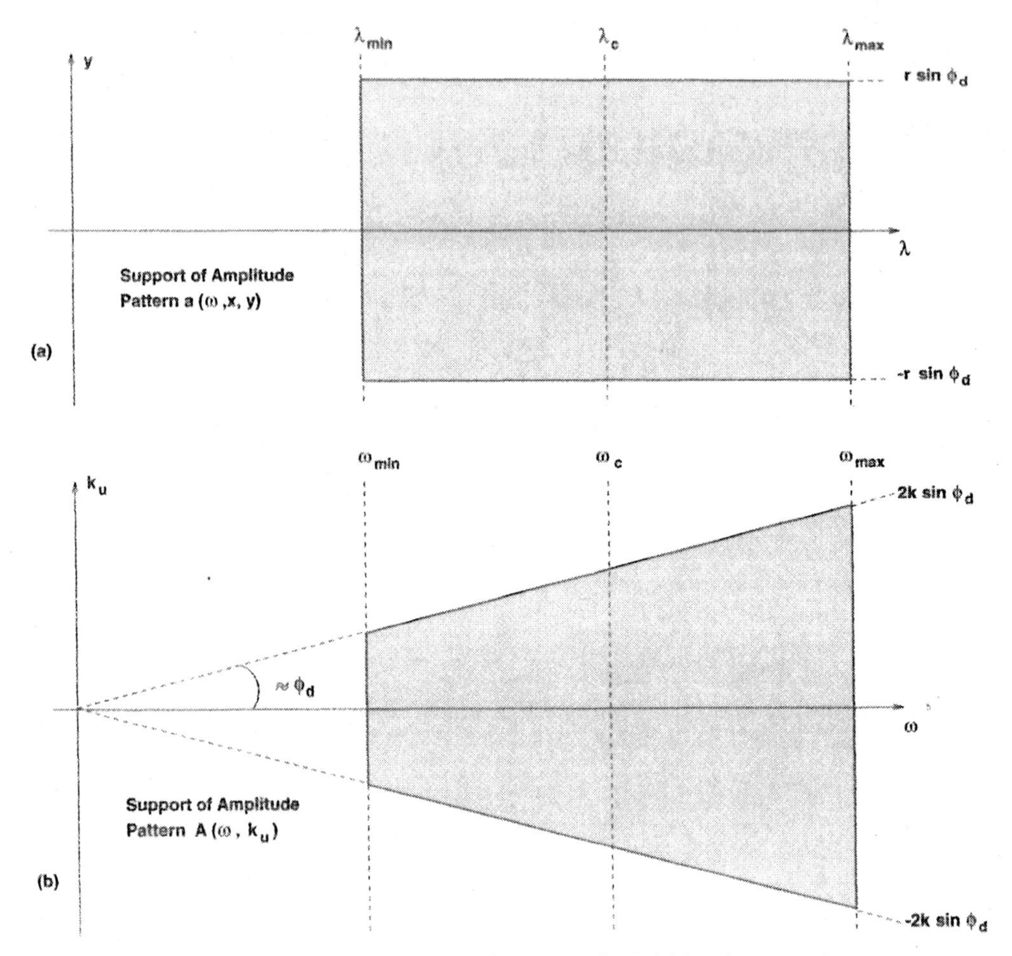

Figure 6.4. Curved radar aperture: (a) radar amplitude pattern in spatial domain $a(\omega, x, y)$ versus wavelength λ and cross-range y; (b) slow-time Doppler amplitude pattern $A(\omega, k_u)$ versus fast-time frequency ω and slow-time Doppler k_u.

the shaded area in Figure 6.3*b* (planar radar) or Figure 6.4*b* (curved radar). Based on the stripmap SAR reconstruction results of Section 6.2, the data within this spectral region are mapped into the target function spatial frequency (spectral) (k_x, k_y) domain via

$$k_x(\omega, k_u) = \sqrt{4k^2 - k_u^2},$$

$$k_y(\omega, k_u) = k_u.$$

Figure 6.5*a* and *b*, respectively, shows this spectral support mapping for a planar and curved radar.

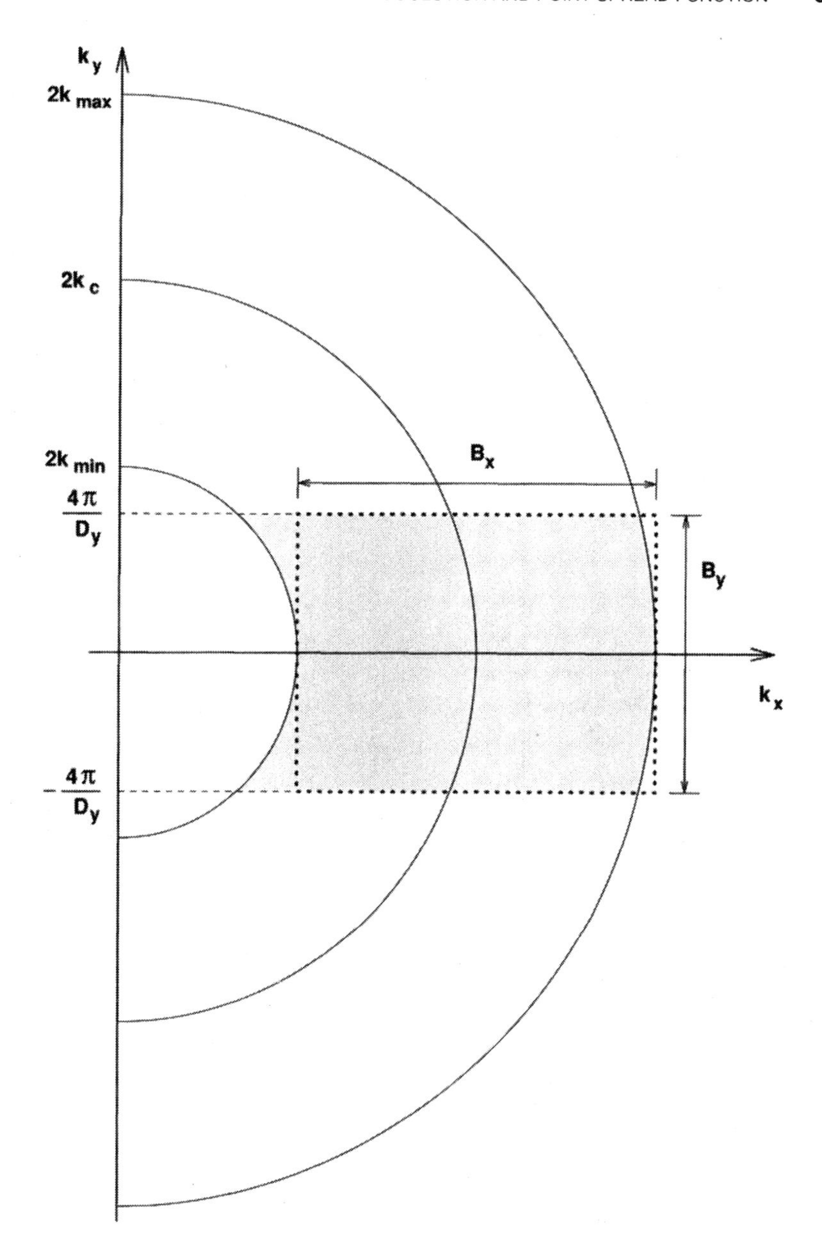

Figure 6.5a. Target spectral support in spatial frequency (k_x, k_y) domain: planar radar aperture.

We should emphasize that Figure 6.5*a* or *b* only represents the *support* region for $F_n(k_x, k_y)$ or the *bandwidth* of the spatial domain target function $f_n(x, y)$; there exist all kinds of phase and amplitude fluctuations within this band which, for example, represent the target location $\exp(-jk_x x_n - jk_y y_n)$, target amplitude pattern $A_n(\omega, k_u)$, and the magnitude-squared of the radar spectrum and amplitude pattern

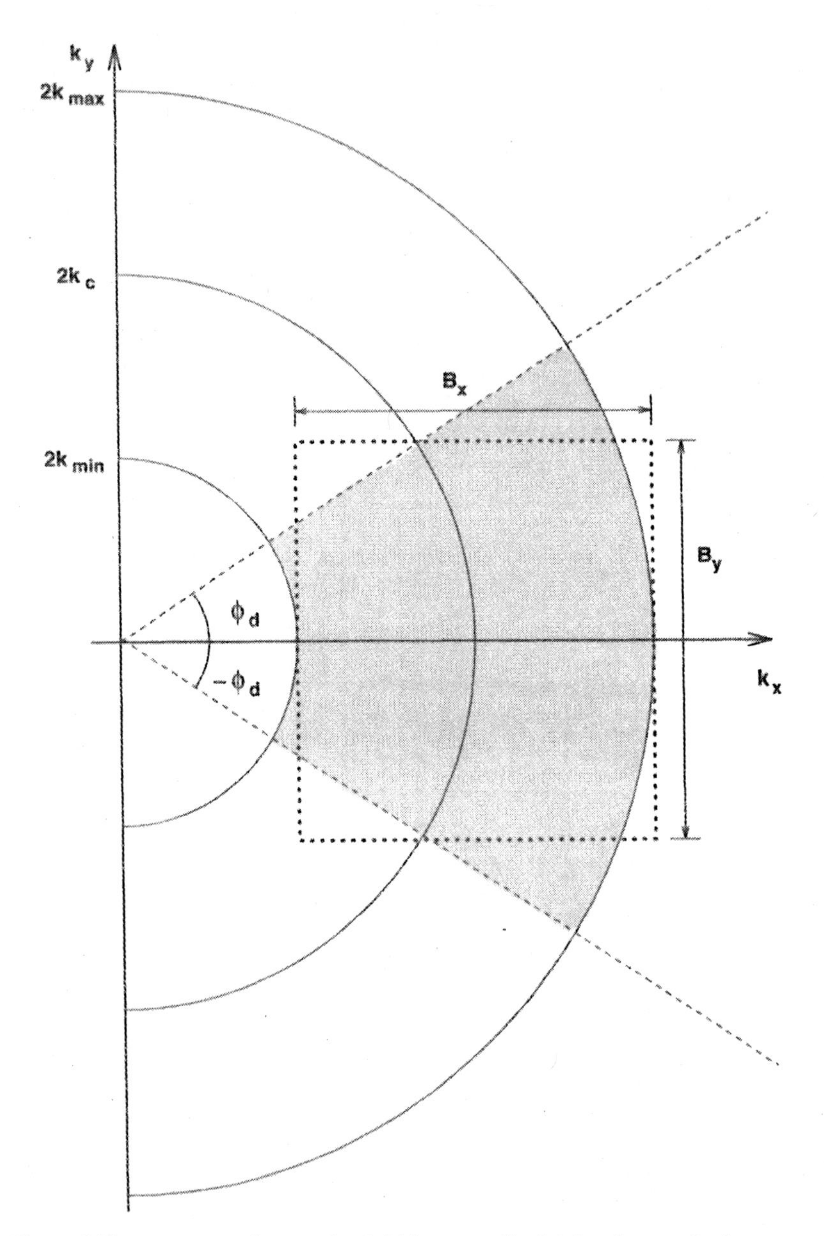

Figure 6.5b. Target spectral support in spatial frequency (k_x, k_y) domain: curved radar aperture.

$|P(\omega)A(\omega, k_u)|^2$ (see the discussion on stripmap SAR reconstruction in Section 6.2). Yet, as was the case of the spotlight SAR, the (k_x, k_y) domain spectral support in Figure 6.5a and b provides a starting point for the user to comprehend the nature of the target point spread function which appears in stripmap SAR images.

The inverse two-dimensional Fourier transform of the shaded area in Figure 6.5*a* and *b* dictates the shape of the point spread function for stripmap SAR systems. To develop an analytical model for the point spread function, one could approximates the target support region in Figure 6.5*a* and *b* via a rectangle in the (k_x, k_y) domain with widths

$$B_x = 2\,(k_{\max} - k_{\min}) \qquad \text{in the } k_x \text{ domain;}$$

$$B_y = \begin{cases} \dfrac{8\pi}{D_y} & \text{for a planar radar} \\ 4k_c \sin \phi_d & \text{for a curved radar} \end{cases} \qquad \text{in the } k_y \text{ domain.}$$

The inverse spatial Fourier transform of the rectangle in Figure 6.5*a* or *b* with respect to (k_x, k_y) is composed of the following separable two-dimensional sinc functions in the (x, y) domain:

$$\operatorname{sinc}\left(\frac{B_x x}{2\pi}\right) \operatorname{sinc}\left(\frac{B_y y}{2\pi}\right).$$

This two-dimensional sinc pattern for the *n*th target is shown as a cross-shaped structure in Figure 6.6.

This two-dimensional sinc pattern represents the point spread function of the stripmap SAR systems for an *ideal* reflector (i.e., the target and radar amplitude pat-

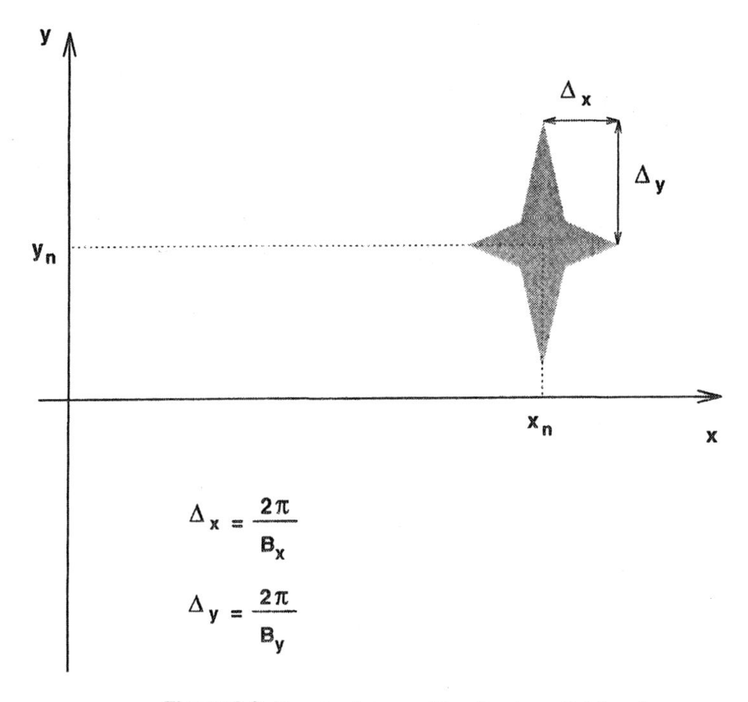

Figure 6.6. Target point spread function in spatial domain.

terns are constants). Unlike the point spread function for the spotlight SAR systems, the point spread function of the stripmap SAR systems is invariant of the coordinates of the target, that is, (x_n, y_n). Thus the ideal point spread function of the stripmap SAR systems is *shift-invariant*.

The main lobes of the two sinc functions in the (x, y) domains are, respectively, within $\pm\Delta_x$ and $\pm\Delta_y$, where

$$\Delta_x = \frac{2\pi}{B_x},$$

$$\Delta_y = \frac{2\pi}{B_y},$$

which after substituting for (B_x, B_y) become

$$\Delta_x = \frac{\pi}{(k_{\max} - k_{\min})} = \frac{\pi c}{2\omega_0},$$

$$\Delta_y = \begin{cases} \dfrac{D_y}{4} & \text{for a planar radar,} \\[2ex] \dfrac{\lambda_c}{4 \sin \phi_d} & \text{for a curved radar.} \end{cases}$$

- The range and cross-range resolution pair in the stripmap SAR systems are invariant in the coordinates of the target.

Based on the discussion of the cross-range resolution of a planar radar in Section 2.9, a more conservative estimate of the cross-range resolution in stripmap SAR that uses a planar radar is

$$\Delta_y = \frac{D_y}{2} \qquad \text{for a planar radar,}$$

which is an increasing function of the radar diameter D_y.

Similarly a more conservative estimate of the cross-range resolution in stripmap SAR which uses a curved radar is

$$\Delta_y = \frac{\lambda_c}{2 \sin \phi_d} \qquad \text{for a curved radar.}$$

Note that in the case of a curved aperture, the divergence angle $\phi_d = \arctan(D_y/2x_f)$ is an increasing function of the radar diameter D_y. Thus, for a curved radar, the cross-range resolution is inversely proportional to the radar diameter D_y.

- For a planar radar, the cross-range resolution Δ_y improves as the radar diameter *decreases*. The opposite is true for a curved radar.

As in the case of spotlight SAR (Section 5.4, Figure 5.6), the above analysis of the point spread function and resolution in the stripmap SAR systems was based on approximating the target spectral support in Figure 6.5*a* or *b* via a rectangular region. This is a good approximation for narrowband stripmap SAR systems where the radar carrier frequency is much larger than its baseband frequency ($\omega_0 \ll \omega_c$), for example, an X-band SAR.

However, in the wide band stripmap SAR systems, approximating the support band in Figure 6.5*a* or *b* via a rectangular region is not a valid approximation. In this case the point spread function cannot be approximated by a two-dimensional sinc pattern, and does not resemble a cross-shaped structure. In fact, for the wideband UHF stripmap SAR systems, the point spread function looks like a *funnel*. One could analytically construct the point spread function or simulate the point spread function numerically via, for example, the Matlab code for the stripmap SAR systems which is discussed at the end of this chapter.

One final note before we close this discussion on the point spread function and resolution of the stripmap SAR systems. We chose the radar beamwidth $2B$ to be the length of the main lobe of the radar radiation pattern. This result may be viewed as too optimistic, since the radar radiation pattern is not a constant (flat) within its main lobe. The actual point spread function of the imaging system in the cross-range domain is related to the inverse Fourier transform of the window function associated with the matched filtering in the k_u domain, that is, $|A(\omega, k_u)|^2$. Clearly the cross-range resolution improves as the support band of $A(\omega, k_u)$ increases in the slow-time Doppler k_u domain.

Our analysis of planar, parabolic, and circular physical radars in Chapter 3 indicated that the slow-time Doppler support band of the amplitude function $A(\omega, k_u)$ increases as the physical radar diameter D_y decreases. In fact we showed earlier that the cross-range resolution improves as the radar diameter D_y decreases, no matter which type of radar (planar or curved) is used.

However, one faces a lower signal-to-noise power ratio as the radar diameter D_y decreases. The choice of D_y is limited by the SNR desired by the user under the environment where the measurements are made. One way to improve the SNR is to utilize *large bandwidth-continuous wave*, e.g., FM-CW, signaling [s94; s97n; s98j]. (The FM-CW radars have their own difficulties, for example, coupling of the transmitter and receiver lines.)

6.5 DATA ACQUISITION AND SIGNAL PROCESSING

We now examine the constraints on the manner in which the stripmap SAR data are collected based on the radar system parameters. This analysis benefits from our discussion of the two-dimensional spectral properties of the stripmap SAR signal in the previous section. Our study also includes preprocessing methods, for example, digital spotlighting to suppress slow-time aliasing caused by the radar side lobes in order to improve the quality of the reconstructed stripmap SAR images.

The reader will notice that some of the principles that govern the stripmap SAR signal processing are identical to those of the spotlight SAR signal in Section 5.5. These are repeated here to provide a complete and independent analysis of the stripmap SAR digital signal processing.

Fast-time Domain Sampling and Processing

Fast-time Interval of Sampling The fast-time samples should be collected over the fast-time interval that covers the returns from all of the reflectors within the radar swath. Since there is no analog spotlighting, the radar swath remains unchanged during the slow-time (synthetic aperture) data acquisition.

Suppose that the closest and farthest radial range distances of the radar swath are $r_{\min}$ and $r_{\max}$, respectively. Then the first echoed signal arrives at the receiving radar at the fast-time

$$T_s = \frac{2r_{\min}}{c}.$$

Moreover the echoed signal from the farthest reflector in the range swath terminates at

$$T_f = \frac{2r_{\max}}{c} + T_p,$$

where T_p is the duration of the transmitted pulsed radar signal $p(t)$.

Thus the fast-time measurements of the stripmap SAR signal $\mathbf{s}(t, u)$ should be made within the following fast-time interval:

$$t \in [T_s, T_f].$$

The above fast-time interval for sampling the SAR signal, which corresponds to the echoed signals from the points within the radial range $[r_{\min}, r_{\max}]$, is also called the *range gate* of the SAR system A/D fast-time sampler.

One can determine the range gate $[r_{\min}, r_{\max}]$ from the knowledge of the stripmap SAR system parameters and, in particular, the beamwidths of the radar in the range and cross-range domains. However, if a planar radar is used, these beamwidths vary with the radar fast-time frequency ω. (Note that the range beamwidth actually corresponds to the radar beamwidth in the slant range domain; see Chapter 3.)

To present a general expression for the range gate, we denote the largest range swath (which occurs at the smallest fast-time frequency $\omega_{\min}$) by

$$[X_c - X_{0\max}, X_c + X_{0\max}],$$

where X_c is the midpoint of the range swath and $2X_{0\max}$ is the maximum size of the range swath. For a planar radar with *effective* diameter D_x in the range domain, the half-width of the maximum range swath is

$$X_{0\max} \approx \frac{X_c \lambda_{\max}}{D_x}.$$

Moreover we denote the largest half-beamwidth in the cross range domain via

$$[-B_{\max}, B_{\max}].$$

For instance, for a planar radar with diameter D_y in the cross-range, the maximum half-width cross-range beamwidth is

$$B_{\max} \approx \frac{(X_c + X_{0\max})\,\lambda_{\max}}{D_y}.$$

We can now use the maximum radar footprint size to determine the range gate. The closest radial distance of the radar to the illuminated target area is

$$r_{\min} = X_c - X_{0\max}.$$

The farthest radial distance of the radar to this target area is

$$r_{\max} = \sqrt{(X_c + X_{0\max})^2 + B_{\max}^2}.$$

(The above expression for $r_{\max}$ is a conservative one. The actual $r_{\max}$ would be less, because of the curvature of the radar radiation pattern.)

Figure 6.7*a* is the (t, u) domain data for the FOPEN stripmap SAR system (the first run data) within a $u \in [-200, 200]$ m synthetic aperture interval (Figures 6.1 and 6.2). We examined this database within a smaller synthetic aperture interval, $u \in [-50, 50]$, in Chapter 5; this was done to show that a stripmap SAR signal within a relatively small synthetic aperture exhibits the functional properties of a spotlight SAR signal (i.e., SAR signal of Type 1).

The current stripmap SAR database with a larger synthetic aperture is now being investigated to show the properties of Type 2 SAR signals. The (t, u) domain SAR data in Figure 6.7*a* are fast-time matched filtered and digitally spotlighted (via the algorithm to be discussed) in order to provide a clearer picture of the stripmap SAR signatures of the targets in the imaging scene.

We should point out that the stripmap SAR data correspond to the first-run data of the radar-carrying aircraft. In our future discussion we will also provide the results of processing the stripmap SAR data which were obtained at the second run of the radar-carrying aircraft over the synthetic aperture interval of $u \in [-391, 391]$ m. The (t, u) domain data for the second run are shown in Figure 6.7*b* after applying the digital spotlighting method discussed below. An interesting feature of the second-run database in Figure 6.7*b* is that it contains what appear to be random streaks (noise) in the fast-time domain; these streaks appear more severe in the original data (prior to digital spotlighting and upsampling) which are not shown. The source of these streaks are believed to be either radio frequency interference or electronic problems with the A/D converter.

Fast-time Sample Spacing Since the baseband bandwidth of the radar signal is $\pm\omega_0$, the A/D fast-time sample spacing should satisfy the following Nyquist criterion:

$$\Delta_t \leq \frac{\pi}{\omega_0}.$$

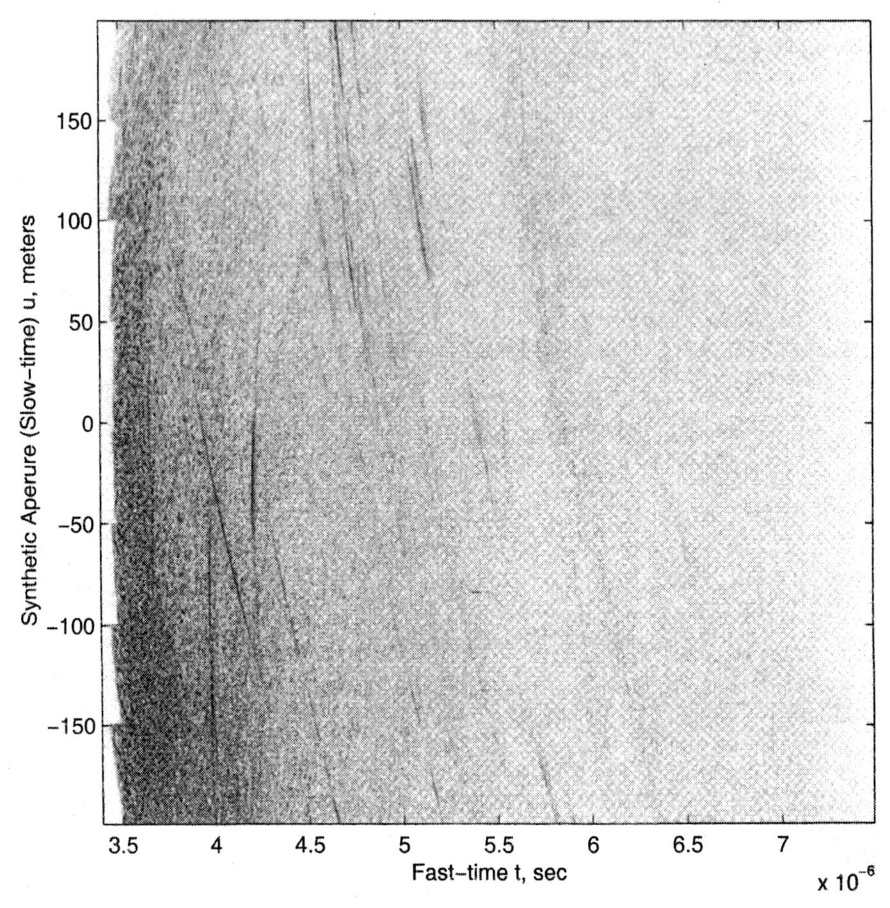

Figure 6.7a. Measured UHF band stripmap SAR data of foliage area and human-made targets in (t, u) domain (after digital spotlighting): first run.

As we mentioned in Section 1.6, in the case of a chirp radar signal, the Nyquist constraint for fast-time domain sampling could be less restrictive for the deramped echoed signal. In such scenarios the upsampling scheme, which was described in Section 1.6, could be utilized to recover the alias-free echoed signal.

Finally the required number of fast-time samples within the fast-time gate $t \in [T_s, T_f]$ is

$$N = 2\left\lceil \frac{T_f - T_s}{2\Delta_t} \right\rceil,$$

where $\lceil a \rceil$ denotes the smallest integer which is larger than a. Since the above operation may increase the duration of the fast-time gate, one should fix the starting point

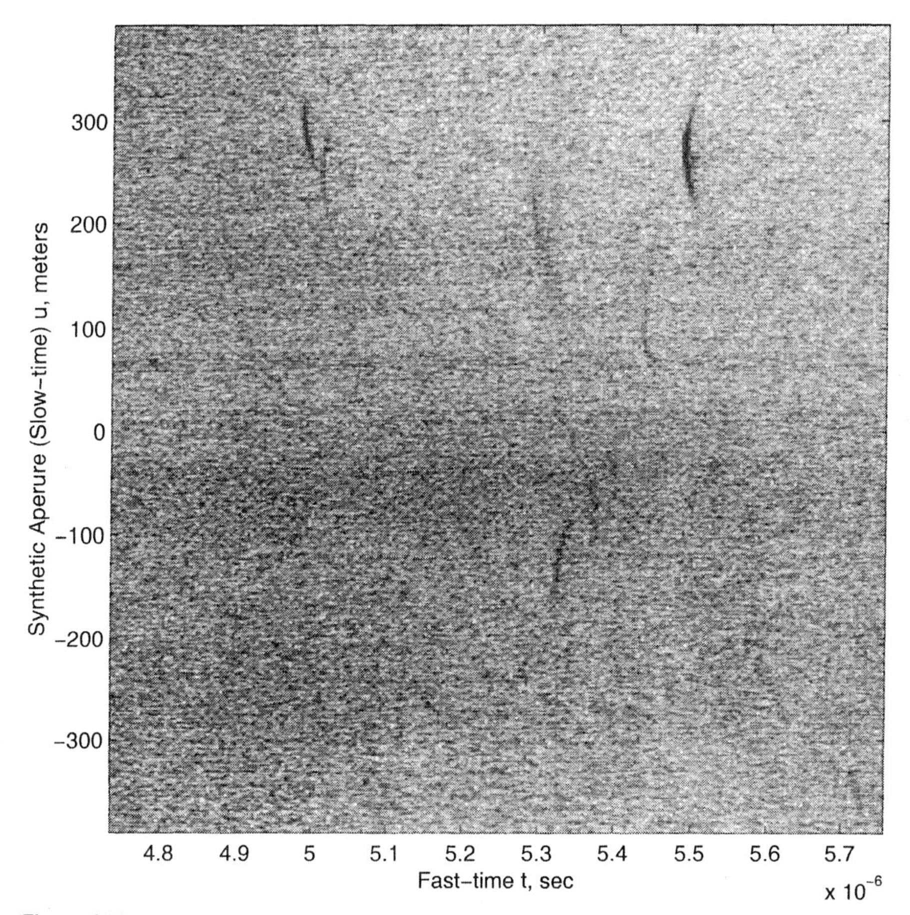

Figure 6.7b. Measured UHF band stripmap SAR data of foliage area and human-made targets in (t, u) domain (after digital spotlighting): second run.

of the time gate T_s and redefine the ending point of the time gate via

$$T_f = T_s + (N - 1)\Delta_t.$$

The other option is to fix the ending point of the time gate T_f and redefine the starting point of the time gate via

$$T_s = T_f - (N - 1)\Delta_t.$$

Fast-time Reference Point for Matched Filtering In our discussion of matched filtering for range imaging in Chapter 1, we pointed out that the matched filter is usually selected based on the echoed signal from the center of the target area. In SAR imaging the fast-time matched filter is commonly chosen to be the echoed signal from the center of the range swath [i.e., $(x, y) = (X_c, 0)$]. Moreover the distance of the radar (at any synthetic aperture position u) from the center of the range swath is X_c; in the stripmap SAR this distance is invariant in the coordinates of the radar $(0, u)$.

The echoed signal from a unit reflector at the center of the range swath is

$$p_0(t) = p(t - T_c),$$

where

$$T_c = \frac{2X_c}{c}$$

is called the reference fast-time point. In this case, the fast-time matched-filtered SAR signal is formed via

$$\mathbf{s}_M(t, u) = \mathbf{s}(t, u) * p_0^*(-t),$$

where $*$ denotes convolution in the fast-time domain, and $p_0^*(-t)$ is the complex conjugate of $p_0(-t)$.

The user, however, may select other values for the reference fast-time point T_c; the choice usually depends on the relative values of the range swath and the pulse duration. For instance, the user could choose the reference fast-time point to be the midpoint of the fast-time gate; that is,

$$T_c = \frac{T_f - T_s}{2}.$$

After choosing a reference fast-time point T_c and performing the above fast-time matched filtering with $p_0^*(-t)$, the fast-time sample number $(N/2) + 1$ (i.e., the sample in the middle of the fast-time array) in the matched-filtered signal $\mathbf{s}_M(t, u)$ corresponds to the fast-time point $t = T_c$. We will examine the implication of this in the stripmap SAR digital reconstruction algorithms.

Slow-time Domain Sampling and Processing

In Section 6.3 we showed that the slow-time Doppler support of the stripmap SAR signal $s(\omega, u)$ at a given fast-time frequency is

$$k_u \in [-2k \sin \phi_d, 2k \sin \phi_d];$$

this was also demonstrated in Figure 6.2. We used this to obtain the two-dimensional spectral supports in Figures 6.3*b* and 6.4*b* for the planar and curved radar apertures, respectively.

Thus, at a given fast-time frequency ω, the Nyquist sampling constraint in the synthetic aperture (slow-time) u domain of the stripmap SAR signal $s(\omega, u)$ is

$$\Delta_u \le \frac{2\pi}{4k \sin \phi_d}$$
$$= \frac{\lambda}{4 \sin \phi_d}.$$

- For an omni-directional radar (i.e., $D_y = 0$) the divergence angle is $\phi_d = 90°$. In this case we have

$$\Delta_u \le \frac{\lambda}{4},$$

which is a familiar constraint that is used in most active array processing systems. Clearly, the *ideal* omni-directional radar, which has the cross-range resolution of $\Delta_y = \lambda/4$, with a *nonzero* power, does not exist in practice.

To select a slow-time sample spacing which is suitable at all the available fast-time frequencies $\omega \in [\omega_{\min}, \omega_{\max}]$, the above constraint must be satisfied for the worst case. This issue is discussed next for the planar and curved radar apertures.

Planar Radar Aperture In the case of a planar radar aperture where

$$\phi_d = \arcsin\left(\frac{\lambda}{D_y}\right),$$

the slow-time Nyquist sampling criterion becomes

$$\Delta_u \le \frac{\lambda}{4\lambda/D_y}$$
$$= \frac{D_y}{4}.$$

Note that this constraint is invariant in the radar fast-time frequency; this fact is also evident from the two-dimensional spectral support of the stripmap SAR signal for the planar radars in Figure 6.3*b*.

- Reminder (Chapter 5): In spotlight SAR, the Nyquist sample spacing for the slow-time compressed spotlight SAR signal $s_c(\omega, u)$ is

$$\Delta_{uc} \le \frac{X_c\lambda}{4Y_0}$$
$$= \frac{D_y}{4} \qquad \text{for a planar radar,}$$

where $[-Y_0, Y_0]$ is the spotlighted target area in the cross-range domain. Clearly this spotlighted area is the beamwidth of the radar, that is, $Y_0 = B$. Thus the slow-time Nyquist sampling rate for the *slow-time compressed spotlight SAR signal* $s_c(\omega, u)$ is the same as the slow-time Nyquist rate for the *stripmap SAR signal* $s(\omega, u)$ for the planar radar apertures.

Curved Radar Aperture For a curved radar aperture the divergence angle, that is,

$$\phi_d = \arctan\left(\frac{D_y}{2x_f}\right),$$

is invariant in the radar fast-time frequency. Thus the worst case for the slow-time sample spacing occurs at the highest fast-time frequency, $\omega = \omega_{\max}$; see Figure 6.4*b*. In this case the Nyquist constraint in the slow-time domain of the stripmap SAR signal of a curved radar is

$$\begin{aligned}\Delta_u &\le \frac{2\pi}{4k_{\max}\sin\phi_d} \\ &= \frac{\lambda_{\min}}{4\sin\phi_d}.\end{aligned}$$

Finally we note that the radar *pulse repetition interval* (PRI) is governed by

$$\text{PRI} = \frac{\Delta_u}{v_r},$$

where v_r is the speed of the radar-carrying aircraft.

Baseband Slow-time Processing The stripmap SAR signal is a lowpass signal in the slow-time Doppler k_u domain in the case of both planar and curved radar apertures; see Figures 6.3*b* and 6.4*b*. Thus there is no need for baseband conversion in stripmap SAR signal processing.

Figure 6.8 shows the two-dimensional spectrum of the stripmap SAR data of Figure 6.7*a*. The spectrum does not appear to possess a guard band (a valley) at the absolute highest slow-time Doppler $|k_u|$; the spectrum is fairly uniform for $k_u \in [-8, 8]$ rad/m. This indicates that this stripmap SAR database is aliased in the slow-time Doppler k_u domain. We will examine methods to reduce some of the slow-time Doppler aliasing in this stripmap SAR database.

Slow-time Compression and Processing

Full Synthetic Aperture Data $u \in [-L, L]$ Consider the slow-time compressed version of the stripmap SAR signal

$$s_c(\omega, u) = s(\omega, u)s_0^*(\omega, u),$$

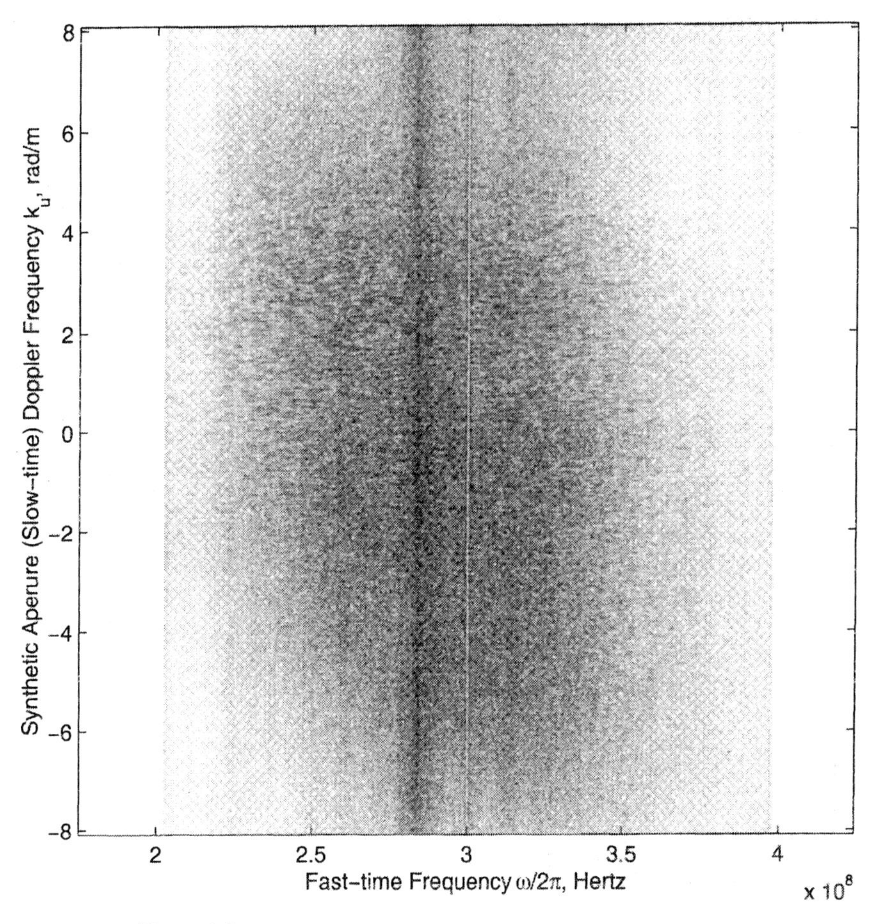

Figure 6.8. Measured SAR data (first run) spectrum in (ω, k_u) domain.

where

$$s_0(\omega, u) = \exp\left(-j2k\sqrt{X_c^2 + u^2}\right)$$

is the reference SAR signal in the (ω, u) domain. In Chapter 5 we used the slow-time compressed spotlight SAR signal to reduce the pulse repetition frequency (PRF) of the radar. In fact we showed that the slow-time Nyquist sample spacing for $s_c(\omega, u)$ is less restrictive than the slow-time Nyquist sample spacing for $s(\omega, u)$ in the spotlight SAR systems; that is,

$$\Delta_{uc} < \Delta_u \qquad \text{in spotlight SAR.}$$

This signal also had utility in digital spotlighting the measured data in the spotlight SAR systems to reduce slow-time Doppler aliasing which is caused by the radar radiation pattern side lobes.

The slow-time compressed SAR signal in the stripmap SAR systems may also be used for digital spotlighting and enhancement of the reconstructed images. However, the slow-time compressed SAR signal in the stripmap SAR systems has different functional properties than the ones we developed for the spotlight SAR systems. Thus, before we study digital spotlighting in stripmap SAR systems, we consider the spectral properties of $s_c(\omega, u)$ in these SAR systems.

Using the results of Section 2.7, in stripmap SAR systems, the two-dimensional spectral domain representation of the slow-time compressed SAR signal is

$$S_c(\omega, k_u) \approx \sum_n \sigma_n \exp[-j2k(r_n - X_c)] \operatorname{sinc}\left[\frac{B_n}{\pi}\left(k_u - \frac{2ky_n}{r_n}\right)\right],$$

where $r_n = \sqrt{x_n^2 + y_n^2}$. Note that in the above we used the fact that the nth target is observable to the radar (at a given fast-time frequency ω) within the following synthetic aperture interval

$$u \in [y_n - B_n, y_n + B_n],$$

where $B_n = x_n \tan \phi_d$.

- The above approximation is valid for $|u| \ll X_c$. This is not the case for the wideband stripmap SAR systems. As we will show later in our discussion, we will utilize *subaperture* processing to deal with this problem.

Moreover, in Section 6.1, we showed that the effective radiated target area in the cross-range domain is $[-Y_e, Y_e]$, where (see Figure 6.1)

$$\begin{aligned} Y_e &= L + Y_0 \\ &= 2B + Y_0, \end{aligned}$$

where $B = x \tan \phi_d$ is the half-beamwidth at a given range x that may also vary with the fast-time frequency ω.

One may present a fairly accurate analysis of the spectral properties of the slow-time compressed SAR signal based on the type of radar used. However, this would involve introducing a whole set of new subscripts for the above parameters, that is, B, Y_e, and so on. One can show that a conservative measure of the slow-time Doppler bandwidth of $S_c(\omega, k_u)$ in the stripmap SAR systems is

$$k_u \in [-4k \sin \phi_d - 2k \sin \theta_{0\,\max}, 4k \sin \phi_d + 2k \sin \theta_{0\,\max}],$$

where

$$\theta_{0\,\max} = \arctan\left(\frac{Y_0}{x_{\min}}\right)$$

is the maximum aspect angle of the desired target area when the radar is at $u = 0$. The above slow-time Doppler band is larger than the slow-time Doppler bandwidth of the SAR signal $S(\omega, k_u)$ which was shown to be (see Section 6.3)

$$k_u \in [-2k \sin \phi_d, 2k \sin \phi_d] .$$

- In spotlight SAR, slow-time compression results in the bandwidth reduction of the SAR signal in the slow-time Doppler k_u domain. The same is not true in stripmap SAR.

At a given fast-time frequency ω, the Nyquist sampling rate for the slow-time compressed stripmap SAR signal is

$$\begin{aligned} \Delta_{uc} &= \frac{\pi}{4k \sin \phi_d + 2k \sin \theta_{0\max}} \\ &= \frac{\lambda}{8 \sin \phi_d + 4 \sin \theta_{0\max}}, \end{aligned}$$

which is more restrictive than the Nyquist slow-time sample spacing of the stripmap SAR signal which is

$$\begin{aligned} \Delta_u &= \frac{\pi}{2k \sin \phi_d} \\ &= \frac{\lambda}{4 \sin \phi_d}. \end{aligned}$$

- Slow-time compression is used in spotlight SAR to achieve a *less restrictive sampling rate* in acquiring the SAR signal in the slow-time domain; this is not the case in the stripmap SAR systems. We will use the slow-time compression in the stripmap SAR systems as a signal processing tool to *digitally spotlight* a desired target area.
- The slow-time Nyquist sample spacing which is used to *measure* the SAR signal in both the spotlight SAR and stripmap SAR systems is dictated by the radar beamwidth; this sample spacing is

  $$\Delta_u \le \frac{\lambda_{\min}}{4 \sin \phi_d}.$$

  The resultant measured data in the spotlight SAR systems are *aliased* and must be upsampled prior to its processing for image formation. In the stripmap SAR systems the measured data are *not aliased*; however, the data must be *upsampled* prior to slow-time compression.

Figure 6.9 shows the two-dimensional spectrum of the slow-time compressed SAR signal $S_c(\omega, k_u)$ for the stripmap SAR data of Figure 6.7a. Similar to the two-

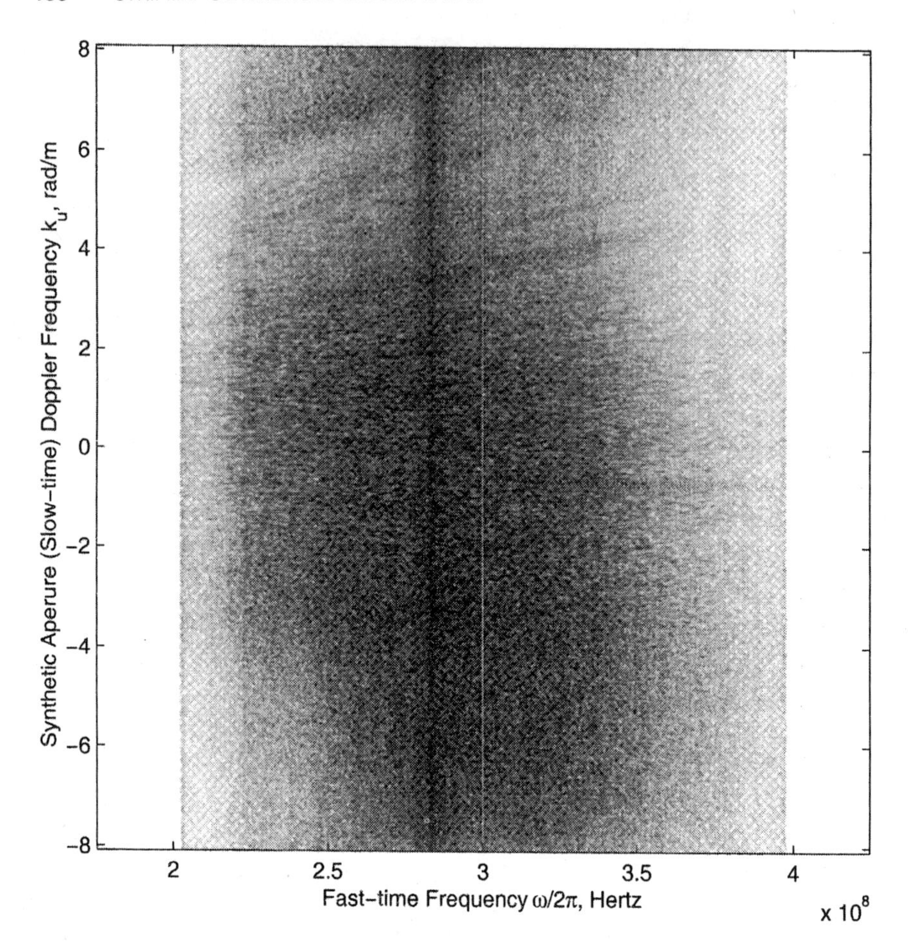

Figure 6.9. Slow-time compressed SAR data (first run) in (ω, k_u) domain.

dimensional SAR signal spectrum $S(\omega, k_u)$ of Figure 6.8, the spectrum in Figure 6.9 is also aliased in the slow-time Doppler k_u domain. We will use the digital spotlighting method to reduce some of the aliasing in this database.

Synthetic Subaperture Data In the spotlight SAR discussion of Chapter 5, we mentioned that in the case of a relatively large synthetic aperture $[-L, L]$, it is advisable to divide the synthetic aperture $[-L, L]$ into smaller subapertures and perform digital spotlighting on the slow-time compressed SAR signal $s_c(\omega, u)$ within these subapertures. Subaperture processing is also a useful tool in the stripmap SAR systems. Subaperture processing has two utilities in the stripmap SAR systems. First, subaperture processing provides a more accurate digital spotlight filter; this was also the reason for using subaperture processing in the spotlight SAR systems.

The second reason for using subaperture processing is the less restrictive Nyquist rate in the slow-time domain for a slow-time compressed stripmap SAR signal within a subaperture. Suppose there are N_s subapertures with length $2L_s$ within $u \in [-L, L]$; these subapertures are

$$[Y_i - L_s, Y_i + L_s], \quad i = 1, \ldots, N_s,$$

where

$$Y_i = (2i - 1)L_s - L.$$

Provided that $L_s \ll L$, then the Nyquist sample spacing in the slow-time domain of the slow-time compressed signal becomes

$$\begin{aligned} \Delta_{uc} &\approx \Delta_u \\ &= \frac{\lambda_{\min}}{4 \sin \phi_d}. \end{aligned}$$

Thus there is no need for upsampling the synthetic aperture data with a subaperture prior to slow-time compression.

Subaperture Digital Spotlighting

Next we examine the problem of digital spotlighting in the stripmap SAR systems. In Section 5.5 (processing of spotlight SAR signal) we provided a discussion on the advantages of digital spotlighting the measured SAR data to reduce the side lobe effects of the radar radiation pattern. Similar issues may also be dealt with via digital spotlighting in the stripmap SAR systems. Digital spotlighting also provides a tool for extracting the stripmap SAR signature of a *finite* target area; this is a crucial issue for the stripmap SAR digital reconstruction methods which should form an *alias-free* target function in the spatial frequency domain.

In Section 5.5 we examined various methods for digital spotlighting (*cross-range gating*, *range and cross-range gating*, *subaperture digital spotlighting*, etc.) and their properties and merits. Our conclusion was that subaperture digital spotlighting of the polar format processed SAR image provides the best tool for extracting the SAR signature of a finite target area. Similar principles also hold in the case of the stripmap SAR systems. In this section we briefly review subaperture digital spotlighting for the stripmap SAR systems.

Suppose that the desired target area is within

$$x \in [X_c - X_0, X_c + X_0] \quad \text{and} \quad y \in [-Y_0, Y_0].$$

(One may be interested in imaging a *squint* target area with respect to the center of the synthetic aperture $u = 0$; for this scenario, see the general treatment of digital spotlighting in the squint case in Section 5.5.) The stripmap SAR data in the ith subaperture can be viewed as the SAR data from a SAR system with squint range and cross-range parameters

$$(X_c, Y_i)$$

and the squint angle

$$\theta_{ci} = \arctan\left(\frac{Y_i}{X_c}\right).$$

The fast-time of arrival of the nth target signature in the ith subaperture can now be approximated via

$$t_{ni} \approx \frac{2\sqrt{x_n^2 + (y_n + Y_i)^2}}{c}.$$

The polar format processed image of the ith subaperture is formed by

$$\mathcal{F}_{(\omega,u)}\big[s_c(\omega, u);\ u \in [Y_i - L_s, Y_i + L_s]\big].$$

For digital spotlighting within the ith subaperture, the resultant digital spotlight filter which is used on the polar format processed image of the ith subaperture becomes

$$W_{di}(t, k_u) = \begin{cases} 1 & \text{for } |ct/2\cos(\phi + \theta_{ci}) - X_c| < X_0 \\ & \text{and } |ct/2\sin(\phi + \theta_{ci}) - Y_i| < Y_0, \\ 0 & \text{otherwise.} \end{cases}$$

The digitally spotlighted stripmap SAR data of the N_s subapertures are appended to each other to form the stripmap SAR data which will be used for image formation.

Figure 6.10 shows the full aperture digital spotlight algorithm for the stripmap SAR systems. For subaperture processing, the full aperture digital spotlight filter, that is,

$$W_d(t, k_u) = \begin{cases} 1 & \text{for } |ct/2\cos\phi - X_c| < X_0 \text{ and } |ct/2\sin\phi| < Y_0, \\ 0 & \text{otherwise,} \end{cases}$$

is replaced with $W_{di}(t, k_u)$. For proper reference fast-time processing, the reference SAR signal is modified to be

$$s_0(\omega, u) = \exp\left[-j2k\sqrt{X_c^2 + u^2} + j2kX_c\right];$$

the addition of the phase term $2kX_c$ to the reference signal ensures that the reference fast-time point T_c is unchanged.

Note that the algorithm includes an option for slow-time upsampling from Δ_u to Δ_{uc} prior to the slow-time compression; for subaperture processing this option is not essential in most practical stripmap SAR systems. After digital spotlighting, the resultant digitally spotlighted SAR signal can be down-sampled from Δ_{uc} to Δ_u after the slow-time decompression.

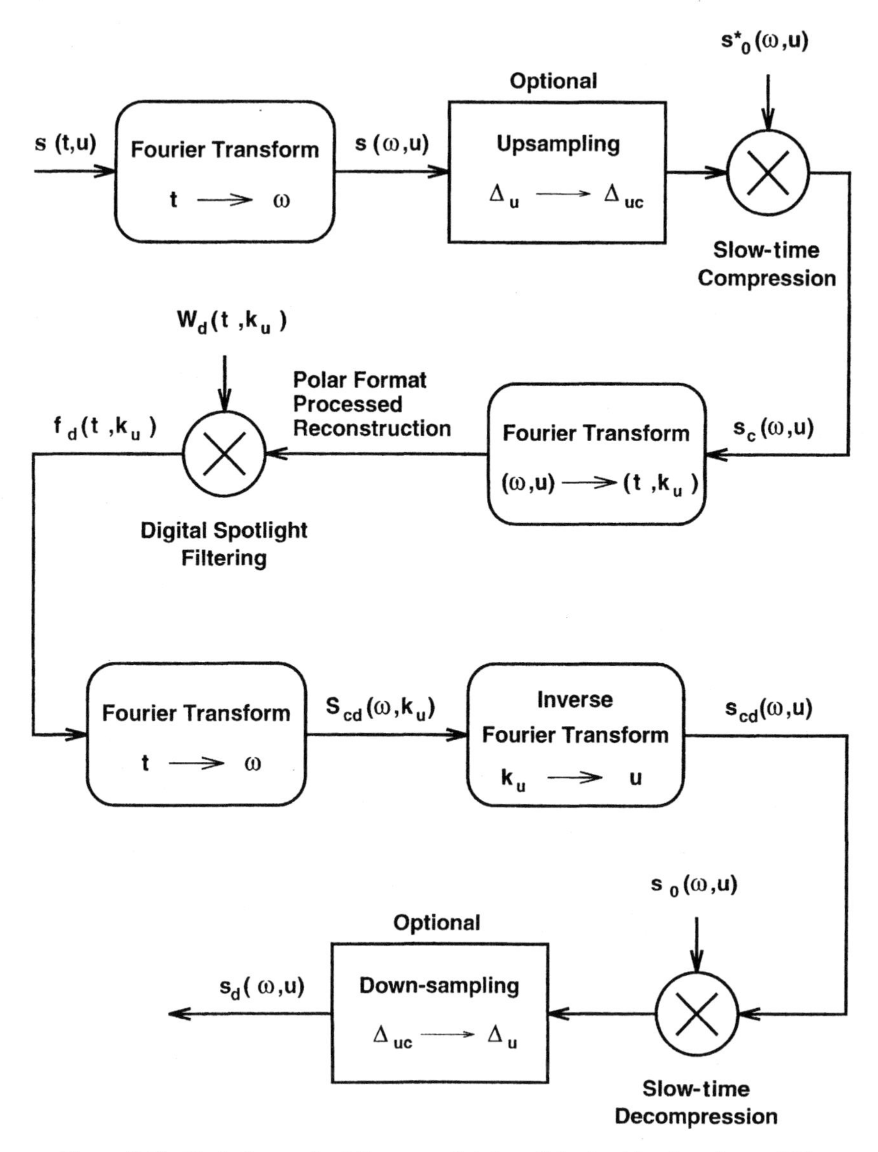

Figure 6.10. Block diagram for full aperture digital spotlight algorithm for stripmap SAR.

Subaperture Size It is desirable to use the size of the subaperture $2L_s$ to be as small as possible to improve the estimate of the fast-time arrival of the targets in the imaging scene. However, if L_s is chosen to be too small, then the rectangular window spreading of the subaperture in the k_u domain, that is, its main lobe $\pm\pi/L_s$, becomes more dominant than the theoretical bandwidth of the subaperture slow-time compressed SAR signal in the k_u domain after digital spotlighting, which is

$$\left[\frac{-2kY_0}{X_c}, \frac{2kY_0}{X_c}\right].$$

A simple rule of thumb is to choose L_s such that the rectangular window spreading of the subapertures is an order of magnitude smaller than (1/100 of) the theoretical slow-time Doppler band of $s_c(\omega, u)$; that is,

$$\frac{\pi}{L_s} \approx \frac{2k_{\min} Y_0}{100 X_c},$$

which yields

$$\begin{aligned} L_s &\approx \frac{25 X_c \lambda_{\max}}{Y_0} \\ &\approx 25 D_y. \end{aligned}$$

In most practical SAR systems this corresponds to approximately a few hundred slow-time samples. For instance, for a planar radar aperture, where $\Delta_u = D_y/4$, the number of the slow-time samples within a subaperture is 200.

Finally subaperture digital spotlight filtering results in some discontinuity in the resultant data at the boundaries of the subapertures. The effects of these discontinuities in the reconstructed images can be seen near the edges of the imaged target area; the effects are relatively weak. One could remove these discontinuities by choosing the subapertures to slightly overlap (e.g., 5% overlap), and discarding the overlap data after digital spotlighting each subaperture data.

Figure 6.11*a* shows the full aperture polar format processed reconstruction, that is, the (t, k_u) domain data, for the stripmap SAR signal of Figure 6.7*a* (first run); $(X_0, Y_0) = (300, 450)$ m are used for the digital spotlight filter. The full aperture digital spotlight filter $W_d(t, k_u)$ is superimposed on this image (the shaded area). Using subaperture digital spotlighting with 8 subapertures on the stripmap SAR signal of Figure 6.7*a* yields slightly different results for the human-made targets in the imaging scene.

To show the merits of subaperture digital spotlighting, we consider the stripmap SAR data which were obtained in the second run of the radar-carrying aircraft. As we noted in the introductory section of this book, this stripmap SAR database contains 2048 slow-time samples over the synthetic aperture interval of $u \in [-391, 391]$ m. The full aperture digitally spotlighted signal for this database is shown in Figure 6.11*b*; $(X_0, Y_0) = (100, 550)$ m are used for the digital spotlight filter. However, it turns out that for the second-run data, full aperture digital spotlight filtering results in the removal of the signature of some of the targets near the edges of the imaging scene.

We apply the subaperture digital spotlight filtering with 16 subapertures on the stripmap SAR data of the second run. The digitally spotlighted signal $s_{cd}(\omega, u)$ is formed by appending the outcomes of each digital spotlight subaperture processing.

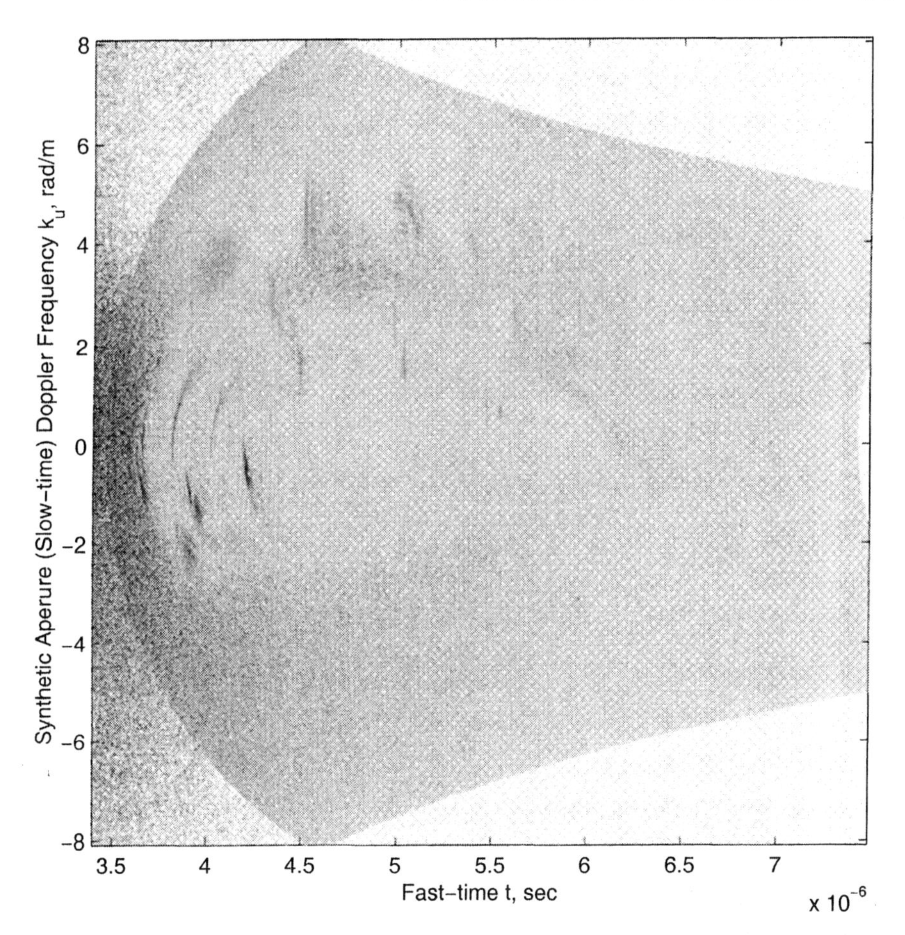

Figure 6.11a. Polar format processed (t, k_u) domain data with *full aperture* digital spotlight filter (shaded area) superimposed, UHF band SAR data of foliage area and human-made targets: first run.

Figure 6.11*c* shows the resultant polar format reconstruction which is obtained via

$$f_d(t, k_u) = \mathcal{F}_{(\omega,u)}\big[s_{cd}(\omega, u)\big].$$

Note the support of the resultant signal $f_d(t, k_u)$ does not resemble the support of the full aperture digital spotlight filter of Figure 6.11*b*. In fact the support region in Figure 6.11*c* is a union of 16 different squint as well as broadside digital spotlight filters (Figure P6.3). Earlier we showed the resultant digitally spotlighted SAR data of the second run in Figure 6.7*b* (Figure P6.4).

Consider the digitally spotlighted (t, k_u) domain SAR data of the first run in Figure 6.11*a*. This database contains the alias-free as well as aliased SAR signature of various targets in the imaging scene. The spotlight filter for $(X_0, Y_0) = (300, 450)$ m

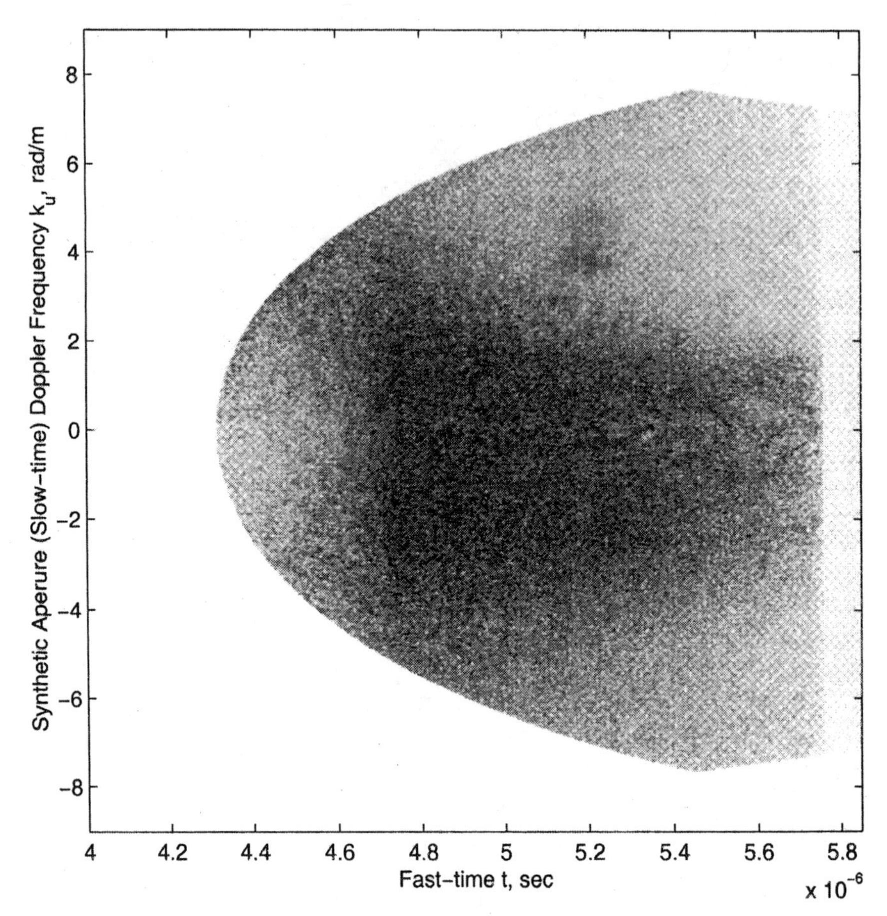

Figure 6.11b. Polar format processed (t, k_u) domain data with *full aperture* digital spotlight filter (shaded area) superimposed, UHF band SAR data of foliage area and human-made targets: second run.

is chosen to be just within the available slow-time Doppler band of $k_u \in [-8, 8]$ rad/m. In this case the signature of the targets within the spotlight filter are almost alias-free. The signature of the other targets, depending on their coordinates, contains some degree of aliasing.

As we showed in Chapter 2, the slow-time compressed signature of a target shows an *inward* behavior toward the zero Doppler frequency $k_u = 0$. Thus the signature of the targets outside the desired target area spill (bleed) into the spotlight filter of Figure 6.11*a*. However, in the final reconstructed SAR image, these inward spills (whether they are aliased or alias-free in the slow-time Doppler domain) yield *outward* signatures, that is, structures outside the desired target area.

From the point of view of one who uses the backprojection method for reconstruction, the above-mentioned outward signature poses no problem. However, for the

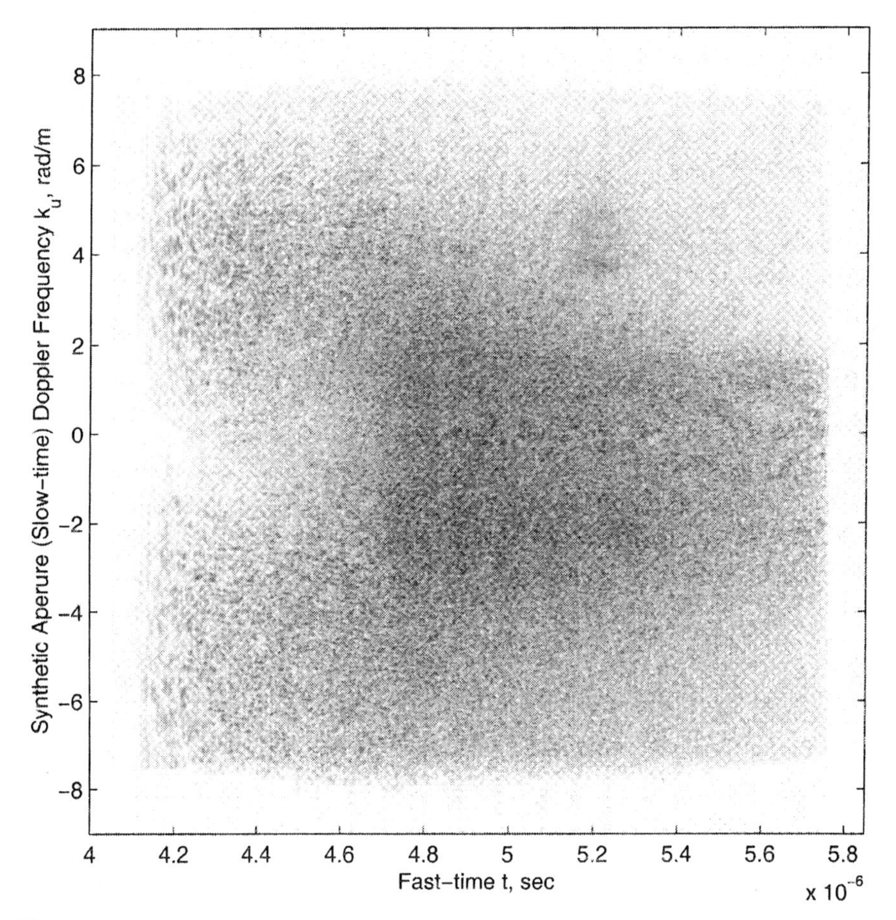

Figure 6.11c. Polar format processed (t, k_u) domain data after applying *subaperture* digital spotlight filters for second run.

digital reconstruction that requires spatial frequency interpolation, the user should employ a finer sample spacing in the k_y domain (i.e., $\Delta_y < \pi / Y_0$) to ensure that the outward signature does not result in the *spatial* domain aliasing. As we will see in our discussion of stripmap SAR reconstruction, this simply translates into lesser subsampling of the SAR data $S_d(\omega, k_u)$ in the k_u domain (Figure P6.5).

Reducing Side Lobes Doppler Aliasing via Slow-time Upsampling

There is another issue associated with the digitally spotlighted stripmap SAR data in Figure 6.11*a* (or 6.11*c*) that has an adverse impact on any SAR reconstruction method. The SAR data of Figure 6.11*a* are almost alias-free for the targets in the region of interest. However, *what will happen after decompression*? In theory, as long as the slow-time Nyquist sample spacing is satisfied, that is,

$$\Delta_u = \frac{\lambda_{\min}}{4 \sin \phi_d},$$

there is no need for concern. However, the divergence angle ϕ_d is an *approximate* measure of the half-beamwidth angle of the radar; the aspect angle interval $[-\phi_d, \phi_d]$ corresponds to the main lobe of the radar radiation pattern. In practice, the radar beamwidth angle exceeds this interval due to its side lobes. Thus the stripmap SAR data obtained based on the above slow-time sample spacing Δ_u contain slow-time Doppler aliasing due to the side lobes of the radar radiation pattern. (Recall that we posed a similar problem with the side lobe effects in Chapter 5.)

To encounter the slow-time Doppler aliasing of the side lobes of the radar radiation pattern, one could simply use the same slow-time sampling scheme as what we outlined for the spotlight SAR systems (see Section 5.5). For the cited stripmap SAR data, this is achieved via the following steps: First, we obtain the one-dimensional Fourier transform of the (t, k_u) domain data of Figure 6.11a (or 6.11c) with respect to the fast-time variable t. The resultant data are the samples of the digitally spotlighted slow-time compressed signal $S_{cd}(\omega, k_u)$.

Next we zero-pad these samples with sufficient zeros to double (or whatever factor the user knows to be sufficient, based on the physical radar properties) the slow-time Doppler k_u domain bandwidth; that is, the zero-padded slow-time Doppler support band for $S_{cd}(\omega, k_u)$ becomes

$$k_u \in [-16, 16] \text{ rad/m}.$$

The inverse Fourier transform of these zero-padded data with respect to the slow-time Doppler frequency k_u yields the samples of $s_{cd}(\omega, u)$ at one-half of the original slow-time sample spacing Δ_u. As we mentioned in Chapter 5, for the stripmap SAR data from the first run of the radar-carrying aircraft, this corresponds to a new slow-time sample spacing of $\Delta_u = 0.39/2 = 0.195$ m.

Slow-time decompression is performed on these upsampled data to yield the samples of

$$s_d(\omega, u) = s_{cd}(\omega, u) s_0(\omega, u),$$

where $s_0(\omega, u)$ is the SAR reference signal. Figure 6.12 shows the resultant data in the spectral (k_u, ω) domain for the SAR data from the first run. Note that the slow-time Doppler support band for this database, in which $u \in [-200, 200]$ m is used, is within

$$k_u \in [-10, 10] \text{ rad/m}.$$

Thus the slow-time upsampling was effective in encountering slow-time Doppler aliasing. (Recall that the *aliased* measured SAR data resided within a smaller slow-time Doppler band of $k_u \in [-8, 8]$ rad/m.)

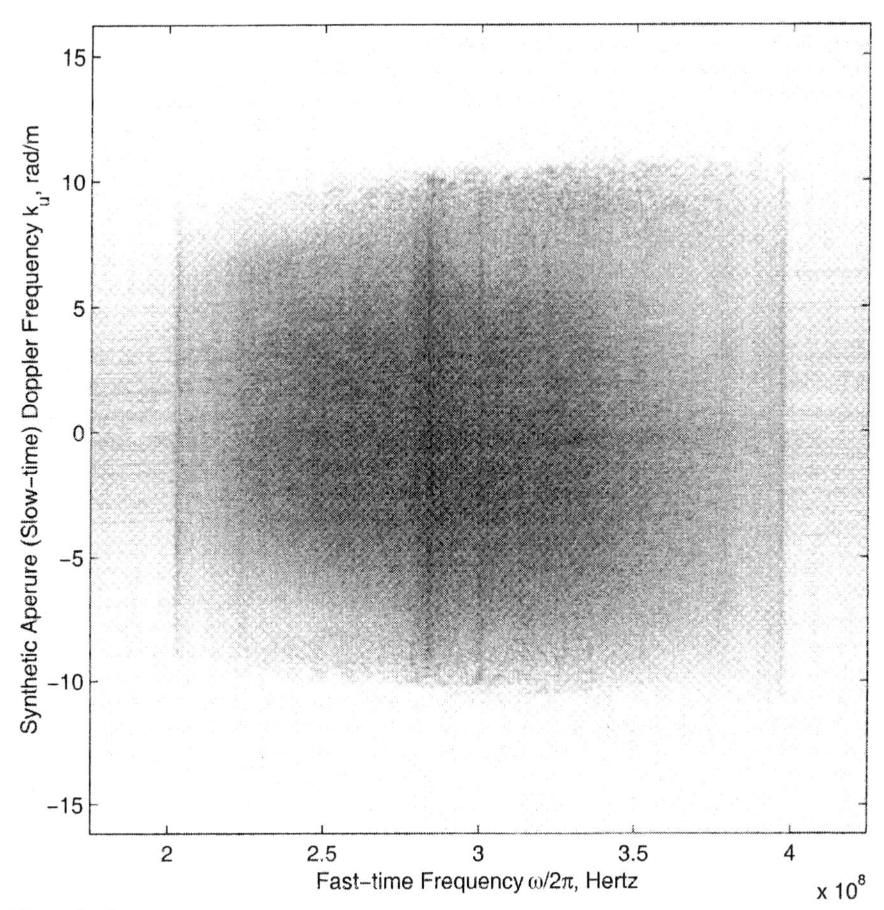

Figure 6.12. SAR data (first run) spectrum in (ω, k_u) domain after digital spotlighting and upsampling.

We showed in Section 5.5 that after digital spotlighting using $Y_0 = 400$ m, the first-run SAR data within $u \in [-50, 50]$ m is within the slow-time Doppler band of $k_u \in [-8, 8]$ rad/m. *Why did the slow-time Doppler band increase for the larger synthetic aperture $u \in [-200, 200]$ m and a larger half cross-range $Y_0 = 450$ m?* The reason is that the beamwidth angle of this stripmap SAR system is relatively large. In fact the SAR signal spectrum in Figure 6.12 (or the spectral support of the individual targets in the imaging scene, which will be shown in the next section) indicates that the beamwidth angle is around $\pm 45°$; for example, the beamwidth at the closest range (500 m) is about $[-500, 500]$ m which is greater than the synthetic aperture interval of $u \in [-200, 200]$ m.

In this case the *digitally spotlighted* stripmap SAR data within any finite synthetic aperture, which is smaller than the beamwidth, are effectively numerically generated *spotlight* SAR data. Thus such a database exhibits certain spectral properties of the spotlight SAR signal, that is, *some* dependence on the size of the target area and length of the synthetic aperture (see Section 5.5). This fact will become more evident in our study of the SAR signature of the individual targets in the imaging scene in the following section.

6.6 RECONSTRUCTION ALGORITHMS AND SAR IMAGE PROCESSING

Most of the issues associated with digital image formation in stripmap SAR are similar to the ones that were outlined in Section 5.6 for spotlight SAR. In this section we briefly outline these issues in addition to other items that are exclusive to stripmap SAR.

Digital Reconstruction via Spatial Frequency Interpolation

Two-dimensional Frequency Domain Matched Filtering As we noted in our discussion of generic SAR reconstruction in Chapter 4, digital reconstruction via spatial frequency interpolation requires the following two-dimensional frequency domain matched filtering of the measured SAR signal with the reference signal, that is,

$$S(\omega, k_u) S_0^*(\omega, k_u),$$

where the two-dimensional frequency domain matched filter

$$S_0^*(\omega, k_u) = \underbrace{P^*(\omega)}_{\text{Fast-time matched filter}} \quad \underbrace{\exp\left(j\sqrt{4k^2 - k_u^2}\, X_c\right)}_{\text{Target area baseband conversion}}$$

is made up of two components: $P^*(\omega)$ which is the fast-time domain matched filter; and the phase function $\exp(j\sqrt{4k^2 - k_u^2} X_c)$ which brings the origin to the center of the target area in the spatial domain.

In the stripmap SAR systems the physical radar radiation pattern introduces an additional amplitude function $A(\omega, k_u)$ in the two-dimensional frequency (ω, k_u) domain; see Sections 6.1 and 6.2. As we stated in Section 6.2 (on reconstruction), an additional matched filtering with the amplitude pattern of the physical radar, that is, $A * (\omega, k_u)$, should be performed. If this amplitude pattern is not known, then the phase of $A(\omega, k_u)$ would cause additional smearing in the point spread function of the stripmap SAR system. For the cited stripmap SAR data in this book, the amplitude pattern $A(\omega, k_u)$ is not known.

Zero-padding in Fast-time Domain For the same reasons that we stated for the spotlight SAR systems in Section 5.6, the minimum fast-time interval for processing the stripmap SAR data should be

$$T_{\min} = \max\left[T_f - T_s, \frac{4X_0}{c \cos\theta_{\text{ax}}}\right],$$

where θ_{ax} is the maximum *absolute* aspect angle of the target area with respect to the radar synthetic aperture coordinates. The above constraint for the minimum interval of fast-time processing could be met by *zero-padding* the measured SAR data in the fast-time domain prior to its Fourier transformation into the ω domain.

In a stripmap SAR system the maximum absolute aspect angle θ_{ax} is the largest divergence angle ϕ_d of the side-looking radar at all of the available fast-time frequencies. Thus we have

$$\theta_{ax} = \begin{cases} \arcsin\left(\dfrac{\lambda_{max}}{D_y}\right), & \text{planar radar,} \\ \arctan\left(\dfrac{D_y}{2x_f}\right), & \text{curved radar.} \end{cases}$$

Zero-padding in Synthetic Aperture Domain This issue was also discussed in Section 5.6 for the spotlight SAR systems. If the size of the target area is greater than the length of the synthetic aperture in which stripmap data are available, the stripmap SAR data in the synthetic aperture u domain should be zero-padded to create an effective aperture of $[-Y_0, Y_0]$ prior to the DFT processing with respect to the discrete samples of u. In the general case ($Y_0 > L$ or $Y_0 \leq L$), the minimum half-length of the processed (zero-padded) synthetic aperture is chosen to be

$$L_{min} = \max(L, Y_0).$$

Algorithm The implementation of the reconstruction algorithm based on the spatial frequency interpolation of the above-mentioned matched-filtered database also requires the knowledge of the support of the stripmap SAR signal in various domains which were described in Sections 6.3 to 6.5. Most of the steps are identical to the algorithm for the spotlight SAR systems. The stripmap version of the reconstruction algorithm is summarized in the following:

Step 1. Perform the (discrete) fast-time matched filtering

$$\mathbf{s}_M(t, u) = \mathbf{s}(t, u) * p_0^*(-t),$$

where

$$p_0(t) = p(t - T_c)$$

and T_c is the user-prescribed reference fast-time point (e.g., $T_c = 2X_c/c$).

Step 2. Perform the fast-time and slow-time processing (digital spotlighting; if it is required, upsampling of the data from Δ_u to Δ_{uc} in the slow-time domain before slow-time compression; etc.) which was described in Section 6.5. After digital spotlighting, down-sample the slow-time domain data from Δ_{uc} to Δ_u.

Step 3. If $T_f - T_s$ does not provide the sufficient sampling density in the k_x domain, zero-pad the SAR data in the fast-time domain according to the constraint on T_{min}. Also, if $Y_0 > L$, zero-pad the data in the slow-time u domain to achieve the minimum length synthetic aperture to the extent of $u \in [-L_{min}, L_{min}]$, where $L_{min} = \max(L, Y_0)$.

Step 4. Obtain the two-dimensional DFT of the digitally spotlighted, upsampled, and zero-padded SAR signal; the result is the samples $S_d(\omega_n, k_{um})$.

Step 5. If $Y_0 < L$, *that is, the size of the target area in the cross-range domain is greater than the length of the synthetic aperture, perform slow-time Doppler domain subsampling on the available* $S_d(\omega_n, k_{um})$ *data; this option, which was also used for the spotlight SAR systems, will be discussed later.*

Step 6. The DFT algorithm assumes that the fast-time sample number $(N/2)+1$ *(i.e., the sample in the middle of the fast-time array) is at* $t = 0$; *however, this sample corresponds to the actual fast-time value of* $t - T_c$. *Thus the two-dimensional DFT of the upsampled and digitally spotlighted SAR signal should be modified by the following linear phase function (discrete) multiplication:*

$$S_d(\omega, k_u)\exp(-j\omega T_c),$$

which moves the fast-time origin back to the actual fast-time zero, that is, $t = 0$. *The above multiplication with the linear phase function of* ω *is crucial for all the SAR digital reconstruction algorithms which involve processing of measured data in the fast-time frequency domain.*

Step 7. Perform (discrete) baseband conversion of target area via the addition of the phase function of $(X_c, 0)$ *to form the target function*

$$F(k_x, k_y) = S_d(\omega, k_u)\exp(-j\omega T_c)\exp\left(j\sqrt{4k^2 - k_u^2}\, X_c\right),$$

where

$$k_x = \sqrt{4k^2 - k_u^2},$$

$$k_y = k_u.$$

For the available discrete samples of SAR data in the (ω, k_u) *domain, labeled* (ω_n, k_{um}), *this results in a set of unevenly spaced samples of the target function in the spatial frequency* (k_x, k_y) *domain at (see Section 4.5; Figure P6.6)*

$$k_{xmn} = \sqrt{4k_n^2 - k_{um}^2},$$

$$k_{ymn} = k_{um}.$$

Step 8. Identify the support of the coverage of SAR data in the spatial frequency (k_x, k_y) *domain, labeled*

$$k_x \in [k_{x\min}, k_{x\max}] \quad \textit{and} \quad k_y \in [k_{y\min}, k_{y\max}],$$

where

$$k_{x\min} = \min[k_{xmn}],$$

$$k_{x\max} = \max[k_{xmn}],$$

and

$$k_{y\min} = \min[k_{ymn}],$$
$$k_{y\max} = \max[k_{ymn}].$$

Identify a grid in this region with the following sample spacings:

$$\Delta_{k_x} = \frac{\pi}{X_0},$$
$$\Delta_{k_y} = \frac{\pi}{Y_0}.$$

The resultant number of samples in the grid are

$$N_x = 2\left\lceil \frac{k_{x\max} - k_{x\min}}{2\Delta_{k_x}} \right\rceil \quad \text{in the } k_x \text{ domain}$$

$$N_y = 2\left\lceil \frac{k_{y\max} - k_{y\min}}{2\Delta_{k_y}} \right\rceil \quad \text{in the } k_y \text{ domain.}$$

Step 9. Interpolate the samples of the target function spectrum $F(k_x, k_y)$ on the uniform grid from its available unevenly spaced data $F(k_{xmn}, k_{ymn})$.

Step 10. Obtain the two-dimensional inverse DFT of the evenly spaced samples of the target function spectrum $F(k_x, k_y)$ on the uniform grid. This yields (N_x, N_y) evenly spaced samples of the spatial domain target function $f(x, y)$ on a uniform grid with the following sample spacings:

$$\Delta_x = \frac{2\pi}{N_x \Delta_{k_x}} = \frac{2X_0}{N_x},$$
$$\Delta_y = \frac{2\pi}{N_y \Delta_{k_y}} = \frac{2Y_0}{N_y}.$$

Step 11. Reduce the bandwidth of the reconstructed image and remove some of the clutter effects by SAR image compression; this is a spotlight SAR carryover which will be discussed below.

Figure 6.13*a* and *b* are the reconstruction of the target function spectrum $F(k_x, k_y)$ from the digitally spotlighted SAR data of Figure 6.11*a* (first run) and Figure 6.11*c* (second run); the target spectrum $F(k_x, k_y)$ is the output of the algorithm after Step 9 (Figure P6.7). Two lines in the (k_x, k_y) domain with the following equations:

$$k_y = \pm \tan(45°) k_x,$$

are superimposed on these target spectra. The support of the two target spectra fall approximately within these two lines. This is indicative of the fact that the beamwidth of the physical radar is approximately ±45° (total of 90°).

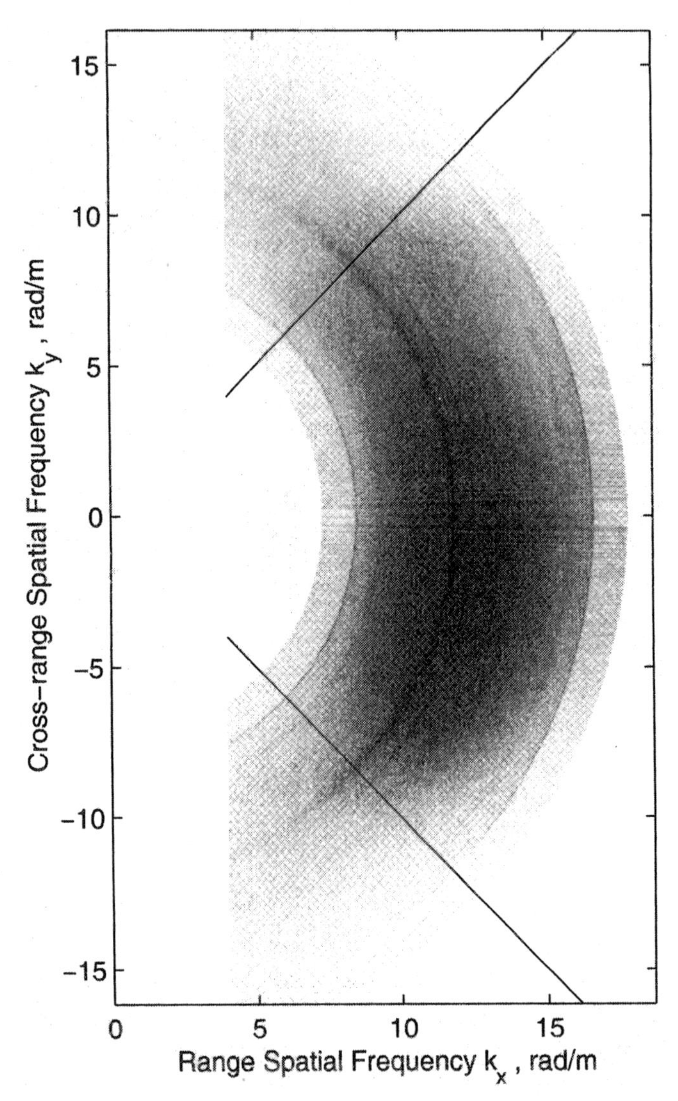

Figure 6.13a. Spectrum of wavefront reconstruction of target function $F(k_x, k_y)$, UHF band SAR data of foliage area and human-made targets: first run.

Figure 6.14*a* is the two-dimensional inverse Fourier transform of the spectrum in Figure 6.13*a*, that is, the spatial domain target function $f(x, y)$ (Figure P6.8) for the first run of the radar-carrying aircraft. This target area, which corresponds to the synthetic aperture data $u \in [-200, 200]$ m, is the same as the one that was shown in Figure 5.18*a* for the (smaller) synthetic aperture data $u \in [-50, 50]$ m. In Figure 6.14*a* one can also see the foliage area at near range on the left side of the image ($x < 600$ m), the lower side of the image (cross-range values $y < -200$ m),

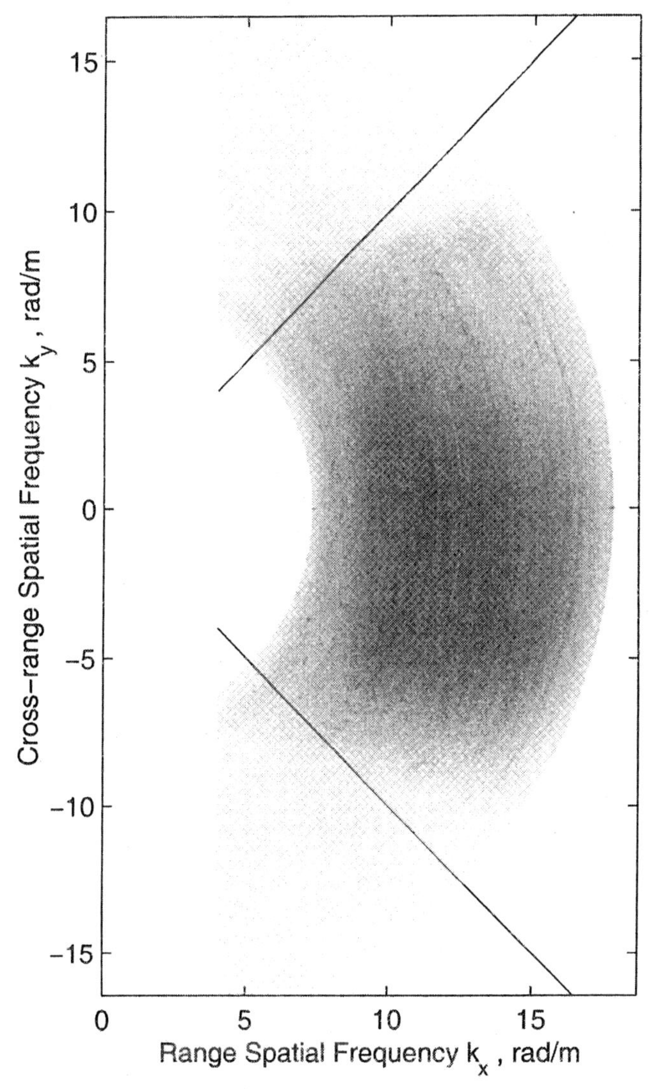

Figure 6.13b. Spectrum of wavefront reconstruction of target function $F(k_x, k_y)$, UHF band SAR data of foliage area and human-made targets: second run.

and the far range at the upper-right side of the image (relatively weak due to the far range).

Figure 6.14*b* and *c* shows close-ups of the SAR reconstruction in Figure 6.14*a*; to compare with the close-ups of the reconstruction with a smaller synthetic aperture, see Figure 5.18*b* and *c*. Figure 6.14*b* contains a portion of the near range and lower cross-range foliage area, and trucks and corner reflectors. Figure 6.14*c* shows a few corner reflectors (note the one at $y = 300$ m at the upper left corner of the image), and a farm area in its upper right corner. (There are several moving targets at or near

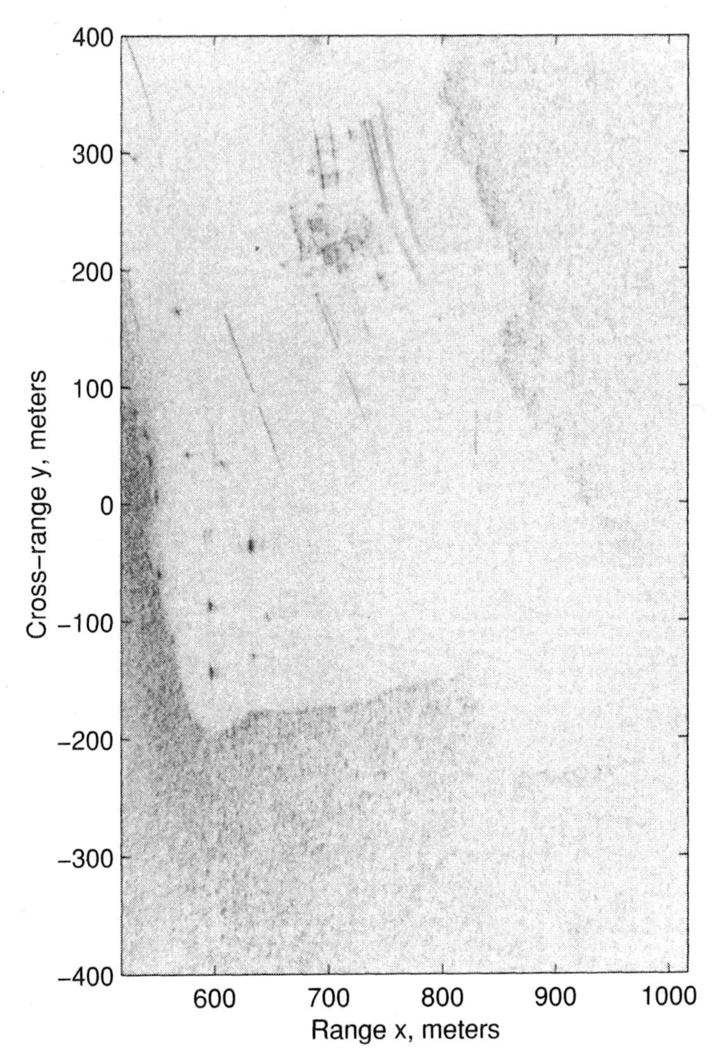

Figure 6.14a. Wavefront reconstruction of target function $f(x, y)$ from UHF band SAR data, first run: target area.

the farm area that appear as smeared and streaking targets.) Note that the streaking signature of the moving targets is more apparent in Figure 6.14*c* than in Figure 5.18*c*.

Consider the target function spectrum for the second run of the radar-carrying aircraft in Figure 6.13*b*. Figure 6.15 shows the two-dimensional inverse Fourier transform of this spectrum at four different target regions. The human-made targets and foliage can be seen in these images (at a different reference range and cross-range as compared with the reconstruction for the first run in Figure 6.14).

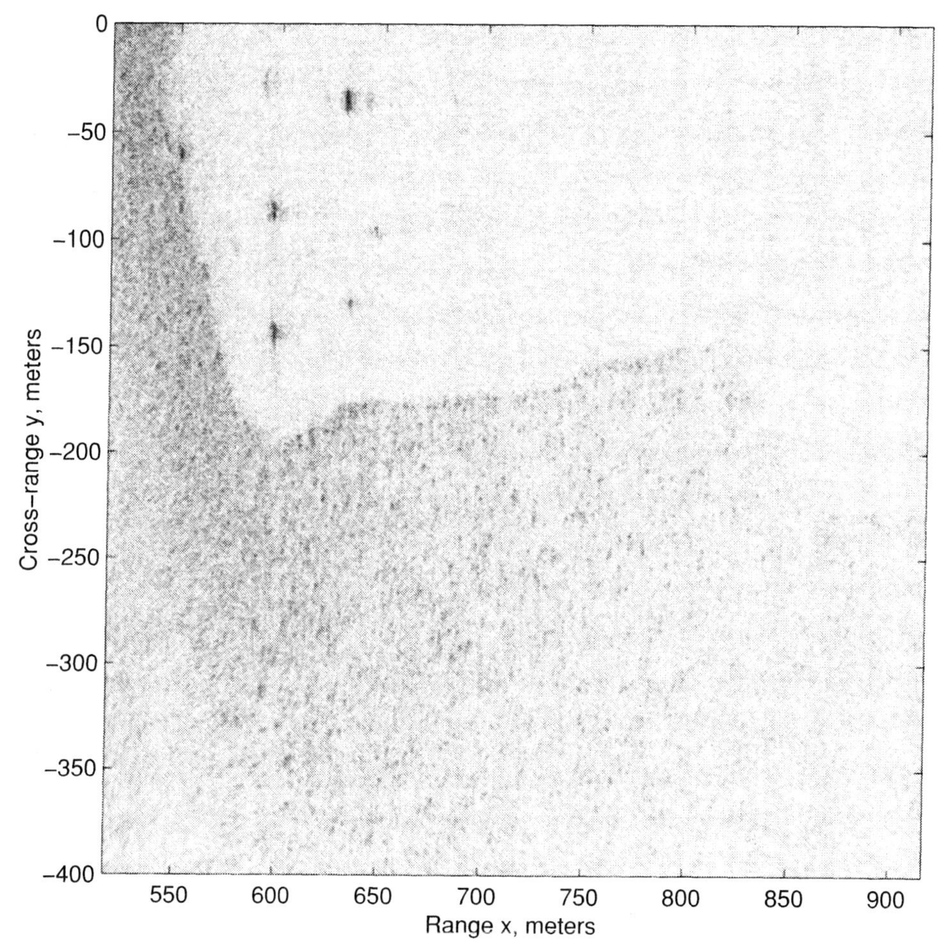

Figure 6.14b. Wavefront reconstruction of target function $f(x, y)$ from UHF band SAR data, first run: close-up of lower target area.

We will examine specific targets in this SAR scene and their spectral properties after discussing SAR image compression.

Slow-time Doppler Domain Subsampling

This issue was discussed in detail in Section 5.6 for the spotlight SAR systems. In the stripmap SAR systems, the size of the aperture $2L$ is larger than the cross-range size of the desired target area $2Y_0$ since $L = B + Y_0$ (B is the half-beamwidth); see Section 6.1. After performing the above-mentioned SAR reconstruction in the spatial frequency domain, the resultant target function in the k_y domain has a sample

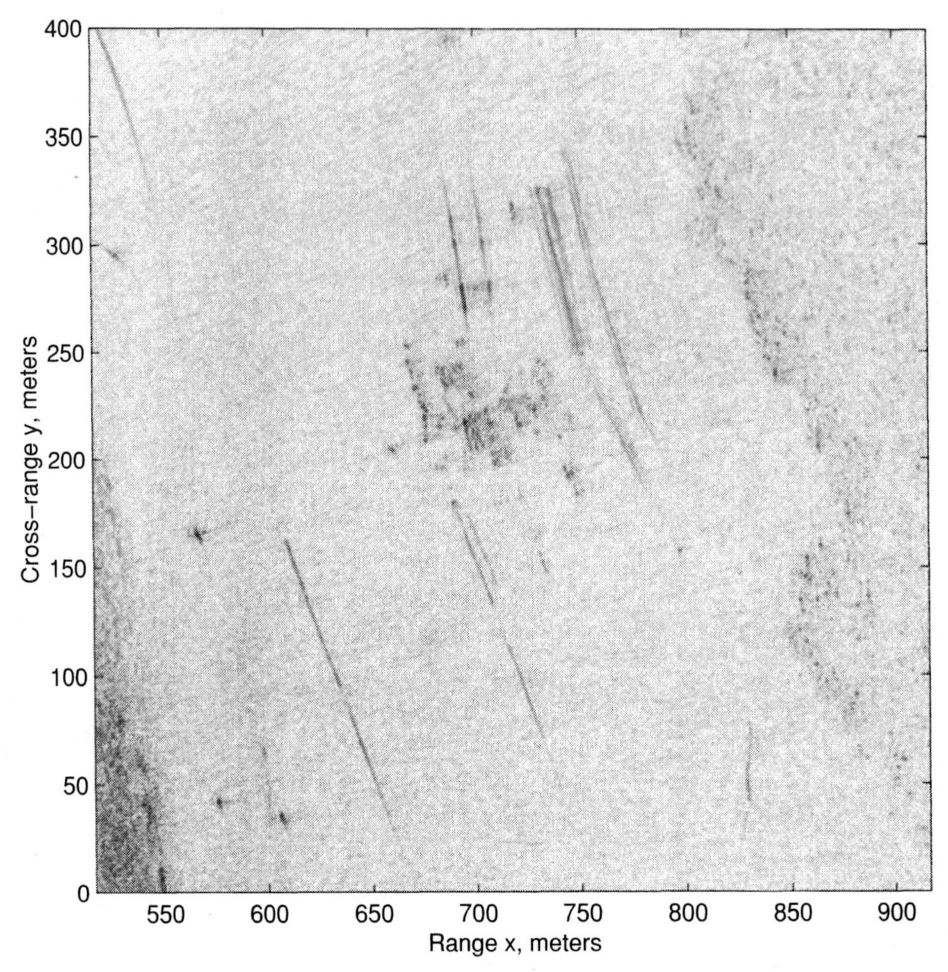

Figure 6.14c. Wavefront reconstruction of target function $f(x, y)$ from UHF band SAR data, first run: close-up of upper target area.

spacing of

$$\Delta_{k_y} = \frac{\pi}{L}.$$

However, the Nyquist sample spacing for the cross-range target function is

$$\Delta_{k_y} \leq \frac{\pi}{Y_0}.$$

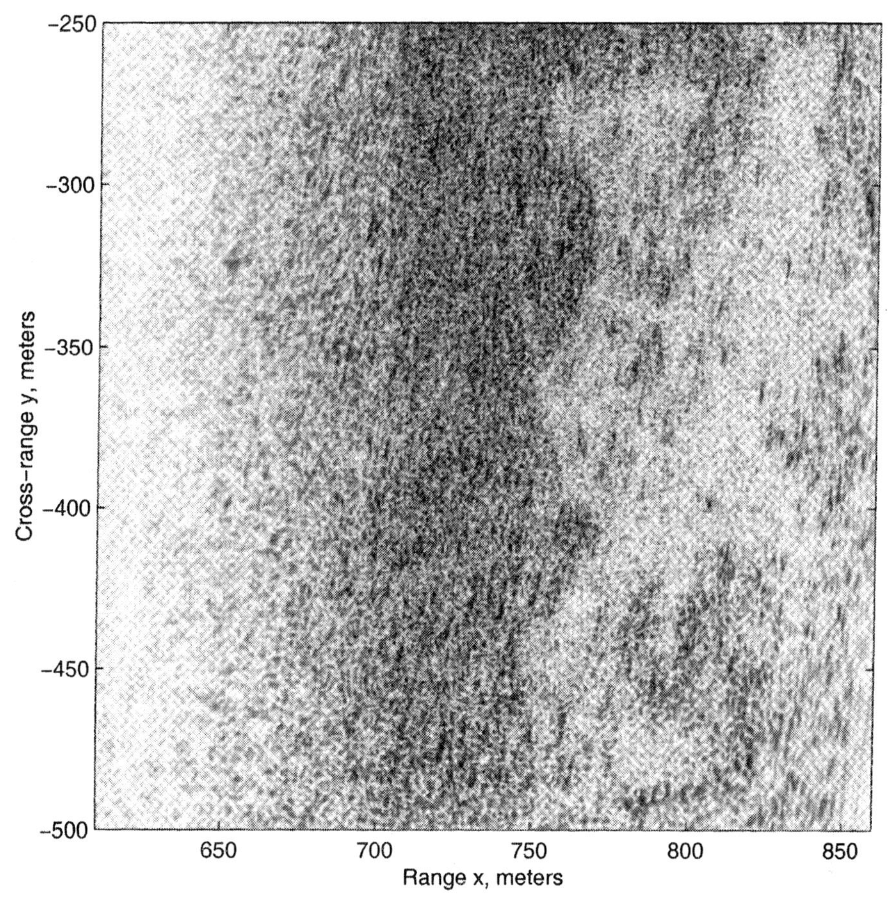

Figure 6.15a. Wavefront reconstruction of target function $f(x, y)$ from UHF band SAR data, second run: its close-ups.

Since L is greater than Y_0, we satisfy the Nyquist criterion for $F(k_y)$ in the k_y domain. In fact the available samples $F(k_{xmn}, k_{ymn})$ contain redundancy (even before interpolation in the k_x domain).

To reduce the computational load and array sizes, one could subsample $F(k_{xmn}, k_{ymn})$ in the k_y domain. Sampling rate reduction can be achieved via interpolation; this, however, increases the computational load. A simple approach is to skip certain samples. The example used in Section 2.8 was as follows: If $L = 3.8Y_0$, then the user could reduce the sampling rate by 3.8 via interpolation, or skip every two samples. In this case there is still some redundancy in the data which provides a guard band in the cross-range y domain [that reduces inverse DFT errors in obtaining $f(x, y)$ from $F(k_x, k_y)$].

The user should be aware of the k_{um} value which corresponds to the middle sample also. Moreover, if the number of subsampled data is an odd number, the

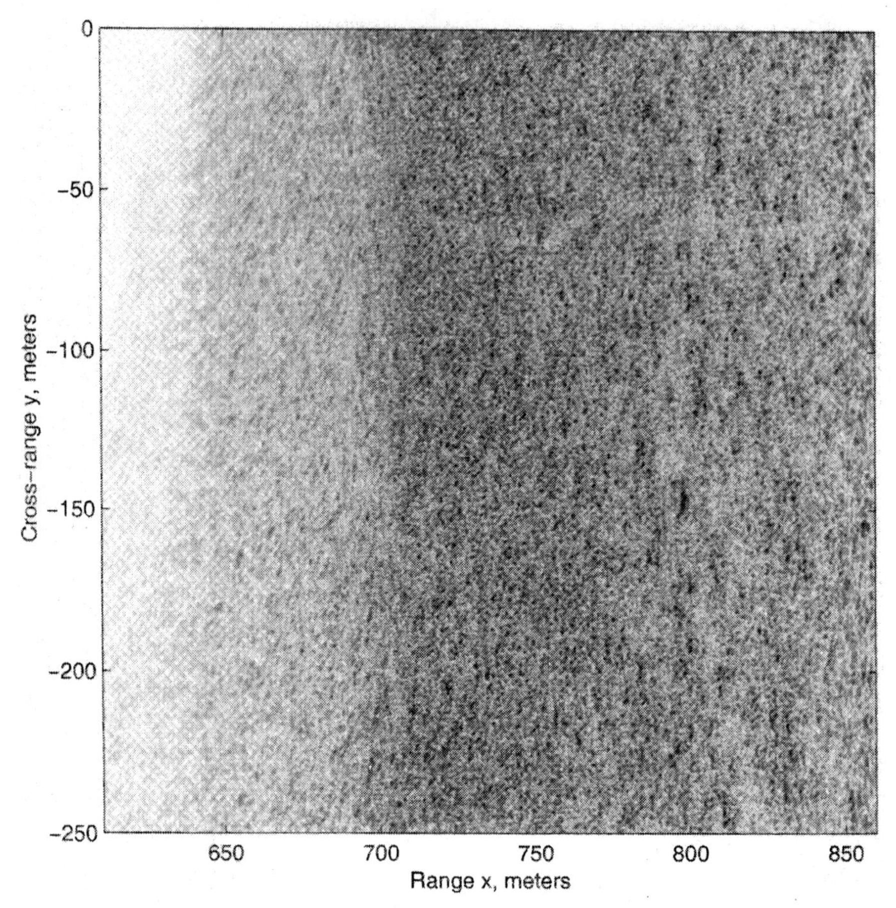

Figure 6.15b. Wavefront reconstruction of target function $f(x, y)$ from UHF band SAR data, second run: its close-ups.

$F(k_{xmn}, k_{ymn})$ array in the k_y domain should be appended with a zero to make the size of the array even (for inverse DFT processing).

Finally we should point out that once the samples of the digitally spotlighted SAR spectrum $S_d(\omega_n, k_{um})$ are formed in Step 4 of the algorithm, then subsampling in the k_u domain can be performed on this database. Thus there is no need to perform two-dimensional frequency-domain matched filtering on the redundant data to form the samples $F(k_{xmn}, k_{ymn})$.

Reducing Bandwidth of Reconstructed Image

For the spotlight SAR systems, compression of the reconstructed image for a broadside target area was achieved via (see Section 5.6)

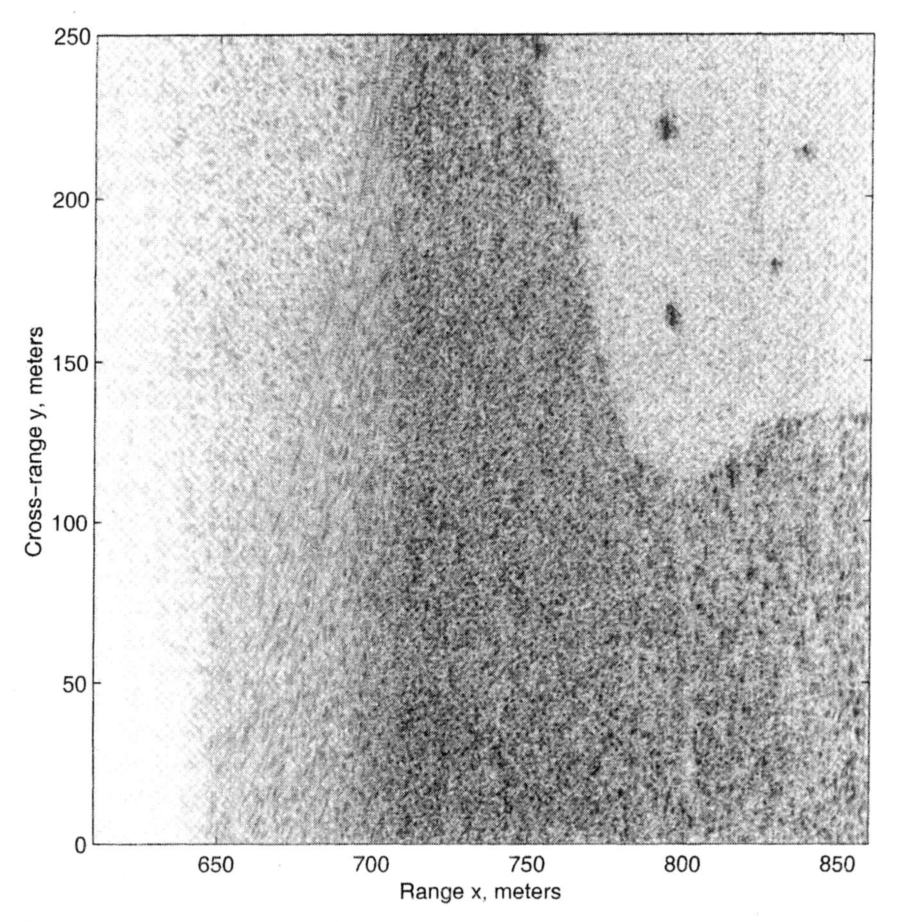

Figure 6.15c. Wavefront reconstruction of target function $f(x, y)$ from UHF band SAR data, second run: its close-ups.

$$f_c(x, y) = f(x, y) \exp\left(j2k_c x - j2k_c\sqrt{x^2 + y^2} \right).$$

With the shift of origin to the center of the target area $(X_c, 0)$, the spotlight SAR image compression for a broadside target area is performed via

$$f_c(x, y) = f(x, y) \exp\left[j2k_c(X_c + x) - j2k_c\sqrt{(X_c + x)^2 + y^2} \right].$$

For a *true* stripmap SAR system, where the interval of the processed synthetic aperture is $2L$ with $L = B + Y_0$, the above operation can be shown to *expand* (not *compress*) the bandwidth of the reconstructed stripmap SAR image. This is due to the fact that the spectral support band of the stripmap SAR signature of a target is

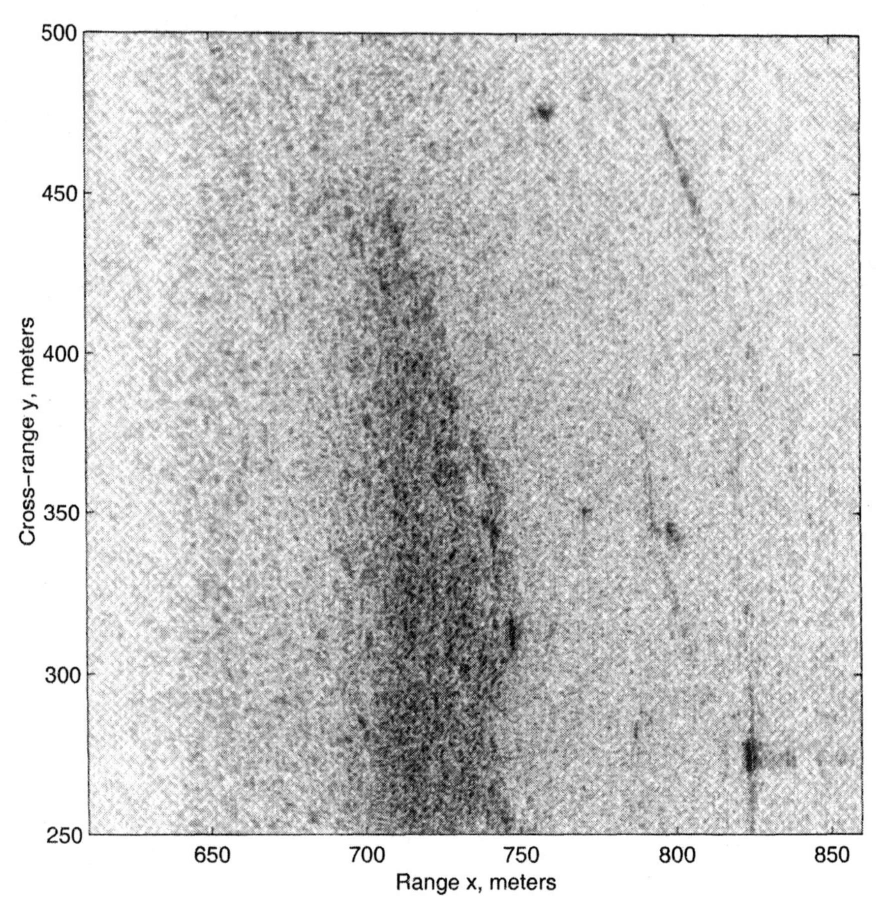

Figure 6.15d. Wavefront reconstruction of target function $f(x, y)$ from UHF band SAR data, second run: its close-ups.

within a *lowpass* slow-time Doppler region and is *invariant* in its coordinates; see Sections 6.3 to 6.4, and Figure 6.5*a* and *b*. Thus any *modulation* of a target stripmap SAR signature (e.g., the above-mentioned compression operation of the spotlight SAR systems) would shift the stripmap Doppler spectrum into a *bandpass* region; this would *increase* the bandwidth of the resultant image.

However, in the case of our cited stripmap SAR data, as we mentioned on various occasions, the beamwidth $2B$ is larger than the interval of synthetic aperture measurements $2L$. Thus we should expect to see the spotlight SAR image compression effect in these stripmap SAR cases.

Compressed SAR image spectra for the two runs of the radar-carrying aircraft are shown in Figure 6.16*a* (first run) and *b* (second run). Both spectra possess strong energy within the following two lines (the lines are not shown in Figure 6.16*a* and *b*):

$$k_y = \tan(25°)k_x,$$

that is, within the angular beamwidth of $[-25°, 25°]$ which is *smaller* than the angular beamwidth of the target function spectra $F(k_x, k_y)$ $[-45°, 45°]$; see Figure 6.13*a* and *b*.

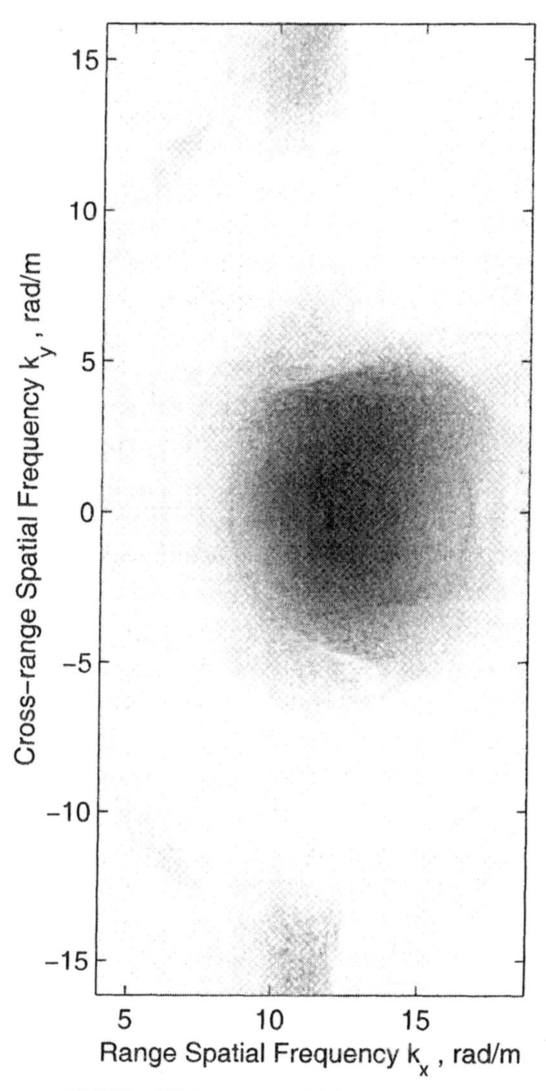

Figure 6.16a. Compressed UHF band SAR spectrum of foliage area and human-made targets: first run.

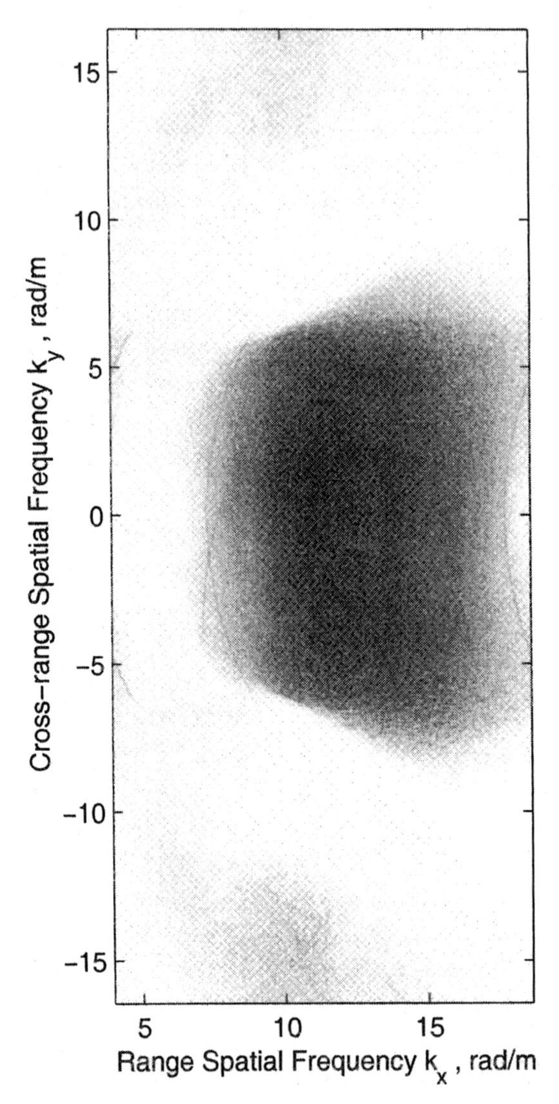

Figure 6.16b. Compressed UHF band SAR spectrum of foliage area and human-made targets: second run.

Consider the broadside target (foliage) area in the image of the first run which we studied for a smaller synthetic aperture $u \in [-50, 50]$ m in Chapter 5 (see Figure 5.24*a* and *b*); this area is centered around $(x_n, y_n) = (526, 0)$ m. Figure 6.17*a* is the reconstruction of this broadside foliage region with $u \in [-200, 200]$ m, and Figure 6.17*b* and *c*, respectively, shows the spectrum and compressed spectrum of the image in Figure 6.17*a*. Note that the spectrum of the broadside foliage area in Figure 6.17*b* also indicates a clear and strong signature within the angular beamwidth of $[-25°, 25°]$.

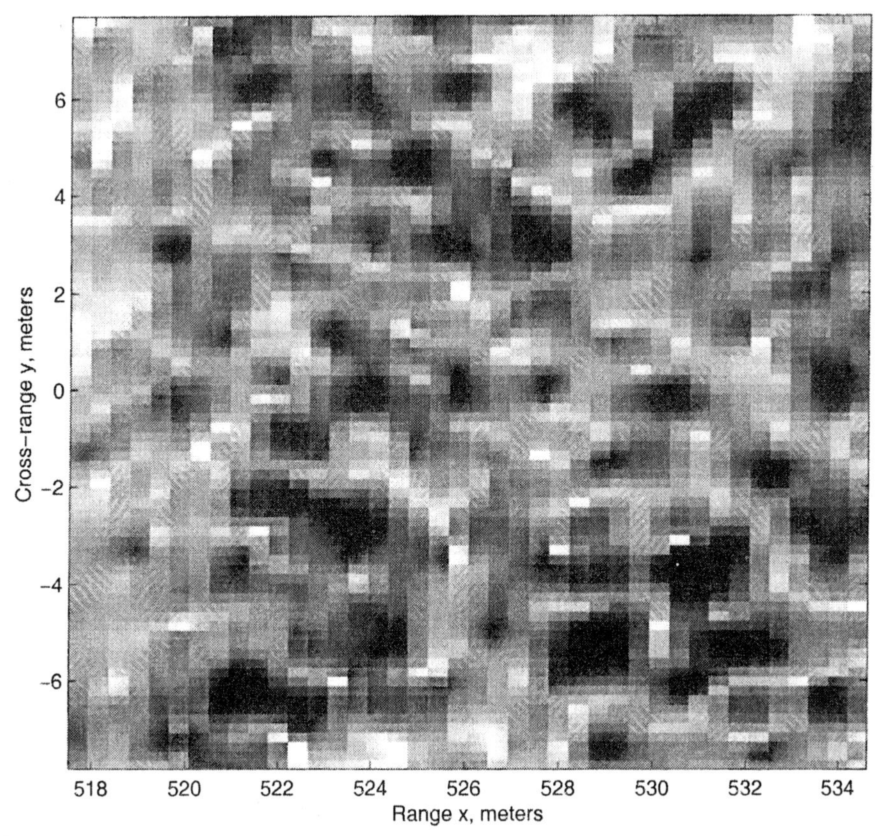

Figure 6.17a. Wavefront reconstruction of broadside target area (first run): spatial domain reconstruction.

Where does the angular beamwidth of $[-25°, 25°]$ *come from?* As we mentioned before, for both stripmap SAR databases the beamwidth is greater than the size of the synthetic aperture ($2B > 2L$). Thus we should anticipate certain spectral properties of the spotlight SAR systems in processing these stripmap SAR databases. In fact the dominant target area, which is the near-range foliage area, is seen within an aspect angle interval of

$$\left[-\arctan\left(\frac{L}{r_{\min}}\right), \arctan\left(\frac{L}{r_{\min}}\right)\right] \approx [-25°, 25°],$$

for *both* runs of the aircraft. The squint foliage area at near range is also seen approximately within the above aspect angle interval, though it is shifted somewhat. The SAR image compression aligns all these in the Doppler domain. The result is a dominant spectrum within the angular beamwidth of $[-25°, 25°]$ which is seen in Figure 6.16*a* and *b*.

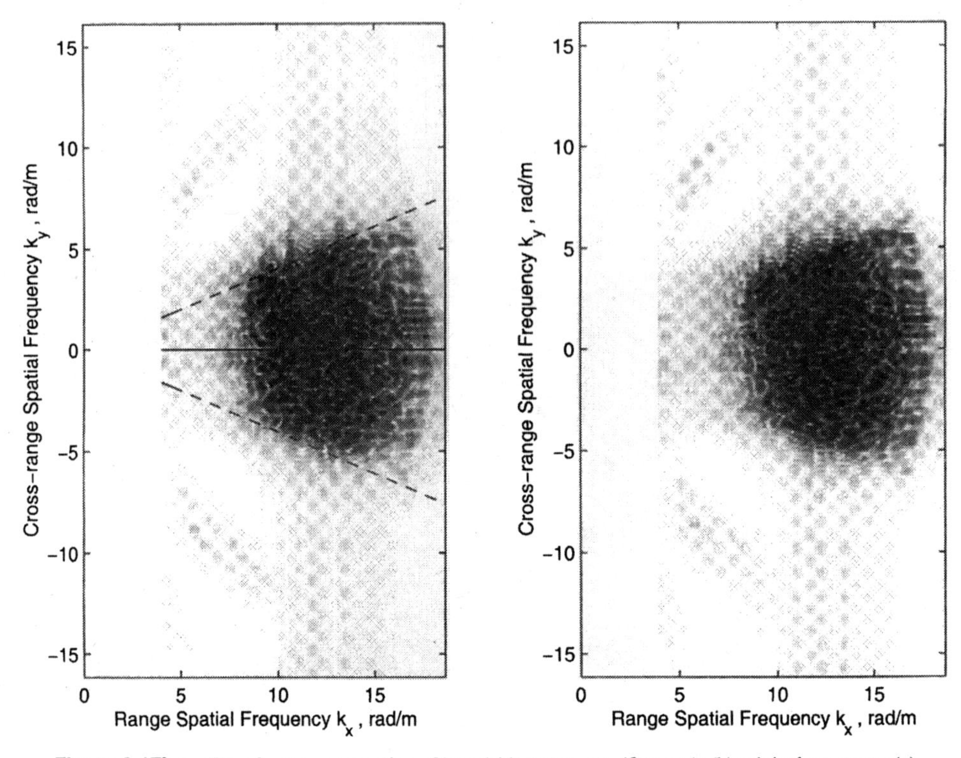

Figure 6.17b-c. Wavefront reconstruction of broadside target area (first run): (b) original spectrum; (c) compressed spectrum.

For the 16 m by 16 m broadside foliage area of Figure 6.17*a*, the aspect angle interval over which this region is observed by the radar is

$$\left[\arctan\left(\frac{y_n - L - 8}{x_n - 8}\right), \arctan\left(\frac{y_n + L + 8}{x_n - 8}\right)\right] = [-22°, 22°].$$

In Figure 6.17*b* the lines

$$k_y = \pm \tan 22° k_x$$

are shown via dashed lines which are superimposed on the target spectrum; these two lines correspond to the theoretical spectral support of this broadside foliage area. The equation of the continuous line in this figure is

$$k_y = \tan \theta_n(0) k_x,$$

where $\theta_n(0) = 0$ for the broadside foliage area.

Consider the targets which were studied in Figures 5.25–5.28, that is, the squint foliage area, the HEMTT truck, the broadside corner reflector, and the slanted corner reflector. Figures 6.18 to 6.21 show the reconstructions, the spectra, and the compressed spectra of these targets.

In the case of the spectrum of the squint foliage area in Figure 6.18*b*, the dashed lines, which identify the theoretical spectral support of this area for an *omni-directional* radar, correspond to the aspect angle interval

$$\left[\arctan\left(\frac{y_n - L - 8}{x_n - 8}\right), \arctan\left(\frac{y_n + L + 8}{x_n + 8}\right)\right] = [-49^\circ, -18.6^\circ].$$

However, as we mentioned earlier, the radar beamwidth angle, which dictates the support of the amplitude pattern $A(\omega, k_u)$ (that is another determining factor on the support of a target spectrum) is approximately within $\pm 45^\circ$. The above theoretical spectral band at -49° for the squint foliage area is beyond this support.

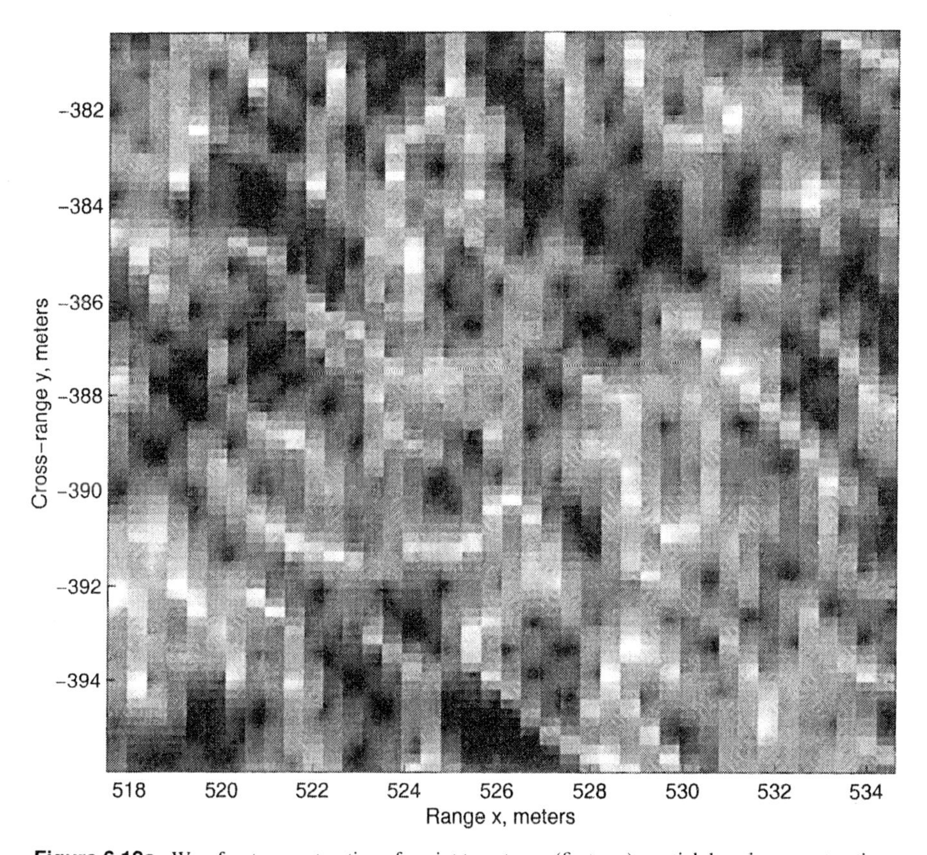

Figure 6.18a. Wavefront reconstruction of squint target area (first run): spatial domain reconstruction.

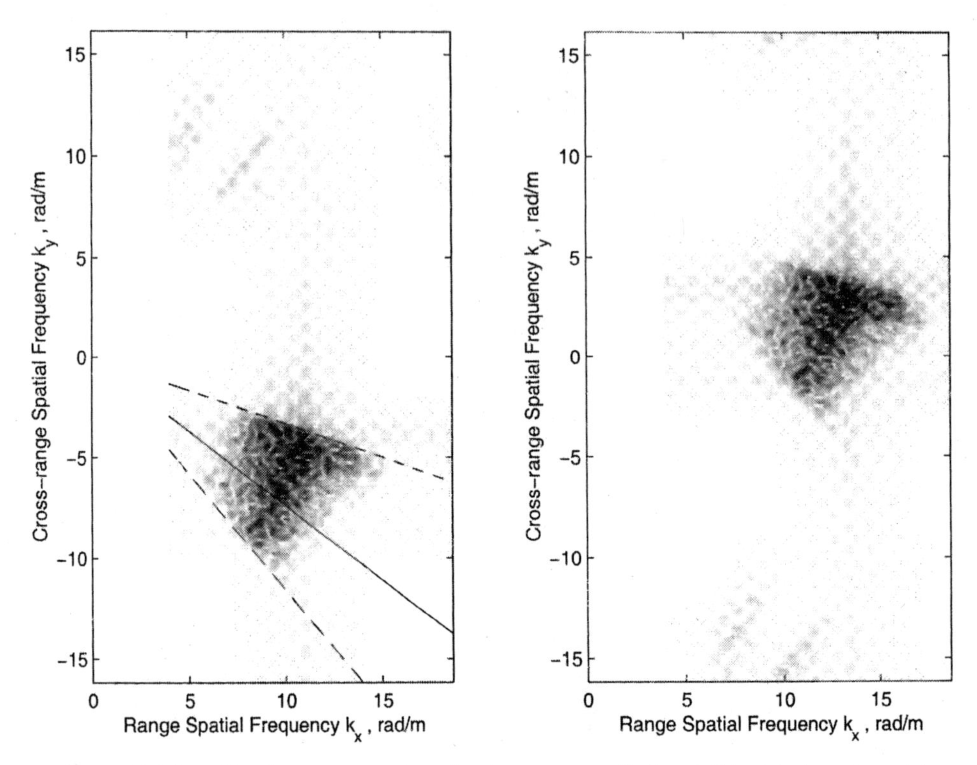

Figure 6.18b-c. Wavefront reconstruction of squint target area (first run): (b) original spectrum; (c) compressed spectrum.

As was the case in the spectrum of this squint foliage with $u \in [-50, 50]$ m in Figure 5.18*b*, the spectrum of this squint target area at $(x_n, y_n) = (526, -388)$ m, which is close to the boundary of the digital spotlight filter at $Y_0 = -450$ m, appears filtered out for

$$\sqrt{k_x^2 + k_y^2} > 2k_c.$$

This is the result of using the *narrow-bandwidth* assumption in digital spotlight filtering. As we noted earlier, this causes filtering of the signature of the near-boundary targets (i.e., $y_n \approx Y_0$) at fast-time frequencies greater than the carrier frequency.

The HEMTT truck exhibits a fairly narrow spectrum in Figure 6.19*b*. This corresponds to the fact that the amplitude pattern of the truck $a_n(\omega, x_n, y_n - u)$ has significant energy only within a small aspect angle interval near the broadside (zero aspect angle); that is, the truck is observable to the radar over a relatively small synthetic aperture interval. In fact the reconstruction of the truck in Figure 5.26*a* with $u \in [-50, 50]$ m is similar to its reconstruction in Figure 6.19*a* with $u \in [-200, 200]$ m.

The broadside corner reflector shows a fairly uniform amplitude pattern $A_n(\omega, k_u)$ within its spectral support in Figure 6.20*b*; this spectrum is centered along the line

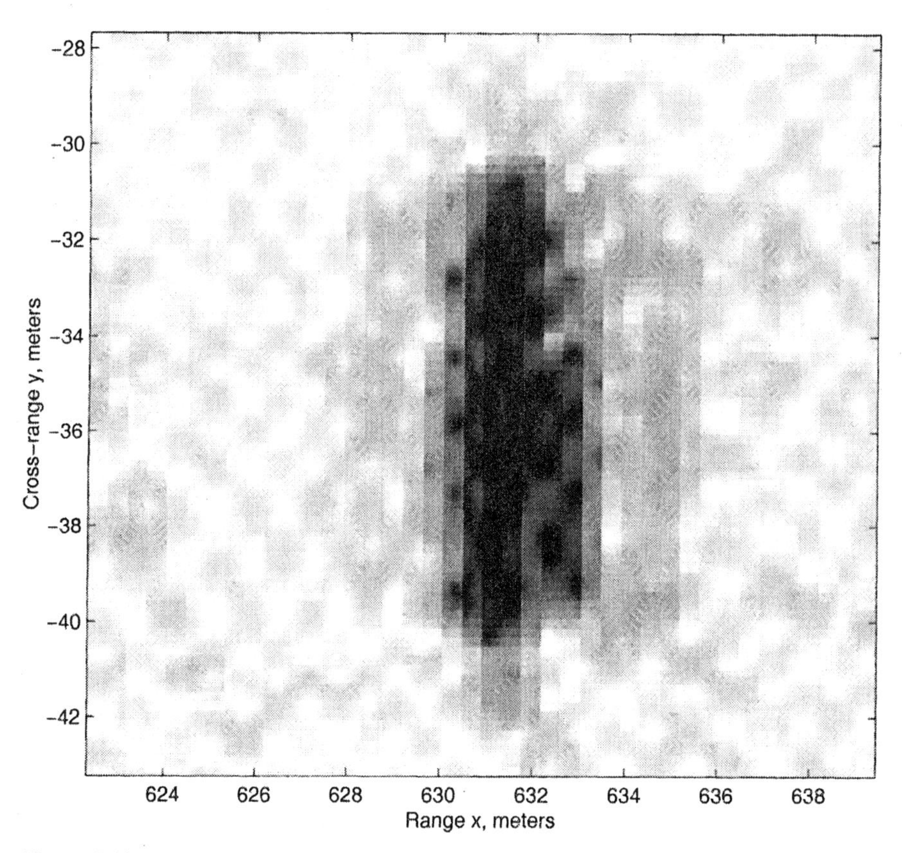

Figure 6.19a. Wavefront reconstruction of HEMTT truck (first run): spatial domain reconstruction.

$k_y = \tan\theta_n(0)k_x$, with $\theta_n(0) = 4.2°$. As we will see in the numerical results with CSAR data in Chapter 7, such a corner reflector shows a uniform amplitude pattern over a $\pm 45°$ aspect angle interval.

The same cannot be seen in the spectrum of the slanted corner reflector in Figure 6.21*b*. The big *flash* of this corner reflector occurs at an aspect angle which is greater than its $\theta_n(0) = 16.3°$. This could be due to a *slant* (off-broadside rotation) in the orientation of the corner reflector or to a defect that has altered the uniform amplitude pattern of the corner reflector. Figure 6.22 shows the results for the third corner reflector.

Next we consider the structure identified as a *moving* target in Figure 5.30 for $u \in [-50, 50]$ m. The results for these targets with $u \in [-200, 200]$ m are shown in Figure 6.23. With a larger synthetic aperture, this region exhibits distinct *streak*-looking signature in Figure 6.23*a*. By using what is referred to as *SAR ambiguity function* [s95], it can be shown that these are in fact signatures of moving targets and not extended stationary targets, for example, walls or power lines; see the discussion on moving target detection and SAR ambiguity function in Section 6.7.

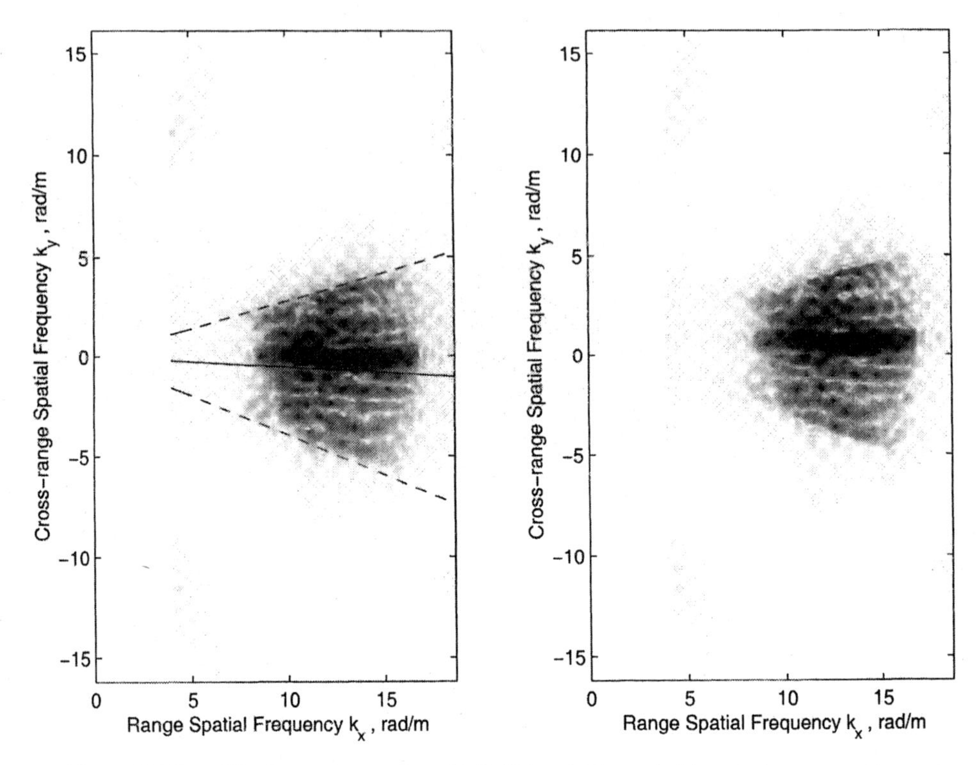

Figure 6.19b-c. Wavefront reconstruction of HEMTT truck (first run): (b) original spectrum; (c) compressed spectrum.

The uniform spectrum seen in Figure 6.23*b* is mainly due to the echoed signals from the ground near the moving targets (i.e., the ground clutter). The *narrow* yet strong signature within the ground clutter spectrum is the spectrum of the moving target. As we noted earlier, the spectrum of the HEMTT truck in Figure 6.19*b* is also fairly narrow (due to its narrow beamwidth angle of observability). Thus it is plausible that one could see a similar narrow-width spectrum for a moving truck or automobile. However, we observe that this narrow-width spectrum is *rotated* in the (k_x, k_y) domain; each moving target spectrum has a different rotation value.

This behavior of moving targets in a SAR scene is well known. In fact the SAR signature of a moving target in the slow-time Doppler domain is shifted with respect to the SAR signature of the stationary targets [s94, ch. 5; yan]; the amount of the shift in the slow-time Doppler k_u domain at the fast-time frequency ω is

$$2k\frac{v_x}{v_r},$$

where v_r is the along-track speed of the radar-carrying aircraft (in the cross-range y domain) and v_x is the speed of the moving target in the range x domain.

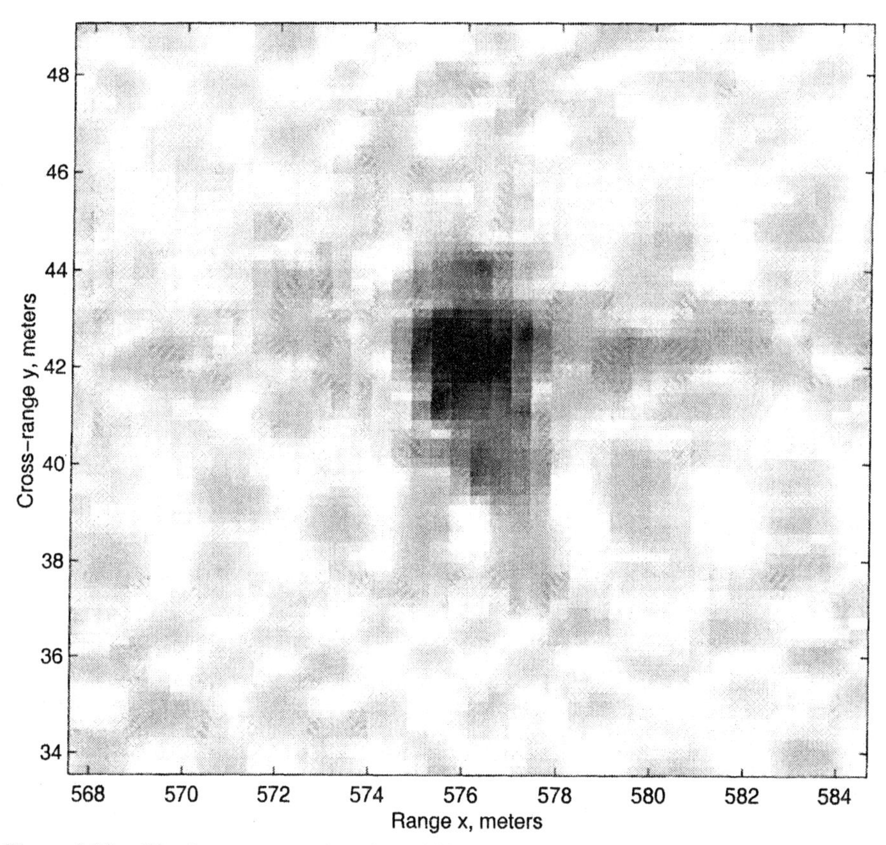

Figure 6.20a. Wavefront reconstruction of broadside corner reflector (first run): spatial domain reconstruction.

Note that the *spotlight* SAR signature of a stationary target located at (x_n, y_n) is centered at $2k \sin \theta_n(0)$ at the fast-time frequency ω. (Note that the *finite* aperture stripmap SAR data exhibits spotlight SAR signal properties.) Thus the SAR signature of the moving target appears centered around

$$2k \sin \theta_n(0) + 2k \frac{v_x}{v_r}.$$

The slow-time Doppler k_u domain shift of the moving target signature translates into the following shift in the angular slow-time Doppler ϕ domain (at all of the available fast-time frequencies):

$$\arcsin\left(\frac{v_x}{v_r}\right).$$

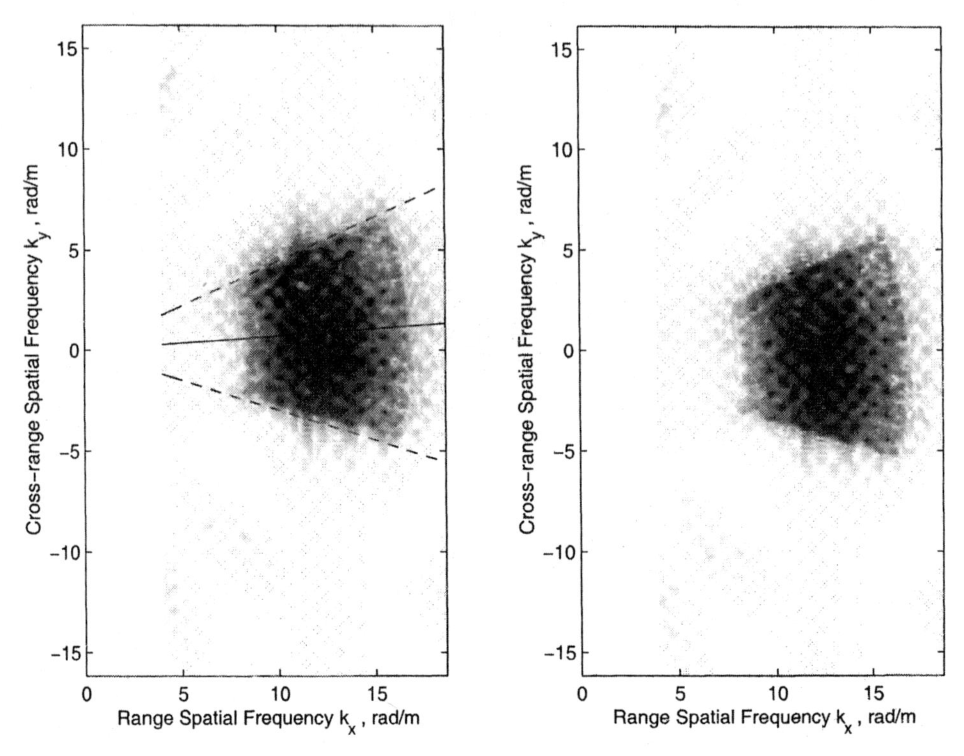

Figure 6.20b-c. Wavefront reconstruction of broadside corner reflector (first run): (b) original spectrum; (c) compressed spectrum.

Using the SAR mapping from the measurement spectral domain (ω, k_u) to the target spectral (k_x, k_y) domain, one can see that the spectrum of a moving target appears rotated by the above angular slow-time Doppler domain shift, that is, $\arcsin(v_x/v_r)$ from the line

$$k_y = \tan\theta_n(0)k_x.$$

Note that this angular rotation in Figure 6.23*b* is *positive*. This is due to the fact that the rotation angle depends on the polarity of v_x, that is, the target speed in the range domain.

Next consider the second-run data of the UHF band stripmap SAR system. As we noted in the introductory chapter, there are 256 fast-time samples in this database which are separated by $\Delta_t = 4$ ns. Since the starting point of fast-time sampling is $T_s = 4.74$ μs, the end point of fast-time sampling is at $T_f = 5.76$ μs.

The reason for bringing up this issue is that in addition to the limited synthetic aperture of $u \in [-391, 391]$ m, the end point of fast-time sampling placed another restriction on the range of aspect angles over which a target is observable to the

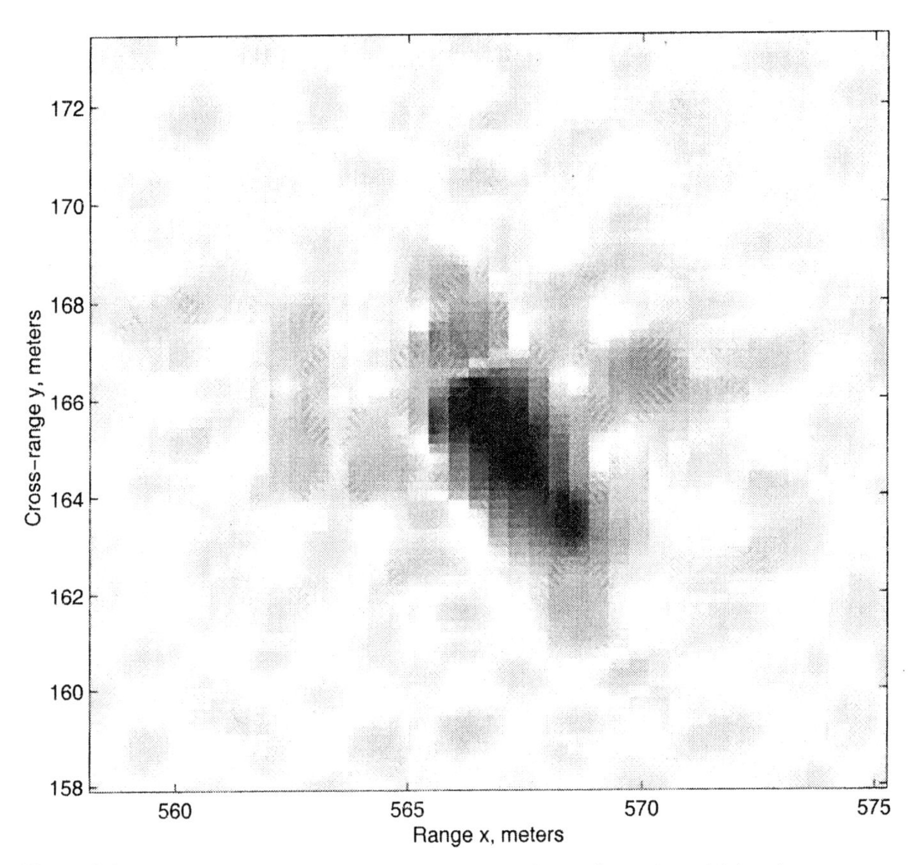

Figure 6.21a. Wavefront reconstruction of slanted corner reflector (first run): spatial domain reconstruction.

radar. For the nth target which is located at (x_n, y_n), the synthetic aperture position, labeled u_{nf}, at which the echoed signal of the nth target arrives at the fast-time T_f is governed by

$$T_f = \frac{2\sqrt{x_n^2 + (y_n - u_{nf})^2}}{c}.$$

This yields two solutions for u_{nf}, which are

$$u_{nf1} = y_n - \sqrt{\left(\frac{cT_f}{2}\right)^2 - x_n^2},$$

$$u_{nf2} = y_n + \sqrt{\left(\frac{cT_f}{2}\right)^2 - x_n^2}.$$

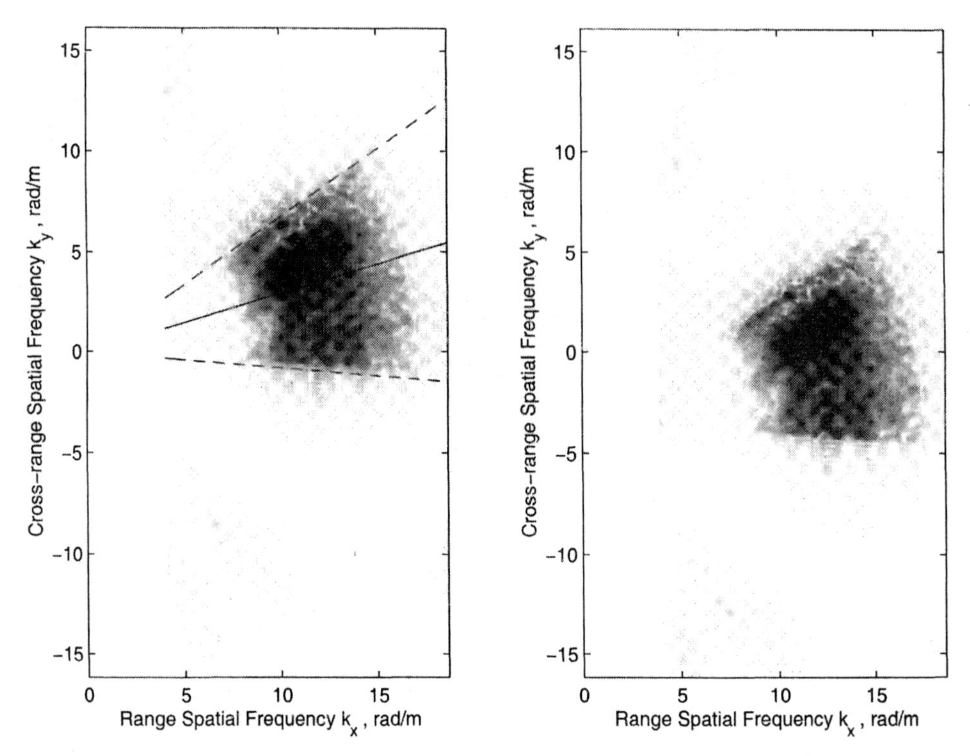

Figure 6.21b-c. Wavefront reconstruction of slanted corner reflector (first run): (b) original spectrum; (c) compressed spectrum.

The region of the observability of the nth target is limited not only by the aspect angle interval imposed by the size of the synthetic aperture (see Section 2.4), that is,

$$[\theta_n(L), \theta_n(-L)] = \left[\arctan\left(\frac{y_n - L}{x_n}\right), \arctan\left(\frac{y_n + L}{x_n}\right)\right],$$

but also for the aspect angle interval corresponding to the pair (u_{nf1}, u_{nf2}), which is

$$\left[\arctan\left(\frac{y_n - u_{nf2}}{x_n}\right), \arctan\left(\frac{y_n - u_{nf1}}{x_n}\right)\right].$$

Figure 6.24a and b show the spectra of the reconstruction of the HEMTT truck and a 60 m by 100 m surrounding ground area (from the second-run data) without any preprocessing and with digital spotlighting and upsampling, respectively; the truck is located at $(x_n, y_n) = (823, 274)$ m in Figure 6.15d. The thin solid line in these figures corresponds to the line $k_y = \tan\theta_n(0)k_x$; the two dashed lines identify the theoretical spectral support of the 60 m by 100 m area. The thick crown-shaped contour is the spectral support band of this area which is dictated by the aspect angles

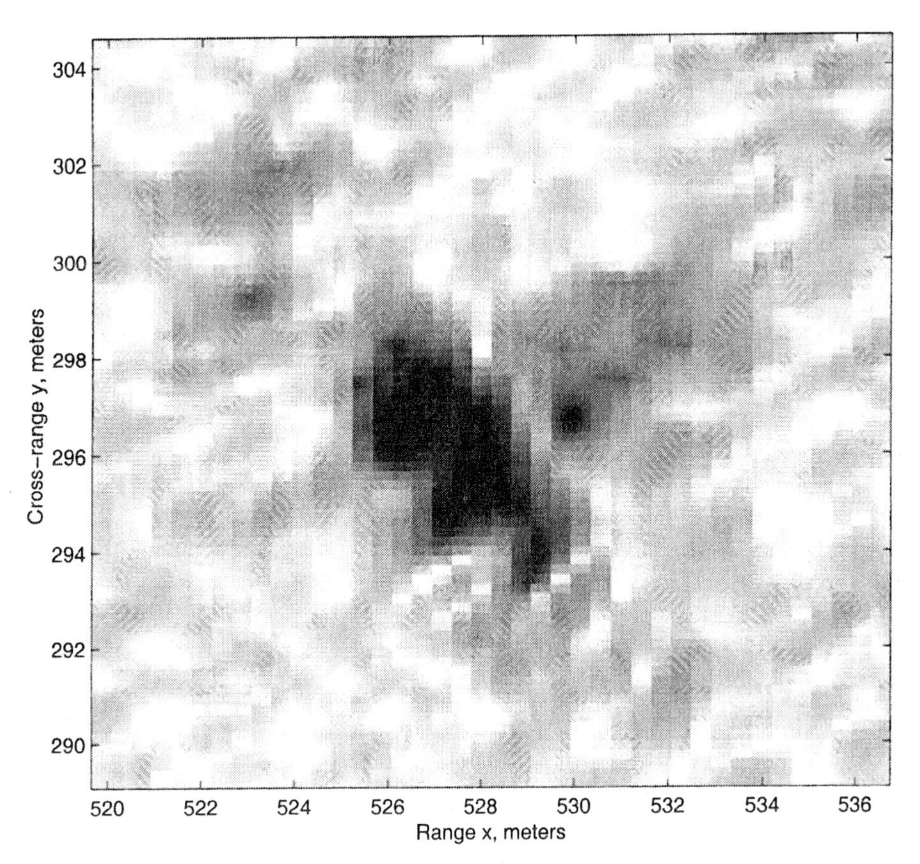

Figure 6.22a. Wavefront reconstruction of corner reflector 3 (wide squint angle) (first run): spatial domain reconstruction.

of the pair (u_{nf1}, u_{nf2}) (for all the targets within the 60 m by 100 m area) and the fast-time bandwidth of the radar which imposes the restriction

$$2k_{\min} \leq \sqrt{k_x^2 + k_y^2} \leq 2k_{\max}.$$

The spectrum in Figure 6.24*b*, which resulted from digital spotlighting and upsampling the SAR data, shows the signature of the ground (clutter) within the crown-shaped region; the truck signature can also be seen here. However, there are other signatures outside the crown-shaped region. In Figure 6.24*a* the target spectrum, which is formed without digital spotlighting and upsampling of the SAR data, contains stronger versions of these undesirable signatures. These signatures are due to various factors, such as the side lobes of the radar radiation pattern and radar instability, which result in slow-time Doppler aliasing.

If these are slow-time Doppler aliasing, why aren't these seen in Figure 6.24a as wrap around *errors?* The answer was found in our discussion on the effect of slow-

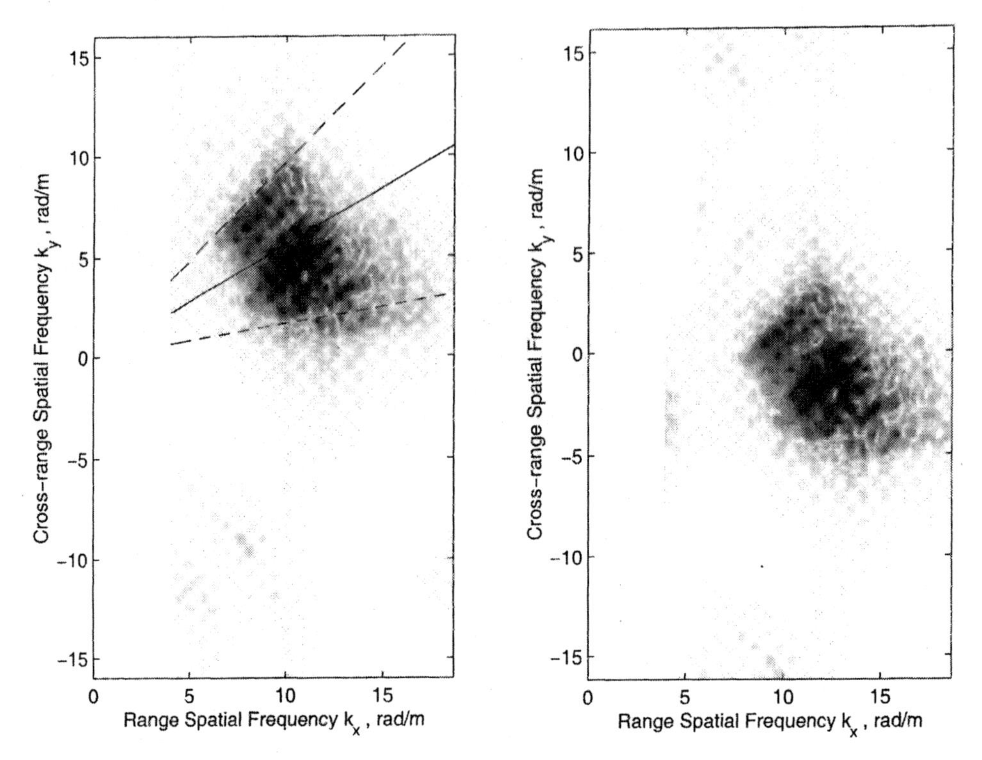

Figure 6.22b-c. Wavefront reconstruction of corner reflector 3 (wide squint angle) (first run): (b) original spectrum; (c) compressed spectrum.

time Doppler aliasing in cross-range imaging in Chapter 2. The wraparound data of the aliased target signature would not appear in the vicinity of the target area in the spatial domain, that is, (x_n, y_n); this is one of the unique functional properties of the spherical PM signal which can be analytically demonstrated. The wraparound aliased data are hidden somewhere in the reconstructed image in the spatial domain. In fact the aliased signatures seen outside the crown-shaped area and even the data within the crown-shaped area contain wraparound aliasing components of the targets that reside somewhere outside the 60 m by 100 m area.

Note that the spectrum in Figure 6.24*b* still contains some aliasing. As we noted earlier, digital spotlighting and upsampling of this stripmap SAR database could reduce some but not all the aliasing artifacts. It is interesting to observe that the aspect angle interval limitation imposed by the pair (u_{nf1}, u_{nf2}) does isolate the target signature from a portion of the aliasing. Thus one can apply a filter (with passband identified by the crown-shaped area) to remove these isolated aliasing components. Note that the crown-shaped area varies with the coordinates of the target area; thus the filter is a shift-varying one.

Figures 6.25 to 6.28 show similar results for 60 m by 100 m areas centered around, respectively, a corner reflector (CR) at (771, 349) m, a *near-range* broadside foliage

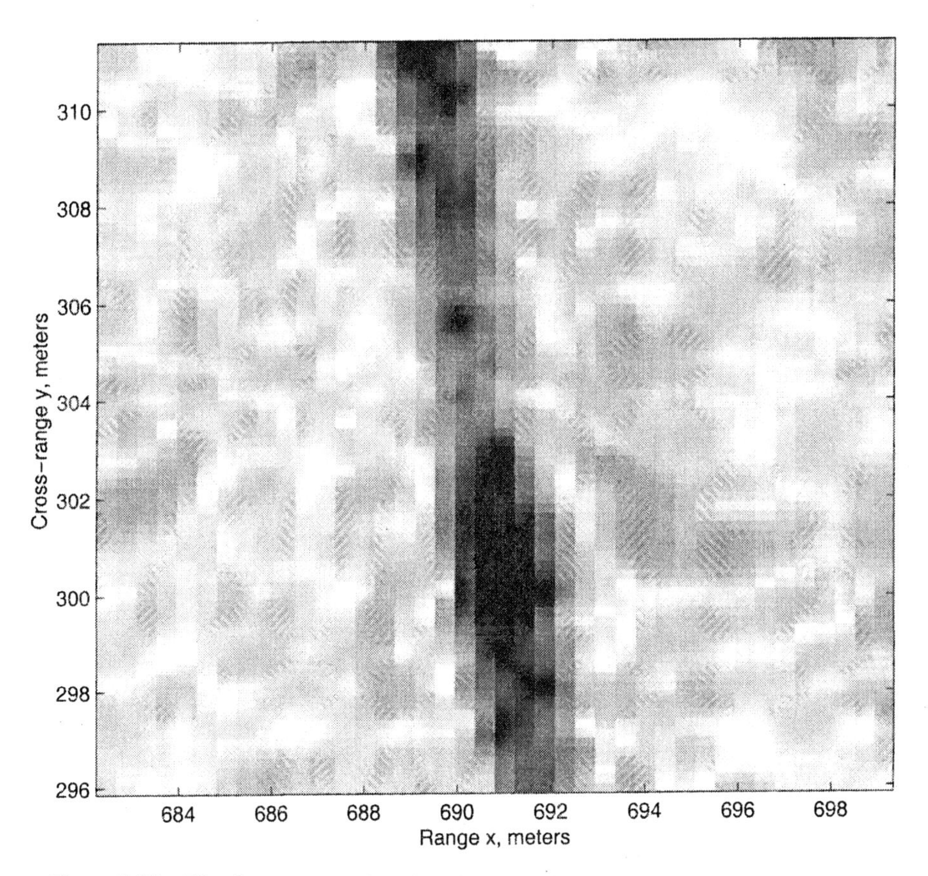

Figure 6.23a. Wavefront reconstruction of moving target (first run): spatial domain reconstruction.

area at (736, 0) m, a *far-range* broadside foliage area at (822, 0) m, and a corner reflector at (759, 476) m; the spatial domain images of these targets appeared in Figure 6.15. These figures exhibit phenomena similar to the ones for the truck in Figure 6.24. Note that the worst aliasing effect is seen in the spectrum of the squint foliage area in Figure 6.28*a*.

Digital Reconstruction via Range Stacking

The stripmap digital reconstruction via range stacking is also very similar to the one for the spotlight SAR systems (Figures P6.9 and P6.10). The main difference is the preprocessing of the stripmap SAR data which was outlined for the stripmap digital reconstruction via spatial frequency interpolation in this section. These are not repeated.

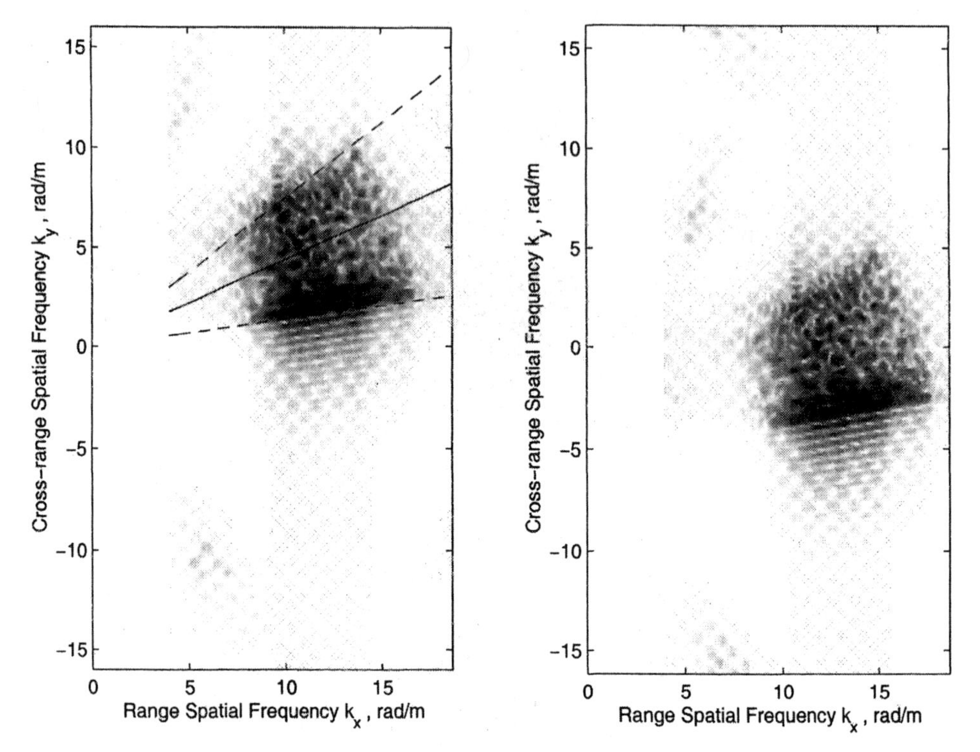

Figure 6.23b-c. Wavefront reconstruction of moving target (first run): (b) original spectrum; (c) compressed spectrum.

Digital Reconstruction via Time Domain Correlation and Backprojection

The time domain correlation (TDC) and backprojection algorithms of Section 4.7 can be used for digital reconstruction in the stripmap SAR systems (Figures P6.11–P6.14). The only modification involves incorporating the radar amplitude pattern $a(\omega, x, y)$ as a two-dimensional matched filter; see the discussion on the stripmap SAR signal model in Section 6.1. This matched filtering can be performed in the spectral (ω, k_u) domain via (see Section 6.2)

$$S_M(\omega, k_u) = P^*(\omega) A^*(\omega, k_u) S(\omega, k_u),$$

where $A(\omega, k_u)$ is the radar amplitude pattern in the spectral domain.

In this case the TDC reconstruction equation for the stripmap SAR becomes

$$f(x_i, y_j) = \int_{-\infty}^{\infty} \int_{\omega_{\min}}^{\omega_{\max}} s(\omega, u) a^*(\omega, x_i, y_i - u) P^*(\omega) \exp[j\omega t_{ij}(u)] \, d\omega \, du$$

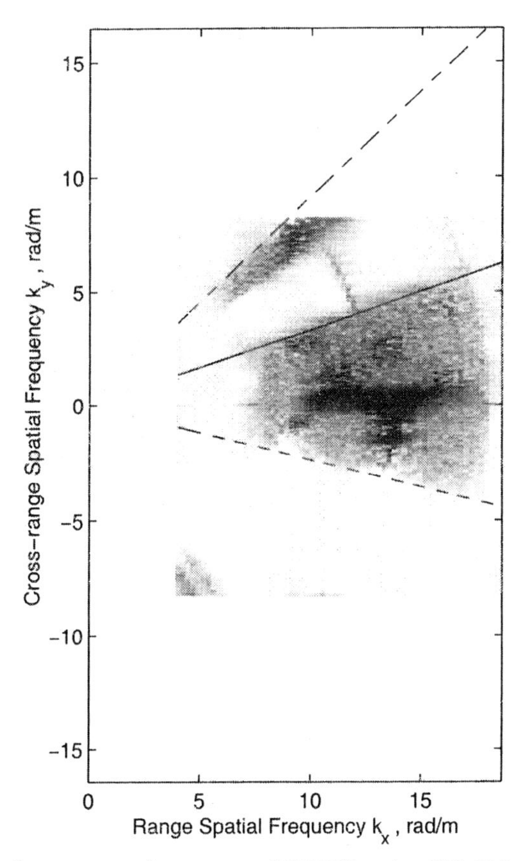

Figure 6.24a. Wavefront reconstruction spectrum of HEMTT truck at (823, 274) m (second run): without preprocessing.

for $T_s \leq t_{ij}(u) \leq T_f$; otherwise, the integrand is zero. In the above expression

$$t_{ij}(u) = \frac{2\sqrt{x_i^2 + (y_j - u)^2}}{c}$$

is the round-trip delay of the echoed signal for the target at (x_i, y_j) when the radar is at $(0, u)$.

The radar amplitude pattern $a(\omega, x_i, y_i - u)$ is nonzero only within its beamwidth, that is, for $|y_j - u| \leq B_i(\omega)$, where the half-beamwidth is defined via

$$B_i(\omega) = x_i \tan \phi_d.$$

(Note that the divergence angle ϕ_d varies with the fast-time frequency ω for a planar radar aperture.) Then the TDC reconstruction equation becomes

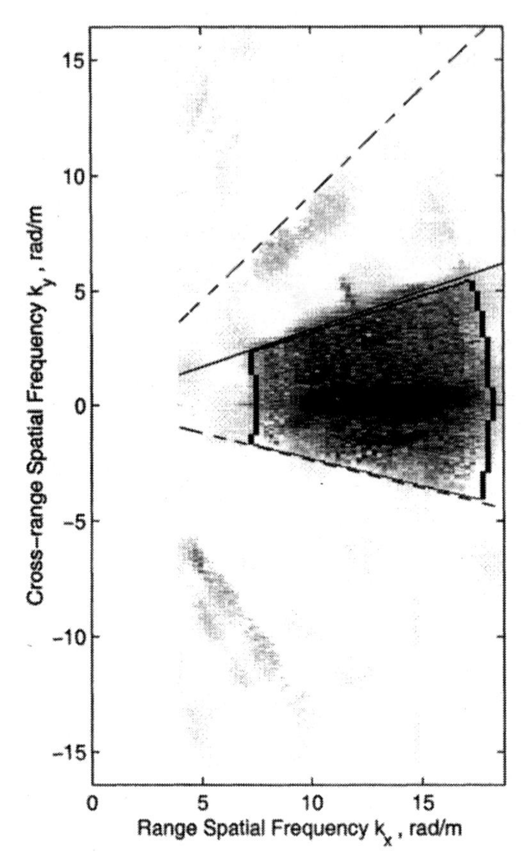

Figure 6.24b. Wavefront reconstruction spectrum of HEMTT truck at (823, 274) m (second run): with digital spotlighting and upsampling.

$$f(x_i, y_j) = \int_{\omega_{\min}}^{\omega_{\max}} \int_{y_j - B_i(\omega)}^{y_j + B_i(\omega)} s(\omega, u) a^*(\omega, x_i, y_i - u) P^*(\omega) \exp[j\omega t_{ij}(u)]\, d\omega\, du$$

$$= \int_{\omega_{\min}}^{\omega_{\max}} \int_{y_j - B_i(\omega)}^{y_j + B_i(\omega)} s_M(\omega, u) \exp[j\omega t_{ij}(u)]\, d\omega\, du.$$

The backprojection algorithm for this scenario becomes

$$f(x_i, y_j) = \int_{y_j - B_i(\omega_{\min})}^{y_j + B_i(\omega_{\min})} \mathbf{s}_M[t_{ij}(u), u]\, du;$$

note that the u domain integral (slow-time coherent integration) must be performed over the *largest* beamwidth of the radar signal $B_i(\omega_{\min})$ which is at the lowest fast-time frequency.

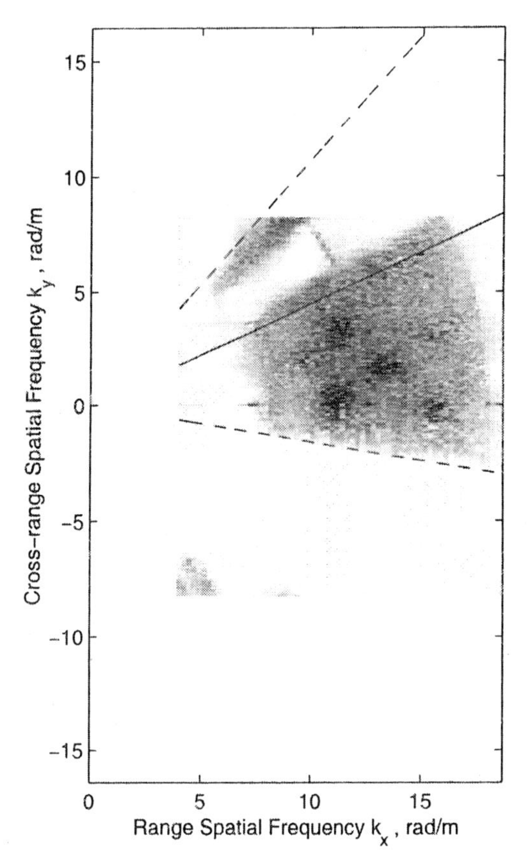

Figure 6.25a. Wavefront reconstruction spectrum of corner reflector at (771, 349) m (second run): without preprocessing.

- Slow-time Doppler subsampling cannot and should not be performed for the TDC and backprojection algorithms.

When the radar amplitude pattern is not known, it is approximated to be a constant within the radar divergence angle, for example,

$$a(\omega, x_i, y_i - u) = \begin{cases} 1, & |y_i - u| \leq B_i, \\ 0, & \text{otherwise.} \end{cases}$$

The steps to implement the TDC and backprojection algorithms in the stripmap SAR are similar to those outlined in Section 5.6; they are not repeated here.

The digital signal processing methods described in Section 6.5 for stripmap SAR data could enhance the quality of images formed via TDC. However, as we noted earlier, most users of the TDC and backprojection algorithms are not aware of this digital signal processing issues; see the discussion on TDC and backprojection in Section 5.6.

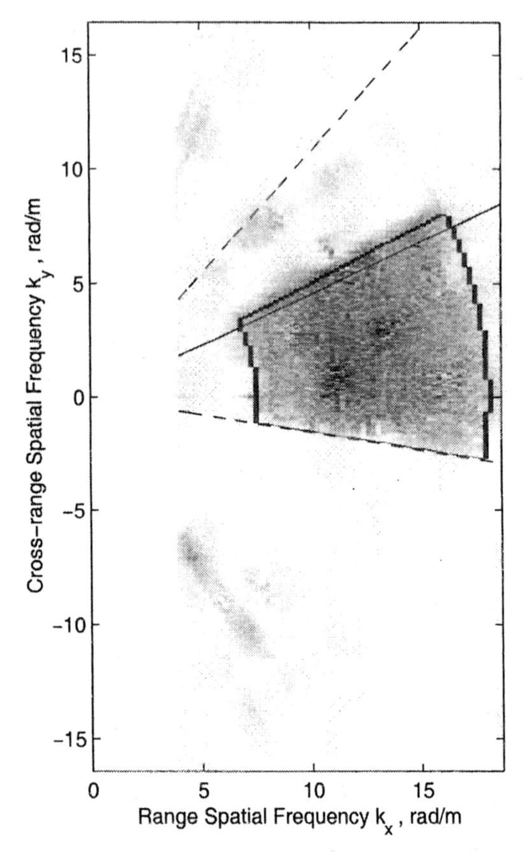

Figure 6.25b. Wavefront reconstruction spectrum of corner reflector at (771, 349) m (second run): with digital spotlighting and upsampling.

We present backprojection reconstruction results for the first-run data of Figure 6.7*a*. To prevent any adverse effects which our Fourier digital spotlighting and upsampling might have on the backprojection method, we use the original measured SAR data (i.e., data not digitally spotlighted and upsampled). Figure 6.29*a* and *b*, respectively, shows the backprojection reconstruction and its spectrum for HEMTT 4 (as a reference, see Figure 6.19*a* and *b*). The truck backprojection image in Figure 6.29*a* contains more clutter than the wavefront reconstruction of the same target in Figure 6.19*a*. By using the digitally spotlighted and upsampled data, the backprojection reconstruction of the truck would look as clean as its wavefront reconstruction in Figure 6.19*a*.

Figure 6.30*a* and *b*, respectively, shows similar results for the moving target; the wavefront reconstruction and spectrum of this target appeared in Figure 6.23*a* and *b*. The signature of this target also contains some slow-time Doppler aliasing.

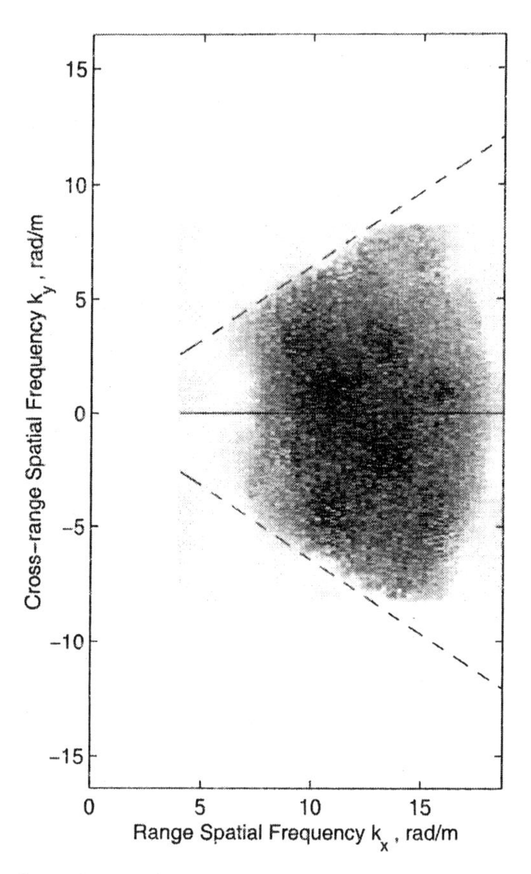

Figure 6.26a. Wavefront reconstruction spectrum of near-range broadside foliage area at (736, 0) m (second run): without preprocessing.

Effect of Beamwidth (Slow-time Doppler) Filtering

We also mentioned in Section 5.6 that the higher slow-time Doppler frequency SAR data are commonly filtered out by the users of TDC and backprojection methods to reduce the heavy computational burden of these reconstruction algorithms. In particular, for the wide-beamwidth UHF SAR systems, for which the radar beamwidth angle is over 80° (i.e., ±40°), the beamwidth angle is assumed to be about 30° (i.e., ±15°). Thus, at a given fast-time frequency ω, the SAR data in the slow-time Doppler frequencies

$$|k_u| > 2k \sin 15^\circ$$

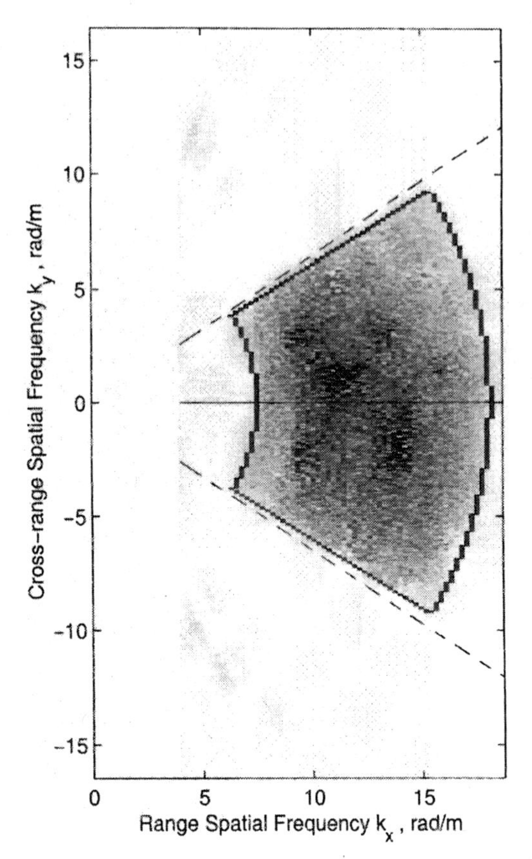

Figure 6.26b. Wavefront reconstruction spectrum of near-range broadside foliage area at (736, 0) m (second run): with digital spotlighting and upsampling.

or the slow-time Doppler angles

$$|\phi| = \left| \arcsin\left(\frac{k_u}{2k}\right) \right| > 15^\circ$$

are discarded.

This slow-time Doppler filtering significantly reduces the computational load of the TDC and backprojection SAR reconstruction methods. This, however, is at the cost of the following in the stripmap SAR systems:

1. Degradations caused by the increase of the power of the side lobes
2. Loss of slanted targets
3. Loss of moving targets

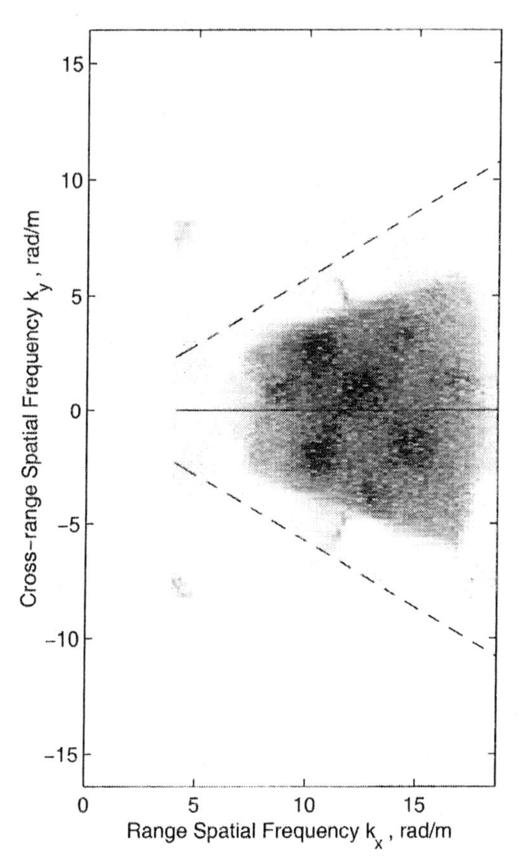

Figure 6.27a. Wavefront reconstruction spectrum of far-range broadside foliage area at (822, 0) m (second run): without preprocessing.

(For the discussion and details see TDC and backprojection in Section 5.6.) Figure 6.31 shows the two-dimensional spectrum of the reconstructed stripmap SAR image which is formed with the backprojection method from the first-run SAR data when the beamwidth angle is limited to the interval $[-15, 15]$ degrees. Figure 6.32*a* is the resultant reconstructed SAR image in the spatial domain; the close-ups are shown in Figure 6.32*b* and *c*. Note the loss of some of the moving targets in Figure 6.32*c*. (For comparison, see the full beamwidth reconstruction in Figures 6.14.)

Effect of Motion Errors in Slow-time Doppler Spectrum

The user should be aware of a source of error that gives SAR data the *appearance* of having a relatively narrow beamwidth angle; this error source is motion errors, which we examine next. We discussed a FOPEN wide-beamwidth and wide-bandwidth ([30, 90] MHz) stripmap database in Section 4.9, on *motion compensation using the global positioning system*. Figure 4.9*a* showed the simulated stripmap data, without

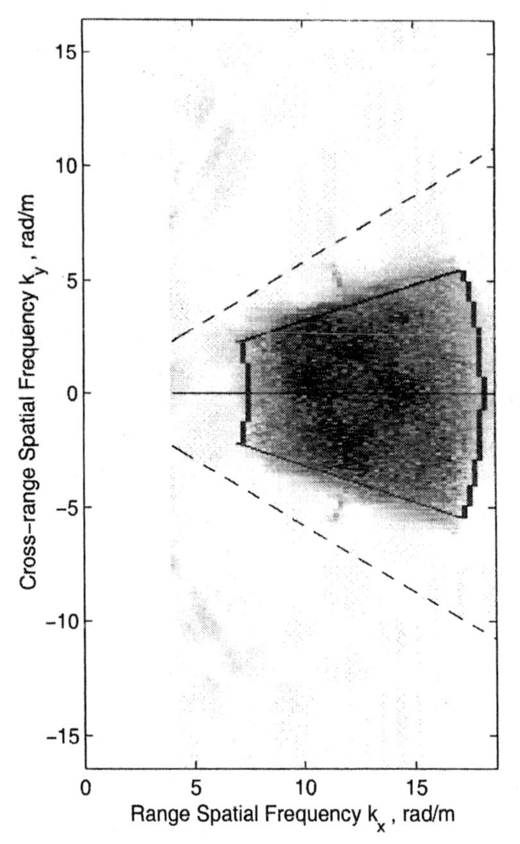

Figure 6.27b. Wavefront reconstruction spectrum of far-range broadside foliage area at (822, 0) m (second run): with digital spotlighting and upsampling.

motion errors, for seven unit reflectors in this SAR system; the Matlab code which is discussed in Section 6.8 of this chapter for stripmap SAR system with a planar radar aperture was used to simulate that data.

Figure 6.33*a* shows the stripmap SAR spectrum $S(\omega, k_u)$ for the SAR data in Figure 4.9*a*. Note that the spectrum shows a slow-time Doppler frequency k_u spread that is invariant in the radar fast-time frequency; this is due to the fact that the radar aperture is planar. Next we consider the SAR data with the motion errors in Figure 4.10*a* of Chapter 4. Figure 6.33*b* is the two-dimensional spectrum of that data. Note that this SAR spectrum appears to be bandlimited in the slow-time Doppler k_u domain with a smaller beamwidth angle as compared with the spectrum of motion error-free SAR data in Figure 6.33*a*.

However, it can be shown that motion phase errors, which correspond to a *multiplicative* noise source, results in the bandwidth expansion of stripmap SAR data in the slow-time Doppler k_u domain [s92a], that is, the opposite of what is seen in

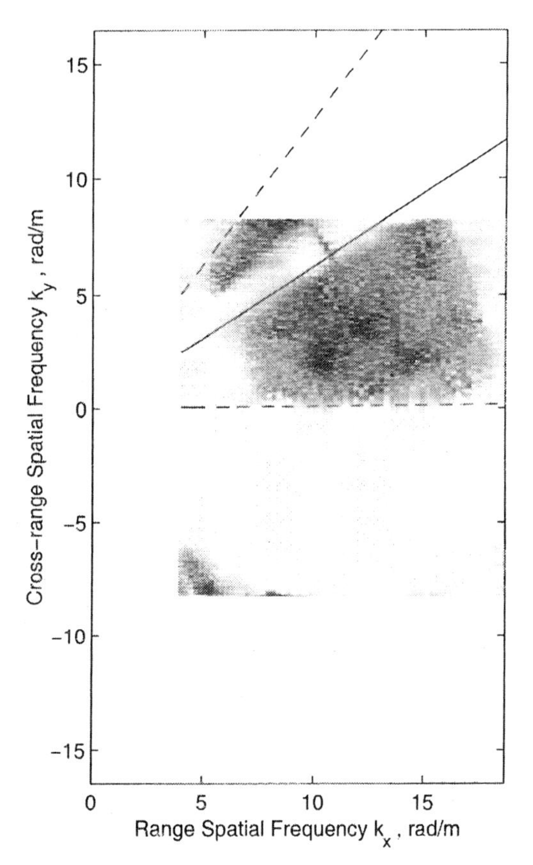

Figure 6.28a. Wavefront reconstruction spectrum of corner reflector at (759, 476) m (second run): without preprocessing.

Figure 6.33*b*. One can show that in fact the phenomenon seen in Figure 6.33*b* is due to wrap around aliasing properties of PM signals.

The main conclusion of this discussion is that the user should not employ the spectrum of a SAR database with motion errors, particularly in wide-beamwidth and wide-bandwidth FOPEN SAR systems, to examine the slow-time Doppler properties of the SAR data. As we noted in Section 4.9, the user should first use narrow-beamwidth motion compensation (using GPS data) which removes the gross motion phase errors from the SAR data.

Figure 6.33*c* shows the spectrum of the narrow-beamwidth motion compensated data for the above-cited stripmap SAR problem; this is the two-dimensional spectrum of the data that appeared in Figure 4.11*b* of Chapter 4. This spectrum has a similar support and even *coherent* properties of the SAR data spectrum without the motion errors of Figure 6.33*a*; this is due to the fact that residual motion errors after narrow-beamwidth compensation result in *localized* smearing and degradations of the SAR data.

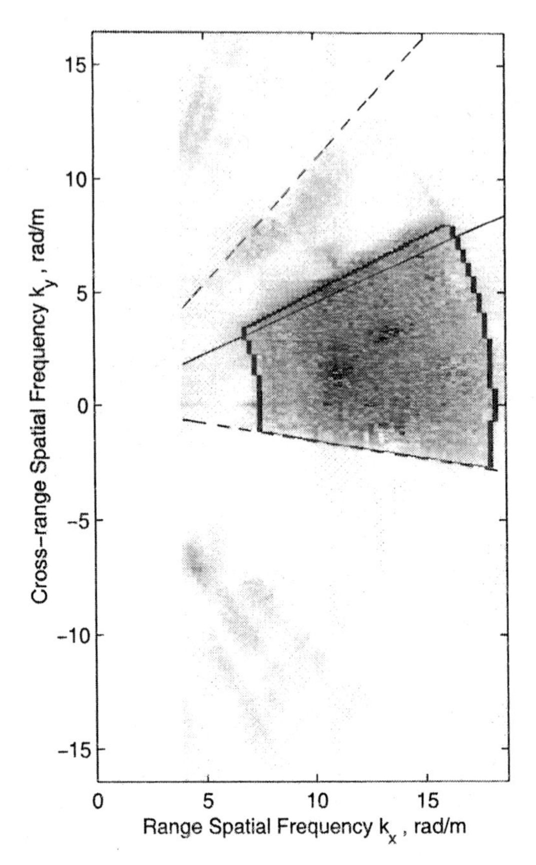

Figure 6.28b. Wavefront reconstruction spectrum of corner reflector at (759, 476) m (second run): with digital spotlighting and upsampling.

In fact the narrow-beamwidth motion compensated stripmap SAR data can be used in the SAR data preprocessing algorithms such as subaperture digital spotlighting. Finally, after forming the SAR image from the processed narrow-beamwidth motion compensated SAR data, the shift-varying filter for wide-beamwidth motion compensation (see Section 4.9) is used to further focus the reconstructed stripmap SAR image.

Subpatch "Mosaic" Digital Reconstruction with Subaperture Data

We have studied four digital image formation methods for stripmap (and spotlight) SAR systems: spatial frequency domain-based interpolation, range stacking, time-domain correlation (TDC), and backprojection. The four methods, provided that they are implemented correctly, should yield comparable results. However, the choice of the digital reconstruction method in a SAR problem depends on the computational resources and time constraints of the user. For instance, a researcher who is studying

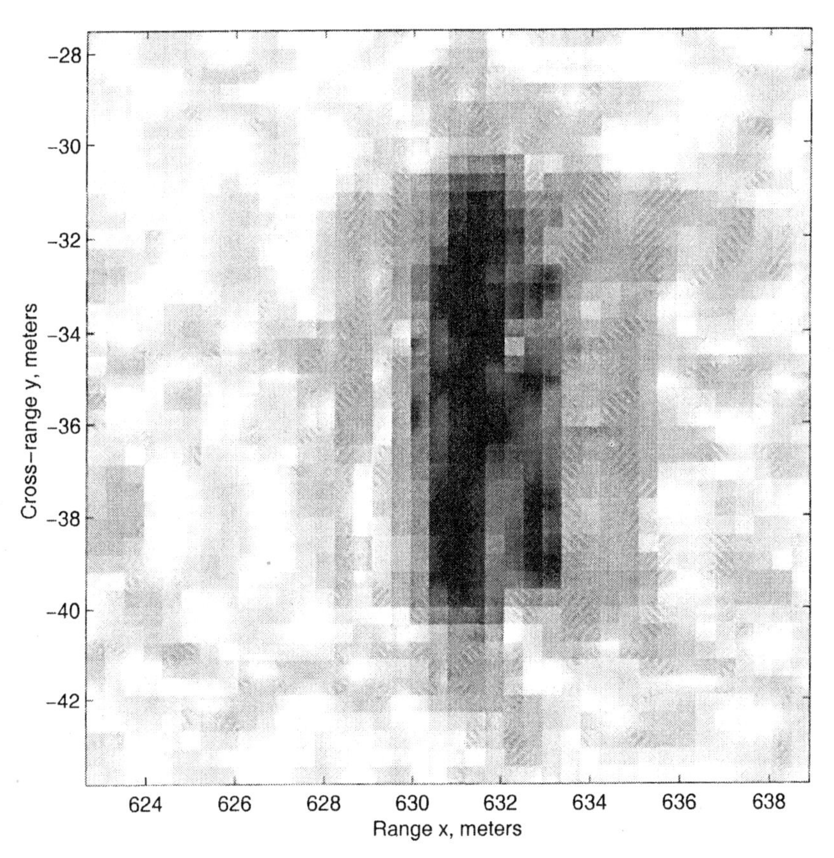

Figure 6.29a. Backprojection reconstruction of HEMTT truck (first run; without digital spotlighting and upsampling): spatial domain reconstruction.

SAR signal/data in an environment with a common computer would be better off with one of the faster digital reconstruction algorithms. On the other hand, a government agency that is collecting SAR data on board an aircraft equipped with a supercomputer may not have the same constraints.

Various investigators have suggested the use of forming subpatches of a terrain from its SAR data and combining the resultant reconstructed subpatches to create a *Mosaic* image of the terrain area [per; s94n]. This approach has two utilities. First, a multiprocessor computer can be programmed to form these subpatches in a parallel form; this would greatly reduce the computational time for image formation. The other positive property of subpatch image formation is in preprocessing of the SAR data to reduce side lobe effects and slow-time Doppler aliasing. In fact the subaperture digital spotlighting of a smaller subpatch area would be more effective and accurate than trying to process the entire target area.

The main criticism of subpath "mosaic" SAR reconstruction is that it generates undesirable-looking (discontinuous) boundaries between the two adjacent sub-

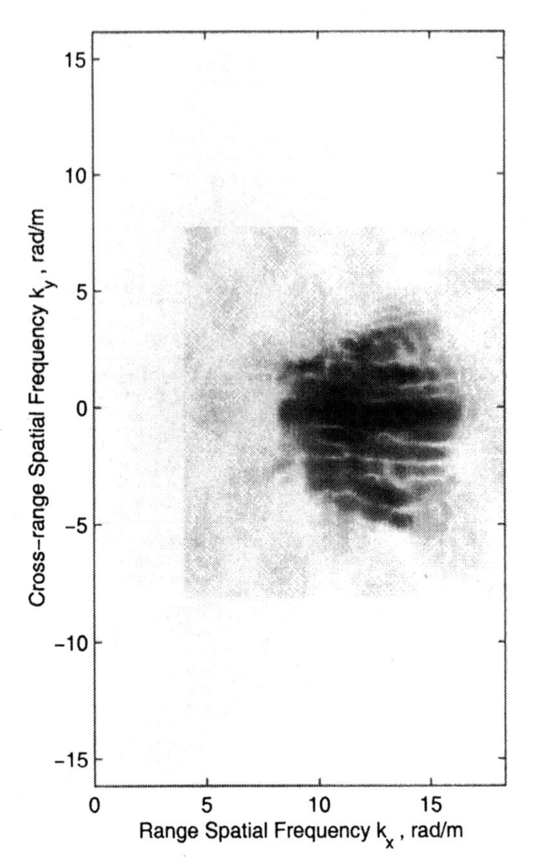

Figure 6.29b. Backprojection reconstruction of HEMTT truck (first run; without digital spotlighting and upsampling): spectrum.

patches. This problem can be easily circumvented by choosing subpatches that slightly overlap with their neighboring subpatches. After forming each subpatch, the user should discard the outer half of its overlapping (four) boundaries.

What is a suitable size of a subpatch? In the case of TDC and backprojection, the primary reason for subpatch SAR imaging is to reduce the time of computation; this is based on certain approximations within a subpatch when it is viewed from a given synthetic *subaperture* interval. It turns out that the size of the subpatch and the subaperture become dependent on the type of approximation which is used. These subpatch/subaperture approximations for TDC and backprojection are beginning to appear in the literature.

For digital SAR imaging via spatial frequency domain interpolation and range stacking, the main merit of subpatch SAR imaging is in its rapid image formation via parallel processing. For this, the most efficient size of a subpatch in the cross-

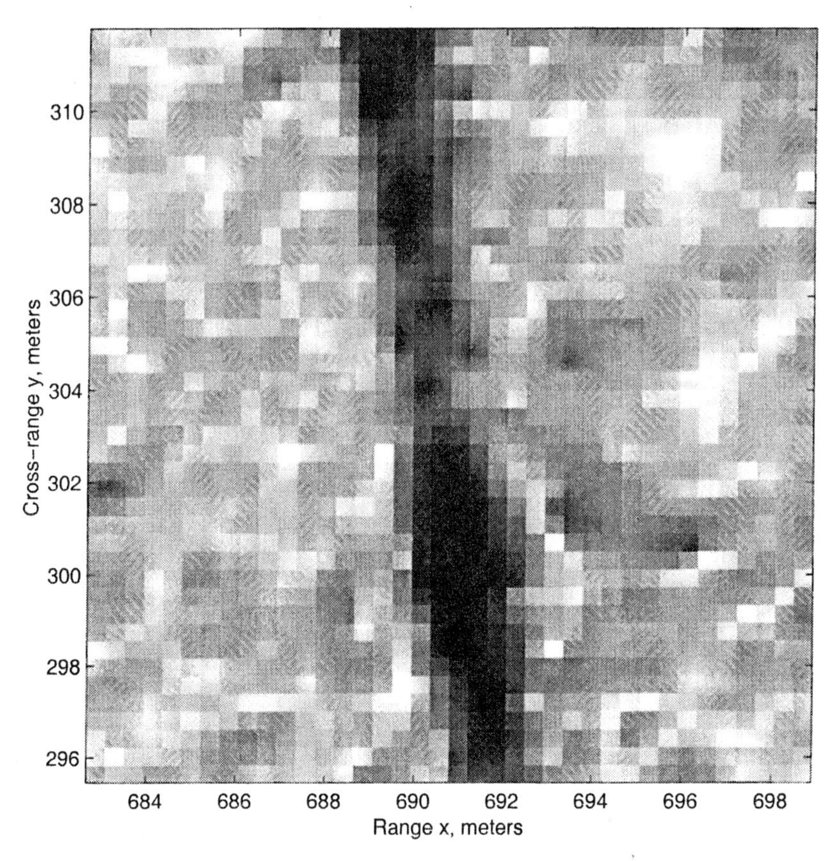

Figure 6.30a. Backprojection reconstruction of moving target (first run; without digital spotlighting and upsampling): spatial domain reconstruction.

range domain is equal to the size of a subaperture; that is,

$$Y_0 = L_s.$$

(It might be a good idea to provide a guard band; e.g., choose $Y_0 = 1.1L_s$.) This choice allows the user to develop a parallel processor for *coherent subaperture-subpatch* SAR image formation. This processor takes the data in each subaperture, digitally spotlights the data for a given subpatch, and forms a preliminary spatial frequency (k_x, k_y) domain [or (x_i, k_y) domain for range stacking] reconstruction of that subpatch.

The parallel processor performs this operation *simultaneously* for all preselected subapertures and subpatches. Note that depending on the size of the radar beamwidth, the user can determine whether or not the data within a subaperture contain any contribution from a subpatch; this should be done to prevent unnecessary computation by the processor. The outcomes of all subaperture frequency domain

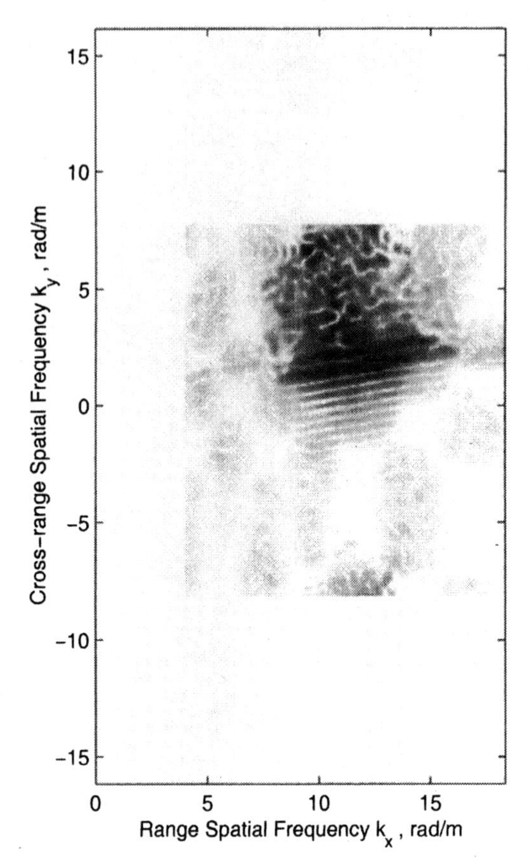

Figure 6.30b. Backprojection reconstruction of moving target (first run; without digital spotlighting and upsampling): spectrum.

reconstructions of each subpatch are *coherently* added; the result is then converted to the spatial domain SAR image of the subpatch via inverse DFT.

The range size of a subpatch, that is, $2X_0$, comes into the picture in the approximations that we use for digital spotlighting. The use of a relatively large X_0 would invalidate some of the approximations that we used to design the digital spotlight filter. It turns out that for comparable range and cross-range resolutions (i.e., $\Delta_x \approx \Delta_y$), the user could choose $X_0 \approx Y_0$.

There exists another factor that could improve the quality of SAR images via filtering clutter, slow-time Doppler aliasing, and, perhaps, radar's faulty transmissions. This factor revealed itself in our study of the stripmap SAR data of the second run which had a relatively *small fast-time interval* for processing. We showed in Figures 6.25 to 6.29 that the small interval of fast-time processing provided another slow-time Doppler filter to isolate the target signature from the noise terms. This corresponds to a shift-varying filter that could enhance the quality of the reconstructed

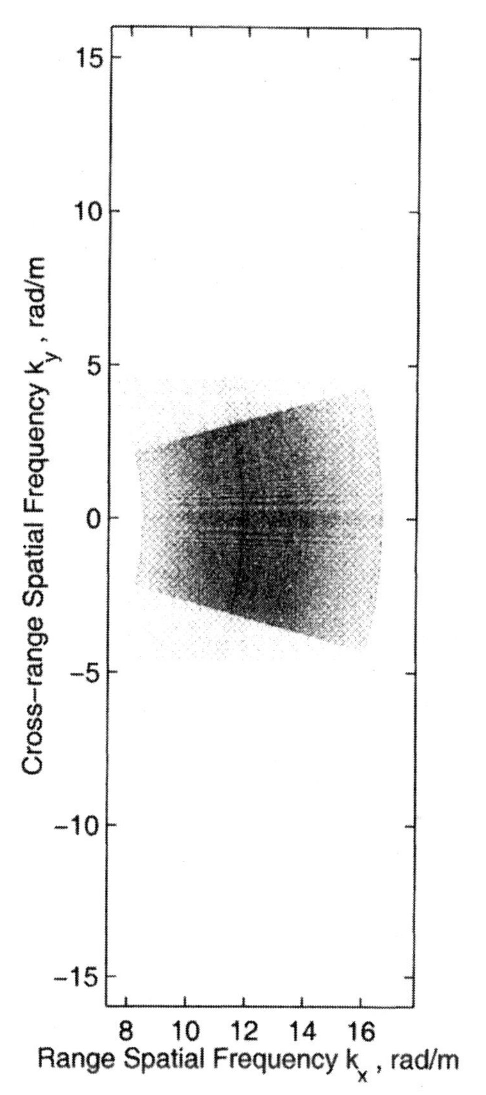

Figure 6.31. Spectrum of backprojection reconstruction with beamwidth angle (Doppler) of [−15, 15] degrees (first run).

image. Note that fast-time gating and processing of the SAR data within each of the fast-time gates is analogous to the slow-time subaperture processing.

Parallel Subaperture-Subpatch Digital Image Formation Algorithm To summarize the main conclusions of this section, the user should seek an approach for preprocessing of the SAR data and image formation that yields fairly clutter-free and alias-free data that could be passed through a parallel processor for fast image formation. Such an approach could be based on the following steps.

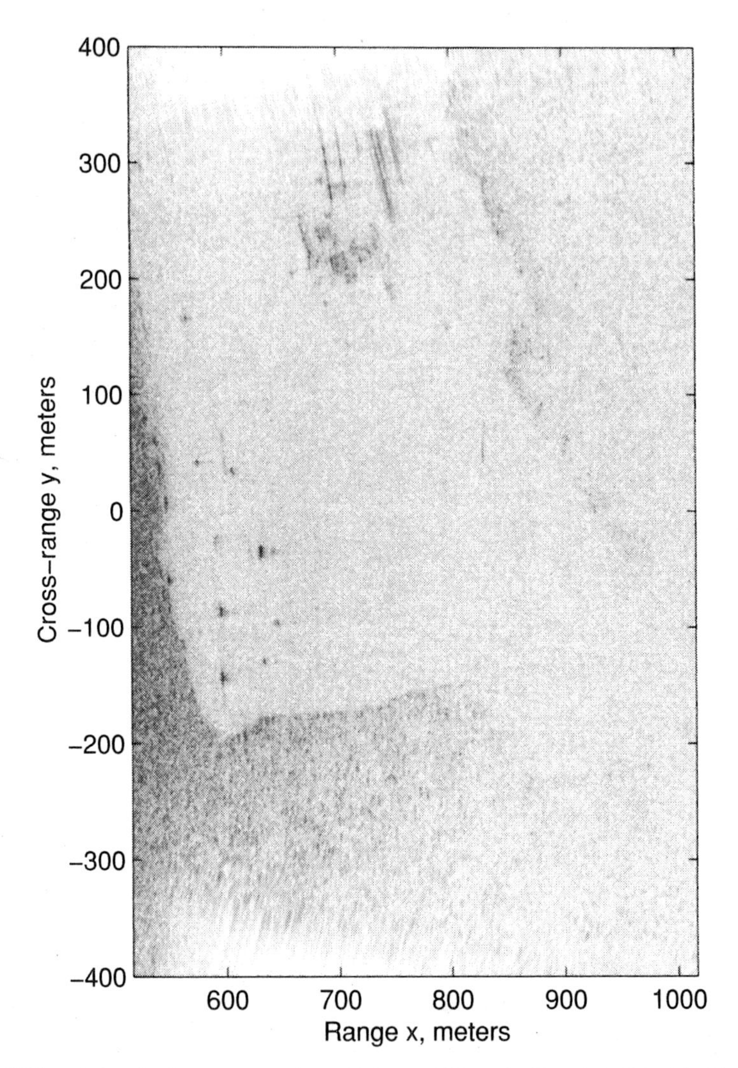

Figure 6.32a. Backprojection reconstruction with beamwidth angle (Doppler) of [−15, 15] degrees (first run): target area.

Step 1. Partition the measured SAR data in both the fast-time and slow-time (t, u) *domains.*

Step 2. Digitally spotlight the SAR data within each partition in the (t, u) *domain for a set of preselected subpatches in the spatial* (x, y) *domain.*

Step 3. Reconstruct the SAR image of each subpatch from its digitally spotlighted data of the partitioned SAR data in the (t, u) *domain.*

Step 4. Coherently add the reconstructions of a subpatch from all the partitioned (t, u) *domain SAR data sets.*

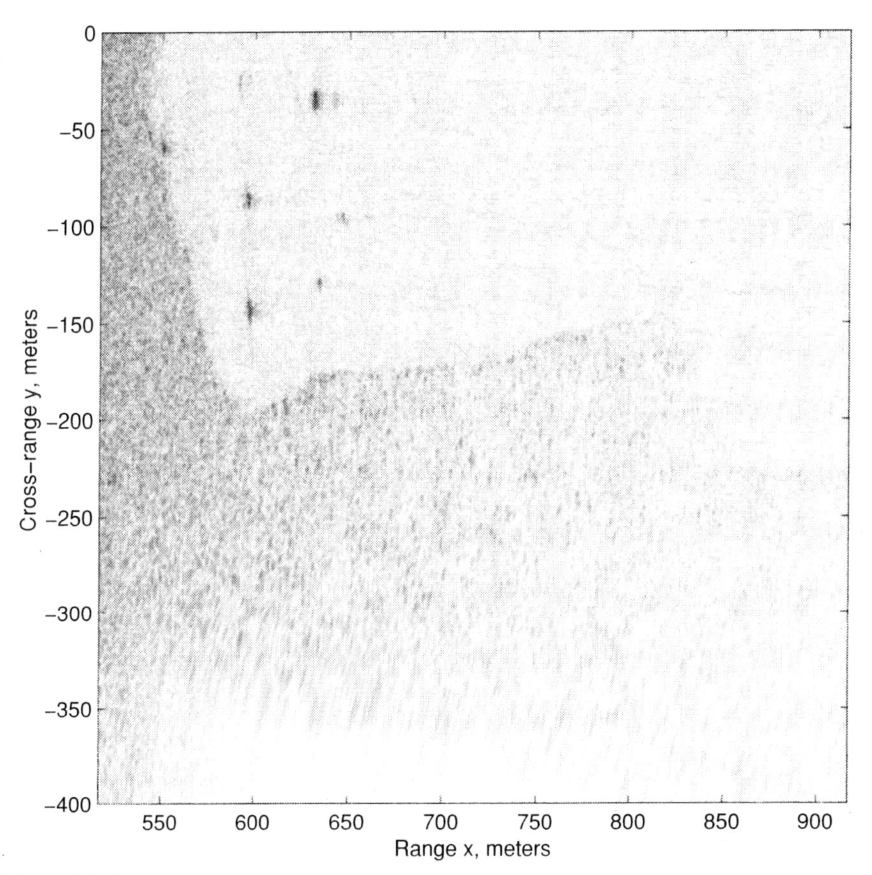

Figure 6.32b. Backprojection reconstruction with beamwidth angle (Doppler) of [−15, 15] degrees (first run): close-up of lower target area.

Depending on the choice of the SAR reconstruction algorithm (i.e., spatial frequency interpolation, range stacking, TDC, or backprojection), the user should make corresponding adjustments in Step 3 of the above algorithm.

Consider the 200 m by 200 m subpatch which is centered at $(x, y) = (750, 400)$ m in the spatial domain for the first run data of Figure 6.7*a*; this subpatch contains a portion of the farm area. Figure 6.34*a* and *b*, respectively, shows the reconstruction spectrum and spatial domain reconstruction of this subpatch; the boundary of the subpatch is identified by a solid line in Figure 6.34*b*. Note that this subpatch is at a nonzero squint angle with respect to the synthetic aperture $u \in [-200, 200]$ m; the squint angle is 28 degrees. The subaperture digital spotlighting and bandpass slow-time Doppler processing of squint SAR data were discussed in Chapter 5.

Next we consider the subpatch wavefront reconstruction of a broadside terrain area which is centered at $(x, y) = (900, 0)$ m; the reconstructed spectrum and spatial domain reconstruction are shown in Figure 6.35*a* and *b*, respectively. (The sub-

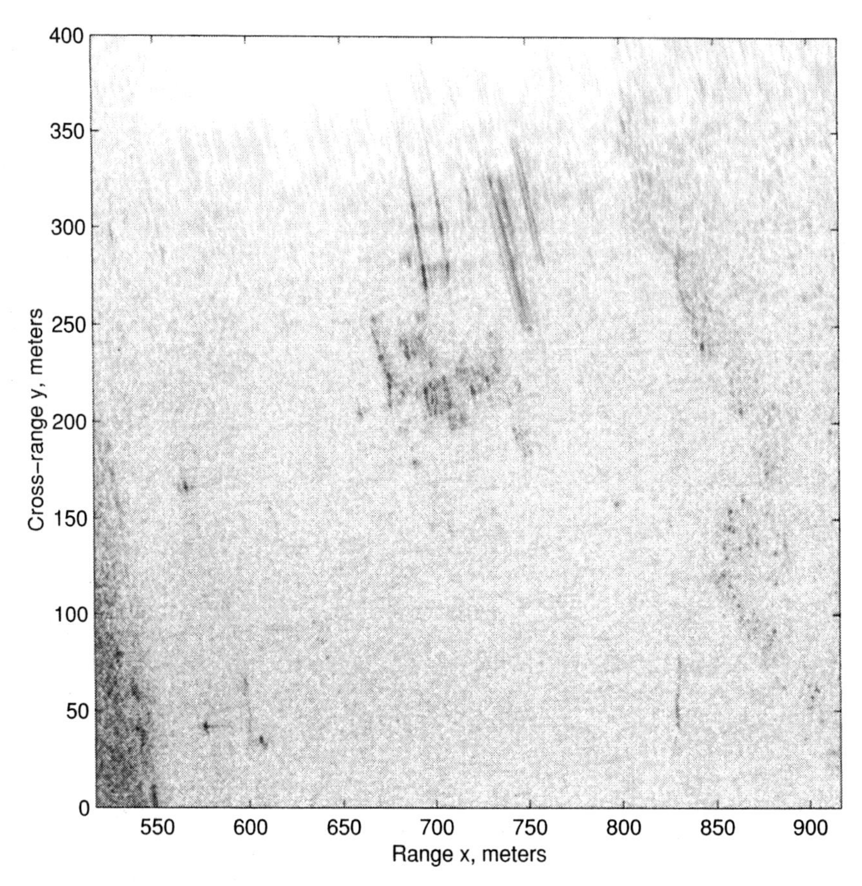

Figure 6.32c. Backprojection reconstruction with beamwidth angle (Doppler) of [−15, 15] degrees (first run): close-up of upper target area.

patch also contains some foliage.) Due to the symmetry of the synthetic aperture $u \in [-200, 200]$ m with respect to the center of this subpatch, the targets in this terrain area exhibit the SAR spectral properties and point spread function of a broadside spotlight SAR or a stripmap SAR system.

Figure 6.36*a* and *b* shows similar results for a subpatch centered at $(x, y) = (750, -550)$; the squint angle of this area is −36.25 degrees. This target area is a foliage region. We should point out that the reconstructions of the two squint subpatches in Figures 6.34*b* and 6.36*b* with the original measured SAR data would show severe clutter-looking aliasing. The *squint* digital spotlighting, upsampling, and bandpass slow-time Doppler processing provide a focused SAR images of the targets in these subpatches in Figures 6.34*b* and 6.36*b*. However, these images contain slow-time Doppler (leakage) aliasing from some other parts of the target area.

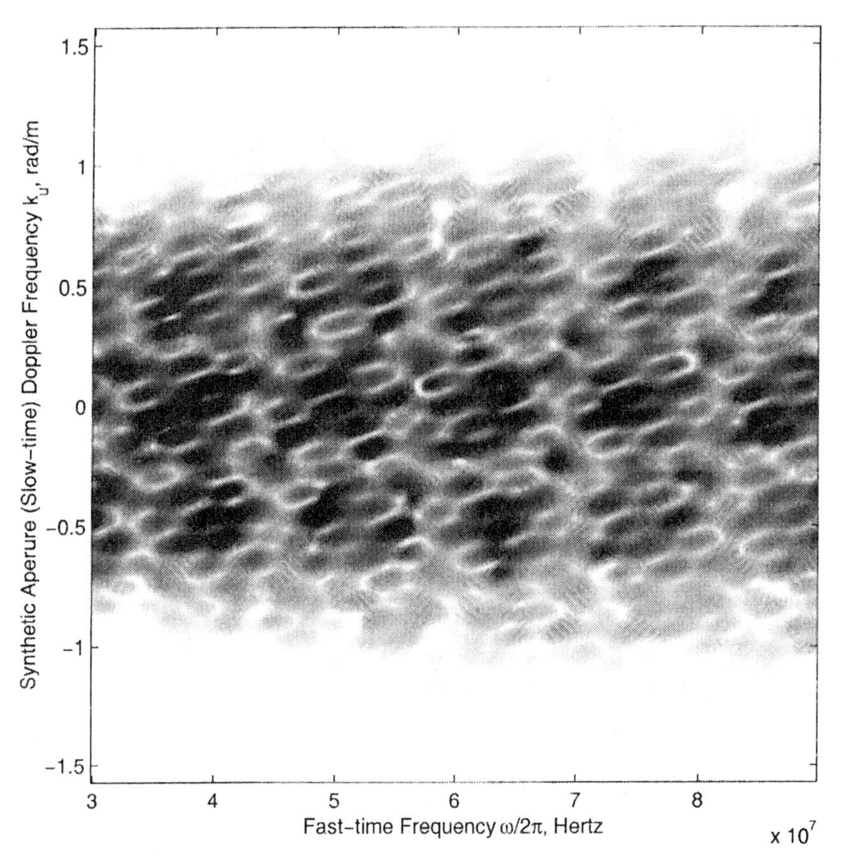

Figure 6.33a. Two-dimensional spectrum of simulated SAR data in (ω, k_u) domain: with no motion errors.

6.7 MOVING TARGET DETECTION AND IMAGING

Next we examine the SAR signal properties of a moving target in a stationary target scene which is being radiated by a SAR system. This analysis provides us with tools for developing SAR signal processing methods for moving target detection and imaging, and the merits of these methods in a realistic SAR system.

The classical radar problem of detecting and estimating the range and speed of a moving target in the one-dimensional spatial domain is based on a two-dimensional reconstructed image or ambiguity function in the (x, v_x) domain [edd; sko]; x and v_x, respectively, are the target range and speed in the range domain. The basic approach is to develop a two-dimensional statistic, which is called the *ambiguity function*, to determine the range coordinates and speed values of moving targets in the irradiated target scene.

The analogous problem which we encounter in SAR is to detect and estimate the location and velocity of a moving target in the two-dimensional (slant-range and

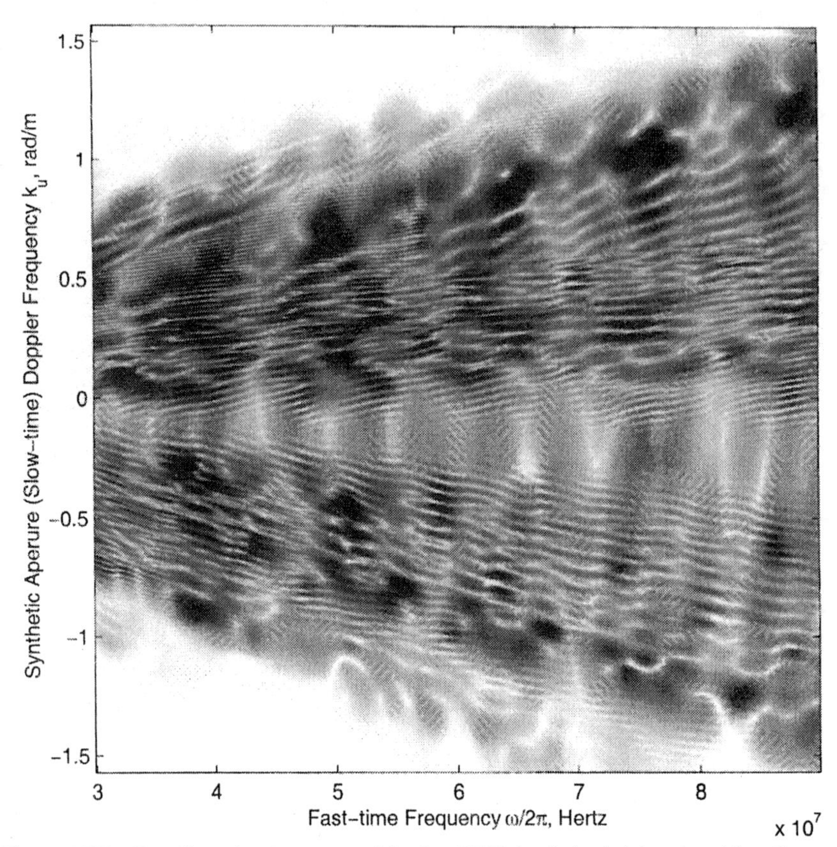

Figure 6.33b. Two-dimensional spectrum of simulated SAR data in (ω, k_u) domain: with motion errors.

cross-range) domain. We consider the *simplest* scenario which involves moving targets with constant vector velocity values as a function of the slow-time; the case of moving targets with nonlinear motion can be examined in a similar fashion [s94, ch. 5; s95].

One may intuitively rationalize that this SAR problem requires a four-dimensional reconstructed image or ambiguity function in the (x, y, v_x, v_y) domain. However, whether a target in a SAR scene is stationary or moving, its *relative* motion trajectory with respect to the radar can be modeled by at most three parameters [s94], which we refer to as the $(\mathbf{X}, \mathbf{Y}, \alpha)$ parameters in our discussion; parameters $(\mathbf{X}, \mathbf{Y})$ are called the *motion-transformed* spatial domain, and α the *relative speed* of a target with respect to the radar.

Based on this model, we show that a three-dimensional image of a moving target scene in the $(\mathbf{X}, \mathbf{Y}, \alpha)$ domain can be formed via the wavefront SAR reconstruction algorithm. We then develop a statistic for moving target detection in a SAR problem. We close this section with the results with the realistic FOPEN stripmap SAR data.

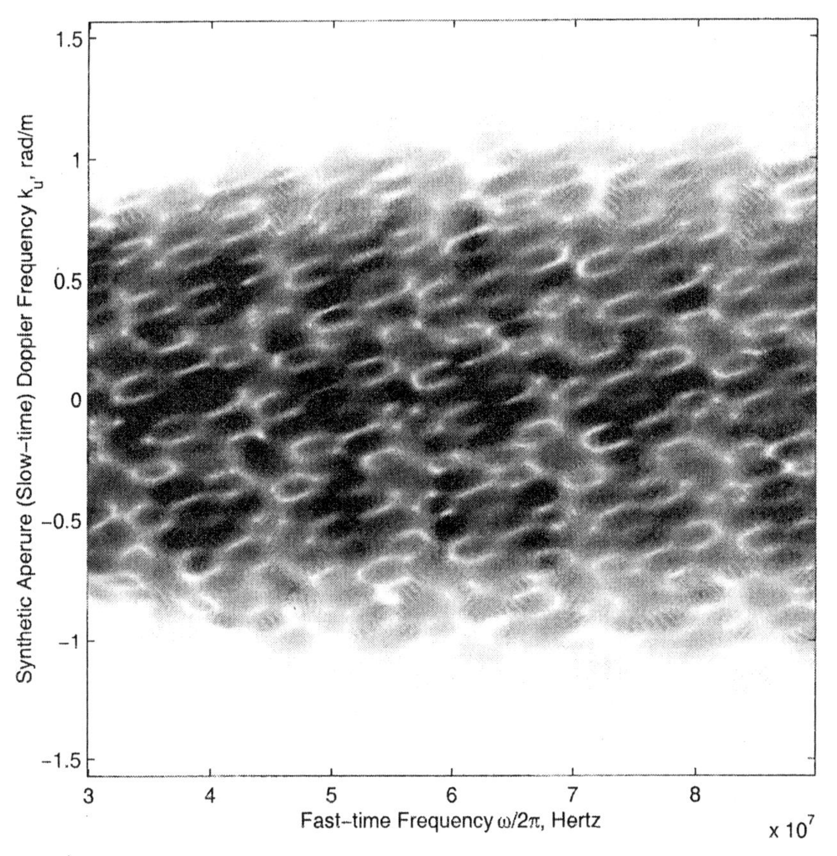

Figure 6.33c. Two-dimensional spectrum of simulated SAR data in (ω, k_u) domain: after narrow-beamwidth motion compensation.

SAR Signal Model for a Moving Target with a Constant Velocity

We denote the speed of the airborne aircraft that carries the radar with v_r and the slow-time domain by

$$\tau = \frac{u}{v_r},$$

where u is the synthetic aperture domain. Suppose that the velocity vector for the nth target is

$$(v_{xn}, v_{yn})$$

in the spatial (x, y) domain which is unknown. Thus the distance of the target from the radar at the slow-time τ is

$$\sqrt{(x_n - v_{xn}\tau)^2 + [y_n - v_{yn}\tau - u]^2} = \sqrt{(x_n - v_{xn}\tau)^2 + [y_n - v_{yn}\tau - v_r\tau]^2}.$$

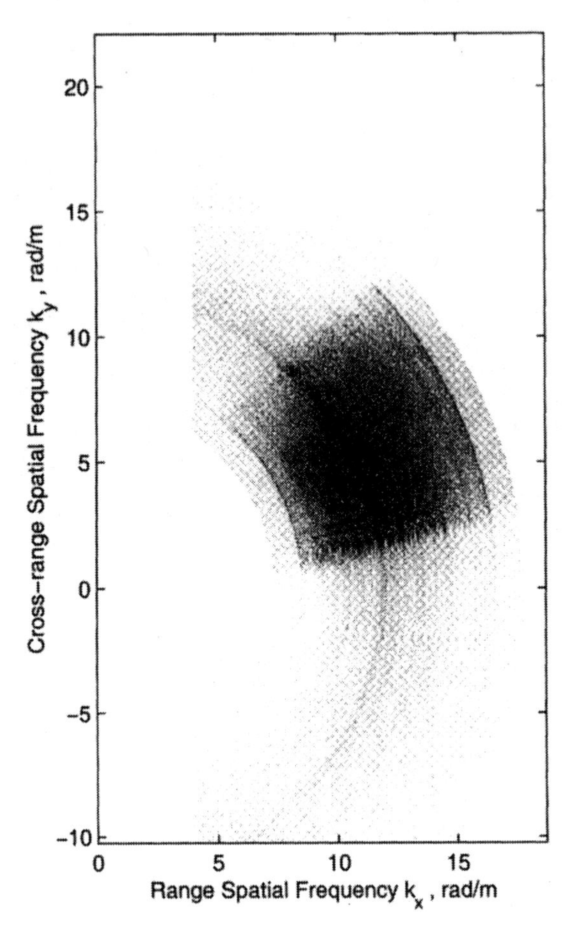

Figure 6.34a. Subpatch wavefront reconstruction of farm area centered at $(x, y) = (750, 400)$ m (first run): spectrum.

In this case the SAR signature of the nth target in the (ω, u) domain is [for notational simplicity, the amplitude patterns $a(\cdot)a_n(\cdot)$ and $P(\omega)$ are suppressed]

$$s_n(\omega, u) = \exp\left[-j2k\sqrt{(x_n - v_{xn}\tau)^2 + (y_n - v_{yn}\tau - v_r\tau)^2}\right]$$

$$= \exp\left[-j2k\sqrt{r_n^2 - 2\left[v_{xn}x_n + (v_{yn} + v_r)y_n\right]\tau + (v_{xn}^2 + v_{yn}^2 + v_r^2)\tau^2}\right]$$

where $r_n = \sqrt{x_n^2 + y_n^2}$.

Note that the motion trajectory for the nth target can be uniquely identified via three parameters: r_n, $v_{xn}x_n + (v_{yn} + v_r)y_n$, and $v_{xn}^2 + (v_{yn} + v_r)^2$. In fact the generalized SAR target motion trajectory with respect to the radar can be expressed via

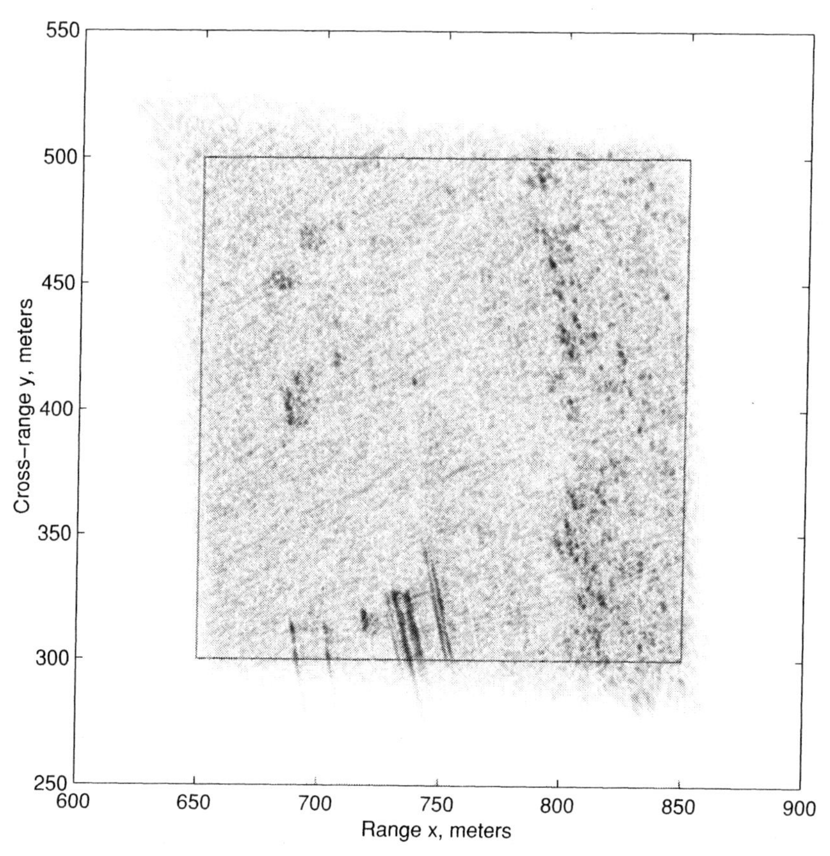

Figure 6.34b. Subpatch wavefront reconstruction of farm area centered at $(x, y) = (750, 400)$ m (first run): spatial domain reconstruction.

the following:

$$\sqrt{\mathbf{X}_n^2 + (\mathbf{Y}_n - \alpha_n u)^2} = \sqrt{\mathbf{X}_n^2 + \mathbf{Y}_n^2 - 2\alpha_n \mathbf{Y}_n u + \alpha_n^2 u^2},$$

and the target SAR signature is

$$\exp\left[-j2k\sqrt{\mathbf{X}_n^2 + (\mathbf{Y}_n - \alpha_n u)^2} = \sqrt{\mathbf{X}_n^2 + \mathbf{Y}_n^2 - 2\alpha_n \mathbf{Y}_n u + \alpha_n^2 u^2}\right].$$

Equating this model to the SAR signal for the nth target yields (note that $u = v_r \tau$)

Radial range	$\sqrt{\mathbf{X}_n^2 + \mathbf{Y}_n^2} = r_n$
Squint cross-range	$\alpha_n \mathbf{Y}_n = v_{xn} x_n + (v_{yn} + v_r) y_n$
Relative speed	$\alpha_n = \dfrac{\sqrt{v_{xn}^2 + (v_{yn} + v_r)^2}}{v_r}$

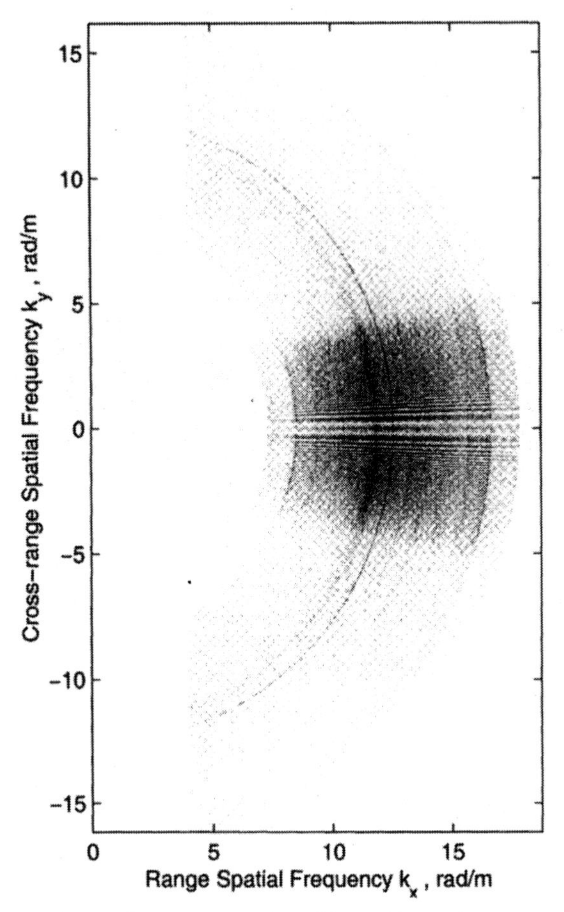

Figure 6.35a. Subpatch wavefront reconstruction of broadside terrain area centered at $(x, y) = (900, 0)$ m (first run): spectrum.

We call $(\mathbf{X}_n, \mathbf{Y}_n)$ the motion-transformed spatial coordinates of the nth target; the parameter α_n is the relative speed of the nth target with respect to the radar. Solving for the motion-transformed coordinates $(\mathbf{X}_n, \mathbf{Y}_n)$ from the above, we obtain

$$\mathbf{X}_n = \frac{(v_{yn} + v_r)x_n - v_{xn}y_n}{\sqrt{v_{xn}^2 + (v_{yn} + v_r)^2}},$$

$$\mathbf{Y}_n = \frac{v_{xn}x_n + (v_{yn} + v_r)y_n}{\sqrt{v_{xn}^2 + (v_{yn} + v_r)^2}},$$

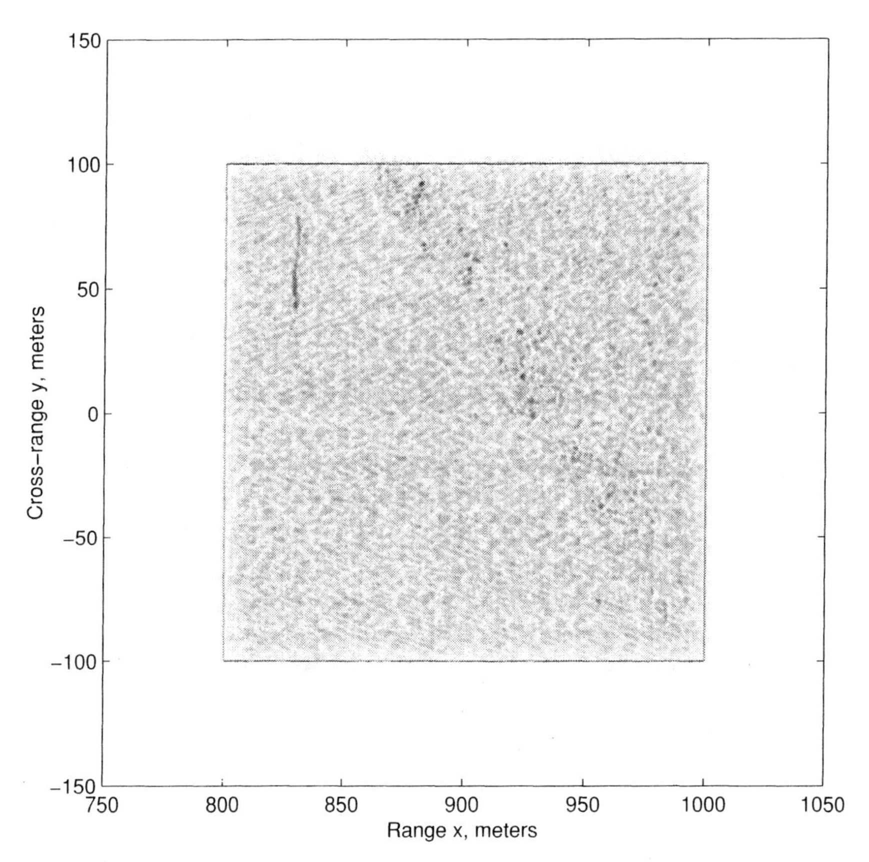

Figure 6.35b. Subpatch wavefront reconstruction of broadside terrain area centered at $(x, y) = (900, 0)$ m (first run): spatial domain reconstruction.

or

$$\begin{bmatrix} \mathbf{X}_n \\ \mathbf{Y}_n \end{bmatrix} = \begin{bmatrix} \dfrac{(v_{yn} + v_r)x_n}{\sqrt{v_{xn}^2 + (v_{yn} + v_r)^2}} & \dfrac{-v_{xn}}{\sqrt{v_{xn}^2 + (v_{yn} + v_r)^2}} \\ \dfrac{v_{xn}}{\sqrt{v_{xn}^2 + (v_{yn} + v_r)^2}} & \dfrac{(v_{yn} + v_r)}{\sqrt{v_{xn}^2 + (v_{yn} + v_r)^2}} \end{bmatrix} \begin{bmatrix} x_n \\ y_n \end{bmatrix}.$$

The above transformation from the (x_n, y_n) domain into the $(\mathbf{X}_n, \mathbf{Y}_n)$ domain is *linear*. Thus it corresponds to a combined rotational and scale transformation of the (x_n, y_n); the parameters of the linear transformation depend on the nth target motion parameters. Note that for a stationary target, that is, $(v_{xn}, v_{yn}) = (0, 0)$, we have $(\mathbf{X}_n, \mathbf{Y}_n) = (x_n, y_n)$.

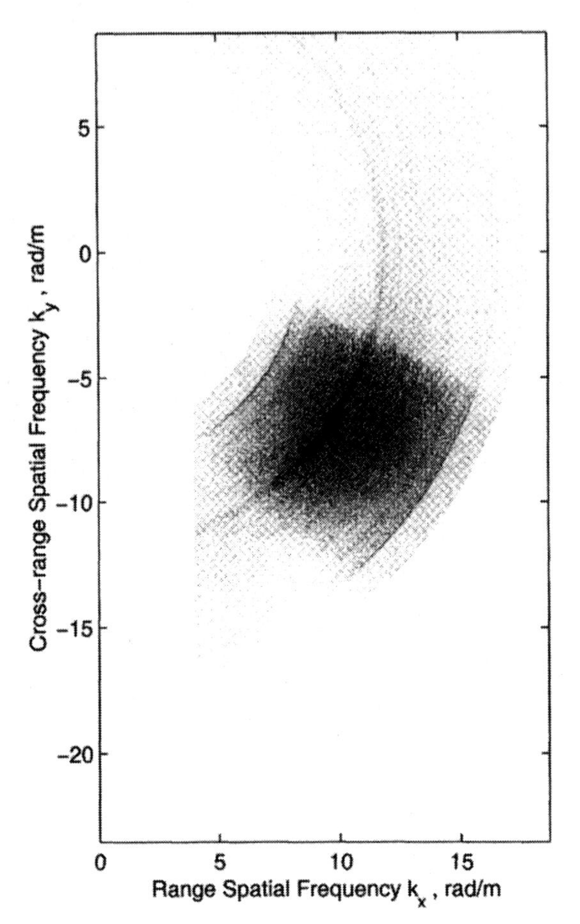

Figure 6.36a. Subpatch wavefront reconstruction of foliage area centered at $(x, y) = (750, -550)$ m (first run): spectrum.

Three-Dimensional Imaging in Motion-Transformed Spatial Domain and Relative Speed Domain

The SAR reconstruction of a moving target appears smeared and shifted if it is treated as a stationary target [s94]. However, the reconstruction of a moving target is focused if the value of its relative speed α is incorporated into the reconstruction algorithm. The SAR imaging of a scene composed of moving targets as well as stationary targets can be formulated as a three-dimensional imaging in the $(\mathbf{X}, \mathbf{Y}, \alpha)$ domain. The three-dimensional reconstruction in what we refer to as the motion-transformed spatial domain $(\mathbf{X}, \mathbf{Y})$ and the relative speed domain α can be achieved via the following:

We denote the two-dimensional Fourier transform of the measured SAR data, that is, $s(t, u)$, with $S(\omega, k_u)$. Then the reconstruction equation in the spatial frequency

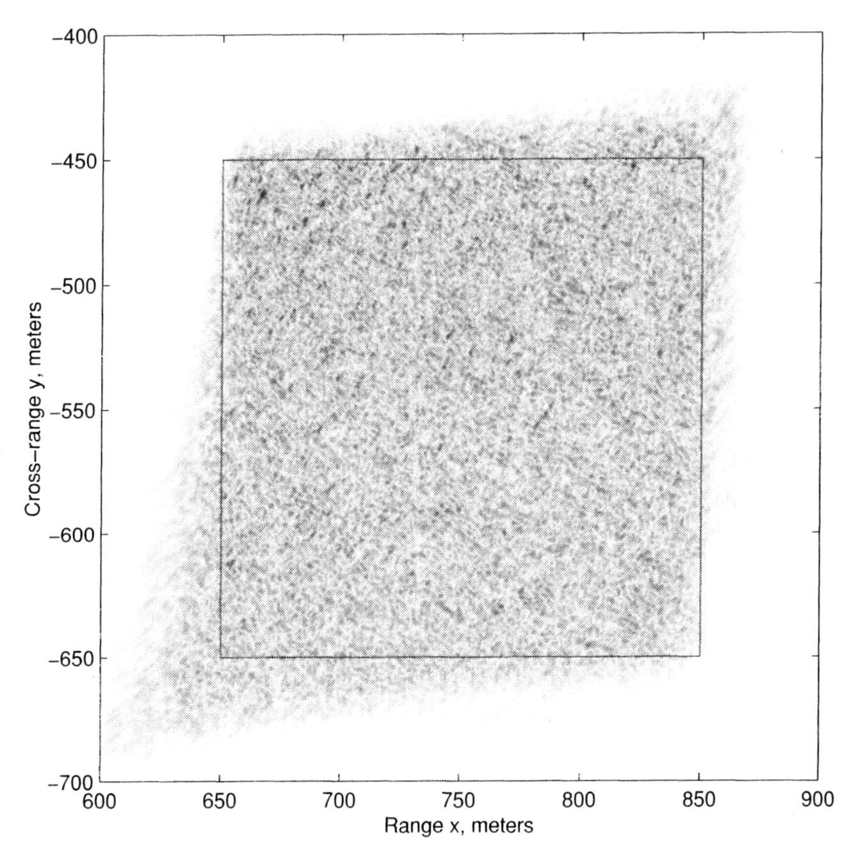

Figure 6.36b. Subpatch wavefront reconstruction of foliage area centered at $(x, y) = (750, -550)$ m (first run): spatial domain reconstruction.

domain of $(\mathbf{X}, \mathbf{Y})$ for targets with relative speed α is [s94]

$$\mathbf{F}(k_{\mathbf{X}}, k_{\mathbf{Y}}, \alpha) = P^*(\omega) S(\omega, k_u)$$

where

$$k_{\mathbf{X}} = \sqrt{4k^2 - \left(\frac{k_u}{\alpha}\right)^2},$$

$$k_{\mathbf{Y}} = \frac{k_u}{\alpha}.$$

The inverse two-dimensional Fourier transform of this signal with respect to $(k_{\mathbf{X}}, k_{\mathbf{Y}})$,

$$\mathbf{f}(\mathbf{X}, \mathbf{Y}, \alpha) = \mathcal{F}^{-1}_{(k_{\mathbf{X}}, k_{\mathbf{Y}})}\big[\mathbf{F}(k_{\mathbf{X}}, k_{\mathbf{Y}}, \alpha)\big],$$

is representative of targets with relative speed α *in focus*.

The signal processing and related issues for this reconstruction are similar to those we developed earlier for SAR imaging of a stationary target scene (i.e., $\alpha = 1$). Considering the practical values for the speed of a moving vehicle on the ground, the values of α (the relative speed) that one should use for the three-dimensional imaging is approximately within [0.75, 1.25].

Moving Target Indicator: SAR Ambiguity Function

One could use the three-dimensional statistic $\mathbf{f}(\mathbf{X}, \mathbf{Y}, \alpha)$, which was developed in the previous section, to search for moving targets. This can be achieved via identifying *focused* structures in the spatial $(\mathbf{X}, \mathbf{Y})$ domain for a given relative speed α. In addition to the computational complexity of such a search, it is difficult to develop a criterion for what is focused and what is not, especially in the case of targets that are surrounded by foliage.

Another difficulty that is encountered by the user stems from the fact that a moving target could possess *unknown* nonlinear motion components; these behave similar to the radar-carrying aircraft (nonlinear) motion errors that result in smearing and loss of focus. In this section we discuss a simpler and more robust statistic that informs the user of the presence of a moving target in the SAR scene, that is, a moving target indicator (MTI).

Digitally Spotlighted SAR Signature in the Reconstructed Image Consider the reconstructed target function for the relative speed of $\alpha = 1$, that is, a stationary SAR scene model

$$f(x, y) = \mathbf{f}(\mathbf{X}, \mathbf{Y}, 1).$$

As we stated earlier, the image of a moving target in this reconstruction would appear smeared. However, since the speed of a moving target (e.g., a truck) on the ground is much smaller than the speed of the radar-carrying aircraft, the moving target would exhibit a relatively *localized* signature (though not focused) in the SAR image $\mathbf{f}(\mathbf{X}, \mathbf{Y}, 1)$.

In this case the SAR signal model of the nth target in the motion-transformed domain $(\mathbf{X}, \mathbf{Y})$, that is,

$$\sqrt{\mathbf{X}_n^2 + (\mathbf{Y}_n - \alpha_n u)^2},$$

shows that the signature of this target in the SAR image $\mathbf{f}(\mathbf{X}, \mathbf{Y}, 1)$ would appear at approximately

$$(\mathbf{X}_{nc}, \mathbf{Y}_{nc}) \approx \left(\mathbf{X}_n, \frac{\mathbf{Y}_n}{\alpha_n}\right).$$

We identify the small region around the above coordinates where the *smeared* signature of the nth target appears as $\mathbf{R}_n$. Thus we can approximately reconstruct the measured SAR signature of the nth target in the (ω, u) domain via the following:

$$s_n(\omega, u) \approx \sum_{(\mathbf{X}_i, \mathbf{Y}_j) \in \mathbf{R}_n} \mathbf{f}(\mathbf{X}_i, \mathbf{Y}_j, 1) \exp\left[-j2k\sqrt{\mathbf{X}_i^2 + (\mathbf{Y}_j - u)^2}\right].$$

Note that $\mathbf{f}(\mathbf{X}_i, \mathbf{Y}_j, 1)$, $(\mathbf{X}_i, \mathbf{Y}_j) \in \mathbf{R}_n$, represent the *coherent* (complex) values of the SAR reconstruction at the region that the nth target signature appears, that is, around the coordinates $(\mathbf{X}_{nc}, \mathbf{Y}_{nc})$. We identify the above operation to reconstruct the target signature $s_n(\omega, u)$ as the *digitally spotlighted SAR signature of the region* $\mathbf{R}_n$ *in the reconstructed image*.

- What we refer to as *digital spotlighting* is a computer (digital signal processing) based process that isolates the signature of a target or a target region in a SAR scene. This operation may be used for imaging, and stationary or moving target detection and identification. A desired target signature may be digitally spotlighted in the original slow-time and fast-time domains of the raw SAR data, or any other one of its transformations which includes the reconstruction domain; the choice of the domain for digital spotlighting depends on the desired application problem.

  Earlier we used digital spotlighting in the polar format processed image in the (t, k_u) domain to isolate the SAR signature of a specific target area, and suppress the side lobes of the radar radiation pattern. To detect moving targets, a suitable domain for digital spotlighting is the reconstruction domain, that is, the (x, y) domain of the coherent reconstructed image $f(x, y)$. In this domain a stationary or moving target signature is fairly focused and thus could be "reasonably" separated from the signatures of the other targets. Once a target signature is spotlighted (extracted) in the spatial (x, y) domain, it can be transformed into any domain of the SAR signal that the user desires for further processing.

Next we show how the digitally spotlighted signature of a target can be used to determine whether the target is stationary or moving.

SAR Ambiguity Function We perform what we referred to as *slow-time compression* on the reconstructed signature $s_n(\omega, u)$ for the region $\mathbf{R}_n$ at the desired relative speed α values (e.g., $\alpha \in [0.75, 1.25]$); this yields

$$\gamma_n(\omega, u, \alpha) = s_n(\omega, u) \exp\left[j2k\sqrt{\mathbf{X}_{nc}^2 + (\mathbf{Y}_{nc} - \alpha u)^2}\right].$$

Note that the reference spatial point for the slow-time compression is chosen to be the center of the region $\mathbf{R}_n$, that is, $(\mathbf{X}_{nc}, \mathbf{Y}_{nc})$.

We also identify the Fourier transform of the above signal with respect to the synthetic aperture u via

$$\Gamma_n(\omega, k_u, \alpha) = \mathcal{F}_{(u)}\left[\gamma_n(\omega, u, \alpha)\right].$$

We call the signal $\Gamma_n(\omega, k_u, \alpha)$ the *SAR ambiguity function*, a statistic that can be used for MTI. The reason for this is described below.

Suppose that the signature appearing at region $\mathbf{R}_n$ belongs to a target with a relative speed of α_n. In this case, at a fixed fast-time frequency ω, the SAR ambiguity function $\Gamma_n(\omega, k_u, \alpha)$ exhibits a sharp peak in the (k_u, α) domain around

$$(k_u, \alpha) = (0, \alpha_n).$$

Moreover, as α moves away from the target relative speed α_n, the SAR ambiguity function $\Gamma_n(\omega, k_u, \alpha)$ becomes weaker and more spread (smeared) [s95].

Note that the SAR ambiguity function $\Gamma_n(\omega, k_u, \alpha)$, similar to the target function at the desired relative speed values $\mathbf{f}(\mathbf{X}, \mathbf{Y}, \alpha)$, is a three-dimensional signal. However, to construct the MTI statistic, one needs only to form the SAR ambiguity function at one of the fast-time frequencies of the radar signal, that is, a fixed ω.

It would have been desirable to form the SAR ambiguity function of the *entire* imaging scene and try to identify its local peaks (maxima) in the (k_u, α) domain at $\alpha \neq 1$ which represent the moving targets. Unfortunately, in reconnaissance problems of SAR such as FOPEN SAR where the suspected moving targets are in a thick foliage area, that is, the static clutter, a moving target signature in the SAR ambiguity function of the entire imaging scene could be completely dominated by the strong and coherent clutter (foliage) signature in the SAR ambiguity function.

The problem encountered with static clutter is well known and well documented in classical one-dimensional MTI and Airborne MTI (AMTI) systems [sko]. Due to the dominance of the foliage power in the collected SAR data, it is not feasible to separate the coherent clutter (foliage) SAR signature from the signatures of the moving targets in the measurement (t, u) domain or its Fourier transformed domain.

As we mentioned earlier in our discussion, the reconstructed image of a moving target in this SAR imaging system appears smeared; the amount of smearing depends on the relative speed of the moving target in the imaging plane and the speed of the radar-carrying aircraft. However, a ground moving target speed is much smaller than the speed of the aircraft. Due to this fact a moving target smeared image appears in a fairly localized region in the reconstructed SAR image.

Thus, as we suggested earlier, a practical use of the SAR ambiguity function for MTI involves constructing this statistic at a relatively small (*digitally spotlighted*) subpatch in the reconstructed image $\mathbf{f}(\mathbf{X}_i, \mathbf{Y}_j, 1)$ which we identified as $\mathbf{R}_n$. This process can then be repeated at the other subpatches in the reconstructed image to determine which subpatches indicate the moving target's signature in their corresponding MTI (SAR ambiguity function).

The subpatches should be chosen to be *overlapping*; this prevents a moving target signature, which is relatively weak, to be split between two subpatches. The size of an individual subpatch $\mathbf{R}_n$, which should be chosen based on the support of the smeared signature of a moving target in the spatial domain, depends on the radar signal bandwidth, the radar beamwidth, and the velocity of the target [s94, ch. 5]. The user should choose the worst-case scenario, which is the maximum size of a smeared signature in the spatial domain, for the extent of a subpatch.

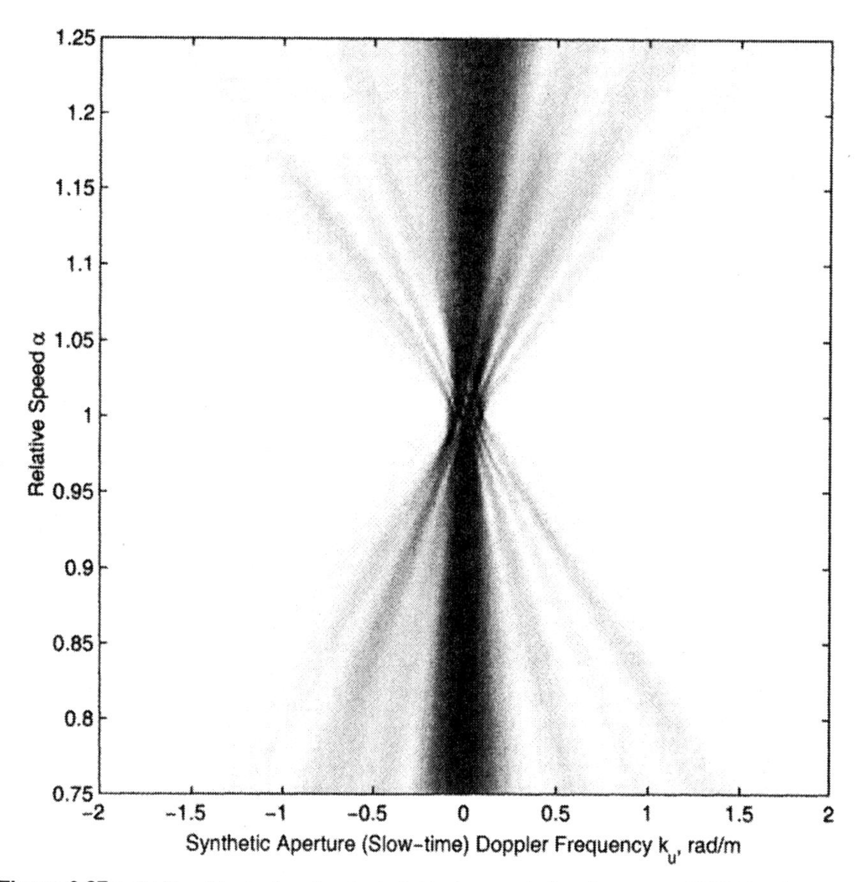

Figure 6.37a. SAR ambiguity function for individual targets in imaging scene of UHF band stripmap SAR data (first run): HEMTT truck.

Consider the first-run stripmap SAR data which we examined earlier. Figure 6.37*a* to *c*, respectively, shows the SAR ambiguity function at 320 MHz for the HEMTT truck (Figure 6.20*a*), the corner reflector at a large squint angle (Figure 6.22*a*), and the moving target (Figure 6.23*a*) For the two stationary targets (HEMTT and corner reflector), Figure 6.37*a* and *b* exhibits SAR ambiguity functions that are fairly symmetric around $(k_u, \alpha) = (0, 1)$; the same is not true for the moving target. For the moving target, Figure 6.37*c* exhibits a chaotic SAR ambiguity function around $\alpha = 1.04$ which is indicative of an accelerating and decelerating moving structure [s95].

6.8 MATLAB ALGORITHMS

The code for stripmap SAR imaging is identified as *pulsed stripmap SAR simulation and reconstruction*. A planar radar with a frequency-dependent amplitude pattern is considered; the amplitude pattern is selected to be a hanning window. The Matlab

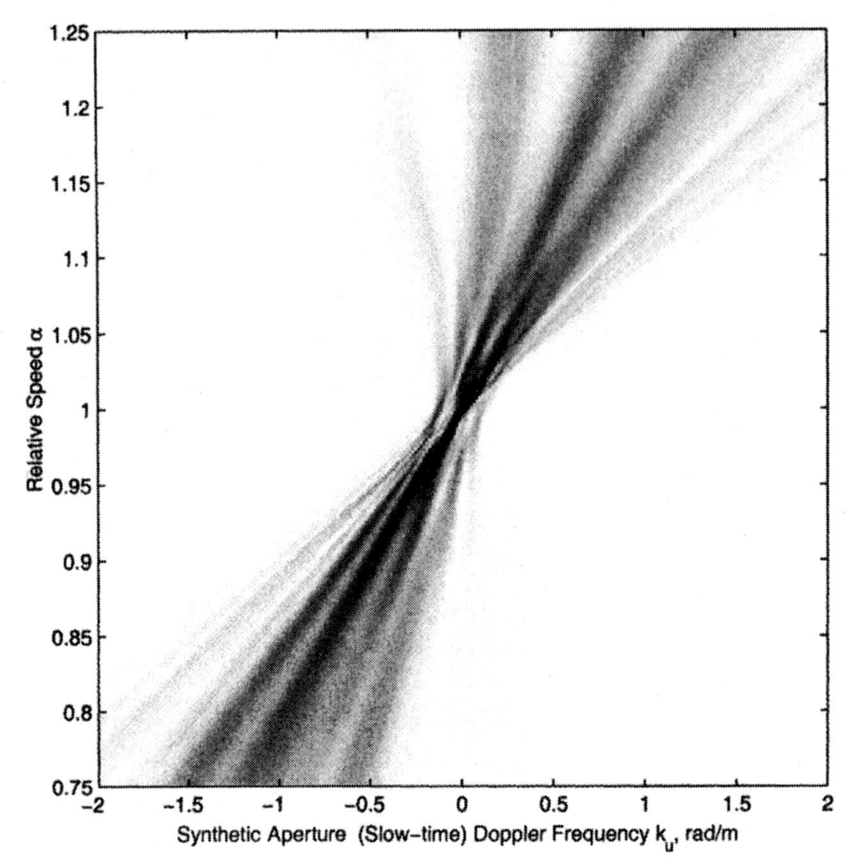

Figure 6.37b. SAR ambiguity function for individual targets in imaging scene of UHF band stripmap SAR data (first run): corner reflector 3.

code provides algorithms for digital reconstruction using the spatial frequency interpolation, range stacking, time-domain correlation, and backprojection. The code also performs slow-time Doppler subsampling when $Y_0 < L$, and subaperture digital spotlighting.

To provide a relatively simple and compact code to demonstrate the main properties of a stripmap SAR system, certain procedures are used that restrict the scope of the code and/or increase its computational load. The user could modify the code to remove these restrictions and increase its speed. These issues are listed in the following:

1. The fast-time sampling is based on the bandwidth of the chirp radar signal, and not the deramped signal. In the case of $T_p > T_x$, use the more efficient fast-time sampling rate of the deramped signal, and convert the resultant samples to alias-free matched-filtered echoed data via the procedure that was described in Section 1.6. After matched filtering, the user could reduce the number of

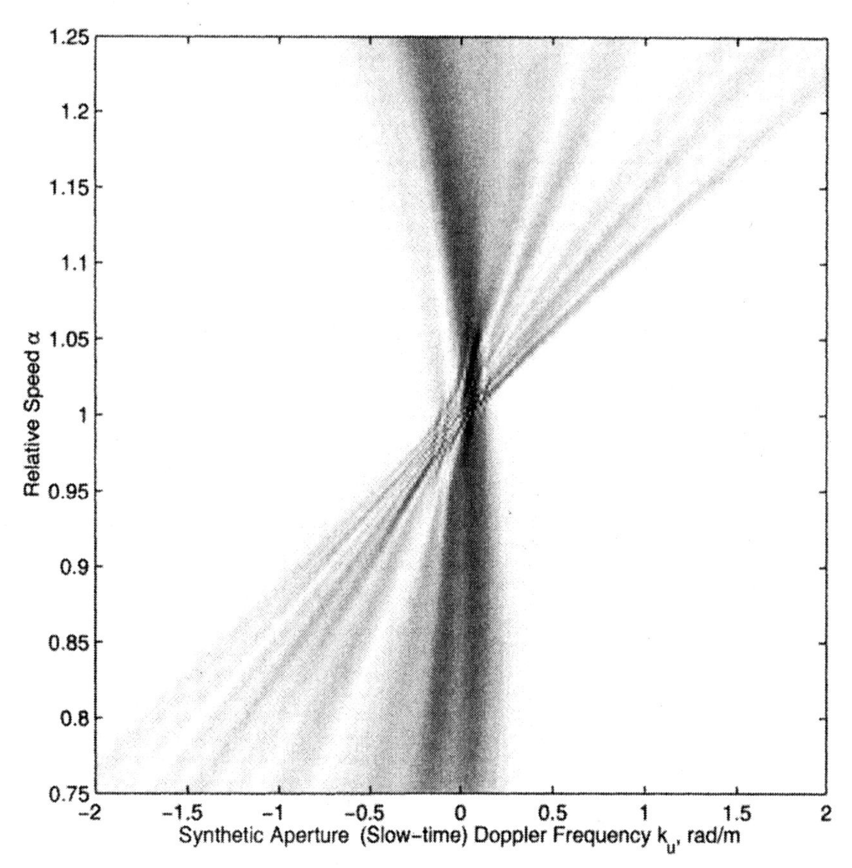

Figure 6.37c. SAR ambiguity function for individual targets in imaging scene of UHF band stripmap SAR data (first run): moving target.

fast-time samples, which are taken over the fast-time interval of $T_f - T_s$, via appropriate fast-time (range) gating; the interval of the resultant gated fast-time interval could be reduced to T_x. This would significantly reduce the number of fast-time samples when $T_p \gg T_x$.

2. Most FOPEN stripmap data contain strong radio frequency interference (RFI) due to the transmission by the local television stations, for example, the [215, 730] MHz P-3 data which were collected by Environmental Research Institute of Michigan (ERIM) with a chirp radar. Figure 6.38*a* shows the deramped SAR data versus the instantaneous frequency of the chirp radar for four consecutive slow-time samples of the P-3 database. As we showed in Section 1.6, the deramped data in Figure 6.38*a* can be interpreted as the SAR signal in the fast-time frequency ω domain. There are certain inaccuracies which are involved in this assumption that are called the residual video phase (RVP) error. One can use an approximation to reduce the RVP error; see Section 1.6 and [carr].

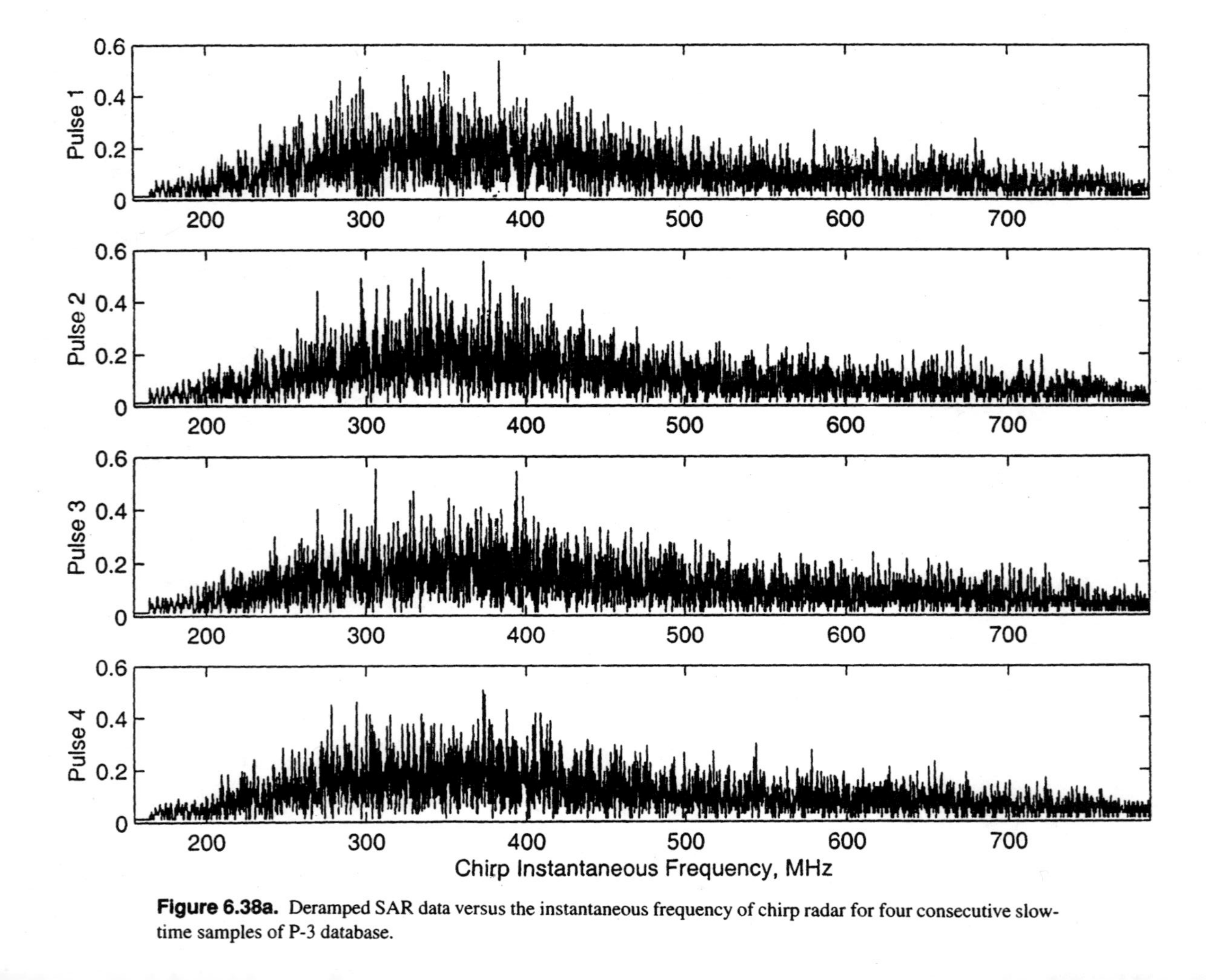

Figure 6.38a. Deramped SAR data versus the instantaneous frequency of chirp radar for four consecutive slow-time samples of P-3 database.

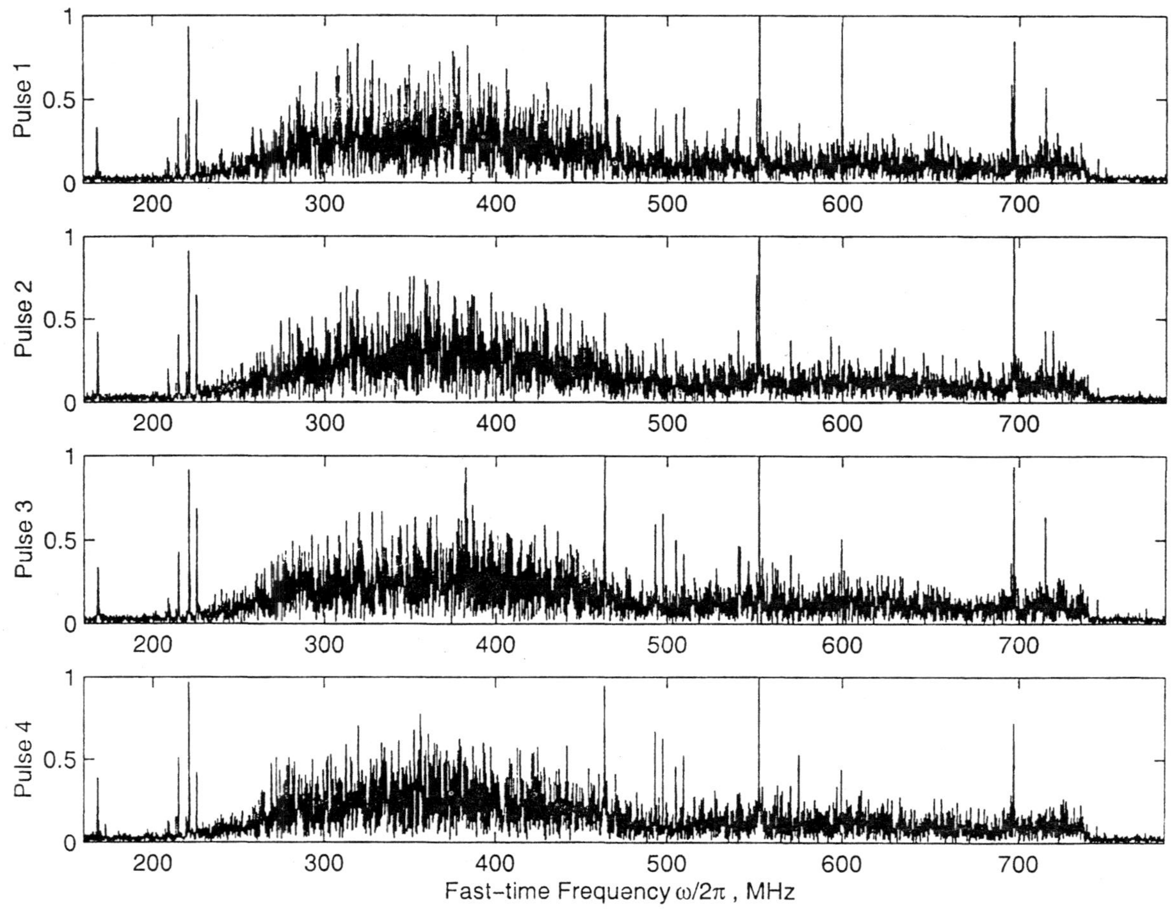

Figure 6.38b. Spectra of echoed signals reconstructed from deramped data in (a).

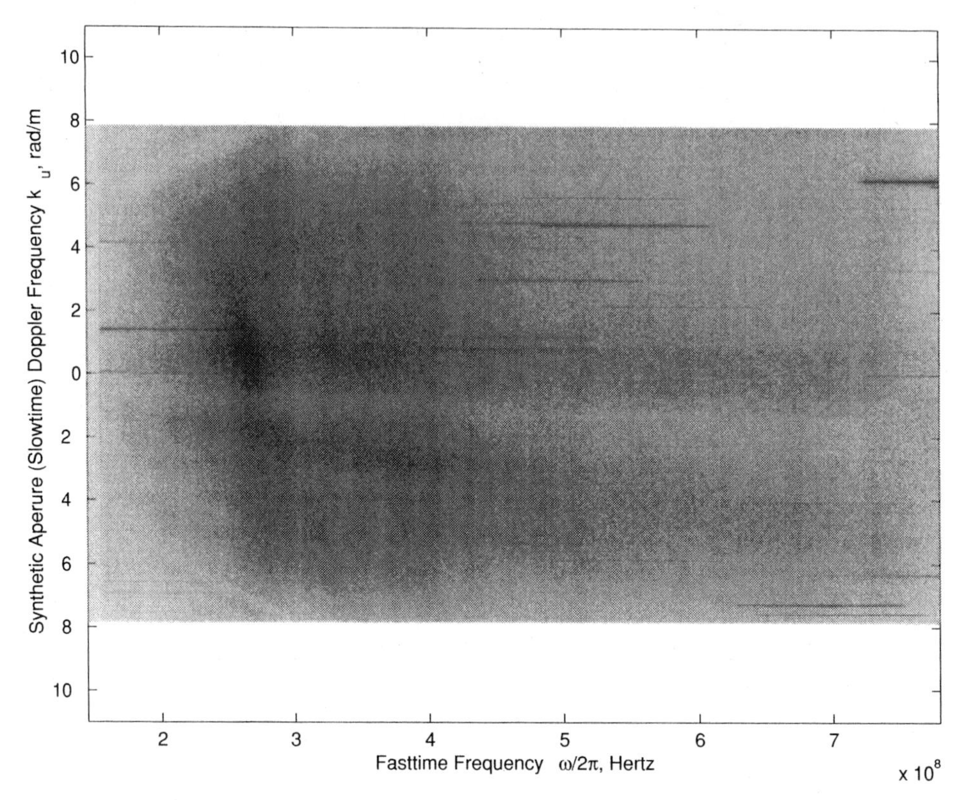

Figure 6.39a. Two-dimensional fast-time and slow-time spectrum of deramped data for P-3 SAR system.

However, the main problem with this processing of the deramped data is not the approximation which is used to remove the RVP error. It turns out that the deramping of the measured signal results in "spectral spreading" of the RFI signals in the fast-time frequency domain. Figure 6.39*a* shows the two-dimensional spectrum (i.e., fast-time frequency and slow-time Doppler frequency) for a portion of the P-3 data. The horizontal lines which appear in this figure are the spectrally-spread RFI signals. A commonly-used method for RFI suppression in deramped databases, such as the P-3 data in Figure 6.39*a*, is based on placing notches in the range equalization filter at the frequency bin locations that are occupied by the strong RFI signals. * However, this would result in degradations in the point spread function (PSF) of the SAR image, particularly in the range domain [s98s]. In addition to degrading the PSF, there is no study on the adverse effects of the spectral spreading of the RFI signals in the reconstructed SAR image. (A preliminary processing of the P-3 data at ERIM had indicated severe RFI degradations in the reconstructed image.)

*L. Bessette, M. Toups, and B. Binder, "Documentation of the Lincoln Laboratory Processing for P-3 UWB imagery," *MIT Lincoln Laboratory Memorandum 47PM-STD-0010*, Contract F19628-95-0002, Boston, August 21, 1996.

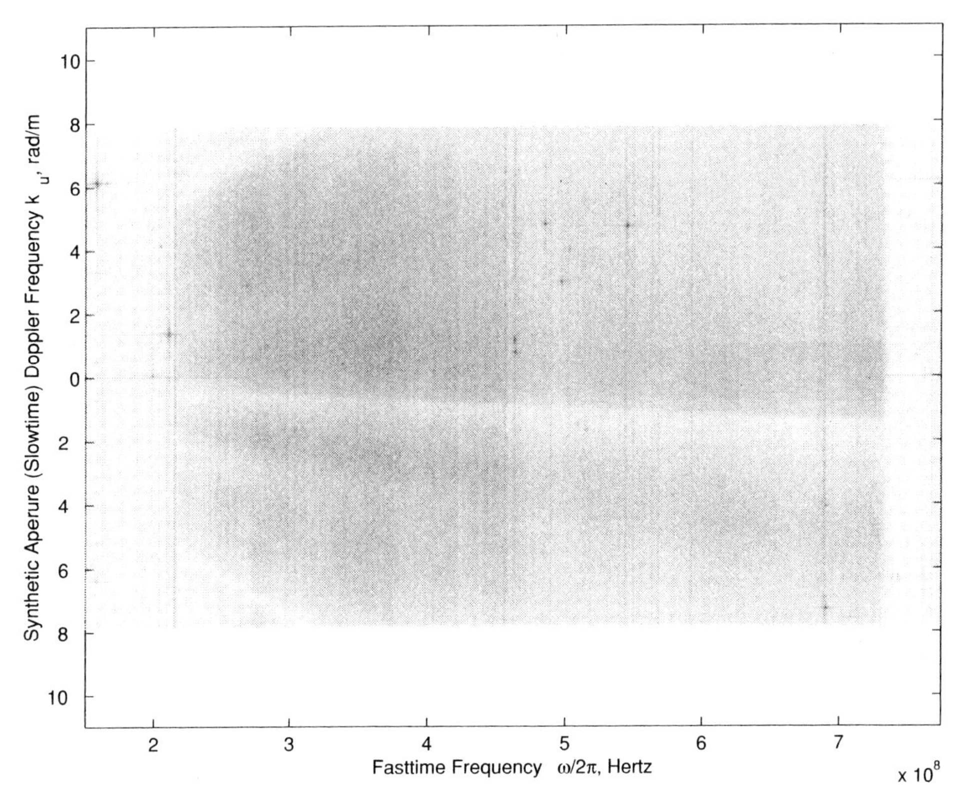

Figure 6.39b. Two-dimensional fast-time and slow-time spectrum of echoed data reconstructed from deramped data in (a).

Figure 6.38*b* shows the fast-time frequency spectra of the echoed signals that are formed from the four deramped P-3 data of Figure 6.38*a* via the method described in Section 1.6. Note that the RFI signals in Figure 6.38*b* appear at narrow and distinct frequency bands (i.e., fairly isolated signatures and not spread) in the fast-time frequency domain; for example, the audio and video signals for Channel 13 appear around 215 MHz. The two-dimensional spectrum of the echoed signal, that is shown in Figure 6.39*b*, exhibits a similar phenomenon. Note that an RFI signal which correspond to a direct-path propagation appears at a specific slow-time Doppler frequency (or "angle of arrival" with respect to the synthetic aperture). The components seen at the other Doppler frequencies are due to multi-path effects as well as slow-time noncoherence in the broadcasted TV signal.

This representation of the RFI signals allows the user to utilize a variety of the array signal processing methods for RFI suppression. However, one may also use simple spectral nulling methods for RFI suppression. For instance, since the direct-path RFI signals pose the main distortion in the reconstructed

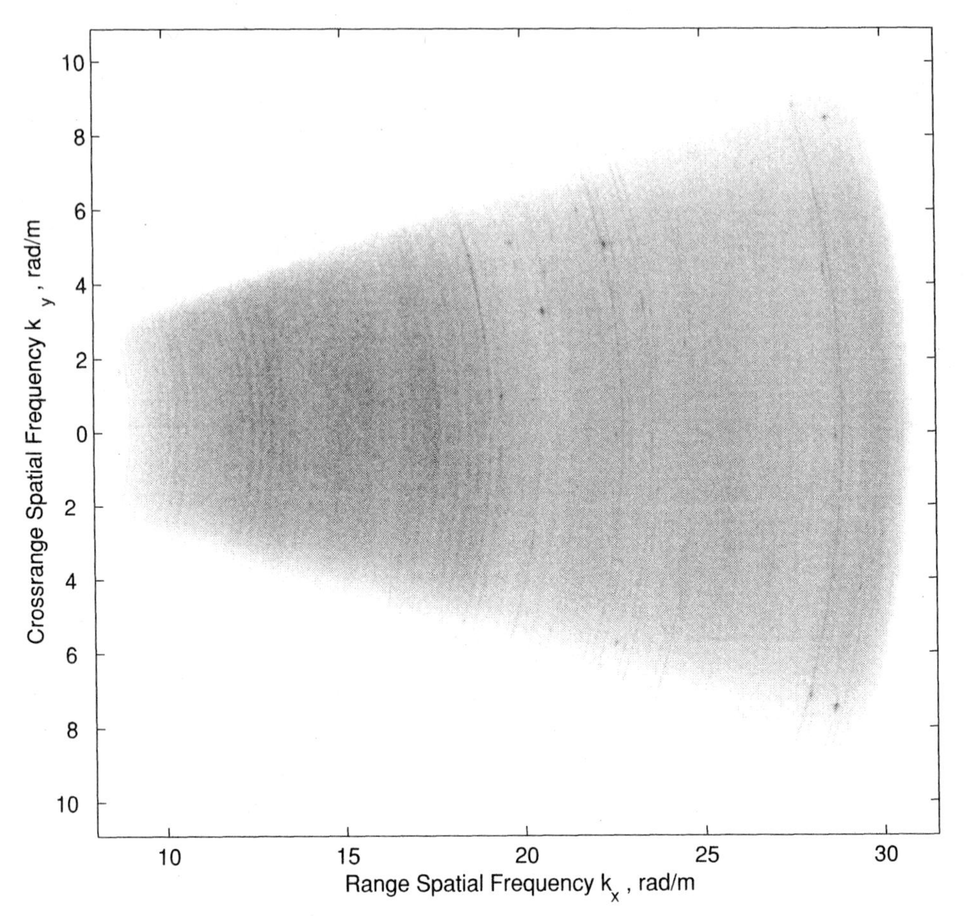

Figure 6.39c. Reconstructed target function spectrum from 8192 pulses of P-3 data (after digital spotlighting and upsampling).

SAR image, the user may apply power equalization on the spectral components of the SAR signal where the direct-path signal is strong.

Note that the SAR wavefront reconstruction is based on a mapping of the spectral domain of the SAR signal into the target function spectral domain; the reconstructed target function spectrum from 8192 pulses of the P-3 data (after digital spotlighting and upsampling [s98s]) is shown in Figure 6.39*c*. Thus, the user may apply a spectral filtering for RFI suppression after forming the SAR image. This allows the user to determine whether such an RFI filtering, which is likely to degrade the PSF of the SAR system, is a necessary operation. In fact, the P-3 results indicate that the strongest RFI signal in the baseband echoed signal results in a background noise which is below 60 dB of the strongest reflector in the scene [s98s].

3. Narrow-bandwidth assumption is used for polar format processing and subaperture digital spotlighting. The code could be modified to remove this assumption. This is a necessary step for wide-bandwidth stripmap SAR data, for example, the [215, 730] MHz P-3 data which were collected by ERIM.
4. For stripmap SAR data which are aliased in the slow-time Doppler domain, for example, the Stanford Research Institute (SRI) data that were cited in this chapter and the P-3 data of ERIM, the user should apply the slow-time upsampling method of Chapter 5 (for spotlight SAR systems). As we mentioned in Section 6.7, the user may avoid processing large upsampled arrays by subpatch image formation. Use subpatches that slightly overlap to avoid mosaic-looking artifacts at the boundaries of the subpatches.
5. The code selects the minimum and maximum range spatial frequency pair $(k_{x\min}, k_{x\max})$ based on all the available discrete k_{xmn} values. However, these two parameters should be chosen from the k_{xmn} values that fall within the two-dimensional bandwidth of the stripmap SAR signal. For instance, one may choose

$$k_{x\min} = 2k_{\min} \cos \theta_{\mathrm{ax}},$$
$$k_{x\max} = 2k_{\max},$$

where θ_{ax} is the maximum absolute aspect angle (the largest divergence angle ϕ_d).

The outputs of the code are:

P6.1 Measured stripmap SAR signal $\mathbf{s}(t, u)$.
P6.2 Stripmap SAR signal after fast-time matched-filtering $\mathbf{s}_M(t, u)$.
P6.3 Polar format SAR reconstruction $f_p(t, k_u)$ with subaperture digital spotlight filter.
P6.4 Stripmap SAR signal after digital spotlighting $\mathbf{s}_d(t, u)$.
P6.5 Stripmap SAR signal spectrum after digital spotlighting $S_d(\omega, k_u)$.
P6.6 Stripmap SAR spatial frequency data coverage
P6.7 Wavefront stripmap SAR reconstruction spectrum
P6.8 Wavefront stripmap SAR reconstruction
P6.9 Range stack stripmap SAR reconstruction
P6.10 Range stack stripmap SAR reconstruction spectrum
P6.11 Time-domain correlation stripmap SAR reconstruction
P6.12 Time-domain correlation stripmap SAR reconstruction spectrum
P6.13 Backprojection stripmap SAR reconstruction
P6.14 Backprojection stripmap SAR reconstruction spectrum

7

CIRCULAR SYNTHETIC APERTURE RADAR

Introduction

Synthetic aperture radar (SAR) research has been mainly focused on imaging and target recognition algorithms when spotlight or stripmap data collection is used. As we showed in Chapters 5 and 6, in these SAR systems a radar-carrying aircraft moves along a line while it illuminates the target scene; we refer to these imaging systems as *linear SAR*. One of the problems with a linear SAR system in the reconnaissance problems of SAR is that a target SAR signature, that is, the amplitude pattern $a_n(\omega, x_n, y_n - u)$ with sufficiently high signal-to-noise power ratio, can be measured only over a limited aspect angle interval.

For example, in a stripmap SAR system with a planar radar aperture having a diameter comparable to the wavelength, the maximum aspect angle of the radar is approximately ± 45 degrees; see Chapter 3. This limitation is mainly caused by the limited divergence angle of the radar radiation pattern $a(\omega, x, y)$. Chapters 5 and 6 provide discussions on this issue and show results where realistic FOPEN stripmap SAR data exhibited this phenomenon. (There are other factors that degrade the reconstructed images, such as commercial radio frequency interference.)

In the linear SAR systems, depending on what the angular orientation of the target is, the user could face a complicated matching algorithm in the aspect angle domain in order to determine the target type and its orientation. The matching algorithm may also be unreliable and produce high false alarm rates or low detection rates (depending on the prescribed threshold level) due to the limited radar look angle data.

This problem can be circumvented by using circular SAR (CSAR) data collection of a spotlighted target region over 360 degrees (full rotation). In reconnaissance with slant plane CSAR, one is faced with comparable values for the altitude of the airborne radar (aircraft), the radius of the aircraft circular path, and the radius of the spotlighted target area. Due to this fact the classical approximation-based SAR imaging algorithms, for example, polar format processing [aus; wal], would fail in these SAR imaging scenarios. This chapter provides a wavefront reconstruction theory-based imaging method for CSAR imaging, and a closely related radar system that utilizes circular and elevation synthetic apertures for three-dimensional imaging.

Outline

This chapter presents a method for imaging from the slant plane CSAR data collected over the full rotation or over a partial segment of a circular flight path. The CSAR system model is described in Section 7.1, and its shift-varying impulse response is identified. A Fourier analysis of the imaging system multidimensional impulse response (the slant plane CSAR *Green's function* [mor]) is provided.

This analysis provides an understanding of the information content of the CSAR signal, and methods to reconstruct the target function from it. In Section 7.2 a method for reconstructing from CSAR data, which exploits the Fourier decomposition of the Green's function, is presented. This scheme is composed of two parts: First the slant plane CSAR data are converted to ground plane CSAR data; then a reconstruction method for ground plane CSAR is formulated.

Section 7.3 provides an analysis of the two-dimensional spectral properties of the CSAR signal. This is then used in Section 7.4 to identify the resolution and point spread function of a CSAR system. The spectral properties of the CSAR signal are also used to outline in Section 7.5 the digital signal processing issues associated with the slant plane CSAR imaging system. Methods for digital image formation in CSAR systems are discussed in Section 7.6.

In Section 7.7 it is shown that the slant plane CSAR, unlike the slant plane linear SAR, has the capability to extract three-dimensional imaging information from a target scene. The theory of slant plane CSAR imaging is accompanied by the real CSAR (turntable) data of a T-72 tank; see the introductory chapter of the book for the parameters of the turntable CSAR data.

A discussion on the classical question of the role of the carrier frequency in the resolution of an imaging system is provided in Section 7.8. In this discussion we develop a procedure, based on work performed in the late 1950s by Howard at the Naval Research Laboratory [sher], to show the resolvability of targets from their single-tone fringe patterns. For this purpose we use the digitally spotlighted CSAR signature of closely spaced targets in the spatial (reconstruction) domain to construct a fringe pattern that enables the user to resolve these targets.

The chapter closes with a discussion on a *true* three-dimensional SAR imaging system that uses synthetic aperture data acquired over a linear elevation synthetic aperture and a circular synthetic aperture in the range and azimuth domains (Section 7.9). This radar imaging system is referred to as an E-CSAR system. A Fourier analysis of the E-CSAR database in its two synthetic aperture domains is provided. This analysis is used for developing a Fourier-based wavefront reconstruction algorithm for the E-CSAR system.

Mathematical Notations and Symbols

The following is a list of mathematical symbols used in this chapter, and their definitions:

c	Wave propagation speed $c = 3 \times 10^8$ m/s for radar waves $c = 1500$ m/s for acoustic waves in water $c = 340$ m/s for acoustic waves in air
D	Diameter of radar
$f(x, y)$	Ground plane CSAR target function in spatial domain
$f_c(x, y)$	Compressed image of ground plane CSAR target function in partial rotation aspect angle measurement
$f_{\text{fp}n}(x, y)$	Single-tone fringe patterns for digitally spotlighted region $\mathbf{R}_n$
$F(k_x, k_y)$	Two-dimensional Fourier transform of ground plane target function
$F_c(k_x, k_y)$	Two-dimensional Fourier transform of compressed CSAR image in partial rotation aspect angle measurement
$F_p(\rho, \phi)$	Two-dimensional Fourier transform of ground plane target function in polar domain
$F_{pn}(\rho, \theta)$	Polar spatial frequency distribution of digitally spotlighted region $\mathbf{R}_n$
$\mathbf{F}_p(\rho, \xi)$	Fourier transform of $F_p(\rho, \theta)$ with respect to θ
$f(x, y, z)$	Target function in three-dimensional spatial domain
$F(k_x, k_y, k_z)$	Three-dimensional Fourier transform of target function
$F_{xy}(k_x, k_y, z)$	Marginal two-dimensional Fourier transform of target function with respect to (x, y)
$g_{xyz}(\omega, \theta, v)$	E-CSAR system shift-varying impulse response
$G_{xyz}(\omega, \theta, k_v)$	One-dimensional Fourier transform of $g_{xyz}(\omega, \theta, v)$ with respect to elevation synthetic aperture v
$g_\theta(\omega, x, y)$	Slant plane CSAR Green's function
$G_\theta(\omega, k_x, k_y)$	Two-dimensional Fourier transform of slant plane CSAR Green's function
$G_{\theta p}(\omega, \rho, \phi)$	Two-dimensional Fourier transform of slant plane CSAR Green's function in polar domain
$H_\xi^{(2)}$	Hankel function of the second kind, ξ order
J_1	Bessel function of the first kind, first order
k	Wavenumber: $k = \omega/c$
k_c	Wavenumber at carrier frequency: $k_c = \omega_c/c$
k_g	Ground plane wavenumber: $k_g = \omega_g/c$
$k_{\max}$	Maximum wavenumber of radar signal

$k_{\min}$	Minimum wavenumber of radar signal
k_r	Spatial frequency or wavenumber domain for radial r domain
k_v	Spatial frequency or wavenumber domain for elevation synthetic aperture v
k_x	Spatial frequency or wavenumber domain for range x
$k_{xn}(\omega, \theta)$	Range spatial frequency trace for nth target
k_y	Spatial frequency or wavenumber domain for cross-range y
$k_{yn}(\omega, \theta)$	Cross-range spatial frequency trace for nth target
k_z	Spatial frequency or wavenumber domain for elevation z
M_θ	Number of slow-time samples
N	Number of fast-time frequency samples
$p(t)$	Transmitted radar signal
$p_0(t)$	Transmitted radar signal shifted by reference fast-time T_c
$P(\omega)$	Fourier transform of transmitted radar signal
r	Ground plane radial distance in spatial domain
$r_{\max}$	Farthest radial distance of radar from spotlighted target area
$r_{\min}$	Closest radial distance of radar from spotlighted target area
r_n	Radial distance of nth target from center of spotlighted target area
R_c	Slant range of center of ground plane target area
R_g	Ground plane radius of radar-carrying aircraft circular flight path
$R_0(\omega)$	Radius of spotlighted target area on ground plane
$\mathbf{R}_n$	Digitally spotlighted region around (x_n, y_n)
$\mathbf{s}(t, \theta)$	Echoed CSAR signal from target area
$\mathbf{s}(t, \theta, v)$	Echoed E-CSAR signal from target area (3D SAR)
$s(\omega, \theta)$	One-dimensional Fourier transform of CSAR signal with respect to fast-time
$s(\omega, \theta, v)$	One-dimensional Fourier transform of E-CSAR signal with respect to fast-time in 3D SAR system
$S(\omega, \xi)$	Two-dimensional Fourier transform of CSAR signal with respect to fast-time and slow-time
$\mathbf{S}(\omega, \xi, k_v)$	Three-dimensional Fourier transform of E-CSAR signal with respect to fast-time and two slow-times
$\mathbf{s}_M(t, \theta)$	Fast-time matched-filtered echoed CSAR signal
$s_M(\omega, \theta)$	One-dimensional Fourier transform of matched-filtered CSAR signal with respect to fast-time
$s_g(\omega_g, \theta)$	One-dimensional Fourier transform of ground plane CSAR signal with respect to fast-time
$S_g(\omega_g, \xi)$	Two-dimensional Fourier transform of ground plane CSAR signal with respect to fast-time and slow-time
$s_{g0}(\omega_g, \theta)$	One-dimensional Fourier transform of reference ground plane signal with respect to fast-time

$S_{g0}(\omega_g, \xi)$	Two-dimensional Fourier transform of reference ground plane CSAR signal with respect to fast-time and slow-time
t	Fast-time domain
$t_{ij}(\theta)$	Round trip delay of echoed signal for target at (x_i, y_j) at slow-time θ
T_c	Reference fast-time point
T_f	Ending point of fast-time sampling for SAR signal
T_p	Radar signal pulse duration
T_s	Starting point of fast-time sampling for SAR signal
u	Azimuthal synthetic aperture (slow-time) domain in linear SAR
v	Synthetic aperture or slow-time domain in elevation
$W_1(\phi)$	Polar angular window support of CSAR Green's function spectrum
$W_2(\omega, \rho)$	Polar radial window support of CSAR Green's function spectrum
x	Range domain
x_f	Focal range of curved radar
x_i	Range bins in reconstruction image
x_n	Range of nth target
x_s	Slant-range in linear SAR
X_c	Mean range swath in linear SAR
y	Cross-range or azimuth domain
y_j	Cross-range bins in reconstructed image
y_n	Cross-range of nth target
z	Elevation or altitude domain
$z(x, y)$	Variations of elevation in spotlighted target area
Z_c	Elevation of radar-carrying aircraft circular flight path
Δ_{k_x}	Sample spacing of reconstructed target function in range spatial frequency domain
Δ_{k_y}	Sample spacing of reconstructed target function in cross-range spatial frequency domain
Δ_t	Sample spacing of CSAR signal in fast-time domain
Δ_x	Sample spacing of reconstructed target function in range domain
Δ_{X_n}	nth target resolution in X_n domain
Δ_y	Sample spacing of reconstructed target function in cross-range domain
Δ_{Y_n}	nth target resolution in Y_n domain
Δ_ϕ	Angular spread of CSAR Green's function spectrum
Δ_ρ	Radial spread of CSAR Green's function spectrum
Δ_θ	Sample spacing of CSAR signal in slow-time domain
$\mathcal{F}$	Forward Fourier transform operator
$\mathcal{F}^{-1}$	Inverse Fourier transform operator

λ	Wavelength: $\lambda = 2\pi c/\omega$
$\lambda_{\max}$	Maximum wavelength: $2\pi c/\omega_{\min}$
$\lambda_{\min}$	Minimum wavelength: $2\pi c/\omega_{\max}$
$\lambda_{n\min}$	Minimum wavelength of radiation experienced by nth target
$\Lambda(\omega, \omega_g)$	Ground plane to slant plane system kernel
$\Lambda^{-1}(\omega_g, \omega)$	Slant plane to ground plane system kernel
ω	Temporal frequency domain for fast-time t
ω_c	Radar signal carrier or center frequency
ω_g	Ground plane temporal frequency domain
$\omega_{g\max}$	Maximum fast-time frequency of radar signal on ground plane
$\omega_{g\min}$	Minimum fast-time frequency of radar signal on ground plane
$\omega_{\max}$	Maximum fast-time frequency of radar signal
$\omega_{\min}$	Minimum fast-time frequency of radar signal
$\omega_{n\min}$	Maximum fast-time frequency of radiation experienced by nth target
ω_0	Radar signal half bandwidth in radians Radar signal baseband bandwidth is $2\omega_0$
ϕ	Polar angle in spatial frequency domain
ϕ_d	Divergence angle of radar
ρ	Polar radius in spatial frequency domain
ρ_c	Polar radius at carrier frequency in spatial frequency domain
$\rho_{\max}$	Maximum polar radius in spatial frequency domain for support of a target at center of spotlighted area
$\rho_{\min}$	Minimum polar radius in spatial frequency domain for support of a target at center of spotlighted area
θ	Slow-time or aspect angle synthetic aperture domain
$\theta_{\max}$	Half-size of aspect angle synthetic aperture
$\theta_n(\theta)$	Aspect angle of nth target at slow-time θ
θ_m	Discrete aspect angles for digitally spotlighted polar data
$\theta_x(\omega)$	Along path target area angle
θ_z	Average depression (elevation/grazing) angle of target area
$\theta_{zn}(\theta)$	Slow-time varying aspect angle of nth target
ξ	Slow-time frequency domain; frequency domain for θ

7.1 SYSTEM MODEL

The CSAR imaging system geometry is shown in Figure 7.1. The radar-carrying aircraft moves along a circular path with radius R_g on the plane $z = Z_c$ with respect to the ground plane. Thus the coordinates of the radar in the spatial domain as a

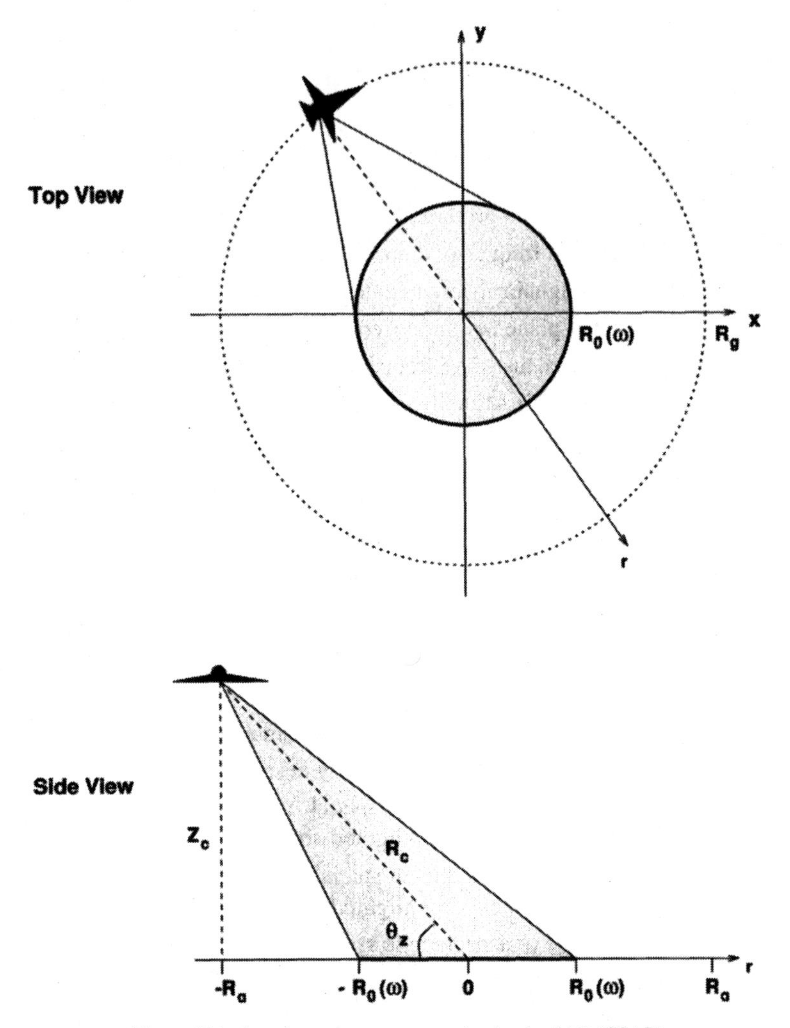

Figure 7.1. Imaging system geometry in circular SAR (CSAR).

function of the slow-time domain are

$$(x, y, z) = (R_g \cos\theta, R_g \sin\theta, Z_c),$$

where $\theta \in [0, 2\pi)$ represents the slow-time or aspect angle synthetic aperture domain. We define the slant range by

$$R_c = \sqrt{R_g^2 + Z_c^2}$$

and the slant *depression* or *grazing* angle via

$$\theta_z = \arctan\left(\frac{Z_c}{R_g}\right).$$

As the radar moves along the circular synthetic aperture, its beam is spotlighted on the disk of radius $R_0(\omega)$ centered at the origin of the spatial (x, y) domain on the ground plane. Note that the radius of the spotlighted $R_0(\omega)$ varies with the fast-time frequency of the radar, that is, ω, in the case of a planar radar aperture. In general, the radius of the spotlighted area is related to the radar parameters by

$$R_0(\omega) = \frac{R_c \tan\phi_d}{\sin\theta_z},$$

where the radar divergence angle is

$$\phi_d = \begin{cases} \arcsin\left(\dfrac{\lambda}{D}\right), & \text{planar radar,} \\ \arctan\left(\dfrac{D}{2x_f}\right), & \text{curved radar.} \end{cases}$$

D is the diameter of the radar, and x_f is the focal range of the curved radar.

- As we will see, whether or not $R_0(\omega)$ varies with the fast-time frequency ω has interesting implications on the shift-varying resolution of a CSAR system.

In our current discussion we assume that the target area is a delta (thin) sheet on the $z = 0$ plane; that is, the target area does not possess altitude variations. The effect of terrain altitude variations from the $z = 0$ plane will be discussed in future sections. Moreover we consider a *continuous* target function model in our discussion of CSAR; this would be more convenient for analysis (the Parseval's identity [car]). We denote the reflectivity function in the target region on the $z = 0$ plane by $f(x, y)$.

We also define the following angle:

$$\theta_x(\omega) = \arcsin\left[\frac{R_0(\omega)}{R_g}\right];$$

for a planar radar aperture, this angle is a *decreasing* function of the fast-time frequency. We call $\pm\theta_x(\omega)$ the *along-track target angles*.

CSAR Signal Model

We denote the transmitted radar signal with $p(t)$. For notational simplicity we assume that the radar radiation pattern is a constant over the spotlighted target area. The measured CSAR signal in the fast-time and slow-time domains (t, θ) can be defined via the following:

$$\mathbf{s}(t,\theta) = \int_y \int_x f(x,y) p \left[t - \frac{2\sqrt{(x - R_g \cos\theta)^2 + (y - R_g \sin\theta)^2 + Z_c^2}}{c} \right] dx\, dy.$$

The Fourier transform of the above model with respect to the fast-time t is

$$\begin{aligned} s(\omega,\theta) &= P(\omega) \int_y \int_x f(x,y) \\ &\quad \times \exp\left[-j2k\sqrt{(x - R_g \cos\theta)^2 + (y - R_g \sin\theta)^2 + Z_c^2} \right] dx\, dy \\ &= P(\omega) \int_y \int_x f(x,y) g_\theta^*(\omega,x,y)\, dx\, dy, \end{aligned}$$

where

$$g_\theta(\omega,x,y) = \exp\left[j2k\sqrt{(x - R_g \cos\theta)^2 + (y - R_g \sin\theta)^2 + Z_c^2} \right],$$

for (x, y) within the disk of radius $R_0(\omega)$ (spotlighted target area), and zero otherwise, is the CSAR imaging system shift-varying impulse response (slant plane Green's function) at the slow-time (aspect angle synthetic aperture) θ and the fast-time frequency ω.

Figures 7.2 and 7.3, respectively, show the CSAR signal $\mathbf{s}(t,\theta)$ and its one-dimensional Fourier transform with respect to the fast-time $s(\omega,\theta)$ for the turntable data of a T-72 tank; this database was described in the introductory chapter of the book. The radar system used for collecting the turntable data was a *stepped frequency* one (see Chapter 1); that is, the radar collected the data shown in Figure 7.3. The data in Figure 7.2 is constructed from the inverse DFT of the data in Figure 7.3 with respect to the discrete values of the fast-time frequency ω.

Fourier Properties of Slant Plane Green's Function

Before examining CSAR reconstruction based on the CSAR signal which was developed in the previous section, we provide a Fourier analysis of the CSAR shift-varying impulse response, that is, the slant plane Green's function; this would facilitate the formulation of the CSAR reconstruction. This Fourier analysis is based on the work [s96a]; we highlight the main conclusion of that work without showing the mathematical derivations.

We denote the rectilinear spatial frequency domain by (k_x, k_y) and its polar transform with (ρ, ϕ) where

$$\phi = \arctan\left(\frac{k_y}{k_x}\right),$$

$$\rho = \sqrt{k_x^2 + k_y^2}.$$

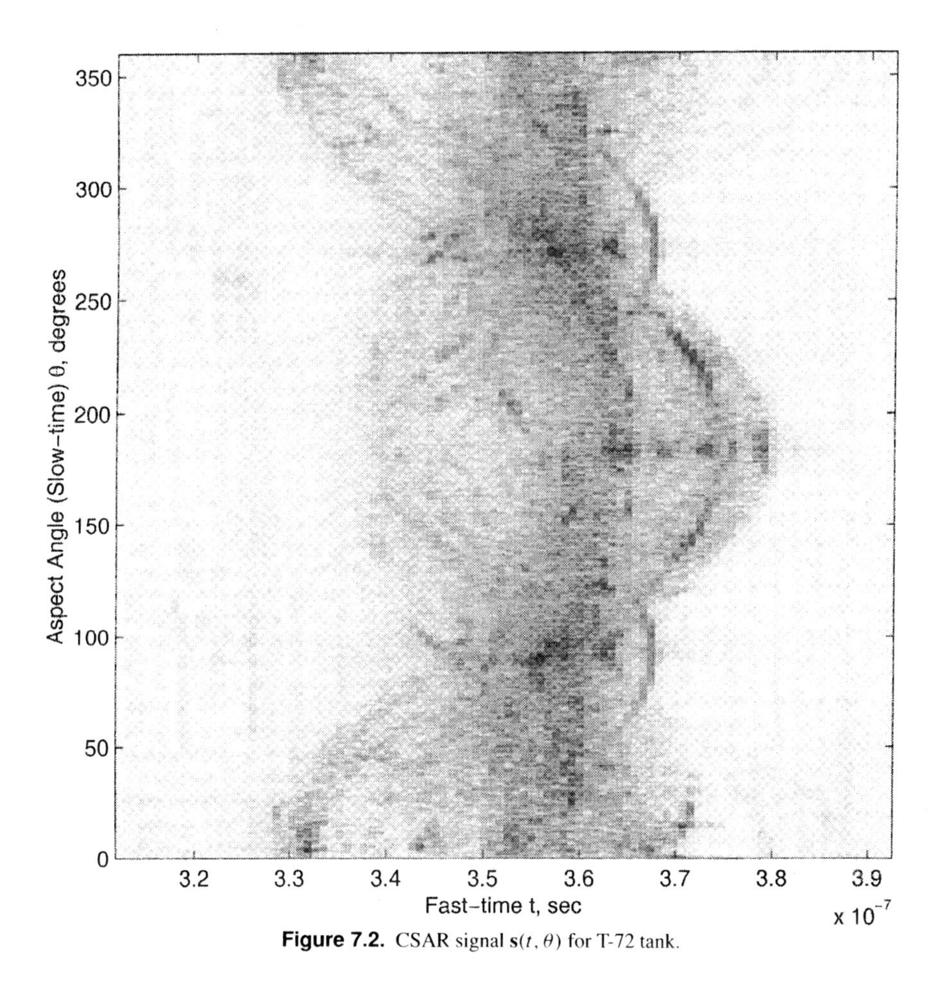

Figure 7.2. CSAR signal $\mathbf{s}(t, \theta)$ for T-72 tank.

We also identify the two-dimensional spatial Fourier transforms of the target function and the slant plane Green's function with

$$F(k_x, k_y) = \mathcal{F}_{(x,y)}[f(x, y)]$$

and

$$G_\theta(\omega, k_x, k_y) = \mathcal{F}_{(x,y)}[g_\theta(\omega, x, y)].$$

We also define the polar coordinate transformation of these two functions in the spatial frequency domain via

$$F_p(\rho, \phi) = F(k_x, k_y)$$

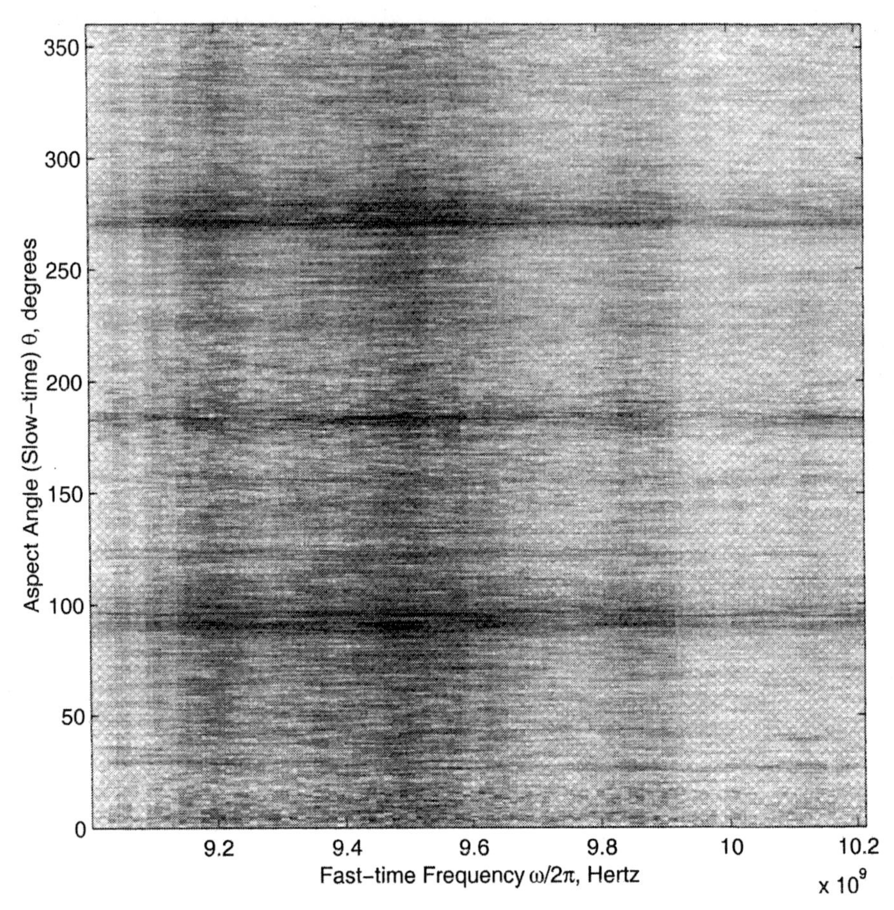

Figure 7.3. Fourier transform of CSAR signal with respect to fast-time, $s(\omega, \theta)$ (stepped frequency data), for T-72 tank.

and

$$G_{\theta p}(\omega, \rho, \phi) = G_\theta(\omega, k_x, k_y).$$

The analysis in [s96a] shows that the Green's function in the polar spatial frequency (ρ, ϕ) domain has the following form:

$$G_{\theta p}(\omega, \rho, \phi) = \underbrace{W_1(\theta - \phi) W_2(\omega, \rho)}_{\text{Window functions}} \times \underbrace{\exp\left[-j\sqrt{4k^2 - \rho^2}\, Z_c - j\rho R_g \cos(\theta - \phi)\right]}_{\text{Phase-modulated signal}};$$

the two window functions are defined via

$$W_1(\phi) = \begin{cases} 1 & \text{for } |\phi| \leq \theta_x(\omega), \\ 0 & \text{otherwise,} \end{cases}$$

and

$$W_2(\omega, \rho) = \begin{cases} 1 & \text{for } |\rho - 2k\cos\theta_z| \leq 2k\sin^2\theta_z \sin\theta_x(\omega), \\ 0 & \text{otherwise.} \end{cases}$$

Examples of the Green's function in the rectilinear spatial frequency domain, that is, $G_\theta(k_x, k_y, \omega)$ for various values of θ_z, are shown in [s96a].

For the ground plane scenario, that is, $\theta_z = 0$ and $\cos\theta_z = 1$, the Green's function spectrum is a pole distribution on the circle of radius $2k$ [mor; s94]. The Green's function possesses a very low energy spread around this circle which corresponds to evanescent waves; the evanescent waves cannot be recorded by a realistic radar system where the range is significantly greater than the wavelength.

For the slant problem, that is, $\theta_z \neq 0$ and $\cos\theta_z < 1$, the Green's function possesses a spatial frequency (Doppler) spread around the circle of radius $\rho = 2k\cos\theta_z$ in the (k_x, k_y) domain; the size of the spread is

$$\Delta_\rho = \pm 2k\sin^2\theta_z \sin\theta_x(\omega),$$

$$\Delta_\phi = \pm\theta_x(\omega),$$

which increases as $\theta_x(\omega)$, or equivalently, $R_0(\omega)$ (the size of the target area) increases.

The physical meaning associated with the Doppler spread for the ground plane case, that is, presence of evanescent waves, should not be associated with the spatial frequency spread for the slant case. The spread in this case is due to the fact that a nonzero slant makes the radar signal experienced by the ground targets to have a wavelength other than $2\pi c/\omega$; this is a well-known fact in classical radar theory [sko]. The effective wavenumber (wavelength) depends on the relative coordinates of the radar and the target; that is, it varies with (x, y, θ). For a fixed θ the spread in the wavenumber domain is due to the variations of (x, y) which yield the spread of Δ_ρ at $\rho = 2k\cos\theta_z$.

7.2 RECONSTRUCTION

Next we utilize the Fourier properties of the slant plane Green's function to develop an analytical solution for CSAR reconstruction. Consider the CSAR signal model in the (ω, θ) domain (see Section 6.1)

$$s(\omega, \theta) = P(\omega) \int_y \int_x f(x, y) g_\theta^*(\omega, x, y)\, dx\, dy.$$

Using the generalized Parseval's theorem, this CSAR system model can be rewritten via

$$s(\omega, \theta) = P(\omega) \int_{k_y} \int_{k_x} F(k_x, k_y) G_\theta^*(k_x, k_y, \omega)\, dk_x\, dk_y.$$

Making variable transformations from the rectilinear spatial frequency (k_x, k_y) domain to the polar spatial frequency (ϕ, θ) domain in the above two-dimensional integral, one obtains

$$s(\omega, \theta) = P(\omega) \int_\phi \int_\rho \rho F_p(\rho, \phi) G_{\theta p}^*(\omega, \rho, \phi)\, d\rho\, d\phi.$$

Using the analytical expression for the slant plane Green's function in the above model yields

$$s(\omega, \theta) = P(\omega) \int_\phi \int_\rho \rho F_p(\rho, \phi) W_1(\theta - \phi) W_2(\omega, \rho)$$
$$\times \exp\left[-j\sqrt{4k^2 - \rho^2}\, Z_c - j\rho R_g \cos(\theta - \phi)\right] d\rho\, d\phi.$$

This CSAR model is the basis for the two-step CSAR reconstruction method which is described next.

Slant Plane to Ground Plane Transformation

The first phase of developing CSAR reconstruction is to convert slant plane CSAR data to a CSAR database of the target region which is collected at the *ground* plane, that is, the zero altitude $z = 0$. For this, we first define the ground plane fast-time frequency by ω_g, and its wavenumber by

$$k_g = \frac{\omega_g}{c}.$$

The ground plane wavenumber can be shown to be related to the target radial spatial frequency via [s96a]

$$\rho = 2k_g.$$

With the help of these ground plane fast-time frequency parameters, we can rewrite the CSAR model developed in the previous section by

$$s(\omega, \theta) = \int_{\omega_g} \Lambda(\omega, \omega_g) s_g(\omega_g, \theta)\, d\omega_g,$$

where

$$\Lambda(\omega, \omega_g) = P(\omega) W_2(\omega, \rho) \exp\left(-j\sqrt{4k^2 - \rho^2}\, Z_c\right)$$

and

$$s_g(\omega_g, \theta) = \rho \int_\phi F_p(\rho, \phi) W_1(\theta - \phi) \exp\bigl[-j\rho R_g \cos(\theta - \phi)\bigr] \, d\phi.$$

We call $s_g(\omega_g, \theta)$ the ground plane CSAR signal.

- Why do we call $s_g(\omega_g, \theta)$ the ground plane CSAR signal? It turns out that the above model which relates $s_g(\omega_g, \theta)$ to the target function $F_p(\rho, \phi)$ is identical to the system model of a CSAR system which radiates the target area at the ground level ($z = 0$) with a radar whose fast-time frequencies are labeled by ω_g [s96a]. In fact the system model for this ground plane CSAR scenario can be obtained from the general system model for a CSAR system in Section 6.1 with $Z_c = 0$, which is

$$s_g(\omega_g, \theta) = \int_y \int_x f(x, y) \times \exp\left[-j2k\sqrt{(x - R_g \cos\theta)^2 + (y - R_g \sin\theta)^2}\right] dx\, dy.$$

Our task is to reconstruct the ground plane CSAR signal $s_g(\omega_g, \theta)$ from the slant plane CSAR signal $s(\omega, \theta)$; the result will be used in the ground plane CSAR reconstruction to obtain the target function.

The relationship between the slant plane CSAR signal $s(\omega, \theta)$ and the ground plane CSAR signal $s_g(\omega_g, \theta)$ is demonstrated in Figure 7.4*a*. This figure shows that

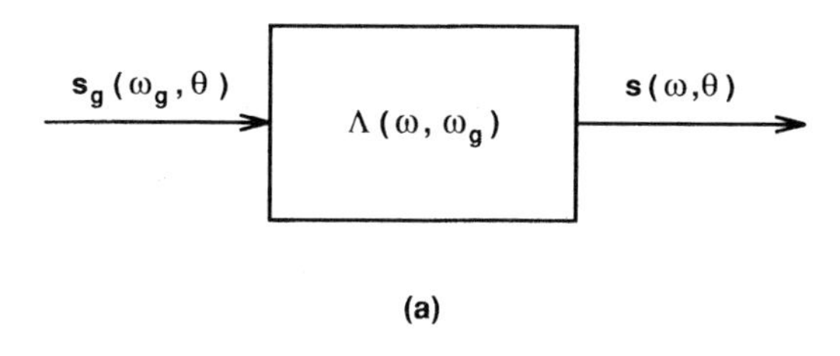

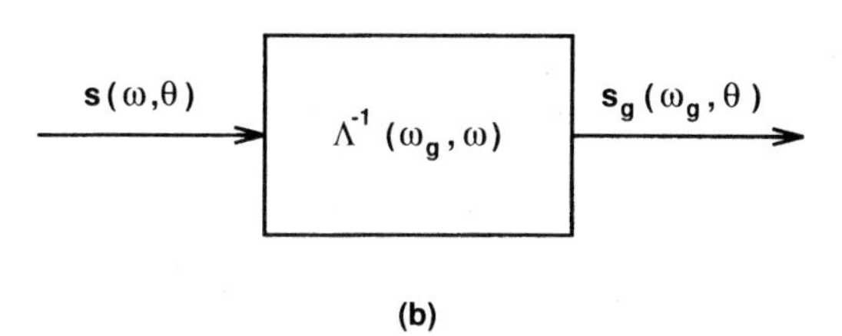

Figure 7.4. (a) Block diagram of linear shift-varying system for generating slant plane CSAR signal $s(\omega, \theta)$ from ground plane CSAR signal $s_g(\omega_g, \theta)$; (b) inverse of linear shift-varying system in (a).

the slant plane CSAR signal $s(\omega, \theta)$ is the output of a linear and *shift-varying* system whose shift-varying impulse response is $\Lambda(\omega, \omega_g)$; the input to this system is the ground plane CSAR signal $s_g(\omega_g, \theta)$. We refer to the shift-varying impulse response $\Lambda(\omega, \omega_g)$ as the linear system kernel; the system kernel is a known analytical signal.

The *inverse* of the linear shift-varying system in Figure 7.4*a* is shown in Figure 7.4*b*. In this system the input is the slant plane CSAR signal, and the output is the ground plane CSAR signal; the shift-varying impulse response of this system is the *inverse* of the system kernel which we denote with $\Lambda^{-1}(\omega_g, \omega)$. The system input/output equation is

$$s_g(\omega_g, \theta) = \int_\omega \Lambda^{-1}(\omega_g, \omega) s(\omega, \theta)\, d\omega.$$

Thus the ground plane CSAR signal can be recovered from the slant plane CSAR signal by constructing the above integral.

In the digital implementation of this method, the inverse system kernel is computed numerically via the pseudo-inverse of the system kernel. Moreover the range of the ground plane fast-time frequencies ω_g is dictated by the support of the window function

$$W_2(\omega, \rho) = \begin{cases} 1 & \text{for } |\rho - 2k\cos\theta_z| \le 2k \sin^2\theta_z \sin\theta_x(\omega), \\ 0 & \text{otherwise.} \end{cases}$$

This yields

$$\omega_g \in \left[\omega_{g\min}, \omega_{g\max}\right],$$

where

$$\omega_{g\min} = \left[\cos\theta_z - \sin^2\theta_z \sin\theta_x(\omega_{\min})\right]\omega_{\min},$$

$$\omega_{g\max} = \left[\cos\theta_z + \sin^2\theta_z \sin\theta_x(\omega_{\max})\right]\omega_{\max},$$

and $\omega \in [\omega_{\min}, \omega_{\max}]$ is the fast-time frequency band of the radar signal.

Figure 7.5*a* and *b*, respectively, shows the system kernel $\Lambda(\omega, \omega_g)$, and the inverse system kernel $\Lambda^{-1}(\omega_g, \omega)$ for the CSAR system of the T-72 turntable data. Figure 7.6*a* shows the ground plane CSAR data $s_g(\omega_g, \theta)$ for the T-72 tank; these are constructed from the slant plane CSAR data $s(\omega, \theta)$ of Figure 7.3 and the inverse kernel $\Lambda^{-1}(\omega_g, \omega)$ of Figure 7.5*b* via the system model of Figure 7.4*b*.

Ground Plane CSAR Reconstruction

The next step is the reconstruction of the target function from the CSAR ground plane signal; as we showed earlier, these two are related via

$$s_g(\omega_g, \theta) = \int_\phi F_p(\rho, \phi) W_1(\theta - \phi) \exp\left[-j\rho R_g \cos(\theta - \phi)\right] d\phi,$$

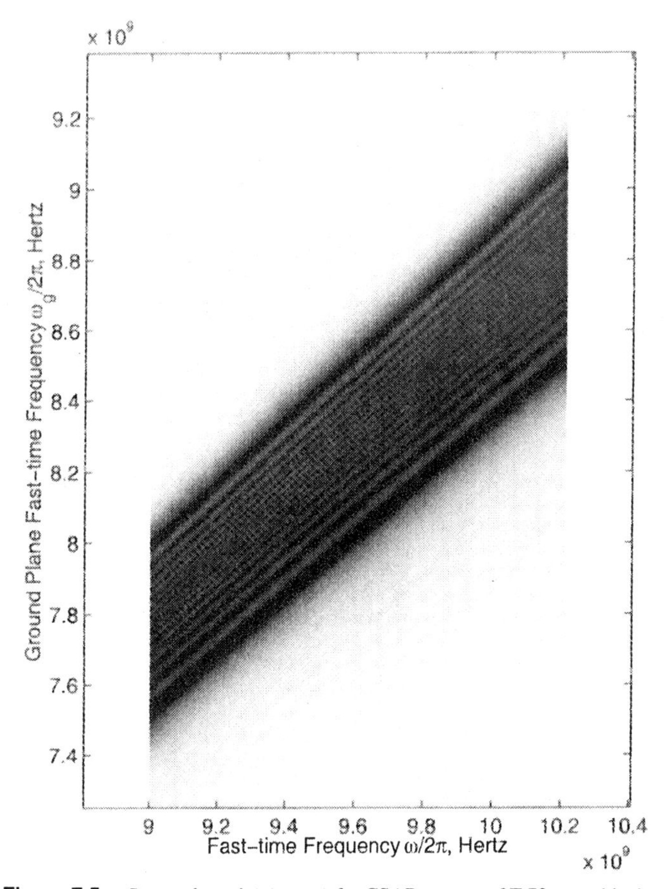

Figure 7.5a. System kernel $\Lambda(\omega, \omega_g)$ for CSAR system of T-72 turntable data.

where $\rho = 2k_g$. The above model can be rewritten in terms of a convolution in the slow-time (aspect angle synthetic aperture) θ domain, which is

$$s_g(\omega_g, \theta) = F_p(\rho, \theta) * \Big[W_1(\theta) \exp\big(- j\rho R_g \cos\theta \big) \Big],$$

where $*$ denotes convolution in the θ domain.

To determine the target function $F_p(\rho, \theta)$ from the above model, we must deconvolve the known *reference* signal

$$s_{g0}(\omega_g, \theta) = W_1(\theta) \exp(-j\rho R_g \cos\theta)$$

(which turns out to be the signature of a unit reflector that is located at the origin) from $s_g(\omega_g, \theta)$; this can be done in the frequency domain of θ. Let ξ represent the frequency domain for the slow-time θ domain. (Note that when $\theta \in (0, 2\pi)$, the ξ domain represents the Fourier series domain, and ξ only takes on integer values.) We

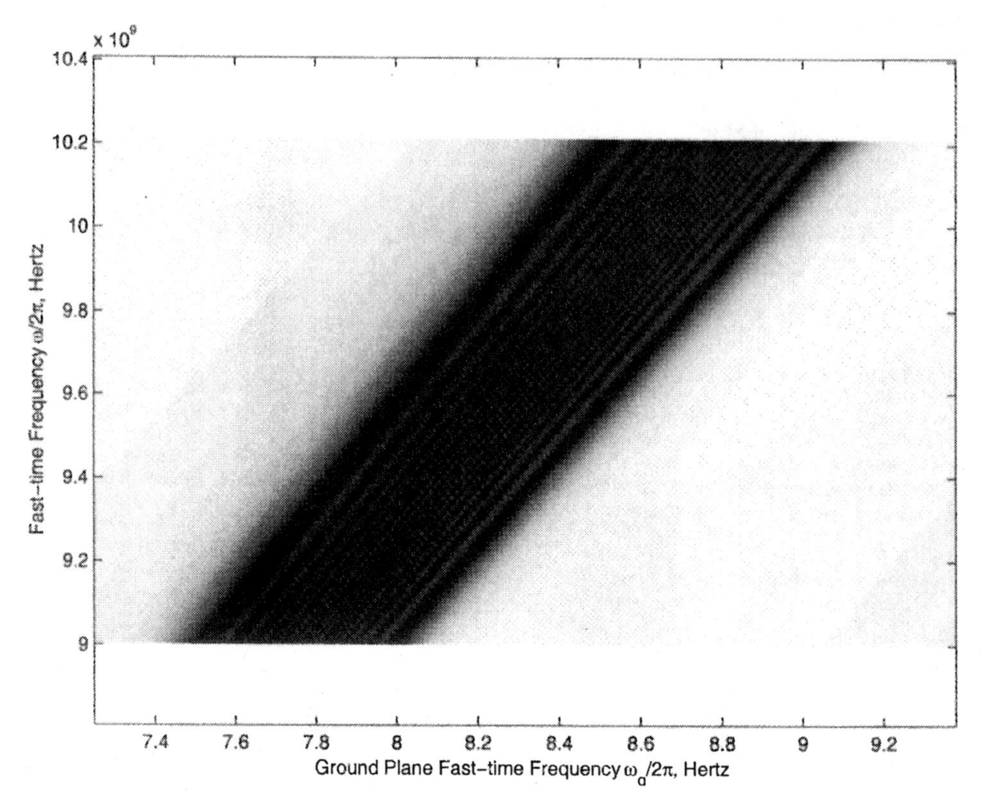

Figure 7.5b. Inverse system kernel $\Lambda^{-1}(\omega_g, \omega)$ for CSAR system of T-72 turntable data.

also define the following Fourier transforms with respect to θ:

$$\mathbf{F}_p(\rho, \xi) = \mathcal{F}_{(\theta)}[F_p(\rho, \theta)]$$

and

$$\begin{aligned} S_g(\omega_g, \xi) &= \mathcal{F}_{(\theta)}[s_g(\omega_g, \theta)], \\ S_{g0}(\omega_g, \xi) &= \mathcal{F}_{(\theta)}[s_{g0}(\omega_g, \theta)]. \end{aligned}$$

Thus, for the deconvolution, we have

$$\mathbf{F}_p(\rho, \xi) = \frac{S_g(\omega_g, \xi)}{S_{g0}(\omega_g, \xi)}.$$

The deconvolution kernel, that is,

$$s_{g0}(\omega_g, \theta) = W_1(\theta) \exp(-j\rho R_g \cos\theta),$$

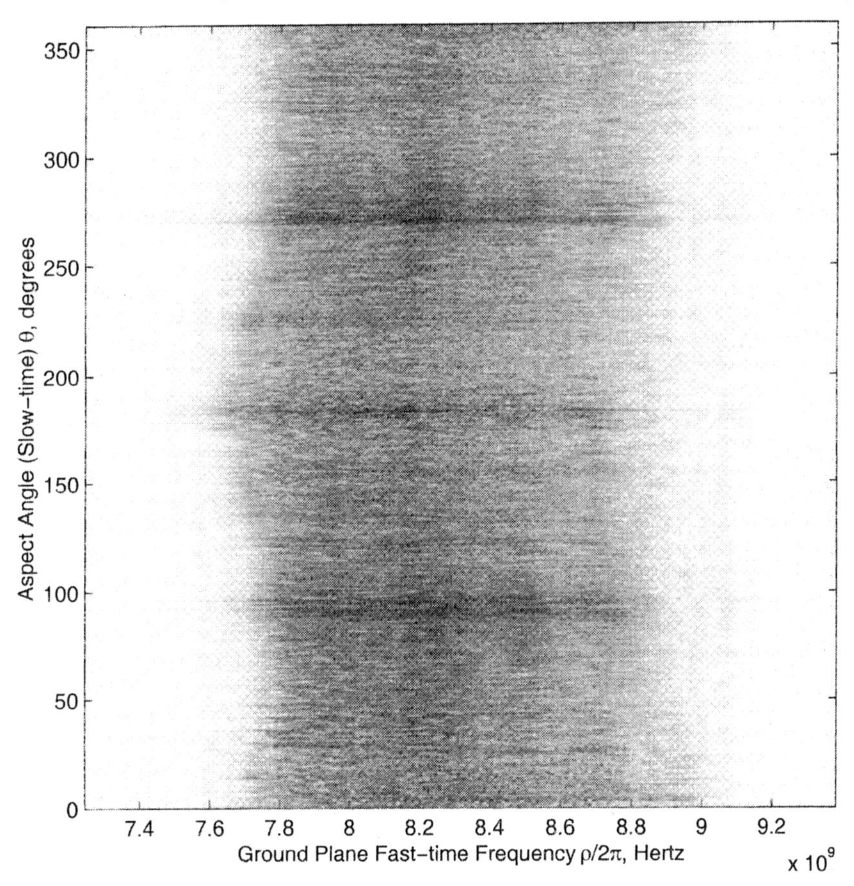

Figure 7.6a. Ground plane CSAR data $s_g(\omega_g, \theta)$ for T-72 tank constructed from slant plane CSAR data $s(\omega, \theta)$ (Figure 7.3) via inverse system of Figure 7.4*b*.

is an AM-PM signal. The PM component has the following one-dimensional Fourier transform with respect to the slow-time θ [s96a]:

$$\mathcal{F}_{(\theta)}\big[\exp(-j\rho R_g \cos\theta)\big] = H_\xi^{(2)}(\rho R_g) \exp\left(-j\frac{\pi\xi}{2}\right),$$

where $H_\xi^{(2)}$ is the Hankel function of the second kind, ξ order [mor; s94]. Using the windowing properties of PM signals (see Chapter 2) and the fact that the support of the window $W_1(\theta)$ is $|\theta| \le \theta_x(\omega)$ (see Section 7.1), the slow-time Fourier transform of the deconvolution kernel is equal to the slow-time Fourier transform of its PM component for

$$|\xi| \le \rho R_g \sin\theta_x(\omega) = \rho R_0(\omega),$$

and zero otherwise.

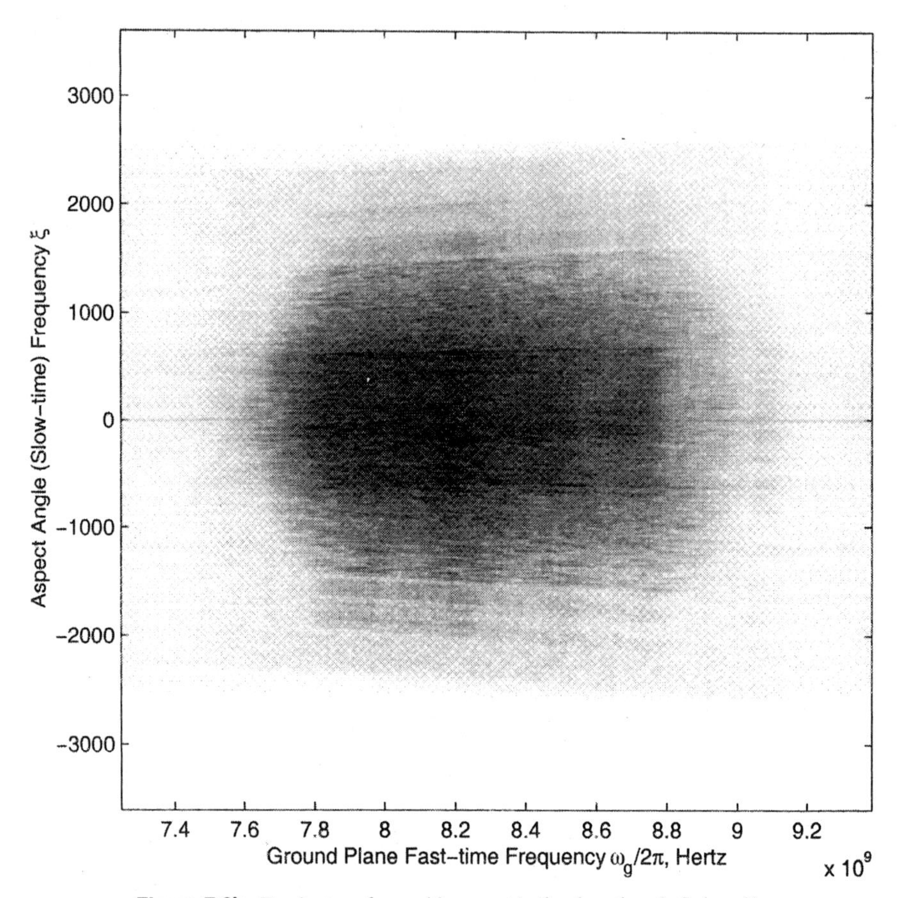

Figure 7.6b. Fourier transform with respect to the slow-time θ, $S_g(\omega_g, \xi)$.

Thus the above deconvolution to reconstruct $\mathbf{F}_p(\rho, \xi)$ is implemented via the following (matched-filtered form):

$$\mathbf{F}_p(\rho, \xi) = S_g(\omega_g, \xi) \underbrace{S_{g0}^*(\omega_g, \xi)}_{\text{Matched filter}},$$

for $|\xi| \leq \rho R_0(\omega)$, where S_{g0}^* is the complex conjugate of S_{g0}.

After reconstructing $\mathbf{F}_p(\rho, \xi)$, the target spectrum in the polar spatial frequency domain, that is, $F_p(\rho, \theta)$, is formed by its inverse Fourier transform with respect to ξ:

$$F_p(\rho, \theta) = \mathcal{F}^{(-1)}{}_{(\xi)}\big[\mathbf{F}_p(\rho, \xi)\big].$$

The target function in the rectilinear spatial frequency domain is obtained by

$$F\big[k_x(\rho, \theta), k_y(\rho, \theta)\big] = F_p(\rho, \theta)$$

where

$$k_x(\omega, \theta) = \rho \cos\theta,$$

$$k_y(\omega, \theta) = \rho \sin\theta.$$

The two-dimensional inverse Fourier transform of this signal is the desired target image $f(x, y)$. This completes the final phase of the reconstruction in slant plane CSAR.

Consider the CSAR system corresponding to the full rotation ($\theta \in [0, 2\pi)$) of T-72 tank turntable data. Figure 7.6*b* shows the two-dimensional spectrum of ground plane CSAR data, $S_g(\omega_g, \xi)$, which is the one-dimensional Fourier transform of $s_g(\omega_g, \theta)$ in Figure 7.6*a* with respect to the slow-time θ. After processing the ground plane CSAR spectral domain data in Figure 7.6*b* in the above-mentioned deconvolution, we obtain the reconstructions shown in Figures 7.7 and 7.8. Figures 7.7 and

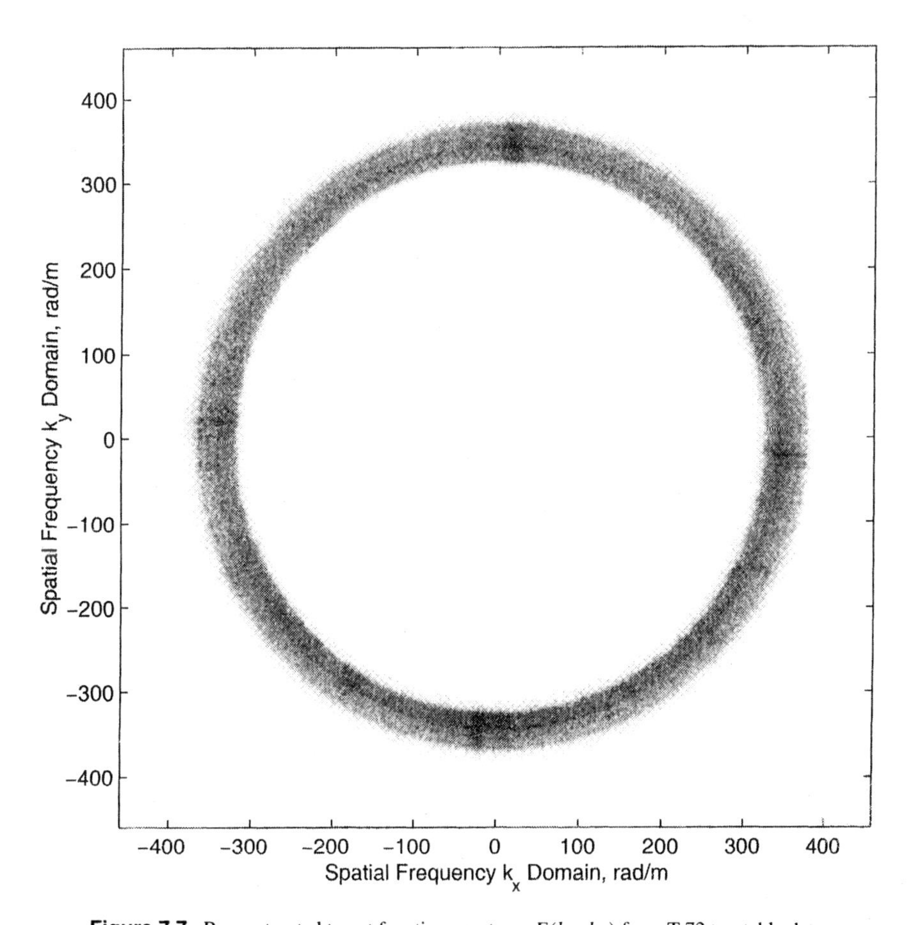

Figure 7.7. Reconstructed target function spectrum $F(k_x, k_y)$ from T-72 turntable data.

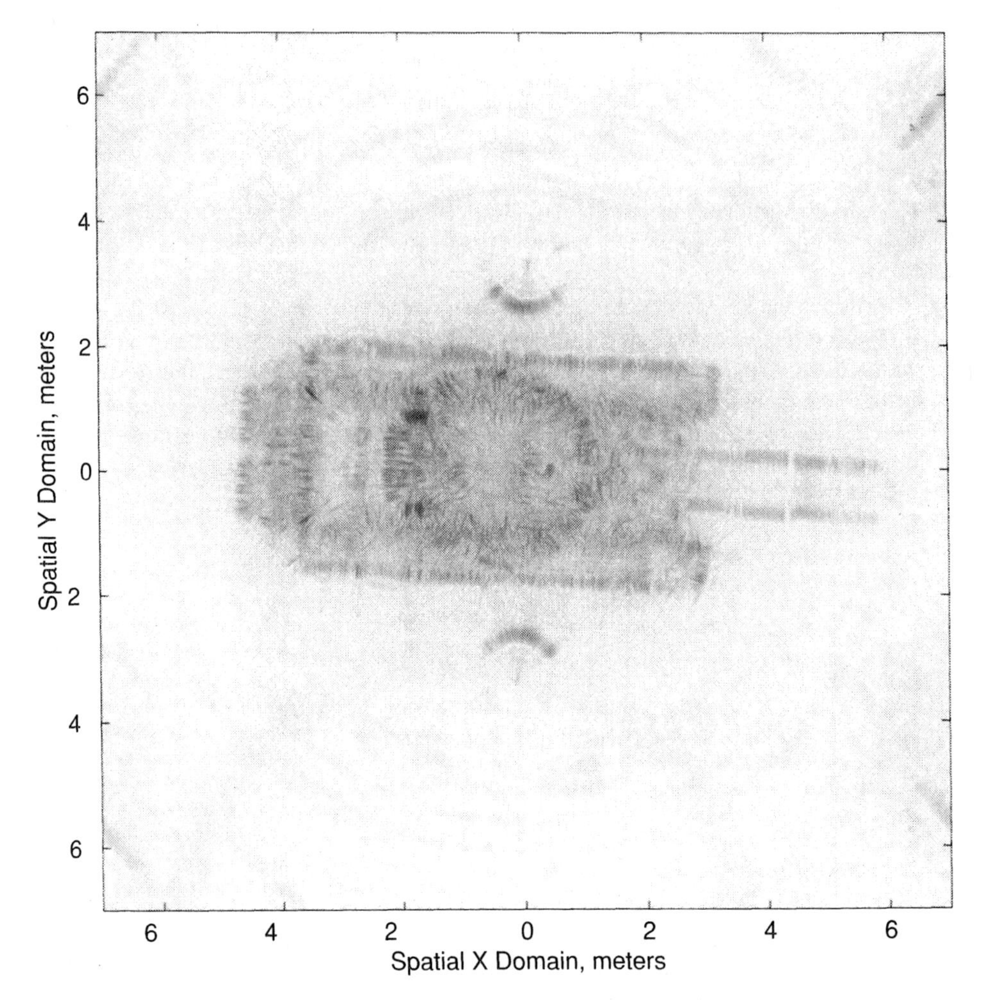

Figure 7.8. Reconstructed target function $f(x, y)$ from T-72 turntable data at elevation $z = 0$.

7.8 are the spectrum of the target function $F(k_x, k_y)$ and the reconstructed target function in the spatial domain $f(x, y)$, respectively.

Note that we have identified the spatial domain target function in Figure 7.8 as $z = 0$. In fact the T-72 tank is not simply a delta (thin) sheet on the ground plane; this target is a three-dimensional structure. Figure 7.8 shows a combination of the reflectivity of various structures on the tank; some reflectors appear to be in focus, and some are not. For example, the cannon of the tank appears to be doubled. There are also two corner reflectors in the imaging scene that are located at the two sides of the tank; both of these corner reflectors appear to be out of focus.

It turns out that slant plane CSAR data provide some form of three-dimensional information about the illuminated target area though the CSAR database is composed of a two-dimensional measurement domain. We will discuss the three-dimensional imaging from slant plane CSAR data later in this chapter. We will develop an imaging algorithm to form a focused image of the target function at a given elevation z value while the targets at the other elevation levels appear out of focus.

7.3 BANDWIDTH OF CSAR SIGNAL

We present the two-dimensional bandwidth of the CSAR signal without the details of its derivation; the mathematical analysis can be found in [s94] and [s96a].

We are interested in the support band of the two-dimensional Fourier transform of the CSAR signal with respect to the fast-time t and slow-time (aspect angle synthetic aperture) θ, that is,

$$S(\omega, \xi) = \mathcal{F}_{(t,\theta)}[\mathbf{s}(t, \theta)].$$

As we stated earlier, the fast-time support band of the radar signal, and thus the CSAR signal is $\omega \in [\omega_{\min}, \omega_{\max}]$.

Next we consider the slow-time frequency support band of the CSAR signal at a given fast-time frequency ω of the radar signal. For this, we should analyze the fluctuations of the phase history CSAR signal for the target which is located at (x, y); that is,

$$\exp\left[-j2k\sqrt{(x - R_g \cos\theta)^2 + (y - R_g \sin\theta)^2 + Z_c^2}\,\right].$$

This is a PM signal in the slow-time θ domain. Thus, to determine its slow-time frequency band, we should determine the minimum and maximum values of instantaneous frequency of the above signal with respect to the slow-time θ.

Using this approach, we can show that at a given fast-time frequency ω, the slow-time bandwidth support of the CSAR signal is approximately within

$$\begin{aligned}
\xi &\in \big[-2k\cos\theta_z R_0(\omega), 2k\cos\theta_z R_0(\omega)\big] \\
&= \begin{cases} \left[\dfrac{-4\pi R_c}{D\tan\theta_z}, \dfrac{4\pi R_c}{D\tan\theta_z}\right], & \text{planar radar,} \\[2ex] \left[\dfrac{-2kR_c D}{2x_f\tan\theta_z}, \dfrac{2kR_c D}{2x_f\tan\theta_z}\right], & \text{curved radar.} \end{cases}
\end{aligned}$$

- For a planar radar aperture the above slow-time frequency band is invariant in the radar fast-time frequency. Recall that we encountered the same phenomenon in the linear stripmap SAR systems; see Section 6.3.

The largest slow-time bandwidth is encountered at the highest fast-time frequency of the radar signal for either the planar or curved radar aperture [though $R_0(\omega_{\max}) \le$

$R_0(\omega_{\min})]$; this is

$$\xi \in \Big[-2k_{\max} \cos\theta_z R_0(\omega_{\max}), 2k_{\max} \cos\theta_z R_0(\omega_{\max})\Big].$$

The above slow-time bandwidth is encountered whether one considers CSAR data within a full rotation of the radar ($\theta \in (0, 2\pi)$), or only a portion of it.

Figure 7.9 shows the two-dimensional spectrum of the T-72 CSAR data, $S(\omega, \xi)$. Note that this spectrum shows a distinct cone-shaped support. The radar used in this experiment effectively had a planar aperture. Moreover the radar bandwidth was relatively narrow. This makes it difficult to see the variations of the CSAR data spectrum with respect to the variations of $R_0(\omega)$ with the fast-time frequency.

The distinct cone-shaped signature in Figure 7.9 belongs to the tank and the corner reflectors, all of which fall within the radar beamwidth at all the available radar fast-time frequencies. We have also performed a fast-time frequency-dependent filtering

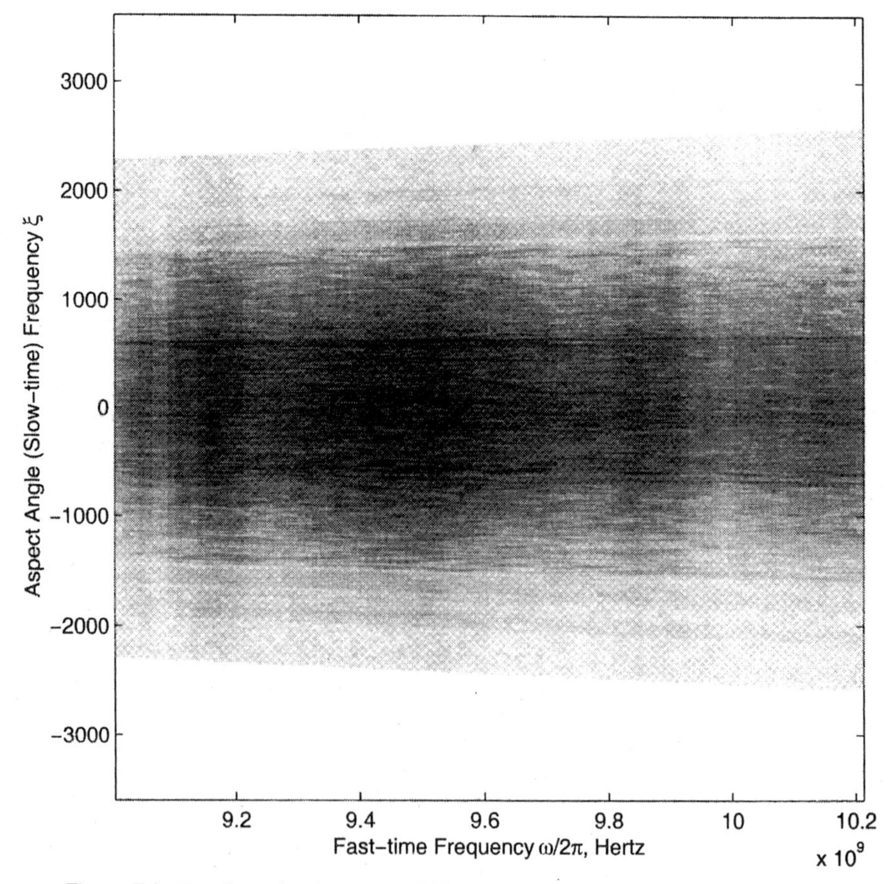

Figure 7.9. Two-dimensional spectrum of CSAR signal $S(\omega, \xi)$ for T-72 turntable data.

in the slow-time frequency ξ; this will be discussed in our analysis of CSAR signal digital spotlighting in Section 7.5.

7.4 RESOLUTION AND POINT SPREAD FUNCTION

Consider a target, labeled the nth target, which is located at the coordinates (x_n, y_n) on the ground plane. We define the slow-time varying depression (grazing) angle of the radar with respect to this target via

$$\theta_{zn}(\theta) = \arctan\left[\frac{Z_c}{\sqrt{(x_n - R_g\cos\theta)^2 + (y_n - R_g\sin\theta)^2}}\right].$$

We also define the ground-plane aspect angle of the radar with respect to the nth target by

$$\theta_n(\theta) = \arctan\left(\frac{y_n - R_g\sin\theta}{x_n - R_g\cos\theta}\right).$$

Using the Fourier properties of AM-PM signals, the following can be stated with regard to the support of the CSAR signature of the nth target in the target spatial frequency (k_x, k_y) domain. At a given fast-time frequency ω, the nth target signature traces the contour identified by

$$k_{xn}(\omega,\theta) = 2k\cos\theta_{zn}(\theta)\cos\theta_n(\theta),$$

$$k_{yn}(\omega,\theta) = 2k\cos\theta_{zn}(\theta)\sin\theta_n(\theta),$$

as the aspect angle (slow-time) θ varies. Next we use the above contour to determine the CSAR spectral support band of a target for both the full rotation and also a partial rotation aspect angle measurement.

Full Rotation Aspect Angle Measurement

Consider the case when the slow-time (aspect angle synthetic aperture) measurement is made over the full rotation, that is, $\theta_n(\theta)$ with $\theta \in (0, 2\pi)$; this also translates into $\theta_n(\theta) \in (0, 2\pi)$. Suppose that the nth target falls within the radar radiation at all the available radar fast-time frequencies. In the case of a curved radar aperture, this occurs when the radial distance of the target from the origin $r_n = \sqrt{x_n^2 + y_n^2}$ is less than the radius of the spotlighted area; that is,

$$r_n \le R_0(\omega) = \frac{R_c\tan\phi_d}{\sin\theta_z};$$

note that $R_0(\omega)$ is invariant in the radar frequency ω for a curved radar aperture. For a planar radar aperture the nth target is observable to the radar at all of the fast-time frequencies of the radar provided that

$$r_n \leq R_0(\omega_{\max}) \approx \frac{R_c \lambda_{\min}}{D \sin \theta_z}.$$

In either case the CSAR support of the nth target in the spatial frequency (k_x, k_y) domain is the region that falls between

$$\left[k_{xn}(\omega_{\min}, \theta), k_{yn}(\omega_{\min}, \theta)\right]$$

and

$$\left[k_{xn}(\omega_{\max}, \theta), k_{yn}(\omega_{\max}, \theta)\right]$$

for $\theta \in (0, 2\pi)$. Figure 7.10*a* shows this spectral band for a target located at the origin (centered), that is, $(x_n, y_n) = (0, 0)$; Figure 7.10*b* represents this spatial frequency support for an off-centered target.

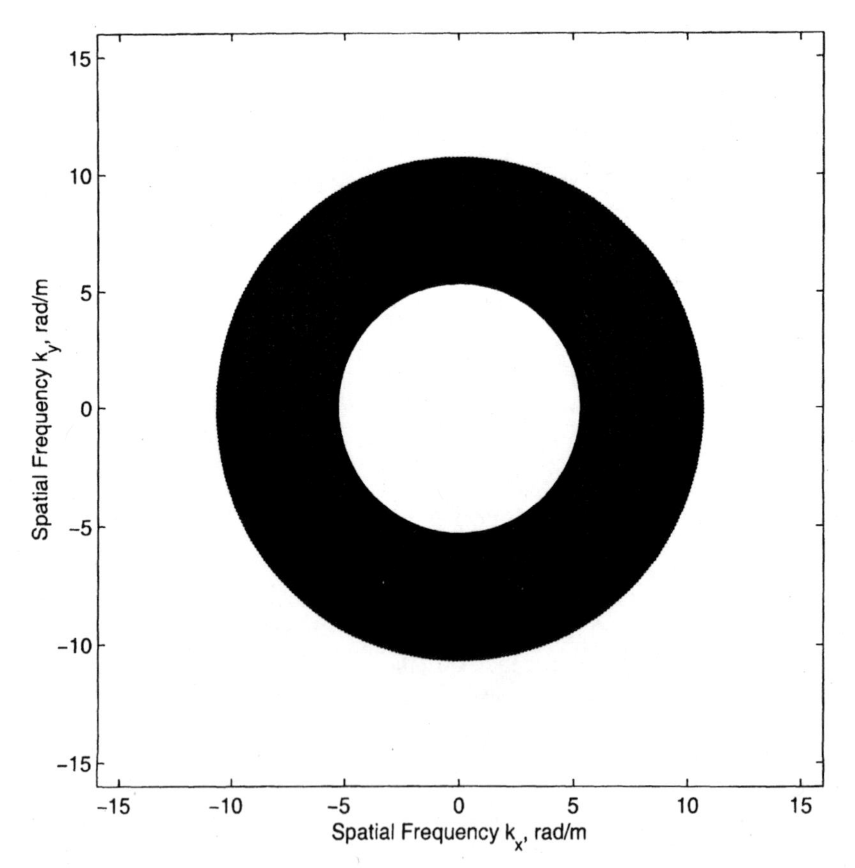

Figure 7.10a. Full rotation, $\theta_{\max} = 180°$ two-dimensional spectral (k_x, k_y) domain support for a target located at origin $(x_n, y_n) = (0, 0)$.

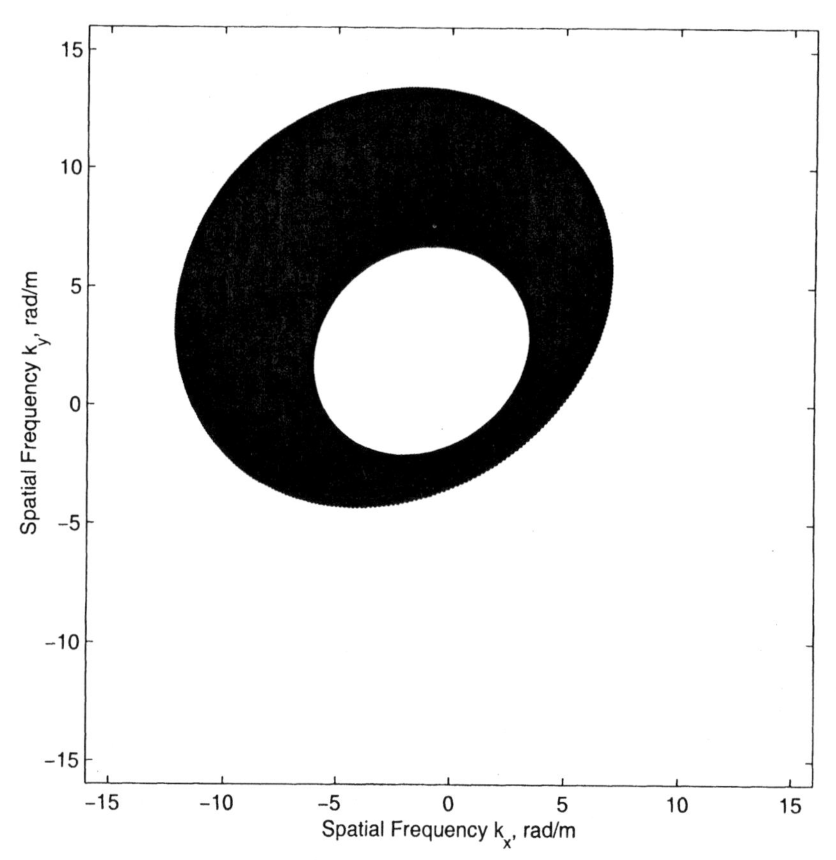

Figure 7.10b. Full rotation, $\theta_{\max} = 180°$ two-dimensional spectral (k_x, k_y) domain support for an off-centered target.

In the case of a planar target, the available fast-time frequencies of the radar may not be observable at all. For instance, suppose that the nth target is radar observable within the wavelengths $[\lambda_{n\min}, \lambda_{\max}]$, where

$$\lambda_{n\min} = \frac{r_n D \sin\theta_z}{R_c}.$$

In other words, the target is not observable to the radar fast-time frequencies whose wavelengths fall within $(\lambda_{\min}, \lambda_{n\min})$. In this case the spectral support of the CSAR signature of the nth target resembles the one which is shown in Figure 7.10*b*; the only difference is that the spectral support band for this target in the fast-time frequency domain is limited to

$$\omega \in [\omega_{\min}, \omega_{n\max}],$$

where $\omega_{n\max} = 2\pi c/\lambda_{n\min}$.

If the nth target is an omni-directional reflector at the origin [i.e., $(x_n, y_n) = (0, 0)$], then its two-dimensional CSAR spectrum $F_n(k_x, k_y)$ is a constant within the donut-shaped band in Figure 7.10a. In this case the point spread function of this target in the spatial domain becomes

$$f_n(x, y) = \rho_{\max} \frac{J_1(r\rho_{\max})}{r} - \rho_{\min} \frac{J_1(r\rho_{\min})}{r},$$

where $r = \sqrt{x^2 + y^2}$, J_1 is the Bessel function of the first kind, first order [mor; s94], and

$$\rho_{\min} = 2k_{\min} \cos\theta_z,$$

$$\rho_{\max} = 2k_{\max} \cos\theta_z.$$

The behavior of this point spread function and its resultant spatial resolution depends on the radar frequency band; an example involving a UHF CSAR system is studied in [s96a].

For an off-centered target with the spectral support of Figure 7.10b, it is difficult to obtain an analytical expression for the target spatial point spread function and its resolution. A viable way for the user to study these is through numerical methods.

For a general target that is not omni-directional, its resolution in a CSAR imaging system depends on the interval of the radar aspect angle over which the target is observable to the radar, that is, the support of its CSAR signature in the θ domain. For instance, suppose that the nth target CSAR signature support is approximately within $[\phi_n - \phi_0, \phi_n + \phi_0]$ in the θ domain, where ϕ_n is a quantity that depends on the target angular orientation in the spatial domain, and typically $\phi_0 < \pi/4$ depends on the physical shape or type of the target. (Most human-made targets, such as trucks and tanks, possess this type of CSAR signature.)

Then the target resolution in the ϕ_n rotated version of the spatial domain, labeled the (X_n, Y_n) domain, can be found via principles that are similar to what we developed for linear spotlight SAR; these yield

$$\Delta_{X_n} \approx \frac{\pi}{\rho_{\max} - \rho_{\min}},$$

$$\Delta_{Y_n} \approx \frac{\pi}{2k_c \cos\theta_z \sin\phi_0}.$$

Partial Rotation Aspect Angle Measurement

Next we consider the case when the aspect angle synthetic aperture measurement is made for $\theta \in [-\theta_{\max}, \theta_{\max}]$ where $\theta_{\max}$ is a constant chosen by the user. The support region of the nth target CSAR signature in the spatial frequency (k_x, k_y) domain is dictated by the area that falls between

$$\left[k_{xn}(\omega_{\min}, \theta), k_{yn}(\omega_{\min}, \theta)\right]$$

and

$$\left[k_{xn}(\omega_{n\max}, \theta), k_{yn}(\omega_{n\max}, \theta)\right]$$

for $\theta \in [-\theta_{\max}, \theta_{\max}]$. Figure 7.11*a* shows this spectral band for a target located at the origin (centered), that is, $(x_n, y_n) = (0, 0)$, with $\theta_{\max} = 10°$; and Figure 7.11*b* is this spatial frequency support for an off-centered target.

Note that the spectral supports in Figures 7.11*a* and *b*, respectively, resemble the spotlight SAR spectrum of a broadside target and an off-broadside target (see Section 5.4 and Figure 5.6). With $\theta_z = 0$ (i.e., zero grazing or depression angle), *the spectral supports of a target in the linear spotlight SAR and CSAR for a given radar (fast-time frequency) band and aspect angle measurement interval are identical.* In fact, if one includes the effect of the depression angle in linear spotlight SAR (this is discussed in [s92ja]), then a target signature spectral support in the linear spot-

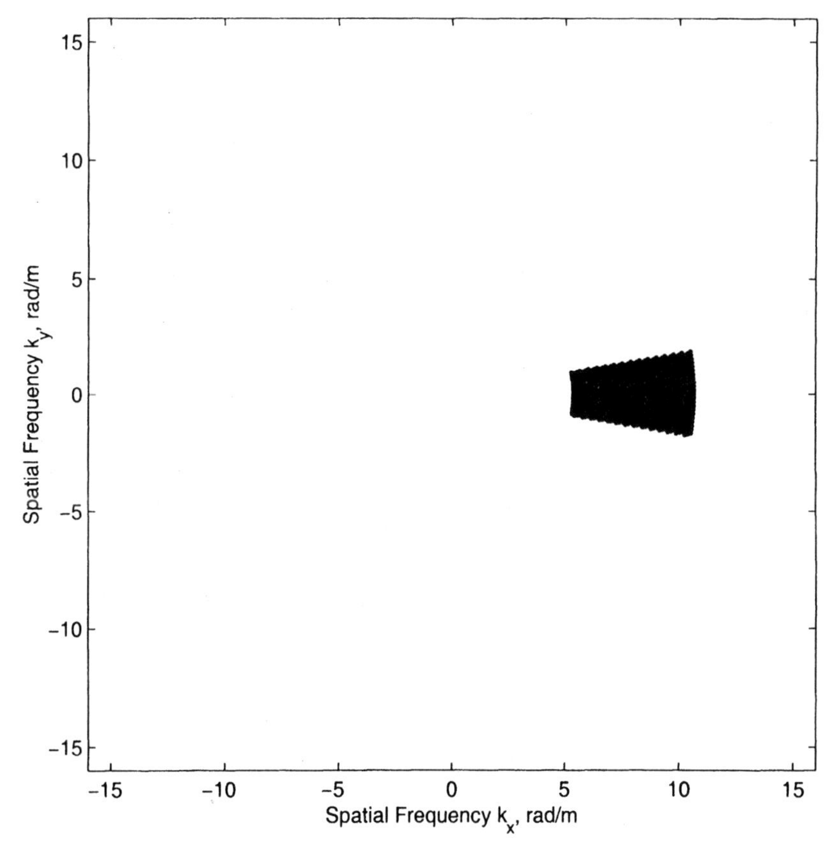

Figure 7.11a. Partial rotation, $\theta_{\max} = 10°$, two-dimensional spectral (k_x, k_y) domain support for a target located at origin $(x_n, y_n) = (0, 0)$.

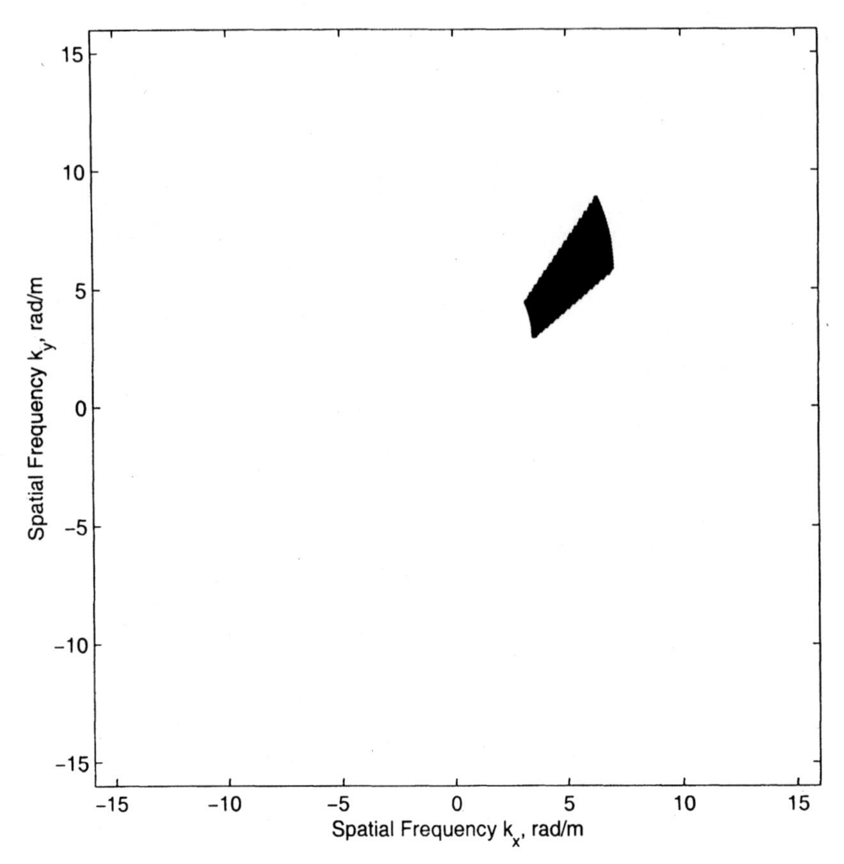

Figure 7.11b. Partial rotation, $\theta_{\max} = 10°$, two-dimensional spectral (k_x, k_y) domain support for an off-centered target.

light SAR and slant plane CSAR can be shown to be the same for any depression angle θ_z.

For an omni-directional target, its resolution in a linear spotlight SAR and CSAR imaging system depends on the interval of the radar aspect angle over which the target is observable to the radar, that is, the support of its signature in the aspect angle domain which is generated by a *linear* or *circular* synthetic aperture trajectory (flight path of the radar-carrying aircraft). The *shift-varying* point spread function of a target in the spatial domain for a linear spotlight SAR was developed in Section 5.4; similar principles also hold for the point spread function of a CSAR system with a partial rotation aspect angle measurement.

7.5 DATA ACQUISITION AND SIGNAL PROCESSING

Next we examine the sampling constraints and processing issues associated with the digital signal processing of the CSAR signal. This analysis benefits from the two-dimensional spectral analysis of the CSAR signal that appeared in Section 7.3.

Fast-time Domain Sampling and Processing

Fast-time Interval of Sampling The distance of the closest reflector on the spotlighted target area from the radar is

$$r_{\min} = \sqrt{\left[R_g - R_0(\omega_{\min})\right]^2 + Z_c^2}.$$

Thus the first echoed signal arrives at the receiving radar at the fast-time

$$T_s = \frac{2r_{\min}}{c}.$$

The distance of the farthest reflector on the spotlighted target area from the radar is

$$r_{\max} = \sqrt{\left[R_g + R_0(\omega_{\min})\right]^2 + Z_c^2}.$$

Thus, if the duration of the pulsed radar signal is T_p, the echoed signal from the farthest reflector lasts until the fast-time

$$T_f = \frac{2r_{\max}}{c} + T_p.$$

To capture the echoed signals from all the reflectors in the spotlighted target area, we must acquire the time samples of the echoed SAR signal $\mathbf{s}(t, \theta)$ in the following time interval:

$$t \in [T_s, T_f].$$

Fast-time Sample Spacing The baseband bandwidth of the radar signal is $\pm\omega_0$; thus the A/D fast-time sample spacing should satisfy the following Nyquist criterion:

$$\Delta_t \le \frac{\pi}{\omega_0}.$$

For chirp radar signals the Nyquist constraint for fast-time domain sampling could be less restrictive for the deramped echoed signal. In such scenarios the upsampling scheme, which was described in Section 1.6, could be utilized to recover the alias-free echoed signal.

The number of fast-time samples within the fast-time gate $t \in [T_s, T_f]$ is

$$N = 2\left\lceil \frac{T_f - T_s}{2\Delta_t} \right\rceil,$$

where $\lceil a \rceil$ denotes the smallest integer larger than a.

Fast-time Reference Point for Matched Filtering The fast-time matched filter (reference signal) is the echoed signal from the center of the spotlighted target area, that is, $(x, y, z) = (0, 0, 0)$. The distance of the radar at any slow-time point θ from the center of the spotlighted area is R_c. Thus the reference echoed signal is

$$p_0(t) = p\big(t - T_c\big),$$

where

$$T_c = \frac{2R_c}{c}$$

is the reference fast-time point. The fast-time matched-filtered CSAR signal is obtained by

$$\mathbf{s}_M(t, \theta) = \mathbf{s}(t, \theta) * p_0^*(-t),$$

where $*$ denotes convolution in the fast-time domain and $p_0^*(-t)$ is the complex conjugate of $p_0(-t)$.

After choosing a reference fast-time point T_c and performing the above fast-time matched filtering with $p_0^*(-t)$, the fast-time sample number $(N/2) + 1$ (i.e., the sample in the middle of the fast-time array) in the matched-filtered signal $\mathbf{s}_M(t, \theta)$ corresponds to the fast-time point $t = T_c$. This will be incorporated in the digital CSAR reconstruction algorithm.

Slow-time Domain Sampling and Processing

Slow-time Sample Spacing We showed that the maximum support of the CSAR signal spectrum $S(\omega, \xi)$ in the slow-time frequency ξ domain is

$$\xi \in \big[-2k_{\max} \cos\theta_z R_0(\omega_{\max}), 2k_{\max} \cos\theta_z R_0(\omega_{\max})\big].$$

Thus the slow-time (aspect angle) sample spacing should satisfy the following:

$$\Delta_\theta \leq \frac{\pi}{2k_{\max} \cos\theta_z R_0(\omega_{\max})}.$$

For a planar radar aperture where

$$R_0(\omega_{\max}) \approx \frac{R_c \lambda_{\min}}{D \sin\theta_z},$$

the slow-time sample spacing constraint becomes

$$\Delta_\theta \leq \frac{D \tan \theta_z}{4R_c},$$

which is invariant in the radar fast-time frequency.

In the case of a curved aperture, with

$$R_0(\omega_{\max}) \approx \frac{R_c D}{2x_f \sin \theta_z},$$

the slow-time Nyquist constraint is

$$\Delta_\theta \leq \frac{\lambda_{\min} x_f \tan \theta_z}{2R_c D}.$$

For slow-time data acquisition over the entire aspect angle interval of $\theta \in [0, 2\pi)$, the value of Δ_θ should be chosen such that the number of the slow-time samples

$$M_\theta = \frac{2\pi}{\Delta_\theta}$$

is an even integer, as well as making sure that the above slow-time Nyquist rate is satisfied.

Digital Spotlighting and Clutter Filtering

The CSAR data usually contain unwanted slow-time frequency information due to the side lobes of the radar radiation pattern. For this reason the user could select a slow-time sample spacing Δ_θ that provides a guard band in the slow-time frequency ξ domain. Then the acquired data will pass through a two-dimensional filter in the spectral (ω, ξ) domain with the following passband:

$$|\xi| \leq 2k \cos \theta_z R_0(\omega).$$

After substitution for $R_0(\omega)$ in terms of the radar parameters, we obtain the following for the above passband:

$$\xi \leq \begin{cases} \dfrac{4\pi R_c}{D \tan \theta_z}, & \text{planar radar,} \\ \dfrac{2k R_c D}{2x_f \tan \theta_z}, & \text{curved radar.} \end{cases}$$

Another source of noise in these systems is a persistent interference-type signal, which is related to the internal circuitry of the radar and is approximately the same at all bursts of the radar (i.e., slow-time values). This noise can be observed in most linear SAR and CSAR databases, for example, the FOPEN stripmap SAR data examined in Chapters 5 and 6 and the T-72 tank CSAR (turntable) data of this chapter.

The effect of this type of radar noise is not obvious in the FOPEN stripmap SAR reconstructions.

The same is not true in a CSAR system with full rotation aspect angle measurement. In full rotation CSAR systems this noise, which is invariant in the slow-time θ, results in ring-shaped artifacts in the reconstructed image; this ring-shaped artifact can be analytically predicted.

The slow-time invariant radar noise in CSAR can be removed by filtering out the zero slow-time frequency, that is, $\xi = 0$, in the measured CSAR signal. This translates into the following:

$$S(\omega, \xi)|_{\xi=0} = S(\omega, 0) = 0$$

at all the available fast-time frequencies. The user may also try a smoother band rejection digital filter that not only removes (or attenuates) the zero slow-time frequency components of the CSAR spectrum $S(\omega, \xi)$ but also attenuates the distribution of $S(\omega, \xi)$ around $\xi = 0$.

We should point out that the FOPEN stripmap SAR data of Chapters 5 and 6 did not show significant improvement with this type of zero slow-time frequency band rejection filter. However, we have examined an X band SAR system that showed visible improvement in its reconstructions with this type of digital band rejection filtering.

7.6 RECONSTRUCTION ALGORITHMS AND CSAR IMAGE PROCESSING

In this section, methods for digital reconstruction from CSAR data are presented. One digital imaging method uses the reconstruction principles developed in Section 7.2. The other two methods are based on the integral algorithms of time-domain correlation and backprojection. Processing issues associated with digital image formation in CSAR are discussed.

Digital Reconstruction via Spatial Frequency Interpolation

Zero-padding in Synthetic Aperture Domain Consider the deconvolution operation for ground plane CSAR reconstruction that is performed via matched filtering in the slow-time frequency ξ domain,

$$\mathbf{F}_p(\rho, \xi) = S_g(\omega_g, \xi) S_{g0}^*(\omega_g, \xi),$$

or via convolution in the slow-time (aspect angle synthetic aperture) domain

$$F_p(\rho, \theta) = s_g(\omega_g, \theta) * s_{g0}^*(\omega_g, -\theta).$$

If the slow-time measurements are made over the full 360° (i.e., $\theta \in [0, 2\pi)$), the digital implementation of the above filtering in the slow-time frequency ξ domain would not be adversely affected by the resultant *circular* convolution.

However, if the slow-time data are obtained over a smaller aspect angle interval, the slow-time frequency filtering will suffer from circular convolution aliasing. To circumvent this problem, the ground plane CSAR data $s_g(\omega_g, \theta)$ should be zero-padded in the slow-time θ domain prior to the above-mentioned matched filtering. The extent of the zero-padding can be shown to be [s96a]

$$\pm \frac{R_0(\omega)}{R_g}.$$

The worst-case scenario (the largest amount of zero-padding) corresponds to

$$\pm \frac{R_0(\omega_{\min})}{R_g}.$$

Algorithm The following steps are used for CSAR image formation based on the reconstruction method of Section 7.2:

Step 1. Perform the (discrete) fast-time matched filtering

$$\mathbf{s}_M(t, \theta) = \mathbf{s}(t, \theta) * p_0^*(-t),$$

where

$$p_0(t) = p(t - T_c)$$

and T_c is the user-prescribed reference fast-time point (e.g., $T_c = 2R_c/c$).

Step 2. Compute the one-dimensional Fourier transform of the matched-filtered CSAR signal with respect to the fast-time t to obtain $s_M(\omega, \theta)$. The DFT algorithm assumes that the fast-time sample number $(N/2) + 1$ (i.e., the sample in the middle of the fast-time array) is at $t = 0$; however, this sample corresponds to the actual fast-time value of $t - T_c$. (The parameter N is the number of the fast-time frequency samples.) Thus the two-dimensional DFT of the upsampled and digitally spotlighted SAR signal should be modified by the following linear phase function (discrete) multiplication:

$$s_M(\omega, \theta) \exp\big(-j\omega T_c\big),$$

which moves the fast-time origin back to the actual fast-time zero, that is, $t = 0$. (Note that Steps (1) and (2) can be combined.) For notational simplicity the resultant data are identified by $s(\omega, \theta)$ in the steps that follow.

Step 3. Compute the samples of the system kernel $\Lambda(\omega, \omega_g)$ for

$$\omega \in [\omega_{\min}, \omega_{\max}]$$

and

$$\omega_g \in [\omega_{g\min}, \omega_{g\max}].$$

Obtain the pseudo-inverse of the resultant matrix to obtain the samples of the inverse system kernel $\Lambda^{-1}(\omega_g, \omega)$.

Step 4. For a fixed radar (aircraft) look angle θ *(i.e., slow-time), convert the slant plane CSAR data* $s(\omega, \theta)$ *into* $s_g(\omega_g, \theta)$ *by constructing the discrete version of the following integral:*

$$s_g(\omega_g, \theta) = \int_\omega \Lambda^{-1}(\omega_g, \omega) s(\omega, \theta)\, d\omega,$$

via the sum of integrand samples over the available fast-time frequencies.

Step 5. Obtain the discrete Fourier transform of the ground plane CSAR data $s_g(\omega_g, \theta)$ *with respect to the available samples of the slow-time* θ*; this yields the samples of two-dimensional spectrum* $S_g(\omega, \xi)$*. In the case of a partial rotation aspect angle measurement, zero-pad the samples of* $s_g(\omega_g, \theta)$ *in the slow-time* θ *domain by*

$$\pm \frac{R_0(\omega_{\min})}{R_g}$$

prior to performing Fourier transform with respect to the slow-time θ.

Step 6. Perform the slow-time frequency domain matched filtering

$$\mathbf{F}_p(\rho, \xi) = S_g(\omega_g, \xi) S_{g0}^*(\omega_g, \xi),$$

where $\rho = 2k_g = 2\omega_g/c$*, to reconstruct the target function in the* (ρ, ξ) *domain.*

Step 7. Obtain the inverse discrete Fourier transform of $\mathbf{F}_p(\rho, \xi)$ *with respect to* ξ*; this yields the polar samples of the target function in the spatial frequency domain, that is, the samples of* $F_p(\rho, \theta)$.

Step 8. For the full rotation aspect angle measurement, the support of the resultant polar samples are within the spatial frequency band

$$\sqrt{k_x^2 + k_y^2} \le 2k_{\max} \cos\theta_z + 2k_{\max} \sin^2\theta_z \sin\theta_x(\omega_{\max}).$$

Identify a rectilinear grid in this region with the following sample spacings:

$$\Delta_{k_x} = \frac{\pi}{R_0(\omega_{\min})},$$

$$\Delta_{k_y} = \frac{\pi}{R_0(\omega_{\min})}.$$

For the partial rotation aspect angle measurement, use the results for the linear spotlight SAR in Chapter 5 to identify the support of the CSAR data in the spatial frequency (k_x, k_y) *domain.*

Step 9. Convert the target function spectrum polar data $F_p(\rho, \theta)$ *into the samples of the target function spectrum* $F(k_x, k_y)$ *on the rectilinear grid, where*

$$k_x(\omega, \theta) = \rho \cos\theta,$$

$$k_y(\omega, \theta) = \rho \sin\theta,$$

via two-dimensional interpolation. This yields evenly-spaced samples of the target function spectrum $F(k_x, k_y)$.

Step 10. Obtain the two-dimensional inverse DFT of the target spectrum $F(k_x, k_y)$ *to reconstruct the samples of the target function* $f(x, y)$ *in the spatial domain. For the full rotation aspect angle measurement, the resultant sample spacing in the spatial domain is*

$$\Delta_x = \frac{\pi}{2k_{\max}\cos\theta_z + 2k_{\max}\sin^2\theta_z \sin\theta_x(\omega_{\max})},$$

and $\Delta_y = \Delta_x$. *For the partial rotation aspect angle measurement, use the results for the linear spotlight SAR in Chapter 5 to identify* (Δ_x, Δ_y) *for the CSAR system.*

Step 11. For the partial rotation aspect angle measurement, perform CSAR image compression; this will be described.

Reducing Bandwidth of Reconstructed Image

The CSAR signal for a partial rotation aspect angle measurement, that is, $\theta \in [-\theta_{\max}, \theta_{\max}]$, has most of the signal properties of the linear spotlight SAR signal of Chapter 5. One of these common properties is the bandwidth compression of the reconstructed image which can be achieved via multiplication of this image with a two-dimensional PM signal in the spatial (x, y) domain. The details of this principle are described in Section 5.6 of Chapter 5.

A similar analysis would yield the following expression for CSAR image compression:

$$f_c(x, y) = f(x, y)\exp\left[j2k_c\cos\theta_z(R_g + x) - j2k_c\cos\theta_z\sqrt{(R_g + x)^2 + y^2}\right].$$

The region of support for the two-dimensional Fourier transform of the compressed image, that is, $F_c(k_x, k_y)$, in the k_y domain is

$$k_y \in \big[-k_x\tan\theta_{\max}, k_x\tan\theta_{\max}\big].$$

However, the support region for the target function spectrum $F(x_x, k_y)$ is approximately dictated by (use the divergence angle at the carrier frequency in the following)

$$k_y \in \big[-k_x\tan\big(\theta_{\max} + \phi_d\big), k_x\big(\tan\theta_{\max} + \phi_d\big)\big].$$

As we stated earlier, these can be justified using the principles that we used in Chapter 5 for the linear spotlight SAR.

Consider the T-72 turntable data for $\theta \in [-2, 2]$ degrees. Figure 7.12*a* and *b*, respectively, shows the reconstructed target function in the spatial frequency (k_x, k_y) domain, $F(k_x, k_y)$, and the spatial (x, y) domain, $f(x, y)$. Figure 7.12*c* is the com-

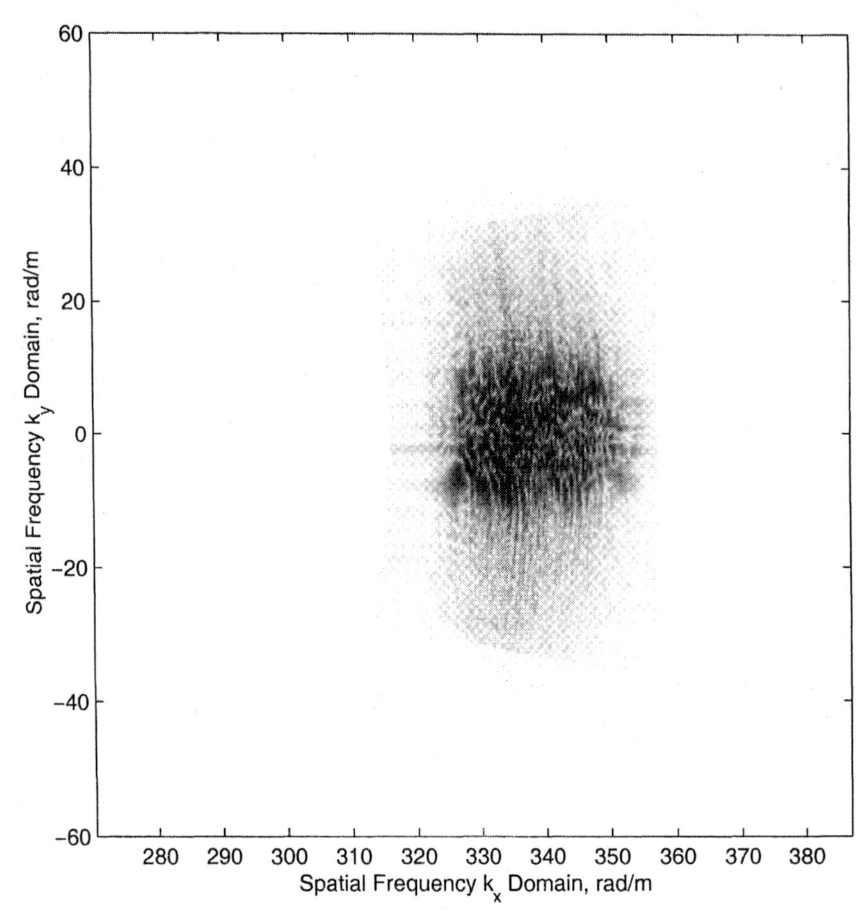

Figure 7.12a. CSAR reconstructions from T-72 turntable data for $\theta \in [-2, 2]$ degrees: target function spectrum $F(k_x, k_y)$.

pressed CSAR spectrum, $F_c(k_x, k_y)$, for this scenario. Similar outputs are shown in Figure 7.13 for $\theta \in [-10, 10]$ degrees.

Digital Reconstruction via Time Domain Correlation and Backprojection

In the same manner as in the linear SAR systems of Chapters 5 and 6, we can correlate the measured CSAR data with the analytical signature of a unit reflector at each pixel point on the desired spatial reconstruction grid to form the CSAR image of the target area. For the time domain correlation (TDC) this is achieved via

$$f(x_i, y_j) = \int_\theta \int_t \mathbf{s}(t, \theta) p^*\big[t - t_{ij}(\theta)\big] dt\, d\theta,$$

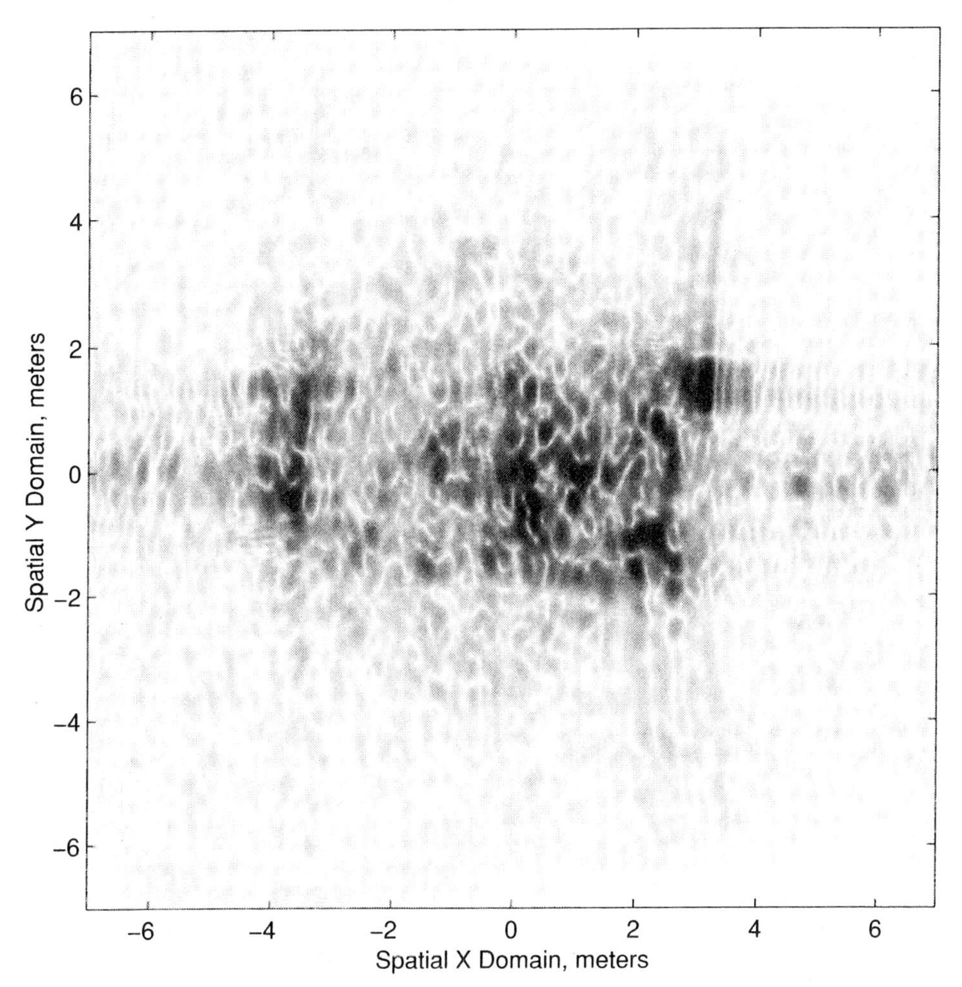

Figure 7.12b. CSAR reconstructions from T-72 turntable data for $\theta \in [-2, 2]$ degrees: target function $f(x, y)$.

where

$$t_{ij}(\theta) = \frac{2\sqrt{(x_i - R_g \cos\theta)^2 + (y_j - R_g \sin\theta)^2 + Z_c^2}}{c}$$

is the round-trip delay of the echoed signal from a reflector at (x_i, y_j) at the slow-time θ. By the generalized Parseval's theorem, the TDC reconstruction equation becomes

$$f(x_i, y_j) = \int_\theta \int_\omega s(\omega, \theta) P^*(\omega) \exp\big[-j2kt_{ij}(\theta)\big]\, d\omega\, d\theta.$$

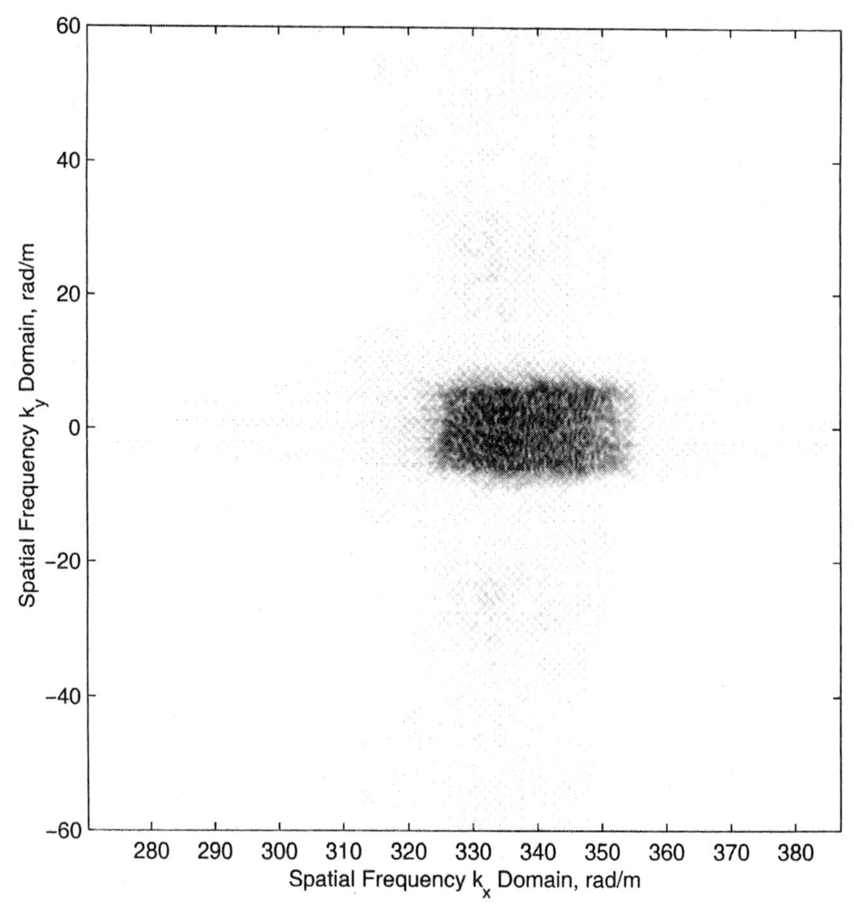

Figure 7.12c. CSAR reconstructions from T-72 turntable data for $\theta \in [-2, 2]$ degrees: compressed target function spectrum $F_c(k_x, k_y)$.

For the backprojection method the reconstruction equation becomes

$$f(x_i, y_j) = \int_\theta \mathbf{s}_M\big[t - t_{ij}(\theta), \theta\big] d\theta,$$

where $\mathbf{s}_M$ is the fast-time matched-filtered CSAR signal.

The algorithmic issues associated with the digital implementation of the TDC and backprojection methods in CSAR are similar to those we outlined in Chapter 5 for the linear spotlight SAR.

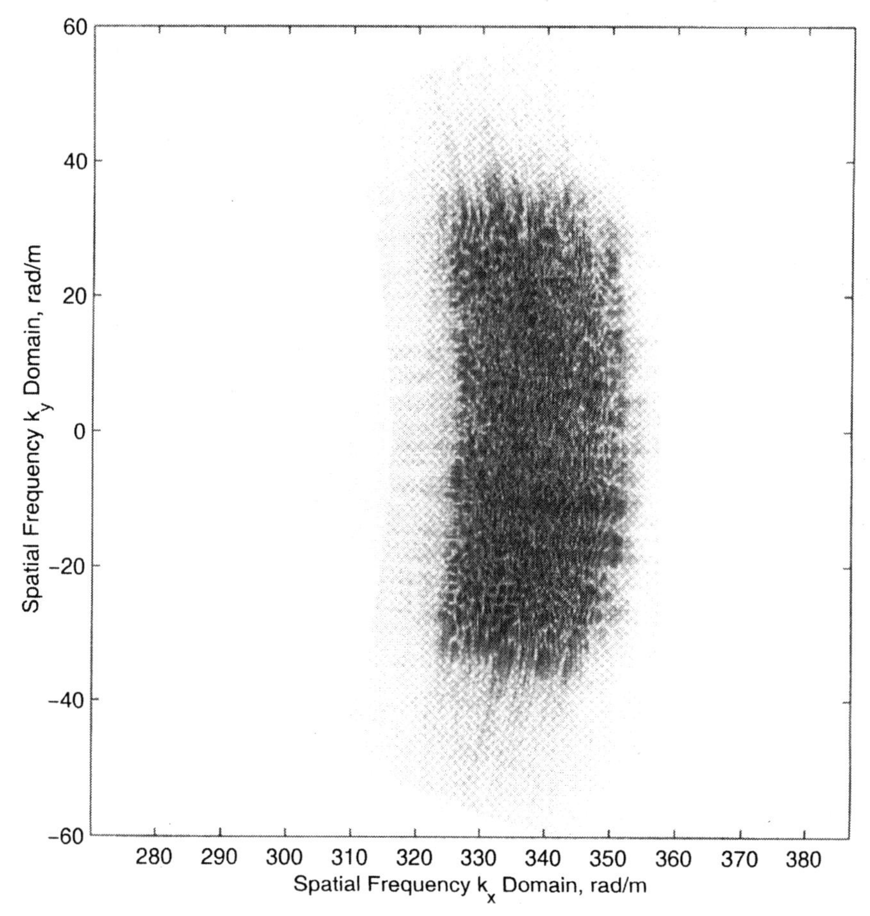

Figure 7.13a. CSAR reconstructions from T-72 turntable data for $\theta \in [-10, 10]$ degrees: target function spectrum $F(k_x, k_y)$.

7.7 THREE-DIMENSIONAL IMAGING

The system model and reconstruction formulated in Sections 7.1 and 7.2 used a target model that was located on the plane $z = 0$. Thus the reconstructed image can be viewed as a three-dimensional target function $f(x, y, z)$ which has nonzero components at $z = 0$; the reconstruction is the two-dimensional distribution of $f(x, y, 0)$.

In practice, we deal with a target region that possesses a varying altitude (z value) at various (x, y) points. We identify this altitude function via $z(x, y)$. The image formed for the plane $z = 0$ contains contributions from the targets that lie at other altitudes. In this section we examine this issue via quantifying the signature of the targets with $z(x, y) \neq 0$ in the reconstructed image $f(x, y, 0)$. Based on this analysis, we investigate the feasibility of three-dimensional imaging with slant plane CSAR data.

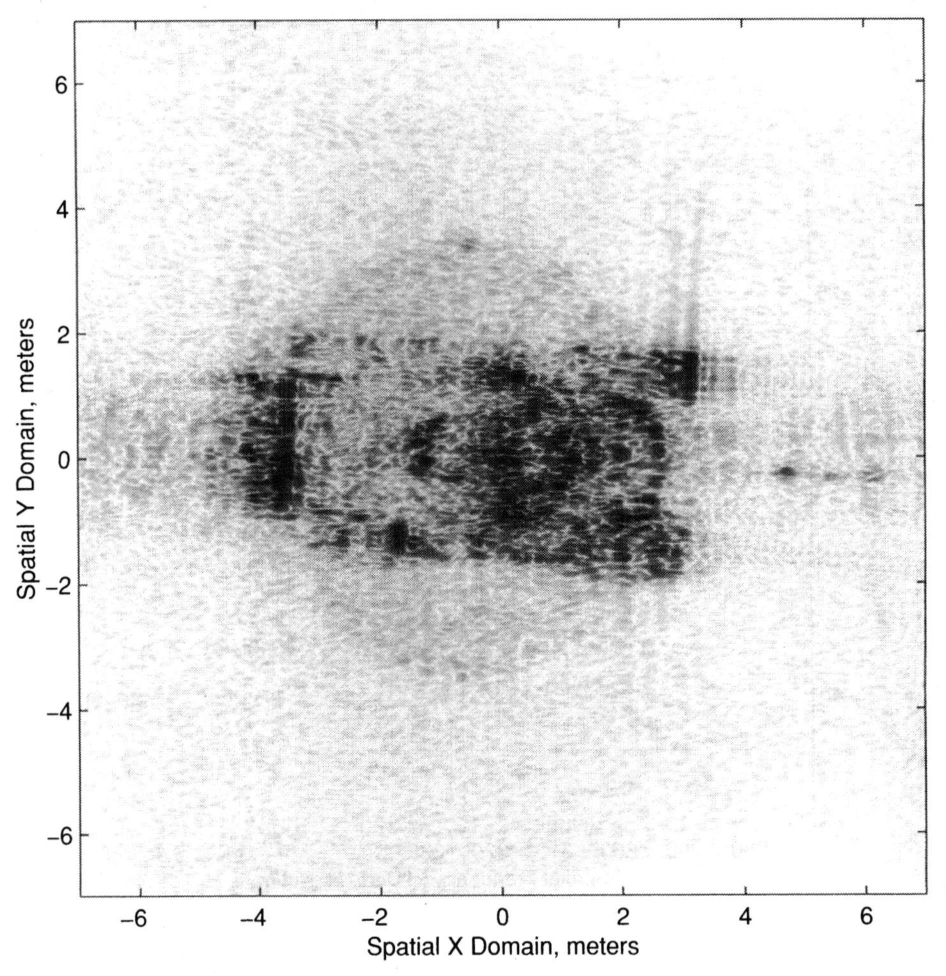

Figure 7.13b. CSAR reconstructions from T-72 turntable data for $\theta \in [-10, 10]$ degrees: target function $f(x, y)$.

Before doing that, we will review the effect of the target area altitude variation in slant plane linear SAR (see Section 3.2). Consider a slant plane linear SAR in which the radar moves along the line $x = X_c$ parallel to the y axis on the $z = Z_c$ plane. Let u represent the linear synthetic aperture (slow-time) domain. The distance of a target located at $[x, y, z(x, y)]$ when the radar is at (X_c, u, Z_c) (i.e., slow-time u) is

$$\sqrt{(x - X_c)^2 + (y - u)^2 + [z(x, y) - Z_c]^2},$$

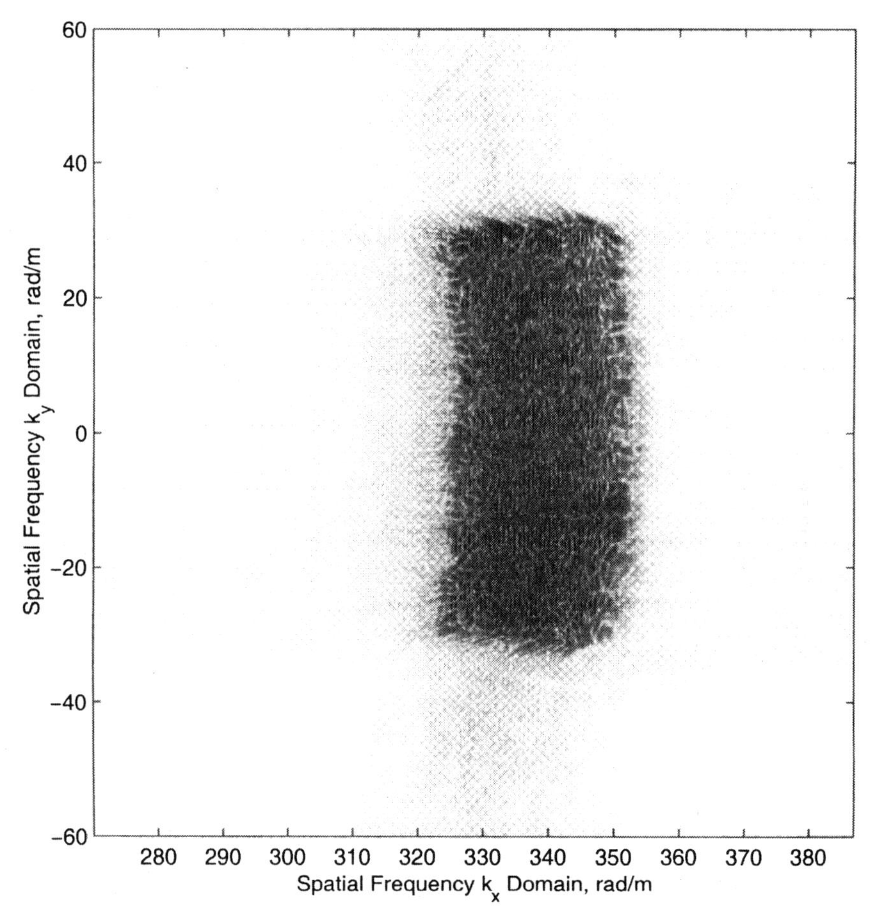

Figure 7.13c. CSAR reconstructions from T-72 turntable data for $\theta \in [-10, 10]$ degrees: compressed target function spectrum $F_c(k_x, k_y)$.

where X_c is the mean swath range of the radar. Thus the target appears at the *fixed* slant-range

$$x_s = \sqrt{(x - X_c)^2 + [z(x, y) - Z_c]^2},$$

as the slow-time varies. The reconstruction algorithm for this SAR system (see Chapters 4–6) performs coherent integration of the data for the slant-range x_s in the slow-time domain irrespective of the target's true altitude (or range); that is, the target altitude is transparent to the algorithm. The linear SAR system does not have the capability to resolve targets in the range-altitude (x, z) domain.

Next we consider the case of slant plane CSAR. For the three-dimensional target scene, the system model developed in Section 7.1 should be modified as follows:

$$\mathbf{s}(t,\theta) = \int_y \int_x f(x,y)$$

$$\times p\left[t - \frac{2\sqrt{(x - R_g\cos\theta)^2 + (y - R_g\sin\theta)^2 + [z(x,y) - Z_c]^2}}{c}\right] dx\,dy.$$

Suppose that there exists only a single reflector at (x, y) with altitude $z(x, y)$ in the target scene. The distance of this target from the radar at the slow-time θ is

$$\sqrt{(x - R_g\cos\theta)^2 + (y - R_g\sin\theta)^2 + [z(x,y) - Z_c]^2}.$$

Thus the target slant-range, which is

$$\sqrt{(x - R_g\cos\theta)^2 + [z(x,y) - Z_c]^2},$$

varies with the slow-time θ. Hence this slow-time dependent slant-range should be incorporated in any coherent processing in the slow-time domain. The reconstruction in Section 7.2 provides such a coherent processing that is sensitive to the target altitude.

It can be shown that the CSAR reconstruction of Section 7.2 provides a *focused* image of the distribution of the three-dimensional target reflectivity function $f(x, y, z)$ at the ground plane $z = 0$; the image of the target function $f(x, y, z)$ at the other altitudes' z values appears *smeared* (out of focus) in the reconstructed image [s96a].

The reconstruction in the two-dimensional spatial frequency (k_x, k_y) domain is an *approximation* to the two-dimensional Fourier transform of the target function $f(x, y, z)$ at $z = 0$ with respect to (x, y); that is,

$$F_{xy}(k_x, k_y, 0) = \mathcal{F}_{(x,y)}[f(x, y, 0)];$$

the function $F_{xy}(k_x, k_y, z)$ is called the *marginal* two-dimensional Fourier transform of the target function $f(x, y, z)$ with respect to (x, y). (For a discussion on the marginal Fourier transforms of a multi-dimensional signal see Section 4.0 of this book and [s94, ch. 2].)

One can obtain the Fourier reconstruction of the target function at an arbitrary altitude z, where $|z| \ll R_c$, via the following [s96a]:

$$F_{xy}(k_x, k_y, z) = F(k_x, k_y, 0)\exp\left(-j\sqrt{k_x^2 + k_y^2}\,\tan\theta_z z\right),$$

where $\theta_z = \arctan(Z_c/R_g)$ is the slant angle for the mid-altitude. The two-dimensional inverse spatial Fourier transform of $F_{xy}(k_x, k_y, z)$ is the desired image $f(x, y, z)$.

Reference [s96a] provides a discussion on the sensitivity (resolution) of this approach in resolving targets in the altitude z domain. Note that this altitude-focusing property of the slant plane CSAR is unique. As we mentioned earlier, the same property is not true for the slant plane linear SAR. One of the applications of this three-dimensional imaging is in ground penetrating (GPEN) CSAR for detection of mines and underground tunnels [s96a].

Next we use the above procedure for three-dimensional CSAR imaging on the T-72 turntable data. Again, our starting point is the reconstructed spectrum in Figure 7.7. First, we multiply this spectrum by the appropriate phase function of (k_x, k_y) for a given elevation z, that is,

$$\exp\left(-j\sqrt{k_x^2 + k_y^2}\,\tan\theta_z z\right).$$

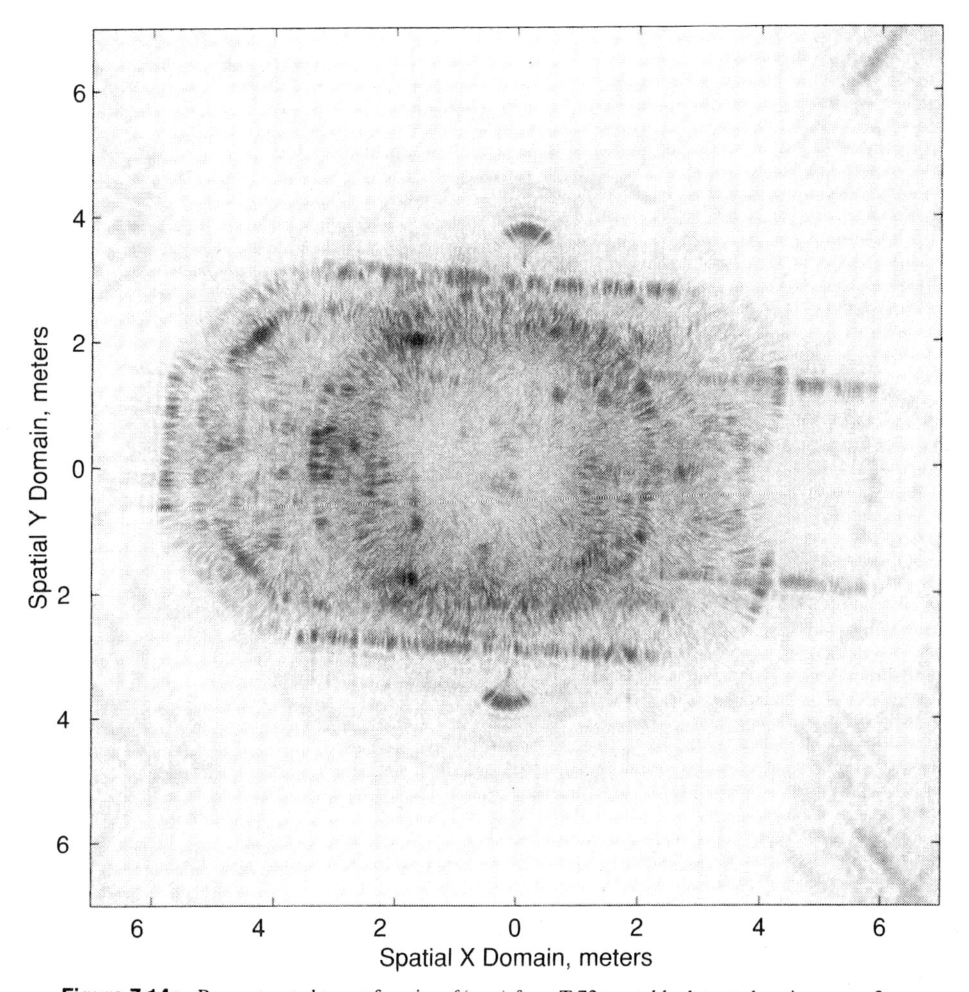

Figure 7.14a. Reconstructed target function $f(x, y)$ from T-72 turntable data at elevation $z = -2$ m.

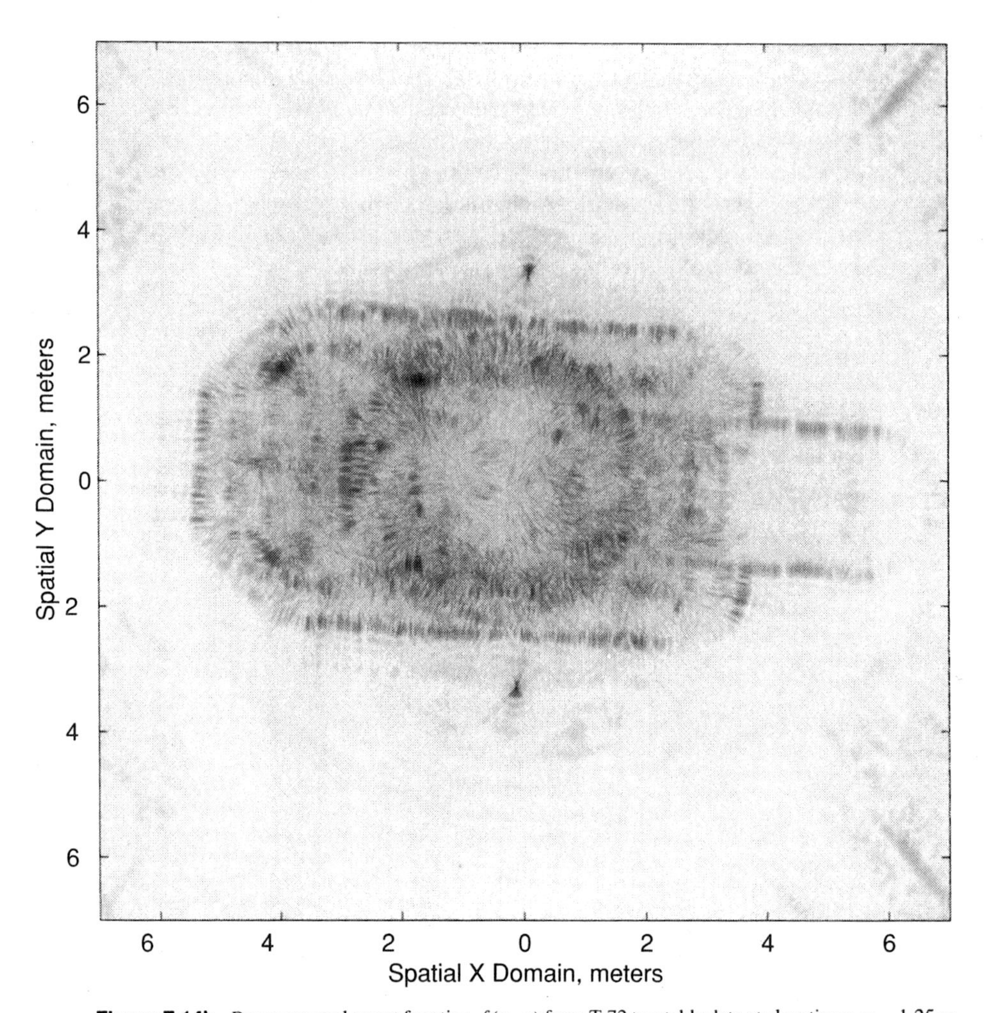

Figure 7.14b. Reconstructed target function $f(x, y)$ from T-72 turntable data at elevation $z = -1.25$ m.

Then we perform an inverse two-dimensional Fourier transform on the resultant data. This yields the slices of the three-dimensional target function $f(x, y, z)$ at the chosen elevation z values. Figure 7.14*a* to *h* shows these slices at $z = -2, -1.25, -0.75, -0.25, 0.25, 0.75, 1.25$, and 2 (all in meters). Note that the two corner reflectors appear to be relatively focused at $z = -1.25$ m (Figure 7.14*b*); the cannon of the tank appears focused at $z = 0.75$ m (Figure 7.14*f*).

By isolating the individual signature of a target in these CSAR reconstructions and obtaining its Fourier spectrum, one can study the aspect angle-dependent as well as fast-time frequency-dependent amplitude pattern of the target (see the discussion on target amplitude pattern in Chapter 2). Consider the CSAR reconstruction in Fig-

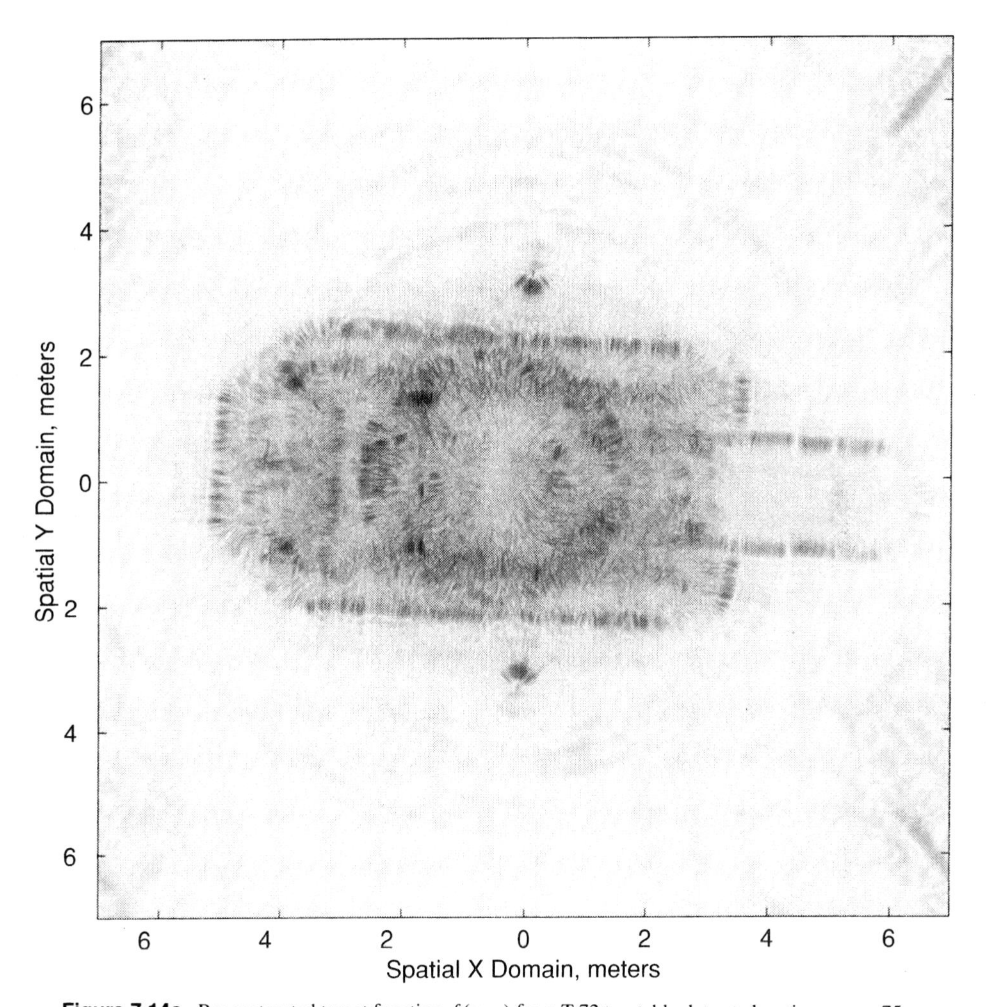

Figure 7.14c. Reconstructed target function $f(x, y)$ from T-72 turntable data at elevation $z = -.75$ m.

ure 7.14*b* ($z = -1.25$ m) at which the two corner reflectors are focused. Figure 7.15*a* shows the close-up image of the bottom corner reflector; Figure 7.15*b* is the two-dimensional Fourier transform of the corner reflector image in Figure 7.15*a*.

The spectrum in Figure 7.15*b* indicates that the corner reflector amplitude pattern is relatively strong and uniform over a contiguous 90-degree interval in the aspect angle domain; at the other aspect angles, the corner reflector amplitude pattern is weak. Thus the corner reflector behaves like an omni-directional reflector over a 90-degree aspect angle interval. Similar results can be observed for the other (top) corner reflector in Figure 7.16*a* to *b*.

Figure 7.17*a* is the close-up of the cannon of the tank in Figure 7.14*f* ($z = 0.75$ m). Note that different partitions of the cannon are observable in this image.

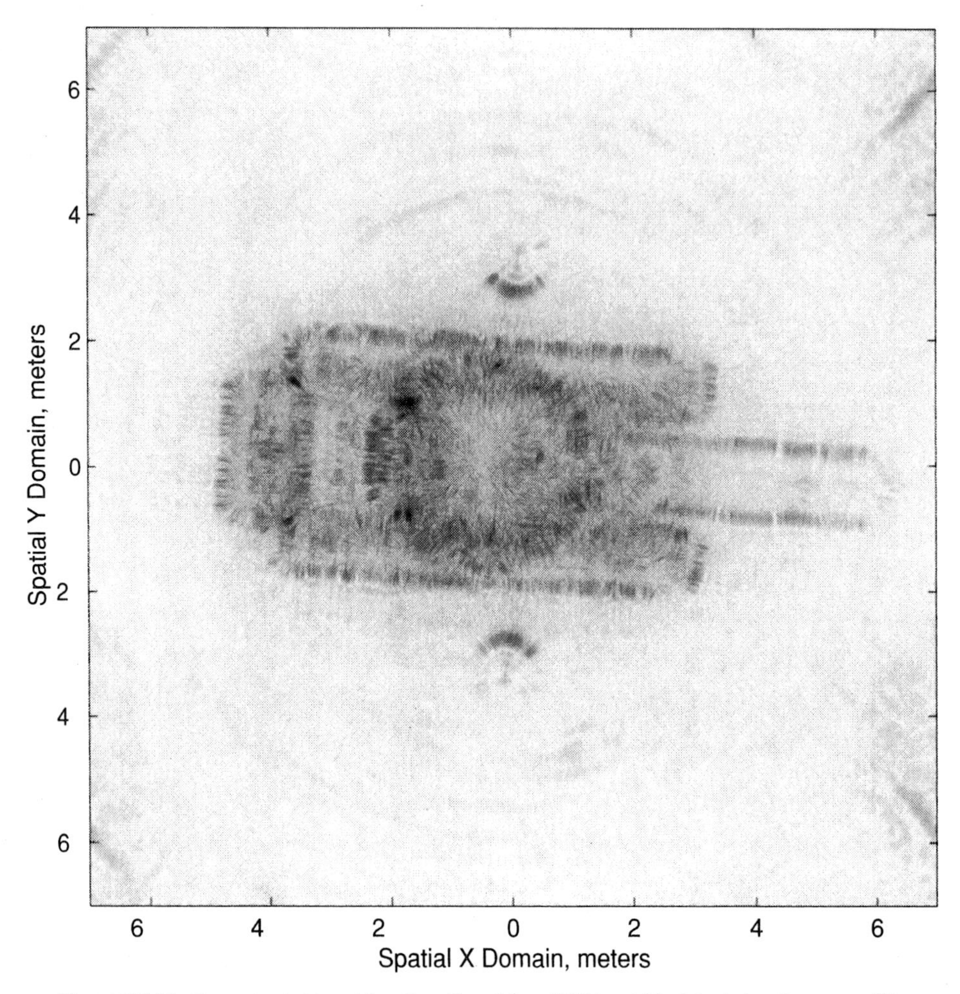

Figure 7.14d. Reconstructed target function $f(x, y)$ from T-72 turntable data at elevation $z = -.25$ m.

The spectrum of the cannon is shown in Figure 7.17*b*. This spectrum shows that the amplitude pattern of the cannon is strong only at two separate and small segments in the aspect angle domain; these are around $\theta \approx 90$ degrees and $\theta \approx 270$ degrees. They can be predicted from the fact that the cannon, unlike a corner reflector, behaves more like a flat surface. Thus the cannon's amplitude pattern is composed of two short *flashes* separated by 180 degrees; one occurs at the front of the flat surface (cannon), and the other occurs at the back of the flat surface.

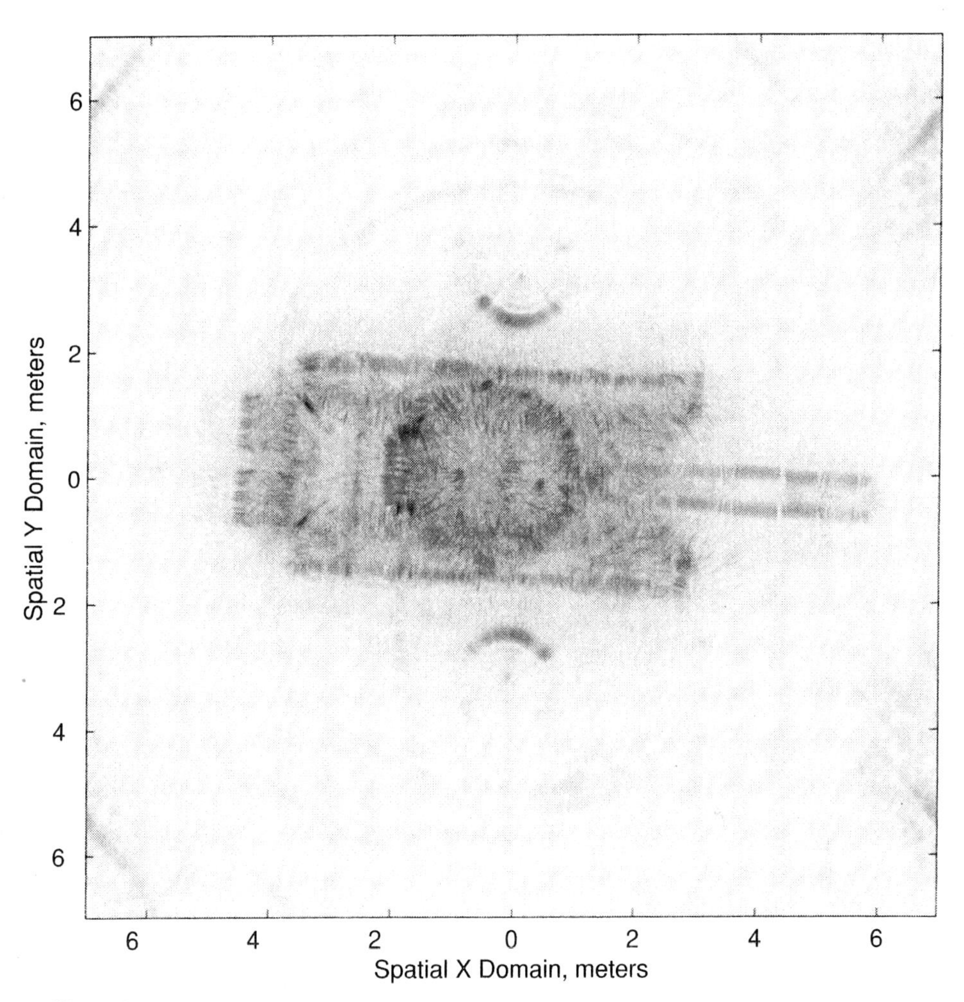

Figure 7.14e. Reconstructed target function $f(x, y)$ from T-72 turntable data at elevation $z = .25$ m.

Target Resolvability from Single-Tone Fringe Patterns

As we showed in Chapter 1, the conventional radar equations relate the range resolution to the baseband bandwidth of the radar signal. However, it has been speculated that the individual frequency values of a radar (e.g., its carrier frequency) could play a role in improving the resolution. In this section we show that two targets that are not resolvable in a SAR image may be individually identified from their SAR signature at a single frequency of the radar.

This approach originates from the work of Howard (1959) at Naval Research Laboratory (NRL) [sher]. In what was referred to as the *ripple tank* experiment, Howard showed that two vibrating single-tone acoustic sources, which were separated by a

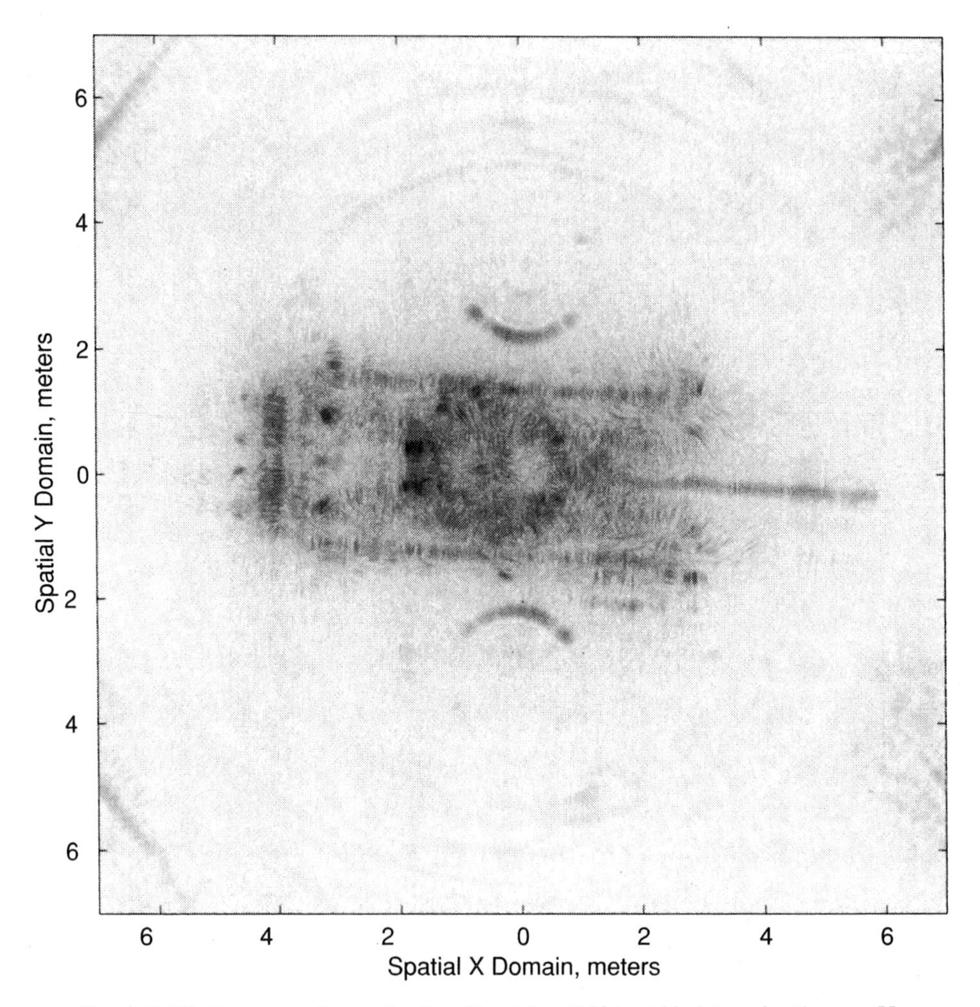

Figure 7.14f. Reconstructed target function $f(x, y)$ from T-72 turntable data at elevation $z = .75$ m.

fraction of the wavelength of their radiation frequency, produced fringe patterns in a water tank that could be used as a measure to resolve them.

In the case of a SAR system with a *wide-bandwidth* radar, the SAR images of two targets separated by a fraction of a wavelength (e.g., consider the wavelength at the carrier frequency) are likely to appear as a single *blip*; this single blip is not suitable for resolving the two targets.

However, the blip signature can be extracted from the SAR image, a process that we refer to as *digital spotlighting* in the reconstructed image. This digitally-spotlighted SAR image can be transformed into the phase history SAR signature at the available radar frequencies. Then the phase history signature at a given tone of

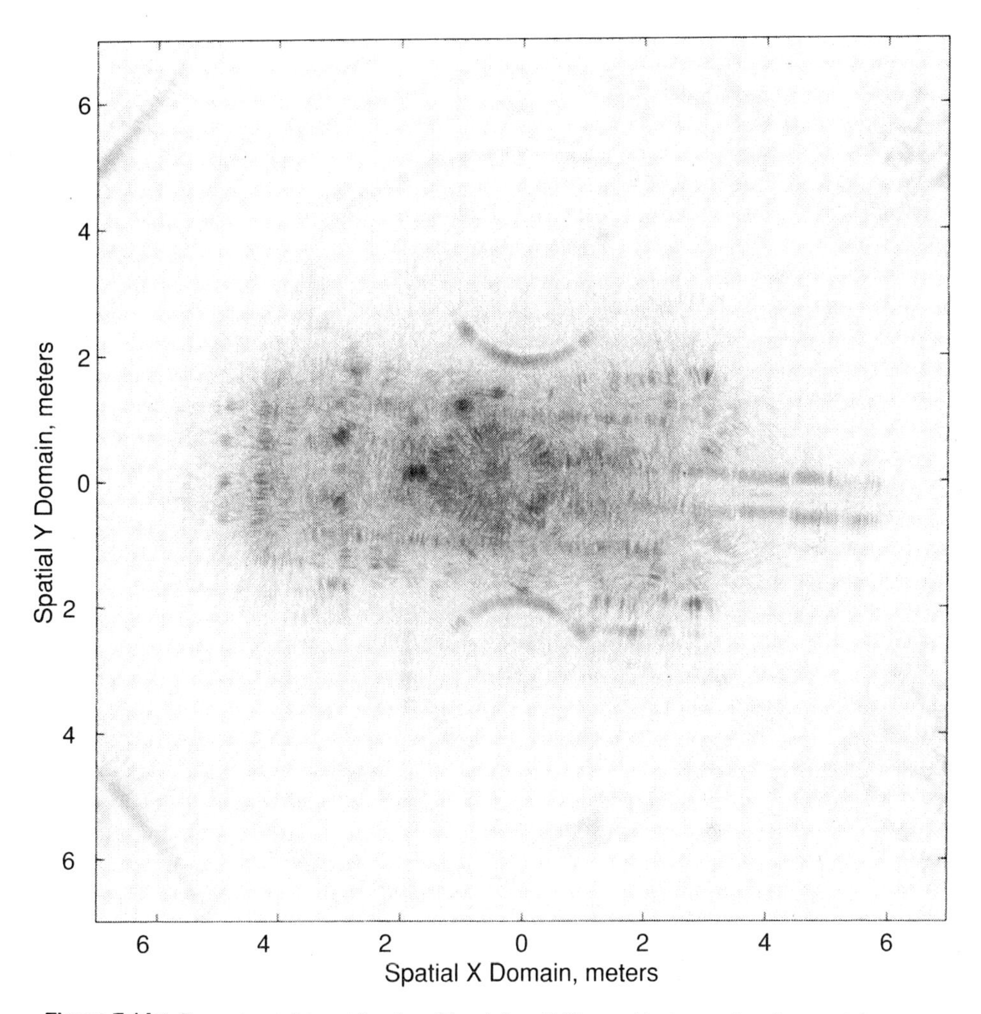

Figure 7.14g. Reconstructed target function $f(x, y)$ from T-72 turntable data at elevation $z = 1.25$ m.

the radar is transformed back into the SAR image domain. This results in fringe patterns in the spatial domain that are mathematically identical to the fringe patterns observed by Howard. The fringe patterns can then be used to resolve the two targets.

This approach is particularly effective in CSAR systems being investigated for mine detection. As we stated earlier, a full rotation CSAR system provides the RCS of a target at all aspect angles. A mine, which behaves similar to an omni-directional reflector, produces single-tone fringe patterns similar to those observed by Howard.

To construct the digital spotlight CSAR signature of a target at a given ground plane fast-time frequency ω_g or target radial spatial frequency $\rho = 2k_g$, the following steps could be used. Suppose that we are interested in an area around the spatial

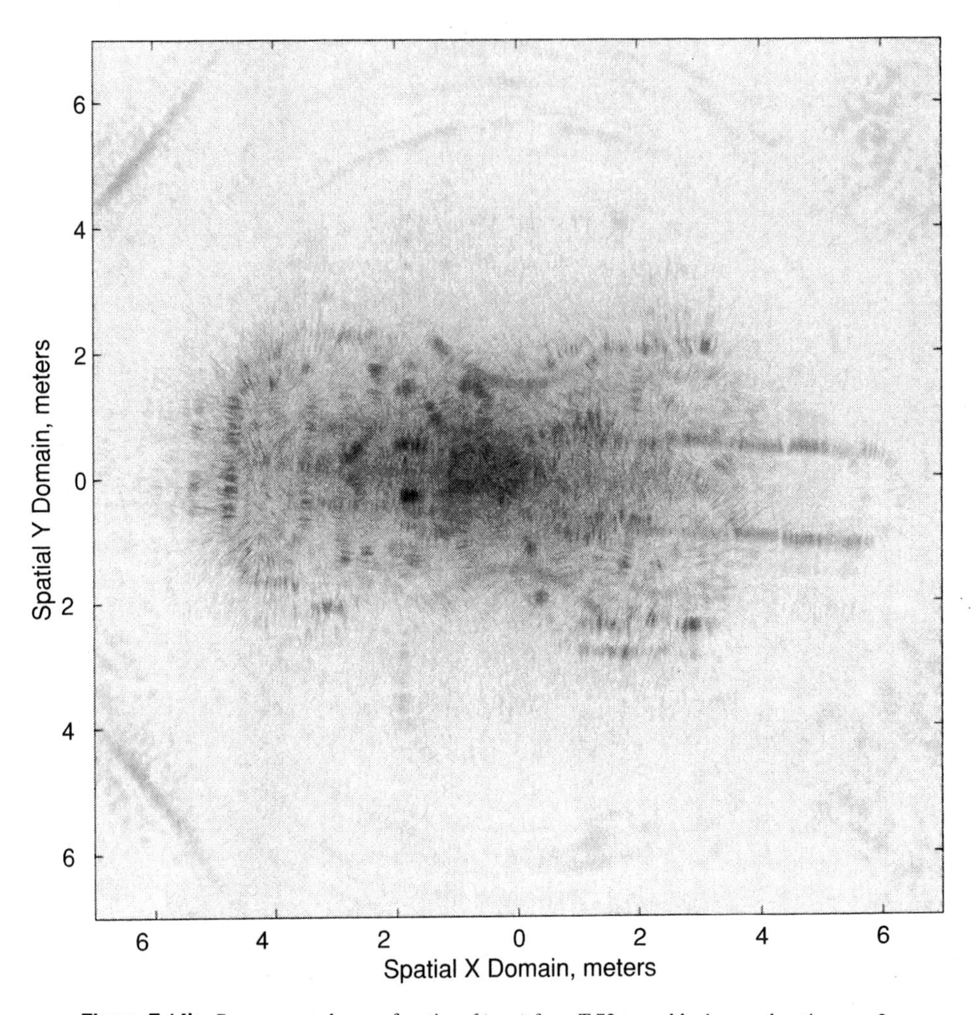

Figure 7.14h. Reconstructed target function $f(x, y)$ from T-72 turntable data at elevation $z = 2$ m.

coordinates (x_n, y_n) in the spatial (x, y) domain, we label this region by $\mathbf{R}_n$. The target spectrum at the polar spatial frequency is formed via

$$F_{pn}(\rho, \theta) \approx \sum_{(x_i, y_j) \in \mathbf{R}_n} f(x_i, y_j) \exp\big(-j\rho \cos\theta x_i - j\rho \sin\theta y_j\big).$$

Since we are interested in the CSAR signature at a single tone (radar frequency), the distribution of $F_{pn}(\rho, \theta)$ could be formed at a fixed ρ, for example,

$$\rho_c = 2k_c \cos\theta_z,$$

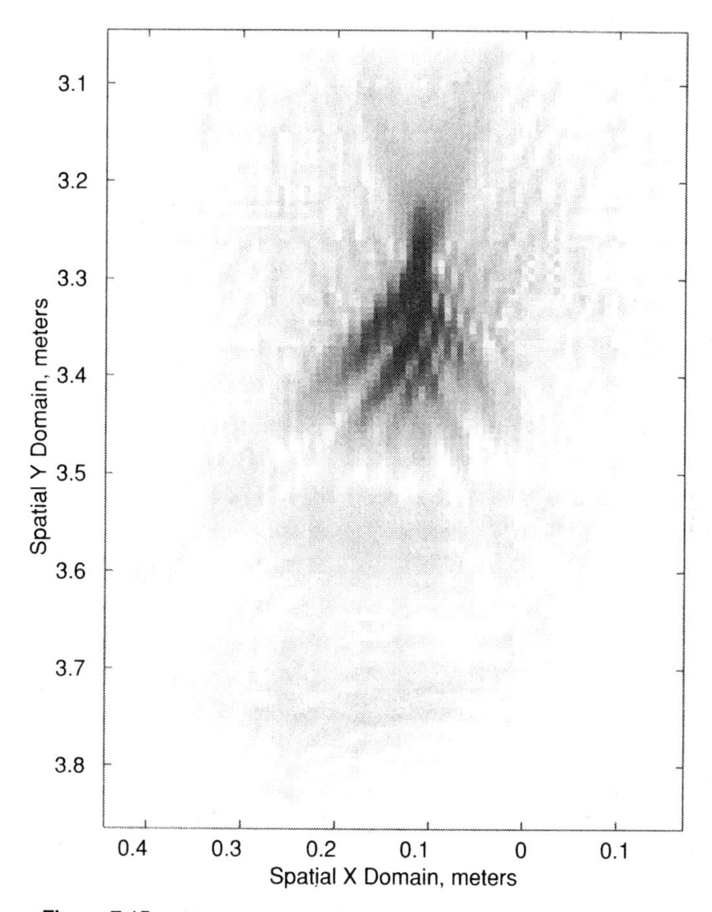

Figure 7.15a. Close-up image of bottom corner reflector at $z = -1.25$ m.

at a set of evenly spaced values of $\theta \in (0, 2\pi)$, for example, θ_m's. For alias-free processing of the resultant database, the Nyquist sampling theorem for polar data would indicate that the number of the discrete aspect angles θ_m's should be $\pi/2$ times the maximum number of pixels in the rows or columns of the region $\mathbf{R}_n$ [s94, ch. 2].

After reconstructing $F_{pn}(\rho_c, \theta_m)$ values, the fringe patterns in the spatial (x, y) domain for the pixels $(x_i, y_j) \in \mathbf{R}_n$ are obtained by

$$f_{\text{fp}n}(x_i, y_j) = \sum_{\theta_m} F_{pn}(\rho_c, \theta_m) \exp\big(j\rho_c \cos\theta_m x_i + j\rho_c \sin\theta_m y_j\big).$$

To show the merits of this approach, we consider the turntable data of the T-72 tank. Figure 7.18 shows a single-tone fringe pattern of the top corner reflector in Figure 7.16*a*. To simulate two closely spaced corner reflectors in the scene, we add the complex image of the top corner reflector to the complex of the same target when

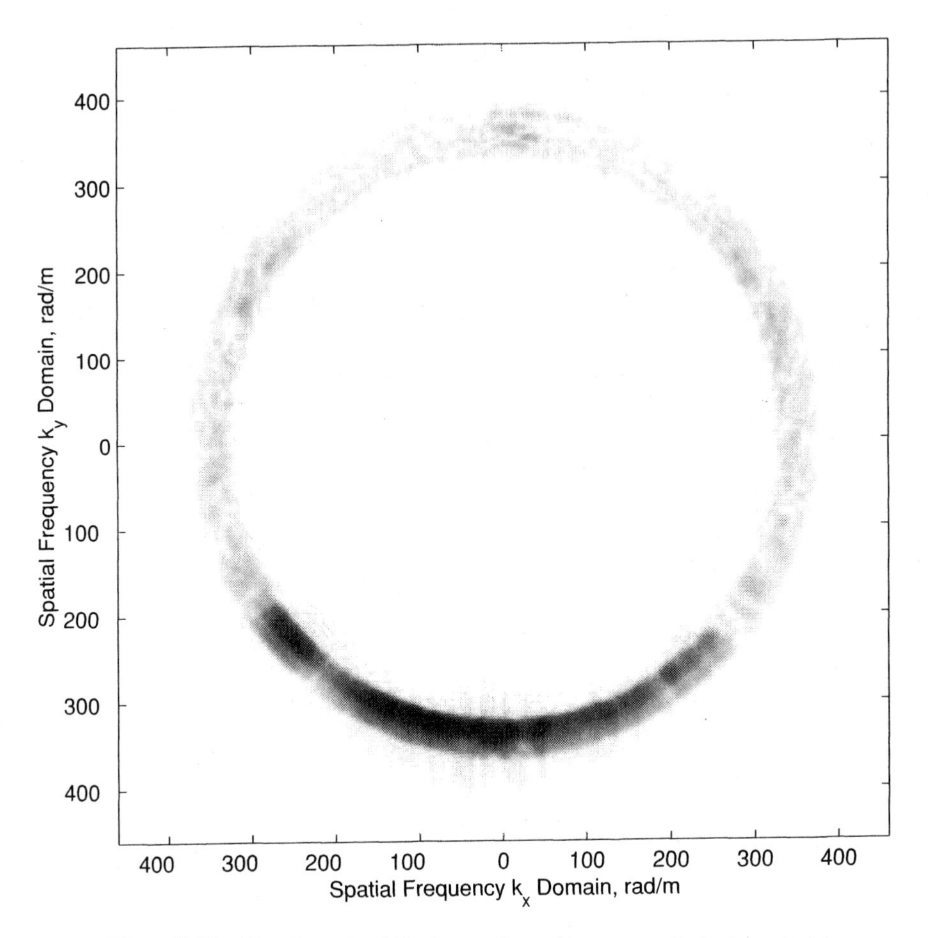

Figure 7.15b. Two-dimensional Fourier transform of top corner reflector image in (a).

it is shifted in the spatial domain by

$$(x_\delta, y_\delta) = (3.25, 0) \text{ cm}.$$

Figure 7.19*a* is the resultant image; due to the close spacing of the two corner reflectors, it is difficult to realize that there are two corner reflectors in the image of Figure 7.19*a*. Figure 7.19*b* shows a single-tone fringe patterns of the image in Figure 7.19*a*. Note that the fringe patterns in Figure 7.19*b* show interference fringe patterns of the two corner reflectors; this fringe pattern distribution is different from the single tone fringe pattern of one corner reflector in Figure 7.18.

Figure 7.20*a* and *b* shows similar results when the two corner reflectors are separated by

$$(x_\delta, y_\delta) = (0, 3.25) \text{ cm}$$

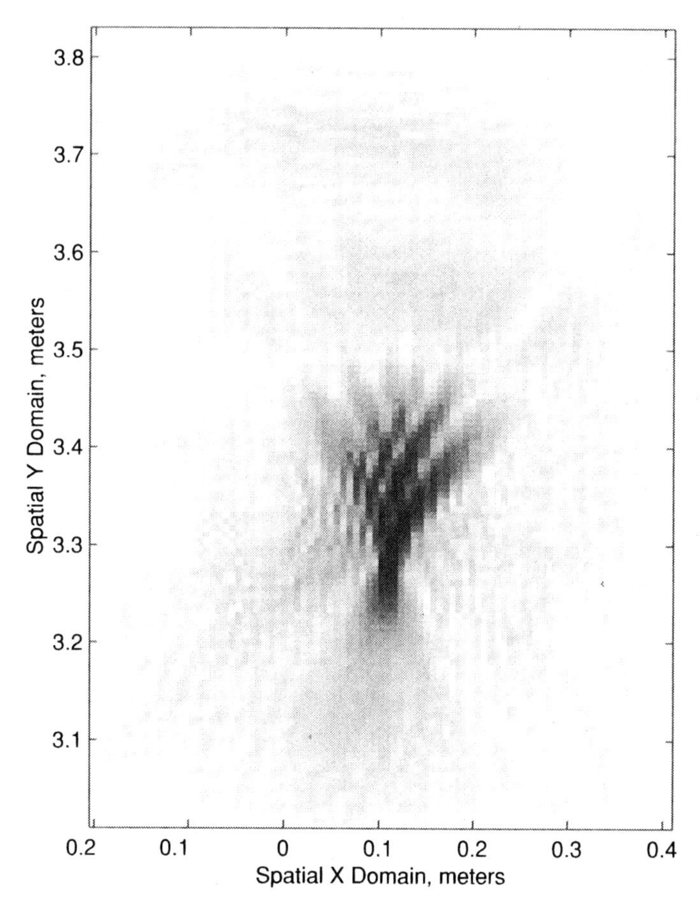

Figure 7.16a. Close-up image of top corner reflector at $z = -1.25$ m.

in the spatial domain. Figure 7.21 shows the single tone fringe patterns for the cannon of the tank; the image of this target was shown in Figure 7.17*a*. Note that a separate set of fringe patterns appears at each partition point of the cannon.

7.8 THREE-DIMENSIONAL IMAGING WITH TWO-DIMENSIONAL CIRCULAR AND ELEVATION SYNTHETIC APERTURES

Current advanced SAR imaging systems provide rich high-resolution coherent information regarding the targets in their imaging scene. To fully utilize and exploit such coherent SAR images, there is a need for a better understanding and characterization of a target's coherent radar cross section (RCS) in the three-dimensional spatial domain. For instance, in reconnaissance with SAR, a priori information on coherent RCS of military vehicles (e.g., trucks and tanks) at different azimuthal and elevation angles are required for automatic target recognition (ATR).

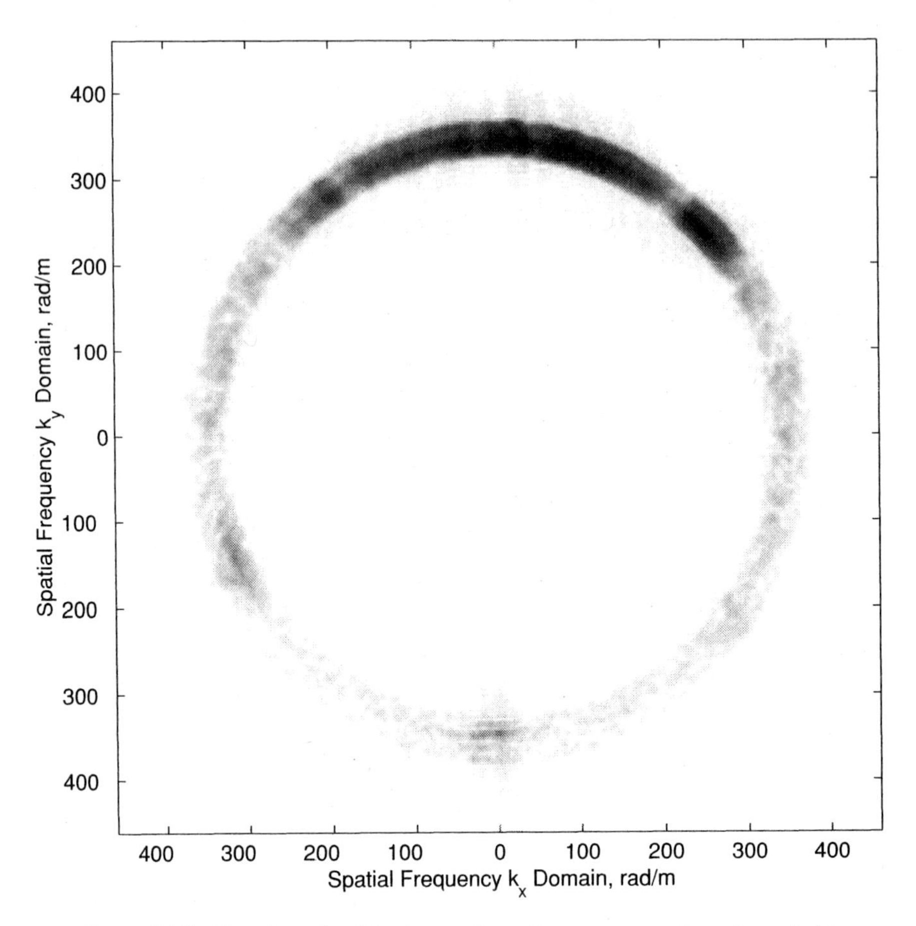

Figure 7.16b. Two-dimensional Fourier transform of bottom corner reflector image in (a).

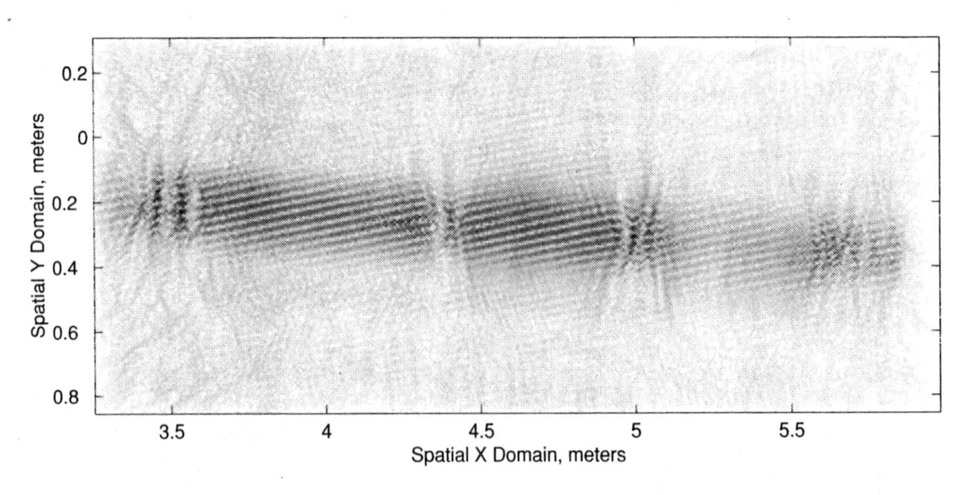

Figure 7.17a. Close-up image of cannon of tank at $z = 0.75$ m.

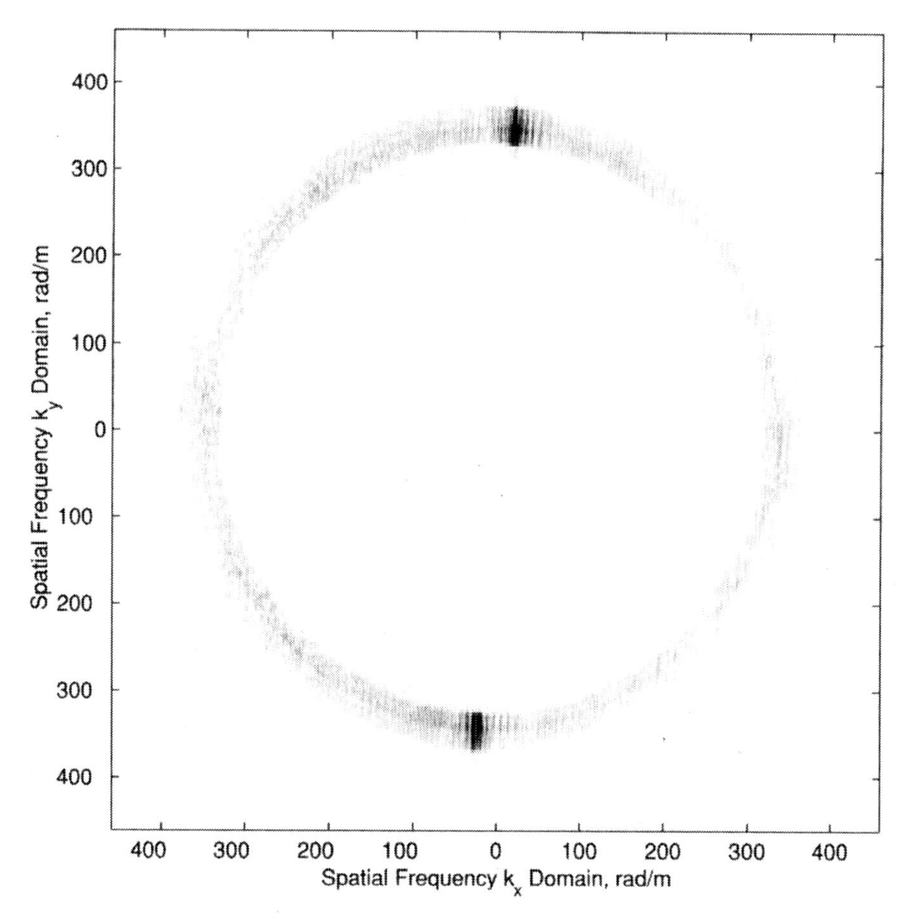

Figure 7.17b. Two-dimensional Fourier transform of cannon image in (a).

This problem has resulted in the construction of three-dimensional SAR systems to measure the coherent RCS of a target as a function of frequency, and azimuthal and elevation (or look) angles. Three-dimensional SAR imaging is also being investigated for interior imaging of buildings for rescue operations in the case of, for example, an earthquake. These SAR systems are currently being constructed and investigated by various radar groups. In fact the CSAR (turntable) data of the T-72 tank, which we examined earlier, were collected at Georgia Tech Research Institute for this purpose.

A particular three-dimensional SAR system of interest performs synthetic aperture measurements via a linear motion of the radar in the elevation domain, and a circular motion of the target in the range and cross-range domains. (Circular synthetic aperture data are also referred to as turntable data.) The radar motion in elevation provides target coherent RCS as a function of the elevation (or depression) angle. The target's circular motion yields the azimuthal look angle information. We call

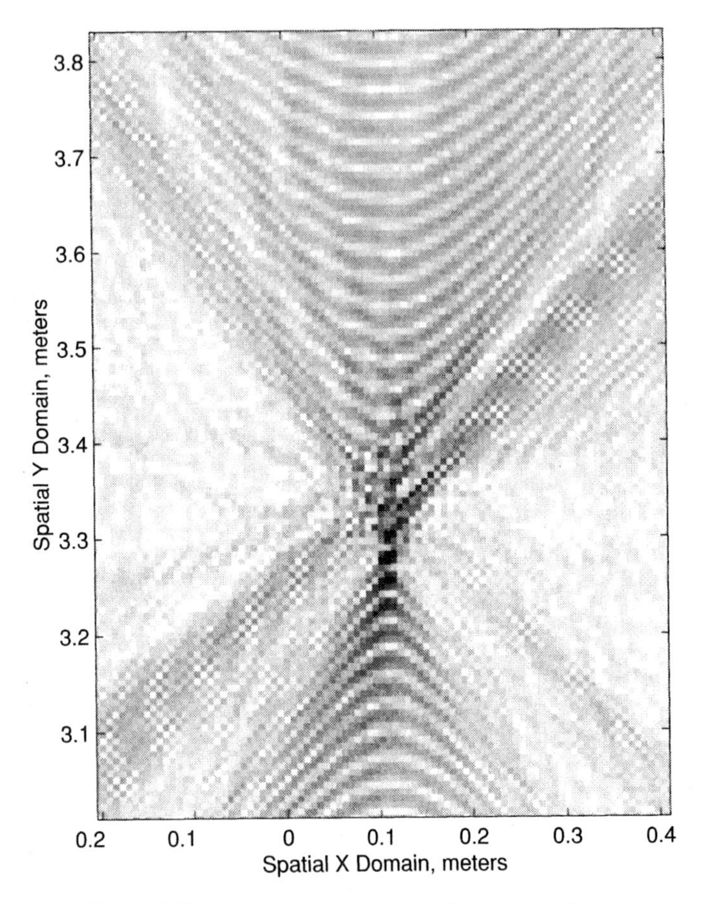

Figure 7.18. Single-tone fringe patterns of top corner reflector.

this synthetic aperture measurement scenario the elevation circular SAR (E-CSAR) system.

To form accurate and coherent three-dimensional reconstructions from the E-CSAR data, the backprojection method has been utilized. Unfortunately, the backprojection algorithm for a typical three-dimensional E-CSAR scenario, for example, imaging a $400 \times 400 \times 200$ volume in the spatial domain, could take a few days. An alternative reconstruction algorithm is the three-dimensional polar format processing of Walker [aus; wal] which requires less computation time. However, this is at the cost of using plane wave approximations, which results in erroneous shifts, smearing, and loss of coherent information in the reconstructed SAR image.

In this section we present a wavefront reconstruction theory-based method for imaging in E-CSAR systems. This method is based on the coherent wavefront reconstruction method for a linear motion SAR in elevation (see Chapter 4) and a circular motion SAR in the range and azimuth domains. These SAR wavefront re-

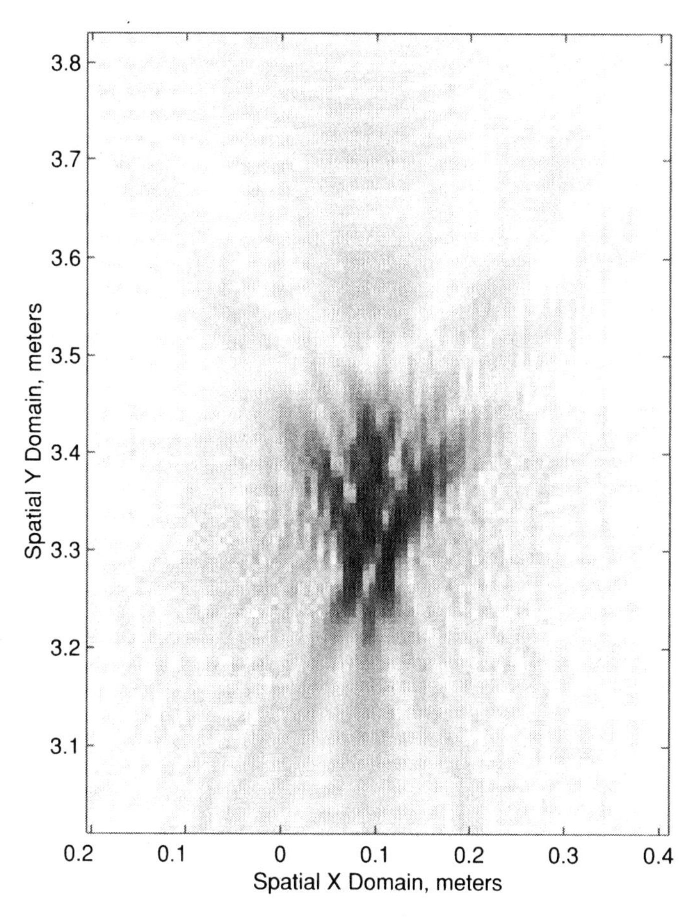

Figure 7.19a. Image of two corner reflectors separated by $(x_\delta, y_\delta) = (3.25, 0)$ cm.

construction methods provide accurate coherent RCS of the targets in the imaging scene.

System Model

The E-CSAR imaging system geometry is shown in Figure 7.22. The radar-carrying aircraft moves along a circular path with radius R_g on the plane $z = Z_c + v$ with respect to the ground plane; Z_c, a known constant, is the squint elevation. The variable $v \in [-L_v, L_v]$, where L_v is a known constant, is the elevation synthetic aperture domain; $2L_v$ is the size of the synthetic aperture in the elevation domain. For notational convenience we associate the rotation, for generating the circular synthetic aperture, with the radar. Thus the coordinates of the radar in the spatial domain along the circular synthetic aperture are $(x, y, z) = (R_g \cos\theta, R_g \sin\theta, Z_c + v)$. The variable $\theta \in [-\theta_{\max}, \theta_{\max}]$ represents the circular synthetic aperture domain; $2\theta_{\max}$, a known constant, is the length of the processed circular synthetic aperture. As the

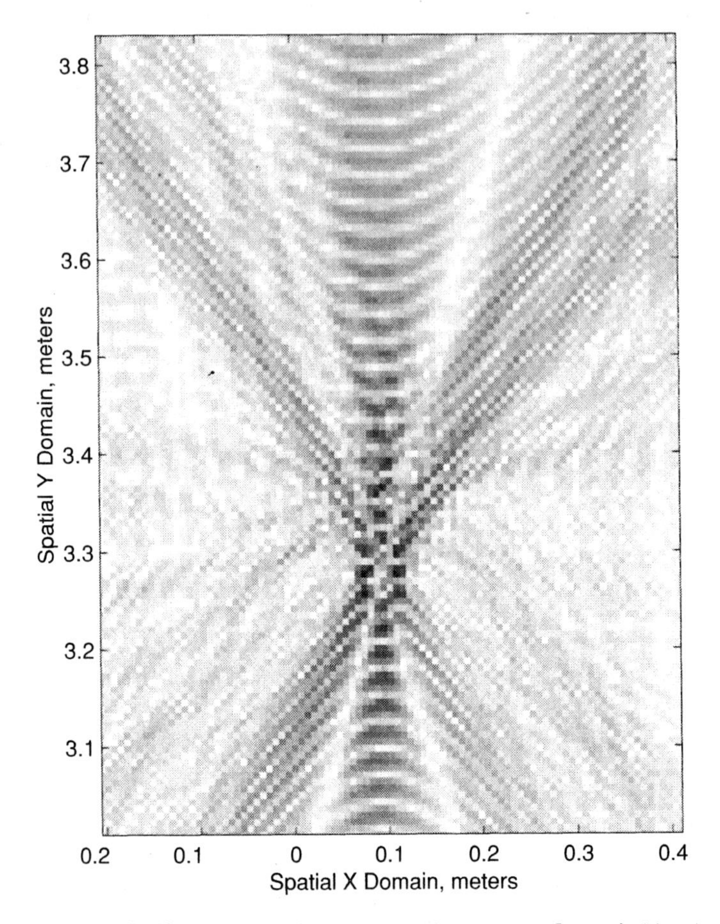

Figure 7.19b. Single-tone fringe patterns of two corner reflectors in (a).

radar moves along the circular and elevation synthetic aperture (θ, v) domains (or *slow-time* domains), its beam is spotlighting the target area centered at the origin of the spatial (x, y) domain on the ground plane.

We denote the reflectivity function in the target region by $f(x, y, z)$. Let $p(t)$ be the transmitted radar signal; t represents the *fast-time* domain. The measured SAR signal when the radar is located at $(x, y, z) = (R_g \cos\theta, R_g \sin\theta, Z_c + v)$ is

$$\mathbf{s}(t, \theta, v) = \int_x \int_y \int_z f(x, y, z)$$

$$\times\, p\left[t - \frac{2\sqrt{(x - R_g\cos\theta)^2 + (y - R_g\sin\theta)^2 + (z - Z_c - v)^2}}{c}\right] dx\, dy\, dz.$$

The Fourier transform of the above model with respect to the fast-time t is

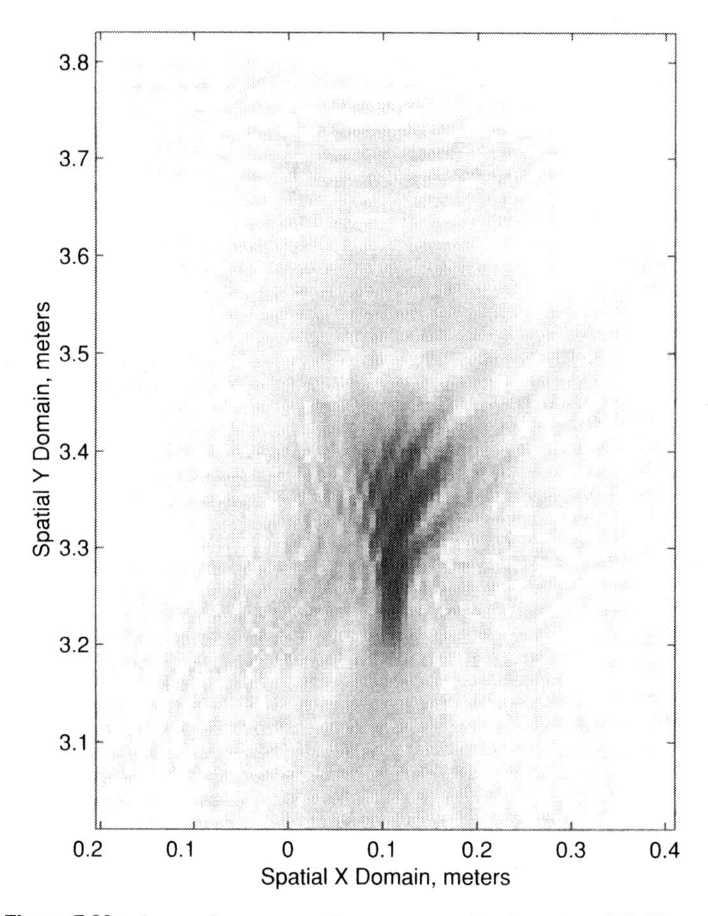

Figure 7.20a. Image of two corner reflectors separated by $(x_\delta, y_\delta) = (0, 3.25)$ cm.

$$s(\omega, \theta, v) = P(\omega) \int_x \int_y \int_z f(x, y, z) g_{xyz}(\omega, \theta, v)\, dx\, dy\, dz,$$

where $k = \omega / c$ is the wavenumber at the fast-time frequency ω and

$$g_{xyz}(\omega, \theta, v) = \exp\left[-j2k\sqrt{(x - R_g \cos\theta)^2 + (y - R_g \sin\theta)^2 + (z - Z_c - v)^2} \right],$$

which we call the E-CSAR system shift-varying impulse response (Green's function).

Reconstruction

The formulation of the reconstruction for the E-CSAR problem depends on the Fourier analysis of its shift-varying impulse response, that is, $g_{xyz}(\omega, \theta, v)$. This involves Fourier decompositions in the elevation synthetic aperture domain and circular synthetic aperture domain.

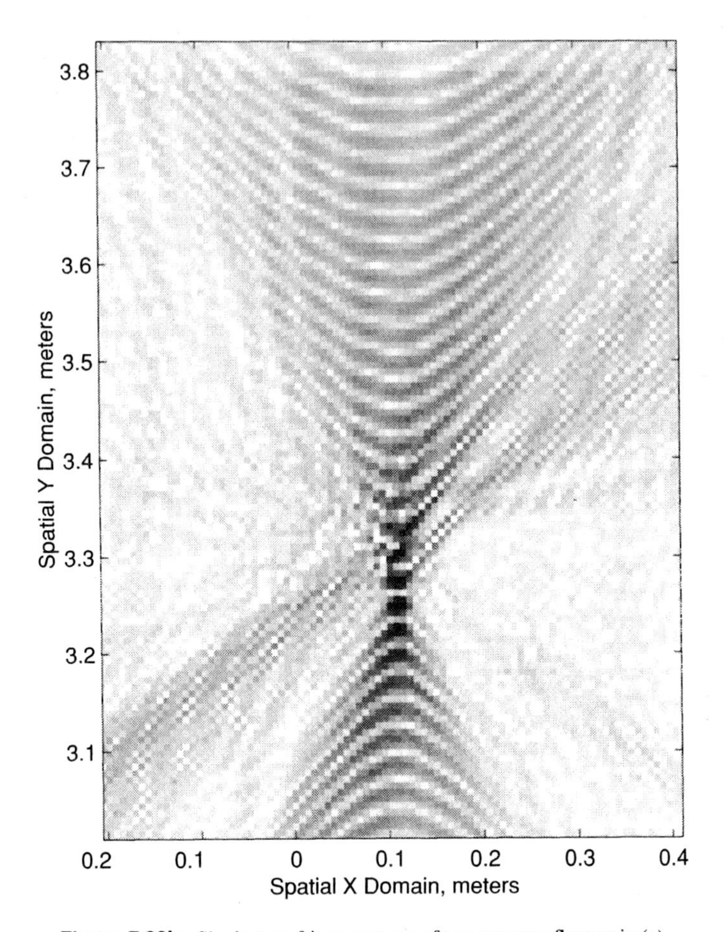

Figure 7.20b. Single-tone fringe patterns of two corner reflectors in (a).

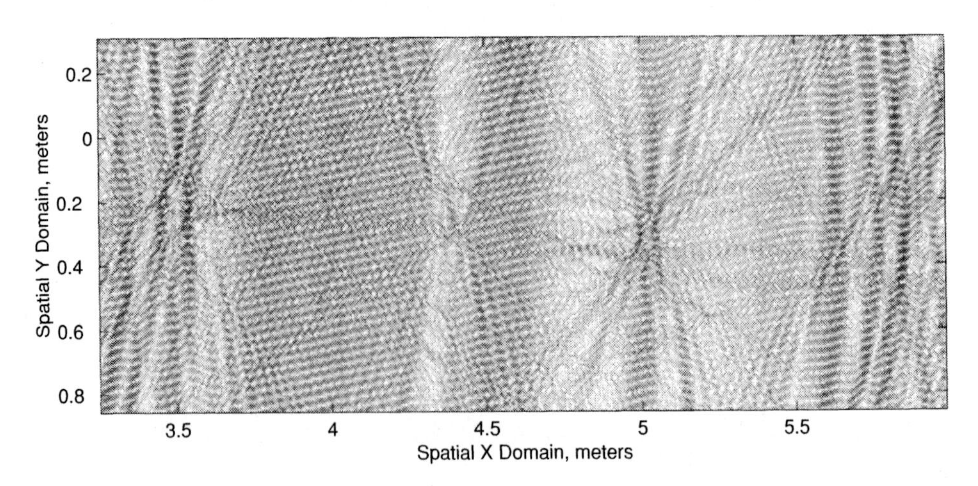

Figure 7.21. Single tone fringe patterns of cannon of tank.

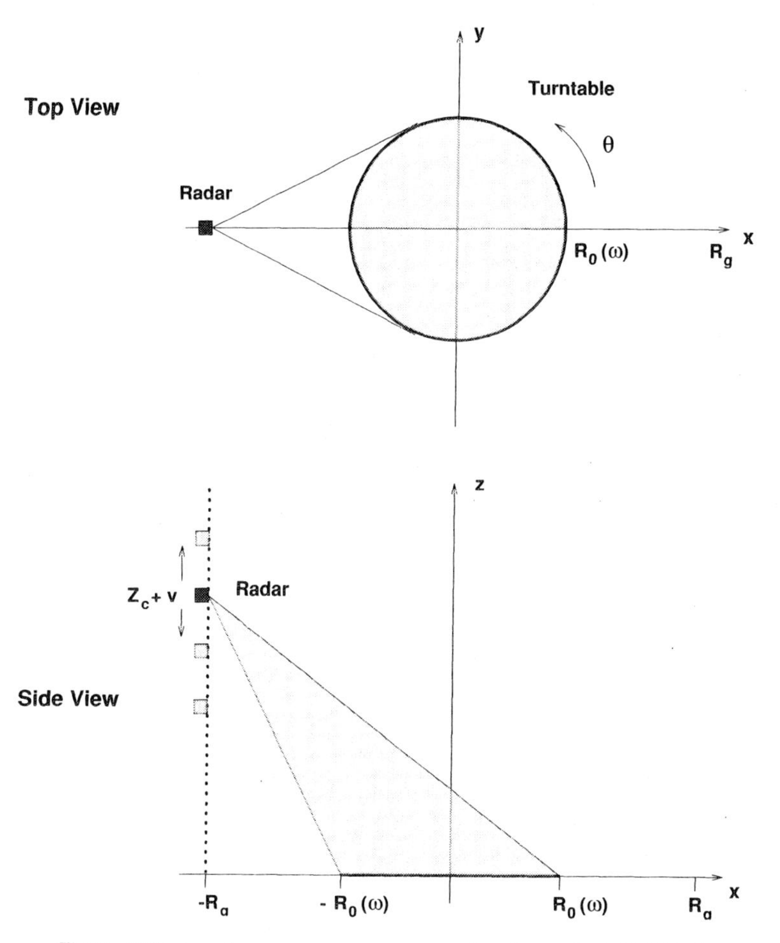

Figure 7.22. Imaging system geometry in elevation circular SAR (E-CSAR).

Elevation Synthetic Aperture Fourier Analysis The following is based on the Fourier analysis of a linear synthetic aperture which we outlined in Chapter 2. Consider the following phase-modulated (PM) signal in the linear elevation synthetic aperture v domain:

$$\exp\left[-j2k\sqrt{r^2-(z-v)^2}\right]$$

which is also called the free space Green's function, or spherical PM signal. The Fourier transform of the spherical PM signal with respect to $v \in (-\infty, \infty)$ is (see Section 2.2)

$$\mathcal{F}_{(v)}\left[\exp\left[-j2k\sqrt{r^2-(z-v)^2}\right]\right] = \exp\big(-jk_r r - jk_v z\big)$$

for $k_v \in [-2k, 2k]$, and zero otherwise, where

$$k_r = \sqrt{4k^2 - k_v^2}.$$

The above Fourier decomposition corresponds to the nonevanescent waves which are recorded in the practical SAR systems. The effect of a finite elevation synthetic aperture, that is, $v \in [-L_v, L_v]$, was discussed in Section 2.4. Note that k_r is a nonlinear mapping of (k_v, k) or (k_v, ω).

Using the above Fourier decomposition in the elevation synthetic aperture v domain, the Fourier transform of the E-CSAR system impulse response with respect to the elevation synthetic aperture can be found to be

$$G_{xyz}(\omega, \theta, k_v) = \exp\left[-jk_r\sqrt{(x - R_g\cos\theta)^2 + (y - R_g\sin\theta)^2} - jk_v(z - Z_c)\right],$$

with $k_r = \sqrt{4k^2 - k_v^2}$. Taking the Fourier transform of both sides of the E-CSAR signal $s(\omega, \theta, v)$ domain with respect to the elevation synthetic aperture v, and with the help of the Fourier expression for $G_{xyz}(\omega, \theta, k_v)$, we can write

$$\begin{aligned} S(\omega, \theta, k_v) &= P(\omega)\int_x\int_y\int_z f(x, y, z) \\ &\quad\times \exp\left[-jk_r\sqrt{(x - R_g\cos\theta)^2 + (y - R_g\sin\theta)^2} - jk_v(z - Z_c)\right] dx\,dy\,dz \\ &= P(\omega)\exp(jk_vZ_c)\int_x\int_y\int_z f(x, y, z)\exp(-jk_vz) \\ &\quad\times \exp\left[-jk_r\sqrt{(x - R_g\cos\theta)^2 + (y - R_g\sin\theta)^2}\right] dx\,dy\,dz. \end{aligned}$$

The z domain integral in the above model is the Fourier transform of the target function $f(x, y, z)$ with respect to z which is evaluated at $k_z = k_v$. Thus we have the following E-CSAR system model in (ω, θ, k_v) domain:

$$\begin{aligned} S(\omega, \theta, k_v) &= P(\omega)\exp(jk_vZ_c)\int_x\int_y F_z(x, y, k_v) \\ &\quad\times \exp\left[-jk_r\sqrt{(x - R_g\cos\theta)^2 + (y - R_g\sin\theta)^2}\right] dx\,dy. \end{aligned}$$

Circular Synthetic Aperture Fourier Analysis The following is based on the Fourier analysis of a circular synthetic aperture which we outlined earlier in this chapter; see Sections 7.1 and 7.2. Consider the following PM signal in the circular synthetic aperture θ domain:

$$\exp\left[-jk_r\sqrt{(x - R_g\cos\theta)^2 + (y - R_g\sin\theta)^2}\right].$$

This PM signal appears on the right side of the E-CSAR system model for $S(\omega, \theta, k_v)$. In Sections 7.1 and 7.2 we used the polar Fourier properties of the free space Green's function to obtain a Fourier decomposition of the above PM signal.

Let $F(k_x, k_y, k_z)$ be the three-dimensional Fourier transform of the target function $f(x, y, z)$. We denote the Fourier transform of the E-CSAR signal $S(\omega, \theta, k_v)$ with respect to the circular synthetic aperture via

$$\mathbf{S}(\omega, \xi, k_v) = \mathcal{F}_{(\theta)}[S(\omega, \theta, k_v)],$$

where ξ is the frequency domain for θ. Then, using the ground plane circular SAR reconstruction in Section 7.2, we obtain the following reconstruction for the E-CSAR problem:

$$F(k_x, k_y, k_z) = \frac{1}{P(\omega)\exp(jk_v Z_c)} \mathcal{F}_{(\xi)}^{(-1)} \left[\frac{\mathbf{S}(\omega, \xi, k_v)}{H_{\xi}^{(2)}(k_r R_g)\exp(j\pi\xi/2)} \right],$$

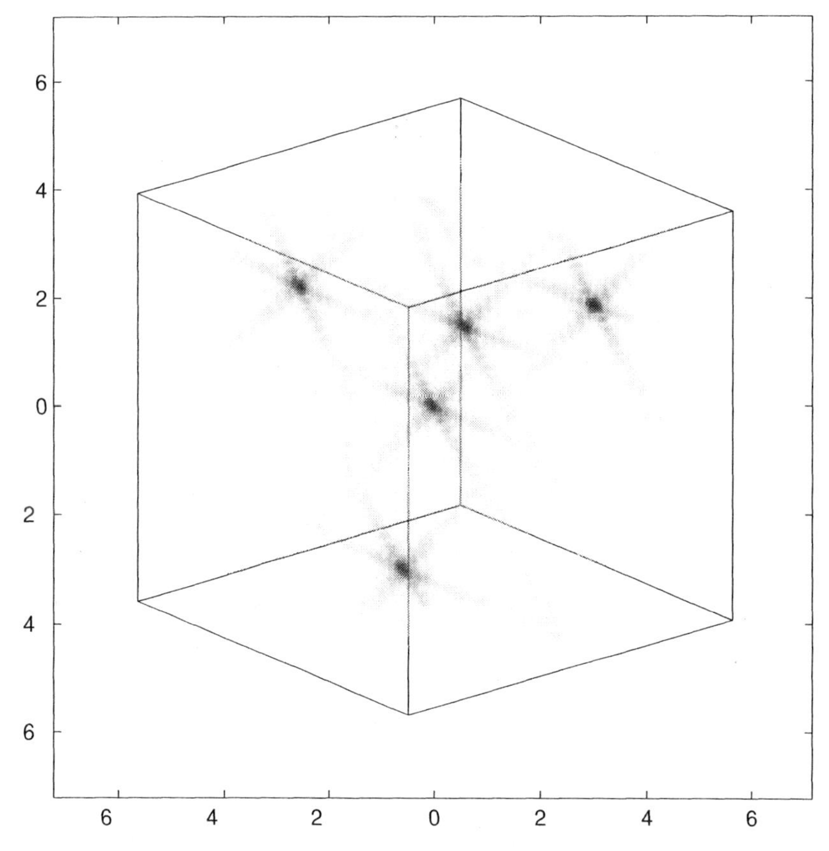

Figure 7.23a. E-CSAR reconstructions of target function $f(x, y, z)$ when viewed at ground plane and elevation angles: (50, 20) degrees.

where $H_\xi^{(2)}$ is the Hankel function of the second kind, ξ order, and

$$k_x(\omega, \theta, k_v) = \sqrt{4k^2 - k_v^2}\cos\theta,$$

$$k_y(\omega, \theta, k_v) = \sqrt{4k^2 - k_v^2}\sin\theta,$$

$$k_z(\omega, \theta, k_v) = k_v.$$

Note that the above three-dimensional mapping from the measurement spectral (ω, θ, k_v) domain into the target spectral (k_x, k_y, k_z) domain is a nonlinear one.

The practical matched-filtered version of the E-CSAR reconstruction, which is not sensitive to the additive noise, is

$$F(k_x, k_y, k_z) = P^*(\omega)\exp(-jk_v Z_c)\mathcal{F}_{(\xi)}^{(-1)}\left[\mathbf{S}(\omega, \xi, k_v) H_\xi^{(2)*}(k_r R_g)\exp\left(-j\frac{\pi\xi}{2}\right)\right].$$

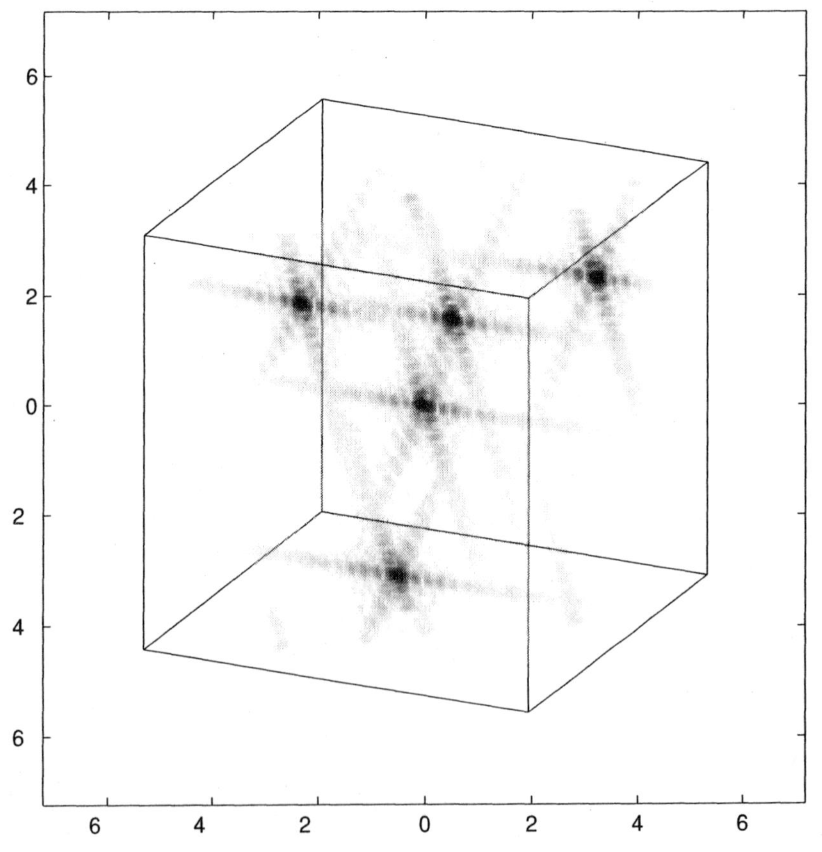

Figure 7.23b. E-CSAR reconstructions of target function $f(x, y, z)$ when viewed at ground plane and elevation angles: (25, 20) degrees.

Digital Reconstruction

In practice, the measured E-CSAR data corresponds to the samples of the signal $\mathbf{s}(t, \theta, v)$. For digital reconstruction, one first performs three-dimensional DFT on this database to obtain samples of the signal $\mathbf{S}(\omega, \xi, k_v)$ at evenly spaced points in the (ω, ξ, k_v) domain. This three-dimensional discrete database is then used in the E-CSAR reconstruction equation. For a given k_v the result is the polar samples of $F(k_x, k_y, k_v)$ at the polar coordinates

$$k_x(\omega, \theta, k_v) = \sqrt{4k^2 - k_v^2}\,\cos\theta,$$

$$k_y(\omega, \theta, k_v) = \sqrt{4k^2 - k_v^2}\,\sin\theta.$$

Note that the location of the polar samples in the (k_x, k_y) domain varies with k_v.

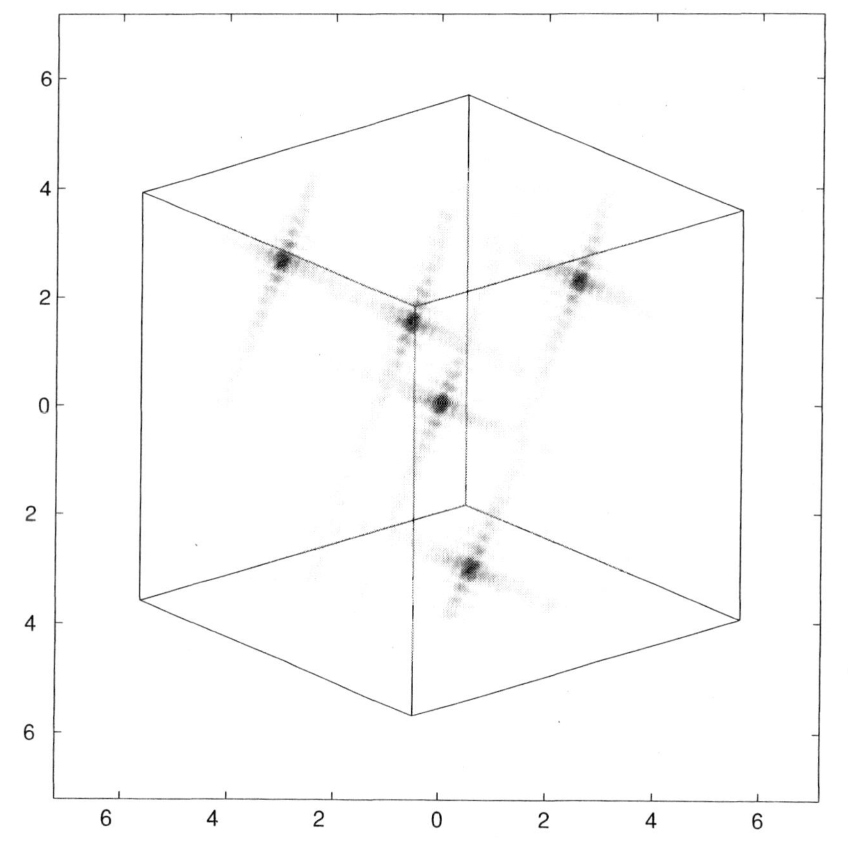

Figure 7.23c. E-CSAR reconstructions of target function $f(x, y, z)$ when viewed at ground plane and elevation angles: (230, 20) degrees.

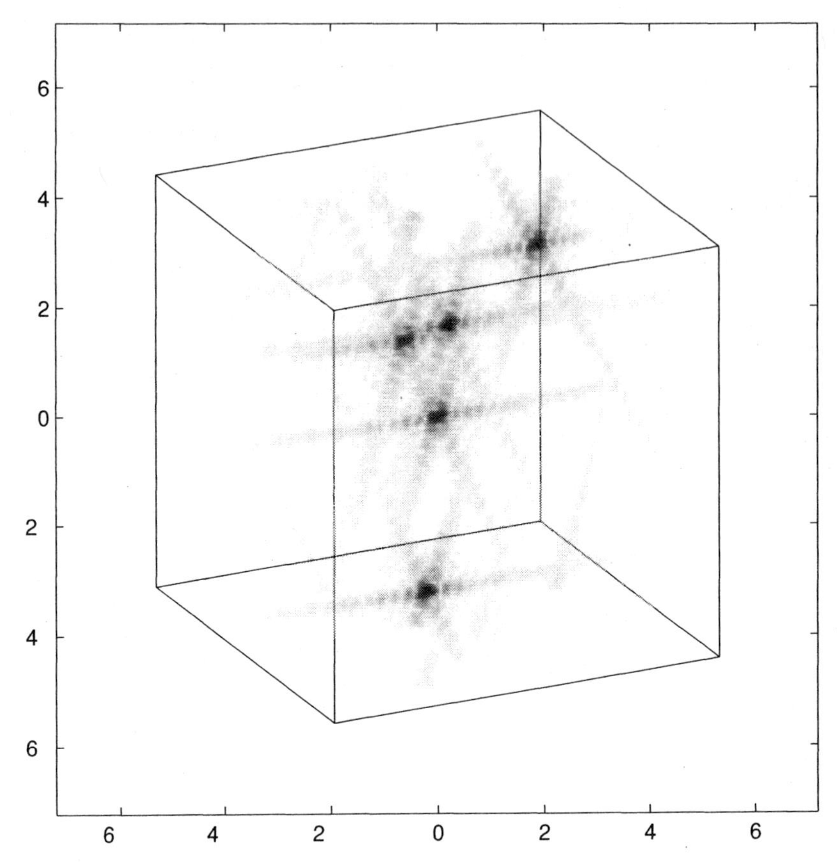

Figure 7.23d. E-CSAR reconstructions of target function $f(x, y, z)$ when viewed at ground plane and elevation angles: $(-25, 20)$ degrees.

The next step is to use an interpolation algorithm to convert the polar samples in the (k_x, k_y) domain into rectilinear samples in this domain. Note that in this step the discrete value of k_v is fixed, and the polar radial lines are at $\sqrt{4k^2 - k_v^2}$, where k (or ω) is a discrete variable. Finally, digital signal processing, image formation, and resolution for E-CSAR systems can be deduced from the analyses in Sections 5.4 to 5.6 and 7.4 to 7.6.

Example Consider an E-CSAR system with the following parameters: fast-time frequency band [9,10.2] GHz, squint elevation $Z_c = 26.44$ m, radius of circular synthetic aperture $R_g = 45.8$ m, $L_v = 2$ m, and $\theta_{\max} = 2$ degrees. (The radar parameters and range are the same as the turntable data of the T-72 tank which were collected at GTRI.) Five unit reflectors are in the target scene with the following (x, y, z) spatial coordinates (all in meters): $(0, 0, 0)$, $(0.4, 0.4, -3.2)$, $(-0.4, -0.4, 1.6)$, $(2, 1.6, 2.4)$, and $(-1.6, -2.8, 2.4)$. Figure 7.23*a* to *d*, respectively, shows the E-CSAR reconstruction when it is viewed at the following ground plane and elevation angles (all in degrees): $(50, 20)$, $(25, 20)$, $(230, 20)$, and $(-25, 20)$.

8

MONOPULSE SYNTHETIC APERTURE RADAR

Introduction

In this chapter we are going to study a more specialized SAR system whose roots can be found in the *dual* sensory human vision system. What is referred to as *stereo* vision is a dual sensory system that provides two maps of a scene from two different aspect angles. These two maps are fused to extract certain types of information that cannot be deduced from a single sensory map. The classical and natural example of a stereo system is the human visual system. The human brain uses the two images that are recorded by the eyes to form *depth* (three-dimensional) information regarding the surrounding objects. In recent years, dual cameras in conjunction with a processor (computer) have been used in a similar manner for artificial (e.g., robot) vision.

The utility of a dual or stereo sensory system in radar has been recognized since its inception. A well-known and practical stereo radar system is called a *monopulse* radar [leo; sher; sko]. This terminology originates from the manner in which the echoed data are collected in these systems. A monopulse radar is composed of a single transmitter that illuminates the target scene with a single bang, that is, a *mono*pulse. The resultant echoed signals are recorded not only by the radar that produced the original bang but also by another radar receiver.

The spatial coordinates of the second radar receiver are different from those of the transmitting/receiving radar. Thus the echoed signals from the target scene, that is, the target scene radar signature, are recorded at two different aspect angles; the monostatic and bistatic measurements of a monopulse radar are analogous to the stereo measurements of the human visual system. The echoed signals measured by the first (transmitting-receiving) radar are called *monostatic* radar data, and the echoed signals recorded by the second (receiving only) radar are called *bistatic* radar data.

Now that we have outlined the basic principle behind monopulse radar data collection, two important questions must be addressed. First, what type of *additional* information can one extract from a monopulse radar system (bistatic as well as monostatic data) that is not available from a single radar (monostatic data)? The second problem is what form of analog or digital processing of monostatic and bistatic data should be used to retrieve that additional information?

The obvious answer to the first question is that we can extract height, that is, elevation, as well as range and azimuth information regarding a terrain via a monopulse radar; although this is not a true three-dimensional SAR imaging system, it has some similarities to the three-dimensional (depth) information which the human visual system produces. But, wait; there is more. Depending on the relative coordinates of the two monostatic and bistatic radars, one can also detect *moving targets* in heavy clutter using a monopulse radar system.

The second question, that is, the *processing* of monopulse radar data, brings us to the topics showcased in this chapter. Our objective here is to study monopulse radar configurations on an airborne SAR system, and multireceiver configurations in an inverse SAR (ISAR) system. Our main task will be to develop a theoretical foundation for quantifying monopulse SAR or ISAR signals, and digital signal processing methods that exploit this theory, for instance, to determine height in a terrain, to detect moving targets in a heavy foliage area, or to track a moving target.

For this, we will examine two different monopulse SAR systems: slant plane and along-track. A slant plane monopulse SAR system has been investigated for topographic terrain mapping [gra; pra; zeb]; an along-track monopulse SAR system has utility in detecting moving targets [s97a]. The roots of both slant plane monopulse SAR and along-track monopulse SAR can be found in the classical radar analog or digital signal processing for height-finding and moving target detection with monopulse physical aperture radars [edd; sko; weh]. An experimental phase-sum-and-difference monopulse radar was developed as early as 1958 in the United States to detect moving targets in vegetative clutter [leo, pp. 340–342]. Some of the extensions of this system are discussed in [sher, ch. 5].

- The main strength of slant plane and along-track monopulse SARs is in the processing of high-resolution formed images to deduce information regarding terrain altitude (slant plane monopulse SAR) or moving targets (along-track monopulse SAR).

Outline

Monopulse SAR is a specialized imaging modality of the SAR systems which we examined in Chapters 5 to 7. Monopulse SAR principles can be formulated in the framework of signal properties of linear spotlight SAR, stripmap SAR, or CSAR. However, most of the operational monopulse SAR systems are the stripmap type. Thus we frame our discussion in terms of monopulse stripmap SAR systems. Nevertheless, we use notations such as R_c and Y_c to provide a more general solution that is also suitable for squint mode spotlight SAR systems.

We begin with a study of along-track monopulse SAR systems, which are used for moving target detection, in Section 8.1. The monostatic and bistatic data collection for this system are outlined; these form the basis for developing a statistic for moving target detection from the SAR images reconstructed from monostatic and bistatic data. This analysis is performed based on the assumption that the two monopulse radars are calibrated and stable. This, however, is not a valid assumption, especially for a relatively large coherent processing slow-time interval (i.e., a large beamwidth SAR system).

In Section 8.2 we examine the calibration problems in along-track monopulse SAR systems. The first task is to quantify the effect of uncalibrated and unstable radars in these systems. This study indicates that these calibration errors can be modeled as an unknown two-dimensional impulse function by which the monostatic and bistatic SAR images are related.

The problem of *blindly* encountering this unknown impulse function is treated in Section 8.3. In this section we identify a general registration problem in which the user wishes to fuse the information in two uncalibrated SAR images. A solution for this problem is provided that is based on constructing a deterministic signal subspace of one image and projecting the other image into this signal subspace. This solution is analogous to what is referred to as a *decorrelator* receiver in communication systems, which is used to handle multipath effects [pro]. Applications of this solution in SAR registration problems for moving target detection (MTD) and automatic target recognition (ATR) are discussed.

Section 8.4 treats the slant plane monopulse SAR system which is used for topographic mapping. First the monostatic and bistatic SAR signals are developed for this system. Then the conventional processing of these signals for extracting height information, which is based on interferometric processing of monostatic and bistatic SAR images, are shown. This method, which is also known as interferometric SAR (IF-SAR), is based on narrow-bandwidth and narrow-beamwidth approximation for the radar system. The general problem, which involves a wide-bandwidth and wide-beamwidth radar, is also discussed.

In Section 8.5 an ISAR system is examined that uses a *multistatic* (multiple) receiver structure. This system can be viewed as a combination of the along-track and slant plane monopulse SAR system. This ISAR system is used not only for three-dimensional imaging of an aircraft but also in tracking its motion. A signal subspace processing algorithm for estimating motion vectors in dual ISAR imagery is presented.

The last section of this chapter, Section 8.6, examines two sets of Matlab codes. The first is an implementation of the signal subspace processing algorithm for registering uncalibrated SAR images for MTD or ATR. The second code is based on another signal subspace processing method in which motion vectors of individual reflectors on an airborne target are estimated from dual ISAR images of the moving target.

Mathematical Notations and Symbols

The following is a list of mathematical symbols used in this chapter, and their definitions:

$a_{\mathrm{b}}(\omega, x, y)$	Bistatic radar amplitude pattern
$a_{\mathrm{m}}(\omega, x, y)$	Monostatic radar amplitude pattern
$a_{\mathrm{b}n}(\omega, x, y)$	Bistatic amplitude pattern for nth target
$a_{\mathrm{m}n}(\omega, x, y)$	Monostatic amplitude pattern for nth target
$A_{\mathrm{b}}(\omega, k_u)$	Slow-time Doppler bistatic radar amplitude pattern
$A_{\mathrm{m}}(\omega, k_u)$	Slow-time Doppler monostatic radar amplitude pattern
$A_{\mathrm{b}n}(\omega, k_u)$	Slow-time Doppler bistatic target amplitude pattern
$A_{\mathrm{m}n}(\omega, k_u)$	Slow-time Doppler monostatic target amplitude pattern
$e_n(\omega, x, y)$	Calibration error amplitude pattern of monopulse radars for nth target
$E_n(\omega, k_u)$	Slow-time Doppler calibration error amplitude pattern of monopulse radars for nth target
c	Wave propagation speed $c = 3 \times 10^8$ m/s for radar waves $c = 1500$ m/s for acoustic waves in water $c = 340$ m/s for acoustic waves in air
c_i	Coefficients for linear (rotation and scaling) transformation of spatial domain
$E_{\hat{h}}$	Energy of estimated coefficients $\hat{h}_{mn}$ or $\hat{h}_m$
$f_1(x, y)$	Output of sensor 1 (reference image)
$f_{1x}(x, y)$	Partial derivative of reference image with respect to x
$f_{1y}(x, y)$	Partial derivative of reference image with respect to y
$f_2(x, y)$	Output of sensor 2 (test image)
$\hat{f}_2(x, y)$	Signal subspace projection of test image
$f_d(x, y)$	Difference image of two sensors
$\hat{f}_d(x, y)$	Signal subspace difference image of two sensors
$f_e(x, y)$	Foreign object (change) image
$f_{1j\ell}$	Projection coefficients of reference image and its shifted versions into signal subspace Ψ
$f_{2\ell}$	Projection coefficients of test image into signal subspace Ψ
$f_{\mathrm{b}}(x, y)$	Bistatic SAR image of target area
$f_{\mathrm{i}}(x, y)$	Interferometric SAR image of target area
$f_{\mathrm{m}}(x, y)$	Monostatic SAR image of target area
$f_{\mathrm{b}n}(x, y)$	Bistatic point spread function of nth target in spatial domain
$f_{\mathrm{d}n}(x, y)$	Monopulse difference image of nth target in spatial domain (shifted to origin)
$f_{\mathrm{i}n}(x, y)$	Monopulse interferometric image of nth target in spatial domain (shifted to origin)
$f_{\mathrm{m}n}(x, y)$	Monostatic point spread function of nth target in spatial domain
$F_{\mathrm{b}n}(k_x, k_y)$	Two-dimensional Fourier transform of bistatic point spread function of nth target

$F_{mn}(k_x, k_y)$	Two-dimensional Fourier transform of monostatic point spread function of nth target
$\mathbf{F}_1$	Matrix containing coefficients $f_{1j\ell}$
$\mathbf{F}_2$	Vector containing coefficients $f_{2\ell}$
$h(x, y)$	Relative calibration impulse (filter) function of two sensors
h_{mn}	Discrete coefficients of filter $h(x, y)$
$\hat{h}_{mn}$	Estimate of discrete filter coefficients h_{mn}
$h_n(x, y)$	Relative calibration impulse function of along-track monopulse SAR images of nth target
$h_n(x_s)$	Relative slant-range shift impulse function of slant plane monopulse SAR images of nth target
h_m	Discrete coefficients of filter $h_n(x_s)$
$\hat{h}_m$	Estimate of discrete filter coefficients h_m
$H_n(k_x, k_y)$	Relative calibration transfer function of monopulse SAR images of nth target
$\mathbf{H}$	Vector containing estimated filter coefficients $\hat{h}_{mn}$
k	Wavenumber: $k = \omega/c$
k_c	Wavenumber at carrier frequency: $k_c = \omega_c/c$
k_u	Spatial frequency or wavenumber domain for azimuthal synthetic aperture u; slow-time Doppler (frequency) domain
k_x	Spatial frequency or wavenumber domain for range x
k_y	Spatial frequency or wavenumber domain for cross-range y
N	Number of basis functions in (size of) signal subspace Ψ
N_x	Size of discrete $h(x, y)$ in x domain; its half-size is n_x
N_y	Size of discrete $h(x, y)$ in y domain; its half-size is n_y
$P(\omega)$	Fourier transform of transmitted radar signal
r_n	Target radial distance from center of synthetic aperture
$r_e(u)$	Nonlinear motion of radar-carrying aircraft
$r_b(x_n, y_n, u)$	Bistatic round trip distance for nth target at slow-time u
$r_m(x_n, y_n, u)$	Monostatic round trip distance for nth target at slow-time u
R_c	Mean radial range of target area
$s_b(\omega, u)$	Bistatic SAR signal of target area
$s_{bn}(\omega, u)$	Bistatic SAR signal for nth target
$s_m(\omega, u)$	Monostatic SAR signal of target area
$s_{mn}(\omega, u)$	Monostatic SAR signal for nth target
$\hat{s}_{mn}(\omega, u)$	Synthesized monostatic SAR signal for nth target from its bistatic SAR signal
u	Synthetic aperture or slow-time domain
v_r	Speed of radar-carrying aircraft in SAR
v_{xn}	Range domain speed of nth target

v_{yn}	Cross-range domain speed of nth target
v_{zn}	Elevation domain speed of nth target
x	Range domain
x_i	Range bins in reconstructed image
x_n	Range of nth target at slow-time $u = 0$
x_s	Slant-range domain
$x_{\mathrm{bs}n}$	Bistatic slant-range of nth target
$x_{\mathrm{ms}n}$	Monostatic slant-range of nth target
$x_{en}(u)$	Nonlinear motion of nth target in range domain also denoted with $x_{en}(\tau)$
$x'_{en}(u)$	Derivative of $x_{en}(u)$ with respect to u
x_ℓ	Range coordinate of ℓth multistatic receiver
X_c	Mean range target area
X_{sc}	Mean slant-range of target area
$\mathbf{X}_{\mathrm{b}n}$	Bistatic range of nth target in motion-transformed spatial domain
$\mathbf{X}_{\mathrm{m}n}$	Monostatic range of nth target in motion-transformed spatial domain
y	Cross-range (azimuth) domain
y_j	Cross-range bins in reconstructed image
y_n	Cross-range of nth target at slow-time $u = 0$
$y_{en}(u)$	Nonlinear motion of nth target in cross-range domain; also denoted with $y_{en}(\tau)$
$y'_{en}(u)$	Derivative of $y_{en}(u)$ with respect to u
y_ℓ	Cross-range coordinate of ℓth multistatic receiver
Y_c	Mean cross-range (azimuth) of target area
$\mathbf{Y}_{\mathrm{b}n}$	Bistatic cross-range of nth target in motion-transformed spatial domain
$\mathbf{Y}_{\mathrm{m}n}$	Monostatic cross-range of nth target in motion-transformed spatial domain
z	Elevation (altitude) domain
z_n	Altitude of nth target at slow-time $u = 0$
$z_{en}(u)$	Nonlinear motion of nth target in altitude domain also denoted with $z_{en}(\tau)$
$z'_{en}(u)$	Derivative of $z_{en}(u)$ with respect to u
z_ℓ	Elevation coordinate of ℓth multistatic receiver
Z_c	Mean altitude (elevation) of target area
α_n	Relative speed of nth target with respect to radar
$\delta(\cdot)$	Delta function
Δ	Half-distance of two monopulse radars; 2Δ is distance between two monopulse radars

Δ_{sn}	Relative shift in slant-range domain of slant plane monopulse SAR images of nth target
$\hat{\Delta}_{sn}$	Estimated slant-range shift for nth target
Δ_{sx}	Relative shift in range domain in multistatic ISAR image
$\hat{\Delta}_{sx}$	Estimated value of Δ_{sx}
Δ_{sy}	Relative shift in cross-range domain in multistatic ISAR image
$\hat{\Delta}_{sy}$	Estimated value of Δ_{sy}
Δ_x	Sample spacing in x domain
Δ_y	Sample spacing in y domain
$\mathcal{F}$	Forward Fourier transform operator
$\mathcal{F}^{-1}$	Inverse Fourier transform operator
ω	Temporal frequency domain for fast-time t
ω_c	Radar signal carrier or center frequency
ω_0	Radar signal half-bandwidth in radians; radar signal baseband bandwidth is $\pm\omega_0$
$\psi_\ell(x_i, y_j)$	Orthogonal basis functions that form signal subspace of reference image and its shifted versions
Ψ	Signal subspace of reference image and its shifted versions
τ	Slow-time domain in seconds
$\theta_n(u)$	Aspect angle of nth target when radar is at $(0, u)$
$\psi_{bn}(\omega, u)$	Bistatic round-trip phase delay for nth target at slow-time u
$\psi_{mn}(\omega, u)$	Monostatic round-trip phase delay for nth target at slow-time u

8.1 ALONG-TRACK MOVING TARGET DETECTOR MONOPULSE SAR

UHF/VHF SAR systems have the ability to penetrate foliage and obtain the SAR signature of concealed targets in foliage [s95]. These radar systems, which are also known as FOPEN (foliage penetrating) SAR, are being investigated for detection of stationary and moving human-made targets in foliage. We examined such a database in Chapter 6.

The task of detecting moving targets in foliage is particularly difficult. This is due to the fact that the image of a moving target in a reconstructed SAR image is smeared and weak in comparison to the SAR image of the surrounding stationary foliage [barb; ran71; s94; yan]. Moreover the foliage possesses a strong coherent signature that overlaps with the target signature in the frequency domain, and thus it cannot be filtered out. The use of SAR ambiguity function as MTI (see Section 6.7) would not be practical in this scenario.

In this section we study a monopulse SAR data collection and processing for moving target detection in foliage as well as clear terrain. The monopulse SAR measurements for moving target detection are made by two radars that possess a common slant range but are separated in the along-track (cross-range) domain; see Figure 8.1.

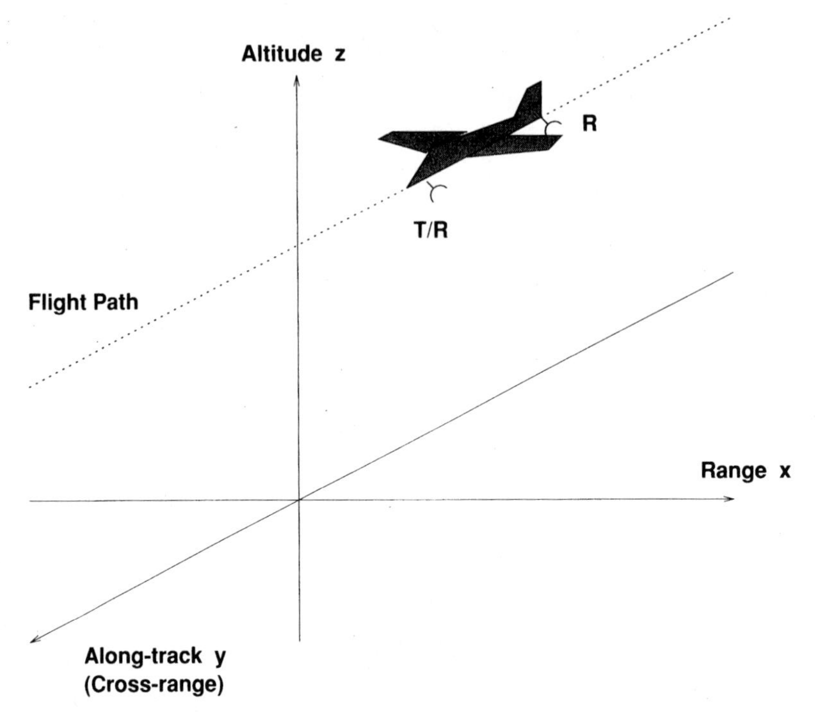

Figure 8.1. Imaging system geometry in along-track monopulse SAR.

(The along-track y domain is reversed in this image for pictorial clarity.) As we will see, the coherent processing of the resultant along-track monopulse SAR images carries information on the presence of a moving target in the imaging scene.

There are unpublished works that utilize the classical monopulse radar processing on SAR data over short slow-time intervals for moving target detection. These methods are not applicable and degrade rapidly when long slow-time intervals (coherent processing time in the slow-time) are used, which is the basis of the method that is outlined in this section.

There are also some common basic points between the along-track monopulse SAR and short-pulse area MTI [weh]. (The original work appears in [can].) In the case of short-pulse area MTI, the echoed signals due to several slow-time bursts (high range resolution) of a stationary monostatic radar are used for clutter filtering via noncoherent or coherent subtraction of successive radar video returns. Similar principles have also been utilized in sonar signal processing to remove ship noise from passive sonar [wid], infrared search and track (IRST) [fri; har], and medical digital subtraction angiography [mac].

For the airborne along-track monopulse SAR, the foliage (clutter) does not appear at the same range at the various slow-time bursts of the moving transmitting radar. We will use a coherent processing of high-resolution formed images in the cross-range (along-track) domain as well as range domain of an along-track monopulse SAR system for moving target detection.

Along-Track Monopulse SAR System Geometry

The along-track monopulse SAR imaging system geometry in the two-dimensional range (slant-range) and cross-range (x, y) domain is depicted in Figure 8.2. We will examine this SAR problem in the three-dimensional spatial domain later. We denote the fast-time domain with t, and the synthetic aperture (slow-time) domain with u. The radar-carrying aircraft moves along the range $x = 0$.

A transmitting/receiving radar (Radar 1) illuminates the target area with a large-bandwidth pulsed signal $p(t)$. The resultant echoed signals are recorded by another receiving radar (Radar 2; bistatic SAR data) as well as Radar 1 (monostatic SAR data). The two radars can be two subapertures of a single phased array, which is also known as a displaced phase center array (DPCA). Both radars have a common range (and altitude), that is, $x = 0$. However, Radar 2 is separated by 2Δ from Radar 1 in the along-track (cross-range) domain. Thus, at a given synthetic aperture position (slow-time) u, the coordinates of Radar 1 in the spatial domain are $(0, u)$, and the coordinates of Radar 2 in the spatial domain are $(0, u - 2\Delta)$.

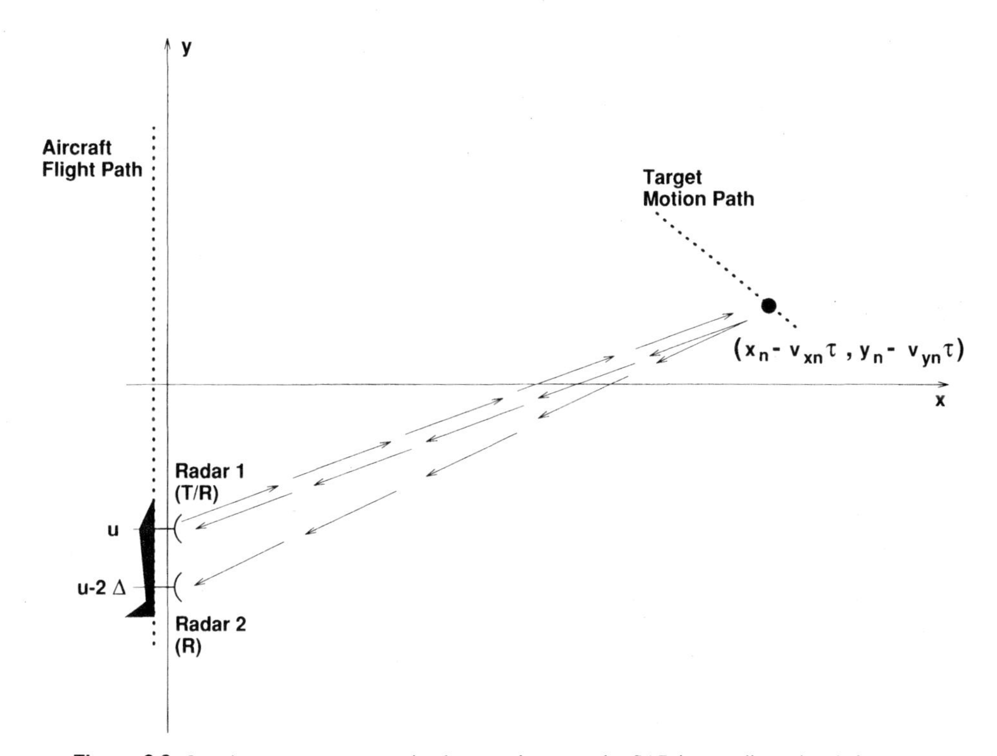

Figure 8.2. Imaging system geometry in along-track monopulse SAR in two-dimensional slant-range and cross-range domain.

We denote the speed of the airborne aircraft which carries the radar with v_r. Suppose that the velocity vector for the nth target is

$$(v_{xn}, v_{yn})$$

in the spatial (x, y) domain, which is unknown. We next develop signal models for monostatic and bistatic measurements of this target signature.

For this, we assume that the radar radiation pattern is omni-directional and that the nth target is a unit reflector whose amplitude pattern is invariant in the radar fast-time frequency and aspect angle. These simplifications are made to bring the reader's attention to the role of monostatic and bistatic SAR *phase history* of the target which is the key component for moving target detection. However, the radar amplitude pattern (whether it is a spotlight or stripmap SAR system) comes into the picture when we discuss an important practical aspect of the along-track monopulse SAR system in Section 8.2.

Monostatic SAR Signal Model

We start by developing the monostatic SAR signal recorded by Radar 1 for the nth target. This model was examined in Section 6.7 on *moving target detection and imaging*; we now provide a brief review of that model. Let the slow-time domain in seconds be τ; that is, the synthetic aperture domain is

$$u = v_r \tau.$$

Let the spatial coordinates of the nth target at the slow-time $u = 0$ be (x_n, y_n); that is, the target motion path is

$$(x_n - v_{xn}\tau, y_n - v_{yn}\tau)$$

as a function of the slow-time domain.

We denote the monostatic SAR signature of the entire target area in the (ω, u) domain by $s_{\text{m}}(\omega, u)$; also we let $s_{\text{m}n}(\omega, u)$ be the nth target monostatic SAR signature. The distance of the nth target from Radar 1 at the slow-time τ is

$$\sqrt{(x_n - v_{xn}\tau)^2 + (y_n - v_{yn}\tau - u)^2} = \sqrt{(x_n - v_{xn}\tau)^2 + (y_n - v_{yn}\tau - v_r\tau)^2}.$$

The round-trip distance from Radar 1 to the nth target, and back from the nth target to Radar 1, is

$$r_{\text{m}}(x_n, y_n, u) = 2\sqrt{(x_n - v_{xn}\tau)^2 + (y_n - v_{yn}\tau - v_r\tau)^2}.$$

Thus the monostatic SAR signature of the nth target is

$$
\begin{aligned}
s_{mn}(\omega, u) &= P(\omega)\exp[-jkr_{m}(x_n, y_n, u)] \\
&= P(\omega)\exp\left[-j2k\sqrt{(x_n - v_{xn}\tau)^2 + (y_n - v_{yn}\tau - v_r\tau)^2}\right],
\end{aligned}
$$

where ω is the fast-time frequency domain, $k = \omega/c$ is the wavenumber, and $P(\omega)$ is the Fourier transform of the transmitted radar signal. For notational simplicity we will not carry $P(\omega)$ in the following discussion. As we stated earlier, we have also suppressed the radar amplitude patterns in the above model; this will be examined in Section 8.2.

The above monostatic SAR signal for the nth target can be rewritten as follows:

$$
\begin{aligned}
s_{mn}(\omega, u) &= \exp\left[-j2k\sqrt{(x_n - v_{xn}\tau)^2 + (y_n - v_{yn}\tau - v_r\tau)^2}\right] \\
&= \exp\left[-j2k\sqrt{r_n^2 - 2\left[v_{xn}x_n + (v_{yn} + v_r)y_n\right]\tau + \left(v_{xn}^2 + v_{yn}^2 + v_r^2\right)\tau^2}\right],
\end{aligned}
$$

where $r_n = \sqrt{x_n^2 + y_n^2}$.

As we showed in Section 6.7, the nth target motion trajectory with respect to the monostatic radar (Radar 1) can be expressed via the following:

$$
\sqrt{\mathbf{X}_{mn}^2 + (\mathbf{Y}_{mn} - \alpha_n u)^2} = \sqrt{\mathbf{X}_{mn}^2 + \mathbf{Y}_{mn}^2 - 2\alpha_n \mathbf{Y}_{mn} u + \alpha_n^2 u^2},
$$

and the nth target monostatic SAR phase signature is

$$
\exp\left[-j2k\sqrt{\mathbf{X}_{mn}^2 + (\mathbf{Y}_{mn} - \alpha_n u)^2} = \sqrt{\mathbf{X}_{mn}^2 + \mathbf{Y}_{mn}^2 - 2\alpha_n \mathbf{Y}_{mn} u + \alpha_n^2 u^2}\right].
$$

After equating this model with the monostatic SAR signature of the nth target, we obtain

$$
\begin{aligned}
&\text{Radial range} && \sqrt{\mathbf{X}_{mn}^2 + \mathbf{Y}_{mn}^2} = r_n \\
&\text{Squint cross-range} && \alpha_n \mathbf{Y}_{mn} = v_{xn}x_n + (v_{yn} + v_r)y_n \\
&\text{Relative speed} && \alpha_n = \frac{\sqrt{v_{xn}^2 + (v_{yn} + v_r)^2}}{v_r}
\end{aligned}
$$

We call $(\mathbf{X}_{mn}, \mathbf{Y}_{mn})$ the monostatic motion-transformed coordinates of the nth target; the parameter α_n is the relative speed of the nth target with respect to the radar. The monostatic motion-transformed coordinates $(\mathbf{X}_{mn}, \mathbf{Y}_{mn})$ are a *linear* transformation of (x_n, y_n); that is,

$$
\mathbf{X}_{mn} = \frac{(v_{yn} + v_r)x_n - v_{xn}y_n}{\sqrt{v_{xn}^2 + (v_{yn} + v_r)^2}},
$$

$$
\mathbf{Y}_{mn} = \frac{v_{xn}x_n + (v_{yn} + v_r)y_n}{\sqrt{v_{xn}^2 + (v_{yn} + v_r)^2}},
$$

or

$$\begin{bmatrix} \mathbf{X}_{mn} \\ \mathbf{Y}_{mn} \end{bmatrix} = \begin{bmatrix} \dfrac{(v_{yn} + v_r)x_n}{\sqrt{v_{xn}^2 + (v_{yn} + v_r)^2}} & \dfrac{-v_{xn}}{\sqrt{v_{xn}^2 + (v_{yn} + v_r)^2}} \\ \dfrac{v_{xn}}{\sqrt{v_{xn}^2 + (v_{yn} + v_r)^2}} & \dfrac{(v_{yn} + v_r)}{\sqrt{v_{xn}^2 + (v_{yn} + v_r)^2}} \end{bmatrix} \begin{bmatrix} x_n \\ y_n \end{bmatrix}.$$

The above transformation corresponds to a combined rotational and scale transformation of (x_n, y_n); the parameters of the transformation depend on the nth target motion parameters. For a stationary target, that is, $(v_{xn}, v_{yn}) = (0, 0)$, we have $(\mathbf{X}_{mn}, \mathbf{Y}_{mn}) = (x_n, y_n)$.

We denote the SAR image of the nth target which is formed by the monostatic SAR data by

$$f_{mn}(x - x_n, y - y_n).$$

As we mentioned before, the speed of a moving target on the ground is much smaller than the radar-carrying aircraft, that is, $|\alpha_n - 1| \ll 1$. In this case, the signature of a moving ground target in the reconstructed monostatic SAR image $f_{mn}(x-x_n, y-y_n)$ appears as a smeared structure around approximately [s94; yan].

$$(x, y) = \left(\mathbf{X}_{mn}, \frac{\mathbf{Y}_{mn}}{\alpha_n} \right).$$

Bistatic SAR Signal Model

The distance of the nth target from Radar 2 at the slow-time τ is

$$\begin{aligned} &\sqrt{(x_n - v_{xn}\tau)^2 + (y_n - v_{yn}\tau - u + 2\Delta)^2} \\ &\quad = \sqrt{(x_n - v_{xn}\tau)^2 + (y_n - v_{yn}\tau - v_r\tau + 2\Delta)^2}. \end{aligned}$$

The round-trip distance from Radar 1 to the nth target, and back from the nth target to Radar 2, is

$$\begin{aligned} r_b(x_n, y_n, u) = &\sqrt{(x_n - v_{xn}\tau)^2 + (y_n - v_{yn}\tau - v_r\tau)^2} \\ &+ \sqrt{(x_n - v_{xn}\tau)^2 + (y_n - v_{yn}\tau - v_r\tau + 2\Delta)^2}. \end{aligned}$$

Thus the bistatic SAR signature of the nth target, which is recorded by Radar 2, is

$$
\begin{aligned}
s_{\mathrm{b}n}(\omega, u) &= P(\omega)\exp\big[-jkr_{\mathrm{b}}(x_n, y_n, u)\big] \\
&= P(\omega)\exp\Bigg[-jk\sqrt{(x_n - v_{xn}\tau)^2 + (y_n - v_{yn}\tau - v_r\tau)^2} \\
&\quad - jk\sqrt{(x_n - v_{xn}\tau)^2 + (y_n - v_{yn}\tau - v_r\tau + 2\Delta)^2}\,\Bigg].
\end{aligned}
$$

For the bistatic SAR signal we will also not carry $P(\omega)$ in our discussion.

Synthesis of Monostatic SAR Signal from Bistatic SAR Signal

Suppose that the nth target is *stationary*, that is, $(v_{xn}, v_{yn}) = (0, 0)$. In this case the monostatic round-trip distance for the nth target

$$
r_{\mathrm{m}}(x_n, y_n, u) = 2\sqrt{x_n^2 + (y_n - v_r\tau)^2},
$$

and its bistatic round-trip distance,

$$
r_{\mathrm{b}}(x_n, y_n, u) = \sqrt{x_n^2 + (y_n - v_r\tau)^2} + \sqrt{x_n^2 + (y_n - v_r\tau + 2\Delta)^2},
$$

can be related via the following approximation:

$$
r_{\mathrm{m}}(x_n, y_n, u) \approx r_{\mathrm{b}}(x_n, y_n, u + \Delta) - \frac{\Delta^2}{2R_c},
$$

where R_c is the mean range of the target area (radar mean range swath) provided that $\Delta \ll R_c$.

- The above approximation is valid for *narrow-beamwidth* SAR systems. We will re-examine the above approximation for the wide-beamwidth SAR systems in Section 8.2.
- The aircraft, which carries the radars, possesses a nonlinear motion component, $r_e(u)$. Provided that $r_e(u)$ is not a highly fluctuating signal (a condition that is met in practice), the above relationship between $r_{\mathrm{m}}(x_n, y_n, u)$ and $r_{\mathrm{b}}(x_n, y_n, u + \Delta)$ is still valid, since $2r_e(u) \approx r_e(u + \Delta) + r_e(u - \Delta)$.

Thus the bistatic SAR measurement can be converted into the monostatic SAR measurement of a transmitting-receiving radar which is located at the midpoint of the line which connects Radar 1 and Radar 2, as shown in the following:

$$
\begin{aligned}
\hat{s}_{\mathrm{m}n}(\omega, u) &= s_{\mathrm{b}n}(u + \Delta, \omega)\exp\left(j\frac{k\Delta^2}{R_c}\right) \\
&\approx s_{\mathrm{m}n}(\omega, u).
\end{aligned}
$$

We refer to $\hat{s}_{mn}(\omega, u)$ as the monostatic SAR signal which is synthesized from the bistatic SAR signal.

- The above synthesis of the monostatic SAR signature of the nth stationary target from its bistatic SAR signature is invariant in the parameters of the nth target. Thus we can perform this synthesis for the signature of all stationary targets within the imaging scene; that is,

$$\hat{s}_{\mathrm{m}}(\omega, u) = s_{\mathrm{b}}(u + \Delta, \omega) \exp\left(j \frac{k\Delta^2}{R_c}\right)$$
$$\approx s_{\mathrm{m}}(\omega, u),$$

  where $s_{\mathrm{m}}(\omega, u)$ and $s_{\mathrm{b}}(\omega, u)$, respectively, are the monostatic and bistatic SAR signatures of all stationary targets in the SAR imaging scene.

Note that in deriving the above synthesis of the monostatic SAR signal from the bistatic SAR signal, we assumed that the nth target is stationary. It turns out that this operation does not work for a moving target. In fact, suppose that we redefine the monostatic and bistatic round-trip distances for a moving nth target via the following:

$$r_{\mathrm{m}}(x_n, y_n, u) = 2\sqrt{(x_n - v_{xn}\tau)^2 + (y_n - v_{yn}\tau - v_r\tau)^2}$$

and

$$r_{\mathrm{b}}(x_n, y_n, u) = \sqrt{(x_n - v_{xn}\tau)^2 + (y_n - v_{yn}\tau - v_r\tau)^2}$$
$$+ \sqrt{(x_n - v_{xn}\tau)^2 + (y_n - v_{yn}\tau - v_r\tau + 2\Delta)^2}.$$

Then, using the fact that $|\alpha_n - 1| \ll 1$ and $\Delta \ll R_c$, we can derive the following approximation:

$$r_{\mathrm{m}}\left(x_n + \frac{v_{xn}\Delta}{v_r}, y_n + \frac{v_{yn}\Delta}{v_r}, u\right) \approx r_{\mathrm{b}}(x_n, y_n, u + \Delta) - \frac{\Delta^2}{2R_c}.$$

Thus the bistatic SAR data of the moving target corresponds to a spatial (x, y) domain shifted version of the monostatic SAR data of the same target, in addition to the shift of Δ in the synthetic aperture domain which is also present for a stationary target. The amount of the shift in the spatial (x, y) domain, that is,

$$\left(\frac{v_{xn}\Delta}{v_r}, \frac{v_{yn}\Delta}{v_r}\right),$$

is related to the target velocity and is unknown.

Moving Target Indicator

If the monostatic SAR signal of a moving target is synthesized from its bistatic SAR signal via the procedure that we described earlier for a stationary target, we obtain the following:

$$\hat{s}_{\text{m}n}(\omega, u) = \exp\left[-j2k\sqrt{\left(x_n + \frac{v_{xn}\Delta}{v_r} - v_{xn}\tau\right)^2 + \left(y_n + \frac{v_{yn}\Delta}{v_r} - v_{yn}\tau - v_r\tau\right)^2}\right].$$

The true monostatic SAR signal $s_{\text{m}n}(\omega, u)$ and the synthesized monostatic SAR signal $\hat{s}_{\text{m}n}(\omega, u)$ differ by the following phase function:

$$s_{\text{m}n}(\omega, u)\hat{s}^*_{\text{m}n}(\omega, u) \approx \exp\left[j2k\frac{\Delta}{R_c v_r}\left[v_{xn}x_n + v_{yn}(y_n - u)\right]\right].$$

Depending on the relative values of the nth target motion parameters and coordinates, and the radar frequency, the above phase function can be either significant or negligible.

The target motion trajectory in the synthesized monostatic SAR signal can be expressed via the following:

$$\sqrt{\mathbf{X}^2_{\text{b}n} + (\mathbf{Y}_{\text{b}n} - \alpha_n u)^2},$$

where

$$\mathbf{X}_{\text{b}n} \approx \mathbf{X}_{\text{m}n} + \frac{v_{xn}\Delta}{v_r},$$

$$\mathbf{Y}_{\text{b}n} \approx \mathbf{Y}_{\text{m}n} + \frac{v_{yn}\Delta}{v_r}.$$

Suppose that we form a SAR image with the synthesized monostatic data from the bistatic data, that is, $\hat{s}_{\text{m}n}(\omega, u)$. We denote the resultant SAR image of the nth target, labeled the bistatic SAR image,

$$f_{\text{b}n}(x - x_n, y - y_n).$$

The signature of the nth moving ground target in the bistatic reconstructed SAR image appears as a smeared structure around approximately

$$(x, y) = \left(\mathbf{X}_{\text{b}n}, \frac{\mathbf{Y}_{\text{b}n}}{\alpha_n}\right).$$

For a stationary target, its coherent (complex) signatures in the two monopulse SAR images, that is, the monostatic image $f_{\text{m}n}(x - x_n, y - y_n)$ and the bistatic image $f_{\text{b}n}(x - x_n, y - y_n)$, are identical. However, in the case of a moving nth target, the coordinates of the nth target in the reconstructed image with the synthesized monostatic data, which are $(\mathbf{X}_{\mathrm{b}n}, \mathbf{Y}_{\mathrm{b}n}/\alpha_n)$, are a shifted version of the coordinates in the true monostatic SAR image in $f_{\mathrm{m}n}(x - x_n, y - y_n)$, that is, $(\mathbf{X}_{\mathrm{m}n}, \mathbf{Y}_{\mathrm{m}n}/\alpha_n)$.

The amount of this shift in the range and cross-range domains, that is,

$$\left(\mathbf{X}_{\mathrm{b}n} - \mathbf{X}_{\mathrm{m}n}, \frac{\mathbf{Y}_{\mathrm{b}n}}{\alpha_n} - \frac{\mathbf{Y}_{\mathrm{m}n}}{\alpha_n}\right) = \left(\frac{v_{xn}\Delta}{v_r}, \frac{v_{yn}\Delta}{\alpha_n v_r}\right),$$

is smaller than the resolution in the range and cross-range domains for a practical SAR system. This is due to the fact that the monopulse radars are mounted on the same aircraft, which implies that $\Delta \ll R_c$, and the speed of a ground moving target is much smaller than the speed of the radar-carrying aircraft, $v_{xn} \ll v_r$ and $v_{yn} \ll v_r$.

Thus, to detect a moving target, a visual inspection of the two monopulse SAR images, which are formed by the monostatic data and synthesized monostatic data (from the bistatic data), would not be practical. Moreover, in the case of a moving target in a foliage area, the moving target signature in the two SAR images may not be visible.

However, the user does not need to rely on visual inspection. As we mentioned earlier, the *coherent* (complex) signatures of a stationary target in the two monopulse SAR images are the same. Thus, by simply subtracting the two monopulse SAR images from each other, that is,

$$f_{\mathrm{d}n}(x - x_n, y - y_n) = f_{\mathrm{b}n}(x - x_n, y - y_n) - f_{\mathrm{m}n}(x - x_n, y - y_n),$$

which we refer to as the difference monopulse SAR image, the signature of all stationary targets in the difference SAR image would disappear. The same is not true for the difference of a moving target signature in the two monopulse SAR images. We call the magnitude of the difference monopulse SAR image, $|f_{\mathrm{d}n}(x - x_n, y - y_n)|$, the monopulse SAR *moving target indicator* (MTI).

In general, it is difficult to quantify the difference monopulse SAR image of a moving target. If the transmitted radar signal is a narrowband one and the available values of u are much smaller than R_c, then the two monopulse SAR images of the nth moving target differ approximately by the following phase function:

$$\exp\left[j2k_c \frac{\Delta}{R_c v_r}(v_{xn}x_n + v_{yn}y_n)\right] = \exp\left(j2k_c \Delta \alpha_n \frac{\mathbf{Y}_{\mathrm{m}n}}{R_c}\right),$$

where k_c is the wavenumber at the carrier frequency. As we mentioned before, depending on the coordinates and motion parameters of the target, and the radar frequency band, the above phase function can be either significant or negligible.

For instance, consider a moving target with

$$v_{xn} = 0 \quad \text{and} \quad |y_n| \ll R_c.$$

In this case the above phase difference is negligible, and the moving target would not be visible in the difference monopulse SAR image. This scenario corresponds to a target that is moving parallel with the aircraft motion. Such a motion causes insignificant phase errors (similar to the small aircraft motion errors) which are not sufficient for MTI. On the other hand, for a target with

$$v_{xn} \neq 0 \quad \text{and} \quad |x_n| \approx R_c,$$

the monopulse SAR MTI would be a useful tool to detect the moving target. Reference [s97a] provides a simulated scenario in which moving targets with different types of motion vectors are studied using the monopulse SAR MTI.

Effect of Variations in Altitude and Nonlinear Motion

Next we consider the three-dimensional spatial domain that is encountered in the practical SAR imaging systems. In this model we also incorporate the effects of maneuvering in a target motion, which is represented as a nonlinear motion model. We use the altitude of the aircraft as the reference $z = 0$ plane; that is, the coordinates of Radar 1 are $(0, u, 0)$, and the coordinates of Radar 2 are $(0, u-2\Delta, 0)$ in the three-dimensional spatial domain.

We identify the three-dimensional target motion model as a function of the slow-time domain, which includes its nonlinear motion as well as variations in the terrain altitude, by the following:

$$\left[x_n - x_{en}(u),\ y_n - y_{en}(u),\ z_n - z_{en}(u)\right],$$

where (x_n, y_n, z_n) are the coordinates of the target at the slow-time $\tau = 0$ in the three-dimensional spatial domain; all the parameters and functions in the above model are unknown. For the model in the previous section, we used $x_{en}(u) = v_{xn}\tau$, $y_{en}(u) = v_{yn}\tau$, and $z_{en}(u) = 0$.

For this motion model, the monostatic and bistatic round-trip distances for a moving target become

$$r_{\text{m}}(x_n, y_n, z_n, u) = 2\sqrt{[x_n - x_{en}(u)]^2 + [y_n - y_{en}(u) - u]^2 + [z - z_{en}(u)]^2}$$

$$r_{\text{b}}(x_n, y_n, z_n, u) = \sqrt{[x_n - x_{en}(u)]^2 + [y_n - y_{en}(u) - u]^2 + [z_n - z_{en}(u)]^2}$$
$$+ \sqrt{[x_n - x_{en}(u)]^2 + [y_n - y_{en}(u) - u + 2\Delta]^2 + [z_n - z_{en}(u)]^2}.$$

Since the speed of a ground moving target is much smaller than the speed of the radar-carrying aircraft, and the terrain altitude variations on the road that the target moves on are unlikely to be rapid, we can write the following for the derivatives of the motion model:

$$|x'_{en}(u)|,\ |y'_{en}(u)|,\ |z'_{en}(u)| \ll 1.$$

By this fact and the fact that $\Delta \ll R_c$, we can obtain the following approximation:

$$r_{\mathrm{m}}\big[x_n + x'_{en}(u)\Delta,\, y_n + y'_{en}(u)\Delta,\, z_n + z'_{en}(u)\Delta,\, u\big] \approx r_{\mathrm{b}}(x_n, y_n, z_n, u+\Delta) - \frac{\Delta^2}{2R_c}.$$

Using the above model, we can show that the true monostatic SAR signal $s_{\mathrm{m}n}(\omega, u)$ and the synthesized monostatic SAR signal $\hat{s}_{\mathrm{m}n}(\omega, u)$ differ by the following phase function:

$$s_{\mathrm{m}n}(\omega, u)\hat{s}^*_{\mathrm{m}n}(\omega, u) \approx \exp\left[j2k\Delta \frac{x'_{en}(u)x_n + y'_{en}(u)(y_n - u) + z'_{en}(u)z_n}{R_c}\right].$$

Note that we used approximations to quantify the above phase difference between the monostatic and bistatic SAR signatures of the target.

However, the fact remains that there is a phase difference between the two (monostatic and bistatic) measurements that depends on the relative values of the target parameters and the radar frequency; this phase function can be either significant or negligible. For the latter case the difference monopulse SAR image can be used to detect moving targets. Reference [s97a] provides a simulated example for this scenario.

8.2 EFFECT OF UNCALIBRATED AND UNSTABLE RADARS

The monopulse SAR MTI, which was developed in the previous section, is a reliable measure for moving target detection in the SAR scene provided that *the two radars are fully calibrated*; that is, there is no relative gain and phase ambiguity in the data collected by the two radars. This idealistic scenario, however, is never encountered in practice. In a realistic monopulse SAR system, the two radars exhibit different amplitude patterns (phase as well as gain) that vary with the radar frequency and the radar position (i.e., the slow-time domain) [s98a].

Moreover these amplitude patterns could vary from one pulse transmission to another due to heat and other uncontrollable natural factors that affect the internal circuitry of the two radars; this is referred to as instability of the radar amplitude pattern. These subtle changes of the amplitude patterns of the radars are difficult to detect and track, so they are unknown to the user.

Another source of calibration error for the monopulse MTI statistic is the *narrow-beamwidth* approximation that we used to develop it. One of the primary application areas of along-track monopulse SAR is in moving target detection in foliage with *wide-beamwidth* FOPEN UHF radars. In these applications the use of the narrowband-beamwidth approximation introduces undesirable phase differences between the two along-track monopulse images of a stationary target.

In the following discussion we show that all of the above three sources of calibration errors (i.e., uncalibrated monopulse radars, radar instability, and narrow-beamwidth approximation) in along-track monopulse SAR can be lumped together; the outcome results in the following relationship between the two monopulse images for the nth stationary target:

$$f_{\mathrm{b}n}(x - x_n, y - y_n) = f_{\mathrm{m}n}(x - x_n, y - y_n) ** h_n(x, y),$$

where $h_n(x, y)$ is an *unknown* two-dimensional impulse function in the spatial (x, y) domain. In Section 8.3 we will present a method for dealing with this unknown impulse function.

The above model, which states that the *point spread function* of the two monopulse radars are different, is intuitively obvious to most radar engineers; a similar problem has been known to exist in the classical monopulse radar signal processing [leo; sher; sko]. The reader may choose to skip the proof of the above model and proceed to Section 8.3 where the calibration problem is treated.

Otherwise, the reader would find the proof to be an interesting application of what we learned about spectral properties of AM-PM signals in Section 2.9, and its utilization to model the SAR radiation pattern (Chapter 3) and SAR motion errors (Sections 4.8–4.10). Our analysis is based on a stripmap SAR system. However, the basic principles that we bring out in this discussion can also be extended to the spotlight SAR systems.

Amplitude Patterns of Monopulse Radars

We did not incorporate the target and radar amplitude patterns in the formulation that was presented in the previous section. We now examine the role of the amplitude patterns of the two radars and the target in monopulse SAR moving target detection.

For this purpose we begin with the signal model for the monostatic and bistatic SAR measurements. We denote the transmit-receive amplitude pattern of the monostatic radar in the spatial (x, y) domain by $a_{\mathrm{m}}(x, y - u, \omega)$ where $(0, u)$ is the radar position on the synthetic aperture and ω is the radar (fast-time) frequency; see Chapter 3. (This amplitude pattern does not include radar *instability* in the slow-time domain; this will be discussed later.)

Thus the monostatic SAR signal for a *stationary* target at (x_n, y_n) is (see Chapters 3 and 6)

$$s_{\mathrm{m}n}(\omega, u) = a_{\mathrm{m}}(\omega, x_n, y_n - u) a_{\mathrm{m}n}(\omega, x_n, y_n - u) \exp\left[j2k\sqrt{x_n^2 + (y_n - u)^2} \right],$$

where $a_{\mathrm{m}n}(\omega, x, y)$ is the target monostatic amplitude pattern. In Chapter 3 we found the radar and target amplitude patterns in the slow-time Doppler domain to be

$$A_{\mathrm{m}}\left[\omega, 2k \sin\theta_n(u)\right] = a_{\mathrm{m}}(\omega, x_n, y_n - u),$$
$$A_{\mathrm{m}n}\left[\omega, 2k \sin\theta_n(u)\right] = a_{\mathrm{m}n}(\omega, x_n, y_n - u),$$

where

$$\theta_n(u) = \arctan\left(\frac{y_n - u}{x_n}\right),$$

is the target aspect angle when the radar is located at $(0, u)$.

Using the wavefront reconstruction algorithm in stripmap SAR for stationary targets (see Section 6.2), the spectrum of the nth target reconstruction from the monostatic SAR data becomes (see Section 6.2)

$$\mathcal{F}_{(x,y)}[f_{\mathrm{m}n}(x,y)] = F_{\mathrm{m}n}(k_x,k_y)\exp\big(jk_x x_n + jk_y y_n\big),$$

where

$$F_{\mathrm{m}n}(k_x,k_y) = A_{\mathrm{m}}(\omega,k_u)A_{\mathrm{m}n}(\omega,k_u),$$

k_u is the spatial frequency domain for u, and

$$k_x = \sqrt{4k^2 - k_u^2},$$
$$k_y = k_u,$$

which we referred to as SAR spectral mapping. The spatial domain monostatic SAR image of the nth target $f_{\mathrm{m}n}(x - x_n, y - y_n)$ is the two-dimensional inverse Fourier transform of $F_{\mathrm{m}n}(k_x,k_y)$. The monostatic SAR image $f_{\mathrm{m}n}(x - x_n, y - y_n)$ is the sum of all these stationary target signatures (as well as the moving target signatures).

A similar analysis can be performed for the bistatic measurement. Suppose that the bistatic amplitude pattern of the nth target is $a_{\mathrm{b}n}(x, y-u, \omega)$ and that the bistatic transmit-receive radar amplitude pattern is $a_{\mathrm{b}}(x, y-u, \omega)$. (As we stated earlier, the radars are assumed to be *stable* in our present discussion.) Then after the phase compensation to convert bistatic SAR data to the equivalent monostatic SAR data under the narrow-beamwidth assumption (see Section 8.1), the synthesized monostatic SAR data from the bistatic SAR data becomes the following SAR signal:

$$\hat{s}_{\mathrm{m}n}(\omega,u) = a_{\mathrm{b}}(\omega,x_n,y_n-u)a_{\mathrm{b}n}(\omega,x_n,y_n-u)\exp\left[j2k\sqrt{x_n^2+(y_n-u)^2}\right].$$

The resultant synthesized monostatic data yields the following target function spectrum, which we referred to as the bistatic target function reconstruction:

$$\mathcal{F}_{(x,y)}[f_{\mathrm{b}n}(x,y)] = F_{\mathrm{b}n}(k_x,k_y)\exp\big(jk_x x_n + jk_y y_n\big),$$

where

$$F_{\mathrm{b}n}(k_x,k_y) = A_{\mathrm{b}}(\omega,k_u)A_{\mathrm{b}n}(\omega,k_u),$$
$$A_{\mathrm{b}}\big[\omega, 2k\sin\theta_n(u)\big] = a_{\mathrm{b}}(\omega,x_n,y_n-u)$$
$$A_{\mathrm{b}n}\big[\omega, 2k\sin\theta_n(u)\big] = a_{\mathrm{b}n}(\omega,x_n,y_n-u),$$

and (k_x,k_y) are related to (ω,k_u) via the SAR spectral mapping. The spatial domain bistatic SAR image of the nth target $f_{\mathrm{b}n}(x - x_n, y - y_n)$ is the two-dimensional inverse Fourier transform of $F_{\mathrm{b}n}(k_x,k_y)$. The bistatic SAR image $f_{\mathrm{b}n}(x - x_n, y - y_n)$ is the sum of all these stationary target signatures (as well as the moving target signatures).

The two radars are considered to be fully calibrated if $a_{\rm b}(\omega, x, y-u) = a_{\rm m}(\omega, x, y-u)$, or

$$A_{\rm b}(\omega, k_u) = A_{\rm m}(\omega, k_u).$$

Moreover, provided that $\Delta \ll R_c$, the monostatic and bistatic amplitude patterns of a target are approximately the same; that is,

$$A_{{\rm b}n}(\omega, k_u) \approx A_{{\rm m}n}(\omega, k_u).$$

In this case the two monopulse SAR images of the nth stationary target are also identical; that is, $F_{{\rm b}n}(k_x, k_y) = F_{{\rm m}n}(k_x, k_y)$, and thus the difference image is a reliable statistic for moving target detection. This, however, is never achieved in practice.

In a practical monopulse SAR system, the two monostatic and bistatic radar and target amplitude patterns are not the same. This difference can be modeled as

$$\begin{aligned} &a_{\rm b}(\omega, x_n, y_n - u, \omega) a_{{\rm b}n}(\omega, x_n, y_n - u, \omega) \\ &\quad = a_{\rm m}(\omega, x_n, y_n - u, \omega) a_{{\rm m}n}(\omega, x_n, y_n - u, \omega)\big[1 + e_n(\omega, x_n, y_n - u, \omega)\big], \end{aligned}$$

where $e_n(\omega, x, y)$ is an unknown amplitude pattern that contains a phase function as well as a gain (magnitude) function. Using the above expression for the bistatic amplitude pattern, the bistatic SAR reconstruction in the spatial frequency domain becomes

$$F_{{\rm b}n}(k_x, k_y) = A_{\rm m}(\omega, k_u) A_{{\rm m}n}(\omega, k_u)\big[1 + E_n(\omega, k_u)\big],$$

where

$$E_n\big[\omega, 2k \sin\theta_n(u)\big] = e_n(\omega, x_n, y_n - u),$$

and (k_x, k_y) are related to (ω, k_u) via the SAR spectral mapping.

Comparing the bistatic and monostatic SAR reconstructions, we can write the following:

$$\begin{aligned} F_{{\rm b}n}(k_x, k_y) &= H_n(k_x, k_y) F_{{\rm m}n}(k_x, k_y) \\ f_{{\rm b}n}(x - x_n, y - y_n) &= f_{{\rm m}n}(x - x_n, y - y_n) ** h_n(x, y) \end{aligned}$$

where

$$H_n(k_x, k_y) = 1 + E_n(\omega, k_u)$$

represents the transfer function of a linear shift-invariant system; $h_n(x, y)$, the impulse response, is the two-dimensional inverse Fourier transform of $H_n(k_x, k_y)$, and $**$ denotes two-dimensional convolution in the spatial domain.

These relationships between the two monopulse SAR images of the *stationary* nth target indicate that the bistatic SAR image is a filtered (smeared and/or shifted) version of the monostatic SAR image; the transfer function of the filter $H_n(k_x, k_y)$

is related to the phase and gain differences of the two uncalibrated radars *and* the coordinates of the target. This can result in gain and phase differences between the monostatic and bistatic SAR images of the stationary target. In this case the difference image is not a reliable statistic for moving target detection.

Instability of Monopulse Radars

In the case of unstable radars, one should associate two separate random amplitude patterns of (ω, u) to the monostatic amplitude pattern $a_{\rm m}(\omega, x, y-u)$ and the bistatic amplitude pattern $a_{\rm b}(\omega, x, y-u)$. In practice, these random amplitude patterns are not highly fluctuating functions. In this case these random amplitude patterns also map into the slow-time Doppler domain via the transformation (recall the spectral properties of AM-PM signals that we used in Section 4.8–4.9)

$$k_u = 2k \sin\theta_n(u).$$

The end result is that these random amplitude patterns introduce additional undesirable and unknown gain and phase variations to the monostatic and bistatic slow-time Doppler amplitude patterns, that is, $A_{\rm m}(\omega, k_u)$ and $A_{\rm b}(\omega, k_u)$. Thus the above model which relates the monostatic and bistatic SAR images of the nth target via an unknown filter function, $h_n(x, y)$, is still valid. Our next task is to seek a procedure to address the adverse effect of this unknown filter function; this is discussed in the next section.

Wide-Beamwidth Monopulse Radars

In the narrow-beamwidth derivation of the MTI statistic for along-track monopulse SAR, we used the following approximation to relate the monostatic and bistatic round-trip distances for the stationary nth target:

$$r_{\rm m}(x_n, y_n, u) \approx r_{\rm b}(x_n, y_n, u+\Delta) - \frac{\Delta^2}{2R_c};$$

as we stated earlier, this approximation is based on the narrow-beamwidth assumption for the two radars.

In the case of wide-beamwidth radars, a better model to relate these two round-trip distances is

$$r_{\rm m}(x_n, y_n, u) \approx r_{\rm b}(x_n, y_n, u+\Delta) - \frac{\Delta^2}{r_{\rm m}(x_n, y_n, u)}.$$

Thus using the narrow-beamwidth assumption results in a distance error (similar to the SAR motion errors in Sections 4.9–4.10) that translates into phase and gain differences (amplitude function) in the spectral domain of the two monopulse SAR images.

This spectral domain error amplitude function adds to the gain and phase variations of the monostatic and bistatic slow-time Doppler amplitude patterns, that is,

$A_{\rm m}(\omega, k_u)$ and $A_{\rm b}(\omega, k_u)$. This again translates into the model which we developed earlier to relate the two monopulse SAR images, which is

$$f_{{\rm b}n}(x - x_n, y - y_n) = f_{{\rm m}n}(x - x_n, y - y_n) ** h_n(x, y).$$

We should point out that, unlike the other two calibration errors, the amplitude error function for the narrow-beamwidth approximation is a *known* signal even though it is shift-varying. Thus the user could encounter its effect via an inverse shift-varying filtering in the spatial domain; see the discussion on GPS-based motion compensation in Section 4.9. However, the processing that we present in the next section deals with all of these three sources of calibration errors, whether they are known or unknown, simultaneously.

8.3 SIGNAL SUBSPACE REGISTRATION OF UNCALIBRATED SAR IMAGES

A classical problem in surveillance with radars or optical sensors, and also diagnostic medicine, involves examining a scene at various time points or with various sensors, which are located at different aspect angles, and fusing the information from these sensors for image registration, or detecting what we refer to as a *change*. For example, in diagnostic medicine, the (X-ray, ultrasonic, etc.) image of a biological structure is acquired at different time points and tested for the presence of irregularities such as a tumor.

Similarly, in surveillance with spaceborne (satellite) or airborne (aircraft) optical devices, the optical images of a scene acquired at different time points are utilized to detect changes in, for example, the environment or the enemy's arsenal. The signals acquired by monopulse radars, which utilize two radars at two different locations (aspect angles), can be used to detect moving targets [leo; sher]. We also used a similar scheme for the SAR system in Section 8.1 to develop a statistic for MTI based on along-track monopulse SAR measurements.

A fundamental problem associated with these systems is that the *stationary background* (e.g., the clutter in radar, or the internal organs of the patient in diagnostic medicine) should exhibit the same behavior (signature) when viewed by different sensory systems or at different time points. We refer to this scenario as perfectly *calibrated* sensors. Unfortunately, perfectly calibrated sensors do not exist in practice.

In the ideal case of perfectly calibrated sensors, the change in two images can be detected by simply subtracting one image from the other. With uncalibrated sensors, the differencing operation is not practical. This is due to the fact that most of these dual sensory systems seek to detect subtle (weak) changes. Unfortunately, the power of the calibration error exceeds the power of a change in most practical scenarios. This is particularly true in along-track monopulse SAR systems; examples of this problem are provided in [s97a].

In this section we examine the problem of fusing or registering information in uncalibrated sensors. The approach is based on manipulating a system model with *un-*

known parameters, which relates the outputs of two uncalibrated sensors, to develop a procedure that *blindly* calibrates the two outputs. This system model is identical to the system model that we developed in Section 8.2 to relate the two uncalibrated monopulse SAR images of a stationary target. Thus the method is also applicable in moving target detection with uncalibrated along-track monopulse SARs. We will show the application of this approach in automatic target recognition (registration) of SAR.

System Model

Consider a sensory system that acquires one-dimensional or multidimensional information (snapshots) of a stationary scene at various time points. Our objective is to detect relative change in any two of these snapshots due to the presence of a foreign object (e.g., a moving target in SAR, nonlinear motion in video, or a tumor in a biological structure). For our discussion we consider two-dimensional snapshots (images). Suppose that the image recorded by one sensor is $f_1(x, y)$, also called the *reference* image (Figure P8.1). Then the recorded image by the other sensor, called the *test* image (Figure P8.2), is modeled via the following:

$$\begin{aligned} f_2(x, y) &= f_1(x, y) ** h(x, y) + f_e(x, y) \\ &= \int_u \int_v h(u, v) f_1(x - u, y - v)\, du\, dv + f_e(x, y), \end{aligned}$$

where $**$ denotes the two-dimensional convolution in the spatial (x, y) domain.

In the above model, $f_e(x, y)$ is the change caused by a foreign object (Figure P8.3), and $h(x, y)$ is an impulse response that represents the relative shift and blurring in the two images due to slight motion and/or change in the point spread function (PSF) of the two sensors; these two signals, that is, $f_e(x, y)$ and $h(x, y)$, are unknown. To provide a general analysis, the signals in the above model are assumed to be complex. A block diagram representation of this model is shown in Figure 8.3.

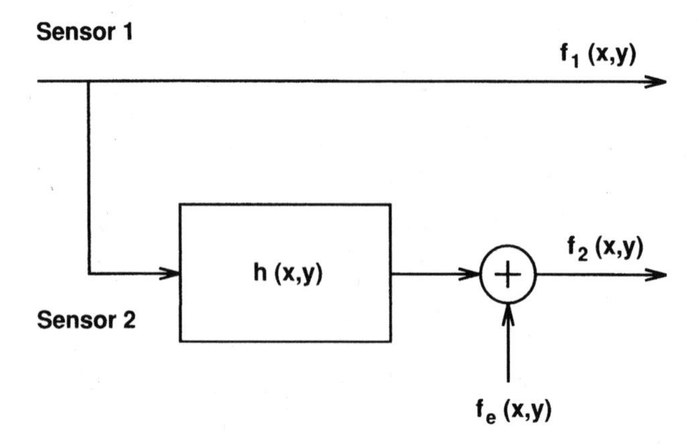

Figure 8.3. Block diagram for generation of uncalibrated images.

The above model states that the output of the second sensor $f_2(x, y)$ is linearly related to the output of the first sensor $f_1(x, y)$ and its shifted versions. There might be cases in which the sensors exhibit nonlinearity over time [s97o]. Also, in some applications, the second snapshot could be a linearly transformed (shifted, rotated, and scaled) version of $f_1(x, y)$. To present the basic concept behind registration of uncalibrated images via signal subspace processing, we begin with the model in Figure 8.3. However, this approach can be applied to the other above-mentioned models (nonlinear sensors, target rotation and scaling, etc.) [s97o]. These will be discussed in the analysis of registering SAR images for automatic target recognition.

If the two sensors were perfectly calibrated, that is, there was no relative shift and/or change in the PSF of the two sensors, then the impulse response in Figure 8.3 would be the two-dimensional delta function

$$h(x, y) = \delta(x, y).$$

In this case one can detect the presence of the foreign object via the difference of the two images (Figure P8.4); that is,

$$f_d(x, y) = f_2(x, y) - f_1(x, y).$$

Our discussion is concerned with scenarios in which the two sensors are not perfectly calibrated. In these problems a slight relative shift and/or blurring in the PSF of the two sensors yields an error signal that can dominate the foreign object signature $f_e(x, y)$. For instance, in along-track monopulse SAR for moving target detection, $f_e(x, y)$ could be an order of magnitude weaker than $f_1(x, y)$.

Adaptive filtering methods have been suggested to solve the above problem in one-dimensional cases [har; wid]. To apply these adaptive filtering methods in the two-dimensional problems, consider the discrete measured data in the (x_i, y_j) domain. The impulse response $h(x, y)$ is modeled by a finite two-dimensional discrete filter h_{mn}; the size of the filter, labeled (N_x, N_y), is chosen by the user based on a priori information. In the following discussion we choose both N_x and N_y to be odd integers and $(n_x, n_y) = (N_x/2 - 0.5, N_y/2 - 0.5)$.

Then the model in Figure 8.3 is rewritten in the following discrete form:

$$f_2(x_i, y_j) = \sum_{m=-n_x}^{n_x} \sum_{n=-n_y}^{n_y} h_{mn} f_1(x_i - m\Delta_x, y_j - n\Delta_y) + f_e(x_i, y_j),$$

where (Δ_x, Δ_y) represent the sensor sample spacing in the (x, y) domain. In the adaptive filtering approach, a solution for the impulse response h_{mn} from the knowledge of $f_1(x_i, y_j)$ and $f_2(x_i, y_j)$ (this operation is called *deconvolution*), labeled $\hat{h}_{mn}$, is obtained via minimizing the following error function:

$$\sum_i \sum_j \left| f_2(x_i, y_j) - \sum_{m=-n_x}^{n_x} \sum_{n=-n_y}^{n_y} \hat{h}_{mn} f_1(x_i - m\Delta_x, y_j - n\Delta_y) \right|^2 .$$

The resultant solution is then used to estimate $f_2(x_i, y_j)$ via the following (Figure P8.5):

$$\hat{f}_2(x_i, y_j) = \sum_{m=-n_x}^{n_x} \sum_{n=-n_y}^{n_y} \hat{h}_{mn} f_1(x_i - m\Delta_x, y_j - n\Delta_y).$$

The statistic used for detecting the foreign object is constructed from the following (Figure P8.6):

$$\hat{f}_d(x_i, y_j) = f_2(x_i, y_j) - \hat{f}_2(x_i, y_j).$$

In the one-dimensional problems the direct solution for $\hat{h}_m$ (based on the normal equations) requires the computation of a large covariance matrix. One may also approach the problem via a more "cost-effective" recursive LMS (least mean square) solution that is based on a gradient descent adaptive algorithm [har]. These methods may be utilized in the two-dimensional problems via, for example, *reshaping* the two-dimensional arrays into one-dimensional arrays. This, however, requires processing very large matrices, especially for the covariance matrix and the reshaped discrete filter.

Signal Subspace Processing

The signal $\hat{f}_2(x_i, y_j)$ is the projection of $f_2(x_i, y_j)$ into the linear subspace which is defined by $f_1(x_i, y_j)$ and $N - 1$, where $N = N_x N_y$, of its shifted versions; that is,

$$\Psi = \left[f_1(x_i - m\Delta_x, y_j - n\Delta_y); m = -n_x, \ldots, n_x, n = -n_y, \ldots, n_y\right].$$

Thus it is sufficient to identify the signal subspace Ψ and then to obtain the projection of $f_2(x_i, y_j)$ into this signal subspace in order to construct $\hat{f}_2(x_i, y_j)$.

Let

$$\psi_\ell(x_i, y_j), \qquad \ell = 1, 2, \ldots, N,$$

be a set of orthogonal basis functions that spans the linear signal subspace of Ψ; that is,

$$\Psi = \left[\psi_\ell(x_i, y_j); \ell = 1, 2, \ldots, N\right],$$

where

$$\langle \psi_\ell, \psi_j \rangle = \sum_i \sum_j \psi_\ell(x_i, y_j)\psi_j^*(x_i, y_j) = \begin{cases} 1 & \text{for } \ell = j, \\ 0 & \text{otherwise.} \end{cases}$$

(In the above, ψ^* is the complex conjugate of ψ.) To generate this signal subspace, one can use a variety of techniques, such as Gram-Schmidt, modified Gram-Schmidt, Householder, Givens orthogonalization procedure, or singular value decomposition.

The size of the signal subspace, that is, N, depends on the user's a priori knowledge of the number of the nonzero coefficients in the discrete model of the impulse response $h(x, y)$, that is, the size of PSF based on the accuracy of the imaging system data acquisition. For instance, if the discrete $h(x, y)$ contains (N_x, N_y) nonzero pixels, then we should select $N = N_x N_y$. (As we pointed out earlier, a similar assignment/model for $h(x, y)$ is used in the adaptive filtering methods [har; wid].)

In practice, the exact value of $N_x N_y$ is not known. In this case an estimate should be used based on the maximum anticipated degree of shift and calibration errors between the two sensors. For instance, in the along-track monopulse SAR for moving target detection, the knowledge of the parameter variations in the electronic circuitry of the two monopulse radars over time could be used.

The projection of the $f_2(x_i, y_j)$ into the basis function $\psi_\ell(x_i, y_j)$, which is identified by the series coefficient $f_{2\ell}$ ($\ell = 1, 2, \ldots, N$), is found via the following:

$$f_{2\ell} = \langle f_2, \psi_\ell \rangle = \sum_i \sum_j f_2(x_i, y_j) \psi_\ell^*(x_i, y_j).$$

The projection of $f_2(x_i, y_j)$ into the signal subspace Ψ is achieved via the following (Figure P8.5):

$$\hat{f}_2(x_i, y_j) = \sum_{\ell=1}^{N} f_{2\ell} \psi_\ell(x_i, y_j).$$

Finally the signal subspace difference image, that is, the statistic for detecting the foreign object, is constructed by (Figure P8.6)

$$\hat{f}_d(x_i, y_j) = f_2(x_i, y_j) - \hat{f}_2(x_i, y_j).$$

Note that both the adaptive filtering method in [har] and the above signal subspace projection seek the same *minimum error energy* solution for the estimate of $f_2(x_i, y_j)$ in the linear subspace of $f_1(x_i, y_j)$ and its shifted versions. However, in the signal subspace method, the object is not to solve the deconvolution problem (i.e., to estimate the impulse response h_{mn}) which is computationally intensive. The signal subspace approach provides a direct solution for $\hat{f}_2(x_i, y_j)$ without the need to solve the deconvolution problem.

Example We examined an X-band ISAR database of an aircraft in Chapter 5. By separately processing the lower and upper portions of the slow-time domain ISAR data, we showed that the aircraft possessed some kind of rolling motion during the data acquisition (see Figure 5.17*b*–*c*). Figure 8.4*a* and *b* shows the squint spatial domain ISAR reconstructions of this aircraft which are formed from the slow-time pulse numbers (PRIs) 89–152 and 105–168. These two overlapping 320-ms slow-time intervals (subapertures) are separated by 80 ms. (The PRF for this ISAR system is 200 Hz.)

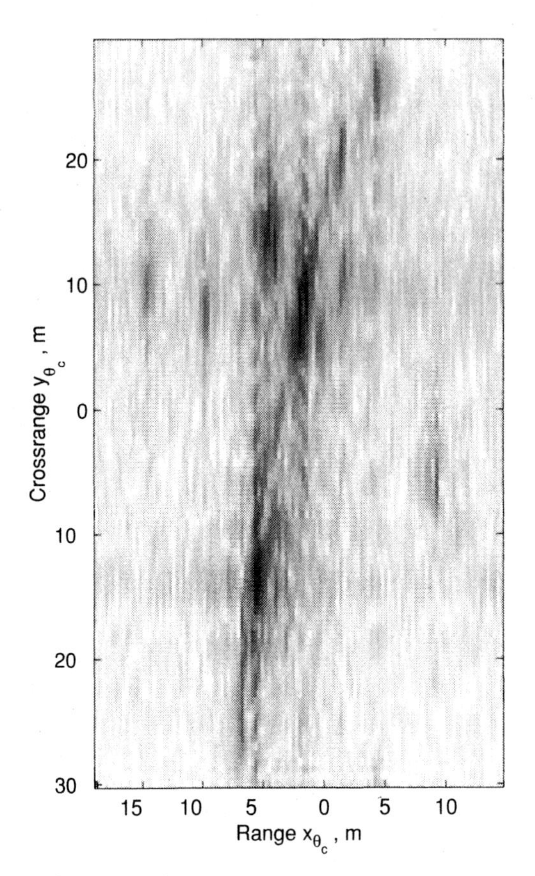

Figure 8.4a. Squint spatial domain ISAR reconstructions of aircraft from its X band ISAR data with slow-time pulse numbers (PRIs): 89–152.

These two images exhibit different point spread functions for each reflector on the aircraft; the variations of the PSF in the two images are also *shift-varying*. These are due to two factors: the roll of the aircraft; and the dependence of the spotlight SAR PSF on the relative coordinates of a reflector with respect to the midpoint of the two subapertures (see Chapter 5). Thus the *magnitude* of the two ISAR images of a reflector on the aircraft exhibit a relative shift as well as change in shape of PSF.

We wish to use the signal subspace processing to register the magnitudes of the two ISAR images in Figure 8.4*a* and *b*. For this we assign the reference image $f_1(x, y)$ and test image $f_2(x, y)$, respectively, to be the ISAR image in Figure 8.4*a* and *b* (Figures P8.1 and P8.2). As we stated earlier, the variations of the two ISAR images are shift-varying. Thus, we cannot apply signal subspace processing, which assumes a common calibration error impulse function $h(x, y)$, throughout these two images.

However, we can assume that this impulse function is approximately shift-invariant within a relatively small area, or *block*, in the reconstructed ISAR image.

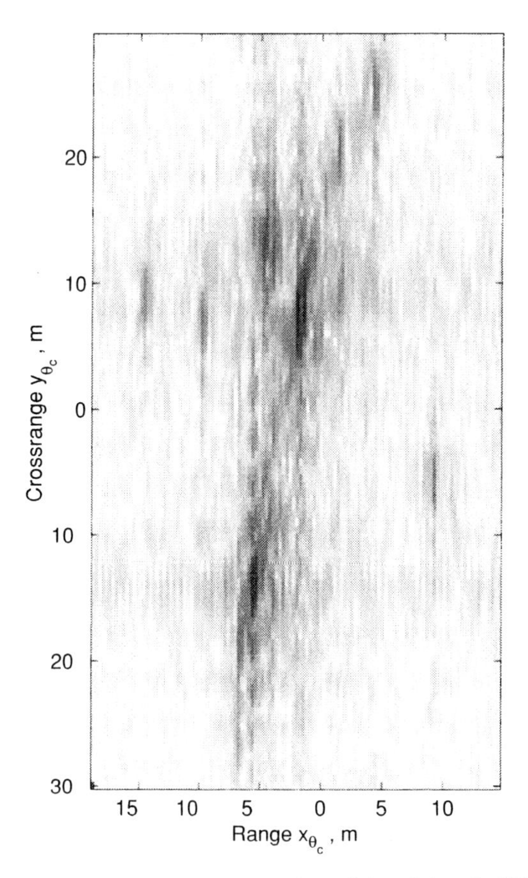

Figure 8.4b. Squint spatial domain ISAR reconstructions of aircraft from its X band ISAR data with slow-time pulse numbers (PRIs): 105–168.

Clearly we cannot choose an arbitrary small size block; the size of the block depends on the extent of a reflector PSF in the ISAR images. For this scenario a block size of 7×15 in the spatial (x, y) domain is suitable.

Note that we also used the reconstructed ISAR images in the *squint* spatial domain. As we stated in Chapter 5, the spotlight SAR PSF of a reflector in the squint spatial domain appears to be a *straight* cross-shaped structure. For block-based processing this is more suitable than the *slanted* PSF in the original spatial domain.

The size of the filter also depends on the size of the pixels relative to the range and cross-range resolutions. For this example the values of $(n_x, n_y) = (2, 4)$ are suitable; these correspond to a filter size of $(2n_x + 1, 2n_y + 1) = (5, 9)$.

Figure 8.4*c* is the difference image, $|f_2(x, y) - f_1(x, y)|$, for this example (Figure P8.3). This figure shows distinct vertical streaks which are due to the relative shift (mismatch) of the PSF of the reflectors caused by the roll of the aircraft. Figure 8.4*d* shows the subspace difference image, $|f_2(x, y) - \hat{f}_2(x, y)|$ (Figure P8.6),

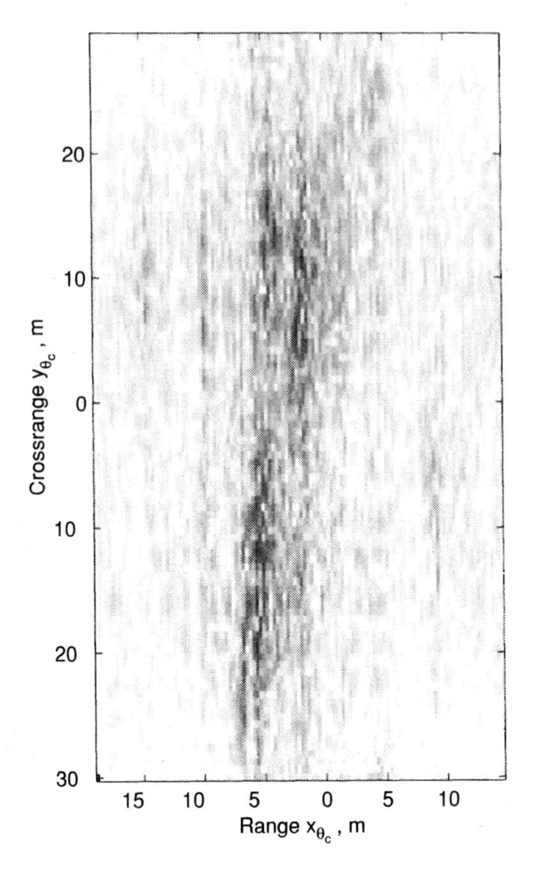

Figure 8.4c. Difference image of (a) and (b).

in which most of those streaks have disappeared (i.e., the ISAR image signatures are registered.) One can also see *patchy looking* areas in the signal subspace difference image. This is caused by the block-based processing which introduces a different measure of similarity in each block; that is, some blocks are matched better than the others.

Finally we did not use *complex* ISAR images in this example. The reason for this is that the overlapping slow-time data of the two sets of PRIs for this experiment (pulse numbers 105–152) correspond to the same complex ISAR data and are perfectly calibrated. However, the coherent ISAR data in the nonoverlapping PRIs (pulse numbers 89–104 and 153–168) map into *disjoint* spectral (k_x, k_y) domain data for a given reflector (i.e., they are orthogonal to each other). In this case these complex data cannot be related by the model in Figure 8.3. The magnitude ISAR reconstructions provide two ISAR point spread functions that have slightly different coordinates and shapes to fit the system model in Figure 8.3.

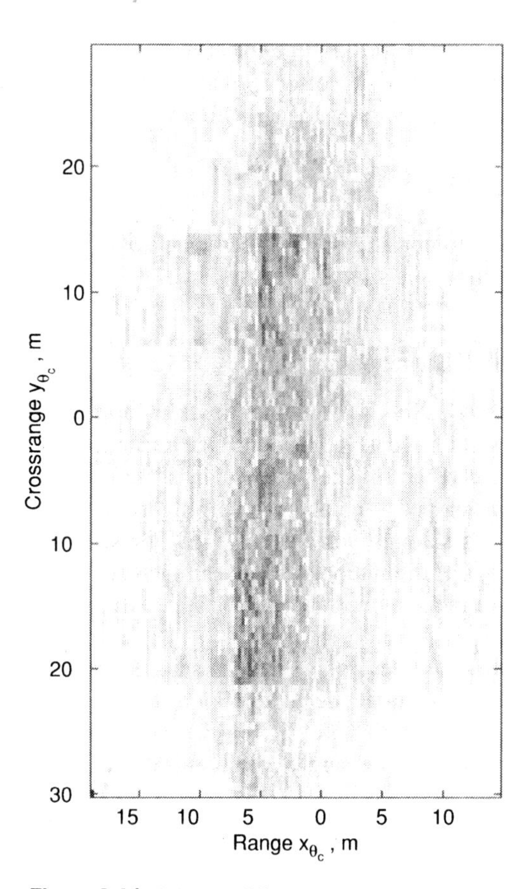

Figure 8.4d. Subspace difference image of (a) and (b).

Estimating Calibration Error Impulse Function

If one desires, the impulse response can be computed via the following procedure. Let $g_j(x_i, y_j)$, $k = 1, \ldots, N$, represent the ordered version of the signals

$$\left[f_1(x_i - m\Delta_x, y_j - n\Delta_y); m = -n_x, \ldots, n_x, n = -n_y, \ldots, n_y\right]$$

in the manner in which the orthogonalization (e.g., Gram-Schmidt) procedure is implemented; the manner in which these shifted versions of $f_1(x_i, y_j)$ are ordered does not significantly alter the outcome.

We define an N by N matrix $\mathbf{F}_1$ whose elements are the projection of $g_j(x_i, y_j)$'s into $\psi_\ell(x_i, y_j)$'s; that is,

$$f_{1j\ell} = \langle g_j, \psi_\ell \rangle = \sum_i \sum_j g_j(x_i, y_j)\psi_\ell^*(x_i, y_j).$$

In the Gram-Schmidt procedure, $\mathbf{F}_1$ is a lower triangular matrix since $f_{1j\ell} = 0$ for $k < \ell$. We also define the 1 by N vector $\mathbf{F}_2$ which is made up of the coefficients $f_{2\ell}$, $\ell = 1, \ldots, N$. Then it can be shown that the following 1 by N vector

$$\mathbf{H} = \mathbf{F}_2 \mathbf{F}_1^{-1}$$

contains the estimated impulse response coefficients $\hat{h}_{mn}$. [The ordering is the same as the ordering of $g_j(x_i, y_j)$'s.]

Application in MTD Monopulse SAR

As we mentioned in Section 8.1, an along-track monopulse SAR imaging system utilizes two radars for its data collection. One radar is used as a transmitter as well as a monostatic receiver. The other radar is used only as a bistatic receiver. In Section 8.1 we developed a signal processing algorithm of the two monostatic and bistatic databases of the along-track monopulse SAR system to obtain two *coherently* identical SAR images of the stationary targets in the scene. While the stationary targets appear the same in the monostatic and bistatic SAR images, the same is not true for moving targets.

This fact is the basis for developing an MTI statistic, which we refer to as the difference image, for moving target detection. For a stationary nth target, if we denote the monostatic SAR image by $f_{mn}(x-x_n, y-y_n)$ and the bistatic image by $f_{bn}(x, y)$, the difference image for moving target detection is defined via the following:

$$f_{dn}(x - x_n, y - y_n) = f_{bn}(x - x_n, y - y_n) - f_{mn}(x - x_n, y - y_n).$$

In Section 8.2 we developed a model for the undesirable variations of the amplitude pattern of uncalibrated monopulse radars. Such a model indicates that the monopulse SAR images are related via

$$f_{bn}(x - x_n, y - y_n) = f_{mn}(x - x_n, y - y_n) ** h_n(x, y),$$

where $h_n(x, y)$ is a function of the calibration error of the two radars. This model is the same as the model in Figure 8.3 of the discussion of uncalibrated images in the previous section with

$$f_1(x, y) = f_{mn}(x - x_n, y - y_n),$$

$$f_2(x, y) = f_{bn}(x - x_n, y - y_n),$$

$$h_n(x, y) = h(x, y),$$

for a stationary nth target, when there is no moving target (foreign object), that is, $f_e(x, y) = 0$.

Thus the signal subspace method described in Section 8.3 can also be applied in the MTD monopulse SAR problem with uncalibrated radars. Note that the calibration error function for the MTD monopulse SAR and, thus, the impulse function $h_n(x, y)$ vary with the coordinates of the nth target. Thus the subspace processing cannot be applied in one step to the entire SAR scene. In this case the SAR image has to be divided into subpatches over which the error function does not vary significantly; this implies that $h_n(x, y)$ remains approximately the same in that subpatch (a condition that is met in practice). The signal subspace registration algorithm can then be applied to each subpatch.

Currently along-track monopulse SAR data are being collected with DPCAs, and the real data will be available soon. Reference [s98a] provides a study of the application of the signal subspace processing in MTD monopulse SAR for a simulated scenario.

Application in Automatic Target Recognition SAR

Next we consider another application of the signal subspace processing registration algorithm in SAR problems. This problem involves registration of two SAR images of a target which are obtained by two different radars or runs of the same radar. After a preliminary processing of the two SAR images to remove gross shifts and rotation (e.g., simple image processing registration methods), the user still faces subtle phase and gain variations in the two *complex* SAR images.

These subtle variations are due to the changes of the point spread function for the two SAR systems, and incremental changes in the relative orientation of the target with respect to the radar (small shift, scaling, and rotation). These physical factors are outlined in the following:

1. The flight path and altitude of the radar-carrying aircraft vary slightly in the two runs. This causes a relative scaling and rotation of targets in the two SAR images; see Chapter 3 and [s94, sec. 4.6].
2. The two radars exhibit calibration errors. This was discussed and modeled in Section 8.2.
3. The two SAR data contain different residual motion errors (even after motion compensation). This results in a fast-time frequency and slow-time (aspect angle) dependent phase error that can be incorporated in the model developed in Section 8.2.

Consider a target function in the spatial domain $f_1(x, y)$. A rotational and scaled version of this function can be expressed via a linear transformation of the (x, y) domain [s94, ch. 2]; that is,

$$f_1(x + c_1 x + c_2 y, y + c_3 x + c_4 y),$$

where c_i's are related to the scale and rotation parameters. If the amount of rotation and scaling is small (i.e., the residual error after, e.g., the moment method is used to estimate the gross rotation and scaling [jai]), then the transformed function in the above model can be approximated via the truncated Taylor series expansion of

$$f_1(x + c_1x + c_2y, y + c_3x + c_4y) \approx f_1(x, y) + c_1xf_{1x}(x, y) + c_2yf_{1x}(x, y) + c_3xf_{1y}(x, y) + c_4yf_{1y}(x, y),$$

where

$$f_{1x}(x, y) = \frac{\partial f_1(x, y)}{\partial x},$$

$$f_{1y}(x, y) = \frac{\partial f_1(x, y)}{\partial y}.$$

A shift and blurring of the function in this approximated model can be identified via convolution with a two-dimensional impulse response $h(x, y)$. However, a more accurate model can be constructed by incorporating the higher derivatives of f_1 in the truncated Taylor series expansion of $f_1(x + c_1x + c_2y, y + c_3x + c_4y)$ and approximating their fluctuations to be relatively small.

Now consider the problem of registering the two SAR images $f_1(x, y)$ and $f_2(x, y)$ where one encounters calibration errors, small rotation, shift, and scaling of the target. Using the above Taylor series approximation, the two SAR images can be related by

$$\begin{aligned} f_2(x, y) = {} & f_1(x, y) ** h_0(x, y) + [xf_{1x}(x, y)] ** h_1(x, y) \\ & + [yf_{1x}(x, y)] ** h_2(x, y) + [xf_{1y}(x, y)] ** h_3(x, y) \\ & + [yf_{1y}(x, y)] ** h_4(x, y) + f_e(x, y), \end{aligned}$$

where $h_k(x, y)$'s are unknown.

For this model the signal subspace is composed of $f_1(x, y)$, $xf_{1x}(x, y)$, $yf_{1x}(x, y)$, $xf_{1y}(x, y)$ $yf_{1y}(x, y)$, and their shifted versions at $(x, y) = (x_i, y_j)$. One must compute the discrete first partial derivatives of the image $f_1(x_i, y_j)$. The discrete model is

$$\begin{aligned} f_2(x_i, y_j) = {} & f_e(x_i, y_j) + \sum_{m=-n_x}^{n_x} \sum_{n=-n_y}^{n_y} h_{0mn} f_1(x_i - m\Delta_x, y_j - n\Delta_y) \\ & + (x_i - m\Delta_x)h_{1mn} f_{1x}(x_i - m\Delta_x, y_j - n\Delta_y) \\ & + (y_j - n\Delta_y)h_{2mn} f_{1x}(x_i - m\Delta_x, y_j - n\Delta_y) \\ & + (x_i - m\Delta_x)h_{3mn} f_{1y}(x_i - m\Delta_x, y_j - n\Delta_y) \\ & + (y_j - n\Delta_y)h_{4mn} f_{1y}(x_i - m\Delta_x, y_j - n\Delta_y). \end{aligned}$$

The size of the resultant signal subspace is $5N$, where $N = N_xN_y$ is the size of the discrete filter h_{kmn}. Examples of applying this approach in ATR problems of SAR are provided in [s98a].

8.4 SLANT PLANE TOPOGRAPHIC MAPPER MONOPULSE SAR

As we stated in the introduction to this chapter, the potential of monopulse radars for *elevation mapping* or *topographic mapping* has been recognized for many years. What is referred to as a *slant plane* monopulse SAR, which was based on the *interferometric* processing of monostatic and bistatic SAR data (this will be discussed later), was developed in the early 1970s for topographic mapping [gra].

The slant plane monopulse SAR system has various commercial and environmental monitoring applications. Topographic data obtained by a slant plane monopulse SAR have been used to correct for geometric distortions in SAR images obtained in future flights [cur]. This monopulse SAR system has also been investigated for detecting changes on the surface of the earth, for example, to monitor volcanoes, faults that cause earthquakes, and glaciers.

The slant plane monopulse SAR measurement for topographic mapping are made by two radars that possess a common azimuth (along-track) coordinates but are separated in the slant range domain; see Figure 8.5. (The along-track y domain is reversed in this image for pictorial clarity.) We will show that the coherent processing of the resultant monostatic and bistatic SAR images carries information on the variations of elevation in the imaging scene.

We should emphasize that slant plane monopulse SAR is *not* a three-dimensional SAR imaging modality. A critical assumption (or *a priori* information) which is

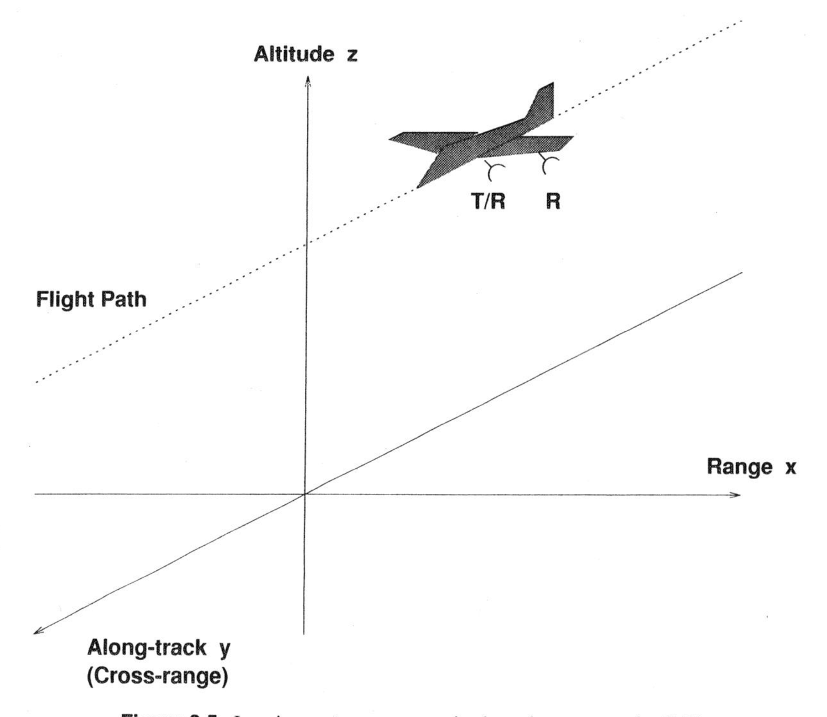

Figure 8.5. Imaging system geometry in slant plane monopulse SAR.

used in this SAR system is that the target area is a scattering *surface* in the three-dimensional spatial domain; this is analogous to modeling the imaging terrain as a *delta* (thin) sheet target function. There are also practical issues in slant plane monopulse SAR, such as foreshortening, layover, and shadow effects, which are discussed in [cur].

In this section we provide a signal theory-based analysis of slant plane monopulse SAR. This includes the processing of the monostatic and bistatic SAR images based on the narrow-bandwidth and narrow-beamwidth assumptions that lead to what is called *interferometric SAR* (IF-SAR), and a signal subspace processing of wide-bandwidth and wide-beamwidth slant plane monopulse SAR data.

Slant Plane Monopulse SAR System Geometry

The slant plane monopulse SAR imaging system geometry in the three-dimensional spatial domain is shown in Figure 8.5. The principles that we will develop for this monopulse SAR system are applicable for any positioning of the receiving radar with respect to the transmitting/receiving radar in the slant plane domain. However, to simplify our notation, we assume that both radars have common range and cross-range values with respect to the center of mass of the target area; we denote these common coordinates for the radars in the range and cross-range domain by $(x, y) = (X_c, Y_c)$.

The coordinate of the transmitting/receiving radar (Radar 1) in the z domain is Z_c; the coordinate of the receiving radar (Radar 2) in the z domain is $Z_c - 2\Delta$. These are shown in the range and altitude (x, z) domain of this SAR system in Figure 8.6. The position of the radar-carrying aircraft on the synthetic aperture is denoted by

$$u = v_r \tau,$$

where v_r is the radar-carrying aircraft speed.

Monostatic and Bistatic SAR Signal Models

Consider a target located at (x_n, y_n, z_n). For a given fast-time frequency ω of the radar signal and synthetic aperture position u of the aircraft, the round-trip phase delay of the echoed signal recorded by Radar 1 (monostatic) and Radar 2 (bistatic) from a reflector at (x_n, y_n, z_n), respectively, are

$$\psi_{\mathrm{m}n}(\omega, u) = 2k\sqrt{(X_c + x_n)^2 + (Y_c + y_n - u)^2 + (Z_c + z_n)^2},$$

$$\psi_{\mathrm{b}n}(\omega, u) = k\sqrt{(X_c + x_n)^2 + (Y_c + y_n - u)^2 + (Z_c + z_n)^2},$$

$$+ k\sqrt{(X_c + x_n)^2 + (Y_c + y_n)^2 + (Z_c + z_n - 2\Delta)^2},$$

where $k = \omega/c$ is the wavenumber at the fast-time frequency ω.

Suppose that the nth target is a unit reflector and that the two monopulse radars are omni-directional. Thus the monostatic and bistatic SAR signatures of the nth target

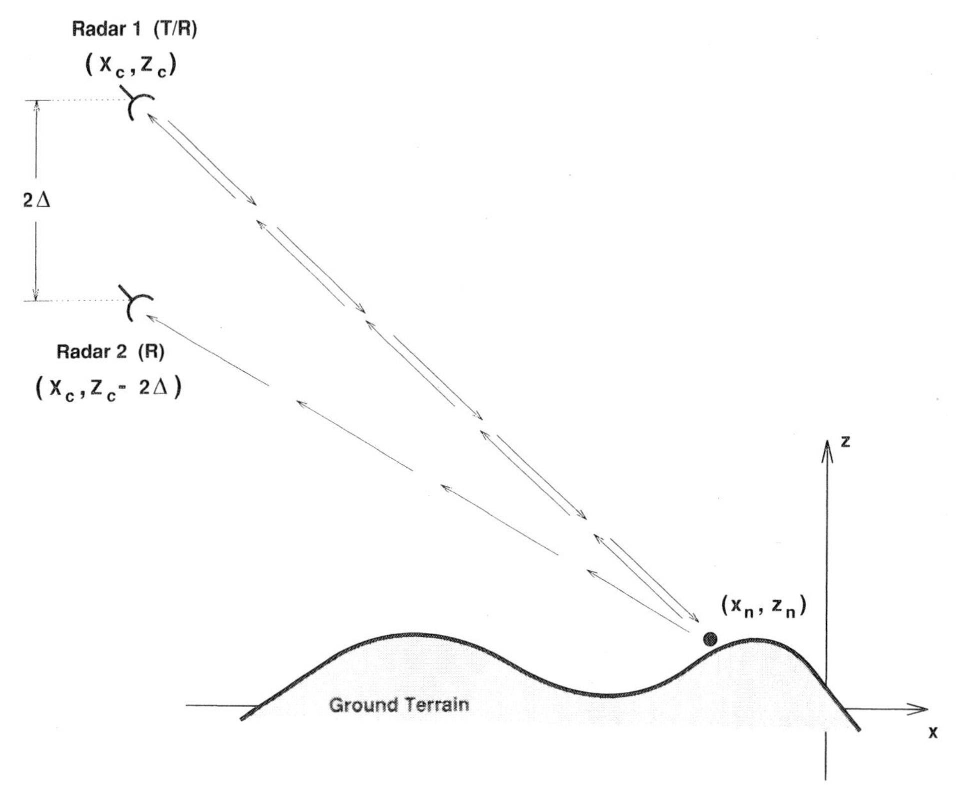

Figure 8.6. Imaging system geometry in slant plane monopulse SAR in two-dimensional range and altitude (x, z) domain.

in the (ω, u), respectively, are

$$s_{mn}(\omega, u) = \exp\left[\psi_{mn}(\omega, u)\right],$$
$$s_{bn}(\omega, u) = \exp\left[\psi_{bn}(\omega, u)\right].$$

Narrow-Bandwidth and Narrow-Beamwidth Approximation: Interferometric SAR (IF-SAR)

Next we use the radar narrow-bandwidth and narrow-beamwidth approximations in the slant plane monopulse SAR signals to develop a procedure to determine variations in the elevation domain for the radiated target area, that is, the elevation mapping. Provided that

$$\Delta \ll R_c = \sqrt{X_c^2 + Y_c^2 + Z_c^2},$$

and under the narrow-beamwidth assumption, that is,

$$|u| \ll R_c,$$

the round-trip phase delay recorded by Radar 2 can be approximated by the following:

$$\psi_{bn}(\omega, u) \approx \psi_{mn}(\omega, u) - \frac{2k\Delta(Z_c - \Delta)}{R_c} - \underbrace{\frac{2k\Delta z_n}{R_c}}_{\text{Target elevation-dependent}}.$$

If the radar signal has a narrow-bandwidth, that is, $|k - k_c| \ll k_c$, the phase term that varies with the target elevation in the above can be approximated by

$$\psi_{bn}(\omega, u) \approx \psi_{mn}(\omega, u) - \frac{2k\Delta(Z_c - \Delta)}{R_c} - \underbrace{\frac{2k_c\Delta z_n}{R_c}}_{\text{Interferometric term}}.$$

The component labeled as the *interferometric term*, that is,

$$\frac{2k_c\Delta z_n}{R_c},$$

is invariant in the (ω, u) domain; this phase term carries information on the altitude of the target z_n. By compensating for the other residual phase term $2k\Delta(Z_c - \Delta)/R_c$, which is known and invariant in the parameters of the target, the only difference between the monostatic and bistatic SAR signals is the interferometric phase term.

Let $f_n(x - x_n, y - y_n, z - z_n)$ be the reflectivity function of the nth target. The SAR reconstruction from the monostatic data recorded by Radar 1 yields the following slant plane target reflectivity function:

$$f_{mn}(x_s - x_{msn}, y - y_n) = f_n(x - x_n, y - y_n, z - z_n),$$

where

$$x_s = \sqrt{(X_c + x)^2 + (Z_c + z)^2} - X_{sc}$$

is the slant-range domain,

$$x_{msn} = \sqrt{(X_c + x_n)^2 + (Z_c + z_n)^2} - X_{sc}$$

is the monostatic slant-range of the nth target, and

$$X_{sc} = \sqrt{X_c^2 + Z_c^2}$$

is the mean slant-range of the target area. After removing the residual phase term $2k\Delta(Z_c - \Delta)/R_c$ from the bistatic data which are recorded by Radar 2, the resultant slant plane SAR reconstruction from this database is

$$f_{bn}(x_s - x_{msn}, y - y_n) = f_n(x - x_n, y - y_n, z - z_n) \exp\left(-j\frac{2k_c \Delta z_n}{R_c}\right).$$

Interferometric processing involves constructing the following target function:

$$\begin{aligned} f_{in}(x_s - x_{msn}, y - y_n) &= f_{mn}(x_s - x_{msn}, y - y_n) f^*_{bn}(x_s - x_{msn}, y - y_n) \\ &= |f_{mn}(x_s - x_{msn}, y - y_n)|^2 \exp\left(j\frac{2k_c \Delta z_n}{R_c}\right), \end{aligned}$$

where * denotes complex conjugation. Thus the interferometric target function contains information on the height variations of the nth target from the $z = 0$ plane.

The formation of the two slant plane monopulse (monostatic and bistatic) SAR images and subsequent interferometric processing are invariant in the coordinates of the nth target. Thus the user could form the two monopulse SAR images for the entire target area, that is, $f_m(x_s, y)$ and $f_b(x_s, y)$. Then the interferometric image of the entire target area is obtained by

$$f_i(x_s, y) = f_m(x_s, y) f^*_b(x_s, y);$$

this is known as the *interferometric* SAR (IF-SAR) image. Provided that there are no overlapping scatterers [cur], the *phase* of the interferometric image contains information on the height (altitude) of individual reflectors in the imaging scene.

- For the success of IF-SAR, the two monopulse radars must be fully calibrated. The blind deconvolution solution of the signal subspace processing of uncalibrated SAR images (see Section 8.3) cannot be used in IF-SAR. This will be discussed later.

An important practical issue associated with deducing the target area height function via the IF-SAR image is the *unwrapping* of the interferometric phase [cur]. The problem here is that the dynamic range of the interferometric phase of the nth target,

$$\frac{2k_c \Delta z_n}{R_c},$$

is not limited to the phase interval $[0, 2\pi)$. However, the phase of the complex interferometric image at $(x_s, y) = (x_{sn}, y_n)$ can be determined only within an integer multiple of 2π. This integer multiple of 2π is unknown to the user and varies in the (x_s, y) domain.

The problem of phase unwrapping appears in many digital signal processing problems. The *two-dimensional* phase unwrapping problem which is encountered in IF-SAR is a difficult one. Various methods have been suggested to solve this problem via exploiting the fact that the height variations of the terrain and thus the IF-SAR phase function, are relatively slowly fluctuating signals [cur; zeb]. In this case the IF-SAR phase unwrapping can be accomplished by adding appropriate integer multiples of 2π to remove rapid discontinuities in the wrapped IF-SAR phase.

Unfortunately, this scheme is susceptible to various sources of noise in a SAR system (even in the absence of calibration errors of the two monopulse radars) [cur; zeb]. Another source of difficulty, which has not been considered in the literature, is the narrow-beamwidth assumption. In the next section we will re-examine the slant plane monopulse SAR under less restrictive assumptions on the radar signal and radiation pattern.

Wide-Bandwidth and Wide-Beamwidth Model

With $\Delta \ll R_c$, one can show that the bistatic measurement of a monopulse SAR system can be converted into the monostatic measurement of a transmitting/receiving radar which is located at the midpoint of the line that connects to Radar 1 and Radar 2; this is true for any monopulse SAR modality (along-track or slant plane), and a wide-bandwidth and wide-beamwidth radar signal. (There are some residual phase errors that are approximately equal to $2k\Delta^2/R_c$ and can be removed from the bistatic SAR image via a spatially varying filter; see the discussion on the effect of wide-beamwidth radiation pattern in Section 8.2).

To give this a mathematical representation, we define the bistatic slant-range of the nth target by

$$x_{\text{bs}n} = \sqrt{(X_c + x_n)^2 + (Z_c + z_n - \Delta)^2} - X_{sc};$$

this is the slant-range of the nth target when it is viewed by a monostatic radar which is located at $(X_c, Y_c, Z_c - \Delta)$, the midpoint of the line that connects the two monopulse radars. Earlier, we also defined the monostatic slant-range of the nth target to be

$$x_{\text{ms}n} = \sqrt{(X_c + x_n)^2 + (Z_c + z_n)^2} - X_{sc}.$$

Note that the relative shift between the two monopulse slant-range values of the nth target is (with $\Delta \ll R_c$)

$$x_{\text{ms}n} - x_{\text{bs}n} \approx \frac{\Delta(Z_c + z_n)}{R_c}.$$

Thus the monostatic and bistatic round-trip phase delay functions can be rewritten as follows:

$$\psi_{\text{m}n}(\omega, u) = 2k\sqrt{(X_{sc} + x_{\text{ms}n})^2 + (Y_c + y_n - u)^2},$$

$$\psi_{\text{b}n}(\omega, u) = 2k\sqrt{(X_{sc} + x_{\text{bs}n})^2 + (Y_c + y_n - u)^2}.$$

Both of the two monopulse SAR databases could be viewed as the monostatic SAR data of the nth target at two different slant-range points. After performing the SAR reconstruction of Section 6.2 on these two monopulse SAR databases, the spectral, that is, the (k_x, k_y) domain, phase of the reconstructed nth target becomes (see Section 6.2)

$$\exp\left(jk_x x_{\mathrm{ms}n} + jk_y y_n\right) \quad \text{for monostatic data}$$

and

$$\exp\left(jk_x x_{\mathrm{bs}n} + jk_y y_n\right) \quad \text{for bistatic data,}$$

where k_u is the spatial frequency domain for u, and

$$k_x = \sqrt{4k^2 - k_u^2},$$
$$k_y = k_u$$

is the SAR spectral mapping.

- Under the *narrow-beamwidth* assumption, the difference of the above two monopulse spectral phase functions is approximated by

$$\begin{aligned}\exp\left(jk_x x_{\mathrm{ms}n} - jk_x x_{\mathrm{bs}n}\right) &= \exp\left(j\sqrt{4k^2 - k_u^2}\,x_{\mathrm{ms}n} - j\sqrt{4k^2 - k_u^2}\,x_{\mathrm{bs}n}\right)\\ &\approx \exp\left(j2kx_{\mathrm{ms}n} - j2kx_{\mathrm{bs}n}\right)\\ &\approx \exp\left[\frac{2k\Delta(Z_c + z_n)}{R_c}\right].\end{aligned}$$

Under the *narrow-bandwidth* assumption, $|k - k_c| \ll k_c$, this phase difference can be further approximated by

$$\exp\left(jk_x x_{\mathrm{ms}n} - jk_x x_{\mathrm{bs}n}\right) \approx \underbrace{\exp\left(\frac{2k\Delta Z_c}{R_c}\right)}_{\text{Residual phase}}\ \underbrace{\exp\left(\frac{2k_c\Delta z_n}{R_c}\right)}_{\text{Interferometric phase}}.$$

The above approximation was shown to be the basis of the IF-SAR processing to determine height variations on the terrain.

The narrow-beamwidth assumption is valid when the target signature is concentrated around the zero slow-time Doppler $k_u = 0$; for example, when $Y_c = 0$, this corresponds to a strong "flash" (target amplitude pattern) when the target is at the boresight (broadside) of the radar. If the target is *slanted*, for example, a mountain area with an oblique angle, then the above narrow-beamwidth assumption is not valid.

Let us continue with the monostatic and bistatic spectral phase models without the narrow-bandwidth and narrow-beamwidth assumptions. These phase functions indicate that the nth target appears at the following slant-range coordinates in the two monopulse SAR images:

$$x_s = x_{msn} \quad \text{in monostatic SAR image,}$$

and

$$x_s = x_{bsn} \quad \text{in bistatic SAR image.}$$

As we stated earlier, the difference between the monostatic and bistatic slant-range values, that is,

$$x_{msn} - x_{bsn} \approx \frac{\Delta(Z_c + z_n)}{R_c},$$

carries information on the altitude of the nth target (the value of z_n). Thus, by determining the relative *shift* in the slant-range domain of the two monopulse images, one can determine the height z_n of the nth reflector.

Unfortunately, this slant-range shift is typically smaller than the slant-range resolution of the SAR images. Determining subpixel shift in dual signals is also a classical problem in digital signal processing, for example, speech processing and radar signal processing. A typical approach involves transferring the (digitally spotlighted) monopulse images, $f_{mn}(x_s - x_{msn}, y - y_n)$ and $f_{bn}(x_s - x_{bsn}, y - y_n)$, into the spectral (k_x, k_y) domain. Then a linear phase function is fitted to the interferometric image of the two spectra. This procedure also has its own phase unwrapping difficulties.

Estimating Slant-range Shift via Signal Subspace Processing

If the two slant plane monopulse SAR images are shifted versions of each other in the slant-range domain, then we can relate the two via

$$f_{bn}(x_s - x_{bsn}, y - y_n) = f_{mn}(x_s - x_{msn}, y - y_n) * h_n(x_s),$$

where $*$ denotes convolution in the slant-range domain,

$$h_n(x_s) = \delta\big(x_s - \Delta_{sn}\big),$$

and the relative shift in the slant-range domain of the two monopulse SAR images is

$$\Delta_{sn} = x_{bsn} - x_{msn}.$$

This model is equivalent to the model in Figure 8.3 of the discussion of uncalibrated images in Section 8.3 with

$$f_1(x, y) = f_{mn}(x_s - x_{bsn}, y - y_n),$$
$$f_2(x, y) = f_{bn}(x_s - x_{bsn}, y - y_n),$$
$$h_n(x_s) = h(x, y),$$

with $f_e(x, y) = 0$, and $x = x_s$. Note that $h(x, y)$ is invariant in y, and thus can be represented as $h(x)$.

As we stated in Section 8.3, the signal subspace algorithm could also be used to determine the unknown impulse response $h(x, y)$. For this purpose the discrete model is

$$f_2(x_i, y_j) = \sum_{m=-n_x}^{n_x} h_m f_1(x_i - m\Delta_{x_s}, y_j),$$

where h_m's are the unknown coefficients which represent the samples of $h(x)$ and Δ_{x_s} is the pixel sample spacing of the two monopulse SAR images in the slant-range domain.

Let $\hat{h}_m$'s be the estimate of the coefficients of the model that are obtained using the method in Section 8.3. Then the relative shift in the slant-range domain can be extracted from the moments of the estimated coefficients; that is,

$$\hat{\Delta}_{sn} = \frac{1}{E_{\hat{h}}} \sum_{m=-n_x}^{n_x} m\Delta_{x_s} \hat{h}_m^2,$$

with

$$E_{\hat{h}} = \sum_{m=-n_x}^{n_x} \hat{h}_m^2,$$

where $\hat{\Delta}_{sn}$ is the estimated slant-range shift for the nth target.

The pixel size in the slant-range domain Δ_{x_s} is typically chosen to be on the order of the slant-range resolution, and as we stated earlier, the relative slant-range shift Δ_{sn} is less than the slant-range resolution. Thus the choice $n_x = 2$, that is, a filter size of $2n_x + 1 = 5$, is reasonable. The size of the digitally spotlighted area around the nth target monopulse SAR images could be chosen to be equal to a few resolution pixels in the (x_s, y) domain.

Finally we should emphasize that both IF-SAR processing and signal subspace processing for determining height variations from slant plane monopulse SAR data use the assumption that the two monopulse radars are *calibrated*. Unlike the along-track monopulse SAR, it is not feasible to use the signal subspace processing method for addressing the calibration errors of the two radars. This is due to the fact that both the calibration error and height information fall into the signal subspace of the shifted versions of $f_1(x, y)$ [or $f_{mn}(x_s - x_{bsn}, y - y_n)$], that is, Ψ.

8.5 MULTISTATIC MONOPULSE ISAR

The slant plane monopulse SAR described in the previous section also has applications in deducing three-dimensional information regarding a moving target in inverse SAR (ISAR) problems. One example involves a slant plane monopulse ISAR system for automatic aircraft landing (AAL) [s96s]. An automatic aircraft landing (AAL) system provides data on the *orientation* of an aircraft approaching a runway. The data can be used by the pilot or an operator (e.g., air traffic controller) to decide whether the aircraft orientation is safe for landing.

Interferometric ISAR (IF-ISAR) with a single transmitter and two receivers in a slant plane can be used to detect the slight angular roll and/or pitch of a landing aircraft. Unlike the SAR image of a terrain region, the ISAR image of an aircraft does not exhibit a contiguous structure [s94, ch. 5; weh]; this was also observed in ISAR images of an aircraft in Chapter 5. The ISAR image of an aircraft is composed of strong (i.e., above the noise level) reflectors at a few corner reflectors of the tail, fuselage, and wings of the aircraft. Thus one cannot utilize the phase unwrapping methods used in IF-SAR imaging of terrain. Despite this fact IF-ISAR contains sufficient phase information to warn the operator of an AAL system of an undesirable roll or pitch of a landing aircraft [s96s].

Multistatic ISAR Model

The above-mentioned approach for AAL may also be extended for the general problem of three-dimensional ISAR imaging and tracking with a single transmitter and multiple receivers; this is called a *multistatic* ISAR system. An example of a multistatic ISAR system is shown in Figure 8.7. Such an ISAR system can be used to retrieve three-dimensional information on spatial and motion parameters of an airborne aircraft. Some of the applications of this system include:

1. Automatic aircraft landing at the civilian and military airports
2. High-resolution three-dimensional air traffic control at the civilian airports
3. Studying vibration of various structures on an airborne aircraft, which is useful for detecting structural flaws on an old aircraft, or designing safer aircrafts

Linear (Constant Velocity) Motion Model Suppose that the transmitter of the multistatic ISAR system is located at the origin in the spatial (x, y, z) domain. Consider a moving reflector whose coordinates at the slow-time $\tau = 0$ is (x_n, y_n, z_n). When the target possesses a constant velocity, for example, (v_{xn}, v_{yn}, v_{zn}), its motion trajectory as a function of the slow-time is

$$\big(x_n - v_{xn}\tau,\ y_n - v_{yn}\tau,\ z_n - v_{zn}\tau\big).$$

The ISAR phase history of this target for a given receiver in the multistatic ISAR system of Figure 8.7 is dictated by its distance from the transmitting radar and that of the receiving radar. For example, for the ℓth receiver located at (x_ℓ, y_ℓ, z_ℓ), this

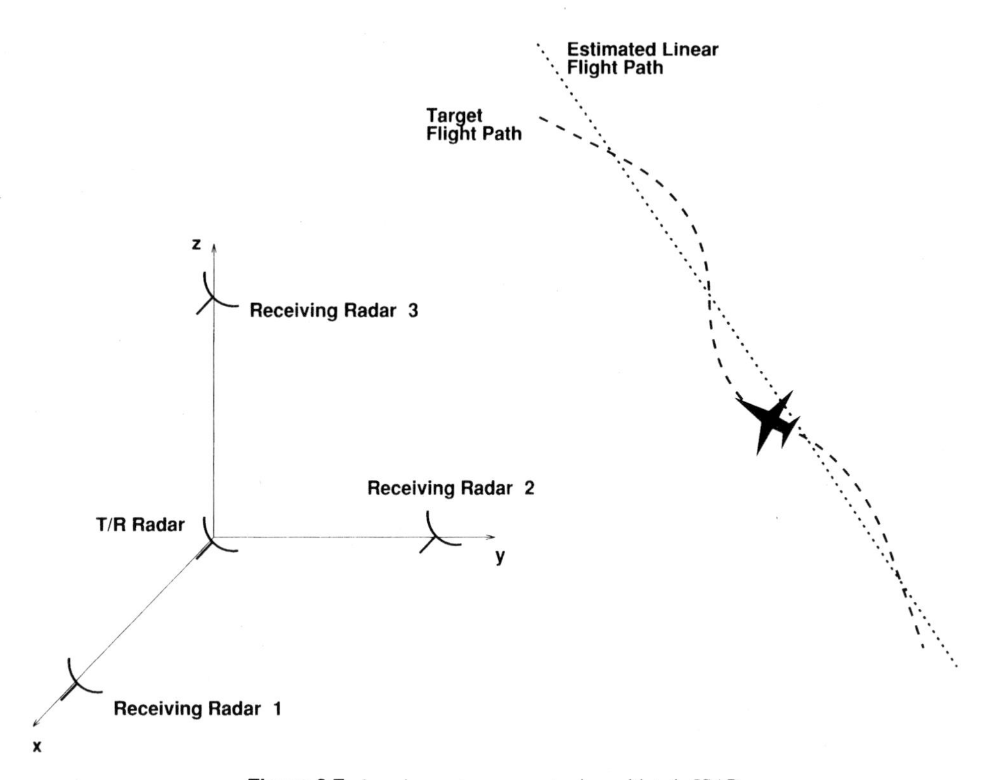

Figure 8.7. Imaging system geometry in multistatic ISAR.

round-trip distance is

$$\sqrt{(x_n - v_{xn}\tau)^2 + (y_n - v_{yn}\tau)^2 + (z_n - v_{zn}\tau)^2} + \sqrt{(x_n - v_{xn}\tau - x_\ell)^2 + (y_n - v_{yn}\tau - y_\ell)^2 + (z_n - v_{zn}\tau - z_\ell)^2}.$$

It can be shown that, in theory, the user needs three receivers to determine the motion parameters of this reflector; for example, this can be achieved by the monostatic ISAR measurement and two bistatic ISAR measurements. The procedure for this involves first estimating the target speed

$$\sqrt{v_{xn}^2 + v_{yn}^2 + v_{zn}^2},$$

from the monostatic ISAR data [carr; s94, ch. 5]. Then, this estimated speed is used to form the monostatic and bistatic ISAR images of the target. (The bistatic ISAR images are formed via first synthesizing monostatic ISAR data for an imaginary transmitting/receiving radar located at the midpoint of the line that connects the transmitting radar and the ℓth receiving radar; see Section 8.4.)

After digital spotlighting the nth target image from its monostatic and bistatic ISAR reconstructions, the target phase history in the (ω, τ) domain can be formed for each image; this can be used to estimate the target parameters. One may also form various IF-ISAR image pairs from these reconstructions which can be translated into the target parameters (under the narrow-bandwidth and narrow-beamwidth assumption); this scheme, however, requires phase unwrapping which is not feasible for an *isolated* reflector (not a contiguous region like a terrain).

- In the case of resonating cavities and corner structures, the reflectors on the moving structure (aircraft) cannot be assumed to possess omni-directional amplitude patterns. In this case each reflector possesses a gain and phase history, which we identified as $a_n(\omega, x_n, y_n - u)$ in our discussion on SAR (see Chapter 3). This phase history varies with the radar frequency and aspect angle, so it corrupts the motion phase history of the reflector. However, provided that the receiving radars are relatively close to the transmitting radar, the target amplitude pattern is the same for the monostatic and bistatic ISAR measurements. Thus the *relative phase difference* of the monostatic and bistatic ISAR images of the reflector still correspond to the reflector motion phase history. In other words, the target amplitude pattern phase history is transparent in *pairwise* processing of multistatic ISAR images; the same is not true in processing the monostatic or a single bistatic ISAR image.

Nonlinear Motion and Maneuvering Target Model In a realistic ISAR problem a moving structure possesses nonlinear motion components in its flight path. Moreover an airborne target, such as an aircraft, may contain maneuvers in its motion. In this case the individual reflectors do not have *parallel* motion trajectories. For example, in the case of the ISAR data examined in Chapter 5, the reconstructions of the aircraft from its lower and upper ISAR data indicated a rotation in its motion.

In such scenarios the motion trajectory of a reflector on the moving structure can be modeled via

$$\left[x_n - x_{en}(\tau), y_n - y_{en}(\tau), z_n - z_{en}(\tau)\right].$$

As we stated in Chapter 4, a practical approach to imaging such a structure is to treat the deviations from the *ideal* (linear) ISAR model as SAR motion errors. In this case the user first estimates a constant speed for the entire structure and forms its monostatic and bistatic ISAR images. Then, after digital spotlighting the prominent reflectors on the target, the monostatic and bistatic phase history of each prominent reflector can be formed and analyzed for motion estimation.

Motion Tracking via Signal Subspace Processing

In Section 8.4, we showed that the signatures of a stationary reflector in the two slant plane monopulse SAR images appear shifted with respect to each other in the slant-range domain; the amount of shift is related to the relative height of the reflector in the slant plane. We used the signal subspace processing for estimating this shift in the slant-range domain.

A similar mathematical formulation can be used to show that for the multistatic ISAR system of Figure 8.7, the signature of a reflector on the moving target in the ℓth bistatic ISAR image appears shifted with respect to its signature in the monostatic ISAR image. The shift is in both the slant-range and cross-range domains*, and it is related to the receiving radar coordinates (x_ℓ, y_ℓ, z_ℓ) and the coordinates of the reflector at the slow-time zero, (x_n, y_n, z_n). By combining these shifts, the user could track the motion of each reflector on the moving structure (aircraft).

Next we address the problem of estimating the relative shift in two images. Our approach is based on an application of the signal subspace processing of Section 8.3; a similar approach was used in estimating the relative slant-range shift in the two slant plane monopulse SAR images (Section 8.4). For this, consider the system model in Figure 8.3. Suppose that the impulse function in this model corresponds to a shift operation, for example,

$$h(x, y) = \delta\big(x - \Delta_{sx}, y - \Delta_{sy}\big),$$

where $(\Delta_{sx}, \Delta_{sy})$ are unknown and the error function $f_e(x, y) = 0$; in this model the variables (x, y) represent the slant-range and cross-range domains. The output of the two sensors is related via

$$f_2(x, y) = f_1\big(x - \Delta_{sx}, y - \Delta_{sy}\big).$$

Our objective is to estimate the shift pair $(\Delta_{sx}, \Delta_{sy})$.

Based on the solution in Section 8.3, we first identify the system in Figure 8.3 by the following discrete model:

$$f_2(x_i, y_j) = \sum_{m=-n_x}^{n_x} \sum_{n=-n_y}^{n_y} h_{mn} f_1(x_i - m\Delta_x, y_j - n\Delta_y),$$

where h_{mn}'s are the unknown coefficients that represent the samples of $h(x, y)$, and (Δ_x, Δ_y) are the pixel sample spacing in the (x, y) domain; that is, $\Delta_x = x_i - xi - 1$ and $\Delta_y = y_j - y_{j-1}$.

Let $\hat{h}_{mn}$'s be the estimate of h_{mn}'s for the above discrete model (see Section 8.3 for this solution). In this case the relative shifts in the (x, y) domain can be determined from the moments of the estimated coefficients as shown in the following:

$$\hat{\Delta}_{sx} = \frac{1}{E_{\hat{h}}} \sum_{m=-n_x}^{n_x} \sum_{n=-n_y}^{n_y} m\Delta_x \hat{h}^2_{mn}, \qquad \hat{\Delta}_{sy} = \frac{1}{E_{\hat{h}}} \sum_{m=-n_x}^{n_x} \sum_{n=-n_y}^{n_y} n\Delta_y \hat{h}^2_{mn},$$

*The slant-range and cross-range domains for a three-dimensional ISAR geometry can be constructed in the same manner as we identified the *motion-transformed* spatial domain for a moving target in a SAR scene; see Sections 6.7 and 8.1.

with

$$E_{\hat{h}} = \sum_{m=-n_x}^{n_x} \sum_{n=-n_y}^{n_y} \hat{h}_{mn}^2,$$

where $(\hat{\Delta}_{sx}, \hat{\Delta}_{sy})$ are the estimated shifts in the (x, y) domain.

Example As we mentioned earlier, monopulse SAR and ISAR systems have recently been built, and their databases are becoming available for investigators. To show the application of the above-mentioned signal subspace processing for motion tracking, we consider the X band ISAR data of an airborne aircraft which was discussed in Chapter 5 in conjunction with spotlight SAR systems. We mentioned in that discussion that this aircraft possesses a rolling motion; this was demonstrated via the aircraft ISAR reconstructions from the lower synthetic aperture data $u \in [-L, 0]$ (see Figure 5.17*b*), and the upper synthetic aperture data $u \in [0, L]$ (see Figure 5.17*c*). Those images indicated that the dominant reflectors on the wing of the aircraft exhibited an inward motion toward the fuselage.

We consider the two ISAR reconstructions of the aircraft that are formed via pulse numbers (PRIs) 81–144 and 89–152 (Figures P8.7 and P8.8); the PRF for this ISAR system is 200 Hz. These two sets of PRIs correspond to two overlapping 320-ms synthetic aperture databases in the slow-time domain which are separated by 40 ms. We assign the two outputs of the sensors in Figure 8.3, $f_1(x, y)$ and $f_2(x, y)$, to be the *magnitude* of these two ISAR reconstructions in the squint spatial domain.

For signal subspace processing, we use a block size of 7×15 in the spatial (x, y) domain with $(n_x, n_y) = (2, 4)$ [i.e., the filter size is $(2n_x + 1, 2n_y + 1) = (5, 9)$]. Moreover the algorithm is applied to a block-neighborhood around every pixel in the reconstructed image, a process that we refer to as *overlapping block-based* signal subspace processing.

The reason for using the ISAR reconstruction in the squint spatial domain is that the point spread function of a reflector in this domain appears to be a *straight* cross-shaped structure (see Chapter 5). As a result we can identify the region of support of a reflector in the ISAR image (i.e., a block that is used for signal subspace processing) as a *rectangle*.

We emphasized throughout this book that the complex (coherent) SAR or ISAR images should be used for target detection, identification, multistatic tracking, and so on. So, why did we use the magnitude of the ISAR image in this experiment? The overlapping slow-time data of the two sets of PRIs for this experiment (i.e., pulse numbers 89–144) correspond to the same ISAR data; there is no relative shift that can be exploited for motion tracking in the same manner as in a multistatic ISAR system. The coherent data in the nonoverlapping PRIs (i.e., pulse numbers 81–88 and 145–152) yield the spatial frequency information on a given reflector in two *disjoint* spectral (k_x, k_y) bands (see Chapter 5); thus these cannot be related by the model in Figure 8.3.

In this case the coherent signal subspace processing of the two ISAR images yields erroneous information. It is possible to perform SAR image compression on these two ISAR images and to apply a spatial frequency domain lowpass (e.g., a rectangular-shaped) filter to force the spectral signatures of a reflector in the two compressed ISAR images to possess a common spatial frequency band. Then one can apply the signal subspace processing on the resultant complex images. This, however, brings up other issues that are beyond the scope of this closing section of this book; we leave this for the reader to pursue.

The magnitude ISAR images provides us with a simple and visible shift in the reconstructions that can be used to demonstrate the utility of the signal subspace processing in motion tracking. Figure 8.8 shows the resultant estimated motion flow for the dominant reflectors on the aircraft; the ISAR image of the aircraft is superimposed on this motion image (Figure P8.9). Note that the estimated motion vectors in Figure 8.8 indicate a slight inward movement of the dominant reflectors on the aircraft wings.

Next we apply the overlapping block-based signal subspace processing to the two ISAR images formed from pulse numbers (PRIs) 89–152 and 105–168. These two

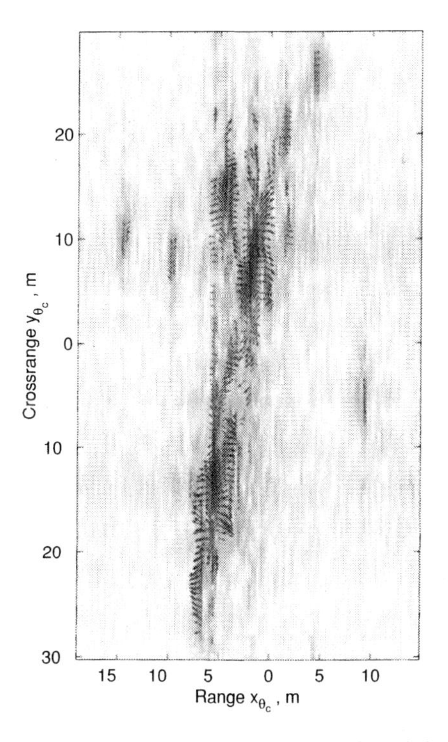

Figure 8.8. Estimated motion flow for dominant reflectors of aircraft; PRIs 81–144 and 89–152 are used to form, respectively, the reference and test ISAR images.

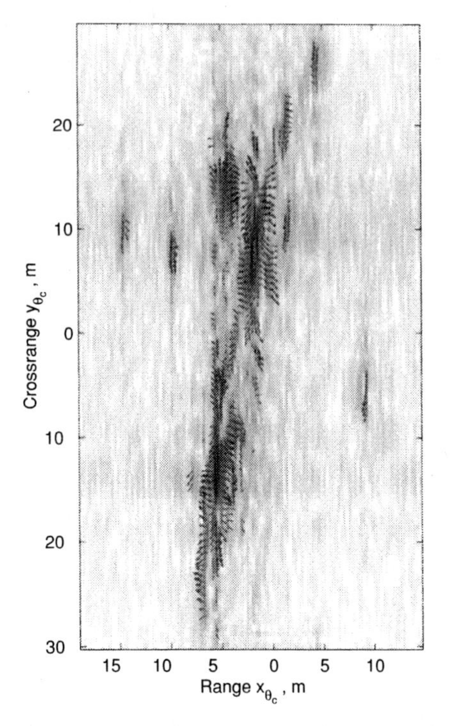

Figure 8.9. Estimated motion flow for dominant reflectors of aircraft; PRIs 89–152 and 105–168 are used to form, respectively, the reference and test ISAR images.

overlapping 320-ms slow-time intervals are separated by 80 ms. Figure 8.9 shows the estimated motion vector for this scenario. Note that the inward motion of the dominant reflectors on the aircraft wings are more prominent in this image. Both Figures 8.8 and 8.9 exhibit some form of flow on the fuselage of the aircraft. The nature of these flow patterns is not clear; some have suggested that these patterns are due to multiple scattering or surface waves.

These and many other interesting synthetic aperture imaging problems are subjects that will be encountered by investigators in the new and advanced SAR, ISAR, and sonar systems. Each of these problems, which requires its own specialized processing and analysis, could benefit from the more basic SAR and ISAR concepts that were presented in this book. Examples of these synthetic aperture-based imaging systems include surveillance with a scanning ISAR [s97m], airborne synthetic aperture acoustic imaging [s97n], and reconnaissance with bistatic SAR imaging from *copied* wide-bandwidth continuous-wave sources [s98j].

However, at some point, this book has to end. As was stated in the classic circuits book by Hayt and Kemmerly, where most of the basic intuition of a radar signal processor comes from, "*. . . we must stop sometime, so let it be now.*"

8.6 MATLAB ALGORITHMS

Two Matlab codes are provided for this chapter. Both codes deal with applications of the signal subspace processing algorithm. The first Matlab code, *block-based signal subspace processing for detecting change or target registration*, is related to the registration issues for MTD-SAR and ATR-SAR (see Section 8.3). The code simulates a complex reference image $f_1(x, y)$, a complex test image $f_2(x, y)$, a complex change image $f_e(x, y)$, and a complex calibration error impulse function $h(x, y)$. It then performs signal subspace processing to register the reference and test images.

The user could remove the simulation part and input his own reference and test images for signal subspace processing. The block size and filter size should also be selected according to the calibration error properties. (The user should know these a priori.)

The following processed signals (outputs) are shown for this code:

P8.1 Reference image $f_1(x, y)$

P8.2 Test image $f_2(x, y)$

P8.3 Change image $f_e(x, y)$

P8.4 Difference image $f_d(x, y)$

P8.5 Projection of test image into the signal subspace $\hat{f}_2(x, y)$

P8.6 Subspace difference image $\hat{f}_d(x, y)$

The second Matlab code, *overlapping block-based signal subspace processing for motion tracking*, is for the tracking and motion flow estimation in ISAR; see Section 8.5. For this code a simulated scenario is also provided in which three targets move with varying motion vectors; the magnitudes of the targets are varied in the reference and test images. The user could remove the simulation, and input the reference and test images. Depending on the type of application and its parameters, the block size and filter size for this code should be chosen accordingly.

The following processed signals (outputs) are shown for this code:

P8.7 Reference image $f_1(x, y)$

P8.8 Test image $f_2(x, y)$

P8.9 Estimated motion vectors in the spatial domain superimposed on the reference image

BIBLIOGRAPHY

[aus] D. Ausherman, A. Kozma, J. Walker, H. Jones, and E. Poggio, "Developments in radar imaging," *IEEE Trans. Aerosp. Electron. Syst.*, **20**: 363, 1984.

[barb] S. Barbarossa, "Detection and imaging of moving objects with synthetic aperture radar, Part 1" *IEE Proc.-F*, **139**: 79–88, 1992.

[ba] B. Barber, "Theory of digital imaging from orbital synthetic aperture radar," *Int. J. Remote Sens.*, **6**: 1009–1057, 1985.

[bar] D. Barrick, "FM/CW radar signal and digital processing," *NOAA Technical Report*, ERL 283-WPL26, July 1973.

[ber80] A. Berkhout, *Seismic Migration-Imaging of Acoustic Energy by Wave Field Extrapolation*, New York: Elsevier/North Holland, 1980.

[ber81] A. Berkhout, "Wavefield extrapolation techniques in seismic migration, a tutorial," *Geophys.*, **46**: 1638–1656, 1981.

[bla] R. Blahut, W. Miller Jr., and C. Wilcox (eds.), *Radar and Sonar, Part I*, New York: Springer Verlag, 1991.

[born] M. Born and E. Wolf, *Principles of Optics*, 6th ed., New York: Pergamon Press, 1983.

[broo] E. Brookner, "Synthetic aperture radar spotlight mapper," in *Radar Technol.*, (E. Brookner, ed.), Norwood, MA: Artech House, 1977, ch. 18.

[br67] W. Brown, "Synthetic aperture radar," *IEEE Trans. Aerosp. Electron. Syst.*, **3**: 217–229, 1967.

[br69] W. Brown and R. Fredricks, "Range-Doppler imaging with motion through resolution cells," *IEEE Trans. Aerosp. Electron. Syst.*, **5**: 98, 1969.

[caf] C. Cafforio, C. Prati, and F. Rocca, "SAR data focusing using seismic migration techniques," *IEEE Trans. Aerosp. Electron. Syst.*, **27**: 194–207, 1991.

[can] B. Cantrell, "A short-pulse area MTI," NRL Report 8162, September 1977.

[car] A. B. Carlson, *Communication Systems*, New York: McGraw-Hill, 1986.

[carr] W. Carrara, R. Goodman, and R. Majewski, *Spotlight Synthetic Aperture Radar*, Norwood, MA: Artech House, 1995, ch. 6.

[che] H. Chen and C. McGillem, "Target motion compensation by spectrum shifting a synthetic aperture radar," *IEEE Trans. Aerosp. Electron. Syst.*, **28**: 895–900, 1992.

[chr] D. Christensen, *Ultrasonic Bioinstrumentation*, New York: Wiley, 1988.

[curl] J. Curlander and R. McDonough, *Synthetic Aperture Radar*, Wiley, 1991.

[curt] J. Curtis, "Synthetic aperture fundamentals," in *Radar Technology*, (E. Brookner, ed.), Norwood, MA: Artech House, 1977, ch. 16.

[cut] L. Cutrona, E. Leith, L. Porcello, and W. Vivian, "On the application of coherent optical processing techniques to synthetic aperture radar," *Proc. IEEE*, **54** (8): 1026–1032, 1966.

[edd] B. Edde, *Radar: Principles, Technology, and Applications,* Englewood Cliffs, NJ: Prentice Hall, 1993.

[fit] J. Fitch, *Synthetic Aperture Radar*, New York: Springer Verlag, 1988.

[fra] L. Franks, *Signal Theory*, Englewood Cliffs, NJ: Prentice Hall, 1969.

[fri] R. Fries, "Three dimensional matched filtering," in *Infrared Systems Signal and Components III*, SPIE, (R. Caswell, ed.) **1050**: 19–27, 1989.

[gab] D. Gabor, "A new microscope principle," *Nature*, **161**: 777, 1948.

[goj] W. Goj, *Synthetic Aperture Radar and Electronic Warfare*, Norwood MA: Artech House, 1989.

[goo] J. Goodman, *Introduction to Fourier Optics*, New York: McGraw-Hill, 1968.

[gou] P. Gough and D. Hawkins, "Imaging algorithms for a stripmap synthetic aperture sonar," *IEEE J. Oceanic Eng.*, **22**: 27–39, 1997.

[gra] L. Graham, "Synthetic aperture radar for topographic imaging," *Proc. IEEE*, **62**: 763–768, 1974.

[gru] F. Grunbaum, M. Bernfeld, and R. Blahut (eds.), *Radar and Sonar, Part II*, New York: Springer Verlag, 1992.

[hag] J. Hagedoorn, "A process of seismic reflection interpretation," *Geophys. Prosp.*, **2**: 85–127, 1954.

[har] M. Hartless and J. Barry, "Shipboard infrared search and track," Final Report of Contract N66001-94-C-6001, NCCOSC, December 1994.

[hel] C. Helstrom, *Elements of Signal Detection and Estimation,* Englewood Cliffs, NJ: Prentice Hall, 1995.

[jai] A. Jain, *Fundamentals of Digital Image Processing,* Englewood Cliffs, NJ: Prentice Hall, 1989.

[jam] R. James and L. Hoff, "Investigation of SAR target detection algorithms using narrowband and ultra wideband sensors," Naval Commmand, Control and Ocean Surveillance Center, Technical Report, San Diego, May 1993.

[kir751] J. Kirk, "Motion compensation for synthetic aperture radar," *IEEE Trans. Aerosp. Electron. Syst.,* **11**: 338–348, 1975.

[kir752] J. Kirk, "A discussion of digital processing in synthetic aperture radar," *IEEE Trans. Aerosp. Electron. Syst.,* **11**: 326–337, 1975.

[kla] J. Klauder, A. Price, S. Darlington, and W. Albershem, "The theory and design of chirp signals," *Bell Syst. Tech. J.,* **39**: 745–808, 1960.

[kov76] J. Kovaly, *Synthetic Aperture Radar*, Norwood, MA: Artech House, 1976.

[kov77] J. Kovaly, "High resolution radar fundamentals (synthetic aperture & pulse compression)," in *Radar Technology*, (E. Brookner, ed.), Norwood, MA: Artech House, 1977, ch. 17.

[lei62] E. Leith and J. Upatnieks, "Reconstructed wavefronts and communication theory," *J. Optical Soc. Am.,* **52**: 1123–1130, 1962.

[lei64] E. Leith and J. Upatnieks, "Wavefront reconstruct with diffused illumination and three-dimensional objects, theory," *J. Optical Soc. Am.,* **54**: 1295–1301, 1964.

[leo] A. Leonov and K. Fomichev, *Monopulse Radar*, trans. by W. Barton, Norwood, MA: Artech House 1986.

[loe] D. Loewnthal, L. Lu, R. Roberson, and J. Sherwood, "The wave equation applied to migration," *Geophys. Prosp.*, **24**: 380–399, 1976.

[mac] A. Macovski, *Medical Imaging Systems,* Englewood Cliffs, NJ: Prentice Hall, 1983.

[mil] A. Milman, "SAR imaging by $\omega - k$ migration," *Int. J. Remote Sens.,* **14**: 1965–1979, 1993.

[mora] P. Mora, "Inversion = migration + tomography," *Geophys.,* **54**: 12, 1575–1586, 1989.

[mor] P. M. Morse and H. Feshbach, *Methods of Theoretical Physics*, New York: McGraw-Hill, 1953, parts 1 and 2.

[mue] R. Mueller, M. Kaveh, and G. Wade, "Reconstructive tomography and applications to ultrasonics," *IEEE Proc.*, **67**: 567–587, 1979.

[nap] H. Naparst, "Dense target signal processing," *IEEE Trans. Inf. Theory,* **37**: 317–327, 1991.

[pap] A. Papoulis, *Systems and Transforms with Applications in Optics,* New York: McGraw-Hill, 1968.

[per] R. Perry, R. DiPietro, B. Johnson, A. Kozma, and J. Vaccaro, "Planar subarray processing for SAR imaging," *Proceedings of IEEE Int. Radar Conf.*, 473–478, Alexandria, VA, May 1995.

[pet] R. Peterson, "Seismography: The writing of earth waves," Technical Report, Pasadena, CA, United Geophysical Corp., 1969.

[pra] C. Prati, F. Rocca, A. Guarnieri, and E. Damonti, "Seismic migration for SAR focusing: interferometrical applications," *IEEE Trans. Geosci. Remote Sens.,* **28**: 627–640, 1990.

[pro] J. Proakis, *Digital Communications*, New York: McGraw-Hill, 1989.

[ran71] R. Raney, "Synthetic aperture imaging radar and moving targets," *IEEE Trans. Aerosp. Electron. Syst.,* **7**: 499–505, 1971.

[ran94] R. Raney, H. Runge, R. Bamler, I. Cumming, and F. Wong, "Precision SAR processing using chirp scaling," *IEEE Trans. Geosci. Remote Sens.,* **32**: 786–799, 1994.

[ras] T. Rastello, D. Vray, and J. Chatillon, "Spatial under-sampling of ultrasoud images using Fourier-based synthetic aperture focusing technique," *Proc. IEEE Ultrason. Symp.*, Toronto, 823–826, October 1997.

[rih] A. Rihaczek, *Principles of High-Resolution Radar*, New York: McGraw-Hill, 1969.

[rob] E. Robinson, "Image reconstruction in exploration geophysics," *IEEE Trans. Sonics Ultrason.*, **31**: 259–270, 1984.

[sco] C. Scott, *The Spectral Domain Method in Electromagnetics,* Norwood, MA: Artech House, 1989.

[she] D. Sheen, N. Malinas, D. Kletzi, T. Lewis, and J. Roman, "Foliage transmission measurements using a ground-based ultrawideband SAR system," *IEEE Trans. Geosci. Remote Sens.*, **32**: 118–130, 1994.

[sher] S. Sherman, *Monopulse Principles and Techniques,* Norwood, MA: Artech House, 1984.

[sko] M.I. Skolnik, *Introduction to Radar Systems,* New York: McGraw-Hill, 1980.

[smi] A. Smith, "A new approach to range-Doppler SAR processing," *Int. J. Remote Sens.,* **12**: 235–251, 1991.

[s88] M. Soumekh, "Band-limited interpolation from unevenly spaced sampled data," *IEEE Trans. Acoust. Speech Signal Process.,* **36**: 110–122, 1988.

[s90] M. Soumekh, "Echo imaging using physical and synthesized arrays," *Optical Eng.*, **29**: 545–554, 1990.

[s91] M. Soumekh, "Bistatic synthetic aperture radar inversion with application in dynamic object imaging," *IEEE Trans. Signal Process.*, **39**: 2044–2055, 1991.

[s92ja] M. Soumekh, "A system model and inversion for synthetic aperture radar imaging," *IEEE Trans. Image Process.*, **1**: 64–76, 1992.

[s92a] M. Soumekh and J. Choi, "Phase and amplitude-phase restoration in synthetic aperture radar imaging," *IEEE Trans. Image Process.*, **1**: 229–242, 1992.

[s92ju] M. Soumekh, "Array imaging with beam-steered data," *IEEE Trans. Image Process.*, **1**: 379–390, 1992.

[s94] M. Soumekh, *Fourier Array Imaging*, Englewood Cliffs, NJ: Prentice Hall, 1994.

[s94n] M. Soumekh, "Digital spotlighting and coherent subaperture image formation for stripmap synthetic aperture radar," *Proc. Int. Conf. Image Process.*, Austin, TX, 476–480, November 1994.

[s95] M. Soumekh, "Reconnaissance with ultra wideband UHF synthetic aperture radar," *IEEE Signal Processing Magazine*, **12**: 21–40, 1995.

[s96a] M. Soumekh, "Reconnaissance with slant plane circular SAR imaging," *IEEE Trans. Image Process.*, **5**: 1252–1265, 1996.

[s96s] M. Soumekh, "Automatic aircraft landing using interferometric inverse synthetic aperture radar imaging," *IEEE Trans. Image Process.*, **5**: 1335–1345, 1996.

[s96n] M. Soumekh, "Super-resolution array processing in SAR," *IEEE Signal Processing Magazine*, **13**: 14–18, 1996.

[s97] M. Soumekh, "Doppler and cinematic phased array imaging via depth-focusing," *Acoustical Imaging*, **23**: 297–302, S. Lees and L. Ferrari (eds.), New York: Plenum, 1997; to appear as "Depth-focused interior echo imaging," in *IEEE Trans. Image Process.*

[s97m] M. Soumekh, "Phased array imaging of moving targets with randomized beam steering and area spotlighting," *IEEE Trans. Image Process.*, **6**: 736–749, 1997.

[s97a] M. Soumekh, "Moving target detection in foliage using along track monopulse synthetic aperture radar imaging," *IEEE Trans. Image Process.*, **6**: 1148–1163, 1997.

[s97o] M. Soumekh, "Signal subspace fusion of uncalibrated sensors with application in SAR, diagnostic medicine and video processing," *Proc. IEEE Int. Conf. Image Process.*, Santa Barbara, CA, 280–283, October 1997; appeared in *IEEE Trans. Image Process.* **8**: 127–137, 1999.

[s97n] M. Soumekh, "Airborne synthetic aperture acoustic imaging," *IEEE Trans. Image Process.*, **6**: 1545–1554, 1997.

[s98a] M. Soumekh, "Moving target detection and automatic target recognition via signal subspace fusion of images," *Proc. SPIE's Annu. Int. Symp. on Aerospace/Defense Sensing, Simulation, and Controls,* 120–131, Orlando, FL, April 1998.

[s98j] M. Soumekh, "Bistatic synthetic aperture radar imaging using wide-bandwidth continuous-wave sources," *Proc. SPIE's Int. Symp. on Optical Science, Engineering and Instrumentation,* San Diego, CA: July 1998.

[s98s] M. Soumekh, "Alias-free processing of P-3 SAR data," Technical Report EE-037, State University of New York, Buffalo, September 1998.

[ste] B.D. Steinberg, *Principles of Aperture and Array System Design*, New York: Wiley, 1976.

[sto] R. Stolt, "Migration by Fourier transform," *Geophys.*, **43**: 23–48, 1978.

[tom] K. Tomiyasu, "Tutorial review of synthetic aperture radar (SAR) with applications to imaging the ocean surface," *Proc. IEEE*, **66**: 563–583, 1978.

[van] H. Van Trees, *Detection, Estimation, and Modulation Theory*, parts I–III, New York: Wiley, 1968.

[wal] J. Walker, "Range-Doppler imaging of rotating objects," *IEEE Trans. Aerosp. Electron. Syst.*, **16**: 23, 1980.

[weh] D. R. Wehner, *High Resolution Radar,* Norwood, MA: Artech House, 1987.

[wid] B. Widrow and S. Stearns, *Adaptive Signal Processing,* Englewood Cliffs, NJ: Prentice Hall, 1985.

[wil] C. Wiley, "Pulsed Doppler radar methods and apparatus," United States Patent, 3,196,436, 1965.

[wol] E. Wolf, "Three-dimensional structure determination of semi-transparent objects from holographic data," *Opt. Commun.*, **1**: 153, 1969.

[woo] P. M. Woodward, *Probability and Information Theory, with Applications to Radar,* Oxford: Pergamon Press, 1964.

[yan] H. Yang and M. Soumekh, "Blind-velocity SAR/ISAR imaging of a moving target in a stationary background," *IEEE Trans. Image Process.*, **2**: 80–95, 1993.

[you] S. Young, N. Nasrabadi, and M. Soumekh, "SAR moving target detection and identification using stochastic gradient techniques," *Proc. ICASSP,* **4**: 2145–2148, Detroit, May 1995.

[zeb] A. Zebker and R. M. Goldstein, "Topographic mapping from interferometric synthetic aperture radar observations," *J. Geophys. Res.,* B5, **91**: 4993–4999, 1986.

INDEX

Aliasing, *see* Sampling
Along the track, *see* Cross-range
Along-track monopulse SAR, *see* Synthetic aperture radar
Along-track target angles, 493
Altitude domain, 160
Ambiguity function, *see* Moving target
Amplitude-modulated-phase-modulated (AM-PM) signal, 97
  slow-time compression, 130, 134
  slow-time Fourier analysis, 98, 124
Amplitude modulation (AM), 97
Amplitude pattern, 96
  beam-steered, 273, 280
  receive mode, 163
  target, 166
  transmit mode, 145
Amplitude pattern in slow-time Doppler
  receive mode, 163
  transmit mode, 143, 144, 147
  transmit-receive mode, 165
Analog spotlighting, 266
Analog to digital (A/D) converter, 2, 20
Angular slow-time Doppler domain, *see* Slow-time Doppler frequency
Aperture, *see* Radar antenna; Synthetic aperture
Array, 47. *See also* Synthetic aperture
  linear, 48, 278
  phased, 48, 278
  physical, 48–49, 277
  synthetic, 48–49
Aspect angle, *see* Synthetic aperture
Aspect angle synthetic aperture domain, *see* Circular SAR; Slow-time
Autocorrelation, 36
Automatic aircraft landing (ALL), 596
Automatic target recognition (ATR), 539. *See also* Signal subspace processing
Azimuth, *see* Cross-range
Azimuthal and elevation synthetic apertures, *see* Synthetic aperture

Backprojection imaging, *see* Reconstruction
Bandpass signal, 5, 69, 298
Bandwidth compression, *see* Compression
Baseband conversion
  in fast-time, 20, 29
  in slow-time, 80, 298
  target function, 353, 356, 359
  target function spectrum, 90, 201, 259, 323, 352, 420
Beam-forming, 48. *See also* Beam steering
Beam steering, 48
  electronic, 277
  mechanical, 270
Beamwidth, 107, 144, 271
  parabolic radar, 149
  planar radar, 145
Beamwidth filtering, *see* Slow-time Doppler filtering
Bistatic SAR data, 561
Boresight angle, 271
Boresight target, *see* Broadside target
Broadside target, 71
Broadside target area, 54, 186

Center of mass, 8, 54
Chirp rate, 23
Chirp signal, 23
  compression, 25
  flat-top, 35
  instantaneous frequency, 23
  phase-modulated, 38
Circular radar, 152
Circular SAR, *see* Synthetic aperture radar
Circular SAR signal, 494
Closest point of approach (CPA), 59, 123, 187

Clutter filtering, *see* Digital spotlighting
Compressed SAR image, 340
  in squint spatial coordinates, 342
Compression
  in fast-time for chirp signal, 25
  in slow-time for AM-PM signal, 130, 134
  in slow-time for polar format processing, 246
  in slow-time for spherical PM signal, 80
  in slow-time for spotlight SAR, 302
  reconstructed image bandwidth, 95, 336, 428, 521
Continuous phase modulation (CPM), 38, 261
Cross-range, 54
Cross-range extended target, 180
Cross-range gating, 88, 307
CSAR data, 494–546
Curved radar, *see* Circular radar; Parabolic radar
  in spotlight SAR, 283, 286
  in stripmap SAR, 391, 404

Decompression
  in fast-time for chirp signal, 32
  in slow-time for spherical PM signal, 88
  in slow-time for spotlight SAR signal, 304
Deconvolution, 577. *See also* Inverse problem
Decorrelator receiver, 555
Depression angle, 161, 228, 493, 541
Deramping, 25
Detecting change, *see* Signal subspace processing; Target detection
Diameter of radar, 103, 132
Digital reconstruction, *see* Reconstruction
Digital spotlight filter
  full aperture, 310, 410
  subaperture, 318, 410
Digital spotlighting
  circular SAR, 517
  moving target indication, 474
  spotlight SAR, 306, 317
  stripmap SAR, 409
  subaperture, 317, 410
Discrete cosine transform (DCT), 87
Discrete Fourier transform (DFT), 18
Displaced phase center array (DPCA), 561
Divergence angle
  circular radar, 154
  parabolic radar, 148
  planar radar, 145
Dynamic focusing, 48

Echoed signal
  circular SAR, 494
  cross-range imaging, 59
  Fourier analysis, 13, 61, 195, 257, 275, 281, 387, 548
  generic SAR, 186
  half-hyperbolic locus, 191
  range imaging, 9, 11, 24
  slow-time compressed, 81, 247, 302, 405
  slow-time Doppler support band, 76, 79, 282
  spotlight SAR, 274, 281
  stripmap SAR, 385
  three-dimensional (azimuthal and elevation) SAR, 257
  three-dimensional E-CSAR, 544
Effective beamwidth, 113, 132
Effective range aperture, 163
Electronic beam steering, *see* Beam steering
Electronic counter-countermeasure (ECCM), 35, 260
Electronic counter measure (ECM), 35, 259
Electronic warfare (EW), 34
Elevation angle, *see* Depression angle
Elevation domain, *see* Altitude domain
Elevation synthetic aperture, *see* Synthetic aperture
Exponential window, *see* Power window

Fast-time, 57, 187
  reference point, 297, 402, 516
Far-field, 146
Far-field approximation, *see* Fraunhofer approximation
Flash point, 169, 367, 437, 593
Flat-top function, 35
Focal point, 152, 153
Focal range, 147
Focusing, 48, 152
Foliage penetrating (FOPEN) SAR
  P-3 data, 479–485
  SRI data, 295–368, 399–479
Forward problem
  cross-range imaging, 60
  range imaging, 12
Fourier transform
  marginal, 178
  one-dimensional, 3
  scaling, 4
  two-dimensional, 178
Fraunhofer approximation, 143
Frequency modulated (FM), 2
Frequency stepping, 9
Fresnel approximation, 81, 86, 143, 251
  narrow-bandwidth and narrow-beamwidth, 255

Fringe patterns, 533
Fusion, *see* Signal subspace processing

Generic SAR, *see* Synthetic aperture radar
Georgia Tech Research Institute (GTRI) data, *see* CSAR data
Global positioning system (GPS), 217
Grazing angle, *see* Depression angle
Green's function, 47, 497
  slant plane, 494
  E-CSAR, 545
Guard band, 19, 459

Hamming window, 204
Hankel function, 503
Hanning amplitude pattern, 110
Hexagonal sampling, 240, 333
Holography, 47, 197
Horizontal polarization, 171
Hyperbolic locus, 191

Ideal source, 174. *See also* Omni-directional radar
Interferometric ISAR (IF-ISAR), 596
Interferometric SAR (IF-SAR), *see* Synthetic aperture radar
Impulse response, *see* Green's function
Inhomogeneous media, 138
Instantaneous frequency, 23, 67, 72, 338
Interpolation. *See also* Reconstruction
  from evenly spaced data, 204
  from unevenly spaced data, 205
Inverse problem
  cross-range imaging, 66, 73
  range imaging, 14
Inverse synthetic aperture radar (ISAR), 244
  maneuvering target, 598
  model, 246, 596
  multistatic monopulse, 596
  polar format processing, 247
ISAR data, 295–343, 579–583, 600–602

Jacobian, 148, 154, 205
Jamming, 35, 259

Linear frequency modulated (LFM) signal, *see* Chirp signal
Linear SAR, *see* Synthetic aperture radar
Look angle, *see* Synthetic aperture

Maneuvering target, *see* Inverse synthetic aperture radar

Matched filtering
  circular SAR, 504
  cross-range imaging, 64, 72, 127
  E-CSAR, 550
  generic SAR, 193
  range imaging, 14, 44
  spotlight SAR, 275, 281, 320
  stripmap SAR, 387, 418
Mean cross-range, 188, 202
Mean elevation, 159
Mean range, 8, 187, 188, 201
Mechanical beam steering, *see* Beam steering
Mine detection, 45
Mixing, 25
Monopulse radar, 553
  instability, 574
Monopulse SAR, *see* Synthetic aperture radar
Monostatic SAR data, 561
Mosaic reconstruction, *see* Reconstruction
Motion compensation
  for polar format processing, 243
  in-scene target, 230
  narrow-bandwidth, 219, 230
  three-dimensional, 226, 238
  wide-bandwidth, 223, 231
Motion errors, 218, 454, 569
  in slow-time Doppler spectrum, 369, 453
Motion tracking, *see* Signal subspace processing
Motion-transformed spatial coordinates, *see* Moving target
Moving target, 328, 348, 368, 437
  ambiguity function, 475
  bistatic SAR signal, 564
  detection, 474, 566
  imaging, 472
  monostatic SAR signal, 562
  motion-transformed spatial coordinates, 470, 563
  motion-transformed target function, 473
  moving target indicator, 476, 568
  relative speed, 469
  signal model, 467
Moving target detection (MTD), *see* Moving target
Moving target indicator (MTI), *see* Moving target
Multiplicative noise, *see* Motion errors
Multistatic monopulse ISAR, *see* Synthetic aperture radar

Narrow-bandwidth, *see* Radar signal
Narrow-beamwidth (look angle), 220, 239, 589

Naval Research and Development (NRaD) data, *see* ISAR data
Near-field approximation, *see* Fresnel approximation
Noise equivalent bandwidth, 113, 132
Nonlinear motion, *see* Motion errors
Nyquist criterion, *see* Sampling

Off-broadside target, *see* Squint target
Omni-directional radar, 157, 263, 383, 403
Omni-directional reflector, 385

P-3 data, *see* Foliage penetrating SAR
Parabolic radar, 147
Parseval's theorem, 355
Phased array, 277
Phase history, 562
Phase modulation, 23, 60
Phase unwrapping, 591
Physical radar, *see* Radar antenna
Planar radar, 144
  in spotlight SAR, 285, 287
  in stripmap SAR, 389, 403
Plane wave approximation, 237. *See also* Polar format processing
  narrow-bandwidth and narrow-beamwidth, 239
  narrow-beamwidth, 238
  wavefront curvature compensation, 242
Point spread function
  circular SAR, 512
  cross-range imaging, 74, 120, 129, 131
  range imaging, 14, 16, 26, 43
  spotlight SAR, 276, 282, 290
  stripmap SAR, 388, 395
Polar format processing, *see* Reconstruction
Polarization, 171
Power equalization, 37
Power window, 37, 102
Pulse diversity, 259
Pulse repetition frequency (PRF), 80, 189, 274
Pulse repetition frequency (PRF) reduction, 80, 302. *See also* Compression
Pulse repetition interval (PRI), 80, 189

Radar antenna, 142
Radar cross section (RCS), 1, 113
Radar dish, 142
Radar footprint, 157, 161, 270
Radar horn, 142
Radar signal
  bandwidth, 14
  carrier (center) frequency, 14
  duration, 9
  FM-CW, 57, 397
  instability, 44
  narrow-bandwidth, 240
  wide-bandwidth, 57
Radar surface, 142
Radar swath, 8
Radar swath echo time period, 29
Radiation pattern
  beam-steered, 273, 280
  receive mode, 163
  side-looking, 382
  slow-time dependence, 156
  slow-time Doppler, 164, 168
  target, 166
  three-dimensional, 157
  transmit mode, 142
  transmit-receive mode, 164
Radio frequency interference (RFI), 399, 479, 486
Ramping, 32
Range bins, 22, 352
Range compression, *see* Compression, in fast-time for chirp signal
Range domain, 4, 8
Range-Doppler imaging, *see* Reconstruction
Range extended target, 43
Range spread target, 43, 180
Range stack reconstruction, *see* Reconstruction
Rayleigh resolution, 75
Reconstruction
  backprojection, 218, 357, 448, 524
  bandwidth reduction, *see* Compression
  in squint spatial coordinates, 329
  parallel implementation, 461
  polar format processing, 234, 249
  range-Doppler imaging, 249
  range stacking, 207, 349, 445
  spatial frequency interpolation, 198, 319, 419, 518, 551
  subaperture, 456
  subpatch (mosaic), 456
  time domain correlation (TDC), 212, 355, 446, 523
Reference fast-time point, *see* Fast-time
Reference signal
  cross-range imaging, 59
  generic SAR, 206
  range imaging, 21
Reflectivity, 8, 54
  frequency-dependent, 40
  frequency and synthetic aperture-dependent, 215
  synthetic aperture-dependent, 95

Registration, *see* Signal subspace processing
Residual video phase (RVP) error, 31
Resolution
  altitude, 528
  circular SAR, 512
  cross-range, 66, 75, 113, 129, 132
  range, 17, 28
  spotlight SAR, 293
  stripmap SAR, 396
Resonating cavity, 598
Round-trip delay, 9, 194, 257, 588
Round-trip distance, 562, 564, 597

Sampled aperture, *see* Array
Sampling
  fast-time, 18, 295
  fast-time compressed chirp, 30
  fast-time frequency, 20
  polar format processing, 241–242
  slow-time, 66, 76, 79, 129, 133, 298, 402, 517
  slow-time compressed SAR signal, 303, 407, 409
  slow-time Doppler frequency, 93
  target function, 325, 421, 521
  target function spectrum, 324
SAR spatial frequency mapping, 197
  circular SAR, 505
  three-dimensional (azimuthal and elevation), 258
  three-dimensional E-CSAR, 550
SAR spectral mapping, *see* SAR spatial frequency mapping
Signal subspace processing
  automatic target recognition, 585
  block-based processing, 580, 600
  calibration error impulse function, 576
  estimated calibration error, 583
  motion tracking, 598
  moving target detection (MTD), 584
  projection, 579
  registration, 575
  shift estimation, 594, 599
  subspace difference image, 579
Sinc function, 16
Slant angle, *see* Depression angle
Slanted target, 367, 437, 593
Slant plane monopulse SAR, *see* Synthetic aperture radar
Slant-range, 162
Slow-time, 51, 57, 187
  aspect angle domain in CSAR, 492
Slow-time Doppler filtering, 362, 451
Slow-time Doppler frequency, 51, 61
  angular, 71, 121, 168, 310
  for CSAR aspect angle domain, 501
  subsampling, 93, 335, 425
Source deconvolution, 14. *See also* Inverse problem
Spatial frequency mapping, *see* SAR spatial frequency mapping
Spherical phase function, *see* Spherical PM signal
Spherical PM signal, 47, 60
  compression, 80
  Fourier analysis, 62, 69
  instantaneous frequency, 67
  phase center, 68
  slow-time angular Doppler representation, 71
  slow-time Doppler support band, 62, 69
  two-dimensional, 257
Spotlight SAR, *see* Synthetic aperture radar
Spotlight SAR signal, 100
  Hamming-windowed, 101
  power-windowed, 102
Squint angle, 222, 271
Squint cross-range, *see* Mean cross-range
Squint spatial coordinates, 249, 272, 329
Squint spatial frequency domain, 330
Squint target, 71
Squint target area, 54, 186
Stanford Research Institute (SRI) data, *see* Foliage penetrating SAR
Stationary phase method, 62, 99
Stripmap SAR, *see* Synthetic aperture radar
Stripmap SAR signal, 103
Subaperture processing, *see* Digital spotlighting; Reconstruction
Subpatch reconstruction, *see* Reconstruction
Synthetic aperture. *See also* Slow-time
  azimuthal synthetic aperture, 57, 257
  aspect angle, 57, 168
  aspect angle domain in CSAR, 492
  elevation synthetic aperture, 257
  finite, 60, 67
  infinite, 61
  look angle, 287
  two-dimensional (azimuthal and elevation), 256
  two-dimensional (circular and elevation), 539
Synthetic aperture frequency, *see* Slow-time Doppler
Synthetic aperture radar (SAR)
  along-track monopulse, 559
  circular, 486
  generic, 181

Synthetic aperture radar (SAR) (*cont.*)
  interferometric (IF-SAR), 588, 591
  inverse synthetic aperture radar (ISAR), 244
  linear, 486
  monopulse, 553
  multistatic monopulse ISAR, 596
  slant plane topographic mapper monopulse, 587
  spotlight, 266–267, 373–374
  stripmap, 266–267, 373–374
  three-dimensional (azimuthal and elevation), 256
  three-dimensional CSAR, 525
  three-dimensional E-CSAR, 539

Target detection, 474, 533, 566, 579, 584
Target function
  bistatic, 567
  ideal, 8, 55, 197
  in squint spatial coordinates, 330
  interferometric, 591
  matched-filtered, 16, 66, 73, 197, 275
  monostatic, 564
  motion-transformed, 473
  spectrum, 17, 55, 65, 73, 275, 504, 528, 549
Target resolvability, *see* Target detection
Three-dimensional imaging, *see* Synthetic aperture radar
Time domain correlation (TDC) imaging, *see* Reconstruction
Topographic mapper monopulse SAR, *see* Synthetic aperture radar
Transition range, 146
Turntable data, *see* CSAR data

Uncalibrated radars, 570
Unstable radars, 570, 574. *See also* Radar signal
Upsampling, 31, 87, 304
Upsweep chirp signal, 23

Vertical polarization, 171

Wave divergence, 13, 186
Wavefront curvature compensation, *see* Plane wave approximation
Wavefront reconstruction, 47, 197. *See also* Reconstruction
Wavelength, 66
Wavenumber, 16, 59
Wavenumber domain, 4
Wavenumber domain mapping, *see* SAR spatial frequency mapping
Wide-bandwidth, *see* Radar signal
Wide-beamwidth, 223, 592

Zero-padding, 34, 87, 91, 320, 322, 418, 419, 518